A GUIDE BOOK OF

CARRY GUN VALUES

VOLUME 1

© 2014 Whitman Publishing, LLC

3101 Clairmont Road - Suite G - Atlanta, GA 30329

THE RED BOOK: A Guide Book of Carry Gun Values, Volume 1

ISBN: 0794842593

EAN: 9780794842598

Correspondence concerning this book may be directed to the publisher via email at contact@RedBookOfGuns.com.

2nd Amendment Media is an imprint of Whitman Publishing, LLC

Printed in the United States of America.

CONTENTS

MANUFACTURERS LISTING (A-Z)

God Made Man, but Samuel Colt Made Them Equal.

EDITORS' LETTER

Dear Reader,

One of the greatest liberties afforded Americans is the right to bear arms. Not only are firearms great fun to own and shoot for hunting or for sport, but they are also wise investments, as they tend to hold and even increase in value. More importantly, a firearm can mean the difference between life and death in dangerous situations. Though firearms can be inherently dangerous, with training and responsible handling, any risk is vastly diminished. That risk is far overshadowed by the fact that in circumstances that require self-defense, firearms ensure that anyone, regardless of physical abilities, can effectively defend himself or herself against any assailant. Thus, there is great truth in the famous saying, "God made man, but Samuel Colt made them equal."

Our right as Americans to own firearms is outlined in the Second Amendment of the Constitution. It is here that our forefathers guaranteed that the government could never impose upon an individual's right to defend themselves and their loved ones. The founders understood that a properly functioning government is solely empowered by its people, as opposed to a privileged few who control all weapons and supplies. Personal firearm ownership is essential to the framework of our nation, as it grants us the ability to discourage and deter political tyranny.

While U.S. citizens enjoy their Second Amendment rights, we, the editors of this book, highly encourage the exercise of extreme caution before purchasing, carrying, and utilizing a firearm. Owning a firearm is a great responsibility. It is our hope that upon reviewing this book and purchasing a firearm, all gun owners would operate their weapons with care, with discretion, and within the confines of the law. Presently, we are faced with heated political debate over the issue of firearms. While one group is heavily in favor of relaxing gun laws, the other is just as eager to ban firearms altogether. Because of these shifting political tides, we encourage those in favor of gun rights to handle their weapons with the utmost care and responsibility, so as not to give credence to anti-gun rhetoric.

We hope this book paints a clear picture of the current handgun market and equips readers with the ability to make an informed decision about which handgun best suits their needs. Furthermore, we would like to thank our readers and encourage them to exercise one of the most cherished rights U.S. citizens are privileged to enjoy.

-THE EDITORS

THE BENEFITS OF THE RED BOOK OF CARRY GUN VALUES

In creating the Red Book of Carry Gun Values, we, the editors and contributors, sought to compile a book that best reflects the current values of handguns suitable for carry. We hope that you, the prospective handgun buyer, can make a well-informed purchase after consulting this book. The editors have no allegiance to any particular company and have remained true to the market when estimating the values of each gun. Many enthusiasts find objectivity difficult when assessing the values of their most beloved firearms, but we've done our best to remain unbiased and to use actual historical sales to guide us as to pricing. We've evaluated the sales history of each model represented in this guide to arrive at real-life estimates, not hypotheticals or ideals. To create an accurate book, we've adhered to the economic principle that a product is worth what a buyer is willing to pay, not what a given expert thinks should be paid.

Along with providing accurate price estimates, we hope this book gives readers an accurate look into the specifications of a handgun, its production history, and other information of value. While a handgun's price is certainly important, its specifications determine which particular model would suit a gun owner's individual needs. Differences among shooters ranging from the extent of firearm experience to a shooter's height or physical strength can significantly influence which handgun best meets that shooter's needs. A gun's specs—ranging from its caliber options, barrel length, and capacity—often determine whether or not it would be optimal for a particular buyer. That's why we've included the specs of each handgun to help educate our readers to make good purchase decisions.

In short, we want you to have a clear picture of the weapon you wish to buy before trying to purchase it through a gun shop, auction house, or individual seller. We want readers to walk away from a purchase feeling secure they've reached a fair deal. To aid our readers in the buying process, in this--the first--edition of The Red Book: A Guide Book of Carry Gun Values, we introduce the Redbook Code™. The Redbook Code is a universal system of organizing all firearms on the secondary market. Many firearms being bought and sold currently pre-date the UPC system that now allows buyers to differentiate between manufacturers and models. Even newer models that do have UPC numbers are more easily organized with our coding system once that arrive on the secondary market, when UPC numbers become irrelevant.

CARRY GUN SPECIFICATIONS

The handguns we've featured range drastically, from pocket-sized derringers to .45 Winchester Magnum M1911s. We wanted to provide readers with the full scope of the carry gun market, so some models may be better characterized as "open carry" guns as opposed to "concealed carry" guns. Additionally, a small number of handguns we've featured arguably wouldn't be used as carry weapons at this time at all, but were included in this collection either due to their historical significance or to their uniqueness to the market. We've included some guns over 100 years old to represent "carry guns" that were popular prior to the invention of smokeless powder. As a result, some guns featured are ideal for concealed carry, some would be better kept in a vehicle glove compartment, and still some others would likely not leave a display case. The specifications given for each gun will help the reader choose wisely for his or her purposes. Our hope is to inform readers so they can feel comfortable and confident about purchasing the right handgun.

When creating and editing this book, we diligently mined several sources to ensure we included accurate specs for each handgun. We consulted manufacturers' records for much of the information as well as a host of online retailers, brick-and-mortar gun shops, auction houses, gun critics and experts, and our panel of contributors. Providing accurate specs for each model was of the utmost importance to us so that we could provide readers with a clear understanding of any gun they consider for purchase. Given the wide array of handguns covered in this book, information for some models remains somewhat obscure or vague. We both welcome and appreciate your feedback if you find an error or oversight within this text. Please help us improve by emailing your comments to contact@RedBookofGuns.com.

HOW WE ESTIMATED VALUES

When creating and editing this book, we consulted many sources to construct our value estimates. We've listed many of these resources in the bibliography. While it would be easy to price a handgun based on our own beliefs about its value, we've refused to ignore the current market and estimated the value of each gun based on actual recent sales. Given the shifting tastes within the gun world, estimates based on recent sales provide the best view of how much one could expect to pay for a particular model.

OUR CONSUMER-FRIENDLY PRICING MODEL

We established our pricing model based on market research, considering market conditions like current gun shop prices, online pricing, and auction sale prices. These values reflect our best efforts to establish average pricing for firearms in the current economy. Bear in mind, however, that the firearms market is volatile, given our country's political climate and changes within the landscape of the firearms industry. Prices are thus subject to change as laws, regulations, media coverage, and manufacturing circumstances influence market values.

Our pricing model provides values for multiple buying and selling scenarios. We include prices for the following:

Dealer-to-Consumer (D2C)

Consumer-to-Consumer (C2C)

Consumer-to-Dealer (Trade-In)

Last Manufacturer's Price (LMP)

Our Dealer-to-Consumer prices will naturally tend to be higher than ones associated with private sales, as "brick and mortar" gun shops must compensate for overhead expenses. Our Consumer-to-Consumer prices reflect fair prices one could typically expect to pay from an individual either locally or from an online auction house like gunbroker.com or rockislandauction.com. Our Trade-In values indicate fair prices that a local gun shop may offer to pay for your firearm. Keep in mind that pricing for a firearm will typically be lower for trade-in than for private sales, as gun dealers who receive trades must factor in a margin for profit that doesn't allow them to offer full value for a firearm. The lower price a seller receives trading in a gun is often considered to be offset by avoiding the hassle of trying to sell locally or the expense of delivering a weapon to a Federal Firearms Licensed dealer for legal delivery to its purchaser.

VARIABLES

The prices we present typically account for the base model of a specific firearm. Upgrades, add-ons, limited editions, modifications, and/or other accessories may bring higher, or even lower, values. Any uncertainty about any aspect of a particular firearm should be addressed by a local firearms expert. Also, for highly collectible firearms, we recommend potential buyers consult an expert before purchase, as miniscule variations can drastically alter a firearm's value. Be aware that certain firearms used in military conflicts, ones owned by famous individuals, or ones used in Hollywood films, etc., should have proper documentation or provenance before purchase.

Also note that you may encounter differences in pricing among states or regions for particular firearms. For example, a long barreled hunting revolver may not carry the same value in Miami, Florida as it would in western Texas or Louisiana.

Additionally, be aware of a fairly recent market trend triggered by modernizations in manufacturing. Gun manufacturers have utilized mass production of parts more and more as demand for their products has increased, and that trend towards mass production has resulted in lower quality control standards. These lower standards have given rise to a corresponding fear that the finished product is compromised, thereby driving up the prices of firearms that pre-date such extensive use of mass manufacturing for their perceived higher quality and workmanship.

EXACT MODELS

Before buying or selling a weapon, buyers should be certain of the exact make, model, and details of the firearm. Small nuances, such as serial number ranges, manufacturer seals or stamps, etc. may significantly impact a firearm's value. If unsure of these factors, one should consult an expert. In addition, many manufacturers maintain useful information related to specific models that one can obtain by providing a gun's specific serial number. Manufacturers may charge a fee for this documentation, but providing this information to potential buyers will typically increase the firearm's value.

nd Auction Company

Serial numbers are a great way to determine a gun's exact model.

OUR FIREARM CONDITION GRADING SCALE

Aside from fundamental aspects like the manufacturer and the model of a firearm, the grade and condition of a gun drastically influences its value. We developed our grading scale based on NRA firearms grades and other industry standards.

Wear is a huge factor in determining a firearm's value. Internally, as the number of rounds fired grows, evidence of wear becomes apparent, and that wear devalues the gun. A qualified gunsmith knows how to determine internal wear, but few gun owners are capable of judging this aspect. The average consumer is left with judging external finish wear in order to discern a gun's value. Unblemished, original factory finish degrades by handling, use, degree of care, and maintenance over time. Through normal use, metal and synthetic surfaces will show evidence of marring. A simple way to grade finish wear on a handgun is to determine the ratio of blemished area to non-blemished area. A rudimentary way to make a quick and dirty judgment about the external wear on a gun is to consider how much of the surface area of the gun is affected. For instance, consider a Colt 1911 with substantial wear on the muzzle extending for about an inch from the tip of the gun; this type of wear originates usually as a result of long-term holster storage. If the rest of the gun is in pristine condition, that wear to the muzzle would constitute about 10% of the surface area of the gun. With that amount of wear to the finish, the gun would rate at a "Very Good +" on our scale.

Our grading scale is as follows:

Courtesy of Rock Island Auction Company

NIB (New in Box) – New from the factory condition. Unused, unfired, includes box and all paperwork and accessories. Must never have been sold before and, if applicable, must have the factory warranty intact.

Courtesy of Rock Island Auction Company

Mint – Same condition as new, but may have been sold from a dealer. Unused, but may have been handled. Includes box, paperwork, and accessories.

Note slight wear

Courtesy of Rock Island Auction Company

Ex (Excellent) – Lightly used and may have very slight signs of wear around the end of the barrel or on the edges of the frame. Minimal to no scratches or dings on wood, polymer, or metal. Mechanically perfect working condition. No rust or corrosion. May or may not include box, paperwork, and accessories. Exposed metal and grip material will only show 2% to 5% of surface wear.

whitman.

Note rust & wear

Note rust & wear

Courtesy of Rock Island Auction Company

VG+ (Very Good Plus) – Used with minor scratches or dings in the wood, polymer, or metal. Must have factory original finish but may show some wear. Mechanically perfect working condition. No rust or corrosion. Exposed metal or grip material will show between 10% and 20% of surface wear.

Note screws on grips removed multiple of times

Note scratches and dents on grips

Courtesy of Rock Island Auction Company

Good – Used with moderate signs of wear and scratches. May contain some replacement parts or may have been refinished. Mechanically safe working condition. Exposed metal and grip material may show up to 45% of surface area wear.

Note rust on slide

Note pitting

Courtesy of Rock Island Auction Company

Fair – Used with significant signs of wear and moderate scratches or dings on wood, polymer, or metal. Mechanically safe working condition, but may need minor repairs or parts soon. May have slight rust or pitting that does not interfere with its mechanics. Half or more of metal and grip material finish may be gone. In some models, only 20% of original finish may be present.

Note the poor finish, pitting, rust and broken hammer. Also assume this gun can no longer fire.

Courtesy of Rock Island Auction Company

Poor – Used with heavy signs of wear. May need full restoration. Needs refinishing and parts replacement. Mechanically inoperable working condition and not safe to fire. May still have 10% of original finish remaining in protected areas.

SAFETY

Firearms safety is of the utmost importance. The following safety rules should always be followed:

1. Treat every weapon as if it were loaded.

2. Never point your weapon at anything you don't intend to shoot.

3. Keep your finger straight and off the trigger until you're ready to fire.

4. Keep your weapon on safe until you intend to fire (if applicable).

5. Know your target and what is beyond/behind your target.

First, when handling any firearm, always check to see if it's loaded. With a revolver, keep the firearm pointed in a safe direction, keep your finger off the trigger, open the cylinder, and check for ammunition. With a semi-automatic, keep the firearm pointed in a safe direction, keep your finger off the trigger, remove the magazine, pull back the slide then lock it to the rear, visually inspect the chamber for a round, and put the weapon on safe (if applicable).

When using a firearm, make sure to use the proper ammunition for your specific model. Also, make sure the firearm is in operating condition. If there's any uncertainty, consult a professional. Always wear ear and eye protection. Know your local laws and regulations.

STORAGE

When storing a firearm, make sure to keep the weapon in a safe and dry location. Use quality gun safes if possible. Keep all firearms away from small children and anyone who is unfamiliar with proper use and care of them. Many firearms may require dehumidifiers or other humidity absorbing products to keep them from rusting. As the saying goes, "Guns have two enemies—rust and politicians."

CLEANING

Before cleaning, go through the previously mentioned safety and handling procedures to make sure the firearm is unloaded. Any firearm you intend to fire must be properly maintained. All firearms have some type of metal parts that are susceptible to rust or corrosion, so properly clean and lubricate any metal or moving part on the firearm. Any type of dirt, debris, or corrosion on a firearm can cause serious malfunctions and potential harm to the shooter. Always use quality cleaners and lubricants specific to firearms. If soaking a firearm without modern coatings, never use water based cleaners for extended periods of time, as this can cause rust and corrosion.

Use extreme caution with collectible firearms. Many collectible firearms should NOT be cleaned at all. You may drastically reduce the value of a collectible weapon by cleaning or removing any rust or patina. If you are unsure, consult an expert.

DISCLAIMER

Given the legal restrictions of selling and purchasing firearms, the editors, publishers, writers, and contributors of this book cannot be held responsible for any legal issues a reader may confront while undertaking the commercial, private, or unlawful sale of firearms. We relinquish all responsibility regarding the purchase, sale, safety, and usage of firearms, and we encourage readers to consult their local law enforcement agency or even professional legal counsel before dealing in firearms in any capacity. This book provides value estimates on firearms only and in no way purports to provide legal guidance of any type or for any purpose.

The editors of this book have researched the firearm prices and specifications with great attention to detail but cannot assume liability for any misinformation within this text. Given the convoluted history and myriad details inherent in the sales of many of these guns, we have tried to ensure the information is as accurate as possible. However, minor discrepancies may exist for which we cannot be held responsible. Also, some figures in this book were rounded or approximated, but these differences are nominal.

CONTRIBUTORS AND EDITORIAL ASSISTANCE

The following people and organizations have, to varying degrees, assisted in the production of this book. Some provided photography, others editorial help, still others their expertise related to firearms. We appreciate their assistance and would like to acknowledge them here for their efforts and contributions.

Kenneth Barlow of North American Arms
Roy Jinks, S&W Historian
Walter McClanahan of WHMC Guns
Gary Cook
Ryan Carlton, Firearms Consultant
Collector's Firearms
Matt Huber of Bud's Gun Shop, at BudsGunShop.com
EAA Corp.
Ronald Eldridge of Sig Sauer
Sig Sauer, Inc.
Bob Grueskin of Les Baer
Smith & Wesson
Frank Harris of Kahr Arms, Auto-Ordnance Corp., and Magnum Research, Inc.
Alea Danielle Ashby, Editorial Consultant
Jacob Perry, Editorial Consultant
Chad Holwerda of Iver Johnson
FNH USA, LLC
icollector.com
Rich Karas and Chip Klaiber of Cimarron Firearms
Major of Rezzguns.com
Beretta USA
Jay Langston, Industry Expert
Dave Miller of CZ-USA
Mo Money Pawn Shop, Pheonix, Arizona
Rock Island Auction Company
Sturm, Ruger & Co., Inc.
Jared Sullivan, Editor of Legally Armed
Rick and Angela Uselton of UA Arms

icollector.com
Online Collectibles Auctions

Firearms & military collectibles, vintage collectibles, coins & paper money, memorabilia, jewelry, prints & paintings, art, antiques, and much more.

---

114 – 1750 Coast Meridian Road,
Port Coquitlam, British Columbia,
V3C 6R8, Canada
1-866-313-0123

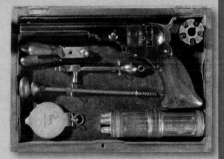

Also by 2nd Amendment Media:

ARMED & SMART

ARMED & SMARTER

LEGALLY ARMED

Look for our upcoming rifle and shotgun value guides.

Arcadia Machine and Tool (AMT)

Founded in Covina, California in the late 1970s, Arcadia Machine and Tool, commonly referred to as AMT, first gained recognition for their series of AutoMag pistols. The AutoMags were chambered in a host of different calibers, ranging from .22 Magnum to .50 AE and became popular with shooting enthusiasts, contributing to their sustained collectability. High Standard Manufacturing of Houston, Texas bought the company in the early 2000s and continues to manufacture and distribute some AMT models.

BACKUP
VERY GOOD+

BACKUP   REDBOOK CODE: RB-AR-H-BACKUP

Black or matte chrome finish, manual thumb safety, High-Standard Manufacturing still produces a variant of the weapon.

Production: discontinued		Caliber: .357 Sig, .40 S&W, .45 ACP, 9mm, .22LR, .380 ACP			Action: DAO, semi-auto		Barrel Length: 2.5"		Capacity: 5, 6, 7		
Magazine: detachable box		Grips: black synthetic									
D2C:	NIB	$ 475	Ex	$ 375	VG+	$ 265	Good	$ 240			
C2C:	Mint	$ 460	Ex	$ 350	VG+	$ 240	Good	$ 215	Fair	$ 110	
Trade-In:	Mint	$ 340	Ex	$ 275	VG+	$ 190	Good	$ 170	Poor	$ 60	

AUTOMAG V   REDBOOK CODE: RB-AR-H-AUTOMG

Slide-mounted safety, matte chrome finish, muzzle brake, less than 3,500 produced.

Production: early 1990s - mid-1990s			Caliber: .50 AE, .440 Cor-Bon								
Action: SA, semi-auto		Barrel Length: 6.5		Sights: adjustable		Capacity: 5					
Magazine: detachable box		Grips: black synthetic									
D2C:	NIB	$ 845	Ex	$ 650	VG+	$ 465	Good	$ 425			
C2C:	Mint	$ 820	Ex	$ 600	VG+	$ 430	Good	$ 385	Fair	$ 195	
Trade-In:	Mint	$ 600	Ex	$ 475	VG+	$ 330	Good	$ 300	Poor	$ 90	

HARDBALLER GOVERNMENT
VERY GOOD+

HARDBALLER GOVERNMENT   REDBOOK CODE: RB-AR-H-HARDBL

Frame-mounted safety, matte stainless finish. Accelerator, Commando, Javelina, and Longslide variants of the gun have differing barrel lengths, calibers, and magazine capacities.

Production: late 1970s - early 2000s			Caliber: .45 ACP		Action: SA, semi-auto						
Barrel Length: 5"		Sights: blade front, adjustable rear			Capacity: 7						
Magazine: detachable box		Grips: black synthetic									
D2C:	NIB	$ 890	Ex	$ 700	VG+	$ 490	Good	$ 445			
C2C:	Mint	$ 860	Ex	$ 625	VG+	$ 450	Good	$ 405	Fair	$ 205	
Trade-In:	Mint	$ 640	Ex	$ 500	VG+	$ 350	Good	$ 315	Poor	$ 90	

ARMALITE

ArmaLite, Inc.

ArmaLite, Inc. began manufacturing firearms in 1954 as a subsidiary of the Fairchild Engine and Airplane Corporation, though the company has since changed hands. The company set out to produce high quality, alloy firearms, seeing a void for such weapons in both commercial and military markets.

Now located in Geneseo, Illinois, the company is best known for their variety of high-power rifles, though their AR-24 pistol has gained attention over the years for its durability and heft. Eugene Stoner, the noted gun designer, famously worked for ArmaLite when he designed the AR-10 and later the AR-15, likely the most successful and popular semi-automatic rifle ever produced.

AR-24
EXCELLENT

AR-24 REDBOOK CODE: RB-AL-H-AR2401
Frame-mounted safety, manganese phosphate finish, six groove rifling.

Production: discontinued Caliber: 9mm Action: DA/SA, semi-auto

Barrel Length: 4.8" OA Length: 8.3" Wt.: 35 oz.

Sights: dovetail, 3-dot Capacity: 15 Magazine: detachable box Grips: black synthetic

D2C:	NIB	$ 525	Ex	$ 400	VG+	$ 290	Good	$ 265	LMP:	$ 550
C2C:	Mint	$ 510	Ex	$ 375	VG+	$ 270	Good	$ 240	Fair	$ 125
Trade-In:	Mint	$ 380	Ex	$ 300	VG+	$ 210	Good	$ 185	Poor	$ 60

AR-24K REDBOOK CODE: RB-AL-H-AR2402
A compact variant of the AR-24.

Production: discontinued Caliber: 9mm Action: DA/SA, semi-auto

Barrel Length: 3.89" Sights: dovetail, 3-dot Capacity: 13

Magazine: detachable box Grips: black synthetic

D2C:	NIB	$ 600	Ex	$ 475	VG+	$ 350	Good	$ 330	LMP:	$ 629
C2C:	Mint	$ 580	Ex	$ 425	VG+	$ 325	Good	$ 240	Fair	$ 300
Trade-In:	Mint	$ 470	Ex	$ 375	VG+	$ 280	Good	$ 210	Poor	$ 155

Arms Tech Ltd.

Incorporated in 1987, Arms Tech Ltd. designs and manufactures an array of firearms and ordnance catering to military and law enforcement standards. The company is currently located in Phoenix, Arizona in a ISO 9001 certified facility.

ARMS TECH

**OSS
HI-STANDARD**
NEW IN BOX

OSS HI-STANDARD REDBOOK CODE: RB-AT-HOSSHI1
Matte steel finish, steel frame and parts, blowback action design, a reproduction of the Hi-Standard pistol used during WWII.

Production: 2011-current Caliber: .22 LR Action: SA, semi-auto

Barrel Length: 4.75", 6.75" Sights: fixed front, adjustable rear Capacity: 10

Magazine: detachable box Grips: checkered walnut

D2C:	NIB	$ 2,200	Ex	$ 1,675	VG+	$ 1,210	Good	$ 1,100	LMP:	$ 2,495
C2C:	Mint	$ 2,120	Ex	$ 1,525	VG+	$ 1,100	Good	$ 990	Fair	$ 510
Trade-In:	Mint	$ 1,570	Ex	$ 1,250	VG+	$ 860	Good	$ 770	Poor	$ 240

Auto-Ordnance Ltd.

Now a division of Saelio Entreprises Inc. (who also owns Kahr Arms), John Blish, John Thompson, and Thomas Ryan founded Auto-Ordnance in 1916 and quickly became a major force in the firearms industry. The company supplied the famed Thompson Sub-Machine Gun for Allied military use in the Second World War and the gun's involvment in organized crime during the prohibition era secured its place in American lore. The company currently produces, among others, a line of quality M1911-style pistols, noted for their affordability and performance.

1911-A1 WWII PARKERIZED PKZSEW
NEW IN BOX

1911-A1 WWII PARKERIZED PKZSEW

REDBOOK CODE: RB-AO-H-191101
Similar to the original "G.I.-style" 1911-A1. Parkerized finish, grips marked "U.S."

Production: current Caliber: .45 ACP Action: SA, semi-auto									
Barrel Length: 5" Wt.: 39 oz. Sights: blade front, drift adjustable rear									
Capacity: 7 Magazine: detachable box Grips: wood, checkered									
D2C:	NIB	$ 505	Ex	$ 400	VG+	$ 280	Good	$ 255	LMP: $ 685
C2C:	Mint	$ 490	Ex	$ 350	VG+	$ 260	Good	$ 230	Fair $ 120
Trade-In:	Mint	$ 360	Ex	$ 300	VG+	$ 200	Good	$ 180	Poor $ 60

1911-A1 WWII PARKERIZED PKZSA
NEW IN BOX

1911-A1 WWII PARKERIZED PKZSA

REDBOOK CODE: RB-AO-H-191102
Similar to the original "G.I.-style" 1911-A1. Parkerized finish, checkered grips.

Production: current Caliber: .45 ACP Action: SA, semi-auto									
Barrel Length: 5" Wt.: 39 oz. Sights: blade front, drift adjustable rear									
Capacity: 7 Magazine: detachable box Grips: wood, checkered									
D2C:	NIB	$ 495	Ex	$ 400	VG+	$ 275	Good	$ 250	LMP: $ 668
C2C:	Mint	$ 480	Ex	$ 350	VG+	$ 250	Good	$ 225	Fair $ 115
Trade-In:	Mint	$ 360	Ex	$ 300	VG+	$ 200	Good	$ 175	Poor $ 60

1911-A1 WWII COMMEMORATIVE
NEW IN BOX

Courtesy of Rock Island Auction Company

1911-A1 WWII COMMEMORATIVE

REDBOOK CODE: RB-AO-H-WWIICO
Only 2,500 produced. Hand finished deluxe walnut grips, etched gold-glit marking inlays, 24k gold plating on controls.

Production: discontinued Caliber: .45 ACP Action: SA, semi-auto									
Barrel Length: 5" Wt.: 39 oz. Sights: blade front, drift adjustable rear Capacity: 7									
Magazine: detachable box Grips: deluxe walnut with medallions									
D2C:	NIB	$1,050	Ex	$ 800	VG+	$ 580	Good	$ 525	
C2C:	Mint	$1,010	Ex	$ 725	VG+	$ 530	Good	$ 475	Fair $ 245
Trade-In:	Mint	$ 750	Ex	$ 600	VG+	$ 410	Good	$ 370	Poor $ 120

1911TC REDBOOK CODE: RB-AO-H-191104

Bright stainless finish, stainless steel frame and slide, grip safety and thumb safety, skeletonized trigger and hammer.

1911TC
NEW IN BOX

Production: current Caliber: .45 ACP Action: SA, semi-auto									
Barrel Length: 5" OA Length: 8.5" Wt.: 39 oz. Sights: fixed, low-profile									
Capacity: 7 Magazine: detachable box Grips: laminate, checkered									
D2C:	NIB	$ 640	Ex	$ 500	VG+	$ 355	Good	$ 320	LMP: $ 813
C2C:	Mint	$ 620	Ex	$ 450	VG+	$ 320	Good	$ 290	Fair $ 150
Trade-In:	Mint	$ 460	Ex	$ 375	VG+	$ 250	Good	$ 225	Poor $ 90

1911 A1 STANDARD 80 REDBOOK CODE: RB-AO-H-A10004
Manual grip safety, blue finish, steel frame.

Production: early 2000s - 2008 Caliber: .45 ACP Action: SA, semi-auto								
Barrel Length: 5" OA Length: 8.5" Sights: blade front, adjustable rear								
Capacity: 7 Magazine: detachable box Grips: synthetic, checkered								
D2C:	Mint	$ 500	Ex $ 400	VG+ $ 275	Good $ 250			
C2C:	NIB	$ 480	Ex $ 350	VG+ $ 250	Good $ 225	Fair $	115	
Trade-In:	Mint	$ 360	Ex $ 300	VG+ $ 200	Good $ 175	Poor $	60	

Beretta

One of the most renowned and famed companies in the history of firearms, Fabbrica d'Armi Pietro Beretta, or simply Beretta, has manufactured guns and weaponry since 1526, making it the oldest firearms manufacturer in the world. The Model 1923 stands as the company's first successful semi-automatic pistol, serving as the standard-issued sidearm of the Italian military until the mid-1940s and helping lead to the development of the modern-era pistol. While the company's presence in the handgun industry waned during the mid-twentieth century, the United States adopted the M9, the service model variant of the Beretta 92, for military use in the 1980s, which led to a resurgence in the company's popularity across the globe. The Beretta 92 remains in the canon of the world's most important pistols, and the company still commands a significant share of the handgun market today. The company is headquartered in Brescia, Italy and is still owned and operated by members of the Beretta family.

MODEL 1915
FAIR

MODEL 1915 REDBOOK CODE: RB-BR-H-M1915X
The first pistol Beretta produced, eventually replaced by the Model 1915/1919. Debate persists regarding its production dates, subsequent variations of the gun saw use in World War I and II.

Production: 1915 - early 1920s Caliber: 7.65mm, 9mm Glisenti								
Action: SA, semi-auto Barrel Length: 3.5" Sights: fixed Capacity: 7								
Magazine: detachable box Grips: walnut								
D2C:	NIB	--	Ex $1,100	VG+ $ 785	Good $ 715			
C2C:	Mint	--	Ex $1,000	VG+ $ 720	Good $ 645	Fair $	330	
Trade-In:	Mint	--	Ex $ 800	VG+ $ 560	Good $ 500	Poor $	150	

MODEL 1915/1919 REDBOOK CODE: RB-BR-H-191519
Similar to the Model 1915, but features an altered barrel design, safety lever, and an extended exposed portion on the slide. Debate persists regarding its production dates.

Production: ~1919 - ~1930 Caliber: 7.65mm Action: SA, semi-auto								
Barrel Length: 3.5" Sights: fixed Capacity: 7 Magazine: detachable box								
Grips: walnut								
D2C:	NIB	--	Ex $ 425	VG+ $ 310	Good $ 280			
C2C:	Mint	--	Ex $ 400	VG+ $ 280	Good $ 250	Fair $	130	
Trade-In:	Mint	--	Ex $ 325	VG+ $ 220	Good $ 195	Poor $	60	

MODEL 1922 REDBOOK CODE: RB-BR-H-1922XX

A variation of the Model 1915, some of these models issued for Navy use and sold at a premium. A precursor to the Model 1923, not chambered for 9mm rounds. Rare, little information exists regarding the model.

Production: 1922 Caliber: 7.65mm Action: SA, semi-auto

Barrel Length: 4" Sights: fixed Capacity: 7 Magazine: detachable box

Grips: walnut, steel

D2C:	NIB	--	Ex	$ 550	VG+ $ 400	Good	$ 360		
C2C:	Mint	--	Ex	$ 500	VG+ $ 360	Good	$ 325	Fair $	170
Trade-In:	Mint	--	Ex	$ 475	VG+ $ 290	Good	$ 255	Poor $	90

MODEL 1923 REDBOOK CODE: RB-BR-H-1923XX

The most popular version of the initial Model 1915 design. Standard service pistol for the Italian Army from 1923 until 1945. Saw extensive use with Italian forces throughout both World Wars. Features an extended opening on top of the slide and frame-mounted safety. Around 10,000 produced.

MODEL 1923

Production: 1923 - 1935 Caliber: 9mm Glisenti Action: SA, semi-auto

Barrel Length: 4" OA Length: 7" Wt.: 28 oz. Sights: fixed Capacity: 7

Magazine: detachable box Grips: steel

D2C:	NIB	--	Ex	$1,450	VG+ $1,045	Good	$ 950		
C2C:	Mint	--	Ex	$1,325	VG+ $ 950	Good	$ 855	Fair $	440
Trade-In:	Mint	--	Ex	$1,075	VG+ $ 750	Good	$ 665	Poor $	210

MODEL 1926 REDBOOK CODE: RB-BR-H-1926XX

Serial numbers range between 187,000 and 198,000. A transition model based off the earlier Model 1919 design, noted for its dolphin-fin front sight, disconnector switch on the frame above the trigger, and circled "PB" emblem in the grip.

Production: 1926 - discontinued Caliber: 9mm Glisenti Action: SA, semi-auto

Barrel Length: 3.5" Sights: fixed, dolphin-fin Capacity: 7

Magazine: detachable box Grips: steel

D2C:	NIB	--	Ex	$ 500	VG+ $ 350	Good	$ 315		
C2C:	Mint	--	Ex	$ 450	VG+ $ 320	Good	$ 285	Fair $	145
Trade-In:	Mint	--	Ex	$ 375	VG+ $ 250	Good	$ 225	Poor $	90

MODEL 1931

MODEL 1931 REDBOOK CODE: RB-BR-H-1931XX

Similar to the earlier Model 1915/1919 design. "RM" emblem in the grip, issued to the Royal Navy, approximately 8,000 units produced. Italian navy models are rare and are priced at a premium.

Production: 1931 Caliber: 7.65mm Action: SA, semi-auto Barrel Length: 3.5"

Sights: fixed Capacity: 8 Magazine: detachable box Grips: walnut

D2C:	NIB	--	Ex	$ 575	VG+ $ 400	Good	$ 365		
C2C:	Mint	--	Ex	$ 525	VG+ $ 370	Good	$ 330	Fair $	170
Trade-In:	Mint	--	Ex	$ 425	VG+ $ 290	Good	$ 255	Poor $	90

MODEL 1934
GOOD

MODEL 1934   REDBOOK CODE: RB-BR-H-1934XX

Saw extensive use in World War II, over a million manufactured. Post-war models have serial numbers beginning at C00001, models marked "RE," "RA," and "RM" have somewhat increased value.

Production: 1934 - late 1950s   Caliber: 9mm Short, .380 ACP   Action: SA, semi-auto

Barrel Length: 3.7"   OA Length: 6"   Sights: fixed   Capacity: 7

Magazine: detachable box   Grips: plastic

D2C:	NIB	--	Ex $ 500	VG+ $ 355	Good $ 325		
C2C:	Mint	--	Ex $ 450	VG+ $ 330	Good $ 295	Fair $ 150
Trade-In:	Mint	--	Ex $ 375	VG+ $ 260	Good $ 230	Poor $ 90

MODEL 1935
EXCELLENT

MODEL 1935   REDBOOK CODE: RB-BR-H-1935XX

Blue finish, simplified design, frame-mounted safety, similar to the Model 1934.

Production: 1935 - late 1950s   Caliber: 7.65mm, .32 ACP   Action: SA, semi-auto

Barrel Length: 3.7"   Sights: fixed   Capacity: 8   Magazine: detachable box

Grips: black plastic

D2C:	NIB	--	Ex $ 475	VG+ $ 340	Good $ 310		
C2C:	Mint	--	Ex $ 425	VG+ $ 310	Good $ 280	Fair $ 145
Trade-In:	Mint	--	Ex $ 350	VG+ $ 240	Good $ 220	Poor $ 90

MODEL 318
GOOD

MODEL 318   REDBOOK CODE: RB-BR-H-M318XX

A modified and smaller version of the earlier Model 1919 design, sometimes called the "Panther." Blue finish, partially shrouded barrel.

Production: 1935 - late 1940s   Caliber: .25 ACP   Action: SA, semi-auto

Barrel Length: 2.5"   Sights: fixed   Capacity: 7   Magazine: detachable box

Grips: black plastic

D2C:	NIB	--	Ex $ 500	VG+ $ 350	Good $ 315		
C2C:	Mint	--	Ex $ 450	VG+ $ 320	Good $ 285	Fair $ 145
Trade-In:	Mint	--	Ex $ 375	VG+ $ 250	Good $ 225	Poor $ 90

MODEL 418
VERY GOOD

MODEL 418   REDBOOK CODE: RB-BR-H-M418XX

Often called the "Bantam," available in several variations. Various finishes, engraved frame and slide, multi-color grips, extended barrel, similar to the Model 318.

Production: late 1950s   Caliber: .25 ACP   Action: SA, semi-auto

Barrel Length: 2.5"   Sights: fixed   Capacity: 7, 8   Magazine: detachable box

Grips: plastic

D2C:	NIB	--	Ex $ 625	VG+ $ 440	Good $ 400		
C2C:	Mint	--	Ex $ 575	VG+ $ 400	Good $ 360	Fair $ 185
Trade-In:	Mint	--	Ex $ 450	VG+ $ 320	Good $ 280	Poor $ 90

MODEL 20
GOOD

MODEL 20   REDBOOK CODE: RB-BR-H-M20XXX

Engraved slide and frame, blue finish, enhanced grip design, frame-mounted safety, compact design.

Production: mid-1980s   Ca3liber: .25 ACP   Action: DA/SA, semi-auto

Barrel Length: 2.5"   Sights: fixed   Capacity: 9   Magazine: detachable box

Grips: plastic, ivory

D2C:	NIB $ 360	Ex $ 275	VG+ $ 200	Good $ 180		
C2C:	Mint $ 350	Ex $ 250	VG+ $ 180	Good $ 165	Fair $ 85
Trade-In:	Mint $ 260	Ex $ 225	VG+ $ 150	Good $ 130	Poor $ 60

MODEL 70
(MODEL 100)
EXCELLENT

MODEL 70 (MODEL 100) REDBOOK CODE: RB-BR-H-M70XXX

Frame-mounted safety, blue finish, alloy frame, available in a host of different variants. Commonly referred to as "Puma."

Production: late 1950s - mid-1980s Caliber: .32 ACP, .380 ACP, .22 LR										
Action: SA, semi-auto Barrel Length: 3.5", 6" Sights: fixed or adjustable										
Capacity: 8 Magazine: detachable box Grips: brown or black plastic										
D2C:	NIB	$ 485	Ex	$ 375	VG+	$ 270	Good	$ 245		
C2C:	Mint	$ 470	Ex	$ 350	VG+	$ 250	Good	$ 220	Fair	$ 115
Trade-In:	Mint	$ 350	Ex	$ 275	VG+	$ 190	Good	$ 170	Poor	$ 60

MODEL 71
VERY GOOD +

MODEL 71 REDBOOK CODE: RB-BR-H-M71XXX

Enhanced and modified grips, alloy frame, frame-mounted safety, cross-bolt safety, some models marked "Jaguar," blue finish.

Production: 1960s - late 1980s Caliber: .22 LR Action: SA, semi-auto										
Barrel Length: 3.5", 4.75", 6" Wt.: 16 oz. Sights: fixed or adjustable										
Capacity: 10 Magazine: detachable box Grips: synthetic										
D2C:	NIB	$ 435	Ex	$ 350	VG+	$ 240	Good	$ 220		
C2C:	Mint	$ 420	Ex	$ 325	VG+	$ 220	Good	$ 200	Fair	$ 105
Trade-In:	Mint	$ 310	Ex	$ 250	VG+	$ 170	Good	$ 155	Poor	$ 60

MODEL 72 REDBOOK CODE: RB-BR-H-M72XXX

Similar to the Model 71, blue finish, frame-mounted safety.

Production: mid-1980s Caliber: .22 LR Action: SA, semi-auto										
Barrel Length: 3.5", 6" Sights: fixed or adjustable Capacity: 10										
Magazine: detachable box Grips: synthetic										
D2C:	NIB	$ 465	Ex	$ 375	VG+	$ 260	Good	$ 235		
C2C:	Mint	$ 450	Ex	$ 325	VG+	$ 240	Good	$ 210	Fair	$ 110
Trade-In:	Mint	$ 340	Ex	$ 275	VG+	$ 190	Good	$ 165	Poor	$ 60

MODEL 74 REDBOOK CODE: RB-BR-H-M74XXX

Similar to the Model 72, but features an adjustable rear sight.

Production: mid-1980s Caliber: .22 LR Action: SA, semi-auto										
Barrel Length: 6" OA Length: 9.25" Wt.: 22.4 oz.										
Sights: fixed front, rear adjustable Capacity: 10 Magazine: detachable box										
Grips: synthetic or wood										
D2C:	NIB	$ 475	Ex	$ 375	VG+	$ 265	Good	$ 240		
C2C:	Mint	$ 460	Ex	$ 350	VG+	$ 240	Good	$ 215	Fair	$ 110
Trade-In:	Mint	$ 340	Ex	$ 275	VG+	$ 190	Good	$ 170	Poor	$ 60

MODEL 76
(MODEL 102)
VERY GOOD +

MODEL 76 (MODEL 102) REDBOOK CODE: RB-BR-H-M76XXX

Aluminum sleeve, blue finish, thumb rest in the grip, frame-mounted safety, skeletonized hammer.

Production: mid-1980s Caliber: .22 LR Action: SA, semi-auto										
Barrel Length: 6" Sights: adjustable Capacity: 10 Magazine: detachable box										
Grips: wood										
D2C:	NIB	$ 545	Ex	$ 425	VG+	$ 300	Good	$ 275		
C2C:	Mint	$ 530	Ex	$ 400	VG+	$ 280	Good	$ 250	Fair	$ 130
Trade-In:	Mint	$ 390	Ex	$ 325	VG+	$ 220	Good	$ 195	Poor	$ 60

MODEL 81/82 (CHEETAH) REDBOOK CODE: RB-BR-H-M8182X

Blue finish, slide-mounted safety, aluminum frame, blowback action, double-stack magazine, rounded trigger guard, available in several different variations. Often referred to by its original model name "Series 81,) or as the "Beretta Cheetah."

Production: late 1970s	Caliber: .32 ACP	Action: DA/SA, semi-auto			
Barrel Length: 3.8"	OA Length: 6.8"	Wt.: 24 oz.	Sights: 3-dot		
Capacity: 10, 12, 13	Magazine: detachable box	Grips: wood or black synthetic			
D2C:	NIB $ 530	Ex $ 425	VG+ $ 295	Good $ 265	
C2C:	Mint $ 510	Ex $ 375	VG+ $ 270	Good $ 240	Fair $ 125
Trade-In:	Mint $ 380	Ex $ 300	VG+ $ 210	Good $ 190	Poor $ 60

MODEL 83/84/85
EXCELLENT

MODEL 83/84/85 (CHEETAH) REDBOOK CODE: RB-BR-H-838485

Included in the Series 81, nickel or blue finish, gold accents, minor variations between models.

Production: 1980s - discontinued	Caliber: .380 ACP				
Action: DA/SA, semi-auto	Barrel Length: 3.8", 4"				
OA Length: 6.7", 7"	Sights: fixed front	Capacity: 8, 13			
Magazine: detachable box	Grips: wood or black synthetic				
D2C:	NIB $ 585	Ex $ 450	VG+ $ 325	Good $ 295	
C2C:	Mint $ 570	Ex $ 425	VG+ $ 300	Good $ 265	Fair $ 135
Trade-In:	Mint $ 420	Ex $ 350	VG+ $ 230	Good $ 205	Poor $ 60

MODEL 86 (CHEETAH) REDBOOK CODE: RB-BR-H-M86XXX

Differs somewhat from the other Series 81 pistols. Hard-chromed barrel, frame-mounted safety, muzzle open breech capabilities, modified barrel design, blue finish.

Production: 1991-1994	Caliber: .380 ACP	Action: DA/SA, semi-auto			
Barrel Length: 4.5"	Sights: 3-dot	Capacity: 8	Magazine: detachable box		
Grips: wood					
D2C:	NIB $ 645	Ex $ 500	VG+ $ 355	Good $ 325	LMP $ 615
C2C:	Mint $ 620	Ex $ 450	VG+ $ 330	Good $ 295	Fair $ 150
Trade-In:	Mint $ 460	Ex $ 375	VG+ $ 260	Good $ 230	Poor $ 90

MODEL 87 (CHEETAH) REDBOOK CODE: RB-BR-H-M87XXX

Hard-chromed barrel, blue finish, frame-mounted safety, blowback action, skeletonized hammer. Target versions of the gun feature a tactical design and larger magazines.

Production: early 2000s - current	Caliber: .22 LR	Action: DA/SA, semi-auto			
Barrel Length: 6"	Sights: 3-dot or adjustable	Capacity: 7, 10			
Magazine: detachable box	Grips: wood or black synthetic				
D2C:	NIB $ 680	Ex $ 525	VG+ $ 375	Good $ 340	LMP $ 845
C2C:	Mint $ 660	Ex $ 475	VG+ $ 340	Good $ 310	Fair $ 160
Trade-In:	Mint $ 490	Ex $ 400	VG+ $ 270	Good $ 240	Poor $ 90

MODEL 87 TARGET
NEW IN BOX

MODEL 87 TARGET REDBOOK CODE: RB-BR-H-87TRGT

Aluminum frame, scope mounts, blue finish, hard-chromed barrel, blowback design, large-bodied slide and frame.

Production: 2000 - current	Caliber: .22 LR	Action: SA, semi-auto			
Barrel Length: 6"	Wt.: 40 oz.	Sights: target	Capacity: 10	Magazine: detachable box	
Grips: synthetic					
D2C:	NIB $ 825	Ex $ 650	VG+ $ 455	Good $ 415	LMP $ 880
C2C:	Mint $ 800	Ex $ 575	VG+ $ 420	Good $ 375	Fair $ 190
Trade-In:	Mint $ 590	Ex $ 475	VG+ $ 330	Good $ 290	Poor $ 90

**MODEL 89
GOLD STANDARD**
EXCELLENT

MODEL 89 GOLD STANDARD REDBOOK CODE: RB-BR-H-M89XXX

Blowback action, aluminum frame, frame-mounted safety, blue finish, considered one of the finest .22 LR pistols ever produced.

Production: late 1980s Caliber: .22 LR Action: DA/SA, semi-auto

Barrel Length: 6" Wt.: 40 oz. Sights: adjustable target Capacity: 10

Magazine: detachable box Grips: wood, competition-style

D2C:	NIB	$ 775	Ex $ 600	VG+ $ 430	Good $ 390	LMP $ 802
C2C:	Mint	$ 750	Ex $ 550	VG+ $ 390	Good $ 350	Fair $ 180
Trade-In:	Mint	$ 560	Ex $ 450	VG+ $ 310	Good $ 275	Poor $ 90

MODEL 948
GOOD

MODEL 948 REDBOOK CODE: RB-BR-H-M948XX

Similar to the Beretta Model 1934/1935, but has limited value.

Production: ~1950s - ~1960s Caliber: .22 LR Action: DA/SA, semi-auto

Barrel Length: 3.5", 6" Wt.: 16 oz. Sights: fixed front, adjustable rear

Capacity: 9 Magazine: detachable box Grips: plastic

D2C:	NIB	$ 315	Ex $ 250	VG+ $ 175	Good $ 160	
C2C:	Mint	$ 310	Ex $ 225	VG+ $ 160	Good $ 145	Fair $ 75
Trade-In:	Mint	$ 230	Ex $ 200	VG+ $ 130	Good $ 115	Poor $ 60

MODEL 950
GOOD

MODEL 950 REDBOOK CODE: RB-BR-H-M950XX

Exists in several different variations, including the Minx, Jetfire, Jetfire Inox, and 950B models, all with similar value. Blue or stainless finish, alloy frame, steel slide and barrel, frame-mounted safety.

Production: mid-1950s - ~2003 Caliber: .22 LR, .25 ACP Action: DA/SA, semi-auto

Barrel Length: 2.25", 2.5", 4" Sights: fixed Capacity: 8 Magazine: detachable box

Grips: synthetic

D2C:	NIB	$ 335	Ex $ 275	VG+ $ 185	Good $ 170	
C2C:	Mint	$ 330	Ex $ 250	VG+ $ 170	Good $ 155	Fair $ 80
Trade-In:	Mint	$ 240	Ex $ 200	VG+ $ 140	Good $ 120	Poor $ 60

MODEL 90
GOOD

Courtesy of Bud's Gun Shop

MODEL 90 REDBOOK CODE: RB-BR-H-M90XXX

6-groove barrel, steel and alloy components, anodized blue finish, frame-mounted safety.

Production: late 1960s - 1983 Caliber: .32 ACP Action: DAO, semi-auto

Barrel Length: 3.5" Sights: fixed Capacity: 8 Magazine: detachable box

Grips: synthetic

D2C:	NIB	$ 470	Ex $ 375	VG+ $ 260	Good $ 235	
C2C:	Mint	$ 460	Ex $ 325	VG+ $ 240	Good $ 215	Fair $ 110
Trade-In:	Mint	$ 340	Ex $ 275	VG+ $ 190	Good $ 165	Poor $ 60

MODEL 951
(BRIGADIER)
GOOD

MODEL 951 (BRIGADIER) REDBOOK CODE: RB-BR-H-M951XX

The first Beretta pistol to feature a locked-breech action, modeled after existing Walther designs. Saw use with Italian military forces. Frame-mounted hammer lock, blue finish. Known as the "Brigadier" on the civilian market.

Production: early 1950s - 1980s Caliber: 9mm, 7.65x22mm Action: SA, semi-auto

Barrel Length: 4.5" OA Length: 8" Wt.: 32 oz.

Sights: fixed front, notch rear Capacity: 8 Magazine: detachable box Grips: black synthetic

D2C:	NIB	$ 470	Ex $ 375	VG+ $ 260	Good $ 235	
C2C:	Mint	$ 460	Ex $ 325	VG+ $ 240	Good $ 215	Fair $ 110
Trade-In:	Mint	$ 340	Ex $ 275	VG+ $ 190	Good $ 165	Poor $ 60

MODEL 8000/8040/ 8045 /8357
(COUGAR SERIES)
NEW IN BOX

MODEL 8000/8040/8045 /8357 (COUGAR SERIES)
REDBOOK CODE: RB-BR-H-8000XX

This is a smaller version of the popular Beretta Model 92. Various models of the gun exist: F (standard version), D (DAO, no manual safety), G (decocker-safety feature), and Inox (stainless slide and barrel). Of these, the Inox has an increased value of $100 or so. Blue or stainless finish, alloy frame, ambidextrous safety (unless already noted), exposed hammer (unless already noted), saw both civilian and law enforcement use, often called the "Cougar."

Production: mid-1950s - 2005   Caliber: 9mm, .357 Sig, .45 ACP, .40 S&W

Action: DA/SA, DAO, semi-auto   Barrel Length: 3.6"   Sights: fixed

Capacity: 10   Magazine: detachable box   Grips: synthetic or wood

D2C:	NIB	$ 585	Ex $ 450	VG+ $ 325	Good $ 295		
C2C:	Mint	$ 570	Ex $ 425	VG+ $ 300	Good $ 265	Fair	$ 135
Trade-In:	Mint	$ 420	Ex $ 350	VG+ $ 230	Good $ 205	Poor	$ 60

MODEL 8000/8040/8045 MINI   REDBOOK CODE: RB-BR-H-8000MM

A smaller variation of the Model 8000 Series. Blue finish, decocker safety, modern design, the Model 8000 fires 9mm rounds, the Model 8045 .45 ACP, Model 8040 .40 S&W, similar values. Referred to as the "Mini Cougar."

Production: late 1990s - 2005   Caliber: .40 S&W, .45 ACP, 9mm

Action: DA/SA, semi-auto   Barrel Length: 3.6", 3.7"   Sights: 3-dot

Capacity: 8 (.40 S&W), 10 (9mm)   Magazine: detachable box   Grips: synthetic or wood

D2C:	NIB	$ 565	Ex $ 450	VG+ $ 315	Good $ 285		
C2C:	Mint	$ 550	Ex $ 400	VG+ $ 290	Good $ 255	Fair	$ 130
Trade-In:	Mint	$ 410	Ex $ 325	VG+ $ 230	Good $ 200	Poor	$ 60

MODEL 92   REDBOOK CODE: RB-BR-H-M92XX1

Serial numbers for these are dated by two letters, not numerically. These guns' serial markings will start with an "A" then a second letter corresponding to their specific year of production, starting with "A." For example: AA=1975, AB=1976. Approximately 50,000 units made and feature blue finish, a frame-mounted safety, and alloy frame.

Production: ~1975 - 1976   Caliber: 9mm   Action: DA/SA, semi-auto

Barrel Length: 5"   OA Length: 8.5"   Wt.: 32 oz.   Sights: fixed

Capacity: 16   Magazine: detachable box   Grips: black synthetic

D2C:	NIB	$ 880	Ex $ 675	VG+ $ 485	Good $ 440		
C2C:	Mint	$ 850	Ex $ 625	VG+ $ 440	Good $ 400	Fair	$ 205
Trade-In:	Mint	$ 630	Ex $ 500	VG+ $ 350	Good $ 310	Poor	$ 90

MODEL 92S
EXCELLENT

MODEL 92S   REDBOOK CODE: RB-BR-H-M92XX2

Serial numbers between AC and AE. Features a slide-mounted safety unlike the original Model 92.

Production: ~1977 - late 1970s   Caliber: 9mm   Action: DA/SA, semi-auto

Barrel Length: 5"   OA Length: 8.5"   Wt.: 32 oz.   Sights: fixed

Capacity: 15   Magazine: detachable box   Grips: black synthetic or wood

D2C:	NIB	$ 835	Ex $ 650	VG+ $ 460	Good $ 420		
C2C:	Mint	$ 810	Ex $ 600	VG+ $ 420	Good $ 380	Fair	$ 195
Trade-In:	Mint	$ 600	Ex $ 475	VG+ $ 330	Good $ 295	Poor	$ 90

BERETTA

MODEL 92SB-P REDBOOK CODE: RB-BR-H-M92SBP

Serial numbers between AF and AN. Firing pin block, ambidextrous safety, magazine release just above the trigger guard, slide-mounted safety, initially produced for military and police use.

Production: 1980 - mid-1980s Caliber: 9mm Action: DA/SA, semi-auto

Barrel Length: 4.3", 5" Sights: fixed Capacity: 15 Magazine: detachable box

Grips: black synthetic or wood

D2C:	NIB	$ 580	Ex	$ 450	VG+	$ 320	Good	$ 290		
C2C:	Mint	$ 560	Ex	$ 425	VG+	$ 290	Good	$ 265	Fair $ 135	
Trade-In:	Mint	$ 420	Ex	$ 325	VG+	$ 230	Good	$ 205	Poor $ 60	

MODEL 92SB-P
GOOD

MODEL 92SB-P COMPACT REDBOOK CODE: RB-BR-H-M92SBC

A smaller version of the standard Model 92SB, models with nickel finish have increased value.

Production: 1980 - 1985 Caliber: 9mm Action: DA/SA, semi-auto

Barrel Length: 4.3" Sights: fixed Capacity: 14 Magazine: detachable box

Grips: black synthetic or wood

D2C:	NIB	$ 510	Ex	$ 400	VG+	$ 285	Good	$ 255	
C2C:	Mint	$ 490	Ex	$ 375	VG+	$ 260	Good	$ 230	Fair $ 120
Trade-In:	Mint	$ 370	Ex	$ 300	VG+	$ 200	Good	$ 180	Poor $ 60

MODEL 92FS SERIES
NEW IN BOX

MODEL 92FS SERIES REDBOOK CODE: RB-BR-H-M92FSS

Enlarged hammer pins, modified disassembly latch, slide-mounted safety, aluminum frame, squared trigger, angled dustcover, adjusted backstrap, easily changeable parts, ambidextrous magazine release, similar to the military issued Beretta M9 pistol.

Production: 1983 - current Caliber: 9mm Action: DA/SA, semi-auto

Barrel Length: 5" OA Length: 8.5" Sights: fixed Capacity: 15

Magazine: detachable box Grips: black synthetic or wood

D2C:	NIB	$ 575	Ex	$ 450	VG+	$ 320	Good	$ 290	
C2C:	Mint	$ 560	Ex	$ 400	VG+	$ 290	Good	$ 260	Fair $ 135
Trade-In:	Mint	$ 410	Ex	$ 325	VG+	$ 230	Good	$ 205	Poor $ 60

MODEL 92F/FS COMPACT
EXCELLENT

MODEL 92F/FS COMPACT REDBOOK CODE: RB-BR-H-M92FSC

A smaller version of the standard Model 92FS. Black finish, enlarged hammer pin, slide-mounted safety, enhanced trigger guard, hard chrome barrel.

Production: early 2000s Caliber: 9mm Action: DA/SA, semi-auto

Barrel Length: 4.3" Wt.:32 oz. Sights: 3-dot Capacity: 13, 14

Magazine: detachable box Grips: black synthetic or wood

D2C:	NIB	$ 555	Ex	$ 425	VG+	$ 310	Good	$ 280	
C2C:	Mint	$ 540	Ex	$ 400	VG+	$ 280	Good	$ 250	Fair $ 130
Trade-In:	Mint	$ 400	Ex	$ 325	VG+	$ 220	Good	$ 195	Poor $ 60

**MODEL 92F/FS
COMPACT "TYPE M"**
VERY GOOD +

MODEL 92FS COMPACT "TYPE M"

REDBOOK CODE: RB-BR-H-M92FSM

A smaller version of the Model 92FS. Slide-mounted safety, alloy frame, black finish. Distinguished by its slim frame and single-stack magazine.

Production: 1993 - 2003 Caliber: 9mm Action: DA/SA, semi-auto

Barrel Length: 4.3" OA Length: 7.8" Wt.: 32 oz. Sights: 3-dot

Capacity: 8 Magazine: detachable box Grips: black synthetic or wood

D2C:	NIB	$ 700	Ex	$ 550	VG+	$ 385	Good	$ 350	LMP $ 691
C2C:	Mint	$ 680	Ex	$ 500	VG+	$ 350	Good	$ 315	Fair $ 165
Trade-In:	Mint	$ 500	Ex	$ 400	VG+	$ 280	Good	$ 245	Poor $ 90

BERETTA

**MODEL 92FS
COMPACT "TYPE M"**
NEW IN BOX

MODEL 92FS BRIGADIER
NEW IN BOX

MODEL 92FS CENTURION   REDBOOK CODE: RB-BR-H-M92FCE

Aluminum frame, slide-mounted safety, black finish.

Production: 1993 - discontinued Caliber: 9mm Action: DA/SA, semi-auto
Barrel Length: 4.3" OA Length: 7.8" Wt.: 33.6 oz. Sights: 3-dot
Capacity: 15 Magazine: detachable box Grips: black synthetic

D2C:	NIB	$ 560	Ex $ 450	VG+ $ 310	Good $ 280	LMP $ 625
C2C:	Mint	$ 540	Ex $ 400	VG+ $ 280	Good $ 255	Fair $ 130
Trade-In:	Mint	$ 400	Ex $ 325	VG+ $ 220	Good $ 200	Poor $ 60

MODEL 92FS BRIGADIER   REDBOOK CODE: RB-BR-H-M92FRB

A variant of the Model 92FS, light-recoil design, black finish, aluminum frame, slide-mounted safety.

Production: late 1990s - discontinued Caliber: 9mm Action: DA/SA, semi-auto
Barrel Length: 5" OA Length: 8.5" Wt.: 35 oz. Sights: 3-dot
Capacity: 15 Magazine: detachable box Grips: black synthetic

D2C:	NIB	$ 650	Ex $ 510	VG+ $ 360	Good $ 325	LMP $ 793
C2C:	Mint	$ 630	Ex $ 450	VG+ $ 330	Good $ 295	Fair $ 150
Trade-In:	Mint	$ 470	Ex $ 375	VG+ $ 260	Good $ 230	Poor $ 90

MODEL 92F   REDBOOK CODE: RB-BR-H-92FXXX

Aluminum frame, short recoil, slide-mounted safety, double-stack magazine, denoted for government testing. Does not include an enlarged hammer pin like the FS models.

Production: ~1983 - discontinued Caliber: 9mm Action: DA/SA, semi-auto
Barrel Length: 5" OA Length: 8.5" Sights: 3-dot Capacity: 15
Magazine: detachable box Grips: black synthetic

D2C:	NIB	$ 565	Ex $ 450	VG+ $ 315	Good $ 285	LMP $ 675
C2C:	Mint	$ 550	Ex $ 400	VG+ $ 290	Good $ 255	Fair $ 130
Trade-In:	Mint	$ 410	Ex $ 325	VG+ $ 230	Good $ 200	Poor $ 60

MODEL 92FS
(U.S.A.)
NEW IN BOX

MODEL 92FS (U.S.A.)   REDBOOK CODE: RB-BR-H-92FSUS

Made in the United States and marked accordingly. Not drastically different than Italian versions of the gun. Debate persists regarding which version is superior. Aluminum frame, short recoil, slide-mounted safety.

Production: current Caliber: 9mm Action: DA/SA, semi-auto
Barrel Length: 5" OA Length: 8.5" Sights: 3-dot Capacity: 15
Magazine: detachable box Grips: black synthetic

D2C:	NIB	$ 575	Ex $ 450	VG+ $ 320	Good $ 290	LMP $ 650
C2C:	Mint	$ 560	Ex $ 400	VG+ $ 290	Good $ 260	Fair $ 135
Trade-In:	Mint	$ 410	Ex $ 325	VG+ $ 230	Good $ 205	Poor $ 60

MODEL 92FS VERTEC   REDBOOK CODE: RB-BR-H-92FSVT

Modified vertical and thinned-grip design, slide-mounted safety, aluminum frame, accessory rail, black finish.

Production: 2002 - discontinued Caliber: 9mm Action: DA/SA, semi-auto
Barrel Length: 4.7" OA Length: 8.3" Sights: fixed, removable front
Capacity: 10 Magazine: detachable box Grips: black synthetic

D2C:	NIB	$ 755	Ex $ 575	VG+ $ 420	Good $ 380	LMP $ 760
C2C:	Mint	$ 730	Ex $ 525	VG+ $ 380	Good $ 340	Fair $ 175
Trade-In:	Mint	$ 540	Ex $ 425	VG+ $ 300	Good $ 265	Poor $ 90

**MODEL 92FS 470TH
ANNIVERSARY**
NEW IN BOX

MODEL 92FS 470TH ANNIVERSARY   REDBOOK CODE: RB-BR-H-92F470

Commemorates Beretta's 470th year of manufacturing. Stainless finish, engraved frame, gold accents, unique emblems, and sold with a walnut display case. Only 470 produced.

Production: 1996 - discontinued Caliber: 9mm Action: DA/SA, semi-auto

Barrel Length: 5" OA Length: 8.5" Sights: fixed front Capacity: 10

Magazine: detachable box Grips: walnut

D2C:	NIB	$2,250	Ex	$1,725	VG+	$1,240	Good	$1,125		
C2C:	Mint	$2,160	Ex	$1,575	VG+	$1,130	Good	$1,015	Fair	$520
Trade-In:	Mint	$1,600	Ex	$1,275	VG+	$880	Good	$790	Poor	$240

MODEL 92EL TEXAS SPECIAL   REDBOOK CODE: RB-BR-H-92ELTX

Gold engraving and accents, black finish, frame-mounted safety, only 50 produced.

Production: discontinued Caliber: 9mm Action: DA/SA, semi-auto

Barrel Length: 5" Sights: fixed front Capacity: 10 Magazine: detachable box

Grips: black synthetic

D2C:	NIB	$2,100	Ex	$1,600	VG+	$1,155	Good	$1,050		
C2C:	Mint	$2,020	Ex	$1,450	VG+	$1,050	Good	$945	Fair	$485
Trade-In:	Mint	$1,500	Ex	$1,200	VG+	$820	Good	$735	Poor	$210

MODEL 92G-SD   REDBOOK CODE: RB-BR-H-M92GXX

Produced for French military and law enforcement use, similar to other Model 92s of the era. No manual safety.

Production: early 2000s - discontinued Caliber: 9mm Action: DA/SA, semi-auto

Barrel Length: 5" OA Length: 8.5" Wt.: 32 oz. Sights: fixed front

Capacity: 15 Magazine: detachable box

D2C:	NIB	$855	Ex	$650	VG+	$475	Good	$430		
C2C:	Mint	$830	Ex	$600	VG+	$430	Good	$385	Fair	$200
Trade-In:	Mint	$610	Ex	$500	VG+	$340	Good	$300	Poor	$90

MODEL 92G ELITE
NEW IN BOX

MODEL 92G ELITE   REDBOOK CODE: RB-BR-H-M92GE1

Produced in the late 1990s, features a flat hammer cap, no manual safety, blue finish. "Brigadier" options change engravings on the slide.

Production: late 1990s - discontinued Caliber: 9mm Action: DA/SA, semi-auto

Barrel Length: 4.7" Wt.: 34 oz. Sights: fixed front Capacity: 15

Magazine: detachable box

D2C:	NIB	$865	Ex	$675	VG+	$480	Good	$435		
C2C:	Mint	$840	Ex	$600	VG+	$440	Good	$390	Fair	$200
Trade-In:	Mint	$620	Ex	$500	VG+	$340	Good	$305	Poor	$90

MODEL 92G ELITE II   REDBOOK CODE: RB-BR-H-M92GE2

Designed for competitive shooting. Black grip and frame, matte stainless slide, slide-mounted safety.

Production: late 1990s - discontinued Caliber: 9mm Action: DA/SA, semi-auto

Barrel Length: 4.7" Wt.: 34 oz. Sights: fixed front Capacity: 10

Magazine: detachable box Grips: black synthetic

D2C:	NIB	$850	Ex	$650	VG+	$470	Good	$425		
C2C:	Mint	$820	Ex	$600	VG+	$430	Good	$385	Fair	$200
Trade-In:	Mint	$610	Ex	$500	VG+	$340	Good	$300	Poor	$90

MODEL 92D   REDBOOK CODE: RB-BR-H-M92DXX

Distinguished by not having a manual safety or exposed hammer. Black finish, aluminum frame.

Production: 1994 - late 1990s   Caliber: 9mm   Action: DAO, semi-auto

Barrel Length: 5"   OA Length: 8.5"   Wt.: 34 oz.   Sights: 3-dot

Capacity: 15   Magazine: detachable box   Grips: black synthetic

D2C:	NIB	$ 580	Ex $ 450	VG+ $ 320	Good $ 290	LMP $ 586
C2C:	Mint	$ 560	Ex $ 425	VG+ $ 290	Good $ 265	Fair $ 135
Trade-In:	Mint	$ 420	Ex $ 325	VG+ $ 230	Good $ 205	Poor $ 60

MODEL 92DS
VERY GOOD +

MODEL 92DS   REDBOOK CODE: RB-BR-H-M92DSX

Unlike the 92D, it features a slide-mounted safety. Black finish, aluminum frame, no external hammer, functions like the Model 92FS.

Production: 1994   Caliber: 9mm   Action: DA/SA, semi-auto   Barrel Length: 5"

OA Length: 8.5"   Sights: fixed   Capacity: 15   Magazine: detachable box

Grips: black synthetic

D2C:	NIB	$ 580	Ex $ 450	VG+ $ 320	Good $ 290	
C2C:	Mint	$ 560	Ex $ 425	VG+ $ 290	Good $ 265	Fair $ 135
Trade-In:	Mint	$ 420	Ex $ 325	VG+ $ 230	Good $ 205	Poor $ 60

MODEL 92A1
NEW IN BOX

MODEL 92A1   REDBOOK CODE: RB-BR-H-M92A1X

Alloy frame, black finish, slide-mounted safety, accessory rail, skeletonized hammer, developed for the U.S. Marine Corps.

Production: 2006 - current   Caliber: 9mm   Action: DA/SA, semi-auto

Barrel Length: 5"   OA Length: 8.5"   Wt.: 32 oz.   Sights: 3-dot

Capacity: 10, 15   Magazine: detachable box   Grips: black synthetic

D2C:	NIB	$ 670	Ex $ 525	VG+ $ 370	Good $ 335	LMP $ 745
C2C:	Mint	$ 650	Ex $ 475	VG+ $ 340	Good $ 305	Fair $ 155
Trade-In:	Mint	$ 480	Ex $ 400	VG+ $ 270	Good $ 235	Poor $ 90

**MODEL 90-TWO
TYPE F**
NEW IN BOX

MODEL 90-TWO TYPE F   REDBOOK CODE: RB-BR-H-M902FX

Slide serrations, updated tactical styling and design, varying grips, accessory rail, skeletonized hammer, slide-mounted safety.

Production: current   Caliber: 9mm   Action: DA/SA, semi-auto

Barrel Length: 5"   OA Length: 8.5"   Wt.: 32 oz.   Sights: removable

Capacity: 17   Magazine: detachable box   Grips: black synthetic, textured

D2C:	NIB	$ 740	Ex $ 575	VG+ $ 410	Good $ 370	LMP $ 795
C2C:	Mint	$ 720	Ex $ 525	VG+ $ 370	Good $ 335	Fair $ 175
Trade-In:	Mint	$ 530	Ex $ 425	VG+ $ 290	Good $ 260	Poor $ 90

MODEL 92 BILLENNIUM
VERY GOOD +

MODEL 92 BILLENNIUM   REDBOOK CODE: RB-BR-H-92BILL

2,000 manufactured to commemorate the millennium. Similar to the Model 92FS Centurion, stainless finish, frame-mounted safety, adjustable trigger, engraved frame and slide.

Production: 2001 - discontinued   Caliber: 9mm   Action: SA, semi-auto

Barrel Length: 5"   OA Length: 8.5"   Sights: adjustable   Capacity: 10, 15

Magazine: detachable box   Grips: carbon fiber grips, checkered

D2C:	NIB	$1,550	Ex $1,200	VG+ $ 855	Good $ 775	
C2C:	Mint	$1,490	Ex $1,075	VG+ $ 780	Good $ 700	Fair $ 360
Trade-In:	Mint	$ 1,110	Ex $ 875	VG+ $ 610	Good $ 545	Poor $ 180

MODEL 96
VERY GOOD +

MODEL 96 COMBAT
VERY GOOD +

MODEL 96   REDBOOK CODE: RB-BR-H-96XXXX

Slide-mounted safety, black finish, light recoil design, similar to the Model 92FS.

Production: 1992 - discontinued Caliber: .40 S&W Action: DA/SA, semi-auto
Barrel Length: 5" OA Length: 8.5" Wt.: 33.6 oz. Sights: fixed Capacity: 10, 11
Magazine: detachable box Grips: black synthetic

D2C:	NIB	$ 675	Ex $ 525	VG+ $ 375	Good $ 340				
C2C:	Mint	$ 650	Ex $ 475	VG+ $ 340	Good $ 305	Fair	$ 160		
Trade-In:	Mint	$ 480	Ex $ 400	VG+ $ 270	Good $ 240	Poor	$ 90		

MODEL 96 COMBAT   REDBOOK CODE: RB-BR-H-96COMP

Designed for competitive shooting. Frame-mounted safety, black finish, extended barrel, adjustable enhanced rear sight.

Production: 1997 - discontinued Caliber: .40 S&W Action: DA/SA, semi-auto
Barrel Length: 5.9" Sights: rear adjustable Capacity: 10 Magazine: detachable box
Grips: black synthetic

D2C:	NIB	$1,785	Ex $1,375	VG+ $ 985	Good $ 895		
C2C:	Mint	$1,720	Ex $1,250	VG+ $ 900	Good $ 805	Fair	$ 415
Trade-In:	Mint	$1,270	Ex $1,000	VG+ $ 700	Good $ 625	Poor	$ 180

MODEL 96 COMPACT TYPE M (REISSUE)
REDBOOK CODE: RB-BR-H-96COMM

A smaller version of the Model 96. Out of production since the early 1990s, but appeared again in 2000. Frame-mounted safety, stainless finish. Distinguished by its slim frame and single-stack magazine.

Production: 2000 - discontinued Caliber: .40 S&W Action: DA/SA, semi-auto
Barrel Length: 4.3" OA Length: 7.8" Sights: 3-dot Capacity: 8, 10
Magazine: detachable box Grips: black synthetic

D2C:	NIB	$ 685	Ex $ 525	VG+ $ 380	Good $ 345		
C2C:	Mint	$ 660	Ex $ 475	VG+ $ 350	Good $ 310	Fair	$ 160
Trade-In:	Mint	$ 490	Ex $ 400	VG+ $ 270	Good $ 240	Poor	$ 90

M9
VERY GOOD +
Courtesy of Bud's Gun Shop

M9   REDBOOK CODE: RB-BR-H-M9XXXX

The military issued version of the Beretta Model 92FS. Aluminum frame, black finish, slide-mounted safety, accessory rail, skeletonized hammer.

Production: 1985 - current Caliber: 9mm Action: DA/SA, semi-auto
Barrel Length: 4.9" OA Length: 8.5" Wt.: 32 oz. Sights: 3-dot
Capacity: 15 Magazine: detachable box Grips: black synthetic

D2C:	NIB	--	Ex $1,125	VG+ $ 800	Good $ 725		
C2C:	Mint	--	Ex $1,025	VG+ $ 730	Good $ 655	Fair	$ 335
Trade-In:	Mint	--	Ex $ 825	VG+ $ 570	Good $ 510	Poor	$ 150

M9 LIMITED EDITION STANDARD   REDBOOK CODE: RB-BR-H-M9LESX

Commemorates the Armed Services' use of the M9 for 10 years. Black finish, frame-mounted safety.

Production: ~1995 - discontinued Caliber: 9mm Action: DA/SA, semi-auto
Barrel Length: 5" OA Length: 8.5" Sights: fixed front Capacity: 15
Magazine: detachable box Grips: black synthetic

D2C:	NIB	$ 650	Ex $500	VG+ $ 360	Good $ 325		
C2C:	Mint	$ 630	Ex $450	VG+ $ 330	Good $ 295	Fair	$ 150
Trade-In:	Mint	$ 470	Ex $375	VG+ $ 260	Good $ 230	Poor	$ 90

BERETTA

M9A1
NEW IN BOX

M9A1 REDBOOK CODE: RB-BR-H-M9A1XX

Aluminum frame, slide-mounted safety, decocker lever, black finish, updated design, partially shrouded barrel.

Production: 2006 - current Caliber: 9mm Action: DA/SA, semi-auto

Barrel Length: 5" OA Length: 8.5" Sights: fixed front, drift rear

Capacity: 10, 15 Magazine: detachable box Grips: black synthetic

D2C:	NIB	$ 670	Ex $ 525	VG+ $ 370	Good $ 335	LMP $ 725
C2C:	Mint	$ 650	Ex $ 475	VG+ $ 340	Good $ 305	Fair $ 155
Trade-In:	Mint	$ 480	Ex $ 400	VG+ $ 270	Good $ 235	Poor $ 90

BU9 NANO
NEW IN BOX

BU9 NANO REDBOOK CODE: RB-BR-H-BUPNNO

Synthetic frame, trigger safety, black finish, compact design.

Production: 2011 - current Caliber: 9mm Action: striker-fire, semi-auto

Barrel Length: 3.07" OA Length: 5.67" Wt.: 17.6 oz. Sights: 3-dot

Capacity: 6 Magazine: detachable box Grips: black synthetic

D2C:	NIB	$ 385	Ex $ 300	VG+ $ 215	Good $ 195	LMP $ 445
C2C:	Mint	$ 370	Ex $ 275	VG+ $ 200	Good $ 175	Fair $ 90
Trade-In:	Mint	$ 280	Ex $ 225	VG+ $ 160	Good $ 135	Poor $ 60

MODEL 21 BOBCAT
NEW IN BOX

MODEL 21 BOBCAT REDBOOK CODE: RB-BR-H-M21BC1

Frame-mounted safety, most feature black finish but available in others. Deluxe and gold plated versions.

Production: ~1984 - current Caliber: .22 LR, .25 ACP Action: DA/SA, semi-auto

Barrel Length: 2.4" OA Length: 5" Sights: fixed front Capacity: 7, 8

Magazine: detachable box Grips: black synthetic

D2C:	NIB	$ 260	Ex $ 200	VG+ $ 145	Good $ 130	
C2C:	Mint	$ 250	Ex $ 200	VG+ $ 130	Good $ 120	Fair $ 60
Trade-In:	Mint	$ 190	Ex $ 150	VG+ $ 110	Good $ 95	Poor $ 30

MODEL 21 LADY BERETTA
REDBOOK CODE: RB-BR-H-M21LB1

Blue finish, gold etched lettering, less than 1,000 produced.

Production: early 1990s - discontinued Caliber: .22 LR Action: DA/SA, semi-auto

Barrel Length: 2.4" OA Length: 5" Sights: fixed front Capacity: 8

Magazine: detachable box Grips: wood

D2C:	NIB	$ 550	Ex $ 425	VG+ $ 305	Good $ 275	
C2C:	Mint	$ 530	Ex $ 400	VG+ $ 280	Good $ 250	Fair $ 130
Trade-In:	Mint	$ 400	Ex $ 325	VG+ $ 220	Good $ 195	Poor $ 60

MODEL 3032 TOMCAT
NEW IN BOX

MODEL 3032 TOMCAT REDBOOK CODE: RB-BR-H-3032TC

Various versions exist, but their values remain similar. Blue, stainless, or titanium finish, aluminum frame, frame-mounted safety, compact design.

Production: late 1990s - current Caliber: .32 ACP Action: DA/SA, semi-auto

Barrel Length: 2.5" OA Length: 5" Wt.: 13 oz. Sights: fixed Capacity: 7

Magazine: detachable box Grips: black synthetic

D2C:	NIB	$ 405	Ex $ 325	VG+ $ 225	Good $ 205	LMP $ 415
C2C:	Mint	$ 390	Ex $ 300	VG+ $ 210	Good $ 185	Fair $ 95
Trade-In:	Mint	$ 290	Ex $ 250	VG+ $ 160	Good $ 145	Poor $ 60

PICO .380
NEW IN BOX

PICO .380   REDBOOK CODE: RB-BR-H-PICO38

Stainless slide, matte finish, polymer frame, blowback design, no manual safety.

Production: current Caliber: .380 ACP Action: DOA, semi-auto											
Barrel Length: 2.7" OA Length: 5.1" Wt.: 11.5 oz. Sights: fixed front, 3-dot											
Capacity: 6 Magazine: detachable box Grips: black synthetic											
D2C:	NIB	$ 385	Ex $ 300	VG+ $ 215	Good $ 195	LMP $ 400
C2C:	Mint	$ 370	Ex $ 275	VG+ $ 200	Good $ 175	Fair $ 90
Trade-In:	Mint	$ 280	Ex $ 225	VG+ $ 160	Good $ 135	Poor $ 60

**PX4 STORM
FULL SIZE**
NEW IN BOX

PX4 STORM, FULL SIZE   REDBOOK CODE: RB-BR-H-PX4ST1

Polymer frame, recoil operated design, slide-mounted safety, black finish. Type C: SAO, no safety. Type D: DAO, no safety. Type F: DA/SA, decocker, manual safety. G: DA/SA, decocker, no safety. These variants have similar values.

Production: 2005 - current Caliber: .45 ACP, .40 S&W, 9mm Action: DA/SA, semi-auto						
Barrel Length: 4" OA Length: 7.6" Sights: 3-dot Capacity: 9, 10, 14, 17						
Magazine: detachable box Grips: black synthetic						
D2C:	NIB	$ 550	Ex $ 425	VG+ $ 305	Good $ 275	LMP $ 575
C2C:	Mint	$ 530	Ex $ 400	VG+ $ 280	Good $ 250	Fair $ 130
Trade-In:	Mint	$ 400	Ex $ 325	VG+ $ 220	Good $ 195	Poor $ 60

**PX4 STORM
SUB-COMPACT**
NEW IN BOX

PX4 STORM SUB-COMPACT   REDBOOK CODE: RB-BR-H-PX4ST2

Tilt-barrel design, slide-mounted safety, polymer frame, skeletonized hammer.

Production: current Caliber: .40 S&W, 9mm Action: DA/SA, semi-auto						
Barrel Length: 3" OA Length: 6.2" Wt.: 25.6 oz. Sights: 3-dot						
Capacity: 10, 13 Magazine: detachable box Grips: black synthetic						
D2C:	NIB	$ 530	Ex $ 425	VG+ $ 295	Good $ 265	LMP $ 575
C2C:	Mint	$ 510	Ex $ 375	VG+ $ 270	Good $ 240	Fair $ 125
Trade-In:	Mint	$ 380	Ex $ 300	VG+ $ 210	Good $ 190	Poor $ 60

**PX4 STORM
COMPACT**
NEW IN BOX

PX4 STORM COMPACT   REDBOOK CODE: RB-BR-H-PX4ST3

Polymer frame, slide-mounted safety, recoil-operated design, skeletonized hammer.

Production: current Caliber: .40 S&W, 9mm Action: DA/SA, semi-auto						
Barrel Length: 3.2" OA Length: 6.8" Sights: 3-dot Capacity: 12, 15						
Magazine: detachable box Grips: black synthetic						
D2C:	NIB	$ 545	Ex $ 425	VG+ $ 300	Good $ 275	LMP $ 575
C2C:	Mint	$ 530	Ex $ 400	VG+ $ 280	Good $ 250	Fair $ 130
Trade-In:	Mint	$ 390	Ex $ 325	VG+ $ 220	Good $ 195	Poor $ 60

**PX4 STORM
SPECIAL DUTY**
NEW IN BOX

PX4 STORM SPECIAL DUTY   REDBOOK CODE: RB-BR-H-PX4ST5

Polymer frame, slide-mounted safety, olive/earth colored frame, black finish slide, comes in a waterproof case.

Production: 2007 - current Caliber: .45 ACP Action: DA/SA, semi-auto						
Barrel Length: 4.6" OA Length: 8.2" Wt.: 28.6 oz. Sights: 3-dot						
Capacity: 9, 10 Magazine: detachable box Grips: olive synthetic						
D2C:	NIB	$1,050	Ex $ 800	VG+ $ 580	Good $ 525	LMP $ 1,145
C2C:	Mint	$1,010	Ex $ 725	VG+ $ 530	Good $ 475	Fair $ 245
Trade-In:	Mint	$ 750	Ex $ 600	VG+ $ 410	Good $ 370	Poor $ 120

BERETTA

U22 NEOS
NEW IN BOX

U22 NEOS   REDBOOK CODE: RB-BR-H-U22NEO

Blue barrel, frame-mounted safety, scope rail, unique design, blowback operated.

Production: 2002 - current   Caliber: .22 LR   Action: SA, semi-auto

Barrel Length: 4.5", 6"   Sights: removable rail   Capacity: 10

Magazine: detachable box   Grips: synthetic, rubber inlay

D2C:	NIB	$ 265	Ex $ 225	VG+ $ 150	Good $ 135	LMP $ 380
C2C:	Mint	$ 260	Ex $ 200	VG+ $ 140	Good $ 120	Fair $ 65
Trade-In:	Mint	$ 190	Ex $ 150	VG+ $ 110	Good $ 95	Poor $ 30

U22 NEOS INOX
NEW IN BOX

U22 NEOS INOX   REDBOOK CODE: RB-BR-H-U22INX

A variant of the U22 Neos, stainless barrel, frame-mounted safety, unique design, blowback operated.

Production: 2002 - current   Caliber: .22 LR   Action: SA, semi-auto

Barrel Length: 6"   Sights: removable   Capacity: 10   Magazine: detachable box

Grips: synthetic, rubber inlay

D2C:	NIB	$ 325	Ex $ 250	VG+ $ 180	Good $ 165	LMP $ 345
C2C:	Mint	$ 320	Ex $ 225	VG+ $ 170	Good $ 150	Fair $ 75
Trade-In:	Mint	$ 240	Ex $ 200	VG+ $ 130	Good $ 115	Poor $ 60

U22 NEOS DLX   REDBOOK CODE: RB-BR-H-U22DLX

An enhanced variant of the U22 Neos with similar components and specs.

Production: 2003   Caliber: .22 LR   Action: SA, semi-auto   Barrel Length: 6", 7.5"

Sights: removable   Capacity: 10   Magazine: detachable box   Grips: synthetic, rubber inlay

D2C:	NIB	$ 415	Ex $ 325	VG+ $ 230	Good $ 210	
C2C:	Mint	$ 400	Ex $ 300	VG+ $ 210	Good $ 190	Fair $ 100
Trade-In:	Mint	$ 300	Ex $ 250	VG+ $ 170	Good $ 150	Poor $ 60

MODEL 9000S
NEW IN BOX

MODEL 9000S   REDBOOK CODE: RB-BR-H-9000SX

Available in several different variations, including DA/SA and DAO. Black finish, ambidextrous frame-mounted safety, modern design.

Production: 2001   Caliber: 9mm, .40 S&W   Action: DA/SA, DAO, semi-auto

Barrel Length: 3.4"   Sights: fixed   Capacity: 10 (9mm), 12 (40 S&W)

Magazine: detachable box   Grips: synthetic

D2C:	NIB	$ 465	Ex $ 375	VG+ $ 260	Good $ 235	
C2C:	Mint	$ 450	Ex $ 325	VG+ $ 240	Good $ 210	Fair $ 110
Trade-In:	Mint	$ 340	Ex $ 275	VG+ $ 190	Good $ 165	Poor $ 60

LARAMIE   REDBOOK CODE: RB-BR-H-LARAMI

Modeled after the Colt Schofield, blue or stainless finish, break-open design, gold medallion grip inserts.

Production: ~2006   Caliber: .38 Special, .45 LC   Action: SA, revolver

Barrel Length: 5", 6.5"   Sights: fixed   Capacity: 6   Grips: walnut

D2C:	NIB	$1,100	Ex $ 850	VG+ $ 605	Good $ 550	
C2C:	Mint	$1,060	Ex $ 775	VG+ $ 550	Good $ 495	Fair $ 255
Trade-In:	Mint	$ 790	Ex $ 625	VG+ $ 430	Good $ 385	Poor $ 120

STAMPEDE SAA
EXCELLENT

*Courtesy of Rock Island
Auction Company*

STAMPEDE SAA   REDBOOK CODE: RB-BR-H-STMP01

Modeled after the Colt Single Action Army. Blue or nickel finish, available in a host of different variations which affect pricing.

Production: ~2003 - 2012   Caliber: .45 LC   Action: SA, revolver

Barrel Length: 3.5", 4.75", 5.5", 7.5"   Sights: fixed   Capacity: 6

Grips: black synthetic or wood

D2C:	NIB	$ 415	Ex	$ 325	VG+	$ 230	Good	$ 210		
C2C:	Mint	$ 400	Ex	$ 300	VG+	$ 210	Good	$ 190	Fair	$ 100
Trade-In:	Mint	$ 300	Ex	$ 250	VG+	$ 170	Good	$ 150	Poor	$ 60

STAMPEDE SAA BISLEY   REDBOOK CODE: RB-BR-H-STMP04

Modeled after the Colt Single Action Army Bisley model, blue or nickel finish.

Production: 2006 - current   Caliber: .357 Magnum, .45 Colt   Action: SA, revolver

Barrel Length: 4.75", 5.5", 7.5"   Sights: fixed   Capacity: 6   Grips: black synthetic

D2C:	NIB	$ 535	Ex	$ 425	VG+	$ 295	Good	$ 270		
C2C:	Mint	$ 520	Ex	$ 375	VG+	$ 270	Good	$ 245	Fair	$ 125
Trade-In:	Mint	$ 380	Ex	$ 300	VG+	$ 210	Good	$ 190	Poor	$ 60

STAMPEDE SAA DELUXE   REDBOOK CODE: RB-BR-H-STMP08

Similar to the standard Stampede, but has enhanced features and an ornate design.

Production: ~2003   Caliber: .45 Colt   Action: SA, revolver

Barrel Length: 4.75", 5.5", 7.5"   Wt.: 38 oz.   Sights: fixed   Capacity: 6   Grips: wood

D2C:	NIB	$ 530	Ex	$ 425	VG+	$ 295	Good	$ 265		
C2C:	Mint	$ 510	Ex	$ 375	VG+	$ 270	Good	$ 240	Fair	$ 125
Trade-In:	Mint	$ 380	Ex	$ 300	VG+	$ 210	Good	$ 190	Poor	$ 60

Bernardelli, Vincenzo

Bernardelli, Vincenzo, headquartered in Brescia, Italy, began manufacturing handguns in 1865 and has remained a fixture in the industry ever since. While perhaps best known for their line of quality, ornate shotguns, their pocket-sized, semi-auto pistols have retained moderate value over the years, given their craftsmanship and early presence in the concealed carry market. Various companies have bought and sold the Bernardelli, Vincenzo trademark over the years, but the standards of their guns have never waned. While the company no longer manufacturers handguns, SAR Arms (or Sarsilmaz), who owns Bernardelli, produces a vest-pocket model indicative of Bernardelli designs.

VEST POCKET MODEL
GOOD

Courtesy of Rock Island Auction Company

VEST POCKET MODEL   REDBOOK CODE: RB-BE-H-VESTPM

Compact design, blue finish, manual safety. Modeled after Walther pistols of the era.

Production: 1945 - late 1940s   Caliber: 6.35mm   Action: SA, semi-auto

Barrel Length: 2.25"   Sights: fixed   Capacity: 5, 8   Magazine: detachable box

Grips: black plastic

D2C:	NIB	$ 280	Ex	$ 225	VG+	$ 155	Good	$ 140		
C2C:	Mint	$ 270	Ex	$ 200	VG+	$ 140	Good	$ 130	Fair	$ 65
Trade-In:	Mint	$ 200	Ex	$ 175	VG+	$ 110	Good	$ 100	Poor	$ 30

BABY
GOOD

BABY   REDBOOK CODE: RB-BE-H-BABYXX

Compact design, blue finish, frame-mounted safety, open-top slide with exposed barrel.

Production: late 1940s - discontinued   Caliber: .22 LR, .22 Short   Action: SA, semi-auto

Barrel Length: 2"   Sights: fixed   Capacity: 6   Magazine: detachable box

Grips: black plastic

D2C:	NIB	$ 330	Ex $ 275	VG+ $ 185	Good $ 165		
C2C:	Mint	$ 320	Ex $ 250	VG+ $ 170	Good $ 150	Fair $	80
Trade-In:	Mint	$ 240	Ex $ 200	VG+ $ 130	Good $ 120	Poor $	60

MODEL 68   REDBOOK CODE: RB-BE-H-M68XXX

Blue finish, compact design, frame-mounted safety, made in Italy, a successor to the VP and Baby models.

Production: 1968 - 1970s   Caliber: .22 LR, .22 Short, .25 ACP   Action: SA, semi-auto

Barrel Length: 2"   Sights: fixed   Capacity: 5   Magazine: detachable box

Grips: black plastic

D2C:	NIB	$ 275	Ex $ 225	VG+ $ 155	Good $ 140		
C2C:	Mint	$ 270	Ex $ 200	VG+ $ 140	Good $ 125	Fair $	65
Trade-In:	Mint	$ 200	Ex $ 175	VG+ $ 110	Good $ 100	Poor $	30

MODEL 80
EXCELLENT

MODEL 80   REDBOOK CODE: RB-BE-H-M80XXX

Blue finish, compact design, frame-mounted safety, made in Italy.

Production: late 1960s - late 1980s   Caliber: .22 LR, .380 ACP   Action: SA, semi-auto

Barrel Length: 3.5"   Sights: adjustable rear   Capacity: 5   Magazine: detachable box

Grips: black plastic

D2C:	NIB	$ 255	Ex $ 200	VG+ $ 145	Good $ 130		
C2C:	Mint	$ 250	Ex $ 200	VG+ $ 130	Good $ 115	Fair $	60
Trade-In:	Mint	$ 190	Ex $ 150	VG+ $ 100	Good $ 90	Poor $	30

Bersa

Imported through Eagle Imports of Wanamassa, New Jersey, Bersa is an Argentinean revolver and pistol manufacturer founded in the late 1950s. The company's Thunder and Firestorm models, both somewhat indicative of the Walther PPK, remain a popular choice for concealed carry use given their dependability, size, and reasonable price. Their BP9/BP40CC model is also of note. Bersa firearms are exclusively sold and distributed by Eagle Imports, who can thus field all concerns and purchase inquiries concerning Bersa.

MODEL 83   REDBOOK CODE: RB-BS-H-M83XXX

Blue or stainless finish, slide-mounted safety, similar to the Bersa Model 383A.

Production: 1988 - early 1990s   Caliber: .380 ACP   Action: DA/SA, semi-auto

Barrel Length: 3.5"   OA Length: 6.6"   Sights: blade, notch fixed   Capacity: 7

Magazine: detachable box   Grips: wood

D2C:	NIB	$ 275	Ex $ 225	VG+ $ 155	Good $ 140	LMP $	288
C2C:	Mint	$ 270	Ex $ 200	VG+ $ 140	Good $ 125	Fair $	65
Trade-In:	Mint	$ 200	Ex $ 175	VG+ $ 110	Good $ 100	Poor $	30

BERNARDELLI

BERSA

THUNDER 9

REDBOOK CODE: RB-BS-H-THUN9X

Alloy frame, frame-mounted safety, blue or stainless finish, ambidextrous design, adjustable trigger.

Production: 1993 - discontinued		Caliber: 9mm		Action: DA/SA, semi-auto						
Barrel Length: 3.5"	OA Length: 6.5"	Wt.: 24 oz.	Sights: fixed front							
Capacity: 15	Magazine: detachable box	Grips: black synthetic								

D2C:	NIB	$ 335	Ex	$ 275	VG+	$ 185	Good	$ 170		
C2C:	Mint	$ 330	Ex	$ 250	VG+	$ 170	Good	$ 155	Fair	$ 80
Trade-In:	Mint	$ 240	Ex	$ 200	VG+	$ 140	Good	$ 120	Poor	$ 60

THUNDER 22
NEW IN BOX
Courtesy of Bud's Gun Shop

THUNDER 22

REDBOOK CODE: RB-BS-H-THN22B

Aluminum-alloy frame, compact design, slide-mounted safety, low-recoil design, black or stainless finish.

Production: current	Caliber: .22 LR	Action: DA/SA, semi-auto	Barrel Length: 3.5"							
OA Length: 6.6"	Wt.: 20 oz.	Sights: dovetail	Capacity: 10	Magazine: detachable box						
Grips: black synthetic										

D2C:	NIB	$ 360	Ex	$ 275	VG+	$ 200	Good	$ 180		
C2C:	Mint	$ 350	Ex	$ 250	VG+	$ 180	Good	$ 165	Fair	$ 85
Trade-In:	Mint	$ 260	Ex	$ 225	VG+	$ 150	Good	$ 130	Poor	$ 60

THUNDER 380
NEW IN BOX
Courtesy of Bud's Gun Shop

THUNDER 380

REDBOOK CODE: RB-BS-H-THN380

Aluminum-alloy frame, slide-mounted safety, compact design, black, nickel or two-toned finish.

Production: mid-1990s - discontinued		Caliber: .380 ACP		Action: DA/SA, semi-auto						
Barrel Length: 3.5"	OA Length: 6.6"	Sights: dovetail front, notched dovetail rear								
Capacity: 7	Magazine: detachable box	Grips: synthetic								

D2C:	NIB	$ 330	Ex	$ 275	VG+	$ 185	Good	$ 165		
C2C:	Mint	$ 320	Ex	$ 250	VG+	$ 170	Good	$ 150	Fair	$ 80
Trade-In:	Mint	$ 240	Ex	$ 200	VG+	$ 130	Good	$ 120	Poor	$ 60

THUNDER 40

REDBOOK CODE: RB-BS-H-THN40X

Matte black finish, aluminum-alloy finish, frame-mounted safety, also produced in high-capacity and compact variants.

Production: early 2000s - discontinued		Caliber: .40 S&W		Action: DA/SA, semi-auto						
Barrel Length: 4.3"	OA Length: 7.5"	Sights: fixed	Capacity: 13							
Magazine: detachable box	Grips: black synthetic									

D2C:	NIB	$ 465	Ex	$ 375	VG+	$ 260	Good	$ 235		
C2C:	Mint	$ 450	Ex	$ 325	VG+	$ 240	Good	$ 210	Fair	$ 110
Trade-In:	Mint	$ 340	Ex	$ 275	VG+	$ 190	Good	$ 165	Poor	$ 60

BERSA

THUNDER 45 ULTRA COMPACT PRO
NEW IN BOX

THUNDER 9MM ULTRA COMPACT PRO
NEW IN BOX

THUNDER 45 ULTRA COMPACT PRO

REDBOOK CODE: RB-BS-H-TH45UC

Various finishes, frame-mounted safety, aluminum-alloy frame, compact design, skeletonized hammer.

Production: 2010 - current Caliber: .45 ACP Action: DA/SA, semi-auto

Barrel Length: 3.6" OA Length: 6.8" Sights: interchangeable front and rear

Capacity: 7 Magazine: detachable box Grips: black synthetic

D2C:	NIB	$ 455	Ex $ 350	VG+ $ 255	Good $ 230	LMP $ 469
C2C:	Mint	$ 440	Ex $ 325	VG+ $ 230	Good $ 205	Fair $ 105
Trade-In:	Mint	$ 330	Ex $ 275	VG+ $ 180	Good $ 160	Poor $ 60

THUNDER 9MM ULTRA COMPACT PRO

REDBOOK CODE: RB-BS-H-TH9UCP

Duotone or matte black finish, decocker, extended slide release.

Production: discontinued Caliber: 9mm Action: DA/SA, semi-auto

Barrel Length: 3.25" Wt.: 23 oz. Sights: 3-dot sight system

Capacity: 10, 12 Magazine: detachable box Grips: checkered black polymer

D2C:	NIB	$ 440	Ex $ 350	VG+ $ 245	Good $ 220	
C2C:	Mint	$ 430	Ex $ 325	VG+ $ 220	Good $ 200	Fair $ 105
Trade-In:	Mint	$ 320	Ex $ 250	VG+ $ 180	Good $ 155	Poor $ 60

FIRESTORM

REDBOOK CODE: RB-BS-H-FRSTRM

Stainless slide, matte black frame/grips, slide-mounted safety, alloy frame, compact design.

Production: ~2004 - current Caliber: .22 LR, .380 ACP Action: DA/SA, semi-auto

Barrel Length: 3.5" Sights: blade front, target rear Capacity: 7, 10

Magazine: detachable box Grips: black synthetic

D2C:	NIB	$ 340	Ex $ 275	VG+ $ 190	Good $ 170	
C2C:	Mint	$ 330	Ex $ 250	VG+ $ 170	Good $ 155	Fair $ 80
Trade-In:	Mint	$ 250	Ex $ 200	VG+ $ 140	Good $ 120	Poor $ 60

BP9/BP40CC
NEW IN BOX

BP9/BP40CC

REDBOOK CODE: RB-BS-H-BP9BP4

Matte black or nickel finish, internal safety, compact design, Picatinny rail, steel slide, polymer frame.

Production: 2011 - current Caliber: 9mm, .40 S&W Action: striker-fired, semi-auto

Barrel Length: 3.3" OA Length: 6.3" Sights: interchangeable front and rear

Capacity: 8 Magazine: detachable box Grips: black synthetic

D2C:	NIB	$ 365	Ex $ 300	VG+ $ 205	Good $ 185	LMP $ 429
C2C:	Mint	$ 360	Ex $ 275	VG+ $ 190	Good $ 165	Fair $ 85
Trade-In:	Mint	$ 260	Ex $ 225	VG+ $ 150	Good $ 130	Poor $ 60

Browning

Founded in Ogden, Utah by John Moses Browning in 1855, Browning Firearms is indisputably one, if not the most influential firearm manufacturer of all time. From the development of the "Potato Digger" to the first semi-automatic shotgun, the company has continuously changed the face of the firearms industry with their constant innovations and improvements. After a notorious dispute with Winchester in 1897, John Moses Browning famously struck a deal with the Belgium firearms manufacturer Fabrique Nationale d'Herstal to produce his .32 caliber semi-automatic pistol in what would become one of the most storied partnerships in the history of American and international firearms. Among their handguns, Browning and FNH produced the BDM, Buck Mark, Hi-Power, and their own version of the M1911, all of which collectors and shooting enthusiasts note for their excellence and dependability. Today, FN owns majority control of Browning and continues to produce many of the guns that made both companies world renowned.

MODEL 1899 (EARLY FN) REDBOOK CODE: RB-BW-H-1899XX
Originally produced by Fabrique Nationale de Herstal of Belgium. Differs from the M1900 with its smaller frame sideplate, absence of safety markings and lanyard ring, different image imprinted on the grip, and studs and nuts instead of screws in the grip. First Browning to use a slide, would lead to the development of the M1911. Approximately 14,400 units made.

Production: 1899	Caliber: .32 ACP	Action: SA, semi-auto					
Barrel Length: 4.8"	Wt.: 22 oz.	Sights: blade front	Capacity: 8				
Magazine: detachable box	Grips: black synthetic						
D2C:	NIB	--	Ex $1,200	VG+ $ 870	Good $ 790		
C2C:	Mint	--	Ex $1,100	VG+ $ 790	Good $ 710	Fair $ 365	
Trade-In: Mint	--	Ex $ 900	VG+ $ 620	Good $ 555	Poor $ 180		

MODEL 1900
(EARLY FN)
GOOD

Courtesy of Rock Island Auction Company

MODEL 1900 (EARLY FN) REDBOOK CODE: RB-BW-H-1900XX
Markings on the gun ("Sur" and "Feu," respectively) distinguish the M1899 and M1900. Similar to the FN M1899, but features a somewhat shorter barrel, larger grips, safety markings, and lanyard ring. Approximately 700,000 units produced. Some feature nickel finish, all include a varied image imprinted on the grip. Models with cases intact have increased value.

Production: 1899 - 1910	Caliber: .32 ACP	Action: SA, semi-auto					
Barrel Length: 4"	Sights: blade front	Capacity: 7	Magazine: detachable box				
Grips: black synthetic							
D2C:	NIB	--	Ex $ 975	VG+ $ 700	Good $ 635		
C2C:	Mint	--	Ex $ 900	VG+ $ 640	Good $ 575	Fair $ 295	
Trade-In: Mint	--	Ex $ 725	VG+ $ 500	Good $ 445	Poor $ 150		

MODEL 1903
(EARLY FN)
GOOD

Courtesy of Rock Island Auction Company

MODEL 1903 (EARLY FN) REDBOOK CODE: RB-BW-H-1903XX
Not the same as the Colt Model 1903, but designed by John Browning specifically for FN. Also known as the Browning No. 2. Similar in design to the Model 1900. Approximately 153,000 units produced.

Production: 1903 - late 1930s	Caliber: 9x20mm Browning Long, .32 ACP						
Action: SA, semi-auto	Barrel Length: 5"	OA Length: 8"	Wt.: 32 oz.				
Sights: fixed	Capacity: 7, 8	Magazine: detachable box	Grips: black synthetic				
D2C:	NIB	--	Ex $1,050	VG+ $ 760	Good $ 690		
C2C:	Mint	--	Ex $ 950	VG+ $ 690	Good $ 620	Fair $ 320	
Trade-In: Mint	--	Ex $ 775	VG+ $ 540	Good $ 485	Poor $ 150		

BROWNING

BABY BROWNING
(EARLY FN)
VERY GOOD +

Courtesy Rock Island Auction Company

BABY BROWNING (EARLY FN) REDBOOK CODE: RB-BW-H-BBYBRW

Models marked with the FN symbol commands a higher premium than those marked with Browning's emblem. Models with nickel finish and pearl grips (circa the 1950s) also have increased value. "Renaissance Models" with engraved grey finish are nearly twice as valuable.

Production: 1931 - early 1980s Caliber: .25 ACP Action: SA, semi-auto

Barrel Length: 2" OA Length: 4.1" Sights: fixed Capacity: 6

Magazine: detachable box Grips: black synthetic, wood, or pearl

D2C:	NIB	--	Ex	$ 475	VG+	$ 330	Good	$ 300		
C2C:	Mint	--	Ex	$ 425	VG+	$ 300	Good	$ 270	Fair	$ 140
Trade-In:	Mint	--	Ex	$ 350	VG+	$ 240	Good	$ 210	Poor	$ 60

MODEL 1905 (EARLY FN) REDBOOK CODE: RB-BW-H-M1905X

Designed in 1905, but entered production in 1906. Smaller than previous FN model semi-auto pistols. Vest Pocket variations exist. Available in nickel finish. Increased value for contract or retailer markings. Those produced after 1908 have diminished value. Approximately 1,000,000 units produced.

Production: 1906 - late 1950s Caliber: .25 ACP Action: SA, semi-auto

Barrel Length: 2" OA Length: 4.5" Wt.: 13 oz. Sights: fixed Capacity: 6

Magazine: detachable box Grips: black synthetic

D2C:	NIB	--	Ex	$ 475	VG+	$ 340	Good	$ 305		
C2C:	Mint	--	Ex	$ 425	VG+	$ 310	Good	$ 275	Fair	$ 145
Trade-In:	Mint	--	Ex	$ 350	VG+	$ 240	Good	$ 215	Poor	$ 90

MODEL 1910
(EARLY FN)
GOOD

Courtesy of Rock Island Auction Company

MODEL 1910 (EARLY FN) REDBOOK CODE: RB-BW-H-M1910X

Used in WWI and WWII, improved modifications on the M1899 and M1900. The FN M1922 is a slight variation of the Model 1910.

Production: ~1910 - early 1950s Caliber: .380 ACP, .32 ACP Action: SA, semi-auto

Barrel Length: 3.5" OA Length: 6" Wt.: 24 oz. Sights: fixed

Capacity: 6, 7, 8, 9 Magazine: detachable box Grips: black synthetic or wood

D2C:	NIB	--	Ex	$ 500	VG+	$ 355	Good	$ 320		
C2C:	Mint	--	Ex	$ 450	VG+	$ 320	Good	$ 290	Fair	$ 150
Trade-In:	Mint	--	Ex	$ 375	VG+	$ 250	Good	$ 225	Poor	$ 90

MODEL 1922
(EARLY FN)
FAIR

Courtesy of Rock Island Auction Company

MODEL 1922 (EARLY FN) REDBOOK CODE: RB-BW-H-M1922X

Very similar to the Model 1910, but features a longer barrel and slide and a larger capacity magazine. Both models saw usage during WWII. Wartime models chambered for .380 ACP have increased value.

Production: 1940 - ~1944 Caliber: .380 ACP, .32 ACP Action: SA, semi-auto

Barrel Length: 4.4" Sights: fixed Capacity: 8, 9 Magazine: detachable box

Grips: black synthetic or wood

D2C:	NIB	--	Ex	$ 425	VG+	$ 295	Good	$ 270		
C2C:	Mint	--	Ex	$ 375	VG+	$ 270	Good	$ 245	Fair	$ 125
Trade-In:	Mint	--	Ex	$ 300	VG+	$ 210	Good	$ 190	Poor	$ 60

**MODEL 1935
COMMERCIAL**
VERY GOOD +

Courtesy of Rock Island Auction Company

MODEL 1935 COMMERCIAL (HI-POWER) (HP)(GP) (PRE-NAZI)

REDBOOK CODE: RB-BW-H-M1935X

An early version of the Hi-Power, commercially available prior to WWII. Features a slot for an attachable shoulder stock. Models with tangent sights or sold in Latvia, Lithuania, and Romania have increased value over other models. Saw use with foreign law enforcement and military forces.

Production: 1930 - 1940 Caliber: 9mm Action: SA, semi-auto Barrel Length: 4.7"

Sights: fixed or adjustable Capacity: 13 Magazine: detachable box

Grips: black synthetic or wood

D2C:	NIB	--	Ex	$3,200	VG+	$2,310	Good	$2,100		
C2C:	Mint	--	Ex	$2,900	VG+	$2,100	Good	$1,890	Fair	$ 970
Trade-In:	Mint	--	Ex	$2,375	VG+	$1,640	Good	$1,470	Poor	$ 420

MODEL 1955
VERY GOOD +

Courtesy Rock Island Auction Company

MODEL 1955   REDBOOK CODE: RB-BW-H-M1955X
A reissue of the original Model 1910, blue finish, frame-mounted safety.

Production: 1954 - late 1960s Caliber: .380 ACP, .32 ACP Action: SA, semi-auto								
Barrel Length: 3.5" OA Length: 7" Sights: fixed Capacity: 6								
Magazine: detachable box Grips: black synthetic								
D2C:	NIB	--	Ex	$ 500	VG+	$ 350	Good	$ 320
C2C:	Mint	--	Ex	$ 450	VG+	$ 320	Good	$ 290	Fair	$ 150
Trade-In:	Mint	--	Ex	$ 375	VG+	$ 250	Good	$ 225	Poor	$ 90

HI-POWER 640   REDBOOK CODE: RB-BW-H-HIPWX1
Produced after the Nazi takeover for German military use. Blue finish is notoriously poor, many fakes in circulation, similar to other Hi-Power's of the era. Some with Nazi markings, some ambiguously marked.

Production: 1940s - discontinued Caliber: 9mm Action: SA, semi-auto								
Barrel Length: 4.7" Sights: fixed Capacity: 13 Magazine: detachable box								
Grips: wood or black synthetic								
D2C:	NIB	$3,100	Ex	$2,375	VG+	$1,705	Good	$1,550
C2C:	Mint	$2,980	Ex	$2,150	VG+	$1,550	Good	$1,395	Fair	$ 715
Trade-In:	Mint	$2,210	Ex	$1,750	VG+	$1,210	Good	$1,085	Poor	$ 330

HI-POWER WAA613 GERMAN FIRST GENERATION COMMERCIAL   REDBOOK CODE: RB-BW-H-HIPWX2
Models produced under German control have significant collector's value. These feature a lanyard ring on the bottom of the grip, tangent sight, and shoulder stock slot in the grip. Serial numbers between approximately 44,000 and 65,300. Many fakes in circulation.

Production: 1940 - 1941 Caliber: 9mm Action: SA, semi-auto								
Barrel Length: 4.7" OA Length: 7.8" Wt.: 2.3 oz. Sights: tangent rear								
Capacity: 13 Magazine: detachable box Grips: wood or black synthetic								
D2C:	NIB	$2,900	Ex	$2,225	VG+	$1,595	Good	$1,450
C2C:	Mint	$2,790	Ex	$2,025	VG+	$1,450	Good	$1,305	Fair	$ 670
Trade-In:	Mint	$2,060	Ex	$1,625	VG+	$1,140	Good	$1,015	Poor	$ 300

**HI-POWER WAA103
GERMAN SECOND
GENERATION
COMMERCIAL**
FAIR

Courtesy of Rock Island Auction Company

HI-POWER WAA103 GERMAN SECOND GENERATION COMMERCIAL   REDBOOK CODE: RB-BW-H-HIPWX3
Models marked WaA 103 don't feature slots for shoulder stocks. Serial numbers typically run between 64,000 and 95,500.

Production: 1941 - 1942 Caliber: 9mm Action: SA, semi-auto								
Barrel Length: 4.7" OA Length: 7.8" Sights: tangent rear								
Capacity: 13 Magazine: detachable box Grips: wood or black synthetic								
D2C:	NIB	$2,870	Ex	$2,200	VG+	$1,580	Good	$1,435
C2C:	Mint	$2,760	Ex	$2,000	VG+	$1,440	Good	$1,295	Fair	$ 665
Trade-In:	Mint	$2,040	Ex	$1,625	VG+	$1,120	Good	$1,005	Poor	$ 300

**HI-POWER WAA140
GERMAN THIRD
GENERATION**
VERY GOOD +

Courtesy of Rock Island Auction Company

HI-POWER WAA140 GERMAN THIRD GENERATION
REDBOOK CODE: RB-BW-H-HIPWX4
No shoulder stock slot, but some may feature tangent rear sights. Serial numbers between approximately 95,000 and B6,350. Those with tangent sights have increased value.

Production: 1941 - 1944 Caliber: 9mm Action: SA, semi-auto								
Barrel Length: 4.7" OA Length: 7.8" Sights: fixed or tangent rear								
Capacity: 13 Magazine: detachable box Grips: wood or black synthetic								
D2C:	NIB	$1,850	Ex	$1,425	VG+	$1,020	Good	$ 925
C2C:	Mint	$1,780	Ex	$1,300	VG+	$ 930	Good	$ 835	Fair	$ 430
Trade-In:	Mint	$1,320	Ex	$1,050	VG+	$ 730	Good	$ 650	Poor	$ 210

BROWNING

HI-POWER POST-WAR COMMERCIAL MODELS
FAIR

Courtesy of Rock Island Auction Company

HI-POWER POST-WAR COMMERCIAL MODELS

REDBOOK CODE: RB-BW-H-HIPWX5

Most feature an "A" or "B" in their serial numbers. Produced prior to 1954, before the company began importing the Hi-Power into U.S. markets.

Production: 1944 - 1954 Caliber: 9mm Action: SA, semi-auto Barrel Length: 4.7"

OA Length: 7.8" Sights: fixed Capacity: 13 Magazine: detachable box

Grips: wood or black synthetic

D2C:	NIB	--	Ex	$ 850	VG+	$ 615	Good	$ 560	
C2C:	Mint	--	Ex	$ 775	VG+	$ 560	Good	$ 505	Fair	$ 260
Trade-In:	Mint	--	Ex	$ 625	VG+	$ 440	Good	$ 395	Poor	$ 120

HI-POWER STANDARD ISSUE

REDBOOK CODE: RB-BW-H-HIPWX6

The gun has remained relatively unchanged since its introduction into U.S. markets. Smaller and lighter than the popular Colt Model 1911. Produced in Belgium, Japan, and Portugal. Over a million produced. Blue finish, checkered grips. Mark I models (1950s and 1980s) have increased value over the Mark II models (early 1980s) and Mark III models (introduced in 1985). Models with rounded hammers given higher premium over models with spur hammers.

HI-POWER STANDARD ISSUE
GOOD

Courtesy of Rock Island Auction Company

Production: 1954 - 2000 Caliber: 9mm, .40 S&W Action: SA, semi-auto

Barrel Length: 4.7" OA Length: 7.8" Sights: fixed front Capacity: 13

Magazine: detachable box Grips: wood or black synthetic

D2C:	NIB	$ 835	Ex	$ 650	VG+	$ 460	Good	$ 420	
C2C:	Mint	$ 810	Ex	$ 600	VG+	$ 420	Good	$ 380	Fair	$ 195
Trade-In:	Mint	$ 600	Ex	$ 475	VG+	$ 330	Good	$ 295	Poor	$ 90

HI-POWER TANGENT SIGHT
VERY GOOD +

Courtesy of Rock Island Auction Company

HI-POWER TANGENT SIGHT

REDBOOK CODE: RB-BW-H-HIPWX7

Noted for its tangent rear sight. Feature a "T" prefix in the serial number. Later models may include spurred hammers and varying serial numbers. Assembled in Portugal. Those with grips slotted for a shoulder stock extension have increased value.

Production: mid-1960s - 1978 Caliber: 9mm, 9x21mm Action: SA, semi-auto

Barrel Length: 4.7" OA Length: 7.8" Sights: tangent Capacity: 13

Magazine: detachable box Grips: checkered wood

D2C:	NIB	$ 900	Ex	$ 700	VG+	$ 495	Good	$ 450	
C2C:	Mint	$ 870	Ex	$ 625	VG+	$ 450	Good	$ 405	Fair	$ 210
Trade-In:	Mint	$ 640	Ex	$ 525	VG+	$ 360	Good	$ 315	Poor	$ 90

HI-POWER 75TH ANNIVERSARY
NEW IN BOX

Courtesy of Rock Island Auction Company

HI-POWER 75TH ANNIVERSARY

REDBOOK CODE: RB-BW-H-HIPWX8

Not markedly different than the Hi-Power Standard. Commemorates the gun's 75th year of production, unique engravings on the slide, sold with an extra magazine.

Production: 2010 - discontinued Caliber: 9mm Action: SA, semi-auto

Barrel Length: 4.7" OA Length: 7.8" Sights: fixed front Capacity: 13

Magazine: detachable box Grips: walnut

D2C:	NIB	$ 915	Ex	$ 700	VG+	$ 505	Good	$ 460	
C2C:	Mint	$ 880	Ex	$ 650	VG+	$ 460	Good	$ 415	Fair	$ 215
Trade-In:	Mint	$ 650	Ex	$ 525	VG+	$ 360	Good	$ 325	Poor	$ 120

**HI-POWER
SILVER CHROME
MODEL**
EXCELLENT

Courtesy of Rock Island Auction Company

HI-POWER SILVER CHROME MODEL

REDBOOK CODE: RB-BW-H-HIPWX9
The first Hi-Power Silver Chrome. Serial numbers should include either PZ, PY, PX, PW, or PV to denote their year of production.

Production: 1981 - mid-1980s Caliber: 9mm, .40 S&W Action: SA, semi-auto

Barrel Length: 4.7" Sights: fixed front Capacity: 13 Magazine: detachable box

Grips: Pachmayr rubber

D2C:	NIB	--	Ex $ 725	VG+ $ 515	Good $ 470				
C2C:	Mint $ 900	Ex $ 650	VG+ $ 470	Good $ 425	Fair $ 220				
Trade-In:	Mint $ 670	Ex $ 525	VG+ $ 370	Good $ 330	Poor $ 120				

HI-POWER SILVER CHROME MODEL (REINTRODUCTION)

REDBOOK CODE: RB-BW-H-HIPW10
A reissue of the original Hi-Power Silver Chrome. Dull metallic finish, black rubber Pachmayr grips, frame-mounted safety.

Production: 1991 - discontinued Caliber: 9mm, .40 S&W Action: SA, semi-auto

Barrel Length: 4.7" OA Length: 7.8" Sights: adjustable Capacity: 13

Magazine: detachable box Grips: Pachmayr rubber

D2C:	NIB $ 920	Ex $ 700	VG+ $ 510	Good $ 460	
C2C:	Mint $ 890	Ex $ 650	VG+ $ 460	Good $ 415	Fair $ 215
Trade-In:	Mint $ 660	Ex $ 525	VG+ $ 360	Good $ 325	Poor $ 120

**HI-POWER
PRACTICAL MODEL**
NEW IN BOX

HI-POWER PRACTICAL MODEL REDBOOK CODE: RB-BW-H-HIPW11

Similar to other Hi-Power Mark III pistols but features a black slide. Silver trigger guard and hammer.

Production: 1993 - 2006 Caliber: 9mm, .40 S&W Action: SA, semi-auto

Barrel Length: 4.7" OA Length: 7.75" Wt.: 38.5 oz. Sights: adjustable

Capacity: 10, 13 Magazine: detachable box Grips: Pachmayr rubber

D2C:	NIB $ 955	Ex $ 750	VG+ $ 530	Good $ 480	
C2C:	Mint $ 920	Ex $ 675	VG+ $ 480	Good $ 430	Fair $ 220
Trade-In:	Mint $ 680	Ex $ 550	VG+ $ 380	Good $ 335	Poor $ 120

HI-POWER NICKEL
EXCELLENT

Courtesy of Rock Island Auction Company

HI-POWER NICKEL REDBOOK CODE: RB-BW-H-HIPW12

Post-dates the later Hi-Power Silver Chrome Model. Nickel or chrome finish.

Production: 1981 - discontinued Caliber: 9mm, .40 S&W Action: SA, semi-auto

Barrel Length: 4.7" OA Length: 7.8" Wt.: 38.5 oz. Sights: fixed front

Capacity: 13 Magazine: detachable box Grips: wood

D2C:	NIB $1,225	Ex $ 950	VG+ $ 675	Good $ 615	
C2C:	Mint $1,180	Ex $ 850	VG+ $ 620	Good $ 555	Fair $ 285
Trade-In:	Mint $ 870	Ex $ 700	VG+ $ 480	Good $ 430	Poor $ 150

HI-POWER GP COMPETITION REDBOOK CODE: RB-BW-H-HIPW13

"GP" stands for "Grande Puissance," which is French for Hi-Power. Has moderate collector's value. Features an extended barrel.

Production: 1987 - late 1980s Caliber: 9mm Action: SA, semi-auto

Barrel Length: 6" Sights: adjustable Capacity: 13 Magazine: detachable box

Grips: Pachmayr grips

D2C:	NIB $ 915	Ex $ 700	VG+ $ 505	Good $ 460	
C2C:	Mint $ 880	Ex $ 650	VG+ $ 460	Good $ 415	Fair $ 215
Trade-In:	Mint $ 650	Ex $ 525	VG+ $ 360	Good $ 325	Poor $ 120

**HI-POWER
RENAISSANCE**
EXCELLENT

Courtesy of Rock Island Auction Company

HI-POWER RENAISSANCE REDBOOK CODE: RB-BW-H-HIPW14

Ornate engravings, silver slide, gold trigger, frame-mounted safety, models with ring hammers or adjustable sights have increased value.

Production: late 1970s Caliber: 9mm, .25 ACP, .380 ACP Action: SA, semi-auto

Barrel Length: 2", 3.4", 4.7" Sights: adjustable Capacity: 13

Magazine: detachable box Grips: pearl

D2C:	NIB	$3,000	Ex	$2,400	VG+	$1,900	Good	$1,475	
C2C:	Mint	$2,840	Ex	$2,050	VG+	$1,480	Good	$1,330	Fair $ 680
Trade-In:	Mint	$2,100	Ex	$1,675	VG+	$1,160	Good	$1,035	Poor $ 300

**HI-POWER
CENTENNIAL MODEL**
(EARLY PRODUCTION)
EXCELLENT

Courtesy of Rock Island Auction Company

HI-POWER CENTENNIAL MODEL (EARLY PRODUCTION)

REDBOOK CODE: RB-BW-H-HIPW16

Less than 4,000 total units produced. Models with serial numbers between 1 and 100 have significant value (almost five times that over the other Centennial models).

Production: 1978 Caliber: 9mm Action: SA, semi-auto Barrel Length: 4.5"

Sights: blade front Capacity: 13 Magazine: detachable box Grips: wood

D2C:	NIB	$1,100	Ex	$ 850	VG+	$ 605	Good	$ 550	
C2C:	Mint	$1,060	Ex	$ 775	VG+	$ 550	Good	$ 495	Fair $ 255
Trade-In:	Mint	$ 790	Ex	$ 625	VG+	$ 430	Good	$ 385	Poor $ 120

**HI-POWER
LOUIS XVI MODEL**
VERY GOOD +

Courtesy of Rock Island Auction Company

HI-POWER LOUIS XVI MODEL REDBOOK CODE: RB-BW-H-HIPW17

Ornate engravings, satin finish, gold trigger, frame-mounted safety, medallion grip. Diamond grip models increase value by 50%.

Production: 1980 - discontinued Caliber: 9mm Action: SA, semi-auto

Barrel Length: 4.6" Sights: blade front Capacity: 13 Grips: wood

D2C:	NIB	$2,220	Ex	$1,700	VG+	$1,225	Good	$1,110	
C2C:	Mint	$2,140	Ex	$1,550	VG+	$1,110	Good	$1,000	Fair $ 515
Trade-In:	Mint	$1,580	Ex	$1,250	VG+	$ 870	Good	$ 780	Poor $ 240

HI-POWER GOLD CLASSIC REDBOOK CODE: RB-BW-H-HIPW18

Ornate engravings, gold trigger, silver finish, frame-mounted safety, approximately 350 made with gold inlays (doubling their value). Less than 6,000 units produced.

Production: 1985 Caliber: 9mm Action: SA, semi-auto Barrel Length: 4.6"

Sights: blade front Capacity: 13 Magazine: detachable box Grips: wood

D2C:	NIB	$1,650	Ex	$1,275	VG+	$ 910	Good	$ 825	
C2C:	Mint	$1,590	Ex	$1,150	VG+	$ 830	Good	$ 745	Fair $ 380
Trade-In:	Mint	$1,180	Ex	$925	VG+	$ 650	Good	$ 580	Poor $ 180

**HI-POWER 125TH
ANNIVERSARY**
EXCELLENT

Courtesy of Rock Island Auction Company

HI-POWER 125TH ANNIVERSARY

REDBOOK CODE: RB-BW-H-HIPW19

Its ornate engravings commemorate the 125th anniversary of John Moses Browning releasing his first firearm, the Model 1885 High Wall.

Production: 2003 Caliber: 9mm Action: SA, semi-auto Barrel Length: 4.7"

Capacity: 10 Magazine: detachable box Grips: ivory

D2C:	NIB	$2,300	Ex	$1,750	VG+	$1,265	Good	$1,150	
C2C:	Mint	$2,210	Ex	$1,600	VG+	$1,150	Good	$1,035	Fair $ 530
Trade-In:	Mint	$1,640	Ex	$1,300	VG+	$ 900	Good	$ 805	Poor $ 240

HI-POWER CAPTAIN
VERY GOOD +

Courtesy of Rock Island Auction Company

HI-POWER CAPTAIN   REDBOOK CODE: RB-BW-H-HIPW22

Blue or stainless finish, ambidextrous frame-mounted safety, plastic carrying case, assembled in Portugal.

Production: 1993 - discontinued   Caliber: 9mm, .40 S&W   Action: SA, semi-auto
Barrel Length: 4.6"   Sights: tangent   Capacity: 10, 13   Magazine: detachable box
Grips: wood

D2C:	NIB	$ 715	Ex $ 550	VG+ $ 395	Good $ 360				
C2C:	Mint	$ 690	Ex $ 500	VG+ $ 360	Good $ 325	Fair $ 165			
Trade-In:	Mint	$ 510	Ex $ 425	VG+ $ 280	Good $ 255	Poor $ 90			

HI-POWER (.30 LUGER)   REDBOOK CODE: RB-BW-H-HIPW23

Somewhat rare with less than 2,000 imported into the U.S. Blue finish, frame-mounted safety.

Production: 1986 - late 1980s   Caliber: .30 Luger   Action: SA, semi-auto
Barrel Length: 4.6"   Sights: fixed   Capacity: 13   Magazine: detachable box
Grips: wood or black synthetic

D2C:	NIB	$ 645	Ex $ 500	VG+ $ 355	Good $ 325		
C2C:	Mint	$ 620	Ex $ 450	VG+ $ 330	Good $ 295	Fair $ 150	
Trade-In:	Mint	$ 460	Ex $ 375	VG+ $ 260	Good $ 230	Poor $ 90	

HI-POWER STANDARD
NEW IN BOX

Courtesy of Rock Island Auction Company

HI-POWER STANDARD   REDBOOK CODE: RB-BW-H-HIPW24

Similar to the original Hi-Powers. Steel barrel, polished blue steel receiver and slide, locked breech, single-action trigger, frame-mounted safety, double-stack magazine.

Production: current   Caliber: 9mm, .40 S&W   Action: SA, semi-auto
Barrel Length: 4.6", 4.7"   Sights: fixed low-profile 3-dot , adjustable target sights
Capacity: 10, 13   Magazine: detachable box   Grips: select walnut, cut checkering

D2C:	NIB $ 990	Ex $ 775	VG+ $ 545	Good $ 495	LMP $1,079	
C2C:	Mint $ 960	Ex $ 700	VG+ $ 500	Good $ 450	Fair $ 230	
Trade-In:	Mint $ 710	Ex $ 575	VG+ $ 390	Good $ 350	Poor $ 120	

HI-POWER MARK III   REDBOOK CODE: RB-BW-H-HIPW25

Steel barrel, matte black frame and slide, locked breech, ambidextrous frame-mounted safety, thumb rest in grip, double stack magazine, modern design.

Production: 1991 - current   Caliber: 9mm, .40 S&W   Action: SA, semi-auto
Barrel Length: 4.6"   Wt.: 32 oz.   Sights: fixed low-profile 3-dot   Capacity: 10, 13
Magazine: detachable box   Grips: composite panels

D2C:	NIB $ 925	Ex $ 725	VG+ $ 510	Good $ 465	LMP $1,069	
C2C:	Mint $ 890	Ex $ 650	VG+ $ 470	Good $ 420	Fair $ 215	
Trade-In:	Mint $ 660	Ex $ 525	VG+ $ 370	Good $ 325	Poor $ 120	

BDM/BRM/BPM-DAO   REDBOOK CODE: RB-BW-H-BDMBRW

The BDM is DA/SA and includes a switch to function as a SA, BRM is DAO (enclosed hammer), and BPM-D is DA/SA with an exposed hammer and no manual safety. The company also produced a BDM Practical Model (matte blue slide, silver frame).

Production: 1991 - late 1990s   Caliber: 9mm   Action: DA/SA, SA, DAO, semi-auto
Barrel Length: 4.75"   Sights: fixed front   Capacity: 10, 15
Magazine: detachable box   Grips: black synthetic

D2C:	NIB $ 515	Ex $ 400	VG+ $ 285	Good $ 260	LMP $ 535	
C2C:	Mint $ 500	Ex $ 375	VG+ $ 260	Good $ 235	Fair $ 120	
Trade-In:	Mint $ 370	Ex $ 300	VG+ $ 210	Good $ 185	Poor $ 60	

BDA
VERY GOOD +

Courtesy of Rock Island Auction Company

BDA   REDBOOK CODE: RB-BW-H-BDAXXX

Manufactured by Sig Sauer and similar to the Sig Model 220, but based on Browning designs. Compact versions of the gun exist. Frame-mounted safety, blue finish. Values vary according to caliber.

Production: 1977 - early 1980s   Caliber: 9mm, .45 ACP, .38 Super, .32 ACP

Action: DA/SA, semi-auto   Barrel Length: 4.6"   OA Length: 8"   Wt.: 32 oz.

Sights: fixed front   Capacity: 9, 14   Magazine: detachable box   Grips: black synthetic

D2C:	NIB	$ 500	Ex $ 400	VG+ $ 275	Good $ 250			
C2C:	Mint	$ 480	Ex $ 350	VG+ $ 250	Good $ 225	Fair $	115
Trade-In:	Mint	$ 360	Ex $ 300	VG+ $ 200	Good $ 175	Poor $	60

BDA-380
VERY GOOD +

Courtesy of Rock Island Auction Company

BDA-380   REDBOOK CODE: RB-BW-H-BDA380

Aluminum frame and slide, nickel or dark blue finish, slide-mounted safety, blowback-style action, Browning emblem in grip, ambidextrous design. Slightly higher premium for nickel finish.

Production: 1977 - late 1990s   Caliber: .380 ACP   Action: DA/SA, semi-auto

Barrel Length: 3.75"   OA Length: 6.8"   Sights: fixed, low profile

Capacity: 14   Magazine: detachable box   Grips: walnut

D2C:	NIB	$ 525	Ex $ 400	VG+ $ 290	Good $ 265		
C2C:	Mint	$ 510	Ex $ 375	VG+ $ 270	Good $ 240	Fair $	125
Trade-In:	Mint	$ 380	Ex $ 300	VG+ $ 210	Good $ 185	Poor $	60

MODEL 10/71   REDBOOK CODE: RB-BW-H-M1071X

A modified version of the Model 1955, marketed and sold as the Model 125. Blue finish, checkered grips, frame-mounted safety.

Production: 1970 - mid-1970s   Caliber: .380 ACP   Action: SA, semi-auto

Barrel Length: 4.5"   Sights: fixed front, adjustable rear   Capacity: 6

Magazine: detachable box   Grips: black synthetic

D2C:	NIB	$ 490	Ex $ 375	VG+ $ 270	Good $ 245		
C2C:	Mint	$ 480	Ex $ 350	VG+ $ 250	Good $ 225	Fair $	115
Trade-In:	Mint	$ 350	Ex $ 275	VG+ $ 200	Good $ 175	Poor $	60

PRO-9
NEW IN BOX

PRO-9/PRO-40   REDBOOK CODE: RB-BW-H-PRO9XX

Frame-mounted safety, composite grips, polymer frame, ambidextrous design, grip back straps, accessory rail in front of the trigger guard. Manufactured at FN's Columbia, South Carolina plant.

Production: 2003 - 2007   Caliber: 9mm, .40 S&W   Action: DA/SA, semi-auto

Barrel Length: 4"   OA Length: 7.25"   Wt.: 29 oz.   Sights: fixed front

Capacity: 10   Magazine: detachable box   Grips: black synthetic

D2C:	NIB	$ 565	Ex $ 450	VG+ $ 315	Good $ 285	LMP $	641
C2C:	Mint	$ 550	Ex $ 400	VG+ $ 290	Good $ 255	Fair $	130
Trade-In:	Mint	$ 410	Ex $ 325	VG+ $ 230	Good $ 200	Poor $	60

NOMAD
EXCELLENT

Courtesy of Rock Island Auction Company

NOMAD   REDBOOK CODE: RB-BW-H-NOMDXX

Blowback action, half-length slide, blue finish, sturdy steel or allow frame. Serial numbers include "2P" indicate the gun was produced between 1962 and 1968. "P" in the serial number indicates production between 1969 and 1974.

Production: 1962 - 1974   Caliber: .22 LR   Action: DAO, semi-auto

Barrel Length: 4.5", 6.75"   Wt.: 32 oz.   Sights: fixed front   Capacity: 10

Magazine: detachable box   Grips: synthetic

D2C:	NIB	$ 435	Ex $ 350	VG+ $ 240	Good $ 220		
C2C:	Mint	$ 420	Ex $ 325	VG+ $ 220	Good $ 200	Fair $	105
Trade-In:	Mint	$ 310	Ex $ 250	VG+ $ 170	Good $ 155	Poor $	60

CHALLENGER
VERY GOOD +

Courtesy of Rock Island Auction Company

**CHALLENGER
RENAISSANCE
MODEL**
EXCELLENT

Courtesy of Rock Island Auction Company

CHALLENGER   REDBOOK CODE: RB-BW-H-CHLGR1

Similar to the Browning Nomad, but features wood grips and gold plated trigger and "U" in the serial number.

Production: early 1960s - 1974		Caliber: .22 LR		Action: DA/SA, semi-auto					
Barrel Length: 4.5", 6.75"		Sights: fixed front		Capacity: 10					
Magazine: detachable box		Grips: wood or synthetic							
D2C:	NIB	$ 525	Ex $ 400	VG+ $ 290	Good $ 265				
C2C:	Mint	$ 510	Ex $ 375	VG+ $ 270	Good $ 240	Fair $ 125			
Trade-In:	Mint	$ 380	Ex $ 300	VG+ $ 210	Good $ 185	Poor $ 60			

CHALLENGER RENAISSANCE MODEL
REDBOOK CODE: RB-BW-H-CHLGR2

The same as the standard Challenger Model, but features ornate engravings on the slide. Nickel finish and unique engravings in the grips.

Production: early 1960s - 1974		Caliber: .22 LR		Action: DA/SA, semi-auto			
Barrel Length: 4.37", 6.75"		Sights: fixed front		Capacity: 10			
Magazine: detachable box		Grips: wood					
D2C:	NIB	--	Ex $2,775	VG+ $2,010	Good $1,825		
C2C:	Mint	$3,510	Ex $2,525	VG+ $1,830	Good $1,645	Fair $ 840	
Trade-In:	Mint	$2,600	Ex $2,050	VG+ $1,430	Good $1,280	Poor $ 390	

CHALLENGER GOLD LINE   REDBOOK CODE: RB-BW-H-CHLGR3

Similar to other Challenger models, but with gold trim around the slide and trigger guard.

Production: 1960s - discontinued		Caliber: .22 LR		Action: DA/SA, semi-auto			
Barrel Length: 4.5", 6.75"		Sights: fixed front		Capacity: 10			
Magazine: detachable box		Grips: wood					
D2C:	NIB	--	Ex $1,675	VG+ $1,210	Good $1,100		
C2C:	Mint	$2,120	Ex $1,525	VG+ $1,100	Good $ 990	Fair $ 510	
Trade-In:	Mint	$1,570	Ex $1,250	VG+ $ 860	Good $ 770	Poor $ 240	

CHALLENGER II   REDBOOK CODE: RB-BW-H-CHLGR4

Gold trigger, rugged steel frame, half-length slide, frame-mounted safety, six groove rifling, and 9.1" sight radius. Serial numbers include a RT, RR, RP, RN, PO, PZ, or PY.

Production: 1976 - 1982		Caliber: .22 LR		Action: DA/SA, semi-auto			
Barrel Length: 6.75"		Wt.: 38 oz.		Sights: fixed front, adjustable rear			
Capacity: 10		Magazine: detachable box		Grips: wood or synthetic			
D2C:	NIB	$ 385	Ex $ 300	VG+ $ 215	Good $ 195		
C2C:	Mint	$ 370	Ex $ 275	VG+ $ 200	Good $ 175	Fair $ 90	
Trade-In:	Mint	$ 280	Ex $ 225	VG+ $ 160	Good $ 135	Poor $ 60	

CHALLENGER III/CHALLENGER III SPORTER
REDBOOK CODE: RB-BW-H-CHLGR5

Gold trigger, rugged steel frame, frame-mounted safety, half-length slide, polished grips. The Sporter Model features a standard barrel, while the standard model features a bull barrel. Serial numbers should a include PY, PX, or PW.

Production: 1982 - 1984		Caliber: .22 LR		Action: DA/SA, semi-auto			
Barrel Length: 5.5", 6.75"		Sights: fixed front, adjustable rear		Capacity: 10, 11			
Magazine: detachable box		Grips: wood					
D2C:	NIB	$ 335	Ex $ 275	VG+ $ 185	Good $ 170		
C2C:	Mint	$ 330	Ex $ 250	VG+ $ 170	Good $ 155	Fair $ 80	
Trade-In:	Mint	$ 240	Ex $ 200	VG+ $ 140	Good $ 120	Poor $ 60	

MEDALIST
EXCELLENT

Courtesy of Rock Island Auction Company

MEDALIST REDBOOK CODE: RB-BW-H-MDLST1

Vent rib barrel, wraparound and checkered grips, gold trigger, blue finish, frame-mounted safety, some models sold in a black Browning case with three interchangeable barrels. A "T" in the serial number can help distinguish it from other Browning .22 LR pistols of the era.

Production: 1962 - 1974 Caliber: .22 LR Action: SA, semi-auto
Barrell Length: 6.75" Wt.: 46 oz. Sights: fixed front, adjustable rear Capacity: 10 Magazine: detachable box Grips: wood

D2C:	NIB	$1,225	Ex	$ 950	VG+	$ 675	Good	$ 615	
C2C:	Mint	$1,180	Ex	$ 850	VG+	$ 620	Good	$ 555	Fair $ 285
Trade-In:	Mint	$ 870	Ex	$ 700	VG+	$ 480	Good	$ 430	Poor $ 150

MEDALIST GOLD LINE REDBOOK CODE: RB-BW-H-MDLST2

A variation of the original Medalist 22 LR, gold trigger, vent rib barrel, some with ornate engraving.

Production: early 1960s - 1974 Caliber: .22 LR Action: SA, semi-auto
Barrel Length: 4.5", 6.75" Capacity: 10 Magazine: detachable box Grips: wood

D2C:	NIB	--	Ex	$2,700	VG+	$1,950	Good	$1,770	
C2C:	Mint	$3,400	Ex	$2,450	VG+	$1,770	Good	$1,595	Fair $815
Trade-In:	Mint	$2,520	Ex	$2,000	VG+	$1,390	Good	$1,240	Poor $360

MEDALIST RENAISSANCE MODEL REDBOOK CODE: RB-BW-H-MDLST3

Chrome finish, gold trigger, shortened slide. Ornate engraved grip, barrel, and trigger guard. Custom models of the gun have increased value.

Production: 1962 - 1974 Caliber: .22 LR Action: SA, semi-auto
Barrel Length: 4.5", 6.75" Sights: fixed front, adjustable rear
Capacity: 10 Magazine: detachable box Grips: wood, engraved

D2C:	NIB	--	Ex	$3,500	VG+	$2,530	Good	$2,300	
C2C:	Mint	$4,420	Ex	$3,175	VG+	$2,300	Good	$2,070	Fair $1,060
Trade-In:	Mint	$3,270	Ex	$2,600	VG+	$1,800	Good	$1,610	Poor $ 480

INTERNATIONAL MEDALIST (EARLY MODEL)

REDBOOK CODE: RB-BW-H-MDLST4
Gold trigger, wraparound and checkered grips, vent rib barrel, frame-mounted trigger. Later model is slightly reduced in value.

Production: late 1970s - 1980 Caliber: .22 LR Action: SA, semi-auto
Barrel Length: 5.8", 5.9" Sights: fixed front, adjustable rear
Capacity: 10 Magazine: detachable box Grips: wood

D2C:	NIB	$1,000	Ex	$ 775	VG+	$ 550	Good	$ 500	
C2C:	Mint	$ 960	Ex	$ 700	VG+	$ 500	Good	$ 450	Fair $ 230
Trade-In:	Mint	$ 710	Ex	$ 575	VG+	$ 390	Good	$ 350	Poor $ 120

BUCK MARK STANDARD
NEW IN BOX

Courtesy Rock Island Auction Company

BUCK MARK STANDARD REDBOOK CODE: RB-BW-H-BCKMX1

Aircraft-grade aluminum frame, chrome or dark blue finish, checkered grips, gold trigger, sleek design, models with stainless finish have a slightly increased value.

Production: 1985 - current Caliber: .22 LR Action: SA, semi-auto
Barrel Length: 5.5" Wt.: 33.6 oz. Sights: fixed front, adjustable rear
Capacity: 10 Magazine: detachable box Grips: Ultragrip RX rubber

D2C:	NIB	$ 390	Ex	$ 300	VG+	$ 215	Good	$ 195	
C2C:	Mint	$ 380	Ex	$ 275	VG+	$ 200	Good	$ 180	Fair $ 90
Trade-In:	Mint	$ 280	Ex	$ 225	VG+	$ 160	Good	$ 140	Poor $ 60

BROWNING

**BUCK MARK
MICRO STANDARD**
NEW IN BOX

BUCK MARK MICRO STANDARD REDBOOK CODE: RB-BW-H-BCKMX2

Ambidextrous and modern design, blue or stainless finish, aircraft-grade aluminum frame, gold trigger, compact size.

Production: 1992 - current Caliber: .22 LR Action: SA, semi-auto

Barrel Length: 4" Wt.: 32 oz. Sights: adjustable Capacity: 10

Magazine: detachable box Grips: Ultragrip RX and Ultragrip DX rubber

D2C:	NIB	$ 385	Ex $ 300	VG+ $ 215	Good $ 195	LMP $ 460
C2C:	Mint	$ 370	Ex $ 275	VG+ $ 200	Good $ 175	Fair $ 90
Trade-In:	Mint	$ 280	Ex $ 225	VG+ $ 160	Good $ 135	Poor $ 60

BUCK MARK MICRO PLUS REDBOOK CODE: RB-BW-H-BCKMX3

Bull barrel, gold trigger, blue finish, Browning emblem in grip.

Production: discontinued Caliber: .22 LR Action: SA, semi-auto

Barrel Length: 4" Sights: adjustable Capacity: 10 Magazine: detachable box

Grips: walnut

D2C:	NIB	$ 390	Ex $ 300	VG+ $ 215	Good $ 195	
C2C:	Mint	$ 380	Ex $ 275	VG+ $ 200	Good $ 180	Fair $ 90
Trade-In:	Mint	$ 280	Ex $ 225	VG+ $ 160	Good $ 140	Poor $ 60

**BUCK MARK
CHALLENGE**
NEW IN BOX

BUCK MARK CHALLENGE REDBOOK CODE: RB-BW-H-BCKMX4

Checkered grips, alloy frame, matte blue finish, gold trigger, shortened slide, tapered barrel.

Production: 1999 - 2011 Caliber: .22 LR Action: SA, semi-auto Barrel Length: 5.5"

OA Length: 9.5" Wt.: 24 oz. Sights: fixed front, adjustable rear Capacity: 10

Magazine: detachable box Grips: walnut

D2C:	NIB	$ 375	Ex $ 300	VG+ $ 210	Good $ 190	LMP $ 430
C2C:	Mint	$ 360	Ex $ 275	VG+ $ 190	Good $ 170	Fair $ 90
Trade-In:	Mint	$ 270	Ex $ 225	VG+ $ 150	Good $ 135	Poor $ 60

BUCK MARK MICRO CHALLENGE

REDBOOK CODE: RB-BW-H-BCKMX5

Checkered grips, alloy frame, matte blue finish, gold trigger, shortened slide, tapered barrel.

Production: discontinued Caliber: .22 LR Action: SA, semi-auto Barrel Length: 4"

OA Length: 8" Sights: fixed front, adjustable rear Capacity: 10

Magazine: detachable box Grips: walnut

D2C:	NIB	$ 345	Ex $ 275	VG+ $ 190	Good $ 175	
C2C:	Mint	$ 340	Ex $ 250	VG+ $ 180	Good $ 160	Fair $ 80
Trade-In:	Mint	$ 250	Ex $ 200	VG+ $ 140	Good $ 125	Poor $ 60

BUCK MARK PLUS
NEW IN BOX

BUCK MARK PLUS REDBOOK CODE: RB-BW-H-BCKMX6

Exists in many variations. Nickel, blue, or stainless finish, gold trigger, shortened slide, ambidextrous design. Different variations of the gun have included Field Plus, rosewood, and Classic Plus models.

Production: 1987 - early 1990s Caliber: .22 LR Action: SA, semi-auto

Barrel Length: 5.5" OA Length: 9.5" Wt.: 2.132 oz.

Sights: fiber-optic front, adjustable rear, some variations Capacity: 10

Magazine: detachable box Grips: laminated wood, rosewood, or UDX synthetic

D2C:	NIB	$ 400	Ex $ 325	VG+ $ 220	Good $ 200	
C2C:	Mint	$ 390	Ex $ 300	VG+ $ 200	Good $ 180	Fair $ 95
Trade-In:	Mint	$ 290	Ex $ 225	VG+ $ 160	Good $ 140	Poor $ 60

BROWNING

**BUCK MARK
BULLSEYE TARGET**
NEW IN BOX

BUCK MARK BULLSEYE TARGET

REDBOOK CODE: RB-BW-H-BCKMX7

Available with polished rosewood grips, frame-mounted safety, fluted barrel, Pro-Target adjustable sights.

Production: 1996 - discontinued   Caliber: .22 LR   Action: SA, semi-auto   Barrel Length: 7.25"   OA Length: 11.3"   Wt.: 43 oz.   Sights: fixed front, Pro-Target adjustable rear   Capacity: 10   Magazine: detachable box   Grips: rosewood or rubber

D2C:	NIB	$ 545	Ex $ 425	VG+ $ 300	Good $ 275	LMP $ 604
C2C:	Mint	$ 530	Ex $ 400	VG+ $ 280	Good $ 250	Fair $ 130
Trade-In:	Mint	$ 390	Ex $ 325	VG+ $ 220	Good $ 195	Poor $ 60

**BUCK MARK BULLSEYE
TARGET STAINLESS**
NEW IN BOX

BUCK MARK BULLSEYE TARGET STAINLESS

REDBOOK CODE: RB-BW-H-BCKMX8

Similar to the Buck Mark Bullseye Target, but features stainless finish.

Production: late 1990s - 2011   Caliber: .22 LR   Action: SA, semi-auto   Barrel Length: 7.25"   OA Length: 11.3"Wt.: 43 oz.   Sights: fixed front, Pro-Target adjustable rear   Capacity: 10   Magazine: detachable box   Grips: rosewood or rubber

D2C:	NIB	$ 570	Ex $ 450	VG+ $ 315	Good $ 285	
C2C:	Mint	$ 550	Ex $ 400	VG+ $ 290	Good $ 260	Fair $ 135
Trade-In:	Mint	$ 410	Ex $ 325	VG+ $ 230	Good $ 200	Poor $ 60

**BUCK MARK BULLSEYE
STANDARD URX**
NEW IN BOX

Courtesy of Rock Island Auction Company

BUCK MARK BULLSEYE STANDARD URX

REDBOOK CODE: RB-BW-H-BCKMX9

Fluted barrel, checkered grips, frame-mounted safety, gold trigger. Some models have URX grips (slightly increasing their value).

Production: 2006 - current   Caliber: .22 LR   Action: SA, semi-auto   Barrel Length: 7.25"   OA Length: 11.3"Wt.: 36.8 oz.   Sights: front fixed   Capacity: 10   Magazine: detachable box   Grips: black synthetic, URX

D2C:	NIB	$ 460	Ex $ 350	VG+ $ 255	Good $ 230	LMP $ 579
C2C:	Mint	$ 450	Ex $ 325	VG+ $ 230	Good $ 210	Fair $ 90
Trade-In:	Mint	$ 330	Ex $ 275	VG+ $ 180	Good $ 165	Poor $ 60

**BUCK MARK
MICRO BULL**
NEW IN BOX

BUCK MARK MICRO BULL   REDBOOK CODE: RB-BW-H-BCKM11

Noted for its bull barrel. Matte blue finish, frame-mounted safety, gold trigger.

Production: 2006 - current   Caliber: .22 LR   Action: SA, semi-auto   Barrel Length: 4"   OA Length: 8"   Wt.: 32 oz.   Sights: fixed front, adjustable rear   Capacity: 10   Magazine: detachable box   Grips: black synthetic

D2C:	NIB	$ 315	Ex $ 250	VG+ $ 175	Good $ 160	LMP $ 339
C2C:	Mint	$ 310	Ex $ 225	VG+ $ 160	Good $ 145	Fair $ 75
Trade-In:	Mint	$ 230	Ex $ 200	VG+ $ 130	Good $ 115	Poor $ 60

BUCK MARK GOLD TARGET   REDBOOK CODE: RB-BW-H-BCKM13

The same as the standard Buck Mark 5.5 Target, but features a gold trigger guard, scope mounts and varying frame color.

Production: discontinued   Caliber: .22 LR   Action: SA, semi-auto   Barrel Length: 5.5"   Wt.: 35 oz.   Sights: hooped   Capacity: 10   Magazine: detachable box   Grips: cocobolo

D2C:	NIB	$ 435	Ex $ 350	VG+ $ 240	Good $ 220	
C2C:	Mint	$ 420	Ex $ 325	VG+ $ 220	Good $ 200	Fair $ 105
Trade-In:	Mint	$ 310	Ex $ 250	VG+ $ 170	Good $ 155	Poor $ 60

BUCK MARK FIELD
NEW IN BOX

BUCK MARK FIELD REDBOOK CODE: RB-BW-H-BCKM14

Alloy frame, frame-mounted safety, matte blue finish, heavy barrel, gold trigger.

Production: 1991 - ~2010 Caliber: .22 LR Action: SA, semi-auto Barrel Length: 5.5"
OA Length: 9.5" Wt.: 35 oz. Sights: Pro-Target adjustable Capacity: 10
Magazine: detachable box Grips: walnut

D2C:	NIB	$ 515	Ex $ 400	VG+ $ 285	Good $ 260	LMP $ 620
C2C:	Mint	$ 500	Ex $ 375	VG+ $ 260	Good $ 235	Fair $ 120
Trade-In:	Mint	$ 370	Ex $ 300	VG+ $ 210	Good $ 185	Poor $ 60

BUCK MARK FIELD (2005) REDBOOK CODE: RB-BW-H-BCKM15

Alloy frame, frame-mounted safety, matte blue finish, heavy barrel, gold trigger, scope mounts, modified grips.

Production: 2005 Caliber: .22 LR Action: SA, semi-auto Barrel Length: 5.5"
Wt.: 35 oz. Sights: scope rail Capacity: 10 Magazine: detachable box Grips: walnut

D2C:	NIB	$ 535	Ex $ 425	VG+ $ 295	Good $ 270	
C2C:	Mint	$ 520	Ex $ 375	VG+ $ 270	Good $ 245	Fair $ 125
Trade-In:	Mint	$ 380	Ex $ 300	VG+ $ 210	Good $ 190	Poor $ 60

BUCK MARK LITE SPLASH URX
NEW IN BOX

BUCK MARK LITE SPLASH URX REDBOOK CODE: RB-BW-H-BCKM17

Similar to the Buck Mark Lite, but features gold splash designs along the barrel.

Production: 2006 Caliber: .22 LR Action: SA, semi-auto Barrel Length: 5.5", 7.25"
Sights: TruGlo front Capacity: 10 Magazine: detachable box Grips: Ultragrip RX

D2C:	NIB	$ 455	Ex $ 350	VG+ $ 255	Good $ 230	
C2C:	Mint	$ 440	Ex $ 325	VG+ $ 230	Good $ 205	Fair $ 105
Trade-In:	Mint	$ 330	Ex $ 275	VG+ $ 180	Good $ 160	Poor $ 60

BUCK MARK VARMINT REDBOOK CODE: RB-BW-H-BCKM19

Similar to other Buck Mark models, black matte finish, extended barrel, frame-mounted safety, gold trigger.

Production: late 1980s - discontinued Caliber: .22 LR Action: SA, semi-auto
Barrel Length: 9.75" Sights: scope mounts Capacity: 10
Magazine: detachable box Grips: wood

D2C:	NIB	$ 400	Ex $ 325	VG+ $ 220	Good $ 200	
C2C:	Mint	$ 390	Ex $ 300	VG+ $ 200	Good $ 180	Fair $ 95
Trade-In:	Mint	$ 290	Ex $ 225	VG+ $ 160	Good $ 140	Poor $ 60

BUCK MARK SILHOUETTE REDBOOK CODE: RB-BW-H-BCKM20

Elongated barrel, wood stock beneath barrel, scope mounts, wood foregrip beneath barrel, gold trigger.

Production: late 1980s - discontinued Caliber: .22 LR Action: SA, semi-auto
Barrel Length: 9.75" Sights: scope mounts Capacity: 10
Magazine: detachable box Grips: wood

D2C:	NIB	--	Ex $ 450	VG+ $ 320	Good $ 290	
C2C:	Mint	$ 560	Ex $ 400	VG+ $ 290	Good $ 260	Fair $ 135
Trade-In:	Mint	$ 410	Ex $ 325	VG+ $ 230	Good $ 205	Poor $ 60

BUCK MARK CAMPER REDBOOK CODE: RB-BW-H-BCKM22

Tapered bull barrel, framed-mounted safety, aluminum-alloy receiver, matte blue finish.

Production: 1999 - current Caliber: .22 LR Action: SA, semi-auto
Barrel Length: 5.5" Wt.: 35 oz. Sights: adjustable Pro-Target
Capacity: 10 Magazine: detachable box Grips: composite black

D2C:	NIB	$ 325	Ex $ 250	VG+ $ 180	Good $ 165	LMP $ 350
C2C:	Mint	$ 320	Ex $ 225	VG+ $ 170	Good $ 150	Fair $ 75
Trade-In:	Mint	$ 240	Ex $ 200	VG+ $ 130	Good $ 115	Poor $ 60

BUCK MARK CAMPER
EXCELLENT

Courtesy of Rock Island Auction Company

BUCK MARK CAMPER STAINLESS
NEW IN BOX

BUCK MARK CAMPER STAINLESS   REDBOOK CODE: RB-BW-H-BCKM23

Tapered stainless bull barrel, aluminum-alloy receiver with matte blue finish, frame-mounted safety, gold trigger.

Production: 2005 - discontinued Caliber: .22 LR Action: SA, semi-auto										
Barrel Length: 5.5" Wt.: 35 oz. Sights: adjustable Pro-Target										
Capacity: 10 Magazine: detachable box Grips: composite black										
D2C:	NIB	$ 345	Ex	$ 275	VG+	$ 190	Good	$ 175	LMP	$ 419
C2C:	Mint	$ 340	Ex	$ 250	VG+	$ 180	Good	$ 160	Fair	$ 80
Trade-In:	Mint	$ 250	Ex	$ 200	VG+	$ 140	Good	$ 125	Poor	$ 60

BUCK MARK CAMPER STAINLESS UFX, FLD
NEW IN BOX

BUCK MARK CAMPER STAINLESS UFX, FLD
REDBOOK CODE: RB-BW-H-BCKM24

Tapered stainless bull barrel, aluminum-alloy receiver with matte blue finish, frame-mounted safety, available only to Browning Full Line and Medallion Dealers.

Caliber: .22 LR Action: SA, semi-auto Barrel Length: 5.5" Wt.: 35 oz.										
Sights: adjustable Pro-Target Capacity: 10 Magazine: detachable box										
Grips: over-molded Ultragrip FX ambidextrous										
D2C:	NIB	$ 355	Ex	$ 275	VG+	$ 200	Good	$ 180	LMP	$ 429
C2C:	Mint	$ 350	Ex	$ 250	VG+	$ 180	Good	$ 160	Fair	$ 85
Trade-In:	Mint	$ 260	Ex	$ 200	VG+	$ 140	Good	$ 125	Poor	$ 60

BUCK MARK CONTOUR URX
NEW IN BOX

BUCK MARK CONTOUR URX   REDBOOK CODE: RB-BW-H-BCKM25

Special contour matte blue barrel, aluminum-alloy receiver with matte blue finish, manual thumb safety, full-length scope base.

Production: 2006 Caliber: .22 LR Action: SA, semi-auto Barrel Length: 5.5", 7.25"										
Wt.: 35 oz. Sights: adjustable Pro-Target Capacity: 10										
Magazine: detachable box Grips: Ultragrip RX ambidextrous										
D2C:	NIB	$ 415	Ex	$ 325	VG+	$ 230	Good	$ 210	LMP	$ 549
C2C:	Mint	$ 400	Ex	$ 300	VG+	$ 210	Good	$ 190	Fair	$ 100
Trade-In:	Mint	$ 300	Ex	$ 250	VG+	$ 170	Good	$ 150	Poor	$ 60

BUCK MARK HUNTER
NEW IN BOX

BUCK MARK HUNTER   REDBOOK CODE: RB-BW-H-BCKM26

Heavy tapered round bull barrel, aluminum-alloy receiver with matte blue finish, manual thumb safety, integral scope base, frame-mounted safety.

Production: 2005 - current Caliber: .22 LR Action: SA, semi-auto										
Barrel Length: 7.25" Wt.: 38 oz. Sights: adjustable Pro-Target rear, Tru-Glo/										
Marble's fiber-optic front Capacity: 10 Magazine: detachable box										
Grips: laminated cocobolo target grips										
D2C:	NIB	$ 385	Ex	$ 300	VG+	$ 215	Good	$ 195	LMP	$ 499
C2C:	Mint	$ 370	Ex	$ 275	VG+	$ 200	Good	$ 175	Fair	$ 90
Trade-In:	Mint	$ 280	Ex	$ 225	VG+	$ 160	Good	$ 135	Poor	$ 60

BUCK MARK LIMITED EDITION 25TH ANNIVERSARY
REDBOOK CODE: RB-BW-H-BCKM27

Commemorates the 25th year of the Buck Mark's production. Ornately engraved grips, blue finish, approximately 1,000 produced.

Production: 2001 - discontinued Caliber: .22 LR Action: SA, semi-auto										
Barrel Length: 6.75" Capacity: 10 Magazine: detachable box Grips: ivory										
D2C:	NIB	$ 625	Ex	$ 475	VG+	$ 345	Good	$ 315		
C2C:	Mint	$ 600	Ex	$ 450	VG+	$ 320	Good	$ 285	Fair	$ 145
Trade-In:	Mint	$ 450	Ex	$ 350	VG+	$ 250	Good	$ 220	Poor	$ 90

**BUCK MARK PLUS
ROSEWOOD UDX, FLD**
NEW IN BOX

BUCK MARK PLUS ROSEWOOD UDX, FLD

REDBOOK CODE: RB-BW-H-BCKM28

Matte blue barrel with polished flats, aluminum-alloy receiver, rosewood-style grips, frame-mounted safety, available only to Browning Full Line and Medallion Dealers.

Production: 2006 - discontinued Caliber: .22 LR Action: SA, semi-auto

Barrel Length: 5.5" Wt.: 34 oz. Sights: adjustable Pro-Target rear,

fiber-optic front Capacity: 10 Magazine: detachable box

Grips: rosewood Ultragrip DX ambidextrous

D2C:	NIB	$ 455	Ex $ 350	VG+ $ 255	Good $ 230	LMP $ 519
C2C:	Mint	$ 440	Ex $ 325	VG+ $ 230	Good $ 205	Fair $ 105
Trade-In:	Mint	$ 330	Ex $ 275	VG+ $ 180	Good $ 160	Poor $ 60

**BUCK MARK PLUS
STAINLESS BLACK
LAMINATED UDX**
NEW IN BOX

BUCK MARK PLUS STAINLESS BLACK LAMINATED UDX

REDBOOK CODE: RB-BW-H-BCKM29

Stainless or blue matte finish, aluminum-alloy frame, frame-mounted safety, gold trigger, flat-sided barrel.

Production: 2007 - current Caliber: .22 LR Action: SA, semi-auto

Barrel Length: 5.5" Sights: adjustable Pro-Target rear, Tru-Glo/Marble's

fiber-optic front Capacity: 10 Magazine: detachable box Grips: black laminated

Ultragrip DX ambidextrous

D2C:	NIB	$ 435	Ex $ 350	VG+ $ 240	Good $ 220	LMP $ 579
C2C:	Mint	$ 420	Ex $ 325	VG+ $ 220	Good $ 200	Fair $ 105
Trade-In:	Mint	$ 310	Ex $ 250	VG+ $ 170	Good $ 155	Poor $ 60

BUCK MARK PLUS STAINLESS UDX

REDBOOK CODE: RB-BW-H-BCKM30

Stainless finish, aluminum-alloy frame, frame-mounted safety, gold trigger, flat-sided barrel.

Production: 2007 - current Caliber: .22 LR Action: SA, semi-auto

Barrel Length: 5.5" Wt.: 34 oz. Sights: adjustable Pro-Target rear, Tru-Glo/

Marble's fiber-optic front Capacity: 10 Magazine: detachable box

Grips: walnut Ultragrip DX ambidextrous

D2C:	NIB	$ 425	Ex $ 325	VG+ $ 235	Good $ 215	LMP $ 539
C2C:	Mint	$ 410	Ex $ 300	VG+ $ 220	Good $ 195	Fair $ 100
Trade-In:	Mint	$ 310	Ex $ 250	VG+ $ 170	Good $ 150	Poor $ 60

**BUCK MARK
PRACTICAL URX**
NEW IN BOX

BUCK MARK PRACTICAL URX

REDBOOK CODE: RB-BW-H-BCKM31

Tapered bull barrel with matte blue finish, aluminum-alloy receiver with matte gray finish, frame-mounted safety, gold trigger, finger-groove grip design.

Production: 2010 - current Caliber: .22 LR Action: SA, semi-auto

Barrel Length: 5.5" Wt.: 34 oz. Sights: adjustable Pro-Target rear, Tru-Glo/

Marble's fiber-optic front Capacity: 10 Magazine: detachable box

Grips: Ultragrip RX ambidextrous

D2C:	NIB	$ 365	Ex $ 300	VG+ $ 205	Good $ 185	LMP $ 439
C2C:	Mint	$ 360	Ex $ 275	VG+ $ 190	Good $ 165	Fair $ 85
Trade-In:	Mint	$ 260	Ex $ 225	VG+ $ 150	Good $ 130	Poor $ 60

BROWNING

1911-22 A1
NEW IN BOX

1911-22 A1   REDBOOK CODE: RB-BW-H-1911XX

Black anodized finish, grip safety, manual sear block, frame-mounted safety, design similar to the popular Colt M1911.

Production: current	Caliber: .22 LR	Action: SA, semi-auto	Barrel Length: 4.25"
Wt.: 15.5 oz.	Sights: fixed	Capacity: 10	Magazine: detachable box
Grips: brown synthetic			

D2C:	NIB	$ 550	Ex $ 425	VG+ $ 305	Good $ 275	LMP $ 600
C2C:	Mint	$ 530	Ex $ 400	VG+ $ 280	Good $ 250	Fair $ 130
Trade-In:	Mint	$ 400	Ex $ 325	VG+ $ 220	Good $ 195	Poor $ 60

1911-22 A1 BLACK LABEL LAMINATE
NEW IN BOX

1911-22 A1 BLACK LABEL LAMINATE

REDBOOK CODE: RB-BW-H-1911X1

Matte black finish, stainless barrel block, frame-mounted safety, ambidextrous design.

Production: current	Caliber: .22 LR	Action: SA, semi-auto	Barrel Length: 4.25"
Sights: fixed	Capacity: 10	Magazine: detachable box	Grips: black laminated, stippled

D2C:	NIB	$ 600	Ex $ 475	VG+ $ 330	Good $ 300	LMP $ 640
C2C:	Mint	$ 580	Ex $ 425	VG+ $ 300	Good $ 270	Fair $ 140
Trade-In:	Mint	$ 430	Ex $ 350	VG+ $ 240	Good $ 210	Poor $ 60

1911-22 A1 BLACK LABEL LAMINATE W/ RAIL
NEW IN BOX

1911-22 A1 BLACK LABEL LAMINATE W/ RAIL

REDBOOK CODE: RB-BW-H-1911X2

Matte black finish, stainless barrel block, frame-mounted safety, ambidextrous design, accessory rail mount.

Production: current	Caliber: .22 LR	Action: SA, semi-auto	Barrel Length: 4.25"
Sights: fixed	Capacity: 10	Magazine: detachable box	Grips: black laminated, stippled

D2C:	NIB	$ 615	Ex $ 475	VG+ $ 340	Good $ 310	LMP $ 670
C2C:	Mint	$ 600	Ex $ 425	VG+ $ 310	Good $ 280	Fair $ 145
Trade-In:	Mint	$ 440	Ex $ 350	VG+ $ 240	Good $ 220	Poor $ 90

1911-22 A1 DESERT TAN
NEW IN BOX

1911-22 A1 DESERT TAN   REDBOOK CODE: RB-BW-H-1911X3

Matte blue slide, tan colored trigger guard and portions of the grip, target crown, machined aluminum frame, frame-mounted safety.

Production: current	Caliber: .22 LR	Action: SA, semi-auto	
Barrel Length: 3.6", 4.25"	Wt.: 14 oz.	Sights: fixed	Capacity: 10
Magazine: detachable box	Grips: black synthetic		

D2C:	NIB	$ 545	Ex $ 425	VG+ $ 300	Good $ 275	LMP $ 580
C2C:	Mint	$ 530	Ex $ 400	VG+ $ 280	Good $ 250	Fair $ 130
Trade-In:	Mint	$ 390	Ex $ 325	VG+ $ 220	Good $ 195	Poor $ 60

1911-22 COMPACT
NEW IN BOX

1911-22 COMPACT   REDBOOK CODE: RB-BW-H-1911X4

Matte blue finish, machined aluminum, target crown, frame-mounted safety, grip safety.

Production: 2011 - current	Caliber: .22 LR	Action: SA, semi-auto	
Barrel Length: 3.6"	OA Length: 6.5"	Sights: fixed	Capacity: 10
Magazine: detachable box	Grips: brown synthetic		

D2C:	NIB	$ 565	Ex $ 450	VG+ $ 315	Good $ 285	LMP $ 600
C2C:	Mint	$ 550	Ex $ 400	VG+ $ 290	Good $ 255	Fair $ 130
Trade-In:	Mint	$ 410	Ex $ 325	VG+ $ 230	Good $ 200	Poor $ 60

1911-22 BCA EDITION   REDBOOK CODE: RB-BW-H-1911X5
Limited edition, less than 150 units produced, gold engravings on the slide, blue finish, frame-mounted safety.

Production: 2011 - 2012		Caliber: .22 LR		Action: SA, semi-auto		Barrel Length: 3.6"	
Sights: fixed	Capacity: 10		Magazine: detachable box		Grips: walnut		
D2C:	NIB	$1,000	Ex $ 775	VG+ $ 550	Good $ 500		
C2C:	Mint	$ 960	Ex $ 700	VG+ $ 500	Good $ 450	Fair $ 230	
Trade-In:	Mint	$ 710	Ex $ 575	VG+ $ 390	Good $ 350	Poor $ 120	

BUL Ltd.

Founded in 1990, BUL Ltd. is located in Tel Aviv, Israel and first gained attention for their M-5 model – an affordable, high-capacity M1911-style pistol. The M-5 is noted for its lightweight, polymer design and modern aesthetics. The company has limited distribution in the United States but has gained a strong following for the M-5 and become a favorite within many handgun carry circles, especially for the Commander and Jet M-5 variants. Testament to the company's craftsmanship, BUL has manufactured pistol frames for a variety of firearms manufacturers, such as Charles Daly and Kimber.

M-5 GOVERNMENT
EXCELLENT

M-5 GOVERNMENT   REDBOOK CODE: RB-BL-H-M5GVRN
Black or stainless finish, frame-mounted and grip safety, polymer frame, similar to the M1911, made in Israel.

Production: mid-1990s - 2003		Caliber: 9mm, 9x21mm, .38 Super, .40 S&W, .45 ACP				
Action: SA, semi-auto		Barrel Length: 5"	OA Length: 8.6"			
Sights: fixed front	Capacity: 13, 17, 18		Magazine: detachable box	Grips: synthetic		
D2C:	NIB	$ 545	Ex $ 425	VG+ $ 300	Good $ 275	
C2C:	Mint	$ 530	Ex $ 400	VG+ $ 280	Good $ 250	Fair $ 130
Trade-In:	Mint	$ 390	Ex $ 325	VG+ $ 220	Good $ 195	Poor $ 60

M-5 COMMANDER   REDBOOK CODE: RB-BL-H-M5CMMD
Frame-mounted and grip safety, blue or nickel finish, skeletonized trigger, made in Israel.

Production: late 1990s - 2003		Caliber: 9mm, 9x21mm, .38 Super, .40 S&W, .45 ACP				
Action: SA, semi-auto		Barrel Length: 4.2"	OA Length: 7.9"			
Sights: fixed front	Capacity: 13, 17, 18		Magazine: detachable box	Grips: synthetic		
D2C:	NIB	$ 530	Ex $ 425	VG+ $ 295	Good $ 265	
C2C:	Mint	$ 510	Ex $ 375	VG+ $ 270	Good $ 240	Fair $ 125
Trade-In:	Mint	$ 380	Ex $ 300	VG+ $ 210	Good $ 190	Poor $ 60

M-5 JET   REDBOOK CODE: RB-BL-H-M5JETX
Black or stainless finish, frame-mounted and grip safety, skeletonized trigger. Discontinued around the turn of the 21st century.

Production: late 1990s - discontinued		Caliber: 9mm, 9x23, .38 Super, .40 S&W, .45 ACP				
Action: SA, semi-auto		Barrel Length: 4.2"	OA Length: 7.8"			
Capacity: 13, 17, 18		Magazine: detachable box	Grips: synthetic			
D2C:	NIB	$ 880	Ex $ 675	VG+ $ 485	Good $ 440	
C2C:	Mint	$ 850	Ex $ 625	VG+ $ 440	Good $ 400	Fair $ 205
Trade-In:	Mint	$ 630	Ex $ 500	VG+ $ 350	Good $ 310	Poor $ 90

CHEROKEE
NEW IN BOX
Courtesy of Bud's Gun Shop

CHEROKEE   REDBOOK CODE: RB-BL-H-CHRKE1
Frame-mounted safety, black finish, polymer frame.

Production: pre-2010 - discontinued   Caliber: 9mm   Action: DA/SA, semi-auto
Barrel Length: 4.25"   OA Length: 8"   Wt.: 24 oz.   Sights: fixed front
Capacity: 17, some variations   Magazine: detachable box   Grips: synthetic

D2C:	NIB	$ 490	Ex $ 375	VG+ $ 270	Good $ 245				
C2C:	Mint	$ 480	Ex $ 350	VG+ $ 250	Good $ 225	Fair $	115		
Trade-In:	Mint	$ 350	Ex $ 275	VG+ $ 200	Good $ 175	Poor $	60		

C.O. Arms

Now defunct, C.O. Arms of Millington, Tennessee (near Memphis) earned a reputation of producing some of the finest, custom made M1911s on the market. Information on the company remains ambiguous, but research indicates the company shut down sometime in 2011. Interest in their products still persists among M1911 enthusiasts and their pistols have retained much of their value.

M1911 SCORPION   REDBOOK CODE: RB-CX-H-1911XX
Blue or stainless finish, frame-mounted safety, skeletonized hammer.

Production: ~2011   Caliber: .45 ACP   Action: SA, semi-auto   Barrel Length: 4.3"
Sights: fixed, dot   Capacity: 8   Magazine: detachable box   Grips: wood

D2C:	NIB	$ 775	Ex $ 600	VG+ $ 430	Good $ 390		
C2C:	Mint	$ 750	Ex $ 550	VG+ $ 390	Good $ 350	Fair $	180
Trade-In:	Mint	$ 560	Ex $ 450	VG+ $ 310	Good $ 275	Poor $	90

M1911 AWP   REDBOOK CODE: RB-CX-H-1911AW
Tactical-style slide, blue or stainless finish, frame-mounted safety, tactical variations also available.

Production: ~2011   Caliber: .45 ACP   Action: SA, semi-auto   Barrel Length: 3.6"
Sights: fixed, dot   Capacity: 8   Magazine: detachable box   Grips: black synthetic or wood

D2C:	NIB	$ 815	Ex $ 625	VG+ $ 450	Good $ 410	LMP	$1,600
C2C:	Mint	$ 790	Ex $ 575	VG+ $ 410	Good $ 370	Fair $	190
Trade-In:	Mint	$ 580	Ex $ 475	VG+ $ 320	Good $ 290	Poor $	90

Calico Light Weapons Systems

Formed in 1985, Calico Light Weapons Systems originally produced .22 LR rifles, but soon transitioned into manufacturing 50 and 100 round helical-fed carbines and handguns. The company remains focused on carbines, but the handgun variants of these guns--though not suitable as carry weapons--are of note for their originality and innovation. The company, now located in Cornelius, Oregon, has become a recognized and trusted manufacturer of large capacity firearms and magazines.

M-110   REDBOOK CODE: RB-C1-H-M110XX
Carbine-style, prime-cast receiver, black matte finish, tactical design, helical feed magazine.

Production: current   Caliber: .22 LR   Action: SA, semi-auto   Barrel Length: 6"
OA Length: 18"   Wt.: 35 oz.   Sights: notch   Capacity: 100   Magazine: drum
Grips: black synthetic

D2C:	NIB	$ 585	Ex $ 450	VG+ $ 325	Good $ 295	LMP $	703
C2C:	Mint	$ 570	Ex $ 425	VG+ $ 300	Good $ 265	Fair $	135
Trade-In:	Mint	$ 420	Ex $ 350	VG+ $ 230	Good $ 205	Poor $	60

M-950 REDBOOK CODE: RB-C1-H-M950XX
Carbine-style, prime-cast receiver, helical feed magazine, black finish.

Production: 1990 - 2010 Caliber: 9mm Action: SA, semi-auto Barrel Length: 6"									
OA Length: 25.5" Wt.: 36 oz. Sights: notch Capacity: 50, 100 Magazine: drum									
Grips: black synthetic									
D2C:	NIB	$ 615	Ex	$ 475	VG+	$ 340	Good	$ 310	LMP	$ 890
C2C:	Mint	$ 600	Ex	$ 425	VG+	$ 310	Good	$ 280	Fair	$ 145
Trade-In:	Mint	$ 440	Ex	$ 350	VG+	$ 240	Good	$ 220	Poor	$ 90

Cabot Guns Co.

Over the last several years, Cabot Guns, of Cabot, Pennsylvania, has made a name for itself by producing a series of high-end M1911 pistols. Known for precision machining, many enthusiasts consider these guns on par or superior to many custom M1911s, given the quality of their design, engineering, and construction. Penn United Technologies, a high-precision, industrial manufacturing company, also of Cabot, PA, formed and operates Cabot Guns, their first venture into the firearms industry. Too few have sold on secondary markets to accurately price at lower grades.

AMERICAN JOE
NEW IN BOX

AMERICAN JOE REDBOOK CODE: RB-C2-H-AMRCJO
M1911 styling, steel frame and slide, flared ejection port, blue finish, ambidextrous design, stainless trigger, grip safety, frame-mounted safety, image of the American flag featured on the grips.

Production: current Caliber: .45 ACP Action: SA, semi-auto Barrel Length: 5"									
Sights: fixed front, adjustable rear Capacity: 8 Magazine: detachable box									
Grips: aluminum									
D2C:	NIB	$7,350	Ex	$5,600	VG+	--	Good	--	LMP	$7,450
C2C:	Mint	$7,060	Ex	$5,075	VG+	--	Good	--	Fair	--
Trade-In:	Mint	$5,220	Ex	$4,125	VG+	--	Good	--	Poor	--

BLACK DIAMOND
NEW IN BOX

BLACK DIAMOND REDBOOK CODE: RB-C2-H-BLCKDM
M1911 styling, steel frame and slide, skeletonized trigger and hammer, flared ejection port, polished black finish, ambidextrous design, stainless trigger, grip safety, frame-mounted safety.

Production: current Caliber: .45 ACP Action: SA, semi-auto Barrel Length: 5"									
Sights: adjustable tritium rear, fixed dot front Capacity: 8 Magazine: detachable box									
Grips: synthetic									
D2C:	NIB	$5,850	Ex	$4,450	VG+	--	Good	--	LMP	$5,950
C2C:	Mint	$5,620	Ex	$4,050	VG+	--	Good	--	Fair	--
Trade-In:	Mint	$4,160	Ex	$3,300	VG+	--	Good	--	Poor	--

JONES 1911
NEW IN BOX

JONES 1911 REDBOOK CODE: RB-C2-H-JNS911
M1911 styling, steel frame and slide, flared ejection port, blue finish, ambidextrous design, stainless trigger, grip safety, frame-mounted safety.

Production: current Caliber: .45 ACP Action: SA, semi-auto Barrel Length: 5"									
Sights: adjustable, 3-dot sights Capacity: 8 Magazine: detachable box Grips: synthetic									
D2C:	NIB	$5,875	Ex	$4,475	VG+	--	Good	--	LMP	$5,950
C2C:	Mint	$5,640	Ex	$4,075	VG+	--	Good	--	Fair	--
Trade-In:	Mint	$4,180	Ex	$3,300	VG+	--	Good	--	Poor	--

JONES DELUXE 1911
NEW IN BOX

JONES DELUXE 1911 REDBOOK CODE: RB-C2-H-JNSDLX

M1911 styling, similar to standard Jones 1911 model, hand fitted, frame-mounted safety, grip safety, polished blue finish, unique custom engravings on the slide.

Production: current Caliber: .45 ACP Action: SA, semi-auto Barrel Length: 5"

Wt.: 40 oz. Sights: Novak, adjustable Capacity: 8 Magazine: detachable box

Grips: wood, elegant design

D2C:	NIB	$9,875	Ex	$7,525	VG+	--	Good	--	LMP $9,950
C2C:	Mint	$9,480	Ex	$6,825	VG+	--	Good	--	Fair --
Trade-In:	Mint	$7,020	Ex	$5,550	VG+	--	Good	--	Poor --

RANGE MASTER
NEW IN BOX

RANGE MASTER REDBOOK CODE: RB-C2-H-RNGMST

M1911 styling, steel frame and slide, flared ejection port, aluminum trigger, polished blue finish, stripped frame near rear sight, extended thumb safety, frame-mounted safety, custom options available.

Production: current Caliber: .45 ACP Action: SA, semi-auto Barrel Length: 5"

Wt.: 40 oz. Sights: Novak adjustable Capacity: 8 Magazine: detachable box

Grips: Pachmayr, black synthetic

D2C:	NIB	$5,000	Ex	$3,800	VG+	--	Good	--	LMP $5,250
C2C:	Mint	$4,800	Ex	$3,450	VG+	--	Good	--	Fair --
Trade-In:	Mint	$3,550	Ex	$2,800	VG+	--	Good	--	Poor --

GI CLASSIC
NEW IN BOX

GI CLASSIC REDBOOK CODE: RB-C2-H-GICLSS

M1911A1 styling, hand-polished blue finish, GI-style, frame-mounted safety, GI-style magazine and grips.

Production: current Caliber: .45 ACP Action: SA, semi-auto Barrel Length: 5"

Wt.: 40 oz. Sights: fixed, dovetail Capacity: 8 Magazine: detachable box

Grips: brown synthetic, walnut

D2C:	NIB	$4,615	Ex	$3,525	VG+	--	Good	--	LMP $4,750
C2C:	Mint	$4,440	Ex	$3,200	VG+	--	Good	--	Fair --
Trade-In:	Mint	$3,280	Ex	$2,600	VG+	--	Good	--	Poor --

CGI CLASSIC
NEW IN BOX

CGI CLASSIC REDBOOK CODE: RB-C2-H-CGICLS

M1911A1 styling, polished blue finish, aluminum trigger, steel frame, frame-mounted safety, GI-style thumb safety.

Production: current Caliber: .45 ACP Action: SA, semi-auto Barrel Length: 5"

Sights: fixed, dovetail Capacity: 8 Magazine: detachable box

Grips: wood, elegant design

D2C:	NIB	$4,850	Ex	$3,700	VG+	--	Good	--	LMP $4,950
C2C:	Mint	$4,660	Ex	$3,350	VG+	--	Good	--	Fair --
Trade-In:	Mint	$3,450	Ex	$2,725	VG+	--	Good	--	Poor --

NATIONAL STANDARD
NEW IN BOX

NATIONAL STANDARD REDBOOK CODE: RB-C2-H-NATSTA

Modern M1911 styling, polished blue or stainless finish, enhanced trigger, rear and top slide serrations, grip safety, frame-mounted safety, skeletonized trigger.

Production: current Caliber: .45 ACP Action: SA, semi-auto Barrel Length: 5"

Sights: Novak adjustable Capacity: 8 Magazine: detachable box

Grips: hand polished, ironwood grips

D2C:	NIB	$6,115	Ex	$4,650	VG+	--	Good	--	LMP $6,250
C2C:	Mint	$5,880	Ex	$4,225	VG+	--	Good	--	Fair --
Trade-In:	Mint	$4,350	Ex	$3,425	VG+	--	Good	--	Poor --

THE SOUTH PAW 1911 REDBOOK CODE: RB-C2-H-STHPAW
Modern M1911 styling, left-handed design, matte gray or blue finish, left-handed frame-mounted safety, grip safety, skeletonized hammer, aluminum trigger, custom options available.

Production: current	Caliber: .45 ACP	Action: SA, semi-auto		Barrel Length: 5"						
Sights: Rozic front sight, fixed or adjustable rear		Capacity: 8	Magazine: detachable box							
Grips: ornate wood										
D2C:	NIB	$5,550	Ex	$4,225	VG+	--	Good	--	LMP	$5,750
C2C:	Mint	$5,330	Ex	$3,850	VG+	--	Good	--	Fair	--
Trade-In:	Mint	$3,950	Ex	$3,125	VG+	--	Good	--	Poor	--

THE SOUTH PAW 1911
NEW IN BOX

Caracal

Formed in Abu Dhabi, United Arab Emirates in 2007, Caracal manufactures a series of striker-fired, semi-automatic pistols suitable for both concealed and open carry. Designed by some of the engineers who helped develop some of the most popular Steyr pistols, the Caracal Model C gained notoriety for its sleek, ergonomic design and affordability. While the company later recalled the Model C for safety concerns, subsequent Caracal models have proven themselves to be reliable and accurate guns, especially given their very recent introduction into the market. Caracal seems primed to become a major force in the firearms industry in the years to come.

CP660 REDBOOK CODE: RB-C3-H-CPXXX1
Ambidextrous design, black finish, frame-mounted safety, improved on the design flaws of earlier Caracal Model C pistols.

Production: 2013 - current	Caliber: 9mm	Action: striker-fired, semi-auto							
Barrel Length: 4"	OA Length: 7.6"	Wt.: 26 oz.	Sights: fixed, dot						
Capacity: 18	Magazine: detachable box	Grips: black synthetic							
D2C:	NIB	$ 515	Ex	$ 400	VG+	$ 285	Good	$ 260	
C2C:	Mint	$ 500	Ex	$ 375	VG+	$ 260	Good	$ 235	Fair $ 120
Trade-In:	Mint	$ 370	Ex	$ 300	VG+	$ 210	Good	$ 185	Poor $ 60

CP661 REDBOOK CODE: RB-C3-H-CPXXX2
Ambidextrous design, black finish, frame-mounted safety, longer barrel than the CP662.

Production: 2013 - current	Caliber: 9mm	Action: striker-fired, semi-auto							
Barrel Length: 3.7"	OA Length: 7.2"	Wt.: 24 oz.	Sights: fixed, dot						
Capacity: 15	Magazine: detachable box	Grips: black synthetic							
D2C:	NIB	$ 495	Ex	$ 400	VG+	$ 275	Good	$ 250	
C2C:	Mint	$ 480	Ex	$ 350	VG+	$ 250	Good	$ 225	Fair $ 115
Trade-In:	Mint	$ 360	Ex	$ 300	VG+	$ 200	Good	$ 175	Poor $ 60

CP662 REDBOOK CODE: RB-C3-H-CPXXX3
Ambidextrous design, black finish, frame-mounted safety, smaller barrel than the rest of the CP series pistols.

Production: 2013 - current	Caliber: 9mm	Action: striker-fired, semi-auto							
Barrel Length: 3.4"	OA Length: 6.6"	Wt.: 22 oz.	Sights: fixed, dot						
Capacity: 13	Magazine: detachable box	Grips: black synthetic							
D2C:	NIB	$ 490	Ex	$ 375	VG+	$ 270	Good	$ 245	
C2C:	Mint	$ 480	Ex	$ 350	VG+	$ 250	Good	$ 225	Fair $ 115
Trade-In:	Mint	$ 350	Ex	$ 275	VG+	$ 200	Good	$ 175	Poor $ 60

CARACAL

CP663 REDBOOK CODE: RB-C3-H-CPXXX4
Ambidextrous design, black finish, frame-mounted safety, longer barrel than that of the other CP series pistols.

Production: 2013 - current	Caliber: 9mm, 9x21mm, .40 S&W								
Action: striker-fired, semi-auto	Barrel Length: 4.7"	OA Length: 8"	Wt.: 35 oz.						
Sights: fixed, dot	Capacity: 18	Magazine: detachable box	Grips: wood						
D2C:	NIB	$ 500	Ex	$ 400	VG+	$ 275	Good	$ 250	
C2C:	Mint	$ 480	Ex	$ 350	VG+	$ 250	Good	$ 225	Fair $ 115
Trade-In:	Mint	$ 360	Ex	$ 300	VG+	$ 200	Good	$ 175	Poor $ 60

MODEL C
VERY GOOD +

MODEL C REDBOOK CODE: RB-C3-H-MCXXXX
A smaller version of Caracal Model F, recalled by the company in 2013, black finish, frame-mounted safety.

Production: 2007 - 2013	Caliber: 9mm, .357 SIG, .40 S&W	Action: striker-fired, semi-auto							
Barrel Length: 3.7", 4"	Sights: fixed, dot	Capacity: 15	Magazine: detachable box						
Grips: black synthetic									
D2C:	NIB	$ 430	Ex	$ 350	VG+	$ 240	Good	$ 215	LMP $ 720
C2C:	Mint	$ 420	Ex	$ 300	VG+	$ 220	Good	$ 195	Fair $ 100
Trade-In:	Mint	$ 310	Ex	$ 250	VG+	$ 170	Good	$ 155	Poor $ 60

MODEL F
VERY GOOD +

MODEL F REDBOOK CODE: RB-C3-H-MFXXXX
One of the handguns first produced by the company, later replaced by the CP660 model.

Production: 2007 - 2013	Caliber: 9mm, 9x23mm, .357 SIG, .40 S&W								
Action: striker-fired, semi-auto	Barrel Length: 4"	OA Length: 7"	Wt.: 26 oz.						
Sights: fixed, dot	Capacity: 18	Magazine: detachable box	Grips: black synthetic						
D2C:	NIB	$ 450	Ex	$ 350	VG+	$ 250	Good	$ 225	LMP $ 720
C2C:	Mint	$ 440	Ex	$ 325	VG+	$ 230	Good	$ 205	Fair $ 105
Trade-In:	Mint	$ 320	Ex	$ 275	VG+	$ 180	Good	$ 160	Poor $ 60

MODEL SC
VERY GOOD +

MODEL SC REDBOOK CODE: RB-C3-H-MSCXXX
Black finish, trigger and firing pin safety, similar to the Caracal Model F.

Production: 2009 - current	Caliber: 9mm, 9x21mm	Action: striker-fired, semi-auto							
Barrel Length: 3.4"	OA Length: 6.3"	Sights: fixed, dot	Capacity: 13						
Magazine: detachable box	Grips: black synthetic								
D2C:	NIB	$ 430	Ex	$ 350	VG+	$ 240	Good	$ 215	
C2C:	Mint	$ 420	Ex	$ 300	VG+	$ 220	Good	$ 195	Fair $ 100
Trade-In:	Mint	$ 310	Ex	$ 250	VG+	$ 170	Good	$ 155	Poor $ 60

Charles Daly

Charles Daly firearms began by contracting foreign manufacturers to produce firearms under the Charles Daly name. They continued importing firearms under their name until 2010. Recently, the company announced they will resume importations. In their current incarnation, the company only produces shotguns. Other manufacturers, such as BUL Ltd. and Armscor, have produced many Charles Daly pistols, and these guns retain moderate interest among handgun carry enthusiasts, given their relative affordability. The company's M1911 Empire ECMT model is of note.

M1911A1 GOVERNMENT FIELD EFS/FS
REDBOOK CODE: RB-C5-H-1911X1
Stainless or matte blue finish, frame-mounted safety, checkered grips, skeletonized hammer and trigger.

Production: late 1990s - 2008	Caliber: .45 ACP	Action: SA, semi-auto			
Barrel Length: 5"	OA Length: 8.7"	Wt.: 40 oz.	Sights: fixed		
Capacity: 8	Magazine: detachable box	Grips: wood			
D2C:	NIB $ 435	Ex $ 350	VG+ $ 240	Good $ 220	LMP $ 590
C2C:	Mint $ 420	Ex $ 325	VG+ $ 220	Good $ 200	Fair $ 105
Trade-In:	Mint $ 310	Ex $ 250	VG+ $ 170	Good $ 155	Poor $ 60

M1911 EMPIRE ECMT   REDBOOK CODE: RB-C5-H-1911X7
Stainless finish, skeletonized hammer and trigger, frame-mounted thumb safety, checkered grips.

Production: 2001 - discontinued	Caliber: .45 ACP	Action: SA, semi-auto			
Barrel Length: 5"	OA Length: 8.7"	Wt.: 40 oz.	Sights: dovetail		
Capacity: 8	Magazine: detachable box	Grips: wood			
D2C:	NIB $ 715	Ex $ 550	VG+ $ 395	Good $ 360	LMP $ 900
C2C:	Mint $ 690	Ex $ 500	VG+ $ 360	Good $ 325	Fair $ 165
Trade-In:	Mint $ 510	Ex $ 425	VG+ $ 280	Good $ 255	Poor $ 90

M1911A1 EMPIRE EFST   REDBOOK CODE: RB-C5-H-1911X8
Skeletonized hammer and trigger, checkered grips, frame-mounted safety, recoil operated, blue finish.

Production: early 2000s - 2007	Caliber: .45 ACP	Action: SA, semi-auto			
Barrel Length: 5"	OA Length: 8.7"	Wt.: 40 oz.	Sights: adjustable		
Capacity: 8	Magazine: detachable box	Grips: wood			
D2C:	NIB $ 635	Ex $ 500	VG+ $ 350	Good $ 320	LMP $ 791
C2C:	Mint $ 610	Ex $ 450	VG+ $ 320	Good $ 290	Fair $ 150
Trade-In:	Mint $ 460	Ex $ 375	VG+ $ 250	Good $ 225	Poor $ 90

M1911A1 FIELD EFST   REDBOOK CODE: RB-C5-H-191110
Skeletonized hammer and trigger, beavertail-style grip safety, matte-blue finish, steel frame.

Production: 2000 - 2007	Caliber: .45 ACP	Action: SA, semi-auto			
Barrel Length: 5"	OA Length: 8.7"	Wt.: 40 oz.	Sights: adjustable		
Capacity: 8	Magazine: detachable box	Grips: wood			
D2C:	NIB $ 510	Ex $ 400	VG+ $ 285	Good $ 255	LMP $ 681
C2C:	Mint $ 490	Ex $ 375	VG+ $ 260	Good $ 230	Fair $ 120
Trade-In:	Mint $ 370	Ex $ 300	VG+ $ 200	Good $ 180	Poor $ 60

Charter Arms

The current incarnation of Charter Arms formed in the late 1990s, though its origins trace back to the 1960s. Douglas McClennahan, a noted gun designer, established Charter after previously working with Colt, High Standard, and Ruger, hoping to make high-quality, affordable handguns. Located in Shelton, Connecticut, the company now makes an extensive array of small-frame and snubbed-nose revolvers, the Bulldog the most noteworthy of these. For the price, these guns stand as a viable option for those looking to carry a concealed weapon in public.

MODEL 79K   REDBOOK CODE: RB-CH-H-MOD79K

An early Charter pistol, stainless finish, slide-mounted safety, similar to the M1911's design.

Production: 1980s	Caliber: .380 ACP	Action: SA, semi-auto								
Barrel Length: 3.2"	OA Length: 6.5"	Sights: fixed	Capacity: 7							
Magazine: detachable box	Grips: wood									
D2C:	NIB	$ 315	Ex	$ 250	VG+	$ 175	Good	$ 160		
C2C:	Mint	$ 310	Ex	$ 225	VG+	$ 160	Good	$ 145	Fair	$ 75
Trade-In:	Mint	$ 230	Ex	$ 200	VG+	$ 130	Good	$ 115	Poor	$ 60

SOUTHPAW   REDBOOK CODE: RB-CH-H-SOUTHP

Hammer block safety, aluminum frame, matte-stainless finish, left-handed design, also offered in a two-tone Pink Lady model.

Production: 2008 - current	Caliber: .38 Special +P	Action: DA/SA, revolver								
Barrel Length: 2"	Wt.: 12 oz.	Sights: fixed	Capacity: 5	Grips: full, black synthetic						
D2C:	NIB	$ 380	Ex	$ 300	VG+	$ 210	Good	$ 190	LMP	$ 428
C2C:	Mint	$ 370	Ex	$ 275	VG+	$ 190	Good	$ 175	Fair	$ 90
Trade-In:	Mint	$ 270	Ex	$ 225	VG+	$ 150	Good	$ 135	Poor	$ 60

PATHFINDER
NEW IN BOX

PATHFINDER (72224)   REDBOOK CODE: RB-CH-H-PTHFND

Stainless finish and frame, compact design, shrouded ejector rod.

Production: 2002 - current	Caliber: .22 LR	Action: DA/SA, revolver								
Barrel Length: 2"	Wt.: 19 oz.	Sights: fixed, notch	Capacity: 6							
Grips: full, black synthetic										
D2C:	NIB	$ 330	Ex	$ 275	VG+	$ 185	Good	$ 165	LMP	$ 362
C2C:	Mint	$ 320	Ex	$ 250	VG+	$ 170	Good	$ 150	Fair	$ 80
Trade-In:	Mint	$ 240	Ex	$ 200	VG+	$ 130	Good	$ 120	Poor	$ 60

TARGET PATHFINDER
NEW IN BOX

TARGET PATHFINDER (72240)   REDBOOK CODE: RB-CH-H-TRGPTH

Stainless finish, steel frame and barrel, noted for its extended barrel and adjustable sights.

Production: 2002 - current	Caliber: .22 LR	Action: DA/SA, revolver								
Barrel Length: 4"	Wt.: 20 oz.	Sights: adjustable target	Capacity: 6							
Grips: full, black synthetic										
D2C:	NIB	$ 345	Ex	$ 275	VG+	$ 190	Good	$ 175	LMP	$405
C2C:	Mint	$ 340	Ex	$ 250	VG+	$ 180	Good	$ 160	Fair	$80
Trade-In:	Mint	$ 250	Ex	$ 200	VG+	$ 140	Good	$ 125	Poor	$60

PATHFINDER .22 MAGNUM
NEW IN BOX

PATHFINDER .22 MAGNUM   REDBOOK CODE: RB-CH-H-PTHF22

Stainless steel frame, compact design, noted for firing .22 Magnum rounds.

Production: current	Caliber: .22 Magnum	Action: DA/SA, revolver								
Barrel Length: 2"	Wt.: 19 oz.	Sights: fixed, notch	Capacity: 6							
Grips: full, black synthetic										
D2C:	NIB	$ 365	Ex	$ 300	VG+	$ 205	Good	$ 185	LMP	$ 363
C2C:	Mint	$ 360	Ex	$ 275	VG+	$ 190	Good	$ 165	Fair	$ 85
Trade-In:	Mint	$ 260	Ex	$ 225	VG+	$ 150	Good	$ 130	Poor	$ 60

TARGET PATHFINDER COMBO REDBOOK CODE: RB-CH-H-TRGPCM
Noted for its interchangeable cylinders. Stainless steel frame, standard hammer.

Production: current Caliber: .22 LR /.22 Magnum Action: DA/SA, revolver									
Barrel Length: 4" Wt.: 20 oz. Sights: adjustable Capacity: 6									
Grips: full, black rubber									
D2C:	NIB	$ 475	Ex $ 375	VG+ $ 265	Good $ 240	LMP $ 548
C2C:	Mint	$ 460	Ex $ 350	VG+ $ 240	Good $ 215	Fair $ 110
Trade-In:	Mint	$ 340	Ex $ 275	VG+ $ 190	Good $ 170	Poor $ 60

**UNDERCOVER
STANDARD**
NEW IN BOX

UNDERCOVER, BLUE STANDARD (13820)
REDBOOK CODE: RB-CH-H-UNDSTD
Blue finish, standard hammer, stainless steel frame, compact design.

Production: current Caliber: .38 Special +P Action: DA/SA, revolver						
Barrel Length: 2" Wt.: 16 oz. Sights: fixed Capacity: 5 Grips: full						
D2C:	NIB	$ 290	Ex $ 225	VG+ $ 160	Good $ 145	LMP $ 342
C2C:	Mint	$ 280	Ex $ 225	VG+ $ 150	Good $ 135	Fair $ 70
Trade-In:	Mint	$ 210	Ex $ 175	VG+ $ 120	Good $ 105	Poor $ 30

UNDERCOVER DAO
NEW IN BOX

UNDERCOVER, BLUE DAO (13811)
REDBOOK CODE: RB-CH-H-UNDDAO
Blue finish, DAO, shaved hammer, compact design.

Production: current Caliber: .38 Special +P Action: DAO, revolver Barrel Length: 2"						
Wt.: 16 oz. Sights: fixed Capacity: 5 Grips: compact, black synthetic						
D2C:	NIB	$ 300	Ex $ 250	VG+ $ 165	Good $ 150	LMP $ 348
C2C:	Mint	$ 290	Ex $ 225	VG+ $ 150	Good $ 135	Fair $ 70
Trade-In:	Mint	$ 220	Ex $ 175	VG+ $ 120	Good $ 105	Poor $ 30

**UNDERCOVER
TIGER**
NEW IN BOX

UNDERCOVER TIGER (13825) REDBOOK CODE: RB-CH-H-UNDTGR
Stainless steel frame, tiger striped and black finish, double/single action hammer, rubber grip, compact design.

Production: current Caliber: .38 Special +P Action: DA/SA, revolver						
Barrel Length: 2" Wt.: 16 oz. Sights: fixed Capacity: 5 Grips: black rubber						
D2C:	NIB	$ 395	Ex $ 325	VG+ $ 220	Good $ 200	LMP $ 400
C2C:	Mint	$ 380	Ex $ 275	VG+ $ 200	Good $ 180	Fair $ 95
Trade-In:	Mint	$ 290	Ex $ 225	VG+ $ 160	Good $ 140	Poor $ 60

**UNDERCOVER
GREEN & BLACK**
NEW IN BOX

UNDERCOVER, OD GREEN & BLACK STANDARD (23820)
REDBOOK CODE: RB-CH-H-UNDGRB
OD Green and black matte finish, coated stainless steel frame, double/single action trigger, compact design.

Production: current Caliber: .38 Special +P Action: DA/SA, revolver						
Barrel Length: 2" Wt.: 16 oz. Sights: fixed Capacity: 5 Grips: black synthetic						
D2C:	NIB	$ 370	Ex $ 300	VG+ $ 205	Good $ 185	LMP $ 379
C2C:	Mint	$ 360	Ex $ 275	VG+ $ 190	Good $ 170	Fair $ 90
Trade-In:	Mint	$ 270	Ex $ 225	VG+ $ 150	Good $ 130	Poor $ 60

**UNDERCOVER
STAINLESS DAO**
NEW IN BOX

UNDERCOVER, STAINLESS DAO (73811)
REDBOOK CODE: RB-CH-H-UNDAO2
Stainless steel frame, DAO hammer, stainless finish, compact design.

Production: late 1990s - current Caliber: .38 Special +P Action: DAO, revolver						
Barrel Length: 2" Wt.: 16 oz. Sights: fixed Capacity: 5 Grips: compact, black synthetic						
D2C:	NIB	$ 315	Ex $ 250	VG+ $ 175	Good $ 160	LMP $ 360
C2C:	Mint	$ 310	Ex $ 225	VG+ $ 160	Good $ 145	Fair $ 75
Trade-In:	Mint	$ 230	Ex $ 200	VG+ $ 130	Good $ 115	Poor $ 60

CHARTER ARMS

UNDERCOVER STAINLESS
NEW IN BOX

UNDERCOVER LITE BLACK STD.
NEW IN BOX

UNDERCOVER LITE BLACK DAO
NEW IN BOX

UNDERCOVER LITE, BRONZE & BLACK STANDARD
NEW IN BOX

UNDERCOVER LITE STANDARD
NEW IN BOX

UNDERCOVER STAINLESS REDBOOK CODE: RB-CH-H-UNDSTN
Stainless steel frame and finish, standard hammer, compact design.

Production: late 1990s - current Caliber: .38 Special +P Action: DA/SA, revolver
Barrel Length: 2" Wt.: 16 oz. Sights: fixed Capacity: 5 Grips: full, black synthetic

D2C:	NIB	$ 295	Ex $ 225	VG+ $ 165	Good $ 150	LMP $ 355
C2C:	Mint	$ 290	Ex $ 225	VG+ $ 150	Good $ 135	Fair $ 70
Trade-In:	Mint	$ 210	Ex $ 175	VG+ $ 120	Good $ 105	Poor $ 30

UNDERCOVER LITE BLACK STD. REDBOOK CODE: RB-CH-H-UNDLBL
Two-tone black and stainless finish, aluminum frame, spurred hammer, compact design.

Production: current Caliber: .38 Special +P Action: DA/SA, revolver
Barrel Length: 2" Wt.: 12 oz. Sights: fixed Capacity: 5 Grips: standard

D2C:	NIB	$ 370	Ex $ 300	VG+ $ 205	Good $ 185	LMP $ 410
C2C:	Mint	$ 360	Ex $ 275	VG+ $ 190	Good $ 170	Fair $ 90
Trade-In:	Mint	$ 270	Ex $ 225	VG+ $ 150	Good $ 130	Poor $ 60

UNDERCOVER LITE, BLACK DAO REDBOOK CODE: RB-CH-H-UNDBDA
Black finish, aluminum frame, DAO hammer, compact design.

Production: current Caliber: .38 Special +P Action: DAO, revolver
Barrel Length: 2" Wt.: 12 oz. Sights: fixed Capacity: 5 Grips: standard

D2C:	NIB	$ 405	Ex $ 325	VG+ $ 225	Good $ 205	LMP $ 415
C2C:	Mint	$ 390	Ex $ 300	VG+ $ 210	Good $ 185	Fair $ 95
Trade-In:	Mint	$ 290	Ex $ 250	VG+ $ 160	Good $ 145	Poor $ 60

UNDERCOVER LITE, BLACK & HI-POLISH STAINLESS STD.
REDBOOK CODE: RB-CH-H-UNDLHP
Black matte aluminum frame, high-polished cylinder and barrel, standard hammer, compact design.

Production: current Caliber: .38 Special +P Action: DA/SA, revolver
Barrel Length: 2" Wt.: 12 oz. Sights: fixed Capacity: 5 Grips: compact

D2C:	NIB	$ 405	Ex $ 325	VG+ $ 225	Good $ 205	LMP $ 428
C2C:	Mint	$ 390	Ex $ 300	VG+ $ 210	Good $ 185	Fair $ 95
Trade-In:	Mint	$ 290	Ex $ 250	VG+ $ 160	Good $ 145	Poor $ 60

UNDERCOVER LITE, BRONZE & BLACK STANDARD
REDBOOK CODE: RB-CH-H-UNDBRZ
Two-tone bronze and black finish, aluminum frame, spurred hammer, fixed sights, compact design.

Production: current Caliber: .38 Special +P Action: DA/SA, revolver
Barrel Length: 2" Wt.: 12 oz. Sights: fixed Capacity: 5 Grips: standard

D2C:	NIB	$ 370	Ex $ 300	VG+ $ 205	Good $ 185	LMP $ 422
C2C:	Mint	$ 360	Ex $ 275	VG+ $ 190	Good $ 170	Fair $ 90
Trade-In:	Mint	$ 270	Ex $ 225	VG+ $ 150	Good $ 130	Poor $ 60

UNDERCOVER LITE STANDARD REDBOOK CODE: RB-CH-H-UNDSTD
Spurred hammer, aluminum frame and finish, fixed sights, compact design.

Production: current Caliber: .38 Special +P Action: DA/SA, revolver
Barrel Length: 2" Wt.: 12 oz. Sights: fixed Capacity: 5 Grips: black synthetic

D2C:	NIB	$ 335	Ex $ 275	VG+ $ 185	Good $ 170	LMP $ 393
C2C:	Mint	$ 330	Ex $ 250	VG+ $ 170	Good $ 155	Fair $ 80
Trade-In:	Mint	$ 240	Ex $ 200	VG+ $ 140	Good $ 120	Poor $ 60

UNDERCOVER LITE, RED & STAINLESS STANDARD
NEW IN BOX

UNDERCOVER LITE, RED & BLACK STANDARD
NEW IN BOX

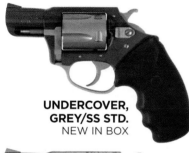

UNDERCOVER, GREY/SS STD.
NEW IN BOX

COUGAR UNDERCOVER LITE, PINK & STAINLESS STD.
NEW IN BOX

SANTA FE UNDERCOVER LITE, TURQUOISE & STAINLESS STD.
NEW IN BOX

CRIMSON UNDERCOVER
NEW IN BOX

UNDERCOVER LITE, RED & STAINLESS STANDARD

(53823) REDBOOK CODE: RB-CH-H-UNDRDS
Two-tone red and stainless finish, aluminum frame, standard hammer, compact design.

Production: current Caliber: .38 Special +P Action: DA/SA, revolver
Barrel Length: 2" Wt.: 12 oz. Sights: fixed Capacity: 5 Grips: standard, rubber

D2C:	NIB	$ 365	Ex $ 300	VG+ $ 205	Good $ 185	LMP $ 410
C2C:	Mint	$ 360	Ex $ 275	VG+ $ 190	Good $ 165	Fair $ 85
Trade-In:	Mint	$ 260	Ex $ 225	VG+ $ 150	Good $ 130	Poor $ 60

UNDERCOVER LITE, RED & BLACK STANDARD

(53824) REDBOOK CODE: RB-CH-H-UNDRBL
Two-tone red and black finish, aluminum frame, standard hammer, compact design.

Production: current Caliber: .38 Special +P Action: DA/SA, revolver
Barrel Length: 2" Wt.: 12 oz. Sights: fixed Capacity: 5 Grips: standard, black rubber

D2C:	NIB	$ 360	Ex $ 275	VG+ $ 200	Good $ 180	LMP $ 410
C2C:	Mint	$ 350	Ex $ 250	VG+ $ 180	Good $ 165	Fair $ 85
Trade-In:	Mint	$ 260	Ex $ 225	VG+ $ 150	Good $ 130	Poor $ 60

UNDERCOVER, GREY/SS STD. REDBOOK CODE: RB-CH-H-UNDGRY

Dark grey and stainless finish, steel frame, standard hammer, compact design.

Production: current Caliber: .38 Special +P Action: DA/SA, revolver
Barrel Length: 2" Wt.: 16 oz. Sights: fixed Capacity: 5 Grips: standard, rubber

D2C:	NIB	$ 360	Ex $ 275	VG+ $ 200	Good $ 180	LMP $ 384
C2C:	Mint	$ 350	Ex $ 250	VG+ $ 180	Good $ 165	Fair $ 85
Trade-In:	Mint	$ 260	Ex $ 225	VG+ $ 150	Good $ 130	Poor $ 60

COUGAR UNDERCOVER LITE, PINK & STAINLESS STANDARD

(53833) REDBOOK CODE: RB-CH-H-UNDLPS
Spurred hammer, two-tone pink and stainless finish, aluminum frame, compact design.

Production: current Caliber: .38 Special +P Action: DA/SA, revolver
Barrel Length: 2" Wt.: 12 oz. Sights: fixed Capacity: 5
Grips: compact, black synthetic

D2C:	NIB	$ 415	Ex $ 325	VG+ $ 230	Good $ 210	LMP $ 429
C2C:	Mint	$ 400	Ex $ 300	VG+ $ 210	Good $ 190	Fair $ 100
Trade-In:	Mint	$ 300	Ex $ 250	VG+ $ 170	Good $ 150	Poor $ 60

SANTA FE UNDERCOVER LITE, TURQUOISE & STAINLESS STANDARD

(53860) REDBOOK CODE: RB-CH-H-UNDSFE
Spurred hammer, two-tone turquoise and stainless finish, aluminum frame, standard hammer, compact design.

Production: current Caliber: .38 Special +P Action: DA/SA, revolver
Barrel Length: 2" Wt.: 12 oz. Sights: fixed Capacity: 5 Grips: standard, synthetic

D2C:	NIB	$ 355	Ex $ 275	VG+ $ 200	Good $ 180	LMP $ 422
C2C:	Mint	$ 350	Ex $ 250	VG+ $ 180	Good $ 160	Fair $ 85
Trade-In:	Mint	$ 260	Ex $ 200	VG+ $ 140	Good $ 125	Poor $ 60

CRIMSON UNDERCOVER (73824) REDBOOK CODE: RB-CH-H-UNDCRM

Stainless steel frame, stainless finish, standard hammer, compact design.

Production: current Caliber: .38 Special +P Action: DA/SA, revolver
Barrel Length: 2" Wt.: 16 oz. Sights: fixed Capacity: 5
Grips: Crimson Trace Lasergrips

D2C:	NIB	$ 535	Ex $ 425	VG+ $ 295	Good $ 270	LMP $ 571
C2C:	Mint	$ 520	Ex $ 375	VG+ $ 270	Good $ 245	Fair $ 125
Trade-In:	Mint	$ 380	Ex $ 300	VG+ $ 210	Good $ 190	Poor $ 60

**POLICE
UNDERCOVER**
NEW IN BOX

POLICE UNDERCOVER (73840) REDBOOK CODE: RB-CH-H-PLCUND
Steel frame, stainless finish, standard hammer, compact design.

Production:	current	Caliber: .38 Special +P		Action: DA/SA, revolver				
Barrel Length: 2"	Wt.: 20 oz.	Sights: fixed	Capacity: 6	Grips: full, rubber				
D2C:	NIB $ 335	Ex $ 275	VG+ $ 185	Good $ 170	LMP $ 390
C2C:	Mint $ 330	Ex $ 250	VG+ $ 170	Good $ 155	Fair $ 80
Trade-In:	Mint $ 240	Ex $ 200	VG+ $ 140	Good $ 120	Poor $ 60

**CRIMSON
UNDERCOVERETTE**
NEW IN BOX

CRIMSON UNDERCOVERETTE REDBOOK CODE: RB-CH-H-CRMUND
Standard hammer, steel frame, crimson finish, compact design, light recoil.

Production:	current	Caliber: .32 H&R Magnum		Action: DA/SA, revolver				
Barrel Length: 2"	Wt.: 16 oz.	Sights: fixed	Capacity: 5					
Grips: Crimson Trace Lasergrips								
D2C:	NIB $ 490	Ex $ 375	VG+ $ 270	Good $ 245	LMP $ 525
C2C:	Mint $ 480	Ex $ 350	VG+ $ 250	Good $ 225	Fair $ 115
Trade-In:	Mint $ 350	Ex $ 275	VG+ $ 200	Good $ 175	Poor $ 60

**UNDERCOVERETTE,
SS STD.**
NEW IN BOX

UNDERCOVERETTE, SS STD. REDBOOK CODE: RB-CH-H-UNDSSS
Standard hammer, stainless steel frame and finish, shrouded ejector rod, compact design.

Production:	current	Caliber: .32 H&R Magnum		Action: DA/SA, revolver				
Barrel Length: 2"	Wt.: 16 oz.	Sights: fixed	Capacity: 5	Grips: black rubber				
D2C:	NIB $ 345	Ex $ 275	VG+ $ 190	Good $ 175	LMP $ 415
C2C:	Mint $ 340	Ex $ 250	VG+ $ 180	Good $ 160	Fair $ 80
Trade-In:	Mint $ 250	Ex $ 200	VG+ $ 140	Good $ 125	Poor $ 60

**PITBULL
9MM RIMLESS**
NEW IN BOX

PITBULL 9MM RIMLESS (79920) REDBOOK CODE: RB-CH-H-PITRML
Stainless steel glass beaded frame, matte finish, spurred hammer.

Production:	2012 - current	Caliber: 9mm	Action: DA/SA, revolver					
Barrel Length: 2.2"	Wt.: 22 oz.	Sights: fixed	Capacity: 6	Grips: Neoprene				
D2C:	NIB $ 375	Ex $ 300	VG+ $ 210	Good $ 190	LMP $ 496
C2C:	Mint $ 360	Ex $ 275	VG+ $ 190	Good $ 170	Fair $ 90
Trade-In:	Mint $ 270	Ex $ 225	VG+ $ 150	Good $ 135	Poor $ 60

PITBULL (74020) REDBOOK CODE: RB-CH-H-PITBLL
Steel frame, matte stainless finish, standard hammer, full grip, DAO hammer optional, compact design, shrouded ejector rod. This model's dual coil spring assembly allows for the use of rimless cartridges without the use of half or full moon clips.

Production:	2011 - current	Caliber: .40 S&W		Action: DA/SA, revolver				
Barrel Length: 2", 2.3"	Wt.: 20 oz.	Sights: fixed	Capacity: 5	Grips: full, synthetic				
D2C:	NIB $ 375	Ex $ 300	VG+ $ 210	Good $ 190	LMP $ 484
C2C:	Mint $ 360	Ex $ 275	VG+ $ 190	Good $ 170	Fair $ 90
Trade-In:	Mint $ 270	Ex $ 225	VG+ $ 150	Good $ 135	Poor $ 60

BULLDOG
NEW IN BOX

BULLDOG (14420) REDBOOK CODE: RB-CH-H-BLLDOG
Blue finish, steel frame, standard hammer, also offered in target model with 4" barrel, compact design, shrouded ejector rod.

Production:	early 1970s - current	Caliber: .44 Special						
Action: DA/SA, revolver	Barrel Length: 2.5"	Wt.: 21 oz.	Sights: fixed					
Capacity: 5	Grips: full, synthetic							
D2C:	NIB $ 345	Ex $ 275	VG+ $ 190	Good $ 175	LMP $ 405
C2C:	Mint $ 340	Ex $ 250	VG+ $ 180	Good $ 160	Fair $ 80
Trade-In:	Mint $ 250	Ex $ 200	VG+ $ 140	Good $ 125	Poor $ 60

CHARTER ARMS

BULLDOG, DAO
NEW IN BOX

BULLDOG, DAO   REDBOOK CODE: RB-CH-H-BLLDAO
Stainless or blue finish, steel frame, target model offered with 4" barrel and adjustable rear sight, shrouded ejector rod, compact design.

Production: current Caliber: .44 Special Action: DAO, revolver Barrel Length: 2.5"
Wt.: 21 oz.   Sights: fixed Capacity: 5 Grips: full, synthetic

D2C:	NIB	$ 375	Ex $ 300	VG+ $ 210	Good $ 190	LMP $ 465
C2C:	Mint	$ 360	Ex $ 275	VG+ $ 190	Good $ 170	Fair $ 90
Trade-In:	Mint	$ 270	Ex $ 225	VG+ $ 150	Good $ 135	Poor $ 60

BULLDOG .44 SPECIAL TIGER   REDBOOK CODE: RB-CH-H-BLLTGR
Black and tiger finish, steel frame, standard hammer, target model 4" barrel, compact design, shrouded ejector rod.

Production: 2011 - current Caliber: .44 Special, .357 Magnum
Action: DA/SA, revolver   Barrel Length: 2.5", 4" Wt.: 21 oz.
Sights: fixed Capacity: 5 Grips: full, synthetic

D2C:	NIB	$ 375	Ex $ 300	VG+ $ 210	Good $ 190	LMP $ 457
C2C:	Mint	$ 360	Ex $ 275	VG+ $ 190	Good $ 170	Fair $ 90
Trade-In:	Mint	$ 270	Ex $ 225	VG+ $ 150	Good $ 135	Poor $ 60

BULLDOG ON DUTY
NEW IN BOX

BULLDOG ON DUTY (74410) REDBOOK CODE: RB-CH-H-BLLSSS
Stainless frame and finish, standard shrouded hammer and ejector rod, black-rubber grip, compact design.

Production: current Caliber: .44 Special Action: DA/SA, revolver
Barrel Length: 2.5" Wt.: 21 oz. Sights: fixed Capacity: 5 Grips: standard, rubber

D2C:	NIB	$ 355	Ex $ 275	VG+ $ 200	Good $ 180	LMP $ 428
C2C:	Mint	$ 350	Ex $ 250	VG+ $ 180	Good $ 160	Fair $ 85
Trade-In:	Mint	$ 260	Ex $ 200	VG+ $ 140	Good $ 125	Poor $ 60

BULLDOG, STAINLESS STD.
NEW IN BOX

BULLDOG, STAINLESS STD. REDBOOK CODE: RB-CH-H-BLLSTS
Standard hammer, steel finish and frame, compact design, shrouded ejector rod, target model offered with 4" barrel and adjustable rear sight.

Production: current Caliber: .44 Special Action: DA/SA, revolver
Barrel Length: 2.5" Wt.: 21 oz. Sights: fixed Capacity: 5 Grips: full, synthetic

D2C:	NIB	$ 345	Ex $ 275	VG+ $ 190	Good $ 175	LMP $ 418
C2C:	Mint	$ 340	Ex $ 250	VG+ $ 180	Good $ 160	Fair $ 80
Trade-In:	Mint	$ 250	Ex $ 200	VG+ $ 140	Good $ 125	Poor $ 60

CRIMSON BULLDOG (74424) REDBOOK CODE: RB-CH-H-CRMBLL
Stainless steel frame and finish, standard hammer, hammer block safety, fixed front sight, notch rear sight.

Production: current Caliber: .44 Special Action: DA/SA, revolver
Barrel Length: 2.5" Wt.: 21 oz. Sights: fixed Capacity: 5
Grips: Crimson Trace Lasergrips

D2C:	NIB	$ 525	Ex $ 400	VG+ $ 290	Good $ 265	LMP $ 683
C2C:	Mint	$ 510	Ex $ 375	VG+ $ 270	Good $ 240	Fair $ 125
Trade-In:	Mint	$ 380	Ex $ 300	VG+ $ 210	Good $ 185	Poor $ 60

TARGET BULLDOG
NEW IN BOX

TARGET BULLDOG   REDBOOK CODE: RB-CH-H-TRGBLL

Steel frame, stainless matte finish, standard hammer, full-size rubber grip, extended barrel, compact design, available in 4" or 5" barrel.

Production: current	Caliber: .44 Special			Action: DA/SA, revolver				
Barrel Length: 4", 5"	Wt.: 21 oz., 23 oz.		Sights: precision		Capacity: 5			
Grips: full, rubber								
D2C:	NIB	$ 380	Ex $ 300	VG+ $ 210	Good $ 190	LMP $	475
C2C:	Mint	$ 370	Ex $ 275	VG+ $ 190	Good $ 175	Fair $	90
Trade-In:	Mint	$ 270	Ex $ 225	VG+ $ 150	Good $ 135	Poor $	60

POLICE BULLDOG (73860)   REDBOOK CODE: RB-CH-H-BLLPLC

Stainless steel finish and frame, full grip, standard hammer, compact design.

Production: current	Caliber: .38 Special +P		Action: DA/SA, revolver				
Barrel Length: 5"	Wt.: 26 oz.	Sights: fixed	Capacity: 6	Grips: full			
D2C:	NIB	$ 370	Ex $ 300	VG+ $ 205	Good $ 185	LMP $	408
C2C:	Mint	$ 360	Ex $ 275	VG+ $ 190	Good $ 170	Fair $	90
Trade-In:	Mint	$ 270	Ex $ 225	VG+ $ 150	Good $ 130	Poor $	60

OFF DUTY
NEW IN BOX

OFF DUTY (53811)   REDBOOK CODE: RB-CH-H-OFFDTY

Aluminum frame and finish, DAO hammer, shrouded ejector rod, hammer safety, compact design.

Production: 2002 - current	Caliber: .38 Special +P		Action: DAO, revolver				
Barrel Length: 2"	Wt.: 12 oz.	Sights: fixed	Capacity: 5	Grips: black synthetic			
D2C:	NIB	$ 345	Ex $ 275	VG+ $ 190	Good $ 175	LMP $	400
C2C:	Mint	$ 340	Ex $ 250	VG+ $ 180	Good $ 160	Fair $	80
Trade-In:	Mint	$ 250	Ex $ 200	VG+ $ 140	Good $ 125	Poor $	60

CRIMSON OFF DUTY
NEW IN BOX

CRIMSON OFF DUTY (53814)   REDBOOK CODE: RB-CH-H-OFFCRM

Stainless steel finish, aluminum frame, DAO hammer, compact design.

Production: current	Caliber: .38 Special +P		Action: DAO, revolver				
Barrel Length: 2"	Wt.: 12 oz.	Sights: fixed	Capacity: 5				
Grips: Crimson Trace Lasergrips							
D2C:	NIB	$ 565	Ex $ 450	VG+ $ 315	Good $ 285	LMP $	651
C2C:	Mint	$ 550	Ex $ 400	VG+ $ 290	Good $ 255	Fair $	130
Trade-In:	Mint	$ 410	Ex $ 325	VG+ $ 230	Good $ 200	Poor $	60

**PINK LADY
UNDERCOVER LITE,
PINK & STAINLESS
STANDARD**
NEW IN BOX

PINK LADY UNDERCOVER LITE, PINK & STAINLESS
STANDARD (53830)   REDBOOK CODE: RB-CH-H-PNKLDY

Standard hammer, two-tone pink and stainless finish, aluminum frame, compact design, light recoil.

Production: current	Caliber: .38 Special +P		Action: DA/SA, revolver				
Barrel Length: 2"	Wt.: 12 oz.	Sights: fixed	Capacity: 5				
Grips: checkered synthetic							
D2C:	NIB	$ 365	Ex $ 300	VG+ $ 205	Good $ 185	LMP $	410
C2C:	Mint	$ 360	Ex $ 275	VG+ $ 190	Good $ 165	Fair $	85
Trade-In:	Mint	$ 260	Ex $ 225	VG+ $ 150	Good $ 130	Poor $	60

PINK LADY WITH CRIMSON TRACE GRIP

REDBOOK CODE: RB-CH-H-PNKCRM

Two-tone pink and stainless finish, standard hammer, black rubber grip, fixed front sight, aluminum frame, compact design, light recoil.

Production: current	Caliber: .38 Special +P								
Action: DA/SA, revolver		Barrel Length: 2"		Wt.: 12 oz.		Sights: fixed		Capacity: 5	
Grips: Crimson Trace Lasergrips									
D2C:	NIB	$ 515	Ex $ 400	VG+ $ 285	Good $ 260	LMP $ 570
C2C:	Mint	$ 500	Ex $ 375	VG+ $ 260	Good $ 235	Fair $ 120
Trade-In:	Mint	$ 370	Ex $ 300	VG+ $ 210	Good $ 185	Poor $ 60

PINK LADY OFF DUTY
NEW IN BOX

PINK LADY OFF DUTY (53851) REDBOOK CODE: RB-CH-H-PNKOFF

Two-tone pink and stainless finish, aluminum frame, DAO hammer, light recoil, compact design.

Production: current	Caliber: .38 Special +P			Action: DAO, revolver				
Barrel Length: 2"		Wt.: 12 oz.	Sights: fixed	Capacity: 5	Grips: checkered synthetic			
D2C:	NIB	$ 365	Ex $ 300	VG+ $ 205	Good $ 185	LMP $ 421
C2C:	Mint	$ 360	Ex $ 275	VG+ $ 190	Good $ 165	Fair $ 85
Trade-In:	Mint	$ 260	Ex $ 225	VG+ $ 150	Good $ 130	Poor $ 60

LAVENDER LADY
NEW IN BOX

LAVENDER LADY (53840) REDBOOK CODE: RB-CH-H-LVNDRL

Two-tone lavender and stainless finish, aluminum frame, standard hammer, compact design.

Production: current	Caliber: .38 Special +P			Action: DA/SA, revolver				
Barrel Length: 2"		Wt.: 12 oz.	Sights: fixed	Capacity: 5	Grips: standard, synthetic			
D2C:	NIB	$ 355	Ex $ 275	VG+ $ 200	Good $ 180	LMP $ 410
C2C:	Mint	$ 350	Ex $ 250	VG+ $ 180	Good $ 160	Fair $ 85
Trade-In:	Mint	$ 260	Ex $ 200	VG+ $ 140	Good $ 125	Poor $ 60

**MAGNUM PUG
BLUE STANDARD**
NEW IN BOX

MAGNUM PUG BLUE STANDARD (13520)

REDBOOK CODE: RB-CH-H-MAGPUG

Blue finish, steel frame, standard hammer, shrouded ejector rod, hammer safety, compact design.

Production: current	Caliber: .357 Magnum			Action: DA/SA, revolver				
Barrel Length: 2.2"		Wt.: 23 oz.	Sights: fixed	Capacity: 5	Grips: full, black rubber			
D2C:	NIB	$ 350	Ex $ 275	VG+ $ 195	Good $ 175	LMP $ 384
C2C:	Mint	$ 340	Ex $ 250	VG+ $ 180	Good $ 160	Fair $ 85
Trade-In:	Mint	$ 250	Ex $ 200	VG+ $ 140	Good $ 125	Poor $ 60

**MAGNUM PUG
STAINLESS**
NEW IN BOX

MAGNUM PUG BLUE STAINLESS (13520)

REDBOOK CODE: RB-CH-H-MGPSSS

Steel frame, spurred hammer, also available in a target model with a 4" barrel and adjustable rear sight, shrouded ejector rod, compact design.

Production: 2001 - current	Caliber: .357 Magnum		Action: DA/SA, revolver				
Barrel Length: 2.2", 4"		Wt.: 23 oz.	Sights: fixed or adjustable	Capacity: 5			
Grips: full, black rubber							
D2C:	NIB	$ 360	Ex $ 275	VG+ $ 200	Good $ 180	LMP $ 390
C2C:	Mint	$ 350	Ex $ 250	VG+ $ 180	Good $ 165	Fair $ 85
Trade-In:	Mint	$ 260	Ex $ 225	VG+ $ 150	Good $ 130	Poor $ 60

CHARTER ARMS

TARGET MAGNUM PUG
NEW IN BOX

ON DUTY
NEW IN BOX

DIXIE DERRINGER
NEW IN BOX
Courtesy of Bud's Gun Shop

PATRIOT
VERY GOOD +

TARGET MAGNUM PUG (73540)
REDBOOK CODE: RB-CH-H-MGTRGT
Standard hammer, steel frame, full-size grips, blue finish, compact design.

Production: 2001 - current	Caliber: .357 Magnum		Action: DA/SA, revolver		
Barrel Length: 4" Wt.: 25 oz.		Sights: adjustable		Capacity: 5	
Grips: full, checkered synthetic					
D2C:	NIB $ 375	Ex $ 300	VG+ $ 210	Good $ 190	LMP $ 479
C2C:	Mint $ 360	Ex $ 275	VG+ $ 190	Good $ 170	Fair $ 90
Trade-In:	Mint $ 270	Ex $ 225	VG+ $ 150	Good $ 135	Poor $ 60

ON DUTY (53810) REDBOOK CODE: RB-CH-H-ONDUTY
Stainless finish, aluminum frame, compact design.

Production: 2009 - current	Caliber: .38 Special +P		Action: DA/SA, revolver		
Barrel Length: 2" Wt.: 12 oz.	Sights: fixed	Capacity: 5	Grips: standard, black rubber		
D2C:	NIB $ 360	Ex $ 275	VG+ $ 200	Good $ 180	LMP $ 398
C2C:	Mint $ 350	Ex $ 250	VG+ $ 180	Good $ 165	Fair $ 85
Trade-In:	Mint $ 260	Ex $ 225	VG+ $ 150	Good $ 130	Poor $ 60

DIXIE DERRINGER REDBOOK CODE: RB-CH-H-DXDRRG
Matte stainless finish, pocket sized, exposed hammer.

Production: early 2000s - discontinued		Caliber: .22 LR, .22 WMR		
Action: SA, single-shot	Barrel Length: 1.1"	Wt.: 6 oz.	Sights: fixed	
Capacity: 5 Grips: black synthetic				
D2C:	NIB $ 185	Ex $ 150	VG+ $ 105	Good $ 95	
C2C:	Mint $ 180	Ex $ 150	VG+ $ 100	Good $ 85	Fair $ 45
Trade-In:	Mint $ 140	Ex $ 125	VG+ $ 80	Good $ 65	Poor $ 30

PATRIOT REDBOOK CODE: RB-CH-H-PATROT
Steel frame, matte stainless finish, rubber full-size grips, shrouded ejector rod.

Production: discontinued	Caliber: .327 Magnum		Action: DA/SA, revolver	
Barrel Length: 2.2", 4" Wt.: 21 oz.	Sights: fixed or adjustable		Capacity: 6	
Grips: full, rubber				
D2C:	NIB $ 400	Ex $ 325	VG+ $ 220	Good $ 200	LMP $ 536
C2C:	Mint $ 390	Ex $ 300	VG+ $ 200	Good $ 180	Fair $ 95
Trade-In:	Mint $ 290	Ex $ 225	VG+ $ 160	Good $ 140	Poor $ 60

GOLDFINGER REDBOOK CODE: RB-CH-H-GLDFNG
Gold-tone and black finish, aluminum frame, hammer safety, compact design, shrouded ejector rod.

Production: current	Caliber: .38 Special +P		Action: DA/SA, revolver	
Barrel Length: 2" Sights: fixed	Capacity: 5	Grips: standard, black synthetic		
D2C:	NIB $ 355	Ex $ 275	VG+ $ 200	Good $ 180	LMP $ 410
C2C:	Mint $ 350	Ex $ 250	VG+ $ 180	Good $ 160	Fair $ 85
Trade-In:	Mint $ 260	Ex $ 200	VG+ $ 140	Good $ 125	Poor $ 60

Chiappa

Ezechiele Chiappa founded Armi Sport in 1958 in what would eventually become the Chiappa Group and Chiappa Firearms. While perhaps best known for their replica rifles, the company, located in Dayton, Ohio, produces a series of semi-auto pistols, similar to the Browning Hi-Power and several modern revolvers. The company remains under the control and leadership of the original Chiappa family.

M9
VERY GOOD +

M9 REDBOOK CODE: RB-CI-H-M9XXXX

Black finish, tactical design, modeled after the original M9, slide-mounted safety.

Production: current   Caliber: 22 LR, 9mm, .40 S&W

Action: DA/SA, semi-auto   Barrel Length: 5"   Wt.: 34 oz.   Sights: drift adjustable rear

Capacity: 10, 15   Magazine: detachable box   Grips: black synthetic

D2C:	NIB	$ 325	Ex $ 250	VG+ $ 180	Good $ 165	LMP $ 527
C2C:	Mint	$ 320	Ex $ 225	VG+ $ 170	Good $ 150	Fair $ 75
Trade-In:	Mint	$ 240	Ex $ 200	VG+ $ 130	Good $ 115	Poor $ 60

M9-22 TACTICAL
NEW IN BOX

M9-22 TACTICAL REDBOOK CODE: RB-CI-H-M922XX

A .22 LR variation of the original M9. Slide-mounted safety, matte black finish, manufactured in Ohio.

Production: 2011 - current   Caliber: .22 LR   Action: DA/SA, semi-auto

Barrel Length: 5"   Wt.: 37 oz.   Sights: Novak-style, fiber-optic front

Capacity: 10   Magazine: detachable box   Grips: black plastic, wooden grips

D2C:	NIB	$ 345	Ex $ 275	VG+ $ 190	Good $ 175	LMP $ 369
C2C:	Mint	$ 340	Ex $ 250	VG+ $ 180	Good $ 160	Fair $ 80
Trade-In:	Mint	$ 250	Ex $ 200	VG+ $ 140	Good $ 125	Poor $ 60

M9 COMPACT
NEW IN BOX

M9 COMPACT REDBOOK CODE: RB-CI-H-M9CMPC

Black finish, tactical design, modeled after the original M9, slide-mounted safety, a smaller version of the M9.

Production: current   Caliber: 9mm, .40 S&W   Action: DA/SA, semi-auto

Barrel Length: 4.3"   Wt.: 29 oz.   Sights: drift adjustable rear

Capacity: 10, 15   Magazine: detachable box   Grips: black synthetic

D2C:	NIB	$ 335	Ex $ 275	VG+ $ 185	Good $ 170	LMP $ 513
C2C:	Mint	$ 330	Ex $ 250	VG+ $ 170	Good $ 155	Fair $ 80
Trade-In:	Mint	$ 240	Ex $ 200	VG+ $ 140	Good $ 120	Poor $ 60

MC27
NEW IN BOX

MC27 REDBOOK CODE: RB-CI-H-MC27XX

Black finish, tactical design, Picatinny rail, double-stack magazine, frame-mounted safety.

Production: 2013 - current   Caliber: 9mm   Action: DA/SA, DOA, semi-auto

Barrel Length: 3.9"   Wt.: 29 oz.   Sights: fixed front, adjustable rear

Capacity: 10, 15   Magazine: detachable box   Grips: black synthetic

D2C:	NIB	$ 415	Ex $ 325	VG+ $ 230	Good $ 210	LMP $ 513
C2C:	Mint	$ 400	Ex $ 300	VG+ $ 210	Good $ 190	Fair $ 100
Trade-In:	Mint	$ 300	Ex $ 250	VG+ $ 170	Good $ 150	Poor $ 60

RHINO 20DS
NEW IN BOX

RHINO 20DS   REDBOOK CODE: RB-CI-H-RHIN20

Matte black or chrome finish, transfer bar safety, compact design.

Production: 2010 - current   Caliber: .357 Magnum   Action: DA/SA, revolver

Barrel Length: 2"   OA Length: 6.5"   Sights: tactical front

Capacity: 6   Grips: black synthetic or walnut

D2C:	NIB	$ 845	Ex $ 650	VG+ $ 465	Good $ 425		
C2C:	Mint	$ 820	Ex $ 600	VG+ $ 430	Good $ 385	Fair $ 195
Trade-In:	Mint	$ 600	Ex $ 475	VG+ $ 330	Good $ 300	Poor $ 90

RHINO 40DS
NEW IN BOX

RHINO 40DS   REDBOOK CODE: RB-CI-H-RHIN40

Transfer bar safety, accessory rail, steel frame, matte black finish.

Production: 2010 - current   Caliber: .40 S&W   Action: DA/SA, revolver

Barrel Length: 4"   OA Length: 8.5"   Capacity: 6

Grips: walnut or synthetic

D2C:	NIB	$ 835	Ex $ 650	VG+ $ 460	Good $ 420		
C2C:	Mint	$ 810	Ex $ 600	VG+ $ 420	Good $ 380	Fair $ 195
Trade-In:	Mint	$ 600	Ex $ 475	VG+ $ 330	Good $ 295	Poor $ 90

RHINO 50DS
NEW IN BOX

RHINO 50DS   REDBOOK CODE: RB-CI-H-RHIN50

Transfer bar safety, accessory rail, steel frame, black or chrome finish, modern design.

Production: 2010 - current   Caliber: .357 Magnum   Action: DA/SA, revolver

Barrel Length: 5"   OA Length: 9.5"   Sights: tactical front   Capacity: 6

Grips: walnut or synthetic

D2C:	NIB	$ 835	Ex $ 650	VG+ $ 460	Good $ 420		
C2C:	Mint	$ 810	Ex $ 600	VG+ $ 420	Good $ 380	Fair $ 195
Trade-In:	Mint	$ 600	Ex $ 475	VG+ $ 330	Good $ 295	Poor $ 90

1911-22
NEW IN BOX

1911-22   REDBOOK CODE: RB-CI-H-91122X

Frame-mounted safety, blowback design, alloy frame, blue or matte stainless finish.

Production: 2010 - current   Caliber: .22 LR   Action: SA, semi-auto

Barrel Length: 5"   OA Length: 8.5"   Sights: fixed combat

Capacity: 10   Magazine: detachable box   Grips: walnut, checkered

D2C:	NIB	$ 265	Ex $ 225	VG+ $ 150	Good $ 135		
C2C:	Mint	$ 260	Ex $ 200	VG+ $ 140	Good $ 120	Fair $ 65
Trade-In:	Mint	$ 190	Ex $ 150	VG+ $ 110	Good $ 95	Poor $ 30

1911-22 TACTICAL
NEW IN BOX

1911-22 TACTICAL   REDBOOK CODE: RB-CI-H-1122TC

Frame-mounted safety, blowback design, alloy frame, skeletonized trigger, blue finish.

Production: 2010 - current   Caliber: .22 LR   Action: SA, semi-auto

Barrel Length: 5"   OA Length: 8.5"   Sights: Novak rear

Capacity: 10   Magazine: detachable box   Grips: walnut, checkered

D2C:	NIB	$ 300	Ex $ 250	VG+ $ 165	Good $ 150		
C2C:	Mint	$ 290	Ex $ 225	VG+ $ 150	Good $ 135	Fair $ 70
Trade-In:	Mint	$ 220	Ex $ 175	VG+ $ 120	Good $ 105	Poor $ 30

Christensen Arms

Founded in the mid-1990s, Christensen Arms developed a series of handguns and rifles around advancements in new materials. Most of their guns feature carbon-fiber barrels, which the company claims has significant advantages over steel with regard to heat dissipation and weight. Their series of M1911 pistols and AR-15 style rifles feature a modern design and command substantial interest among firearms collectors and enthusiasts. The company is located in Gunnison, Utah.

M1911 CUSTOM COMMANDER
NEW IN BOX

M1911 CUSTOM COMMANDER REDBOOK CODE: RB-C6-H-CUSCOM

Hand-fitted stainless slide with tungsten carbide runners, unique slide serrations, titanium frame, match grade barrel, optional threaded barrel, stainless adjustable trigger, frame-mounted safety, grip safety, many optional Cerakote finish combinations, optional proprietary carbon grips.

Production: current Caliber: .45 ACP Action: SA, semi-auto

Barrel Length: 4.3" Wt.: 34 oz. Sights: tritium night sights

Capacity: 8 Magazine: detachable box Grips: VZ G10, optional carbon panels

D2C:	NIB	$3,200	Ex	$2,450	VG+	--	Good	--		
C2C:	Mint	$3,080	Ex	$2,225	VG+	--	Good	--	Fair	--
Trade-In:	Mint	$2,280	Ex	$1,800	VG+	--	Good	--	Poor	--

M1911 GOVERNMENT DAMASCUS
NEW IN BOX

M1911 GOVERNMENT DAMASCUS REDBOOK CODE: RB-C6-H-GOVDAM

Metallic finish, skeletonized trigger and hammer, hand-fit Damascus slide.

Production: 2013 - current Caliber: .45 ACP Action: SA, semi-auto

Barrel Length: 5" Wt.: 35 oz. Sights: fixed, dot Capacity: 7

Magazine: detachable box Grips: carbon fiber

D2C:	NIB	$4,700	Ex	$3,575	VG+	--	Good	--	LMP	$ 4,751
C2C:	Mint	$4,520	Ex	$3,250	VG+	--	Good	--	Fair	--
Trade-In:	Mint	$3,340	Ex	$2,650	VG+	--	Good	--	Poor	--

M1911 GOVERNMENT
NEW IN BOX

M1911 GOVERNMENT REDBOOK CODE: RB-C6-H-GOVERN

Metallic matte finish, skeletonized trigger and hammer, modern tactical design, frame mounted safety.

Production: current Caliber: .45 ACP Action: SA, semi-auto

Barrel Length: 5" Wt.: 35 oz. Sights: fixed Capacity: 8

Magazine: detachable box Grips: VZ G10

D2C:	NIB	$3,750	Ex	$2,850	VG+	--	Good	--		
C2C:	Mint	$3,600	Ex	$2,600	VG+	--	Good	--	Fair	--
Trade-In:	Mint	$2,670	Ex	$2,100	VG+	--	Good	--	Poor	--

M1911 OFFICER
NEW IN BOX

M1911 OFFICER REDBOOK CODE: RB-C6-H-OFFICE

Hand-fitted stainless slide with tungsten carbide runners, custom slide serrations, titanium frame, match grade barrel, adjustable trigger, titanium firing pin and hammer strut, titanium grip safety.

Production: 2013 - current Caliber: .45 ACP Action: SA, semi-auto

Barrel Length: 3.5" Wt.: 30 oz. Sights: tritium night sights Capacity: 6

Magazine: detachable box Grips: VZ G10, optional carbon panels

D2C:	NIB	$3,250	Ex	$2,475	VG+	--	Good	--		
C2C:	Mint	$3,120	Ex	$2,250	VG+	--	Good	--	Fair	--
Trade-In:	Mint	$2,310	Ex	$1,825	VG+	--	Good	--	Poor	--

Cimarron

Mike Harvey established Cimarron in early 1984 and the company almost exclusively produces replica Colt, Remington, and military model rifles and shotguns. Most of these guns cost significantly less than the original versions of these guns and are often used in military and western reenactments and films. The guns themselves are manufactured by ArmiSport, Chiappa, and Uberti, among others, but Cimarron imports and markets them in the U.S. The company is located in Fredricksburg, Texas.

1847 WALKER REDBOOK CODE: RB-CM-H-1874WL

Replicates the Colt 1847 Walker, blue finish, hardened brass components, brass accents.

1847 WALKER
NEW IN BOX

Production: current   Caliber: .44 cal.   Action: SA, revolver   Barrel Length: 9"
OA Length: 15.5"   Sights: fixed   Capacity: 6   Grips: walnut

D2C:	NIB	$ 420	Ex $ 325	VG+ $ 235	Good $ 210	LMP $ 438
C2C:	Mint	$ 410	Ex $ 300	VG+ $ 210	Good $ 190	Fair $ 100
Trade-In:	Mint	$ 300	Ex $ 250	VG+ $ 170	Good $ 150	Poor $ 60

1ST MODEL DRAGOON
REDBOOK CODE: RB-CM-H-1STDRG

Replicates the Colt Dragoon revolver, blue or charcoal finish, brass trigger guard, percussion design.

1ST MODEL DRAGOON
NEW IN BOX

Production: current   Caliber: .44 cal.   Action: SA, revolver   Barrel Length: 7.5"
OA Length: 14.75"   Sights: blade front   Capacity: 6   Grips: walnut

D2C:	NIB	$ 400	Ex $ 325	VG+ $ 220	Good $ 200	LMP $ 432
C2C:	Mint	$ 390	Ex $ 300	VG+ $ 200	Good $ 180	Fair $ 95
Trade-In:	Mint	$ 290	Ex $ 225	VG+ $ 160	Good $ 140	Poor $ 60

M1849 WELLS FARGO REDBOOK CODE: RB-CM-H-1849WLF

Replicates the Colt 1849 Wells Fargo, blue or charcoal finish, hardened brass components, compact design.

M1849 WELLS FARGO
NEW IN BOX

Production: current   Caliber: .31 cal.   Action: SA, revolver   Barrel Length: 4"
Sights: fixed   Capacity: 6   Grips: walnut

D2C:	NIB	$ 295	Ex $ 225	VG+ $ 165	Good $ 150	LMP $ 337
C2C:	Mint	$ 290	Ex $ 225	VG+ $ 150	Good $ 135	Fair $ 70
Trade-In:	Mint	$ 210	Ex $ 175	VG+ $ 120	Good $ 105	Poor $ 30

M1849 POCKET MODEL REDBOOK CODE: RB-CM-H-1849PM

Replicates the Colt 1849 Pocket Model, either blue or charcoal finish, hardened brass components, compact design.

M1849 POCKET MODEL
NEW IN BOX

Production: current   Caliber: .31 cal.   Action: SA, revolver   Barrel Length: 4"
Sights: fixed   Capacity: 6   Grips: walnut

D2C:	NIB	$ 295	Ex $ 225	VG+ $ 165	Good $ 150	LMP $ 347
C2C:	Mint	$ 290	Ex $ 225	VG+ $ 150	Good $ 135	Fair $ 70
Trade-In:	Mint	$ 210	Ex $ 175	VG+ $ 120	Good $ 105	Poor $ 30

M1851 NAVY LONDON

M1851 NAVY LONDON
GOOD

REDBOOK CODE: RB-CM-H-1851NV

Replicates the Colt 1851 Navy London, blue finish, hardened steel components.

Production: current Caliber: .36 cal. Action: SA, revolver Barrel Length: 7.5"
OA Length: 15.5" Sights: fixed Capacity: 6 Grips: walnut

D2C:	NIB	$ 315	Ex $ 250	VG+ $ 175	Good $ 160	LMP $ 350
C2C:	Mint	$ 310	Ex $ 225	VG+ $ 160	Good $ 145	Fair $ 75
Trade-In:	Mint	$ 230	Ex $ 200	VG+ $ 130	Good $ 115	Poor $ 60

M1851 NAVY OVAL TRIGGER GUARD

M1851 NAVY OVAL TRIGGER GUARD
NEW IN BOX

REDBOOK CODE: RB-CM-H-1851NO

Replicates the Colt 1851 Navy Oval Trigger Guard, blue finish, hardened steel components, brass trigger guard.

Production: current Caliber: .36 cal. Action: SA, revolver Barrel Length: 7.5"
OA Length: 15.5" Sights: fixed Capacity: 6 Grips: walnut

D2C:	NIB	$ 290	Ex $ 225	VG+ $ 160	Good $ 145	LMP $ 321
C2C:	Mint	$ 280	Ex $ 225	VG+ $ 150	Good $ 135	Fair $ 70
Trade-In:	Mint	$ 210	Ex $ 175	VG+ $ 120	Good $ 105	Poor $ 30

M1858 REMINGTON NAVY

M1858 REMINGTON NAVY
NEW IN BOX

REDBOOK CODE: RB-CM-H-1858RN

Replicates the Remington 1858 Navy Model revolver, blue or charcoal finish, two piece grips, brass accents, updated percussion cylinder, strap-top frame.

Production: current Caliber: .36 cal. Action: SA, revolver Barrel Length: 7.5"
Sights: fixed Capacity: 6 Grips: walnut

D2C:	NIB	$ 315	Ex $ 250	VG+ $ 175	Good $ 160	LMP $ 350
C2C:	Mint	$ 310	Ex $ 225	VG+ $ 160	Good $ 145	Fair $ 75
Trade-In:	Mint	$ 230	Ex $ 200	VG+ $ 130	Good $ 115	Poor $ 60

M1858 REMINGTON ARMY

M1858 REMINGTON ARMY
NEW IN BOX

REDBOOK CODE: RB-CM-H-1858RA

Replicates the Remington 1858 Army Model revolver, blue or charcoal finish, two piece grips, brass accents, updated percussion cylinder, strap-top frame.

Production: current Caliber: .44 cal. Action: SA, revolver Barrel Length: 5.5", 8"
Sights: fixed Capacity: 6 Grips: walnut

D2C:	NIB	$ 315	Ex $ 250	VG+ $ 175	Good $ 160	LMP $ 350
C2C:	Mint	$ 310	Ex $ 225	VG+ $ 160	Good $ 145	Fair $ 75
Trade-In:	Mint	$ 230	Ex $ 200	VG+ $ 130	Good $ 115	Poor $ 60

M1860 ARMY MILITARY

M1860 ARMY MILITARY
NEW IN BOX

REDBOOK CODE: RB-CM-H-1860AM

Replicates the 1860 Army Model revolver, hardened brass components, blue or charcoal finish, one-piece walnut grip, classic design.

Production: current Caliber: .44 cal. Action: SA, revolver Barrel Length: 8"
Sights: fixed Capacity: 6 Grips: walnut

D2C:	NIB	$ 310	Ex $ 250	VG+ $ 175	Good $ 155	LMP $ 342
C2C:	Mint	$ 300	Ex $ 225	VG+ $ 160	Good $ 140	Fair $ 75
Trade-In:	Mint	$ 230	Ex $ 175	VG+ $ 130	Good $ 110	Poor $ 60

M1860 ARMY CIVILIAN
NEW IN BOX

M1860 ARMY CIVILIAN REDBOOK CODE: RB-CM-H-1860AC
Replicates the 1860 Army Civilian revolver, blue finish, steel frame, brass accents.

Production: current Caliber: .44 cal. Action: SA, revolver Barrel Length: 8"
Sights: fixed Capacity: 6 Grips: walnut

D2C:	NIB	$ 295	Ex $ 225	VG+ $ 165	Good $ 150	LMP $ 337
C2C:	Mint	$ 290	Ex $ 225	VG+ $ 150	Good $ 135	Fair $ 70
Trade-In:	Mint	$ 210	Ex $ 175	VG+ $ 120	Good $ 105	Poor $ 30

M1861 NAVY MILITARY
NEW IN BOX

M1861 NAVY MILITARY
REDBOOK CODE: RB-CM-H-1861NM
Replicates the 1861 Navy Military revolver, blue or charcoal finish, one-piece walnut grip, brass accents.

Production: current Caliber: .36 cal. Action: SA, revolver Barrel Length: 7.5"
Sights: fixed Capacity: 6 Grips: walnut

D2C:	NIB	$ 295	Ex $ 225	VG+ $ 165	Good $ 150	LMP $ 347
C2C:	Mint	$ 290	Ex $ 225	VG+ $ 150	Good $ 135	Fair $ 70
Trade-In:	Mint	$ 210	Ex $ 175	VG+ $ 120	Good $ 105	Poor $ 30

M1861 NAVY CIVILIAN
NEW IN BOX

M1861 NAVY CIVILIAN REDBOOK CODE: RB-CM-H-1861NC
Replicates the 1861 Navy Civilian revolver, blue or charcoal finish, one-piece walnut grip, brass accents.

Production: current Caliber: .36 cal. Action: SA, revolver Barrel Length: 7.5"
Sights: fixed Capacity: 6 Grips: walnut

D2C:	NIB	$ 285	Ex $ 225	VG+ $ 160	Good $ 145	LMP $ 338
C2C:	Mint	$ 280	Ex $ 200	VG+ $ 150	Good $ 130	Fair $ 70
Trade-In:	Mint	$ 210	Ex $ 175	VG+ $ 120	Good $ 100	Poor $ 30

M1858 NEW ARMY
NEW IN BOX

M1858 NEW ARMY REDBOOK CODE: RB-CM-H-1858NA
Replicates the 1858 New Army, blue finish, brass accents, strap-top frame.

Production: current Caliber: .38 Special, .44-40 WCF, .45 LC Action: SA, revolver
Barrel Length: 5.5" Sights: fixed Capacity: 6 Grips: walnut

D2C:	NIB	$ 480	Ex $ 375	VG+ $ 265	Good $ 240	LMP $ 552
C2C:	Mint	$ 470	Ex $ 350	VG+ $ 240	Good $ 220	Fair $ 115
Trade-In:	Mint	$ 350	Ex $ 275	VG+ $ 190	Good $ 170	Poor $ 60

M1875 OUTLAW
NEW IN BOX

M1875 OUTLAW REDBOOK CODE: RB-CM-H-1875OL
Replicates the Colt M1875 Outlaw, blue or original finish.

Production: current Caliber: .357 Special, .44 WCF, .45 LC Action: SA, revolver
Barrel Length: 5.5", 7.5" Sights: blade front Capacity: 6 Grips: walnut

D2C:	NIB	$ 480	Ex $ 375	VG+ $ 265	Good $ 240	
C2C:	Mint	$ 470	Ex $ 350	VG+ $ 240	Good $ 220	Fair $ 115
Trade-In:	Mint	$ 350	Ex $ 275	VG+ $ 190	Good $ 170	Poor $ 60

MODEL P OLD MODEL
NEW IN BOX

MODEL P OLD MODEL REDBOOK CODE: RB-CM-H-MDPOLD
Replicates the Colt 1873 Single-Action, charcoal blue finish, hardened steel frame.

Production: current Caliber: .32 WCF, .38 WCF, .357 Magnum, .44 Special, .44 WCF, .45 LC, .45 ACP Action: SA, revolver Barrel Length: 4.74", 5.5", 7.5"
Sights: fixed Capacity: 6 Grips: walnut

D2C:	NIB	$ 460	Ex $ 350	VG+ $ 255	Good $ 230	
C2C:	Mint	$ 450	Ex $ 325	VG+ $ 230	Good $ 210	Fair $ 110
Trade-In:	Mint	$ 330	Ex $ 275	VG+ $ 180	Good $ 165	Poor $ 60

SAA EVIL ROY
NEW IN BOX

THUNDERER
NEW IN BOX

SAA EVIL ROY   REDBOOK CODE: RB-CM-H-SAEVLR

Replicates the Colt SAA Evil Roy Model P, blue or nickel finish, slim grips, light trigger pull.

Production: current	Caliber: .44-40 WCF, .357 Magnum, .45 LC	Action: SA, revolver						
Barrel Length: 4.75"	Sights: fixed blade	Capacity: 6	Grips: walnut					
D2C:	NIB	$ 585	Ex $ 450	VG+ $ 325	Good $ 295	LMP $ 740		
C2C:	Mint	$ 570	Ex $ 425	VG+ $ 300	Good $ 265	Fair $ 135		
Trade-In:	Mint	$ 420	Ex $ 350	VG+ $ 230	Good $ 205	Poor $ 60		

THUNDERER   REDBOOK CODE: RB-CM-H-THNDXX

Replicates the Colt Thunderer, blue and nickel finish, unique grip design.

Production: current	Caliber: .357 Magnum, .44-40 WCF, .45 LC	Action: SA, revolver				
Barrel Length: 3.5", 4.75", 5.5"	Sights: fixed blade	Capacity: 6	Grips: polished walnut			
D2C:	NIB	$ 525	Ex $ 400	VG+ $ 290	Good $ 265	
C2C:	Mint	$ 510	Ex $ 375	VG+ $ 270	Good $ 240	Fair $ 125
Trade-In:	Mint	$ 380	Ex $ 300	VG+ $ 210	Good $ 185	Poor $ 60

DIABLO DERRINGER   REDBOOK CODE: RB-CM-H-DIABDR

Black or chrome finish, break-top action, pocket-size design, stainless hammer, Cimarron emblem in the grip.

Production: current	Caliber: .22 LR/.22 Magnum, .38 Special/.32 H&R					
Action: SA, derringer	Barrel Length: 2.4"	Sights: fixed	Capacity: 2	Grips: walnut		
D2C:	NIB	$ 200	Ex $ 175	VG+ $ 110	Good $ 100	LMP $ 220
C2C:	Mint	$ 200	Ex $ 150	VG+ $ 100	Good $ 90	Fair $ 50
Trade-In:	Mint	$ 150	Ex $ 125	VG+ $ 80	Good $ 70	Poor $ 30

TITAN DERRINGER
NEW IN BOX

TITAN DERRINGER   REDBOOK CODE: RB-CM-H-TITDRR

Black or stainless finish, enhanced trigger guard, break-top action, pocket-size design.

Production: current	Caliber: .45 LC/.410, 9mm					
Action: SA, derringer	Barrel Length: 3.5"	Sights: fixed	Capacity: 2	Grips: walnut		
D2C:	NIB	$ 385	Ex $ 300	VG+ $ 215	Good $ 195	LMP $ 428
C2C:	Mint	$ 370	Ex $ 275	VG+ $ 200	Good $ 175	Fair $ 90
Trade-In:	Mint	$ 280	Ex $ 225	VG+ $ 160	Good $ 135	Poor $ 60

CIMARRON

Citadel (Legacy Sports)

A trademark of Legacy Sports, Citadel manufactures a series of M1911 pistols noted for their affordability and quality. The company produces both modern and classic M1911 pistols that have begun to garner acclaim since their introduction circa 2009. Legacy Sports can field all questions and sales inquiries regarding Citadel pistols.

M1911
NEW IN BOX

M1911   REDBOOK CODE: RB-C7-H-911XXX

Flared ejection port, full-length guide rod, grooved slide stop, skeletonized hammer and trigger, ambidextrous frame-mounted safety, grip safety, beveled magwell, matte black or polished nickel finish, available with Hogue grips.

Production: current	Caliber: .45 ACP, .38 Super, 9mm	Action: SA, semi-auto				
Barrel Length: 5"	Wt.: 37 oz.	Sights: Novak	Capacity: 8			
Magazine: detachable box	Grips: wood or Hogue synthetic					
D2C:	NIB	$ 480	Ex $ 375	VG+ $ 265	Good $ 240	LMP $ 589
C2C:	Mint	$ 470	Ex $ 350	VG+ $ 240	Good $ 220	Fair $ 115
Trade-In:	Mint	$ 350	Ex $ 275	VG+ $ 190	Good $ 170	Poor $ 60

CITADEL

M1911 COMPACT
NEW IN BOX

M1911 COMPACT   REDBOOK CODE: RB-C7-H-911CMP

Bushingless barrel, flared ejection port, skeletonized hammer and trigger, grip safety, ambidextrous frame-mounted safety, beveled magwell, matte black with brushed nickel or polished nickel finish, available with Hogue grips.

Production: current Caliber: .45 ACP, 9mm Action: SA, semi-auto									
Barrel Length: 3.5" Wt.: 34 oz. Sights: Novak Capacity: 7									
Magazine: detachable box Grips: wood or Hogue synthetic									
D2C:	NIB	$ 490	Ex $ 375	VG+ $ 270	Good $ 245	LMP $ 589
C2C:	Mint	$ 480	Ex $ 350	VG+ $ 250	Good $ 225	Fair $ 115
Trade-In:	Mint	$ 350	Ex $ 275	VG+ $ 200	Good $ 175	Poor $ 60

M1911 .22 TACTICAL
NEW IN BOX

M1911 .22 TACTICAL   REDBOOK CODE: RB-C7-H-911TAC

Rimfire replica of the 1911, fixed barrel design, matte black finish, skeletonized hammer and trigger, slide stop with grooves.

Production: current Caliber: .22 LR Action: SA, semi-auto Barrel Length: 5"									
Wt.: 34 oz. Sights: fiber-optic front Capacity: 10 Magazine: detachable box									
Grips: Hogue grips									
D2C:	NIB	$ 340	Ex $ 275	VG+ $ 190	Good $ 170	LMP $ 349
C2C:	Mint	$ 330	Ex $ 250	VG+ $ 170	Good $ 155	Fair $ 80
Trade-In:	Mint	$ 250	Ex $ 200	VG+ $ 140	Good $ 120	Poor $ 60

M1911 .22 G.I.
NEW IN BOX

M1911 .22 G.I.   REDBOOK CODE: RB-C7-H-911GIX

Rimfire replica of the M1911 Government, matte black finish, G.I. style hammer and trigger, lanyard loop, slide-mounted safety.

Production: current Caliber: .22 LR Action: SA, semi-auto									
Barrel Length: 5" Wt.: 34 oz. Sights: G.I. style, fixed front									
Capacity: 10 Magazine: detachable box Grips: wood, Hogue synthetic									
D2C:	NIB	$ 295	Ex $ 225	VG+ $ 165	Good $ 150	LMP $ 310
C2C:	Mint	$ 290	Ex $ 225	VG+ $ 150	Good $ 135	Fair $ 70
Trade-In:	Mint	$ 210	Ex $ 175	VG+ $ 120	Good $ 105	Poor $ 30

Cobra Firearms (Enterprises)

Founded around the turn of the twenty-first century, Cobra Firearms (sometimes referred to as Cobra Enterprises of Utah, Inc.) manufactures a host of different derringers, compact revolvers, and pocket-sized pistols. Their guns are extremely inexpensive and gained a moderate following over the last several years due to their affordability. The company is based in Salt Lake City, Utah.

STANDARD DERRINGER   REDBOOK CODE: RB-C8-H-STNDRR

Model names differ depending on caliber: C22, C22m, C25, C32. Alloy frame and available in either chrome, black, or nickel finish.

Production: 2001 - current Caliber: .22LR, .22 Magnum, .25 ACP, .32 ACP									
Action: SA, derringer Barrel Length: 2.4" OA Length: 4" Wt.: 9.5 oz.									
Sights: fixed Capacity: 2 Grips: rosewood									
D2C:	NIB	$ 150	Ex $ 125	VG+ $ 85	Good $ 75	LMP $ 170
C2C:	Mint	$ 150	Ex $ 125	VG+ $ 80	Good $ 70	Fair $ 35
Trade-In:	Mint	$ 110	Ex $ 100	VG+ $ 60	Good $ 55	Poor $ 30

CITADEL

COBRA FIREARMS

BIG BORE DERRINGER   REDBOOK CODE: RB-C8-H-BGBRDR

Model names differ depending on caliber: CB22, CB32, CB38, CB380, CB9. Alloy frame and available in black, red, majestic, copper, blue, or purple finish.

Production: 2000 - current   Action: SA, derringer

Caliber: .22 Magnum, .32 H&R Magnum, .38 Special, .380 ACP, 9mm

Barrel Length: 2.75"   OA Length: 4.65"   Wt.: 14 oz.   Sights: fixed

Capacity: 2   Grips: ivory, rosewood, black, pink

D2C:	NIB	$ 175	Ex $ 150	VG+ $ 100	Good $ 90	LMP $ 187
C2C:	Mint	$ 170	Ex $ 125	VG+ $ 90	Good $ 80	Fair $ 45
Trade-In:	Mint	$ 130	Ex $ 100	VG+ $ 70	Good $ 65	Poor $ 30

LONG BORE DERRINGER
NEW IN BOX

LONG BORE DERRINGER   REDBOOK CODE: RB-C8-H-LGBRDR

Model names differ depending on caliber: CLB22, CLB38, CLB9. Alloy frame and available in black, red, majestic, copper, blue, or purple finish.

Production: 2001 - current   Caliber: .22 Magnum, .38 Special, 9mm

Action: SA, derringer   Barrel Length: 3.5"   OA Length: 5.4"   Wt.: 16 oz.

Sights: fixed   Capacity: 2   Grips: ivory, rosewood, black, pink

D2C:	NIB	$ 175	Ex $ 150	VG+ $ 100	Good $ 90	LMP $ 187
C2C:	Mint	$ 170	Ex $ 125	VG+ $ 90	Good $ 80	Fair $ 45
Trade-In:	Mint	$ 130	Ex $ 100	VG+ $ 70	Good $ 65	Poor $ 30

TITAN DERRINGER   REDBOOK CODE: RB-C8-H-TITDRR

Stainless steel frame, rebounding hammer, available in stainless, black, or brushed finish.

Production: 2001 - current   Caliber: .45 LC/.410   Action: SA, derringer

Barrel Length: 3.5"   OA Length: 5.5"   Wt.: 16.5 oz.   Sights: fixed

Capacity: 2   Grips: rosewood

D2C:	NIB	$ 370	Ex $ 300	VG+ $ 205	Good $ 185	LMP $ 400
C2C:	Mint	$ 360	Ex $ 275	VG+ $ 190	Good $ 170	Fair $ 90
Trade-In:	Mint	$ 270	Ex $ 225	VG+ $ 150	Good $ 130	Poor $ 60

CA SERIES
EXCELLENT

CA SERIES   REDBOOK CODE: RB-C8-H-CASRES

Models differ by caliber: CA32 and CA380. Alloy frame and either red, pink, copper, blue, purple, or black finish.

Production: 2001 - current   Caliber: .32 ACP, .380 ACP   Action: SA, semi-auto

Barrel Length: 2.8"   OA Length: 5.4"   Wt.: 22 oz.   Sights: fixed   Capacity: 5, 6

Magazine: detachable box   Grips: black, pink, pearl synthetic

D2C:	NIB	$ 145	Ex $ 125	VG+ $ 80	Good $ 75	
C2C:	Mint	$ 140	Ex $ 125	VG+ $ 80	Good $ 70	Fair $ 35
Trade-In:	Mint	$ 110	Ex $ 100	VG+ $ 60	Good $ 55	Poor $ 30

COLT

Colt

Given Colt's Manufacturing Company's extensive history and significance in the firearms industry, most of their history cannot fit within the context of this book. With that said, Samuel Colt founded the company in 1836 in Paterson, New Jersey, manufacturing revolving handguns and rifles, though this initial incarnation of the company failed and shut down in the early 1840s. With assistance from Samuel Walker and Eli Whitney Jr., Samuel Colt continued developing his early handgun designs and eventually opened another factory in Hartford, Connecticut around 1848. In the years that followed, Colt's Patent Fire Arms Manufacturing Company, as it was called, began to prosper thanks to assembly-line manufacturing techniques and the development of interchangeable parts for all of the company's weapons. Revolvers manufactured during this time include the Model 1849 Army and Navy Pocket, the Model 1860 Army, the Single Action Army, and Lightning and Thunderer. During this time, Samuel Colt grew exceedingly wealthy and fostered the reputation of running a state-of-the-art, yet stringent, factory thus ensuring the quality of his guns. After Samuel Colt's death in the early 1860s, his wife, Elizabeth, ran and managed the company for several more decades until selling the organization to an investment group around the turn of the twentieth century. Around this time, Colt executives began exploring the realities of developing a semi-automatic pistol. Designed by John Moses Browning, the Colt M1900 stands as the first semi-automatic pistol Colt produced to any significance. It, and subsequent models--such as the Model 1902, Model 1903 Hammerless, and the Model 1905--would lead to the development of the Model 1911, the most iconic semi-automatic pistol ever produced.

1851 NAVY REVOLVER
NEW IN BOX

Courtesy of Rock Island Auction Company

1851 NAVY REVOLVER REDBOOK CODE: RB-CL-H-1851NV

One of Colt's most acclaimed and successful revolvers. Fires cap and ball rounds and operates on single-action mechanisms. Highly collectible. Colt manufactured First, Second, Third, Fourth, and Iron Grip strap models of the 1851 Navy, which affect the price of the weapon. First and Second Models demand the highest value.

Production: 1850 - 1873	Caliber: .36 cal.	Action: SA, revolver						
Barrel Length: 6", 7.5"	Sights: bead	Capacity: 6	Grips: walnut					
D2C:	NIB	--	Ex	$16,725	VG+	$12,100	Good	$11,000
C2C:	Mint	--	Ex	$15,200	VG+	$11,000	Good $9,900	Fair $5,060
Trade-In:	Mint	--	Ex	$12,325	VG+	$8,580	Good $7,700	Poor $2,220

1847 WALKER
VERY GOOD +

Courtesy of Rock Island Auction Company

1847 WALKER REDBOOK CODE: RB-CL-H-1847WL

One of the most famed and treasured revolvers to ever enter production. At its initial time of production, it was the most powerful black powder revolver available. Less than 1200 units produced, making it one of the rarest yet most influential guns in the history of American firearms. A civilian model with no markings was produced and commands a smaller premium.

Production: 1847 - 1848	Caliber: .44 Colt	Action: SA, revolver				
Barrel: 9"	OA Length: 15.5"	Wt.: 72 oz.	Sights: blade front			
Capacity: 6	Grips: wood					
D2C:	NIB $ 885,000	Ex $672,600	VG+ $486,750	Good $442,500		
C2C:	Mint $ 849,600	Ex $ 610,650	VG+ $442,500	Good $398,250	Fair $203,550	
Trade-In:	Mint $ 628,350	Ex $495,600	VG+ $ 345,150	Good $309,750	Poor $88,500	

COLT

SINGLE ACTION ARMY STANDARD ISSUE
FIRST GEN. SERIAL NUMBER RANGE: 1-357,000

The "gun that won the west," or the "Peacemaker"--a term coined by Benjamin Kittredge, one of Colt's major distributor's--is perhaps the most iconic and influential revolver to ever enter production. Introduced in 1873, the first generation models of the SAA fired .45 Long Colt cartridges, built on contract to the U.S. Government. Issued to soldiers during the Indian Wars, Spanish-American War, and the Philippine conflict, among others. Colt briefly ceased production of the gun around the beginning of WWII. Original and black powder versions of the gun are highly sought after. Currently, the market is volatile with regard to SAAs. Prices are subjective and continually fluctuate. Any non-original finish, part, or adjustment to any and all SAA revolvers significantly decreases their value. SAAs First Generation were produced between 1873 and 1941 with the exception of 400+ SAAs built from pre-WWII parts and given to company VIPs throughout the 1950s. Last known serial number is approximately 357,850. A factory letter has potential to increase premiums.

Production: 1873 - 1941 Caliber: .45 LC, .44 WCF, .38 WCF, .32 WCF, .41 LC, and 31 other caliber options.

Action: SA, revolver Barrel Length: 4.5", 5.5", 7.5" Wt.: 40 oz.

Sights: blade front Capacity: 6 Grips: wood, plastic, or ivory

SAA PINCH FRAME
GOOD

Courtesy of Rock Island Auction Company

SINGLE ACTION ARMY PINCHED FRAME

(FIRST GEN. 1873) REDBOOK CODE: RB-CL-H-SAXXX1

A part of the first generation of SAA revolvers, noted for their rear sight located 1/2" in front of the hammer notch (a minor detail to the untrained eye). Also features a "V" groove above the cylinder. Serial numbers range between 1 and 160. Non-factory restored or refinished models have a significantly decreased value.

Production: 1873 Caliber: .44 S&W, .45 LC Action: SA, revolver

Barrel Length: 7.5" Wt.: 40 oz. Sights: blade front

Capacity: 6 Grips: wood

D2C:	NIB	--	Ex	$ 55,500	VG+	$ 40,145	Good $ 36,500		
C2C:	Mint	--	Ex	$ 50,375	VG+	$ 36,500	Good $ 32,850	Fair $ 16,790	
Trade-In:	Mint	--	Ex	$ 40,900	VG+	$ 28,470	Good $ 25,550	Poor $ 7,320	

SINGLE ACTION ARMY EARLY CIVILIAN FIRST GEN.

REDBOOK CODE: RB-CL-H-SAXXX3

Similar to military issued versions of the same era, but features no military markings. Models with three or four digit serial numbers have increased value. Serial numbers concurrent to the Military models, ranging between approximately 160 and 22,000.

Production: 1873 - 1876 Caliber: .45 LC Action: SA, revolver Capacity: 6

D2C:	NIB	--	Ex	$72,200	VG+	$52,250	Good	$47,500	
C2C:	Mint	--	Ex	$65,550	VG+	$47,500	Good	$42,750	Fair $21,850
Trade-In:	Mint	--	Ex	$53,200	VG+	$37,050	Good	$33,250	Poor $ 9,510

SAA FIRST GEN.
VERY GOOD +

Courtesy of Rock Island Auction Company

SINGLE ACTION ARMY FIRST GEN.

REDBOOK CODE: RB-CL-H-SAXXX8

Serial numbers range between approximately 1 and 1892. Nickel finish doubles their value.

Production: 1875 - 1880 Caliber: .44 Henry Action: SA, revolver

Barrel Length: 4.75", 5.5" Wt.: 40 oz. Sights: blade front

Capacity: 6 Grips: gutta percha

D2C:	NIB	--	Ex $62,325	VG+ $	45,100	Good	$41,000		
C2C:	Mint $78,720	Ex $56,600	VG+ $41,000	Good $36,900	Fair	$18,860			
Trade-In:	Mint $58,220	Ex $45,925	VG+ $31,980	Good $28,700	Poor	$ 8,220			

SAA PRE-WAR FIRST GEN.
EXCELLENT

Courtesy of Rock Island Auction Company

SINGLE ACTION ARMY FIRST GEN.

REDBOOK CODE: RB-CL-H-SAXX11

Serial numbers range between approximately 53,000 and 129,999.

Production: 1880 - 1890	Caliber: .44-40 WCF	Action: SA, revolver				
Barrel Length: 7.5"	Sights: blade front	Capacity: 6	Grips: wood or rubber			
D2C:	NIB	--	Ex $46,375	VG+ $33,550	Good $30,500	
C2C:	Mint	--	Ex $42,100	VG+ $30,500	Good $27,450	Fair $14,030
Trade-In:	Mint	--	Ex $34,175	VG+ $23,790	Good $21,350	Poor $ 6,120

SAA PRE-WAR FIRST GEN.
EXCELLENT

Courtesy of Rock Island Auction Company

SINGLE ACTION ARMY PRE-WAR FIRST GEN.

REDBOOK CODE: RB-CL-H-SAXX13

Serial numbers between approximately 130,000 to 167,999. Features a horizontal latch cylinder pin instead of a vertical screw.

Production: 1890 - 1896	Caliber: .45 LC, .44-40 WCF, .38 WCF, .32 WCF, etc					
Barrel Length: 5.5"	Action: SA, revolver	Wt.: 40 oz.	Sights: blade front			
Capacity: 6	Grips: wood, plastic, or ivory					
D2C:	NIB	--	Ex $10,650	VG+ $7,700	Good $7,000	
C2C:	Mint	--	Ex $9,675	VG+ $7,000	Good $6,300	Fair $3,220
Trade-In:	Mint	--	Ex $7,850	VG+ $5,460	Good $4,900	Poor $ 1,410

SINGLE ACTION ARMY FIRST GEN.

REDBOOK CODE: RB-CL-H-SAXX14

Features unique blue finish and some include medallions or emblems in the grip. Serial numbers range between approximately 163,000 and 191,999.

Production: 1896 - 1899	Caliber: .45 LC, .44-40 WCF, .38 WCF, .32 WCF, etc					
Action: SA, revolver	Barrel Length: 5.5"	OA Length: 11"	Wt.: 40 oz.			
Sights: blade front	Capacity: 6	Grips: wood, plastic, or ivory				
D2C:	NIB	--	Ex $10,275	VG+ $7,425	Good $6,750	
C2C:	Mint	--	Ex $9,325	VG+ $6,750	Good $6,075	Fair $ 3,105
Trade-In:	Mint	--	Ex $7,575	VG+ $5,270	Good $4,725	Poor $ 1,350

SAA FIRST GEN.
GOOD

Courtesy of Rock Island Auction Company

SINGLE ACTION ARMY FIRST GEN.

REDBOOK CODE: RB-CL-H-SAXX18

Serial numbers range between approximately 182,000 and 307,999. SAAs of this era still have substantial collector's value, but pale in comparison to earlier SAA models.

Production: 1899 - 1908	Caliber: .45 LC, .44-.40 WCF, .38-.40, 32-.20, .38 Colt,					
.38 Special, etc.	Action: SA, revolver	Barrel Length: 5.5", some minor				
variations exist	OA Length: 11"	Wt.: 40 oz.	Sights: blade front			
Capacity: 6	Grips: wood, plastic, or ivory					
D2C:	NIB	--	Ex $8,750	VG+ $6,325	Good $ 5,750	
C2C:	Mint	--	Ex $7,950	VG+ $5,750	Good $ 5,175	Fair $ 2,645
Trade-In:	Mint	--	Ex $6,450	VG+ $4,490	Good $ 4,025	Poor $ 1,170

SAA
EXCELLENT

Courtesy of Rock Island Auction Company

SINGLE ACTION ARMY   REDBOOK CODE: RB-CL-H-SAXX21

Most in blue finish. Made from an assortment of older SAA parts. Distributed to Colt employees. Serial numbers begin around 307,999.

Production: 1920s	Caliber: .30 cal., .357 Magnum, .38 Special,					
.44 Special, .44-.40 WCF, .45 LC	Action: SA, revolver	Barrel Length: 5.5"				
some minor variations exist	OA Length: 11"	Wt.: 40 oz.	Sights: blade front			
Capacity: 6	Grips: wood, plastic, or ivory					
D2C:	NIB	--	Ex $2,750	VG+	$1,980	Good $ 1,800	
C2C:	Mint	--	Ex $2,500	VG+	$ 1,800	Good $1,620	Fair $830
Trade-In:	Mint	--	Ex $2,025	VG+	$ 1,410	Good $1,260	Poor $360

SAA EARLY MILITARY
(FIRST GEN.)
FAIR

Courtesy of Rock Island Auction Company

SINGLE ACTION ARMY EARLY MILITARY (FIRST GEN.)
REDBOOK CODE: RB-CL-H-SAXXX2
Highly collectible. Be sure to consult an expert before purchase, as fakes exist.

roduction: 1873 - 1877 Caliber: .45 LC Action: SA, revolver Barrel Length: 5.5", 7.5", Wt.: 40 oz. Sights: blade front Capacity: 6 Grips: wood, plastic, ivory							
D2C:	NIB	--	Ex $ 72,200	VG+ $	52,250	Good	$47,500
C2C:	Mint	--	Ex $ 65,550	VG+ $	47,500	Good	$42,750 Fair $21,850
Trade-In:	Mint	--	Ex $ 53,200	VG+ $	37,050	Good	$33,250 Poor $9,510

SAA FLATTOP TARGET
EXCELLENT

Courtesy of Rock Island Auction Company

SINGLE ACTION ARMY FLATTOP TARGET
REDBOOK CODE: RB-CL-H-SAXX10
Less than 950 units produced. Serial numbers approximately between 127,000 and 162,000. These models are highly collectible and prices can fluctuate drastically. Be advised before purchasing.

Production: 1888 - 1896 Caliber: .22 LR, .38 Colt, .41 Short, .45 LC, .450 Boxer, .476 Eley, etc. Action: SA, revolver Sights: blade front Capacity: 6								
D2C:	NIB	--	Ex $19,775	VG+	$14,300	Good	$13,000	
C2C:	Mint	--	Ex $17,950	VG+	$13,000	Good	$11,700 Fair	$5,980
Trade-In:	Mint	--	Ex $14,575	VG+	$10,140	Good	$9,100 Poor	$2,610

SAA BISLEY
(FIRST GEN.)
GOOD

Courtesy of Rock Island Auction Company

SINGLE ACTION ARMY BISLEY (FIRST GEN.)
REDBOOK CODE: RB-CL-H-SAXX15
Serial numbers range between 155,00 and 332,000.

Production: 1894 - 1915 Caliber: .32 LC, .455 Eley, .32-30, .38-40, .41, .44-40, .45 LC Action: SA, revolver Sights: blade front Capacity: 6								
Grips: wood, plastic, or ivory								
D2C:	NIB	--	Ex $9,125	VG+	$6,600	Good	$6,000	
C2C:	Mint	--	Ex $8,300	VG+	$6,000	Good	$5,400 Fair	$2,760
Trade-In:	Mint	--	Ex $6,725	VG+	$4,680	Good	$4,200 Poor	$1,200

**SAA FLATTOP BISLEY
MODEL FIRST GEN.**
EXCELLENT

Courtesy of Rock Island Auction Company

SINGLE ACTION ARMY FLATTOP BISLEY
FIRST GEN. REDBOOK CODE: RB-CL-H-SAXX16
Less than 1,000 produced. Noted for its flattop frame and rear dovetail sight.

Production: late 1890s - 1913 Caliber: .32 LC, .455 Eley, .32-30, .38-40, .41, .44-40, .45 LC Action: SA, revolver Barrel Length: 7.5"								
Sights: blade front, dovetail rear Capacity: 6 Grips: wood, plastic, or ivory								
D2C:	NIB	--	Ex $20,150	VG+	$14,575	Good	$13,250	
C2C:	Mint	--	Ex $18,300	VG+	$13,250	Good	$11,925 Fair	$6,095
Trade-In:	Mint	--	Ex $14,850	VG+	$10,340	Good	$9,275 Poor	$2,670

COLT

SAA STANDARD ISSUE SECOND GEN.
VERY GOOD +

Courtesy of Rock Island Auction Company

SINGLE ACTION ARMY STANDARD ISSUE SECOND GEN.
REDBOOK CODE: RB-CL-H-SAXX22
Very similar to the First Generation SAA. Most have serial numbers between approximately 1 and 3,000SA. Chambered for many different calibers (all with similar values). Models chambered for .45 LC have slightly increased value.

Production: 1956 - 1975   Caliber: .45 LC, .44-40 WCF, .38-.40, .32-.20, .38 LC, .38 Special, etc.   Action: SA, revolver   Barrel Length: 5.5", some minor variations exist   OA Length: 11"   Wt.: 40 oz.

Sights: blade front   Capacity: 6   Grips: wood, plastic, or ivory

D2C:	NIB	--	Ex	$2,400	VG+	$1,735	Good	$1,575	
C2C:	Mint	$3,030	Ex	$2,175	VG+	$1,580	Good	$1,420	Fair $725
Trade-In:	Mint	$2,240	Ex	$1,775	VG+	$1,230	Good	$1,105	Poor $330

SAA SECOND GEN. EARLY PRODUCTION
VERY GOOD +

Courtesy of Rock Island Auction Company

SINGLE ACTION ARMY SECOND GEN. EARLY PRODUCTION
REDBOOK CODE: RB-CL-H-SAXX23
Colt began producing the SAA again in 1956. Early guns of this generation have serial numbers between 0001SA and approximately 40,100SA. These guns have moderate value, though drastically less than any First Generation SAA.

Production: 1956 - 1965   Caliber: .45 LC, .44-.40 WCF, .38-.40, .32-.20, .38 LC, .38 Special, etc.   Action: SA, revolver   Barrel Length: 5.5", some minor variations exist   OA Length: approximately 11", varies   Wt.: 40 oz.

Sights: blade front   Capacity: 6   Grips: wood, plastic, or ivory

D2C:	NIB	--	Ex	$2,550	VG+	$1,845	Good	$1,675	
C2C:	Mint	$3,220	Ex	$2,325	VG+	$1,680	Good	$1,510	Fair $775
Trade-In:	Mint	$2,380	Ex	$1,900	VG+	$1,310	Good	$1,175	Poor $360

SAA FRONTIER SCOUT
GOOD

Courtesy of Rock Island Auction Company

SINGLE ACTION ARMY FRONTIER SCOUT
REDBOOK CODE: RB-CL-H-FRNTSC
Serial numbers range between approximately 1000Q and 250,000. Available in several variants, denoted by Q, F, and K suffixes. Models with walnut grips, an extra cylinder, and Buntline variants have increased value of nearly half.   K versions command a slightly high premium.

Production: 1957 - 1971   Caliber: .22 LR, .22 WMR   Action: SA, revolver

Barrel Length: 4.75", 9.5"   Sights: blade front   Capacity: 6

Grips: rubber, wood, or ivory

D2C:	NIB	$550	Ex	$425	VG+	$305	Good	$275	
C2C:	Mint	$530	Ex	$400	VG+	$280	Good	$250	Fair $130
Trade-In:	Mint	$400	Ex	$325	VG+	$220	Good	$195	Poor $60

SAA SECOND GEN. MID PRODUCTION
EXCELLENT

Courtesy of Rock Island Auction Company

SINGLE ACTION ARMY SECOND GEN. MID PRODUCTION
REDBOOK CODE: RB-CL-H-SAXX24
Serial numbers range between approximately 39,500SA to 70,050SA.

Production: 1965 - 1975   Caliber: .45 LC, .44 Special, .357 Magnum

Action: SA, revolver   Barrel Length: 5.5", some minor variations exist

OA Length: 11"   Wt.: 40 oz.   Sights: fixed   Capacity: 6   Grips: wood, plastic, or ivory

D2C:	NIB	--	Ex	$2,000	VG+	$1,430	Good	$1,300	
C2C:	Mint	$2,500	Ex	$1,800	VG+	$1,300	Good	$1,170	Fair $600
Trade-In:	Mint	$1,850	Ex	$1,475	VG+	$1,020	Good	$910	Poor $270

**SAA NEW FRONTIER
SECOND GEN.**
EXCELLENT
Courtesy of Rock Island Auction Company

SINGLE ACTION ARMY NEW FRONTIER SECOND GEN.
REDBOOK CODE: RB-CL-H-SAXX26
Serial numbers feature a "NF" suffix. Variances in pricing exist between different barrel lengths and calibers.

Production: 1961 - 1975 Caliber: .357 Magnum, .45 LC, .44 Special, .38 Special
Action: SA, revolver Barrel Length: 4.75", 5.5", 7.5" Capacity: 6

D2C:	NIB	--	Ex	$1,725	VG+	$1,240	Good	$1,125	
C2C:	Mint	--	Ex	$1,575	VG+	$1,130	Good	$1,015	Fair $520
Trade-In:	Mint	$1,600	Ex	$1,275	VG+	$880	Good	$790	Poor $240

**SAA SECOND GEN. LATER
PRODUCTION**
GOOD

Courtesy of Rock Island Auction Company

SINGLE ACTION ARMY SECOND GEN. LATER PRODUCTION
REDBOOK CODE: RB-CL-H-SAXX27
Serial numbers range between approximately 70,050SA and 73,200SA. After 73,200SA, the "SA" was changed to a prefix, with the first of these models being around SA01000.

Production: 1973 - early 1980s Caliber: .45 LC, .357 Magnum Action: SA, revolver
Barrel Length: 5.5", some minor variations exist OA Length: 11"
Wt.: 40 oz. Sights: fixed Capacity: 6 Grips: wood, plastic, or ivory

D2C:	NIB	--	Ex	$1,600	VG+	$1,155	Good	$1,050	
C2C:	Mint	$2,020	Ex	$1,450	VG+	$1,050	Good	$945	Fair $485
Trade-In:	Mint	$1,500	Ex	$1,200	VG+	$820	Good	$735	Poor $210

SINGLE ACTION ARMY STANDARD ISSUE THIRD GEN.
REDBOOK CODE: RB-CL-H-SAXX28
Features more interchangeable and adjustable parts, a polished hammer, black composite eagle stocks, and second-generation style cylinder finish.

Production: early 1980s - current Caliber: .45 LC, .357 Magnum Action: SA, revolver
Barrel Length: 5.5", some minor variations exist OA Length: 11" Wt.: 40 oz.
Sights: iron Capacity: 6 Grips: wood, plastic, or ivory

D2C:	NIB	$1,000	Ex	$775	VG+	$550	Good	$500	
C2C:	Mint	$960	Ex	$700	VG+	$500	Good	$450	Fair $230
Trade-In:	Mint	$710	Ex	$575	VG+	$390	Good	$350	Poor $120

**SAA STANDARD
ISSUE 3RD AND CURRENT
GENERATION**
VERY GOOD +

Courtesy of Rock Island Auction Company

SAA NEW FRONTIER
EXCELLENT
Courtesy of Rock Island Auction Company

SINGLE ACTION ARMY NEW FRONTIER THIRD GEN.
REDBOOK CODE: RB-CL-H-SAXX29
Five-digit serial numbers, blue finish, one-piece grips, most popular with 7.5" barrel. Models chambered for .44-40 SCF are rare and haev more value. Smaller barrel lengths also command a higher premium.

Production: early 1980s - current Caliber: .357 Magnum, .44-.40 WCF, .44 Special, .45 LC Action: SA, revolver Barrel Length: 4.75", 5.5", 7.5" Sights: blade front
Capacity: 6 Grips: wood, plastic, or ivory

D2C:	NIB	--	Ex	$875	VG+	$635	Good	$575	
C2C:	Mint	$1,110	Ex	$800	VG+	$580	Good	$520	Fair $265
Trade-In:	Mint	$820	Ex	$650	VG+	$450	Good	$405	Poor $120

SINGLE ACTION ARMY COWBOY THIRD GEN.
REDBOOK CODE: RB-CL-H-SAXX31
Reintroduced in 1998, but discontinued again in 2003. Marked Cowboy .45 LC. Limited value.

Production: 1998 - 2003 Caliber: .45 LC Action: SA, revolver
Barrel Length: 4.75", 5.5", 7.5"
Wt.: 40 oz., varies Sights: blade front Capacity: 6 Grips: black hard rubber

D2C:	NIB	$815	Ex	$625	VG+	$450	Good	$410	
C2C:	Mint	$790	Ex	$575	VG+	$410	Good	$370	Fair $190
Trade-In:	Mint	$580	Ex	$475	VG+	$320	Good	$290	Poor $90

SAA COWBOY
EXCELLENT
Courtesy of Rock Island Auction Company

COLT

**SAA SHERIFF'S MODEL
THIRD GEN.**
EXCELLENT
Courtesy of Rock Island Auction Company

SINGLE ACTION ARMY SHERIFF'S MODEL THIRD GEN.

REDBOOK CODE: RB-CL-H-SAXX32
Produced on-and-off between 1980 and 2010. Increase in value if nickel finish and factory ivory grips.

Production: early 1980s - ~2010 Caliber: .44-40 WCF, .45 LC Action: SA, revolver

Barrel Length: 3", 4" OA Length: 7", 8" Wt.: 40 oz., varies Sights: blade front

Capacity: 6 Grips: black synthetic

D2C:	NIB	$1,200	Ex	$ 925	VG+	$ 660	Good	$ 600	
C2C:	Mint	$1,160	Ex	$ 850	VG+	$ 600	Good	$ 540	Fair $ 280
Trade-In:	Mint	$ 860	Ex	$ 675	VG+	$ 470	Good	$ 420	Poor $ 120

**SAA
ENGRAVERS
SAMPLER
THIRD GEN.**
EXCELLENT
Courtesy of Rock Island Auction Company

SINGLE ACTION ARMY ENGRAVERS SAMPLER

THIRD GEN REDBOOK CODE: RB-CL-H-SAXX38
Nickel finish, unique engravings on the frame and cylinder, ornate design.

Production: late 1990s Caliber: .45 LC Action: SA, revolver

Barrel Length: 5.5" OA Length: 11" Wt.: 40 oz., varies

Sights: blade front Capacity: 6 Grips: ivory

D2C:	NIB	$3,100	Ex	$2,375	VG+	$1,705	Good	$1,550	
C2C:	Mint	$2,980	Ex	$2,150	VG+	$1,550	Good	$1,395	Fair $ 715
Trade-In:	Mint	$2,210	Ex	$1,750	VG+	$1,210	Good	$1,085	Poor $ 330

SINGLE ACTION ARMY MODEL P THIRD GEN.

REDBOOK CODE: RB-CL-H-SAXX41
Discontinued in 2010, but since reintroduced. Models marked "B" have twice the value.

Production: current Caliber: .45 LC, .357 Magnum Action: SA, revolver

Barrel Length: 4.75", 5.5", 7.5" Wt.: 40 oz., varies

Sights: fixed Capacity: 6 Grips: black synthetic

D2C:	NIB	$1,600	Ex	$ 1,225	VG+	$ 880	Good	$ 800	
C2C:	Mint	$1,540	Ex	$1,125	VG+	$ 800	Good	$ 720	Fair $ 370
Trade-In:	Mint	$1,140	Ex	$ 900	VG+	$ 630	Good	$ 560	Poor $ 180

NEW LINE .22
EXCELLENT

NEW LINE .22 (1ST AND 2ND MODELS) REDBOOK CODE: RB-CL-H-NL22XX

Compact design. 1st models with short cylinder flutes command higher premium.

Production: 1873 - 1877 Caliber: .22 LR Action: SA, revolver

Barrel Length: 2.25" Sights: blade front Capacity: 5, 7 Grips: rosewood or ivory

D2C:	NIB	--	Ex	$1,125	VG+	$815	Good	$740	
C2C:	Mint	--	Ex	$1,025	VG+	$740	Good	$665	Fair $340
Trade-In:	Mint	--	Ex	$850	VG+	$580	Good	$520	Poor $150

NEW LINE .30 REDBOOK CODE: RB-CL-H-NL30XX

Slightly larger than the .22 LC version. Roughly 11,000 manufactured. Blue models command nearly doubled premium.

Production: 1874 - 1876 Caliber: .30 cal. Action: SA, revolver

Barrel Length: 2.25" Sights: blade front Capacity: 5 Grips: rosewood or ivory

D2C:	NIB	--	Ex	$ 1,225	VG+	$ 880	Good	$ 800	
C2C:	Mint	--	Ex	$ 1,125	VG+	$ 800	Good	$ 720	Fair $ 370
Trade-In:	Mint	--	Ex	$ 900	VG+	$ 630	Good	$ 560	Poor $ 180

NEW LINE .32 GOOD

Courtesy of Rock Island Auction Company

NEW LINE .32 REDBOOK CODE: RB-CL-H-NL32XX

Similar to the other New Line models, sometimes called the "Ladies Colt." 4" barrel model commands a doubled premium.

Production: 1873 - 1884 Caliber: .32 S&W Action: SA, revolver

Barrel Length: 2.25", 4" Sights: blade front Capacity: 5 Grips: rosewood or ivory

D2C:	NIB	--	Ex	$2,225	VG+	$1,600	Good	$1,455		
C2C:	Mint	--	Ex	$2,025	VG+	$1,460	Good	$1,310	Fair	$670
Trade-In:	Mint	--	Ex	$1,650	VG+	$1,140	Good	$1,020	Poor	$300

NEW LINE .38 GOOD

Courtesy of Rock Island Auction Company

NEW LINE .38 REDBOOK CODE: RB-CL-H-NL38XX

Sometimes called the "Pet Colt." Roughly 5,000 manufactured. 4" barrel version worth twice as much.

Production: 1874 - 1880 Caliber: .38 RF, .38 LC, .38 SC Action: SA, revolver

Barrel Length: 2.25", 4" Sights: blade front Capacity: 5 Grips: rosewood or ivory

D2C:	NIB	--	Ex	$1,950	VG+	$1,405	Good	$1,275		
C2C:	Mint	--	Ex	$1,775	VG+	$1,280	Good	$1,150	Fair	$590
Trade-In:	Mint	--	Ex	$1,450	VG+	$1,000	Good	$895	Poor	$270

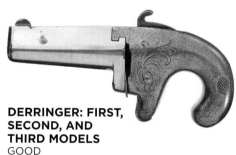

NEW LINE .41 FAIR

Courtesy of Rock Island Auction Company

NEW LINE .41 REDBOOK CODE: RB-CL-H-NL41XX

Sometimes called the "Big Colt." Approximately 7,000 produced. 4" barrel models command nearly a doubled premium.

Production: 1874 - 1879 Caliber: .41 RF/CF, .41 LC Action: SA, revolver

Barrel Length: 2.25", 4" Sights: blade front Capacity: 5 Grips: rosewood or ivory

D2C:	NIB	--	Ex	$2,550	VG+		$1,850	Good	$1,680		
C2C:	Mint	--	Ex	$2,325	VG+		$1,680	Good	$1,510	Fair	$775
Trade-In:	Mint	--	Ex	$1,900	VG+		$1,310	Good	$1,175	Poor	$360

DERRINGER: FIRST, SECOND, AND THIRD MODELS GOOD

Courtesy of Rock Island Auction Company

DERRINGER: FIRST, SECOND, AND THIRD MODELS

REDBOOK CODE: RB-CL-H-DR1234

First and second model Derringers have moderate collectors value. Later models have a decreased value, but are still collectible. Minor variations exist between each model with no great distinctions in price.

Production: 1870 - 1963 Caliber: .41 RF, .41 CF (.41 Cal) Action: SA, derringer

Barrel Length: 2.5", OA Length: 5.5" Sights: blade front Capacity: 1

Grips: ivory or wood

D2C:	NIB	$3,350	Ex	$2,550	VG+	$1,845	Good	$1,675		
C2C:	Mint	$3,220	Ex	$2,325	VG+	$1,680	Good	$1,510	Fair	$775
Trade-In:	Mint	$2,380	Ex	$1,900	VG+	$1,310	Good	$1,175	Poor	$360

LORD AND LADY DERRINGER REDBOOK CODE: RB-CL-H-LRLDDR

Serial numbers range between approximately 1,000 and 60,000 with either DER or LDR suffixes. Cosmetic differences: the Lady features ivory grips and a gold-looking finish, the Lord walnut grips and blue finish. Values nearly doubled if still in their original cases or in a set.

Production: 1970 - 1973 Caliber: .22 short Action: SA, derringer

Barrel Length: 2.5" OA Length: 5.5" Sights: blade front Capacity: 1 Grips: walnut, pearl

D2C:	NIB	$ 225	Ex	$ 175	VG+	$ 125	Good	$ 115		
C2C:	Mint	$ 220	Ex	$ 175	VG+	$ 120	Good	$ 105	Fair	$ 55
Trade-In:	Mint	$ 160	Ex	$ 150	VG+	$ 90	Good	$ 80	Poor	$ 30

LORD AND LADY DERRINGER EXCELLENT

Courtesy of Rock Island Auction Company

COLT

MODEL 1877 LIGHTNING
VERY GOOD +

Courtesy of Rock Island Auction Company

MODEL 1877 LIGHTNING

REDBOOK CODE: RB-CL-H-1877LGT

166,849 produced from January 1877 to 1909. Offered in 3 calibers: the Lightning, .38 LC; the Thunderer, .41 LC; and the Rainmaker .32 Colt. Rainmaker model was offered only in 1877, so has higher value. 1877 nickel-plated or case-hardened frame with blue or nickel-plated barrel and cylinder (premium for blue). Barrel lengths range from 1.5" to 10.5", with ones under 2.5" and over 6" at a 50% premium. Available with or without ejector rod and housing. Thunderer model pricing is about 10% higher than below; Rainmaker models, about 15% higher.

Production: 1877 - 1909 Caliber: .38 LC Action: DA/SA, revolver

Barrel Length: 1.5" - 10 Sights: blade front Capacity: 6 Grips: checkered rosewood, hard rubber, pearl, or black synthetic

D2C:	NIB	--	Ex	$2,900	VG+	$2,090	Good	$1,900	
C2C:	Mint	--	Ex	$2,625	VG+	$1,900	Good	$1,710	Fair $ 875
Trade-In:	Mint	--	Ex	$2,150	VG+	$1,490	Good	$1,330	Poor $ 390

1878 SHERIFF'S MODEL

REDBOOK CODE: RB-CL-H-1878SHR

A snubbed-nose version of the M1878, same set of serial numbers. Models "J.P. Lower" have increased value ($1,500-$2,250).

Production: 1878 - 1905 Caliber: .44-40, .45 LC Action: DA/SA, revolver

Barrel Length: 3", 3.5", 4" Sights: blade front Capacity: 6

Grips: black synthetic, ivory, or walnut

D2C:	NIB	--	Ex	$5,800	VG+	$4,180	Good	$3,800	
C2C:	Mint	--	Ex	$5,250	VG+	$3,800	Good	$3,420	Fair $1,750
Trade-In:	Mint	--	Ex	$4,275	VG+	$2,970	Good	$2,660	Poor $780

MODEL 1878 SHERIFF'S MODEL
FAIR

Courtesy of Rock Island Auction Company

MODEL 1889 NAVY (CIVILIAN MODEL) REDBOOK CODE: RB-CL-H-1889NV

Nickel or blue finish, around 30,000 produced with corresponding serial numbers. Martial models command higher premium.

Production: 1889 - 1894 Caliber: .38 LC, .41 LC Action: DA/SA, revolver

Barrel Length: 3", 4.5", 6" Sights: blade front Capacity: 6 Grips: black synthetic

D2C:	NIB	--	Ex	$2,300	VG+	$1,650	Good	$1,500	
C2C:	Mint	--	Ex	$2,075	VG+	$1,500	Good	$1,350	Fair $ 690
Trade-In:	Mint	--	Ex	$1,700	VG+	$1,170	Good	$1,050	Poor $ 300

MODEL 1889 NAVY
(CIVILIAN MODEL)
GOOD

Courtesy of Rock Island Auction Company

MODEL 1905 MARINE CORPS REDBOOK CODE: RB-CL-H-1905MC

Less than a thousand made. Serial numbers range between 10,001 and 10,926. Most feature a "W" stamped on the heel of the grip.

Production: 1905 - 1909 Caliber: .38 SC, .38 LC Action: DA/SA, revolver

Barrel Length: 4", 4.5", 5", 6" Sights: blade front Capacity: 6 Grips: walnut

D2C:	NIB	--	Ex	$3,500	VG+	$2,530	Good	$2,300	
C2C:	Mint	$4,420	Ex	$3,175	VG+	$2,300	Good	$2,070	Fair $1,060
Trade-In:	Mint	$3,270	Ex	$2,600	VG+	$1,800	Good	$1,610	Poor $ 480

MODEL 1905 MARINE CORPS
GOOD

Courtesy of Rock Island Auction Company

NEW ARMY/NAVY MODEL
GOOD

Courtesy of Rock Island Auction Company

OFFICER'S TARGET MODEL
GOOD

Courtesy of Rock Island Auction Company

NEW POCKET MODEL
GOOD

Courtesy of Rock Island Auction Company

NEW POLICE
EXCELLENT

Courtesy of Rock Island Auction Company

**NEW SERVICE
MODEL COMMERCIAL**
GOOD

Courtesy of Rock Island Auction Company

MODEL 1892 NEW ARMY/NAVY MODEL (CIVILIAN/MARTIAL)
REDBOOK CODE: RB-CL-H-OFFXX1
Serial numbers range approximately between 1 and 300,000. Blue finish, sideplate located on the right. Martially marked models command a premium of nearly 75%.

Production: 1892 - 1908   Caliber: .38 Special, .38 LC   Action: DA/SA, revolver
Barrel Length: 6"   Sights: blade front   Capacity: 6   Grips: synthetic

D2C:	NIB	--	Ex	$1,450	VG+	$1,045	Good	$ 950	
C2C:	Mint	--	Ex	$1,325	VG+	$950	Good	$ 855	Fair $ 440
Trade-In:	Mint	--	Ex	$1,075	VG+	$750	Good	$ 665	Poor $ 210

OFFICER'S TARGET MODEL (FIRST ISSUE)
REDBOOK CODE: RB-CL-H-NWOFM1
Serial numbers range between approximately 225,000 and 292,000.

Production: 1904 - 1908   Caliber: .38 Special   Action: DA/SA, revolver
Barrel Length: 6"   Sights: adjustable   Capacity: 6   Grips: walnut, checkered

D2C:	NIB	--	Ex	$1,375	VG+	$990	Good	$900	
C2C:	Mint	--	Ex	$1,250	VG+	$900	Good	$810	Fair $415
Trade-In:	Mint	--	Ex	$1,025	VG+	$710	Good	$630	Poor $180

NEW POCKET MODEL   REDBOOK CODE: RB-CL-H-NWPCKT
Serial numbers range between approximately 1 and 30,000. Features either blue or nickel finish. Roughly 30,000 produced.

Production: 1893 - 1905   Caliber: .32 Short, .32 LR, .32 Colt New Police
Action: DA/SA, revolver   Barrel Length: 2.5", 3.5", 5", 6"
Sights: blade front   Capacity: 6   Grips: hard rubber

D2C:	NIB	--	Ex	$1,050	VG+	$760	Good	$690	
C2C:	Mint	--	Ex	$950	VG+	$690	Good	$620	Fair $320
Trade-In:	Mint	--	Ex	$775	VG+	$540	Good	$485	Poor $150

NEW POLICE   REDBOOK CODE: RB-CL-H-NWPOLC
Serial numbers range between approximately 1 and 49,000. Nickel finish commands a 25% premium.

Production: 1887 - 1908   Caliber: .32 Colt, .32 Colt New Police, .38 S&W
Action: DA/SA, revolver   Barrel Length: 2.5", 4", 6"
Wt.: 32 oz.   Sights: blade front   Capacity: 6   Grips: hard rubber

D2C:	NIB	--	Ex	$ 600	VG+	$ 430	Good	$ 390	
C2C:	Mint	--	Ex	$ 550	VG+	$ 390	Good	$ 355	Fair $ 180
Trade-In:	Mint	--	Ex	$ 450	VG+	$ 310	Good	$ 275	Poor $ 90

NEW SERVICE MODEL COMMERCIAL (EARLY AND IMPROVED MODELS) (FIRST & SECOND SERIES) REDBOOK CODE: RB-CL-H-NWSCMM
These were the largest swing-out cylinder revolvers ever produced by Colt. Early models command a slightly higher premium. Nickel finish is also slightly more valuable. Ivory or pear grips from the factory can increase premium by half.

Production: 1898 - ~1940   Caliber: .38 Special, .357 Magnum, .38-40 WCF,
.44-40 WCF, .44 Russian, .44 Special, .45 ACP, etc.
Action: DA/SA, revolver   Barrel Length: 4", 4.5", 5", 5.5", 6", 7.5"   Sights: blade front
Capacity: 6   Grips: walnut, black synthetic

D2C:	NIB	--	Ex	$1,300	VG+	$ 935	Good	$ 850	
C2C:	Mint	--	Ex	$1,175	VG+	$ 850	Good	$ 765	Fair $ 395
Trade-In:	Mint	--	Ex	$975	VG+	$ 670	Good	$ 595	Poor $ 180

NEW SERVICE MODEL SHOOTING MASTER

REDBOOK CODE: RB-CL-H-NWMSHM

Less than 4,500 produced, serial numbers ranging from approximately 333,000 to 350,000. Versions chambered in .44 Special, .45 ACP, or .45 LC command slightly higher premium.

Production: 1900 - 1941 Caliber: .38 Special, . 45 LC, .45 ACP, .357 Magnum

Action: DA/SA, revolver Barrel Length: 6" OA Length: 10.8" Wt.: 40 oz.

Sights: blade front Capacity: 6 Grips: walnut

D2C:	NIB	--	Ex	$1,425	VG+	$1,020	Good	$ 925	
C2C:	Mint	$1,780	Ex	$1,300	VG+	$ 930	Good	$ 835	Fair	$ 430
Trade-In:	Mint	$1,320	Ex	$1,050	VG+	$ 730	Good	$ 650	Poor	$ 210

POCKET POSITIVE REDBOOK CODE: RB-CL-H-PCKPOS

Models with nickel finish have increased value. A modified and enhanced version of the New Pocket.

Production: 1905 - 1940 Caliber: .32 Colt, .32 S&W, .32 Colt New Police

Action: DA/SA, revolver Barrel Length: 2", 4", 5", 6", some variations

Sights: blade front Capacity: 6 Grips: hard rubber

D2C:	NIB	--	Ex	$ 575	VG+	$ 410	Good	$ 370	
C2C:	Mint	$ 720	Ex	$ 525	VG+	$ 370	Good	$ 335	Fair	$ 175
Trade-In:	Mint	$ 530	Ex	$ 425	VG+	$ 290	Good	$ 260	Poor	$ 90

POCKET POSITIVE
EXCELLENT

Courtesy of Rock Island Auction Company

POLICE POSITIVE (FIRST, SECOND, SPECIAL, TARGET MODELS)

(FIRST AND SECOND ISSUE) REDBOOK CODE: RB-CL-H-PLPOS1

Police Positive Target models, indicated with a "G" or "C" in the serial number, have nearly doubled value.

Production: 1905 - late 1940s Caliber: .32 Colt, .32 New Police, .38 New Police, .38 S&W Action: DA/SA, revolver Barrel Length: 4", 5", 6" Wt.: 32 oz.

Sights: blade front Capacity: 6 Grips: hard rubber, walnut

D2C:	NIB	--	Ex	$ 600	VG+	$ 420	Good	$ 380	
C2C:	Mint	$ 730	Ex	$ 525	VG+	$ 380	Good	$ 345	Fair	$ 175
Trade-In:	Mint	$ 540	Ex	$ 450	VG+	$ 300	Good	$ 270	Poor	$ 90

POLICE POSITIVE
VERY GOOD +

Courtesy of Rock Island Auction Company

OFFICER'S TARGET MODEL (2ND ISSUE)

REDBOOK CODE: RB-CL-H-NWOFM2

Serial numbers range between approximately 291,000 and 540,000. A part of the second series of New Service models, but designed for target shooting, noted for its flat-style barrel.

Production: 1908 - 1940 Caliber: .22 LR, .32 Police Positive, .38 Special

Action: DA/SA, revolver Barrel Length: 4, 4.5, 5, 6, 7.5 Sights: blade front Barrel Length: 4", 4.5", 5", 6", 7.5" Sights: blade front Capacity: 6 Grips: walnut

D2C:	NIB	--	Ex	$1,000	VG+	$ 715	Good	$ 650	
	Mint	$1,250	Ex	$ 900	VG+	$ 650	Good	$ 585	Fair	$ 300
Trade-In:	Mint	$930	Ex	$ 750	VG+	$ 510	Good	$ 455	Poor	$ 150

NEW ARMY/NAVY OFFICER'S MODEL TARGET
GOOD

Courtesy of Rock Island Auction Company

NEW ARMY 1909
VERY GOOD +

NEW ARMY 1909 REDBOOK CODE: RB-CL-H-NWS09A

More valuable than the Commercial models. Serial numbers range approximately between 30,000 and 50,000.

Production: 1909   Caliber: .38 Special, .357 Magnum, .38-40 WCF, .44-40 WCF, .44 Russian, .44 Special, .45 ACP, etc.   Action: DA/SA, revolver
Barrel Length: 4", 4.5", 5", 5.5", 6", 7.5"   Sights: blade front   Capacity: 6
Grips: black synthetic

D2C:	NIB	--	Ex	$1,325	VG+	$ 950	Good	$ 865	
C2C:	Mint	$1,660	Ex	$1,200	VG+	$ 870	Good	$ 780	Fair	$ 400
Trade-In:	Mint	$1,230	Ex	$ 975	VG+	$ 680	Good	$ 605	Poor	$ 180

NEW NAVY 1909
VERY GOOD +

Courtesy of Rock Island Auction Company

NEW NAVY 1909 REDBOOK CODE: RB-CL-H-NWM09N

Less than 2,000 manufactured. Marked "U.S.N. No. 972" on the bottom of the grip. Serial numbers likely range between 50,000 and 52,000.

Production: 1909   Caliber: .38 Special, .357 Magnum, .38-40 WCF, .44-40 WCF, .44 Russian, .44 Special, .45 ACP, etc.   Action: DA/SA, revolver
Barrel Length: 4", 4.5", 5", 5.5", 6", 7.5"   Sights: blade front   Capacity: 6
Grips: walnut or black synthetic

D2C:	NIB	--	Ex	$4,575	VG+	$3,300	Good	$3,000	
C2C:	Mint	$5,760	Ex	$4,150	VG+	$3,000	Good	$2,700	Fair	$1,380
Trade-In:	Mint	$4,260	Ex	$3,375	VG+	$2,340	Good	$2,100	Poor	$ 600

NEW SERVICE MODEL U.S. MARINE CORPS. 1909
VERY GOOD +

Courtesy of Rock Island Auction Company

NEW SERVICE MODEL U.S. MARINE CORPS. 1909

REDBOOK CODE: RB-CL- H-NWUSM9

Similar to the other New Service models, but marked U.S.M.C. Serial numbers range between 21,000 and 23,000, though less than two thousand manufactured.

Production: 1909 - discontinued   Caliber: .38 Special, .357 Magnum, .38-40 WCF, .44-40 WCF, .44 Russian, .44 Special, .45 ACP, etc. Action: DA/SA, revolver
Barrel Length: 4", 4.5", 5", 5.5", 6", 7.5"   Sights: blade front   Capacity: 6
Grips: black, walnut, or ivory

D2C:	NIB	--	Ex	$4,200	VG+	$3,025	Good	$2,750	
C2C:	Mint	$5,280	Ex	$3,800	VG+	$2,750	Good	$2,475	Fair	$1,265
Trade-In:	Mint	$3,910	Ex	$3,100	VG+	$2,150	Good	$1,925	Poor	$ 570

NEW ARMY 1917
EXCELLENT

Courtesy of Rock Island Auction Company

NEW ARMY 1917 REDBOOK CODE: RB-CL-H-NWM17A

Similar to the M1909 revolver, but only fires .45 ACP rounds. Serial numbers range between approximately 150,000 and 301,000.

Production: 1917 - 1944 Caliber: .45 ACP Action: DA/SA, revolver
Barrel Length: 5.5" OA Length: 10.8" Wt.: 40 oz. Sights: blade front
Capacity: 6 Grips: synthetic or wood

D2C:	NIB	--	Ex	$1,375	VG+	$ 990	Good	$ 900	
C2C:	Mint	$1,730	Ex	$1,250	VG+	$ 900	Good	$ 810	Fair	$ 415
Trade-In:	Mint	$1,280	Ex	$1,025	VG+	$ 710	Good	$ 630	Poor	$ 180

NEW SERVICE COMMERICAL 1917
VERY GOOD +

Courtesy of Rock Island Auction Company

NEW SERVICE CIVILIAN/COMMERCIAL 1917

REDBOOK CODE: RB-CL-H-NWM17C

Similar to the M1909 revolver, but only fires .45 ACP rounds. Serial numbers range between approximately 335,000 and 336,000, though less than 2,000 produced.

Production: 1917 - 1927 Caliber: .45 ACP Action: DA/SA, revolver
Barrel Length: 5.5" OA Length: 10.8" Wt.: 40 oz. Sights: blade front
Capacity: 6 Grips: walnut

D2C:	NIB	--	Ex	$1,225	VG+	$ 880	Good	$ 800	
C2C:	Mint	$1,540	Ex	$1,125	VG+	$ 800	Good	$ 720	Fair	$ 370
Trade-In:	Mint	$1,140	Ex	$ 900	VG+	$ 630	Good	$ 560	Poor	$ 180

COLT

CAMP PERRY MODEL
EXCELLENT

Courtesy of Rock Island Auction Company

CAMP PERRY MODEL REDBOOK CODE: RB-CL-H-CMPRRY

One of the most unique revolvers Colt has ever produced, made in honor of the National Rifle and Pistol Matches. Single-shot action on a revolver frame. Roughly 2,500 produced.

Production: 1920 - early 1940s Caliber: .22 LR Action: DA/SA, revolver, single-shot

Barrel Length: 8", 10" Wt.: 16 oz. Sights: blade front Capacity: 1 Grips: plastic

D2C:	NIB	--	Ex	$2,375	VG+	$1,705	Good	$1,550	
C2C:	Mint	$2,980	Ex	$2,150	VG+	$1,550	Good	$1,395	Fair	$ 715
Trade-In:	Mint	$2,210	Ex	$1,750	VG+	$1,210	Good	$1,085	Poor	$ 330

NEW SERVICE TARGET MODEL

REDBOOK CODE: RB-CL-H-NWMTRG

Checkered trigger and back strap. Round butt version available after 1930. Models made with a 6" barrel or with a nickel finish are rare and command a higher premium.

Production: 1900 - 1940 Caliber: .45 LC, .45 ACP, .44 Special, .44 Russian

Action: DA/SA, revolver   Barrel Length: 6", 7.5"   OA Length: 10.8" Wt.: 40 oz.

Sights: adjustable rear Capacity: 6 Grips: checkered walnut or rubber

D2C:	NIB	--	Ex	$2,300	VG+	$1,650	Good	$1,500	
C2C:	Mint	$2,880	Ex	$2,075	VG+	$1,500	Good	$1,350	Fair	$ 690
Trade-In:	Mint	$2,130	Ex	$1,700	VG+	$1,170	Good	$1,050	Poor	$ 300

NEW SERVICE MODEL TARGET
EXCELLENT

Courtesy of Rock Island Auction Company

OFFICIAL POLICE REDBOOK CODE: RB-CL-H-OFFPLC

Models with a rounded butt and factory lettering are worth twice as much. Snub nose .38 Special models are rare and priced at a premium.

Production: 1927 - 1969 Caliber: .32-20 WCF, .38-200, .41 L, .38 Special, .22 LR

Action: DA/SA, revolver Barrel Length: 4", 6" Wt.: 40 oz. Sights: blade front

Capacity: 6 Grips: walnut

D2C:	NIB	$ 625	Ex	$ 475	VG+	$ 345	Good	$ 315	
C2C:	Mint	$ 600	Ex	$ 450	VG+	$ 320	Good	$ 285	Fair	$ 145
Trade-In:	Mint	$ 450	Ex	$ 350	VG+	$ 250	Good	$ 220	Poor	$ 90

OFFICIAL POLICE
EXCELLENT

Courtesy of Rock Island Auction Company

OFFICIAL POLICE MARK III REDBOOK CODE: RB-CL-H-OFPLMK

Features an updated J frame design, unlike the previous Official Police models. Very slight premium for nickel finish.

Production: late 1960s - 1978 Caliber: .38 Special Action: DA/SA, revolver

Barrel Length: 4", 5", 6" Wt.: 40 oz. Sights: blade front Capacity: 6 Grips: walnut

D2C:	NIB	$ 465	Ex	$ 375	VG+	$ 260	Good	$ 235	
C2C:	Mint	$ 450	Ex	$ 325	VG+	$ 240	Good	$ 210	Fair	$ 110
Trade-In:	Mint	$ 340	Ex	$ 275	VG+	$ 190	Good	$ 165	Poor	$ 60

OFFICIAL POLICE MARK III
EXCELLENT

Courtesy of Rock Island Auction Company

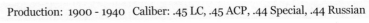

OFFICIAL POLICE MARSHAL MODEL
EXCELLENT

Courtesy of Rock Island Auction Company

OFFICIAL POLICE MARSHAL MODEL

REDBOOK CODE: RB-CL-H-OFFPXX

Serial numbers range between approximately 833,350M and 845,320M. Approximately 10,000 purchased for military and law enforcement use.

Production: 1955 - 1956   Caliber: .38 Special   Action: DA/SA, revolver

Barrel Length: 4", 5", 6"   Sights: blade front   Capacity: 6   Grips: walnut

D2C:	NIB	$1,250	Ex	$ 950	VG+	$ 690	Good	$ 625		
C2C:	Mint	$1,200	Ex	$ 875	VG+	$ 630	Good	$ 565	Fair	$ 290
Trade-In:	Mint	$ 890	Ex	$ 700	VG+	$ 490	Good	$ 440	Poor	$ 150

NEW SERVICE OFFICER'S MODEL MATCH

REDBOOK CODE: RB-CL-H-OFFXX4

Models chambered in .22 LR have increased value as shown below. More common models have a D2C Excellent price near $825.

Production: 1953 - late 1960s   Caliber: .22 LR, .22 WMR, .38 Special, .22 Magnum

Action: DA/SA, revolver   Barrel Length: 4", 4.5", 5", 6"   Wt.: 40 oz.

Sights: Accro   Capacity: 6   Grips: walnut

D2C:	NIB	$1,100	Ex	$ 850	VG+	$ 605	Good	$ 550		
C2C:	Mint	$1,060	Ex	$ 775	VG+	$ 550	Good	$ 495	Fair	$ 255
Trade-In:	Mint	$ 790	Ex	$ 625	VG+	$ 430	Good	$ 385	Poor	$ 120

COBRA (FIRST SERIES)   REDBOOK CODE: RB-CL-H-CBRAX1

Similar to the Colt Detective, but features an alloy frame, some chambered in .22 LR. Later versions of the gun have a shrouded ejection rod and hammer, increasing their value.

Production: 1950 - early 1970s   Caliber: .22 LR, .32 Colt N.P., .38 Special

Action: DA/SA, revolver   Barrel Length: 2", 3", 4"   Wt.: 16 oz.

Sights: blade front   Capacity: 6   Grips: wood or plastic

D2C:	NIB	$ 785	Ex	$ 600	VG+	$ 435	Good	$ 395		
C2C:	Mint	$ 760	Ex	$ 550	VG+	$ 400	Good	$ 355	Fair	$ 185
Trade-In:	Mint	$ 560	Ex	$ 450	VG+	$ 310	Good	$ 275	Poor	$ 90

COBRA
FIRST SERIES
EXCELLENT

Courtesy of Rock Island Auction Company

COBRA
SECOND SERIES
EXCELLENT

Courtesy of Rock Island Auction Company

COBRA (SECOND SERIES)   REDBOOK CODE: RB-CL-H-CBRAX2

The second series of the Cobra model. Nickel models command a higher premium.

Production: 1974 - 1981   Caliber: .38 Special   Action: DA/SA, revolver

Barrel Length: 2", 3", 4", 5"   Sights: blade front   Capacity: 6   Grips: wood or plastic

D2C:	NIB	$ 685	Ex	$ 525	VG+	$ 380	Good	$ 345		
C2C:	Mint	$ 660	Ex	$ 475	VG+	$ 350	Good	$ 310	Fair	$ 160
Trade-In:	Mint	$ 490	Ex	$ 400	VG+	$ 270	Good	$ 240	Poor	$ 90

COMMANDO
VERY GOOD +

Courtesy of Rock Island Auction Company

COMMANDO REDBOOK CODE: RB-CL-H-CMMDOM

Serial numbers range between approximately 1 and 51,000. A variant of the Official Police. Features non-parkerized finish, no checkering on the trigger, hammer, or latch. Approximately 50,000 units produced. 2" barrel commands a slightly higher premium.

Production: 1942 - 1943   Caliber: .38 Special   Action: DA/SA, revolver

Barrel Length: 2", 4", 6"   Sights: blade front   Capacity: 6   Grips: plastic

D2C:	NIB	$1,800	Ex	$1,375	VG+	$ 990	Good	$ 900		
C2C:	Mint	$1,730	Ex	$1,250	VG+	$ 900	Good	$ 810	Fair	$ 415
Trade-In:	Mint	$1,280	Ex	$1,025	VG+	$ 710	Good	$ 630	Poor	$ 180

COMMANDO SPECIAL REDBOOK CODE: RB-CL-H-CMMDSP

A reissued and enhanced version of the Commando. Features a compact design, shrouded ejector rod, and matte finish.

Production: 1984 - late 1980s   Caliber: .38 Special   Action: DA/SA, revolver

Barrel Length: 2"   Wt.: 16 oz.   Sights: blade front   Capacity: 6

Grips: hard rubber

D2C:	NIB	$ 640	Ex	$ 500	VG+	$ 355	Good	$ 320		
C2C:	Mint	$ 620	Ex	$ 450	VG+	$ 320	Good	$ 290	Fair	$ 150
Trade-In:	Mint	$ 460	Ex	$ 375	VG+	$ 250	Good	$ 225	Poor	$ 90

POLICE POSITIVE MARK V
VERY GOOD +

Courtesy of Rock Island Auction Company

POLICE POSITIVE MARK V REDBOOK CODE: RB-CL-H-PLPOS2

Similar to older models, but may feature a shrouded ejector rod and steel frame.

Production: 1994 - discontinued   Caliber: .38 Special +P   Action: DA/SA, revolver

Barrel Length: 4", 5", 6"   Wt.: 32 oz.   Sights: blade front   Capacity: 6

Grips: hard rubber or walnut

D2C:	NIB	$ 525	Ex	$ 400	VG+	$ 290	Good	$ 265		
C2C:	Mint	$ 510	Ex	$ 375	VG+	$ 270	Good	$ 240	Fair	$ 125
Trade-In:	Mint	$ 380	Ex	$ 300	VG+	$ 210	Good	$ 185	Poor	$ 60

COURIER
EXCELLENT

Courtesy of Rock Island Auction Company

COURIER REDBOOK CODE: RB-CL-H-CURIER

Less than 3,500 units produced, a variant of the Cobra.

Production: 1955 - 1956   Caliber: .22 LR, .32 S&W   Action: DA/SA, revolver

Barrel Length: 3"   Wt.: 16 oz.   Sights: blade front   Capacity: 6   Grips: wood

D2C:	NIB	$1,450	Ex	$1,125	VG+	$ 800	Good	$ 725		
C2C:	Mint	$1,400	Ex	$1,025	VG+	$ 730	Good	$ 655	Fair	$ 335
Trade-In:	Mint	$1,030	Ex	$825	VG+	$ 570	Good	$ 510	Poor	$ 150

AIRCREWMAN
VERY GOOD +

Courtesy of Rock Island Auction Company

AIRCREWMAN REDBOOK CODE: RB-CL-H-AIRCRW

Less than 1,500 manufactured. Marked "U.S." or "A.F." to denote its use with the U.S. Air Force. A variant of the Colt Cobra.

Production: 1951   Caliber: .38 Special   Action: DA/SA, revolver

Barrel Length: 2"   OA Length: 7"   Wt.: 16 oz.   Sights: blade front

Capacity: 6   Grips: wood

D2C:	NIB	--	Ex	$3,500	VG+	$2,530	Good	$2,300		
C2C:	Mint	$4,420	Ex	$3,175	VG+	$2,300	Good	$2,070	Fair	$1,060
Trade-In:	Mint	$3,270	Ex	$2,600	VG+	$1,800	Good	$1,610	Poor	$ 480

COLT

BORDER PATROL
(FIRST SERIES)
EXCELLENT
Courtesy of Rock Island Auction Company

BORDER PATROL
EXCELLENT
Courtesy of Rock Island Auction Company

AGENT
VERY GOOD +
Courtesy of Rock Island Auction Company

AGENT
VERY GOOD +
Courtesy of Rock Island Auction Company

DETECTIVE SPECIAL
VERY GOOD +

Courtesy of Rock Island Auction Company

BORDER PATROL (FIRST SERIES) REDBOOK CODE: RB-CL-H-BRDPTR

Similar to the Official Police Model. Serial numbers are in the 610,000 range, although they are very rare with only 400 produced.

Production: 1952 Caliber: .38 Special Action: DA/SA, revolver

Barrel Length: 4" OA Length: 7" Wt.: 32 oz. Sights: blade front

Capacity: 6 Grips: wood or plastic

D2C:	NIB	$2,150	Ex	$1,650	VG+	$1,185	Good	$1,075	
C2C:	Mint	$2,070	Ex	$1,500	VG+	$1,080	Good	$ 970	Fair $ 495
Trade-In:	Mint	$1,530	Ex	$1,225	VG+	$ 840	Good	$ 755	Poor $ 240

BORDER PATROL (SECOND SERIES) REDBOOK CODE: RB-CL-H-BRDPTL

Similar to the Official Police Model. Just over 6,000 units produced. Some have nickel finish.

Production: 1970 - 1975 Caliber: .357 Magnum Action: DA/SA, revolver

Barrel Length: 4" OA Length: 7" Wt.: 32 oz. Sights: blade front

Capacity: 6 Grips: wood or plastic

D2C:	NIB	$2,000	Ex	$1,525	VG+	$1,100	Good	$1,000	
C2C:	Mint	$1,920	Ex	$1,400	VG+	$1,000	Good	$ 900	Fair $ 460
Trade-In:	Mint	$1,420	Ex	$1,125	VG+	$ 780	Good	$ 700	Poor $ 210

AGENT (FIRST SERIES) REDBOOK CODE: RB-CL-H-AGNTX1

A smaller variant of the Colt Cobra. Later versions of the gun have matte finish and an unshrouded ejector rod.

Production: 1955 - early 1970s Caliber: .22 LR, .32 Colt N.P., .38 Special

Action: DA/SA, revolver Sights: blade front Capacity: 6 Grips: wood or plastic

D2C:	NIB	$ 675	Ex	$ 525	VG+	$ 375	Good	$ 340	
C2C:	Mint	$ 650	Ex	$ 475	VG+	$ 340	Good	$ 305	Fair $ 160
Trade-In:	Mint	$ 480	Ex	$ 400	VG+	$ 270	Good	$ 240	Poor $ 90

AGENT (SECOND SERIES) REDBOOK CODE: RB-CL-H-AGNTX2

A smaller variant of the Colt Cobra. Later versions of the gun have matte finish and a shrouded ejector rod.

Production: 1973 - early 1980s Caliber: .22 LR, .32 Colt N.P., .38 Special

Action: DA/SA, revolver Sights: blade front Capacity: 6 Grips: wood or plastic

D2C:	NIB	$ 570	Ex	$ 450	VG+	$ 315	Good	$ 285	
C2C:	Mint	$ 550	Ex	$ 400	VG+	$ 290	Good	$ 260	Fair $ 135
Trade-In:	Mint	$ 410	Ex	$ 325	VG+	$ 230	Good	$ 200	Poor $ 60

DETECTIVE SPECIAL (FIRST ISSUES)
REDBOOK CODE: RB-CL-H-DTCTX3

A shorter variant of the Police Positive model. Nickel finish commands a slightly higher premium.

Production: 1927 - late 1940s Caliber: .38 Special, .32 New Police, .38 New Police

Action: DA/SA, revolver Barrel Length: 2" OA Length: 7" Wt.: 16 oz.

Sights: blade front Capacity: 6 Grips: wood or plastic

D2C:	NIB	$1,150	Ex	$ 875	VG+	$ 635	Good	$ 575	
C2C:	Mint	$ 1,110	Ex	$ 800	VG+	$ 580	Good	$ 520	Fair $ 265
Trade-In:	Mint	$ 820	Ex	$ 650	VG+	$ 450	Good	$ 405	Poor $ 120

COLT

DETECTIVE SPECIAL
EXCELLENT
Courtesy of Rock Island Auction Company

DETECTIVE SPECIAL (SECOND ISSUES)

REDBOOK CODE: RB-CL-H-DTCTX4

A shorter variant of the Police Positive model. Features a larger frame than previous Detective Special model.

Production: 1947 - 1972   Caliber: .38 Special   Action: DA/SA, revolver

Barrel Length: 2", 3"   Wt.: 16 oz.   Sights: blade front   Capacity: 6

Grips: wood or plastic

D2C:	NIB	$1,000	Ex $	775	VG+ $	550	Good $	500		
C2C:	Mint	$ 960	Ex $	700	VG+ $	500	Good $	450	Fair $	230
Trade-In:	Mint	$ 710	Ex $	575	VG+ $	390	Good $	350	Poor $	120

DETECTIVE SPECIAL
EXCELLENT
Courtesy of Rock Island Auction Company

DETECTIVE SPECIAL (THIRD ISSUE)

REDBOOK CODE: RB-CL-H-DTCTX5

Serial numbers should begin around F01001.

Production: 1973 - late 1980s   Caliber: .38 Special   Action: DA/SA, revolver

Barrel Length: 2", 3"   Wt.: 16 oz.   Sights: blade front   Capacity: 6

Grips: wood, wrapped

D2C:	NIB	$ 825	Ex $	650	VG+ $	455	Good $	415		
C2C:	Mint	$ 800	Ex $	575	VG+ $	420	Good $	375	Fair $	190
Trade-In:	Mint	$ 590	Ex $	475	VG+ $	330	Good $	290	Poor $	90

DETECTIVE SPECIAL
EXCELLENT
Courtesy of Rock Island Auction Company

DETECTIVE SPECIAL (FOURTH ISSUE)

REDBOOK CODE: RB-CL-H-DTCTX6

Features gold Colt emblems in the grips and an enhanced safety device. Replaced by the SF-VI model.

Production: early 1990s - 1995   Caliber: .38 Special   Action: DA/SA, revolver

Barrel Length: 2"   Wt.: 21 oz.   Sights: blade front

Capacity: 6   Grips: black synthetic or wood

D2C:	NIB	$ 700	Ex $	550	VG+ $	385	Good $	350		
C2C:	Mint	$ 680	Ex $	500	VG+ $	350	Good $	315	Fair $	165
Trade-In:	Mint	$ 500	Ex $	400	VG+ $	280	Good $	245	Poor $	90

DETECTIVE SPECIAL II   REDBOOK CODE: RB-CL-H-DTCTX7

Features an updated safety mechanism and modified action. Replaced by the Colt Magnum Carry.

Production: late 1990s - discontinued   Caliber: .38 Special   Action: DA/SA, revolver

Barrel Length: 2"   Wt.: 21 oz.   Sights: blade front with night sights   Capacity: 6

Grips: rubber

D2C:	NIB	$ 670	Ex $	525	VG+ $	370	Good $	335		
C2C:	Mint	$ 650	Ex $	475	VG+ $	340	Good $	305	Fair $	155
Trade-In:	Mint	$ 480	Ex $	400	VG+ $	270	Good $	235	Poor $	90

MAGNUM CARRY
EXCELLENT
Courtesy of Rock Island Auction Company

MAGNUM CARRY   REDBOOK CODE: RB-CL-H-MGNCRY

Features a stainless steel finish. Early models stamped "2nd Edition" and "Magnum Carry" on the barrel.

Production: 1999   Caliber: .357 Magnum   Action: DA/SA, revolver

Barrel Length: 2.1"   Wt.: 21 oz.   Sights: ramp front   Capacity: 6   Grips: black rubber

D2C:	NIB	$ 730	Ex $	575	VG+ $	405	Good $	365		
C2C:	Mint	$ 710	Ex $	525	VG+ $	370	Good $	330	Fair $	170
Trade-In:	Mint	$ 520	Ex $	425	VG+ $	290	Good $	260	Poor $	90

COLT

DIAMONDBACK 2.5"
EXCELLENT
Courtesy of Rock Island Auction Company

DIAMONDBACK 4"
EXCELLENT

Courtesy of Rock Island Auction Company

DIAMONDBACK 6"
EXCELLENT

Courtesy of Rock Island Auction Company

DIAMONDBACK
(.38 SPECIAL)
EXCELLENT

Courtesy of Rock Island Auction Company

.357 MAGNUM
VERY GOOD +

Courtesy of Rock Island Auction Company

DIAMONDBACK (2.5" BARREL) REDBOOK CODE: RB-CL-H-DIAMNX

Serial numbers range between approximately D1,000 and S66,000. Models in .22 (especially .22 Magnum) with 2.5" barrels have substantially increased value compared to other Diamondback models. Similar to the Detective Special.

Production: 1966 - late 1980s   Caliber: .22 LR, .22 Magnum   Action: DA/SA, revolver
Barrel Length: 2.5"   Wt.: 16 oz.   Sights: blade front   Capacity: 6   Grips: walnut

D2C:	NIB	$1,475	Ex	$1,125	VG+	$ 815	Good	$ 740	
C2C:	Mint	$1,420	Ex	$1,025	VG+	$ 740	Good	$ 665	Fair $ 340
Trade-In:	Mint	$1,050	Ex	$ 850	VG+	$ 580	Good	$ 520	Poor $ 150

DIAMONDBACK (4" BARREL) REDBOOK CODE: RB-CL-H-DIAMN1

Serial numbers range between approximately D1,000 and S66,000. Diamondbacks in .22 with 4" barrels have less value than those with 2.5" barrels. Similar to the Detective Special.

Production: 1966 - late 1980s   Caliber: .22 LR, .22 Magnum   Action: DA/SA, revolver
Barrel Length: 4"   Wt.: 32 oz.   Sights: blade front   Capacity: 6   Grips: walnut

D2C:	NIB	$1,200	Ex	$ 925	VG+	$ 660	Good	$ 600	
C2C:	Mint	$1,160	Ex	$ 850	VG+	$ 600	Good	$ 540	Fair $ 280
Trade-In:	Mint	$ 860	Ex	$ 675	VG+	$ 470	Good	$ 420	Poor $ 120

DIAMONDBACK (6" BARREL) REDBOOK CODE: RB-CL-H-DIAMN2

Serial numbers range between approximately D1,000 and S66,000. Diamondbacks in .22 with 6" barrels are not quite as valuable as those with 2.5" or 4". Similar to the Detective Special.

Production: 1966 - late 1980s   Caliber: .22 LR, .22 Magnum   Action: DA/SA, revolver
Barrel Length: 6"   Wt.: 32 oz.   Sights: blade front   Capacity: 6   Grips: walnut

D2C:	NIB	$1,400	Ex	$1,075	VG+	$ 770	Good	$ 700	
C2C:	Mint	$1,350	Ex	$975	VG+	$ 700	Good	$ 630	Fair $ 325
Trade-In:	Mint	$1,000	Ex	$800	VG+	$ 550	Good	$ 490	Poor $ 150

DIAMONDBACK (.38 SPECIAL) REDBOOK CODE: RB-CL-H-DIAMN3

Serial numbers range between approximately D1,000 and S66,000.

Production: 1966 - late 1980s   Caliber: .38 Special   Action: DA/SA, revolver
Barrel Length: 2.5", 4", 6"   Wt.: 40 oz.   Sights: blade front
Capacity: 6   Grips: walnut

D2C:	NIB	$1,100	Ex	$ 850	VG+	$ 605	Good	$ 550	
C2C:	Mint	$1,060	Ex	$ 775	VG+	$ 550	Good	$ 495	Fair $ 255
Trade-In:	Mint	$ 790	Ex	$ 625	VG+	$ 430	Good	$ 385	Poor $ 120

.357 MAGNUM REDBOOK CODE: RB-CL-H-357TRP

Serial numbers range between approximately 1 and 85,000, though less than 20,000 manufactured. The deluxe variant of the Colt Trooper. Models with a wide hammer and grip have a slightly increased value.

Production: 1954 - late 1960s   Caliber: .357 Magnum   Action: DA/SA, revolver
Barrel Length: 4", 6"   Sights: fixed front, adjustable rear   Capacity: 6
Grips: checkered walnut

D2C:	NIB	--	Ex	$ 750	VG+	$ 540	Good	$ 490	
C2C:	Mint	$ 940	Ex	$ 675	VG+	$ 490	Good	$ 440	Fair $ 225
Trade-In:	Mint	$ 700	Ex	$ 550	VG+	$ 390	Good	$ 345	Poor $ 120

TROOPER
EXCELLENT

Courtesy of Rock Island Auction Company

TROOPER REDBOOK CODE: RB-CL-H-TROPRX

Serial numbers for the early model Troopers range from approximately 900,000 to 940,000. Considered one of the best revolvers of the last century. Similar design to the Colt Officer's Model. Models with a wide hammer and grip are slightly more valuable than those without.

Production: 1953 - late 1960s   Caliber: .22 LR, .357 Magnum, .38 Special

Action: DA/SA, revolver   Barrel Length: 4", 6"   Sights: fixed   Capacity: 6

Grips: black synthetic or wood

D2C:	NIB	$ 925	Ex $ 725	VG+ $ 510	Good $ 465		
C2C:	Mint	$ 890	Ex $ 650	VG+ $ 470	Good $ 420	Fair	$ 215
Trade-In:	Mint	$ 660	Ex $ 525	VG+ $ 370	Good $ 325	Poor	$ 120

TROOPER MARK III
EXCELLENT

Courtesy of Rock Island Auction Company

TROOPER MARK III REDBOOK CODE: RB-CL-H-TROPR1

Serial numbers range between approximately J1000 and 1000L. Features a modified lock, but some think this compromises its quality compared to earlier models. Slight premium for nickel finish.

Production: 1969 - 1982   Caliber: .357 Magnum   Action: DA/SA, revolver

Barrel Length: 2", 4"   Sights: fixed   Capacity: 6   Grips: black synthetic or wood

D2C:	NIB	$ 770	Ex $ 600	VG+ $ 425	Good $ 385		
C2C:	Mint	$ 740	Ex $ 550	VG+ $ 390	Good $ 350	Fair	$ 180
Trade-In:	Mint	$ 550	Ex $ 450	VG+ $ 310	Good $ 270	Poor	$ 90

TROOPER MARK V
EXCELLENT

Courtesy of Rock Island Auction Company

TROOPER MARK V REDBOOK CODE: RB-CL-H-TROPR2

Features an improved trigger and lock and redesigned V frame, similar to that of the Lawman Mark V. Slight premium for nickel finish.

Production: 1982 - 1985   Caliber: .357 Magnum   Action: DA/SA, revolver

Barrel Length: 4", 6", 8"   Sights: fixed   Capacity: 6   Grips: black synthetic or wood

D2C:	NIB	$ 760	Ex $ 600	VG+ $ 420	Good $ 380		
C2C:	Mint	$ 730	Ex $ 525	VG+ $ 380	Good $ 345	Fair	$ 175
Trade-In:	Mint	$ 540	Ex $ 450	VG+ $ 300	Good $ 270	Poor	$ 90

LAWMAN MARK III
EXCELLENT

Courtesy of Rock Island Auction Company

LAWMAN MARK III REDBOOK CODE: RB-CL-H-LWMXX1

Serial numbers range between approximately J1000 and 1000L. Some models feature a shrouded ejector rod. Models with nickel finish have slightly increased value.

Production: 1969 - 1982   Caliber: .357 Magnum   Action: DA/SA, revolver

Barrel Length: 2", 4"   Wt.: 42 oz.   Sights: fixed   Capacity: 6   Grips: walnut

D2C:	NIB	$ 580	Ex $ 450	VG+ $ 320	Good $ 290		
C2C:	Mint	$ 560	Ex $ 425	VG+ $ 290	Good $ 265	Fair	$ 135
Trade-In:	Mint	$ 420	Ex $ 325	VG+ $ 230	Good $ 205	Poor	$ 60

LAWMAN MARK V REDBOOK CODE: RB-CL-H-LWMXX2

A part of the Trooper Mark V series. Models with nickel finish have increased value.

Production: 1982 - early 1990s   Caliber: .357 Magnum   Action: DA/SA, revolver

Barrel Length: 2", 4"   Wt.: 42 oz.   Sights: fixed   Capacity: 6   Grips: walnut

D2C:	NIB	$ 565	Ex $ 450	VG+ $ 315	Good $ 285		
C2C:	Mint	$ 550	Ex $ 400	VG+ $ 290	Good $ 255	Fair	$ 130
Trade-In:	Mint	$ 410	Ex $ 325	VG+ $ 230	Good $ 200	Poor	$ 60

PEACEKEEPER
VERY GOOD +

Courtesy of Rock Island Auction Company

PEACEKEEPER REDBOOK CODE: RB-CL-H-PEACKP

Replaced the Lawman Mark V. Features a Python-style barrel, matte blue finish, and shrouded ejector rod.

Production: 1985 - 1987 Caliber: .357 Magnum Action: DA/SA, revolver

Barrel Length: 4", 6" Wt.: 42 oz. Sights: adjustable Capacity: 6 Grips: rubber

D2C:	NIB	$ 700	Ex	$ 550	VG+	$ 385	Good	$ 350	
C2C:	Mint	$ 680	Ex	$ 500	VG+	$ 350	Good	$ 315	Fair $ 165
Trade-In:	Mint	$ 500	Ex	$ 400	VG+	$ 280	Good	$ 245	Poor $ 90

BOA
EXCELLENT

Courtesy of Rock Island Auction Company

BOA REDBOOK CODE: RB-CL-H-BOAXXX

Features a Python-style frame, a deluxe version of the Trooper. Less than 1,500 manufactured. Some accounts claim these were distributed exclusively by the Lew Horton Company of Massachusetts.

Production: mid-1980s Caliber: .357 Magnum Action: DA/SA, revolver

Barrel Length: 4", 6" Wt.: 42 oz. Sights: fixed Capacity: 6 Grips: walnut

D2C:	NIB	$1,550	Ex	$1,200	VG+	$ 855	Good	$ 775	
C2C:	Mint	$1,490	Ex	$1,075	VG+	$ 780	Good	$ 700	Fair $ 360
Trade-In:	Mint	$ 1,110	Ex	$ 875	VG+	$ 610	Good	$ 545	Poor $ 180

COMBAT COBRA
EXCELLENT

Courtesy of Rock Island Auction Company

COMBAT COBRA REDBOOK CODE: RB-CL-H-CMBCBR

A high-polished variant of the Cobra, only available with a 2.5" barrel.

Production: 1987 Caliber: .357 Magnum Action: DA/SA, revolver

Barrel Length: 2.5" Wt.: 21 oz. Sights: fixed Capacity: 6 Grips: wood

D2C:	NIB	$1,400	Ex	$1,075	VG+	$770	Good	$700	
C2C:	Mint	$1,350	Ex	$975	VG+	$700	Good	$630	Fair $ 325
Trade-In:	Mint	$1,000	Ex	$800	VG+	$550	Good	$490	Poor $ 150

KING COBRA
EXCELLENT

Courtesy of Rock Island Auction Company

KING COBRA REDBOOK CODE: RB-CL-H-KGCBRX

Features high-grade carbon components and bright or royal blue finish. Considered one of Colt's best double-action revolvers. Slight premium for 2.5" barrel.

Production: 1986 - 1998 Caliber: .357 Magnum, .38 Special Action: DA/SA, revolver

Barrel Length: 2.5", 4", 6" Wt.: 42 oz. Sights: fixed Capacity: 6

Grips: black synthetic, walnut

D2C:	NIB	$1,200	Ex	$ 925	VG+	$ 660	Good	$ 600	
C2C:	Mint	$1,160	Ex	$ 850	VG+	$ 600	Good	$ 540	Fair $ 280
Trade-In:	Mint	$ 860	Ex	$ 675	VG+	$ 470	Good	$ 420	Poor $ 120

KING COBRA ULTIMATE BRIGHT STAINLESS
EXCELLENT

Courtesy of Rock Island Auction Company

KING COBRA ULTIMATE BRIGHT STAINLESS

REDBOOK CODE: RB-CL-H-KGCBR1

Bright stainless finish, gold medallions in the grip, shrouded ejector rod.

Production: 1988 - early 1990s Caliber: .357 Magnum, .38 Special

Action: DA/SA, revolver Barrel Length: 2.5", 4", 6", 8" Wt.: 42 oz.

Sights: fixed Capacity: 6 Grips: black synthetic, walnut

D2C:	NIB	$1,250	Ex	$ 950	VG+	$ 690	Good	$ 625	
C2C:	Mint	$1,200	Ex	$ 875	VG+	$ 630	Good	$ 565	Fair $ 290
Trade-In:	Mint	$ 890	Ex	$ 700	VG+	$ 490	Good	$ 440	Poor $ 150

COLT

VIPER
EXCELLENT

Courtesy of Rock Island Auction Company

PYTHON
EXCELLENT

Courtesy of Rock Island Auction Company

**PYTHON LATER
PRODUCTION**
EXCELLENT

Courtesy of Rock Island Auction Company

VIPER REDBOOK CODE: RB-CL-H-VIPRXX

Similar to the Police Positive Special and Cobra, serial numbers falling within the same range. The gun didn't sell well and was quickly discontinued.

Production: 1977 - late 1970s   Caliber: .38 Special   Action: DA/SA, revolver

Barrel Length: 4"   Wt.: 32 oz.   Sights: blade front   Capacity: 6

Grips: wood or plastic

D2C:	NIB	$1,275	Ex	$ 975	VG+	$ 705	Good	$ 640	
C2C:	Mint	$1,230	Ex	$ 900	VG+	$ 640	Good	$ 575	Fair $ 295
Trade-In:	Mint	$ 910	Ex	$ 725	VG+	$ 500	Good	$ 450	Poor $ 150

PYTHON (1950S-1960S) REDBOOK CODE: RB-CL-H-PYNTHX

Widely revered and noted for its hand-fitted components. Early models of the gun are moderately more valuable than later models. Models with 3-inch-barrels and non-varnished grips are worth approximately $1000 more than other models. Fake models of the Python exist, use discretion before purchasing.

Production: 1955 - late 1960s   Caliber: .357 Magnum   Action: DA/SA, revolver

Barrel Length: 2.5", 3", 4", 6", 8"   Sights: fixed front, adjustable rear

Capacity: 6   Grips: black synthetic or walnut

D2C:	NIB	$2,500	Ex	$1,900	VG+	$1,375	Good	$1,250	
C2C:	Mint	$2,400	Ex	$1,725	VG+	$1,250	Good	$1,125	Fair $ 575
Trade-In:	Mint	$1,780	Ex	$1,400	VG+	$ 980	Good	$ 875	Poor $ 270

PYTHON (LATER PRODUCTION) REDBOOK CODE: RB-CL-H-PYNTH1

Available in several variants with nominal differences between them. Highly revered among collectors.

Production: 1970 - 1996   Caliber: .357 Magnum, .41 Magnum,
.256 Win. Magnum, .44 Special   Action: DA/SA, revolver

Barrel Length: 3", 4", 6", 8"Sights: fixed front, adjustable rear   Capacity: 6

Grips: black synthetic, walnut

D2C:	NIB	$ 2,100	Ex	$1,600	VG+	$ 1,155	Good	$1,050	
C2C:	Mint	$ 2,020	Ex	$1,450	VG+	$1,050	Good	$ 945	Fair $ 485
Trade-In:	Mint	$ 1,500	Ex	$1,200	VG+	$ 820	Good	$ 735	Poor $ 210

PYTHON .38 SPECIAL REDBOOK CODE: RB-CL-H-PYNT10

Limited production, available only in blue finish.

Production: mid-1950s - 1956   Caliber: .38 Special   Action: DA/SA, revolver

Barrel Length: 8"   Sights: fixed front, adjustable rear   Capacity: 6   Grips: walnut

D2C:	NIB	$1,900	Ex	$1,450	VG+	$1,045	Good	$ 950	
C2C:	Mint	$1,830	Ex	$1,325	VG+	$950	Good	$855	Fair $ 440
Trade-In:	Mint	$1,350	Ex	$1,075	VG+	$750	Good	$665	Poor $ 210

PYTHON (NICKEL) REDBOOK CODE: RB-CL-H-PYNTH2

Noted for its polished nickel finish. Serial numbers should correspond with those of the standard Python.

Production: mid-1980s   Caliber: .357 Magnum   Action: DA/SA, revolver

Barrel Length: 2.5", 3", 4", 6", 8"   Wt.: 37 oz.   Sights: fixed front, adjustable rear

Capacity: 6   Grips: walnut

D2C:	NIB	$2,550	Ex	$1,950	VG+	$1,405	Good	$1,275	
C2C:	Mint	$2,450	Ex	$1,775	VG+	$1,280	Good	$1,150	Fair $ 590
Trade-In:	Mint	$1,820	Ex	$1,450	VG+	$1,000	Good	$ 895	Poor $ 270

**PYTHON
STAINLESS STEEL**
EXCELLENT

Courtesy of Rock Island Auction Company

PYTHON ULTIMATE
EXCELLENT

Courtesy of Rock Island Auction Company

COLT

PYTHON (STAINLESS STEEL) REDBOOK CODE: RB-CL-H-PYNTH3

A stainless steel variant of the standard Python. Serial numbers should correspond with those of the standard Python.

Production: early 1980s - late 1990s   Caliber: .357 Magnum

Action: DA/SA, revolver   Barrel Length: 4", 6", 8"   Wt.: 37 oz.

Sights: fixed front, adjustable rear   Capacity: 6   Grips: black synthetic

D2C:	NIB	$2,350	Ex	$1,800	VG+	$1,295	Good	$1,175		
C2C:	Mint	$2,260	Ex	$1,625	VG+	$1,180	Good	$1,060	Fair	$ 545
Trade-In:	Mint	$1,670	Ex	$1,325	VG+	$920	Good	$825	Poor	$ 240

PYTHON ULTIMATE (BRIGHT STAINLESS) (1970-1985)

REDBOOK CODE: RB-CL-H-PYNTH4

Considered a deluxe version of the gun. Features high-polished steel and vent rib barrel. Serial numbers should correspond with those of the standard Python. Premium nearly doubles for 2.5" barrel.

Production: 1970 - mid-1980s   Caliber: .357 Magnum   Action: DA/SA, revolver

Barrel Length: 2.5", 4", 6", 8"   Wt.: 37 oz.   Sights: fixed front, adjustable rear

Capacity: 6   Grips: black synthetic or walnut

D2C:	NIB	$2,450	Ex	$1,875	VG+	$1,350	Good	$1,225		
C2C:	Mint	$2,360	Ex	$1,700	VG+	$1,230	Good	$1,105	Fair	$565
Trade-In:	Mint	$1,740	Ex	$1,375	VG+	$ 960	Good	$ 860	Poor	$270

PYTHON ELITE REDBOOK CODE: RB-CL-H-PYNTH5

The deluxe model of the Python made by the Colt Custom Shop. Little to no difference between this model and the standard Python model.

Production: 1997 - 1999   Caliber: .357 Magnum   Action: DA/SA, revolver

Barrel Length: 4", 6"   Wt.: 37 oz.   Sights: plastic front   Capacity: 6

Grips: black synthetic or walnut

D2C:	NIB	$2,400	Ex	$1,825	VG+	$1,320	Good	$1,200		
C2C:	Mint	$2,310	Ex	$1,675	VG+	$1,200	Good	$1,080	Fair	$ 555
Trade-In:	Mint	$1,710	Ex	$1,350	VG+	$ 940	Good	$ 840	Poor	$ 240

PYTHON ELITE (REINTRODUCTION) REDBOOK CODE: RB-CL-H-PYNTH6

Reintroduced after a break in production. The deluxe model of the Python made by the Colt Custom Shop.

Production: 2001 - ~2003   Caliber: .357 Magnum   Action: DA/SA, revolver

Barrel Length: 4", 6"   Wt.: 37 oz.   Sights: fixed front, rear adjustable

Capacity: 6   Grips: black synthetic or walnut

D2C:	NIB	$ 2,300	Ex	$1,750	VG+	$1,265	Good	$ 1,150		
C2C:	Mint	$ 2,210	Ex	$1,600	VG+	$1,150	Good	$ 1,035	Fair	$ 530
Trade-In:	Mint	$ 1,640	Ex	$1,300	VG+	$ 900	Good	$ 805	Poor	$ 240

PYTHON ELITE
REINTRODUCTION
EXCELLENT

Courtesy of Rock Island Auction Company

COLT

PYTHON HUNTER
EXCELLENT

Courtesy of Rock Island Auction Company

PYTHON SILHOUETTE
VERY GOOD +

Courtesy of Rock Island Auction Company

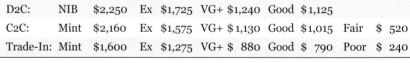

GRIZZLY
EXCELLENT

Courtesy of Rock Island Auction Company

PYTHON HUNTER REDBOOK CODE: RB-CL-H-PYNTH8
Noted for its extended barrel. Sold with a Halliburton carrying case and Leupold scope.

Production: 1981 Caliber: .357 Magnum Action: DA/SA, revolver Barrel
Length: 8" Sights: Leupold brand scope Capacity: 6 Grips: black synthetic or walnut

D2C:	NIB	$2,800	Ex	$2,150	VG+	$1,540	Good	$1,400	
C2C:	Mint	$2,690	Ex	$1,950	VG+	$1,400	Good	$1,260	Fair	$645
Trade-In:	Mint	$1,990	Ex	$1,575	VG+	$1,100	Good	$980	Poor	$300

PYTHON SILHOUETTE REDBOOK CODE: RB-CL-H-PYNTH9
Sold with a Leupold scope and black carrying case.

Production: 1981 - discontinued Caliber: .357 Magnum Action: DA/SA, revolver Barrel
Length: 8" Sights: Leupold brand scope Capacity: 6 Grips: black synthetic or walnut

D2C:	NIB	$2,600	Ex	$2,000	VG+	$1,430	Good	$1,300	
C2C:	Mint	$2,500	Ex	$1,800	VG+	$1,300	Good	$1,170	Fair	$600
Trade-In:	Mint	$1,850	Ex	$1,475	VG+	$1,020	Good	$910	Poor	$270

PYTHON TEN POINTER REDBOOK CODE: RB-CL-H-PYNT11
Less than 300 manufactured. Sold with a Burris scope.

Production: late 1980s   Caliber: .357 Magnum Action: DA/SA, revolver
Barrel Length: 8" Sights: Burris brand scope Capacity: 6 Grips: black synthetic or walnut

D2C:	NIB	$2,300	Ex	$1,750	VG+	$1,265	Good	$1,150	
C2C:	Mint	$2,210	Ex	$1,600	VG+	$1,150	Good	$1,035	Fair	$530
Trade-In:	Mint	$1,640	Ex	$1,300	VG+	$900	Good	$805	Poor	$240

GRIZZLY REDBOOK CODE: RB-CL-H-GRZZLY
A variant of the Python. Features matte finish and only available with a 6-inch barrel. The Colt Custom Shop produced only 2,000 of these weapons.

Production: 1994 Caliber: .357 Magnum Action: DA/SA, revolver Barrel Length: 6"
Sights: fixed front, adjustable rear Capacity: 6 Grips: black synthetic or walnut

D2C:	NIB	$2,250	Ex	$1,725	VG+	$1,240	Good	$1,125	
C2C:	Mint	$2,160	Ex	$1,575	VG+	$1,130	Good	$1,015	Fair	$520
Trade-In:	Mint	$1,600	Ex	$1,275	VG+	$880	Good	$790	Poor	$240

WHITETAILER REDBOOK CODE: RB-CL-H-WHTTLX
Based on the Colt Trooper MK V design. Features a Burris scope and matte finish.

Production: 1982 - 1985 Caliber: .357 Magnum Action: DA/SA, revolver Barrel
Length: 8 Sights: Burris brand scope, iron Capacity: 6 Grips: black synthetic

D2C:	NIB	$1,780	Ex	$1,375	VG+	$980	Good	$890	
C2C:	Mint	$1,710	Ex	$1,250	VG+	$890	Good	$805	Fair	$410
Trade-In:	Mint	$1,270	Ex	$1,000	VG+	$700	Good	$625	Poor	$180

WHITETAILER II REDBOOK CODE: RB-CL-H-WHTTL1
Similar to the Colt Whitetailer. Features a Burris scope and high polish finish.

Production: 1990 - 1999 Caliber: .357 Magnum Action: DA/SA, revolver
Barrel Length: 8 Sights: Burris brand scope, fixed Capacity: 6
Grips: black synthetic

D2C:	NIB	$1,650	Ex	$1,275	VG+	$910	Good	$825	
C2C:	Mint	$1,590	Ex	$1,150	VG+	$830	Good	$745	Fair	$380
Trade-In:	Mint	$1,180	Ex	$925	VG+	$650	Good	$580	Poor	$180

KODIAK
EXCELLENT
Courtesy of Rock Island Auction Company

KODIAK REDBOOK CODE: RB-CL-H-KODIAK

Features an Anaconda frame, non-fluted cylinder, and ported barrel. Only 2,000 units produced. Higher premium for .357 Magnum model.

Production: 1994 Caliber: .44 Magnum Action: DA/SA, revolver
Barrel Length: 4", 6" Sights: fixed front, rear adjustable Capacity: 6
Grips: Pachmayr

D2C:	NIB	$ 2,075	Ex	$ 1,600	VG+	$ 1,145	Good	$ 1,040	
C2C:	Mint	$ 2,000	Ex	$ 1,450	VG+	$1,040	Good	$ 935	Fair $ 480
Trade-In:	Mint	$ 1,480	Ex	$ 1,175	VG+	$ 810	Good	$ 730	Poor $ 210

ANACONDA REDBOOK CODE: RB-CL-H-ANACX1

A hybrid between the Colt King Cobra and Python models. Features a big frame and vent rib barrel. Less than 2,500 manufactured. 5" barrel models are valued substantially higher. Variant chambered for .45 caliber offered in 1993.

Production: 1990 - 2003 Caliber: .44 Magnum, .45 LC Action: DA/SA, revolver
Barrel Length: 4", 5", 6", 8" Sights: ramp front, adjustable rear Capacity: 6
Grips: black synthetic, walnut

D2C:	NIB	$1,400	Ex	$1,075	VG+	$ 770	Good	$ 700	
C2C:	Mint	$1,350	Ex	$975	VG+	$ 700	Good	$ 630	Fair $ 325
Trade-In:	Mint	$1,000	Ex	$ 800	VG+	$ 550	Good	$ 490	Poor $ 150

ANACONDA
EXCELLENT
Courtesy of Rock Island Auction Company

ANACONDA HUNTER REDBOOK CODE: RB-CL-H-ANACX3

Produced briefly in the early 1990s. Features a Leupold scope.

Production: early 1990s Caliber: .44 Magnum, .45 LC
Action: DA/SA, revolver Barrel Length: 8" Sights: Leupold scope
Capacity: 6 Grips: walnut or rubber

D2C:	NIB	$1,300	Ex	$1,000	VG+	$ 715	Good	$ 650	
C2C:	Mint	$1,250	Ex	$ 900	VG+	$ 650	Good	$ 585	Fair $ 300
Trade-In:	Mint	$ 930	Ex	$ 750	VG+	$ 510	Good	$ 455	Poor $ 150

MODEL 1900
VERY GOOD +

Courtesy of Rock Island Auction Company

MODEL 1900 (MILITARY/STANDARD) REDBOOK CODE: RB-CL-H-M1900X

Designed by John Moses Browning. One of the first handguns with a full-length slide. Led to the development of M1911, though its design was largely unsuccessful. High premium for military models.

Production: 1900 - 1903 Caliber: .38 ACP Action: SA, semi-auto
Barrel Length: 6" OA Length: 9" Wt.: 35 oz. Sights: blade front
Capacity: 7 Magazine: detachable box Grips: walnut or rubber

D2C:	NIB	--	Ex	$ 12,175	VG+	$8,800	Good	$8,000	
C2C:	Mint	$ 15,360	Ex	$ 11,050	VG+	$8,000	Good	$ 7,200	Fair $3,680
Trade-In:	Mint	$ 11,360	Ex	$ 8,975	VG+	$ 6,240	Good	$ 5,600	Poor $1,620

MODEL 1902 MILITARY/STANDARD

REDBOOK CODE: RB-CL-H-M1902X

Based on the M1900's design. Both Military and Standard models (similar values) produced. Military models feature a longer barrel and squared receiver. Sporting serial numbers: ~3,300-31,000. Military serial numbers: ~150,000-30,500. Those marked "Military-U.S. Army" with serial numbers between 150,000 and 150,300 have increased value. Military and Standard models have similar values.

MODEL 1902
MILITARY/STANDARD
EXCELLENT

Courtesy of Rock Island Auction Company

Production: 1902 - 1908 Caliber: .38 ACP Action: SA, semi-auto
Barrel Length: 6" OA Length: 9" Wt.: 35 oz. Sights: blade front
Capacity: 7, 8 Magazine: detachable box Grips: rubber

D2C:	NIB	--	Ex	$ 4,325	VG+	$ 3,120	Good	$2,835	
C2C:	Mint	$ 5,450	Ex	$ 3,925	VG+	$ 2,840	Good	$ 2,555	Fair $ 1,305
Trade-In:	Mint	$ 4,030	Ex	$ 3,200	VG+	$ 2,220	Good	$ 1,985	Poor $ 570

COLT

MODEL 1903 POCKET
EXCELLENT

Courtesy of Rock Island Auction Company

MODEL 1903 POCKET REDBOOK CODE: RB-CL-H-M1903P

Serial numbers range between approximately 19,999 and 4722F, though less than 35,000 produced. Based on the Model 1900's design.

Production: 1903 - 1947 Caliber: .38 ACP Action: SA, semi-auto

Barrel Length: 4.5" OA Length: 7" Wt.: 32 oz. Sights: blade front

Capacity: 7 Magazine: detachable box Grips: black synthetic or wood

D2C:	NIB	--	Ex $ 1,750	VG+ $ 1,265	Good $ 1,150			
C2C:	Mint $ 2,210	Ex $ 1,600	VG+ $ 1,150	Good $ 1,035	Fair $ 530		
Trade-In: Mint $ 1,640	Ex $ 1,300	VG+ $ 900	Good $ 805	Poor $ 240		

**MODEL 1903
POCKET HAMMERLESS**
EXCELLENT

Courtesy of Rock Island Auction Company

MODEL 1903 POCKET HAMMERLESS

REDBOOK CODE: RB-CL-H-M1903H

A hammerless version of the Model 1903 Pocket. Those marked with an "M" prefix and "U.S. Property" have increased value. Grip safety and blue or parkerized finish.

Production: 1903 - 1945 Caliber: .32 ACP, .380 ACP Action: SA, semi-auto

Barrel Length: 3.75", 4" Sights: blade front, adjustable rear Capacity: 7, 8

Magazine: detachable box Grips: black synthetic or wood

D2C:	NIB --	Ex $ 925	VG+ $ 660	Good $ 600
C2C:	Mint $1,160	Ex $ 850	VG+ $ 600	Good $ 540 Fair $ 280
Trade-In:	Mint $ 860	Ex $ 675	VG+ $ 470	Good $ 420 Poor $ 120

MODEL 1905 CIVILIAN REDBOOK CODE: RB-CL-H-M1905M

The first pistol to fire .45 ACP rounds, similar to the M1903. Less than 6,500 produced. Military models have serial numbers between approximately 1 and 200 are worth roughly $16,000.

Production: 1905 - 1911 Caliber: .45 ACP Action: SA, semi-auto

Barrel Length: 5" OA Length: 9" Sights: blade front Capacity: 7, 8

Magazine: detachable box Grips: black synthetic or wood

D2C:	NIB --	Ex $5,800	VG+ $4,180	Good $3,800
C2C:	Mint $ 7,300	Ex $ 5,250	VG+ $3,800	Good $ 3,420 Fair $ 1,750
Trade-In:	Mint $ 5,400	Ex $ 4,275	VG+ $ 2,970	Good $ 2,660 Poor $ 780

MODEL 1905 MILITARY
EXCELLENT

Courtesy of Rock Island Auction Company

MODEL 1907 REDBOOK CODE: RB-CL-H-M1907X

Less than 220 manufactured for military testing. Blue finish and marked "K.M." Actual field test models command nearly a doubled premium.

Production: 1907 - 1908 Caliber: .45 ACP Action: SA, semi-auto Barrel Length: 5"

Sights: blade front Capacity: 7 Magazine: detachable box Grips: walnut

D2C:	NIB --	Ex $15,500	VG+ $ 8,415	Good $ 7,650
C2C:	Mint --	Ex $14,075	VG+ $ 7,650	Good $ 6,885 Fair $ 1,265
Trade-In:	Mint --	Ex $11,425	VG+ $ 5,970	Good $ 5,355 Poor $ 570

MODEL 1907
GOOD

Courtesy of Rock Island Auction Company

MODEL 1908 POCKET REDBOOK CODE: RB-CL-H-1908XP

One of the first Colts to fire .380 ACP. Models with nickel finish and pearl grips have increased value. Model 1908s marked "U.S. Property" and General Officer models have increased value.

Production: 1908 - 1945 Caliber: .25 ACP, .380 ACP Action: SA, semi-auto Barrel

Length: 2" Sights: blade front Capacity: 7 Magazine: detachable box Grips: rubber

D2C:	NIB --	Ex $1,150	VG+ $ 825	Good $ 750
C2C:	Mint $1,440	Ex $1,050	VG+ $ 750	Good $ 675 Fair $ 345
Trade-In:	Mint $1,070	Ex $ 850	VG+ $ 590	Good $ 525 Poor $ 150

**MODEL 1908
POCKET**
GOOD

Courtesy of Rock Island Auction Company

**MODEL 1908
POCKET
HAMMERLESS**
GOOD

Courtesy of Rock Island Auction Company

MODEL 1908 POCKET HAMMERLESS (CIVILIAN)

REDBOOK CODE: RB-CL-H-1908PH

One of the first concealed pistols in .25 ACP. Models with pearl grips and/or nickel finish have increased value. Military models command a higher premium.

Production: 1908 - 1941 Caliber: .25 ACP, .380 ACP Action: SA, semi-auto

Barrel Length: 2" OA Length: 4.5" Sights: fixed Capacity: 7

Magazine: detachable box Grips: rubber or walnut

D2C:	NIB	$ 825	Ex	$ 650	VG+	$ 455	Good	$ 415		
C2C:	Mint	$ 800	Ex	$ 575	VG+	$ 420	Good	$ 375	Fair	$ 190
Trade-In:	Mint	$ 590	Ex	$ 475	VG+	$ 330	Good	$ 290	Poor	$ 90

MODEL 1909 REDBOOK CODE: RB-CL-H-1909XX

An extremely rare gun, less than 25 produced with corresponding serial numbers. Most feature a lanyard loop and marked "United States Property." Too rare to price.

Production: 1909 - discontinued Caliber: .45 ACP Action: SA, semi-auto Barrel

Length: 5" Sights: fixed Capacity: 5 Magazine: detachable box Grips: walnut

D2C:	NIB	--	Ex	--	VG+	--	Good	--		
C2C:	Mint	--	Ex	--	VG+	--	Good	--	Fair	--
Trade-In:	Mint	--	Ex	--	VG+	--	Good	--	Poor	--

MODEL 1910 REDBOOK CODE: RB-CL-H-1910XX

A test and trial version of the M1911. Less than 15 known to exist, making it one of the rarest, most valuable Colts to ever produced. Features minor details modified and changed for the M1911. Too rare to price.

Production: 1910 Caliber: .45 ACP Action: SA, semi-auto Barrel Length: 5"

OA Length: 8.5" Sights: fixed Capacity: 7 Magazine: detachable box Grips: walnut

D2C:	NIB	--	Ex	--	VG+	--	Good	--		
C2C:	Mint	--	Ex	--	VG+	--	Good	--	Fair	--
Trade-In:	Mint	--	Ex	--	VG+	--	Good	--	Poor	--

**MODEL 1911
COMMERCIAL
GOVERNMENT**
FAIR

Courtesy of Rock Island Auction Company

MODEL 1911 COMMERCIAL GOVERNMENT

(EARLY PRODUCTION) REDBOOK CODE: RB-CL-H-911XX1

Serial numbers range between approximately C1 and C138,600. Produced in both regular and high-polish finish. Feature a "C" on their serial numbers, denoting their commercial use. The earliest models of these guns have serial numbers between C1 and C4,500 and have nearly doubled premium. Minor variations exist between models.

Production: 1911 - 1924 Caliber: .45 ACP Action: SA, semi-auto Barrel Length: 5"

OA Length: 8.25" Sights: fixed Capacity: 7 Grips: synthetic or wood

D2C:	NIB	--	Ex	$ 5,875	VG+	$ 4,235	Good	$ 3,850		
C2C:	Mint	--	Ex	$ 5,325	VG+	$ 3,850	Good	$ 3,465	Fair	$ 1,775
Trade-In:	Mint	--	Ex	$ 4,325	VG+	$ 3,010	Good	$ 2,695	Poor	$ 780

**MODEL 1911 MILITARY
ISSUED** (SERIALS 1-99)
VERY GOOD +

Courtesy of Rock Island Auction Company

MODEL 1911 MILITARY ISSUED (SERIALS 1-99)

REDBOOK CODE: RB-CL-H-911100

Models with serial numbers between approximately 1 and 99 have substantial value. Marked "United States Property." Bright-polish finish. Frame-mounted and grip safety. These models are extremely rare. Prices are speculative and subject to change.

Production: 1911 Caliber: .45 ACP Action: SA, semi-auto Barrel Length: 5"

OA Length: 8.25" Sights: fixed Capacity: 7 Magazine: detachable box Grips: wood

D2C:	NIB	--	Ex	$109,250	VG+	--	Good	--		
C2C:	Mint	--	Ex	$99,200	VG+	--	Good	--	Fair	--
Trade-In:	Mint	--	Ex	$80,500	VG+	--	Good	--	Poor	--

COLT

**MODEL 1911
MILITARY ISSUED**
EXCELLENT

Courtesy of Rock Island Auction Company

**MODEL 1911
MILITARY ISSUED**
FAIR

Courtesy of Rock Island Auction Company

**MODEL 1911
MILITARY ISSUED**
GOOD

Courtesy of Rock Island Auction Company

**MODEL 1911
MILITARY ISSUED**
EXCELLENT

Courtesy of Rock Island Auction Company

MODEL 1911 MILITARY ISSUED (SERIALS 100-999)
REDBOOK CODE: RB-CL-H-911STX
Serial numbers range between approximately 100 and 999. Marked "United States Property." High-polish blue finish, frame-mounted and grip safety.

Production: 1911 - 1912 Caliber: .45 ACP Action: SA, semi-auto Barrel Length: 5"
Sights: fixed Capacity: 7 Magazine: detachable box Grips: black synthetic or wood

D2C:	NIB	--	Ex	$19,775	VG+	$14,300	Good	$ 5,050	
C2C:	Mint	--	Ex	$17,950	VG+	$13,000	Good	$ 4,545	Fair $ 767
Trade-In:	Mint	--	Ex	$14,575	VG+	$10,140	Good	$ 3,535	Poor $ 336

MODEL 1911 MILITARY ISSUED (FOUR-DIGIT SERIAL NUMBERS)
REDBOOK CODE: RB-CL-H-911XX5
Serial numbers range between 1,000 and 9,999. Marked "United States Property." Used by the U.S. Army, Marine Corps, and Navy, marked accordingly.

Production: 1912 Caliber: .45 ACP Action: SA, semi-auto Barrel Length: 5"
Sights: blade front Capacity: 7 Magazine: detachable box Grips: black synthetic or wood

D2C:	NIB	--	Ex	$ 5,325	VG+	$3,850	Good	$3,500	
C2C:	Mint	--	Ex	$ 4,850	VG+	$3,500	Good	$ 3,150	Fair $ 1,610
Trade-In:	Mint	--	Ex	$ 3,925	VG+	$ 2,730	Good	$ 2,450	Poor $ 720

MODEL 1911 MILITARY ISSUED (FIVE-DIGIT SERIAL NUMBERS)
REDBOOK CODE: RB-CL-H-911XX6
Serial numbers range between 10,000 and 99,999. Marked "United States Property." Standard blue finish.

Production: 1912 - 1914 Caliber: .45 ACP Action: SA, semi-auto
Barrel Length: 5" Sights: blade front Capacity: 7
Magazine: detachable box Grips: black synthetic or wood

D2C:	NIB	--	Ex	$ 5,875	VG+	$4,235	Good	$3,850	
C2C:	Mint	--	Ex	$ 5,325	VG+	$3,850	Good	$ 3,465	Fair $ 1,775
Trade-In:	Mint	--	Ex	$ 4,325	VG+	$ 3,010	Good	$ 2,695	Poor $ 780

MODEL 1911 MILITARY ISSUED (SIX-DIGIT SERIAL NUMBER)
REDBOOK CODE: RB-CL-H-911X16
Serial numbers range between approximately 100,000 and 999,999. Produced by Colt and several other manufacturers for use in WWII.

Production: 1914 - 1942 Caliber: .45 ACP Action: SA, semi-auto
Barrel Length: 5" Sights: blade front Capacity: 7 Magazine: detachable box
Grips: black synthetic or wood

D2C:	NIB	--	Ex	$ 5,700	VG+	$ 4,125	Good	$ 1,240	
C2C:	Mint	--	Ex	$ 5,175	VG+	$3,750	Good	$ 1,120	Fair $ 190
Trade-In:	Mint	--	Ex	$ 4,200	VG+	$2,930	Good	$ 870	Poor $ 90

MODEL 1911 (ARGENTINE) REDBOOK CODE: RB-CL-H-911X27
Serial numbers range between approximately C6,200 and C21,100. Engraved with the Argentine crest and a "C" in the serial number.

Production: 1914 - early 1920s Caliber: .45 ACP Action: SA, semi-auto
Barrel Length: 5" OA Length: 8.25" Wt.: 40 oz. Sights: fixed
Capacity: 7 Magazine: detachable box Grips: wood

D2C:	NIB	--	Ex	$ 850	VG+	$ 605	Good	$ 550	
C2C:	Mint	--	Ex	$ 775	VG+	$ 550	Good	$ 495	Fair $ 255
Trade-In:	Mint	--	Ex	$ 625	VG+	$ 430	Good	$ 385	Poor $ 120

MODEL 1911 (RUSSIAN)
EXCELLENT

Courtesy of Rock Island Auction Company

MODEL 1911 (RUSSIAN) REDBOOK CODE: RB-CL-H-911X28

Serial numbers range between approximately C21,000 and C89,000. Less than 52,000 made for the Russian Government. Similar to other M1911s of the era, frame-mounted safety, blue finish.

Production: 1915 - 1917 Caliber: .45 ACP Action: SA, semi-auto Barrel Length: 5"
OA Length: 8.25" Wt.: 40 oz. Sights: fixed Capacity: 7 Magazine: detachable box
Grips: wood

D2C:	NIB	$ 9,000	Ex	$ 6,850	VG+	$4,950	Good	$ 4,500		
C2C:	Mint	$ 8,640	Ex	$ 6,225	VG+	$4,500	Good	$ 4,050	Fair	$2,070
Trade-In:	Mint	$ 6,390	Ex	$ 5,050	VG+	$ 3,510	Good	$ 3,150	Poor	$ 900

MODEL 1911 (NORWEGIAN KONGSBERG VAPENFABRIK) (1912)

REDBOOK CODE: RB-CL-H-911X31

Less than 100 produced with corresponding serial numbers. Made in Norway for Norwegian military forces. Experienced substantial collector's interest over the last decade or so.

Production: ~1912 Caliber: .45 ACP Action: SA, semi-auto Barrel
Length: 5" OA Length: 8.25" Sights: fixed Capacity: 7 Magazine: detachable box

D2C:	NIB	--	Ex	$ 2,900	VG+	$2,090	Good	$ 1,900		
C2C:	Mint	--	Ex	$ 2,625	VG+	$1,900	Good	$ 1,710	Fair	$ 875
Trade-In:	Mint	--	Ex	$ 2,150	VG+	$1,490	Good	$ 1,330	Poor	$ 390

MODEL 1911 (NORWEGIAN KONGSBERG VAPENFABRIK) (1914)

REDBOOK CODE: RB-CL-H-911X32

Less than 33,000 made with corresponding serial numbers. Made in Norway for Norwegian military forces. Substantial collector's interest in this model has arisen over the last decade or so.

Production: ~1914 Caliber: .45 ACP Action: SA, semi-auto Barrel
Length: 5" OA Length: 8.25" Sights: fixed Capacity: 7 Magazine: detachable box

D2C:	NIB	--	Ex	$ 850	VG+	$ 605	Good	$ 550		
C2C:	Mint	--	Ex	$ 775	VG+	$ 550	Good	$ 495	Fair	$ 255
Trade-In:	Mint	--	Ex	$ 625	VG+	$ 430	Good	$ 385	Poor	$ 120

MODEL 1911 (CANADIAN)
VERY GOOD +

Courtesy of Rock Island Auction Company

MODEL 1911 (CANADIAN) REDBOOK CODE: RB-CL-H-911X34

Serial numbers range between approximately C3,000 and C140,000. Approximately 5,100 manufactured for the Canadian Military. These guns can be difficult to identify, given their minimal markings.

Production: ~1914 Caliber: .45 ACP Action: SA, semi-auto Barrel Length: 5"
OA Length: 8.25" Sights: fixed Capacity: 7 Magazine: detachable box Grips: wood

D2C:	NIB	--	Ex	$ 1,825	VG+	$ 1,320	Good	$ 1,200		
C2C:	Mint	--	Ex	$ 1,675	VG+	$ 1,200	Good	$ 1,080	Fair	$ 555
Trade-In:	Mint	--	Ex	$ 1,350	VG+	$ 940	Good	$ 840	Poor	$ 240

MODEL 1911 NORTH AMERICAN ARMS COMPANY
FAIR

Courtesy of Rock Island Auction Company

MODEL 1911 NORTH AMERICAN ARMS COMPANY

REDBOOK CODE: RB-CL-H-911X36

North American Arms produced M1911 models on contract from Colt for use during WWI. Collectors consider some of these models the most valuable versions of the M1911 in existence. Produced in Canada, noted for their poor finish, marked "North American Arms." Very rare.

Production: 1918 - 1920s Caliber: .45 ACP Action: SA, semi-auto Barrel Length: 5"
Wt.: 40 oz. Sights: fixed Capacity: 7 Magazine: detachable box Grips: wood

D2C:	NIB	--	Ex	$28,125	VG+	$20,350	Good	$18,500		
C2C:	Mint	--	Ex	$25,550	VG+	$18,500	Good	$16,650	Fair	$ 8,510
Trade-In:	Mint	--	Ex	$20,725	VG+	$14,430	Good	$12,950	Poor	$3,720

**MODEL 1911
REMINGTON UMC,
SPRINGFIELD ARMORY,
NAVY, MARINE CORPS**
EXCELLENT

Courtesy of Rock Island Auction Company

MODEL 1911 REMINGTON UMC, SPRINGFIELD ARMORY, NAVY, MARINE CORPS REDBOOK CODE: RB-CL-H-911X37

Based on the Colt design, but manufactured for the WWI effort. Guns produced by these companies have similar value, often determined by condition. Markings should correspond with their place of manufacture.

Production: 1918 - ~ 1919 Caliber: .45 ACP Action: SA, semi-auto Barrel Length: 5"
OA Length: 8.25" Wt.: 40 oz. Sights: fixed Capacity: 7 Magazine: detachable box
Grips: wood

D2C:	NIB	--	Ex	$ 1,800	VG+	$ 1,295	Good	$ 1,175	
C2C:	Mint	--	Ex	$ 1,625	VG+	$ 1,180	Good	$ 1,060	Fair $ 545
Trade-In:	Mint	--	Ex	$ 1,325	VG+	$ 920	Good	$ 825	Poor $ 240

MODEL 1911 SUPER MATCH .38 REDBOOK CODE: RB-CL-H-11SPMX

Serial numbers range approximately between 14,000 and 37,000, approximately 4,000 units produced. Models with adjustable sights are typically priced $1,000 more valuable than fixed sight models.

Production: 1935 - 1947 Caliber: .38 Super Action: SA, semi-auto
Barrel Length: 5" OA Length: 8.25" Wt.: 40 oz. Sights: fixed
Capacity: 7 Magazine: detachable box Grips: synthetic or wood

D2C:	NIB	--	Ex	$ 6,500	VG+	$ 4,705	Good	$ 4,275	
C2C:	Mint	$ 8,210	Ex	$ 5,900	VG+	$ 4,280	Good	$ 3,850	Fair $ 1,970
Trade-In:	Mint	$ 6,080	Ex	$ 4,800	VG+	$ 3,340	Good	$ 2,995	Poor $ 870

MODEL 1911 COMMANDER
EXCELLENT
Courtesy of Rock Island Auction Company

MODEL 1911 COMMANDER (PRE-1970) REDBOOK CODE: RB-CL-H-11CMM7

Similar to other M1911s of the era, but built with a more compact design.

Production: 1949 - late 1960s Caliber: 9mm, .38 Super, .45 ACP Action: SA, semi-auto
Barrel Length: 4.25" Sights: fixed Capacity: 7, 9 Magazine: detachable box Grips: synthetic

D2C:	NIB	$ 2,200	Ex	$ 1,675	VG+	$ 1,210	Good	$ 1,100	
C2C:	Mint	$ 2,120	Ex	$ 1,525	VG+	$ 1,100	Good	$ 990	Fair $ 510
Trade-In:	Mint	$ 1,570	Ex	$ 1,250	VG+	$ 860	Good	$ 770	Poor $ 240

**MODEL 1911A1
COMMERCIAL**
EXCELLENT

Courtesy of Rock Island Auction Company

MODEL 1911A1 COMMERCIAL REDBOOK CODE: RB-CL-H-A1XXXX

Serial numbers range between approximately C135,500 and C215,100. Features several improvements to the original M1911 design: modified butt and grip, shortened hammer, improved safety, and square-notch rear sight, among others. Commercially sold both abroad and domestically.

Production: 1924 - 1940s Caliber: .45 ACP Action: SA, semi-auto
Barrel Length: 5" OA Length: 8.25" Wt.: 40 oz. Sights: fixed
Capacity: 7 Magazine: detachable box Grips: synthetic or wood

D2C:	NIB	--	Ex	$ 3,650	VG+	$ 2,640	Good	$ 2,400	
C2C:	Mint	$ 4,610	Ex	$ 3,325	VG+	$ 2,400	Good	$ 2,160	Fair $ 1,105
Trade-In:	Mint	$ 3,410	Ex	$ 2,700	VG+	$ 1,880	Good	$ 1,680	Poor $ 480

**MODEL 1911A1
MILITARY ISSUED**
VERY GOOD +

Courtesy of Rock Island Auction Company

MODEL 1911A1 MILITARY ISSUED (TRANSITION)
REDBOOK CODE: RB-CL-H-A1X15X

Serial numbers range between approximately 700,000 and 710,200. Limited distribution and sales.

Production: ~ 1924 Caliber: .45 ACP Action: SA, semi-auto Barrel Length: 5", 7"

D2C:	NIB	--	Ex	$ 5,950	VG+	$ 4,290	Good	$ 3,900	
C2C:	Mint	$ 12,650	Ex	$ 5,400	VG+	$ 3,900	Good	$ 3,510	Fair $ 1,795
Trade-In:	Mint	$ 9,360	Ex	$ 4,375	VG+	$ 3,050	Good	$ 2,730	Poor $ 780

MODEL 1911A1 COMMERCIAL (NEW SAFETY DEVICE)
REDBOOK CODE: RB-CL-H-A1XX3X

Serial numbers corresponding to commercial M1911A1s of the era. Sometimes referred to as the Swartz Safety models. Features a modified firing pin block safety and "C" prefix in their serial numbers, post-war models have a "C" suffix. The modified safety has a piston in the slide that can block the firing pin when engaged. Models with fixed sights are slightly less valuable than those with adjustable.

Production: 1930s - early 1940s Caliber: .45 ACP Action: SA, semi-auto

Barrel Length: 5" Magazine: detachable box

D2C:	NIB	--	Ex	$ 5,525	VG+	$3,990	Good	$3,625	
C2C:	Mint	--	Ex	$ 5,025	VG+	$3,630	Good	$ 3,265	Fair	$ 1,670
Trade-In:	Mint	--	Ex	$ 4,075	VG+	$2,830	Good	$ 2,540	Poor	$ 750

MODEL 1911A1 GOVERNMENT EXCELLENT +

Courtesy of Rock Island Auction Company

MODEL 1911A1 GOVERNMENT (PRE-WWII)
REDBOOK CODE: RB-CL-H-A1XX1X

Very similar to the early production M1911A1 pistols. Distributed both domestically and abroad for military use. Blue finish, frame-mounted safety. Serial number range includes models produced by Colt, Ithaca, and Springfield, among others. Most marked "Government Model."

Production: 1930s - 1942 Caliber: .45 ACP Action: SA, semi-auto

Barrel Length: 5" OA Length: 8.25" Wt.: 40 oz. Sights: fixed front

Capacity: 7 Magazine: detachable box Grips: synthetic

D2C:	NIB	--	Ex	$4,600	VG+	$3,330	Good	$3,025	
C2C:	Mint	--	Ex	$4,175	VG+	$3,030	Good	$2,725	Fair	$1,395
Trade-In:	Mint	--	Ex	$3,400	VG+	$2,360	Good	$2,120	Poor	$ 630

MODEL 1911A1 MILITARY ISSUED (FIRST TRANSITION)
REDBOOK CODE: RB-CL-H-A1X16X

Approximate serial numbers: 710,000-711,600.

Production: 1937 - discontinued Caliber: .45 ACP Action: SA, semi-auto Barrel Length: 5" Sights: fixed front Capacity: 7 Magazine: detachable box Grips: synthetic

D2C:	NIB	--	Ex	$ 9,350	VG+	$ 6,765	Good	$ 6,150	
C2C:	Mint	$13,500	Ex	$ 8,500	VG+	$ 6,150	Good	$ 5,535	Fair	$2,830
Trade-In:	Mint	$ 9,990	Ex	$ 6,900	VG+	$4,800	Good	$ 4,305	Poor	$1,230

MODEL 1911A1 MILITARY ISSUED (SECOND TRANSITION)
REDBOOK CODE: RB-CL-H-A1X17X

Approximate serial numbers: 711,600-712,400.

Production: 1937 - discontinued Caliber: .45 ACP Action: SA, semi-auto Barrel Length: 5" Sights: fixed front Capacity: 7 Magazine: detachable box Grips: synthetic

D2C:	NIB	--	Ex	$ 8,175	VG+	$5,915	Good	$ 5,375	
C2C:	Mint	--	Ex	$ 7,425	VG+	$5,380	Good	$4,840	Fair	$ 2,475
Trade-In:	Mint	--	Ex	$ 6,025	VG+	$4,200	Good	$ 3,765	Poor	$1,080

MODEL 1911A1 MILITARY ISSUED
REDBOOK CODE: RB-CL-H-A1X18X Approximate serial numbers: 712,400-713,700.

Production: 1938 - discontinued Caliber: .45 ACP Action: SA, semi-auto

Barrel Length: 5" Sights: fixed front Capacity: 7 Magazine: detachable box

Grips: synthetic

D2C:	NIB	--	Ex	$ 9,500	VG+	$6,875	Good	$ 6,250	
C2C:	Mint	--	Ex	$ 8,625	VG+	$6,250	Good	$ 5,675	Fair	$ 2,875
Trade-In:	Mint	--	Ex	$ 7,000	VG+	$4,880	Good	$ 4,375	Poor	$ 1,260

MODEL 1911A1 MILITARY ISSUED NAVY
FAIR
Courtesy of Rock Island Auction Company

MODEL 1911A1 MILITARY ISSUED
VERY GOOD +
Courtesy of Rock Island Auction Company

MODEL 1911A1 MILITARY ISSUED
VERY GOOD +
Courtesy of Rock Island Auction Company

MODEL 1911A1 MILITARY ISSUED
EXCELLENT
Courtesy of Rock Island Auction Company

MODEL 1911A1 MILITARY ISSUED
VERY GOOD +
Courtesy of Rock Island Auction Company

MODEL 1911A1 MILITARY ISSUED (NAVY)

REDBOOK CODE: RB-CL-H-A1X19X Serial numbers: 716,004-716,873.

Production: 1939 - discontinued Caliber: .45 ACP Action: SA, semi-auto Barrel Length: 5 Sights: fixed front Capacity: 7 Grips: synthetic

D2C:	NIB	--	Ex	$ 6,250	VG+	$ 4,310	Good	$ 4,100	
C2C:	Mint	--	Ex	$ 5,675	VG+	$ 4,100	Good	$ 3,690	Fair $ 1,890
Trade-In:	Mint	--	Ex	$ 4,600	VG+	$ 3,200	Good	$ 2,870	Poor $ 840

MODEL 1911A1 MILITARY ISSUED REDBOOK CODE: RB-CL-H-A1X20X

Serial numbers: 718,092-721,200.

Production: 1940 - discontinued Caliber: .45 ACP Action: SA, semi-auto Barrel Length: 5 Sights: fixed front Capacity: 7 Grips: synthetic

D2C:	NIB	--	Ex	$ 4,950	VG+	$ 3,575	Good	$ 3,250	
C2C:	Mint	--	Ex	$ 4,500	VG+	$ 3,250	Good	$ 2,925	Fair $ 1,495
Trade-In:	Mint	--	Ex	$ 3,650	VG+	$ 2,540	Good	$ 2,275	Poor $ 660

MODEL 1911A1 MILITARY ISSUED REDBOOK CODE: RB-CL-H-A1X21X

Serial numbers: 722,645-736,632.

Production: 1941 - discontinued Caliber: .45 ACP Action: SA, semi-auto Barrel Length: 5" Sights: fixed front Capacity: 7 Grips: synthetic

D2C:	NIB	--	Ex	$ 5,175	VG+	$ 3,740	Good	$ 3,400	
C2C:	Mint	--	Ex	$ 4,700	VG+	$ 3,400	Good	$ 3,060	Fair $ 1,565
Trade-In:	Mint	--	Ex	$ 3,825	VG+	$ 2,660	Good	$ 2,380	Poor $ 690

MODEL 1911A1 MILITARY ISSUED REDBOOK CODE: RB-CL-H-A1X22X

Serial numbers: 770,588- 825,604.

Production: 1942 - discontinued Caliber: .45 ACP Action: SA, semi-auto Barrel Length: 5" Sights: fixed front Capacity: 7 Grips: synthetic

D2C:	NIB	--	Ex	$ 3,550	VG+	$ 2,570	Good	$ 2,335	
C2C:	Mint	--	Ex	$ 3,225	VG+	$ 2,340	Good	$ 2,105	Fair $ 1,075
Trade-In:	Mint	--	Ex	$ 2,625	VG+	$ 1,830	Good	$ 1,635	Poor $ 480

MODEL 1911A1 MILITARY ISSUED REDBOOK CODE: RB-CL-H-A1X29X

Serial numbers: 879,706- 1,121,296.

Production: 1943 - 1944 Caliber: .45 ACP Action: SA, semi-auto Barrel Length: 5" Sights: fixed front Capacity: 7 Grips: synthetic

D2C:	NIB	--	Ex	$ 3,325	VG+	$ 2,395	Good	$ 2,175	
C2C:	Mint	--	Ex	$ 3,025	VG+	$ 2,180	Good	$ 1,960	Fair $ 1,005
Trade-In:	Mint	--	Ex	$ 2,450	VG+	$ 1,700	Good	$ 1,525	Poor $ 450

MODEL 1911A1 MILITARY ISSUED REDBOOK CODE: RB-CL-H-A1X30X

Serial numbers: 2,244,804- 2,368,718.

Production: 1945 - discontinued Caliber: .45 ACP Action: SA, semi-auto Barrel Length: 5" Sights: fixed front Capacity: 7 Grips: synthetic

D2C:	NIB	--	Ex	$ 3,225	VG+	$ 2,320	Good	$ 2,110	
C2C:	Mint	--	Ex	$ 2,925	VG+	$ 2,110	Good	$ 1,900	Fair $ 970
Trade-In:	Mint	--	Ex	$ 2,375	VG+	$ 1,650	Good	$ 1,480	Poor $ 450

**MODEL 1911A1
MILITARY ISSUED
(ITHACA)**
EXCELLENT

Courtesy of Rock Island Auction Company

MODEL 1911A1 MILITARY ISSUED (ITHACA)

REDBOOK CODE: RB-CL-H-A1X32X
Serial numbers: 856,500-916,500, 1,209,000-1,280,000, 1,442,000-1,472,000, 1,817,000-1,891,000, 2,075,100-2,135,00, 2,619,000-2,694,000.

Production: 1943 - 1945 Caliber: .45 ACP Action: SA, semi-auto Barrel Length: 5"								
Sights: fixed front Capacity: 7 Magazine: detachable box Grips: synthetic								
D2C:	NIB	--	Ex	$ 1,825	VG+	$ 1,320	Good	$ 1,200
C2C:	Mint	--	Ex	$ 1,675	VG+	$ 1,200	Good $ 1,080	Fair	$ 555
Trade-In: Mint	--		Ex	$ 1,350	VG+	$ 940	Good $ 840	Poor	$ 240

**MODEL 1911A1
MILITARY ISSUED
REMINGTON RAND**
VERY GOOD +

Courtesy of Rock Island Auction Company

MODEL 1911A1 MILITARY ISSUED REMINGTON RAND

REDBOOK CODE: RB-CL-H-A1X33X
Serial numbers: 916,405-1,041,404; 1,279,699-1,363,700. Marked "M1911A1 U.S. Army" and "United States Army" on the frame.

Production: 1942 - 1943 Caliber: .45 ACP Action: SA, semi-auto Barrel Length: 5"								
Sights: fixed front Capacity: 7 Magazine: detachable box Grips: synthetic								
D2C:	NIB	--	Ex	$ 1,525	VG+	$ 1,100	Good	$ 1,000
C2C:	Mint	--	Ex	$ 1,400	VG+	$ 1,000	Good $ 900	Fair	$ 460
Trade-In: Mint	--		Ex	$ 1,125	VG+	$ 780	Good $ 700	Poor	$ 210

**MODEL 1911A1
MILITARY ISSUED
REMINGTON RAND**
VERY GOOD +

Courtesy of Rock Island Auction Company

MODEL 1911A1 MILITARY ISSUED REMINGTON RAND

REDBOOK CODE: RB-CL-H-A1X35X
Marked "M1911A1 U.S. Army" and "United States Property" on the frame. Serial numbers: 1,471,431-1,609,528; 1,743,847-1,816,861; 1,890,504-2,075,103; 2,134,404-2,244,803; 2,380,014-2,465,139.

Production: 1944 - 1945 Caliber: .45 ACP Action: SA, semi-auto Barrel Length: 5"								
Sights: fixed front Capacity: 7 Magazine: detachable box Grips: synthetic								
D2C:	NIB	$ 1,900	Ex	$ 1,450	VG+	$ 1,045	Good	$ 950
C2C:	Mint	$ 1,830	Ex	$ 1,325	VG+	$ 950	Good $ 855	Fair	$ 440
Trade-In: Mint	$ 1,350	Ex	$ 1,075	VG+	$ 750	Good $ 665	Poor	$ 210

MODEL 1911A1 MILITARY ISSUED UNION SWITCH

SIGNAL CO. (BATCH 1) REDBOOK CODE: RB-CL-H-A1X37X
Union Switch Signal Co. manufactured approximately 55,000 units. Similar to other M1911A1s of the era. Early versions marked with a "P" on slide.

Production: 1943 Caliber: .45 ACP Action: SA, semi-auto Barrel Length: 5"								
Sights: fixed front Capacity: 7 Magazine: detachable box Grips: synthetic								
D2C:	NIB	--	Ex	$ 4,250	VG+	$ 3,070	Good	$ 2,790
C2C:	Mint	--	Ex	$ 3,850	VG+	$ 2,790	Good $ 2,510	Fair	$ 1,285
Trade-In: Mint	--		Ex	$ 3,125	VG+	$ 2,180	Good $ 1,955	Poor	$ 570

MODEL 1911A1 MILITARY ISSUED UNION SWITCH

SIGNAL CO. (BATCH 2) REDBOOK CODE: RB-CL-H-A1X38X
A part of the 55,000 models produced by US&S. Serial numbers range between approximately 141,405 and 1,096,404. Marked "P" on the slide's edge and "HS" on the barrel.

Production: 1943 Caliber: .45 ACP Action: SA, semi-auto Barrel Length: 5"								
Sights: fixed front Capacity: 7 Magazine: detachable box Grips: synthetic								
D2C:	NIB	--	Ex	$ 4,100	VG+	$ 2,960	Good	$ 2,690
C2C:	Mint	--	Ex	$ 3,725	VG+	$ 2,690	Good $ 2,420	Fair	$ 1,240
Trade-In: Mint	--		Ex	$ 3,025	VG+	$ 2,100	Good $ 1,885	Poor	$ 540

COLT

MODEL 1911A1 MILITARY ISSUED UNION SWITCH SIGNAL CO.
VERY GOOD +

Courtesy of Rock Island Auction Company

MODEL 1911A1 MILITARY ISSUED UNION SWITCH SIGNAL CO. (BATCH 3) REDBOOK CODE: RB-CL-H-A1X39X

A part of the 55,000 models produced by US&S. Serial numbers range between approximately 1,080,001 and 1,096,000. Marked "P" on the frame and slide and "HS" on the barrel.

Production: 1943   Caliber: .45 ACP   Action: SA, semi-auto   Barrel Length: 5"

Sights: fixed front   Capacity: 7   Magazine: detachable box   Grips: synthetic

D2C:	NIB	--	Ex	$4,000	VG+	$2,890	Good	$2,625	
C2C:	Mint	--	Ex	$3,625	VG+	$2,630	Good	$2,365	Fair $1,210
Trade-In:	Mint	--	Ex	$2,950	VG+	$2,050	Good	$1,840	Poor $540

MODEL 1911A1 MILITARY ISSUED STATE OF NEW YORK GOVERNMENT REDBOOK CODE: RB-CL-H-A1X40X

Marked Property of the State of New York. Less than 300 units produced. Little information exists pertaining to these models.

Production: ~1960s - discontinued   Caliber: .45 ACP   Action: SA, semi-auto

Barrel Length: 5"   Sights: fixed front   Capacity: 7   Grips: synthetic

D2C:	NIB	--	Ex	$2,025	VG+	$1,460	Good	$1,325	
C2C:	Mint	$2,550	Ex	$1,850	VG+	$1,330	Good	$1,195	Fair $610
Trade-In:	Mint	$1,890	Ex	$1,500	VG+	$1,040	Good	$930	Poor $270

MODEL 1911A1 ACE
VERY GOOD +

Courtesy of Rock Island Auction Company

MODEL 1911A1 ACE   REDBOOK CODE: RB-CL-H-11ACEX

Serial numbers range between approximately 1 and 10,800 with as many manufactured. Similar to the M1911A1 Government model. Noted for firing .22 LR rounds.

Production: 1931 - 1947   Caliber: .22LR   Action: SA, semi-auto

Barrel Length: 4.75"   Sights: fixed front, adjustable rear   Capacity: 10

Magazine: detachable box   Grips: synthetic

D2C:	NIB	--	Ex	--	VG+	$3,400	Good	$3,090	
C2C:	Mint	--	Ex	$4,275	VG+	$3,090	Good	$2,780	Fair $1,425
Trade-In:	Mint	--	Ex	$3,475	VG+	$2,410	Good	$2,165	Poor $630

MODEL 1911A1 SERVICE MODEL ACE
EXCELLENT

Courtesy of Rock Island Auction Company

MODEL 1911A1 SERVICE MODEL ACE (PRE-WAR)

REDBOOK CODE: RB-CL-H-11SERX

An improved variation of the original Ace model. Features a modified chamber and "SM" prefixes in the serial number. Roughly 13,800 manufactured. Parkerized or blue (slight premium) finish.

Production: 1937 - 1945   Caliber: .22LR   Action: SA, semi-auto

Barrel Length: 4.75"   Sights: fixed front, adjustable rear   Capacity: 10

Magazine: detachable box   Grips: synthetic

D2C:	NIB	--	Ex	$4,650	VG+	$3,355	Good	$3,050	
C2C:	Mint	--	Ex	$4,225	VG+	$3,050	Good	$2,745	Fair $1,405
Trade-In:	Mint	--	Ex	$3,475	VG+	$2,380	Good	$2,135	Poor $630

MODEL 1911A1 SERVICE MODEL ACE (POST-WAR)

REDBOOK CODE: RB-CL-H-11SER1

A later production of the Service Model Ace. Converted models can greatly increase premium.

Production: 1978 - 1982   Caliber: .22LR   Action: SA, semi-auto

Barrel Length: 5"   Capacity: 10   Magazine: detachable box   Grips: synthetic

D2C:	NIB	--	Ex	$1,000	VG+	$715	Good	$650	
C2C:	Mint	$1,250	Ex	$900	VG+	$650	Good	$585	Fair $300
Trade-In:	Mint	$930	Ex	$750	VG+	$510	Good	$455	Poor $150

**MODEL 1911A1
NATIONAL MATCH**
GOOD

Courtesy of Rock Island Auction Company

MODEL 1911A1 NATIONAL MATCH REDBOOK CODE: RB-CL-H-A1XX2

Less than 6,000 produced. Match grade barrel, blue or nickel finish. Some models may have a Swartz safety, increasing their value (+$450-$2,500). Less valuable second generation models denoted by "NM" in the serial number.

Production: 1932 - early 1940s   Caliber: .45 ACP   Action: SA, semi-auto

Barrel Length: 5"   Sights: fixed or adjustable   Capacity: 7

Magazine: detachable box   Grips: synthetic or wood

D2C:	NIB	--	Ex	$ 5,475	VG+	$3,960	Good	$ 3,600		
C2C:	Mint	$ 6,920	Ex	$ 4,975	VG+	$3,600	Good	$ 3,240	Fair	$ 1,660
Trade-In:	Mint	$ 5,120	Ex	$4,050	VG+	$ 2,810	Good	$ 2,520	Poor	$ 720

MODEL 1911A1 COMMERCIAL (POST-WAR, DOMESTIC)

REDBOOK CODE: RB-CL-H-A1XX4

Serial numbers range between approximately C220,000 to 340,000C. Blue or stainless finish. Later models will feature Government Model markings.

Production: 1946 - 1969   Caliber: .45 ACP   Action: SA, semi-auto

Barrel Length: 5"   OA Length: 8.25"   Wt.: 40 oz.   Sights: fixed

Capacity: 7   Magazine: detachable box   Grips: synthetic or wood

D2C:	NIB	$ 2,100	Ex	$ 1,600	VG+	$ 1,155	Good	$ 1,050		
C2C:	Mint	$ 2,020	Ex	$ 1,450	VG+	$ 1,050	Good	$ 945	Fair	$ 485
Trade-In:	Mint	$ 1,500	Ex	$ 1,200	VG+	$ 820	Good	$ 735	Poor	$ 210

**MODEL 1911A1
SUPER .38 AUTOMATIC**
VERY GOOD +

Courtesy of Rock Island Auction Company

MODEL 1911A1 .38 SUPER AUTOMATIC REDBOOK CODE: RB-CL-H-A1XX5

Marked Super .38 Automatic. Similar to the Government Model 1911A1 but fires .38 Super instead of .45 ACP. The U.S. Government bought less than 500 units near the end of WWII, each engraved "GHD." Due to a large span of production dates earlier models will bring a higher premium.

Production: 1929 - 1971   Caliber: .38 Super   Action: SA, semi-auto

Barrel Length: 5"   OA Length: 8.25"   Wt.: 40 oz.   Sights: fixed

Capacity: 7   Magazine: detachable box   Grips: synthetic or wood

D2C:	NIB	--	Ex	$4,150	VG+	$3,060	Good	$ 2,725		
C2C:	Mint	--	Ex	$3,775	VG+	$ 2,730	Good	$ 2,455	Fair	$ 1,255
Trade-In:	Mint	--	Ex	$3,075	VG+	$ 2,130	Good	$ 1,910	Poor	$ 570

MODEL 1911A1 GOVERNMENT .22 CAL REDBOOK CODE: RB-CL-H-A1XX7

Manufactured by Umarex. Noted for firing .22 LR rounds instead of .45 ACP.

Production: 2011 - current   Caliber: .22 LR   Action: SA, semi-auto

Barrel Length: 5"   OA Length: 8.25"   Wt.: 40 oz.   Sights: fixed

Capacity: 10, 12   Magazine: detachable box   Grips: wood

D2C:	NIB	$ 400	Ex	$ 325	VG+	$ 220	Good	$ 200		
C2C:	Mint	$ 390	Ex	$ 300	VG+	$ 200	Good	$ 180	Fair	$ 95
Trade-In:	Mint	$ 290	Ex	$ 225	VG+	$ 160	Good	$ 140	Poor	$ 60

**MODEL 1911A1 GOLD CUP NATIONAL
MATCH**
Courtesy of Rock Island Auction Company VERY GOOD +

MODEL 1911A1 GOLD CUP NATIONAL MATCH (PRE-1971)

REDBOOK CODE: RB-CL-H-A1XX8

Serial numbers include an "NM" prefix. Blue or stainless finish and deluxe fitting. Released prior to the Series 70 line.

Production: late 1950s - 1971   Caliber: .45 ACP   Action: SA, semi-auto   Barrel Length: 5"

Sights: adjustable   Capacity: 7, 8   Magazine: detachable box   Grips: black rubber, wood

D2C:	NIB	--	Ex	$ 1,000	VG+	$ 715	Good	$ 650		
C2C:	Mint	$ 1,250	Ex	$ 900	VG+	$ 650	Good	$ 585	Fair	$ 300
Trade-In:	Mint	$ 930	Ex	$ 750	VG+	$ 510	Good	$ 455	Poor	$ 150

MODEL 1911A1 GOLD CUP MKIII NATIONAL MATCH

REDBOOK CODE: RB-CL-H-A1XX9

Similar to the Gold Cup National Match, but chambered for .38 Special WC. Serial numbers should correspond to other M1911A1s of the era.

Production: early 1960s - 1974		Caliber: .38 Special WC		Action: SA, semi-auto						
Barrel Length: 5"		Sights: 3-dot		Capacity: 9		Magazine: detachable box				
D2C:	NIB	$1,600	Ex	$1,225	VG+	$ 880	Good	$ 800		
C2C:	Mint	$1,540	Ex	$1,125	VG+	$ 800	Good	$ 720	Fair	$ 370
Trade-In:	Mint	$1,140	Ex	$ 900	VG+	$ 630	Good	$ 560	Poor	$ 180

MODEL 1911A1 COMBAT COMMANDER
VERY GOOD +
Courtesy of Rock Island Auction Company

MODEL 1911A1 COMBAT COMMANDER

REDBOOK CODE: RB-CL-H-A1X12

Several variants produced. Gold Cup versions of the gun are more valuable than the standard, stainless steel, Officer's ACP, Model ACP Lightweight, and nickel versions of the gun.

Production: late 1980s - 1997		Caliber: 9mm, .45 ACP, .38 Super								
Action: SA, semi-auto		Barrel Length: 4.25"		OA Length: 8.25"		Wt.: 35 oz.				
Sights: fixed		Capacity: 8		Magazine: detachable box		Grips: synthetic				
D2C:	NIB	$ 815	Ex	$ 625	VG+	$ 450	Good	$ 410		
C2C:	Mint	$ 790	Ex	$ 575	VG+	$ 410	Good	$ 370	Fair	$ 190
Trade-In:	Mint	$ 580	Ex	$ 475	VG+	$ 320	Good	$ 290	Poor	$ 90

MODEL 1911 GOVERNMENT MKIV SERIES 70

REDBOOK CODE: RB-CL-H-11GS70

Marked "70G," if produced between 1970 and 1976, "G70" if produced between 1976 and 1980, or 70B if produced between 1980 and 1983. Blue or nickel finish, available in a host of different calibers. Does not feature a firing pin block safety, unlike the M1911 Series 80. Collectors typically consider the Series 70 superior to the Series 80 for their smoother trigger.

Production: 1970 - 1983		Caliber: .45 ACP		Action: SA, semi-auto						
Barrel Length: 5"		Capacity: 7		Magazine: detachable box						
D2C:	NIB	$1,150	Ex	$ 875	VG+	$ 635	Good	$ 575		
C2C:	Mint	$1,110	Ex	$ 800	VG+	$ 580	Good	$ 520	Fair	$ 265
Trade-In:	Mint	$ 820	Ex	$ 650	VG+	$ 450	Good	$ 405	Poor	$ 120

MODEL 1911 GOVERNMENT MKIV SERIES 70
EXCELLENT
Courtesy of Rock Island Auction Company

MODEL 1911 COMMANDER SERIES 70

REDBOOK CODE: RB-CL-H-11CMME

Serial numbers begin around 70G01000. Features an enhanced hammer, extended safety lock, beavertail grip, and full length guide rod. Similar to the Colt M1911 Government.

Production: 1970 - 1983		Caliber: .45 ACP		Action: SA, semi-auto						
Barrel Length: 4.25"		Wt. 36 oz.		Sights: Novak low mount		Capacity: 8				
Magazine: detachable box		Grips: synthetic or wood								
D2C:	NIB	$1,100	Ex	$ 850	VG+	$ 605	Good	$ 550		
C2C:	Mint	$1,060	Ex	$ 775	VG+	$ 550	Good	$ 495	Fair	$ 255
Trade-In:	Mint	$ 790	Ex	$ 625	VG+	$ 430	Good	$ 385	Poor	$ 120

MODEL 1911 COMMANDER SERIES 70
EXCELLENT
Courtesy of Rock Island Auction Company

MODEL 1911 MODEL 1918 REPLICA SERIES 70

REDBOOK CODE: RB-CL-H-191870

Produced by the Colt Custom Shop between 2008 and 2009. Similar to the M1911 models issued to U.S. soldiers in WWI. A facotry recall concerning parts replacement was issued for models sold after 2007.

Production: 2008 - 2009		Caliber: .45 ACP		Action: SA, semi-auto						
Barrel Length: 5"		OA Length: 8.25"		Wt.: 40 oz.		Sights: fixed				
Capacity: 7		Magazine: detachable box		Grips: synthetic or wood						
D2C:	NIB	$ 815	Ex	$ 625	VG+	$ 450	Good	$ 410		
C2C:	Mint	$ 790	Ex	$ 575	VG+	$ 410	Good	$ 370	Fair	$ 190
Trade-In:	Mint	$ 580	Ex	$ 475	VG+	$ 320	Good	$ 290	Poor	$ 90

MODEL 1911 GOVERNMENT MKIV SERIES 80
NEW IN BOX

MODEL 1911 GOVERNMENT MKIV SERIES 80

REDBOOK CODE: RB-CL-H-11GV80

The successor to the Series 70. Features a modified action and firing pin block safety, which some feel compromises its quality, but decreases the risk of an accidental misfire if dropped. Available in blue, nickel, stainless, or polished stainless.

Production: 1983 - late 1980s   Caliber: .45 ACP, .22 LR   Action: SA, semi-auto

Barrel Length: 5"   Capacity: 7   Magazine: detachable box

D2C:	NIB	$1,300	Ex	$1,000	VG+	$ 715	Good	$ 650	
C2C:	Mint	$1,250	Ex	$ 900	VG+	$ 650	Good	$ 585	Fair	$ 300
Trade-In:	Mint	$ 930	Ex	$ 750	VG+	$ 510	Good	$ 455	Poor	$ 150

MODEL 1911 COMMANDER MKIV SERIES 80
EXCELLENT
Courtesy of Rock Island Auction Company

MODEL 1911 COMMANDER MKIV SERIES 80

REDBOOK CODE: RB-CL-H-CMMK80

Very similar to the M1911 Government Model Series 80. Alloy frame, steel slide, and frame-mounted and grip safety. Nickel plated and stainless steel models command a slightly higher premium.

Production: early 1980s - late 1980s   Caliber: .45 ACP, 9mm, .38 Super

Action: SA, semi-auto   Barrel Length: 4.25"   Sights: 3-dot   Capacity: 7, 8

Magazine: detachable box

D2C:	NIB	$1,250	Ex	$ 950	VG+	$ 690	Good	$ 625	
C2C:	Mint	$1,200	Ex	$ 875	VG+	$ 630	Good	$ 565	Fair	$ 290
Trade-In:	Mint	$ 890	Ex	$ 700	VG+	$ 490	Good	$ 440	Poor	$ 150

MODEL 1911 GOLD CUP NATIONAL MATCH MKIV SERIES 70
EXCELLENT
Courtesy of Rock Island Auction Company

MODEL 1911 GOLD CUP NATIONAL MATCH
MKIV SERIES 70   REDBOOK CODE: RB-CL-H-GLDC70

Serial numbers range between 70G01000 and 70B11300. An improved variant of the original M1911 Gold Cup National Match, noted for its aluminum frame.

Production: 1970 - early 1980s   Caliber: .45 ACP   Action: SA, semi-auto

Barrel Length: 5"   Capacity: 7   Magazine: detachable box

D2C:	NIB	$1,725	Ex	$1,325	VG+	$ 950	Good	$ 865	
C2C:	Mint	$1,660	Ex	$1,200	VG+	$ 870	Good	$ 780	Fair	$ 400
Trade-In:	Mint	$1,230	Ex	$ 975	VG+	$ 680	Good	$ 605	Poor	$ 180

MODEL 1911A1 DELTA ELITE SERIES 80
EXCELLENT
Courtesy of Rock Island Auction Company

MODEL 1911A1 DELTA ELITE SERIES 80

REDBOOK CODE: RB-CL-H-11DE80

Available in both blue and stainless finish, which have similar values. A deluxe model of the M1911A1. The first 600 units of the gun are worth several hundred dollars more in value.

Production: 1986 - 1996   Caliber: 10mm   Action: SA, semi-auto

Barrel Length: 5"   OA Length: 8.25"   Wt.: 38 oz.   Sights: fixed

Capacity: 8   Magazine: detachable box   Grips: neoprene

D2C:	NIB	$1,000	Ex	$ 775	VG+	$ 550	Good	$ 500	
C2C:	Mint	$ 960	Ex	$ 700	VG+	$ 500	Good	$ 450	Fair	$ 230
Trade-In:	Mint	$ 710	Ex	$ 575	VG+	$ 390	Good	$ 350	Poor	$ 120

MODEL 1911A1 COMBAT ELITE
NEW IN BOX
Courtesy of Rock Island Auction Company

MODEL 1911A1 COMBAT ELITE   REDBOOK CODE: RB-CL-H-11CMBE

Features an enhanced hammer, blue or stainless finish, beavertail grip and frame-mounted safety, flared ejection port, and full-length guide rod.

Production: 1987 - current   Caliber: .45 ACP, .38 Super   Action: SA, semi-auto

Barrel Length: 5"   OA Length: 8.5"   Wt.: 38 oz.   Sights: Novak Low Mount sights

Capacity: 8, 9   Magazine: detachable box   Grips: synthetic or wood

D2C:	NIB	$ 880	Ex	$ 675	VG+	$ 485	Good	$ 440	
C2C:	Mint	$ 850	Ex	$ 625	VG+	$ 440	Good	$ 400	Fair	$ 205
Trade-In:	Mint	$ 630	Ex	$ 500	VG+	$ 350	Good	$ 310	Poor	$ 90

COLT

**MODEL 1911A1 DELTA GOLD CUP
SERIES 80**
EXCELLENT
Courtesy of Rock Island Auction Company

**MODEL 1911
SERIES 70**
NEW IN BOX
Courtesy of Rock Island Auction Company

**MODEL 1911 GOLD CUP
NATIONAL MATCH
MKIV SERIES 80**
EXCELLENT
Courtesy of Rock Island Auction Company

**MODEL 1911 CONCEALED CARRY
OFFICER'S MODEL O**
EXCELLENT

MODEL 1911A1 DELTA GOLD CUP SERIES 80
REDBOOK CODE: RB-CL-H-11DEGC
Similar to the M1911A1 Delta Elite. Noted for firing 10mm rounds. A deluxe model of the M1911A1. Stainless steel valued at $100-$250 more than those with blue finish.

Production: 1992 - 1996 Caliber: 10mm Action: SA, semi-auto
Barrel Length: 5" OA Length: 8.25" Wt.: 36 oz. Sights: Accro sight Capacity: 8
Magazine: detachable box Grips: black rubber, wrap-around

D2C:	NIB	$1,300	Ex	$1,000	VG+	$ 715	Good	$ 650	
C2C:	Mint	$1,250	Ex	$ 900	VG+	$ 650	Good	$ 585	Fair $ 300
Trade-In:	Mint	$ 930	Ex	$ 750	VG+	$ 510	Good	$ 455	Poor $ 150

MODEL 1911 SERIES 70 (CURRENT ISSUE) REDBOOK CODE: RB-CL-H-11S70C
A reintroduction of the M1911 Series 70. Features a special hammer, safety lock, grip safety, short steel trigger, and standard ejection port.

Production: 2011 - current Caliber: .45 ACP Action: SA, semi-auto
Barrel Length: 5" OA Length: 8.5" Wt.: 38 oz. Sights: high-profile Capacity: 7
Magazine: detachable box Grips: checkered walnut

D2C:	NIB	$1,000	Ex	$ 775	VG+	$ 550	Good	$ 500	
C2C:	Mint	$ 960	Ex	$ 700	VG+	$ 500	Good	$ 450	Fair $ 230
Trade-In:	Mint	$ 710	Ex	$ 575	VG+	$ 390	Good	$ 350	Poor $ 120

MODEL 1911 GOLD CUP NATIONAL MATCH
MKIV SERIES 80 REDBOOK CODE: RB-CL-H-GLDCNM
Similar to the Series 70 Gold Cup, but features a modified firing pin, which some think compromises their quality. Available in blue, stainless, or polished stainless, but all have similar values.

Production: early 1980s - late 1980s Caliber: .45 ACP Action: SA, semi-auto
Magazine: detachable box

D2C:	NIB	$1,450	Ex	$1,125	VG+	$ 800	Good	$ 725	
C2C:	Mint	$1,400	Ex	$1,025	VG+	$ 730	Good	$ 655	Fair $ 335
Trade-In:	Mint	$1,030	Ex	$ 825	VG+	$ 570	Good	$ 510	Poor $ 150

MODEL 1911 GOVERNMENT MODEL O
REDBOOK CODE: RB-CL-H-911GMO
Features serrations on the front and rear of the slide, enhanced trigger, stainless or blue finish, and an extended magazine. All pistols marked "Model O" feature Novak rear and dovetail front sights, a modified grip safety, de-horned frame, and enhanced finished. Produced by the Colt Custom Shop.

Production: 1999 Caliber: .45 ACP Action: SA, semi-auto
Barrel Length: 5" OA Length: 8.5" Wt.: 36 oz. Sights: Novak rear, dovetail front
Capacity: 8 Magazine: detachable box Grips: black synthetic or wood, checkered

D2C:	NIB	$ 1,050	Ex	$ 800	VG+	$ 580	Good	$ 525	
C2C:	Mint	$ 1,010	Ex	$ 725	VG+	$ 530	Good	$ 475	Fair $ 245
Trade-In:	Mint	$ 750	Ex	$ 600	VG+	$ 410	Good	$ 370	Poor $ 120

MODEL 1911 CONCEALED CARRY OFFICER'S MODEL O
REDBOOK CODE: RB-CL-H-911CCO
Features serrations on the front and rear of the slide, enhanced trigger, and carbon or blue finish.

Production: 1999 Caliber: .45 ACP Action: SA, semi-auto
Barrel Length: 4.25" Sights: Novak rear, dovetail front Capacity: 7
Magazine: detachable box Grips: black synthetic or wood, checkered

D2C:	NIB	$1,150	Ex	$ 875	VG+	$ 635	Good	$ 575	
C2C:	Mint	$ 1,110	Ex	$ 800	VG+	$ 580	Good	$ 520	Fair $ 265
Trade-In:	Mint	$ 820	Ex	$ 650	VG+	$ 450	Good	$ 405	Poor $ 120

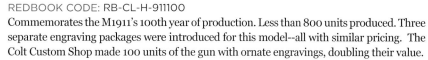

COLT

MODEL 1911 COMMANDER MODEL O
NEW IN BOX

MODEL 1911 COMMANDER ("O" PACKAGE)
REDBOOK CODE: RB-CL-H-911CMO
Features serrations on the front and rear of the slide, enhanced trigger, and carbon or blue finish. Ambidextrous design.

Production: 1999 - discontinued Caliber: .45 ACP Action: SA, semi-auto										
Barrel Length: 4.25" Sights: Novak rear, dovetail front Capacity: 8										
Magazine: detachable box Grips: black synthetic or wood, checkered										
D2C:	NIB	$ 960	Ex	$ 750	VG+	$ 530	Good	$ 480		
C2C:	Mint	$ 930	Ex	$ 675	VG+	$ 480	Good	$ 435	Fair	$ 225
Trade-In:	Mint	$ 690	Ex	$ 550	VG+	$ 380	Good	$ 340	Poor	$ 120

MODEL 1911 100TH ANNIVERSARY
REDBOOK CODE: RB-CL-H-911100
Commemorates the M1911's 100th year of production. Less than 800 units produced. Three separate engraving packages were introduced for this model--all with similar pricing. The Colt Custom Shop made 100 units of the gun with ornate engravings, doubling their value.

Production: 2011 Caliber: .45 ACP Action: SA, semi-auto										
Barrel Length: 5" OA Length: 8.25" Wt.: 40 oz. Sights: fixed Capacity: 7										
Magazine: detachable box Grips: cocobolo, ivory, or walnut										
D2C:	NIB	$ 4,225	Ex	$ 3,225	VG+	$ 2,325	Good	$ 2,115		
C2C:	Mint	$ 4,060	Ex	$ 2,925	VG+	$ 2,120	Good	$ 1,905	Fair	$ 975
Trade-In:	Mint	$ 3,000	Ex	$ 2,375	VG+	$ 1,650	Good	$ 1,480	Poor	$ 450

**MODEL 1911
100TH
ANNIVERSARY**
EXCELLENT
Courtesy of Rock Island Auction Company

MODEL 1911 GOLD CUP TROPHY ("O" PACKAGE)
REDBOOK CODE: RB-CL-H-11GDNT
National Match barrel, high-polish stainless finish, round-top slide, aluminum trigger, frame-mounted and grip safety, gold Colt medallions in the grip.

Production: late 1990s - current Caliber: .45 ACP Action: SA, semi-auto										
Barrel Length: 5" OA Length: 8.5" Wt.: 40 oz. Sights: adjustable rear, fixed front										
Capacity: 7 Magazine: detachable box Grips: checkered walnut										
D2C:	NIB	$1,000	Ex	$ 775	VG+	$ 550	Good	$ 500		
C2C:	Mint	$ 960	Ex	$ 700	VG+	$ 500	Good	$ 450	Fair	$ 230
Trade-In:	Mint	$ 710	Ex	$ 575	VG+	$ 390	Good	$ 350	Poor	$ 120

**MODEL 1911 GOLD CUP TROPHY
MODEL O**
EXCELLENT
Courtesy of Rock Island Auction Company

MODEL 1911 DELTA ELITE (CURRENT ISSUE)
REDBOOK CODE: RB-CL-H-11DELE
M1911 Government-style design, skeletonized hammer, lowered ejection port, frame-mounted and grip safety, blue or stainless finish.

Production: current Caliber: 10mm Action: SA, semi-auto										
Barrel Length: 5" OA Length: 8.25" Wt.: 36 oz. Sights: high profile, dot										
Capacity: 8 Magazine: detachable box Grips: black synthetic										
D2C:	NIB	$ 1,250	Ex	$ 950	VG+	$690	Good	$ 625		
C2C:	Mint	$ 1,200	Ex	$ 875	VG+	$630	Good	$ 565	Fair	$ 290
Trade-In:	Mint	$ 890	Ex	$ 700	VG+	$490	Good	$ 440	Poor	$ 150

**MODEL 1911
DELTA ELITE**
NEW IN BOX

NEW AGENT   REDBOOK CODE: RB-CL-H-NEWAGT
Frame-mounted safety, enhanced hammer, grip safety, flared ejection port, compact design. Several variations, including special editions with different accessories and a DAO model.

Production: 2007 - current Caliber: .45 ACP, 9mm Action: SA, semi-auto										
Barrel Length: 3" OA Length: 6.75" Wt.: 23 oz. Sights: fixed, trench-style										
Capacity: 7 Magazine: detachable box Grips: Crimson Trace brand										
D2C:	NIB	$ 970	Ex	$ 750	VG+	$ 535	Good	$ 485		
C2C:	Mint	$ 940	Ex	$ 675	VG+	$ 490	Good	$ 440	Fair	$ 225
Trade-In:	Mint	$ 690	Ex	$ 550	VG+	$ 380	Good	$ 340	Poor	$ 120

M1911 NEW AGENT
NEW IN BOX

COLT

SERIES 1991 SERIES 80
EXCELLENT

Courtesy of Rock Island Auction Company

**SERIES 1991
GOVERNMENT**
NEW IN BOX

SERIES 1991 SERIES 80 (CURRENT ISSUE)
REDBOOK CODE: RB-CL-H-91CURR
Almost identical to the M1911 pistol. Stainless steel, blue, or green army finish, all with similar values. Features a solid aluminum trigger, lowered ejection port, frame-mounted and grip safety, and Series 80s firing mechanisms.

Production: 1991 - current Caliber: 9mm, .45 ACP, .38 Super Action: SA, semi-auto
Barrel Length: 4.25", 5" Wt.: 38 oz. Sights: fixed, high-profile Capacity: 7, 8, 9
Magazine: detachable box Grips: synthetic or wood, checkered

D2C:	NIB	$ 980	Ex $ 750	VG+ $ 540	Good $ 490				
C2C:	Mint	$ 950	Ex $ 700	VG+ $ 490	Good $ 445	Fair $ 230			
Trade-In:	Mint	$ 700	Ex $ 550	VG+ $ 390	Good $ 345	Poor $ 120			

SERIES 1991 GOVERNMENT REDBOOK CODE: RB-CL-H-9180GV
Stainless steel or parkerized matte finish. Design indicative of the M1991A1 series.

Production: 1992 - current Caliber: .45 ACP, 9x23mm Action: SA, semi-auto
Barrel Length: 5" Wt.: 38 oz. Sights: fixed
Capacity: 7 Magazine: detachable box Grips: synthetic, tactical style

D2C:	NIB	$ 985	Ex $ 750	VG+ $ 545	Good $ 495		
C2C:	Mint	$ 950	Ex $ 700	VG+ $ 500	Good $ 445	Fair $ 230	
Trade-In:	Mint	$ 700	Ex $ 575	VG+ $ 390	Good $ 345	Poor $ 120	

SERIES 1991 COMMANDER REDBOOK CODE: RB-CL-H-91A1CM
Produced between 1993 and 1995 and reintroduced again in 1997. Models with stainless finish are slightly more valuable than blue finish models. Design much like the M1991 series.

Production: 1993 - current Caliber: .45 ACP, 9mm Action: SA, semi-auto
Barrel Length: 4.25" Sights: fixed Capacity: 7 Magazine: detachable box
Grips: synthetic or wood

D2C:	NIB	$ 960	Ex $ 750	VG+ $ 530	Good $ 480	LMP $ 993	
C2C:	Mint	$ 930	Ex $ 675	VG+ $ 480	Good $ 435	Fair $ 225	
Trade-In:	Mint	$ 690	Ex $ 550	VG+ $ 380	Good $ 340	Poor $ 120	

**SERIES 1991
COMMANDER**
(STAINLESS)
NEW IN BOX

SERIES 1991 COMMANDER (STAINLESS)
REDBOOK CODE: RB-CL-H-91A1CB
Discontinued in 2005, but reintroduced in 2007.

Production: 1997 - current Caliber: .45 ACP, 9mm Action: SA, semi-auto
Barrel Length: 4.25" Sights: fixed Capacity: 7 Magazine: detachable box
Grips: synthetic or wood

D2C:	NIB	$ 990	Ex $ 775	VG+ $ 545	Good $ 495	LMP $ 1,058	
C2C:	Mint	$ 960	Ex $ 700	VG+ $ 500	Good $ 450	Fair $ 230	
Trade-In:	Mint	$ 710	Ex $ 575	VG+ $ 390	Good $ 350	Poor $ 120	

SERIES 1991 OFFICER'S REDBOOK CODE: RB-CL-H-91OFFX
Blue, matte blue, nickel, or stainless finish, all with differing values. Stainless and nickel finishes command a slightly higher premium with a NIB price around $1,000. Noted for its compact design.

Production: 1985 - mid-1990s Caliber: .45 ACP Action: SA, semi-auto
Barrel Length: 3" Sights: fixed Capacity: 6 Magazine: detachable box
Grips: synthetic or wood

D2C:	NIB	$ 770	Ex $ 600	VG+ $ 425	Good $ 385		
C2C:	Mint	$ 740	Ex $ 550	VG+ $ 390	Good $ 350	Fair $ 180	
Trade-In:	Mint	$ 550	Ex $ 450	VG+ $ 310	Good $ 270	Poor $ 90	

COLT

MODEL 1991 GOVERNMENT ("O" PACKAGE)

REDBOOK CODE: RB-CL-H-91GVM2

The same as the M1991A1, but features Novak rear and dovetail front sights, a modified grip safety, de-horned frame, and enhanced finish. Produced by the Colt Custom Shop.

Production: early 1990s   Caliber: .45 ACP   Action: SA, semi-auto

Barrel Length: 5"   OA Length: 8.5"   Sights: Novak rear, dovetail front   Capacity: 7

Magazine: detachable box   Grips: synthetic or wood

D2C:	NIB	$1,100	Ex	$ 850	VG+	$ 605	Good	$ 550	
C2C:	Mint	$1,060	Ex	$ 775	VG+	$ 550	Good	$ 495	Fair $ 255
Trade-In:	Mint	$ 790	Ex	$ 625	VG+	$ 430	Good	$ 385	Poor $ 120

MODEL 1991 COMMANDER ("O" PACKAGE)

REDBOOK CODE: RB-CL-H-91CMMO

Design similar to previously manufactured Commander models. Features Novak rear and dovetail front sights, a modified grip safety, de-horned frame, and enhanced finish. Produced by the Colt Custom Shop.

Production: early 1990s   Caliber: .45 ACP   Action: SA, semi-auto

Barrel Length: 4.25"   Sights: Novak rear, dovetail front   Capacity: 7

Magazine: detachable box   Grips: synthetic or wood

D2C:	NIB	$1,050	Ex	$ 800	VG+	$ 580	Good	$ 525	
C2C:	Mint	$1,010	Ex	$ 725	VG+	$ 530	Good	$ 475	Fair $ 245
Trade-In:	Mint	$ 750	Ex	$ 600	VG+	$ 410	Good	$ 370	Poor $ 120

MODEL 1991 GOLD CUP ("O" PACKAGE)

REDBOOK CODE: RB-CL-H-91GCMO

True to the M1991 series design but features Novak rear and dovetail front sights, a modified grip safety, de-horned frame, and enhanced finish. Produced by the Colt Custom Shop.

Production: 1999   Caliber: .45 ACP   Action: SA, semi-auto

Barrel Length: 5"   Sights: Novak rear, dovetail front   Capacity: 8

Magazine: detachable box   Grips: synthetic or wood

D2C:	NIB	$1,200	Ex	$ 925	VG+	$ 660	Good	$ 600	
C2C:	Mint	$1,160	Ex	$ 850	VG+	$ 600	Good	$ 540	Fair $ 280
Trade-In:	Mint	$ 860	Ex	$ 675	VG+	$ 470	Good	$ 420	Poor $ 120

MODEL 1991 DEFENDER ("O" PACKAGE)

REDBOOK CODE: RB-CL-H-91DFNO

An updated variant of the M1991 Defender. Features Novak rear and dovetail front sights, a modified grip safety, de-horned frame, and enhanced finished.

Production: 2000   Caliber: .45 ACP   Action: SA, semi-auto

Barrel Length: 3"   Sights: Novak rear, dovetail front   Capacity: 7

Magazine: detachable box   Grips: synthetic or wood

D2C:	NIB	$ 950	Ex	$ 725	VG+	$ 525	Good	$ 475	
C2C:	Mint	$ 920	Ex	$ 675	VG+	$ 480	Good	$ 430	Fair $ 220
Trade-In:	Mint	$ 680	Ex	$ 550	VG+	$ 380	Good	$ 335	Poor $ 120

**JUNIOR COLT
POCKET MODEL**
VERY GOOD +

Courtesy of Rock Island Auction Company

JUNIOR COLT POCKET MODEL REDBOOK CODE: RB-CL-H-JNRPCK

Manufactured by the Astra company of Spain. Blue finish, compact design.

Production: 1958 - early 1970s Caliber: .22 Short, .25 ACP Action: DA/SA, semi-auto

Barrel Length: 2.25" OA Length: 4.5" Sights: notch

Capacity: 6 Magazine: detachable box Grips: black synthetic

D2C:	NIB	$ 475	Ex	$ 375	VG+	$ 265	Good	$ 240	
C2C:	Mint	$ 460	Ex	$ 350	VG+	$ 240	Good	$ 215	Fair $ 110
Trade-In:	Mint	$ 340	Ex	$ 275	VG+	$ 190	Good	$ 170	Poor $ 60

DOUBLE EAGLE
EXCELLENT

Courtesy of Rock Island Auction Company

DOUBLE EAGLE REDBOOK CODE: RB-CL-H-91DE90

Several variations exist: I, II, Combat Commander, Officer's Model, and Officer's Lightweight Model, all of which are similar in value. Features a decocking lever, steel frame, blue or usually stainless finish, and frame-mounted safety.

Production: 1989 - 2000 Caliber: .45 ACP, .38 Super, 9mm, 10mm

Action: DA/SA, semi-auto Barrel Length: 5" OA Length: 8.5" Wt.: 40 oz.

Sights: 3-dot Capacity: 8 Magazine: detachable box Grips: black synthetic

D2C:	NIB	$1,100	Ex	$ 850	VG+	$ 605	Good	$ 550	
C2C:	Mint	$1,060	Ex	$ 775	VG+	$ 550	Good	$ 495	Fair $ 255
Trade-In:	Mint	$ 790	Ex	$ 625	VG+	$ 430	Good	$ 385	Poor $ 120

2000 ALL AMERICAN
EXCELLENT
Courtesy of Rock Island Auction Company

2000 ALL AMERICAN REDBOOK CODE: RB-CL-H-912000

Designed for law enforcement use, but never sold well. Firearm experts and critics note that inaccuracy and unreliability plagued the model. Aluminum and polymer versions of the gun exist. Features a polymer frame, steel barrel and slide, and frame-mounted safety.

Production: early 1990s - 1994 Caliber: 9mm Action: DAO, semi-auto

Barrel Length: 3.75" OA Length: 7.5" Sights: fixed, 3-dot Capacity: 15

Magazine: detachable box Grips: black synthetic or wooden

D2C:	NIB	$ 800	Ex	$ 625	VG+	$ 440	Good	$ 400	
C2C:	Mint	$ 770	Ex	$ 575	VG+	$ 400	Good	$ 360	Fair $ 185
Trade-In:	Mint	$ 570	Ex	$ 450	VG+	$ 320	Good	$ 280	Poor $ 90

PONY/PONY POCKETLITE
GOOD
Courtesy of Rock Island Auction Company

PONY/PONY POCKETLITE REDBOOK CODE: RB-CL-H-PONYPC

Blue or stainless finish, frame-mounted safety, compact design. The Pocketlite version is made of stainless steel and more lightweight components.

Production: 1997 - late 1990s Caliber: .380 ACP Action: DAO, semi-auto

Barrel Length: 3.75" OA Length: 5.5" Sights: fixed

Capacity: 6 Magazine: detachable box Grips: black synthetic

D2C:	NIB	$ 715	Ex	$ 550	VG+	$ 395	Good	$ 360	
C2C:	Mint	$ 690	Ex	$ 500	VG+	$ 360	Good	$ 325	Fair $ 165
Trade-In:	Mint	$ 510	Ex	$ 425	VG+	$ 280	Good	$ 255	Poor $ 90

POCKET NINE
VERY GOOD +
Courtesy of Rock Island Auction Company

POCKET NINE REDBOOK CODE: RB-CL-H-PCKNIN

Aluminum-alloy frame, matte stainless finish, wraparound grips, frame-mounted safety.

Production: 1999 - 2001 Caliber: 9mm Action: DAO, semi-auto

Barrel Length: 2.75" OA Length: 5.5" Sights: fixed

Capacity: 6 Magazine: detachable box Grips: black rubber

D2C:	NIB	$1,225	Ex	$ 950	VG+	$ 675	Good	$ 615	
C2C:	Mint	$1,180	Ex	$ 850	VG+	$ 620	Good	$ 555	Fair $ 285
Trade-In:	Mint	$ 870	Ex	$ 700	VG+	$ 480	Good	$ 430	Poor $ 150

MUSTANG REDBOOK CODE: RB-CL-H-MUSTNG
Features a frame-mounted safety, compact design, and blue, nickel, or stainless finish.

Production: 1987 - 2000 Caliber: .380 ACP Action: SA, semi-auto

Barrel Length: 2.75" OA Length: 5.5" Wt.: 19 oz.

Sights: fixed Capacity: 6 Magazine: detachable box Grips: synthetic

D2C:	NIB	$ 860	Ex	$ 675	VG+	$ 475	Good	$ 430	
C2C:	Mint	$ 830	Ex	$ 600	VG+	$ 430	Good	$ 390	Fair	$ 200
Trade-In:	Mint	$ 620	Ex	$ 500	VG+	$ 340	Good	$ 305	Poor	$ 90

MUSTANG POCKETLITE
NEW IN BOX

MUSTANG POCKETLITE REDBOOK CODE: RB-CL-H-MUSTX4
Discontinued in 2000, but reintroduced in 2011. Features a firing pin safety block, lowered ejection port, solid aluminum trigger, and high profile sights. Available in both black matte and nickel finishes.

Production: 1987 - current Caliber: .380 ACP Action: SA, semi-auto

Barrel Length: 2.75" OA Length: 5.5" Wt.: 20 oz.

Sights: fixed, high-profile Capacity: 6 Magazine: detachable box Grips: synthetic

D2C:	NIB	$ 725	Ex	$ 575	VG+	$ 400	Good	$ 365	
C2C:	Mint	$ 700	Ex	$ 525	VG+	$ 370	Good	$ 330	Fair	$ 170
Trade-In:	Mint	$ 520	Ex	$ 425	VG+	$ 290	Good	$ 255	Poor	$ 90

MUSTANG XSP
NEW IN BOX

Courtesy of Rock Island Auction Company

MUSTANG XSP
REDBOOK CODE: RB-CL-H-MSTXSP
Stainless steel slide, polymer frame, accessory rail, commander style hammer, thumb safety.

Production: 2013 - current Caliber: .380 ACP Action: SA, semi-auto

Barrel Length: 2.75" Sights: high profile fixed

Capacity: 6 Magazine: detachable box Grips: polymer grip frame

D2C:	NIB	$ 615	Ex	$ 475	VG+	$ 340	Good	$ 310	
C2C:	Mint	$ 600	Ex	$ 425	VG+	$ 310	Good	$ 280	Fair	$ 145
Trade-In:	Mint	$ 440	Ex	$ 350	VG+	$ 240	Good	$ 220	Poor	$ 90

PRE-WOODSMAN
EXCELLENT

Courtesy of Rock Island Auction Company

PRE-WOODSMAN REDBOOK CODE: RB-CL-H-PRWOOD
Approximately 55,000 units made with corresponding serial numbers. Factory engraved models are more valuable than those engraved by a third party. Noted for their thin barrels. There are no official "Woodsman" markings on this model.

Production: 1915 - 1927 Caliber: .22 LR Action: SA, semi-auto

Barrel Length: 6.5" Wt.: 29 oz. Sights: adjustable front and rear

Capacity: 10 Magazine: detachable box Grips: wood

D2C:	NIB	$ 1,400	Ex	$ 1,075	VG+	$ 770	Good	$ 700	
C2C:	Mint	$ 1,350	Ex	$ 975	VG+	$ 700	Good	$ 630	Fair	$ 325
Trade-In:	Mint	$ 1,000	Ex	$ 800	VG+	$ 550	Good	$ 490	Poor	$ 150

WOODSMAN TARGET MODEL
GOOD

Courtesy of Rock Island Auction Company

WOODSMAN TARGET MODEL (FIRST SERIES)
REDBOOK CODE: RB-CL-H-WMTMFS
Serial numbers range between approximately 84,000 and 200,000. Features a heavier barrel than the Pre-Woodsman and Woodsman Specialist models. The novelist Ernest Hemingway famously used and praised the Woodsman.

Production: 1934 - 1947 Caliber: .22 LR Action: SA, semi-auto

Barrel Length: 6.6" Wt.: 29 oz. Sights: adjustable front and rear

Capacity: 10 Magazine: detachable box Grips: walnut

D2C:	NIB	$ 1,625	Ex	$ 1,250	VG+	$ 895	Good	$ 815	
C2C:	Mint	$ 1,560	Ex	$ 1,125	VG+	$ 820	Good	$ 735	Fair	$ 375
Trade-In:	Mint	$ 1,160	Ex	$ 925	VG+	$ 640	Good	$ 465	Poor	$ 180

WOODSMAN MATCH (FIRST SERIES)
REDBOOK CODE: RB-CL-H-WMMTTP
Noted for its larger, wrap-around walnut grip. Serial numbers range from MT1 to MT16611.

Production: late 1930s - 1942 Caliber: .22 LR Action: SA, semi-auto						
Barrel Length: 6.6" Sights: adjustable front and rear						
Capacity: 10 Magazine: detachable box Grips: walnut, wrap-around						
D2C:	NIB	$1,650	Ex $1,275	VG+ $ 910	Good $ 825	
C2C:	Mint	$1,590	Ex $1,150	VG+ $ 830	Good $ 745	Fair $ 380
Trade-In:	Mint	$1,180	Ex $ 925	VG+ $ 650	Good $ 580	Poor $ 180

WOODSMAN SPORT
EXCELLENT

Courtesy of Rock Island Auction Company

WOODSMAN SPORT (FIRST SERIES)
REDBOOK CODE: RB-CL-H-WMSPFS
Serial numbers range from approximately 85,000 to 158,000. Very similar to the Target model but features a shortened barrel and fixed front sight. Less than 150,000 produced of this and the Target model.

Production: 1933 - 1942 Caliber: .22 LR Action: SA, semi-auto Barrel Length: 4.5"						
Sights: fixed front Capacity: 10 Magazine: detachable box Grips: walnut						
D2C:	NIB	$1,475	Ex $1,125	VG+ $ 815	Good $ 740	
C2C:	Mint	$1,420	Ex $1,025	VG+ $ 740	Good $ 665	Fair $ 340
Trade-In:	Mint	$1,050	Ex $850	VG+ $ 580	Good $ 520	Poor $ 150

WOODSMAN CHALLENGER, MATCH, SPORT, AND TARGET MODELS, (SECOND SERIES)
EXCELLENT

Courtesy of Rock Island Auction Company

WOODSMAN CHALLENGER MATCH, SPORT, AND TARGET MODELS (SECOND SERIES)
REDBOOK CODE: RB-CL-H-WMCHMT
These models have similar values, except the Challenger, which is worth almost half as much. These guns, except the Challenger, are denoted by an "S" in their serial number. The Woodsman Target and Sport models fetch an NIB premium near $1,700, with the 6" barrel being slightly higher. The Challenger 4.5" barrel is priced below, with the 6" barrell being priced slightly lower.

Production: 1947 - 1955 Caliber: .22 LR Action: SA, semi-auto						
Barrel Length: 4.5", 6" Sights: fixed or adjustable						
Capacity: 10 Magazine: detachable box Grips: plastic or wood						
D2C:	NIB	$ 985	Ex $ 750	VG+ $ 545	Good $ 495	
C2C:	Mint	$ 950	Ex $ 700	VG+ $ 500	Good $ 445	Fair $ 230
Trade-In:	Mint	$ 700	Ex $ 575	VG+ $ 390	Good $ 345	Poor $ 120

WOODSMAN HUNTSMAN, HUNTSMAN MODEL S MASTER, SPORT, TARGET, TARGETSMAN MODELS (THIRD SERIES)
EXCELLENT

Courtesy of Rock Island Auction Company

WOODSMAN HUNTSMAN, HUNTSMAN MODEL S MASTER, SPORT, TARGET, TARGETSMAN MODELS (THIRD SERIES) REDBOOK CODE: RB-CL-H-WMHUNT
These guns differ from the Second Series Woodsman with their magazine release located on the bottom of the grip. All similar in value, except for the Model S Master, which features gold engraving. Colt produced less than 450 units of the Model S Master, doubling their value. Woodsman Target and Sport models are priced below. Target Match models have an NIB premium of nearly $1,000, with the 6" barrel being slightly higher. Huntsman and Targetsman models are the cheapest in the series and are priced NIB around $725.

Production: 1955 - 1977 Caliber: .22 LR Action: SA, semi-auto						
Barrel Length: 4.5", 6" Sights: fixed or adjustable Capacity: 10						
Magazine: detachable box Grips: plastic or wood						
D2C:	NIB	$ 915	Ex $ 700	VG+ $ 505	Good $ 460	
C2C:	Mint	$ 880	Ex $ 650	VG+ $ 460	Good $ 415	Fair $ 215
Trade-In:	Mint	$ 650	Ex $ 525	VG+ $ 360	Good $ 325	Poor $ 120

**CADET .22/
COLT .22**
EXCELLENT

Courtesy of Rock Island Auction Company

CADET .22/COLT .22 REDBOOK CODE: RB-CL-H-0

Standard and Target model both produced. Features a vent rib and matte stainless barrel. Marketed as an improved and new version of the Woodsman.

Production: 1994 - late 1990s Caliber: .22 LR Action: SA, semi-auto

Barrel Length: 4.5" Sights: fixed, grooved-sight rib

Capacity: 11 Magazine: detachable box Grips: rubber

D2C:	NIB	$ 400	Ex $ 325	VG+ $ 220	Good $ 200		
C2C:	Mint	$ 390	Ex $ 300	VG+ $ 200	Good $ 180	Fair $	95
Trade-In:	Mint	$ 290	Ex $ 225	VG+ $ 160	Good $ 140	Poor $	60

Comanche

Imported by Eagle Imports Inc. of Township, New Jersey, Comanche Firearms is an Argentinean-based handgun manufacturer, best known for their Super Comanche single-shot pistol and Comanche II-A revolver. Since their introduction into U.S. markets, the handguns have gained favorable reviews, especially considering their affordability. Eagle Imports can field all questions and sales inquiries regarding Comanche Firearms.

**SUPER COMANCHE
SINGLE-SHOT**
NEW IN BOX

SUPER COMANCHE SINGLE-SHOT REDBOOK CODE: RB-C9-H-COMMXX

Blue or nickel finish, noted for its one round capacity and break-top action.

Production: 2002 - current Caliber: .45 LC/.410 gauge, .22 Magnum

Action: SA, single-shot Barrel Length: 6", 10" Wt.: 48 oz. Sights: adjustable

Capacity: 1 Grips: black rubber, white

D2C:	NIB	$ 200	Ex $ 175	VG+ $ 110	Good $ 100	LMP $	220
C2C:	Mint	$ 200	Ex $ 150	VG+ $ 100	Good $ 90	Fair $	50
Trade-In:	Mint	$ 150	Ex $ 125	VG+ $ 80	Good $ 70	Poor $	30

COMANCHE II-A
NEW IN BOX

COMANCHE II-A REDBOOK CODE: RB-C9-H-COMMX2

Noted as an extremely reliable pistol. Features blue or nickel finish, alloy frame, and lightweight design.

Production: 2002 - current Caliber: .38 Special Action: DA/SA, revolver

Barrel Length: 3", 4" OA Length: 8", 8.5" Wt.: 33 oz., 35 oz.

Sights: fixed Capacity: 6 Grips: black rubber

D2C:	NIB	$ 215	Ex $ 175	VG+ $ 120	Good $ 110		
C2C:	Mint	$ 210	Ex $ 150	VG+ $ 110	Good $ 100	Fair $	50
Trade-In:	Mint	$ 160	Ex $ 125	VG+ $ 90	Good $ 80	Poor $	30

Connecticut Valley Arms

Established in 1971, Connecticut Valley Arms of Duluth, Georgia is perhaps best known for their muzzleloader rifles, but their single-shot, black powder pistols remain significant, given the company's presence in the black-powder industry. CVA voluntarily recalled one gun in 1997 and has since gone to extensive lengths to ensure their pistols and rifles remain of the utmost quality and safety for use. The company is a subsidiary of BPI Outdoors.

SCOUT V2 PISTOL
NEW IN BOX

SCOUT V2 PISTOL REDBOOK CODE: RB-CX-H-SCOUTX

Similar to the original Scout pistol, noted for firing large caliber rounds and its single shot design. Stainless finish, scope rail.

Production: 2010 - current   Action: SA, single-shot

Caliber: .357 Magnum, .44 Magnum, .243 Win, 300 blackout

Barrel Length: 14" Sights: fixed Capacity: 1 Grips: black synthetic

D2C:	NIB	$ 355	Ex $ 275	VG+ $ 200	Good $ 180	LMP $ 410
C2C:	Mint	$ 350	Ex $ 250	VG+ $ 180	Good $ 160	Fair $ 85
Trade-In:	Mint	$ 260	Ex $ 200	VG+ $ 140	Good $ 125	Poor $ 60

Coonan Arms/Coonan Inc.

While Coonan Arms and Coonan Inc. were separate entities, Dan Coonan, a noted gun designer, founded and managed both companies, producing high-end M1911s. Coonan Arms, formed in the early 1980s, dissolved in the late 1990s. Coonan Inc. was established around 2009 and currently manufactures pistols from Blaine, Minnesota. Coonan M1911s fire .357 Magnum rounds, a rarity for a M1911, and retain much of their value.

M1911 CLASSIC .357 MAGNUM (NEW MODEL)
REDBOOK CODE: RB-CO-H-1911CL

Features an enhanced barrel, recoil operated design, available sight upgrades, and in camo, stainless, or blue finish. Manufactured by the revamped Coonan Firearms.

Production: 2009 - current Caliber: .357 Magnum Action: SA, semi-auto

Barrel Length: 5" OA Length: 8.3" Wt.: 42 oz. Sights: fixed, dovetail

Capacity: 7 Magazine: detachable box Grips: walnut

D2C:	NIB	$1,300	Ex $1,000	VG+ $ 715	Good $ 650	LMP $ 1,375
C2C:	Mint	$1,250	Ex $ 900	VG+ $ 650	Good $ 585	Fair $ 300
Trade-In:	Mint	$ 930	Ex $ 750	VG+ $ 510	Good $ 455	Poor $ 150

**M1911 .357
MAGNUM MODEL B**
EXCELLENT

M1911 .357 MAGNUM MODEL B REDBOOK CODE: RB-CO-H-911357

Classic 1911 styling. Later models have a modified safety, stainless finish, frame-mounted and grip safety.

Production: 1983 - late 1990s Caliber: .357 Magnum Action: SA, semi-auto

Barrel Length: 5" OA Length: 8.3" Wt.: 42 oz. Sights: blade or adjustable

Capacity: 7 Magazine: detachable box Grips: walnut

D2C:	NIB	$ 825	Ex $ 650	VG+ $ 455	Good $ 415	LMP $ 875
C2C:	Mint	$ 800	Ex $ 575	VG+ $ 420	Good $ 375	Fair $ 190
Trade-In:	Mint	$ 590	Ex $ 475	VG+ $ 330	Good $ 290	Poor $ 90

M1911 CADET REDBOOK CODE: RB-CO-H-911CDT
A smaller version of Coonan M1911 .357 Magnum, stainless finish.

Production: 1993 - 1998 Caliber: .357 Magnum Action: SA, semi-auto

Barrel Length: 4" Sights: blade or adjustable Capacity: 6

Magazine: detachable box Grips: walnut

D2C:	NIB	$1,050	Ex	$ 800	VG+ $ 580	Good	$ 525		
C2C:	Mint	$1,010	Ex	$ 725	VG+ $ 530	Good	$ 475	Fair $	245
Trade-In:	Mint	$ 750	Ex	$ 600	VG+ $ 410	Good	$ 370	Poor $	120

CZ-U.S.A.

Formed in 1936, Česká Zbrojovka, or C.Z., is a Czech firearms manufacturer with a significant presence in the U.S. markets due to their American subsidiary, C.Z. U.S.A., which they established in the late 1990s. The company produces a wide array of pistols, the C.Z. 75 and its variants being the most popular. Most of the company's pistols feature an ergonomic design, matte black finish, and polymer grip frame. C.Z.'s typically sell at reasonable prices and remain a good option for concealed carry use. C.Z. U.S.A. is located in Kansas City, Kansas.

CZ 1922 REDBOOK CODE: RB-CZ-H-CZ1922
Designed for military use, somewhat similar to Mauser models of the era.

Production: 1923 Caliber: 6.35mm (.25 ACP) Action: SA, semi-auto

Barrel Length: 3.4" Sights: fixed Capacity: 8 Magazine: detachable box

Grips: wood

D2C:	NIB	--	Ex	$ 575	VG+ $ 405	Good	$ 370		
C2C:	Mint	$ 710	Ex	$ 525	VG+ $ 370	Good	$ 335	Fair $	170
Trade-In:	Mint	$ 530	Ex	$ 425	VG+ $ 290	Good	$ 260	Poor $	90

CZ 1924 STANDARD
VERY GOOD +

CZ 1924 STANDARD REDBOOK CODE: RB-CZ-H-CZ1924
An updated version of the CZ 22 Model, designed for military use with further modifications in later models. Approximately 200,000 units produced.

Production: 1924 - discontinued Caliber: .380 ACP (9mm Short)

Action: SA, semi-auto Barrel Length: 3.5" Sights: fixed Capacity: 8

Magazine: detachable box Grips: wood

D2C:	NIB	--	Ex	$ 525	VG+ $ 375	Good	$ 340		
C2C:	Mint	$ 660	Ex	$ 475	VG+ $ 340	Good	$ 310	Fair $	160
Trade-In:	Mint	$ 490	Ex	$ 400	VG+ $ 270	Good	$ 240	Poor $	90

CZ 1927
VERY GOOD +

CZ 1927 REDBOOK CODE: RB-CZ-H-CZ1927
Less than 550,000 units produced. Features a blowback action, exposed hammer, and blue finish.

Production: 1939 - 1945 Caliber: .32 ACP (7.65mm) Action: SA, semi-auto

Barrel Length: 3.85" Sights: fixed Capacity: 9 Magazine: detachable box

Grips: wood

D2C:	NIB	--	Ex	$ 500	VG+ $ 355	Good	$ 320		
C2C:	Mint	$ 620	Ex	$ 450	VG+ $ 320	Good	$ 290	Fair $	150
Trade-In:	Mint	$ 460	Ex	$ 375	VG+ $ 250	Good	$ 225	Poor $	90

CZ 1928   REDBOOK CODE: RB-CZ-H-CZ1928

Produced for military use in WWII. Similar to other CZ and Mauser pistols of the era. Blue or nickel finish. Marked "CZ" on the slide.

Production: 1915 - 1919	Caliber: .32 ACP (7.65mm)		Action: SA, semi-auto		
Barrel Length: 3.4"	Sights: fixed	Capacity: 7	Magazine: detachable box		
Grips: wood					
D2C:	NIB --	Ex $ 475	VG+ $ 330	Good $ 300	
C2C:	Mint $ 580	Ex $ 425	VG+ $ 300	Good $ 270	Fair $ 140
Trade-In:	Mint $ 430	Ex $ 350	VG+ $ 240	Good $ 210	Poor $ 60

CZ 1936   REDBOOK CODE: RB-CZ-H-CZ1936

A precursor to the more popular CZ 1945. Pocket-size design, blowback operated, blue finish.

Production: late 1930s - 1940	Caliber: .25 ACP (6.35mm)		Action: DAO, semi-auto		
Barrel Length: 2"	OA Length: 5"	Sights: fixed	Capacity: 8		
Magazine: detachable box	Grips: synthetic				
D2C:	NIB --	Ex $ 475	VG+ $ 340	Good $ 310	
C2C:	Mint $ 600	Ex $ 425	VG+ $ 310	Good $ 280	Fair $ 145
Trade-In:	Mint $ 440	Ex $ 350	VG+ $ 240	Good $ 220	Poor $ 90

CZ 1938
VERY GOOD +

CZ 1938   REDBOOK CODE: RB-CZ-H-CZ1938

Similar to other CZ pistols of the era, saw use with the German law enforcement officers. Some feature manual safeties.

Production: late 1930s - 1945	Caliber: .380 ACP (9mm Short)		Action: SA, semi-auto		
Barrel Length: 4.6"	OA Length: 8.125"	Sights: fixed	Capacity: 9		
Magazine: detachable box	Grips: wood				
D2C:	NIB --	Ex $ 450	VG+ $ 325	Good $ 295	
C2C:	Mint $ 570	Ex $ 425	VG+ $ 300	Good $ 270	Fair $ 140
Trade-In:	Mint $ 420	Ex $ 350	VG+ $ 240	Good $ 210	Poor $ 60

CZ 1945   REDBOOK CODE: RB-CZ-H-CZ1945

Pocket-sized design, blue finish.

Production: 1945 - discontinued	Caliber: .25 ACP (6.35mm)		Action: SA, semi-auto		
Barrel Length: 2"	OA Length: 5"	Sights: fixed	Capacity: 8		
Magazine: detachable box	Grips: synthetic				
D2C:	NIB --	Ex $ 450	VG+ $ 315	Good $ 285	
C2C:	Mint $ 550	Ex $ 400	VG+ $ 290	Good $ 260	Fair $ 135
Trade-In:	Mint $ 410	Ex $ 325	VG+ $ 230	Good $ 200	Poor $ 60

POCKET AUTOMATIC   REDBOOK CODE: RB-CZ-H-PCKATM
Similar to other M1910 pocket models.

Production: 1940s - discontinued		Caliber: .25 ACP, .380 ACP		Action: SA, semi-auto		
Barrel Length: 2"	OA Length: 5"	Sights: fixed	Capacity: 6			
Magazine: detachable box	Grips: synthetic					
D2C:	NIB	--	Ex $ 250	VG+ $ 165	Good $ 150	
C2C:	Mint $ 290	Ex $ 225	VG+ $ 150	Good $ 135	Fair $ 70	
Trade-In:	Mint $ 220	Ex $ 175	VG+ $ 120	Good $ 105	Poor $ 30	

CZ 75
EXCELLENT

Courtesy of Rock Island Auction Company

CZ 75   REDBOOK CODE: RB-CZ-H-CZ75ST
Widely known and highly revered in the semi-auto world. Many variants followed the original model, which was copied in many countries. Political climates in early production made importation sporadic in the U.S., so early pricing was quite high. Chambered for 9mm Parabellum cartridge. Frame-mounted thumb safety, spurred hammer. Finish is polished or matte blue, with dual-tone, nickel, or stainless at a higher premium.

Production: 1975 - discontinued		Caliber: 9mm	Action: DA/SA, semi-auto		
Barrel Length: 4.72"	OA Length: 8"	Wt.: 35 oz.	Sights: fixed		
Capacity: 15	Magazine: detachable box	Grips: black synthetic			
D2C:	NIB $ 515	Ex $ 400	VG+ $ 285	Good $ 260	LMP $ 528
C2C:	Mint $ 500	Ex $ 375	VG+ $ 260	Good $ 235	Fair $ 120
Trade-In:	Mint $ 370	Ex $ 300	VG+ $ 210	Good $ 185	Poor $ 60

CZ 75 P-01
NEW IN BOX

CZ 75 P-01   REDBOOK CODE: RB-CZ-H-75P01X
Frame-mounted safety, aluminum-alloy frame, accessory rail, black finish, compact design. Manufactured for law enforcement use.

Production: 2001 - current		Caliber: 9mm	Action: DA/SA, semi-auto		
Barrel Length: 3.84"	OA Length: 7.2"	Wt.: 28 oz.	Sights: fixed		
Capacity: 14	Magazine: detachable box	Grips: rubber			
D2C:	NIB $ 550	Ex $ 425	VG+ $ 305	Good $ 275	LMP $ 660
C2C:	Mint $ 530	Ex $ 400	VG+ $ 280	Good $ 250	Fair $ 130
Trade-In:	Mint $ 400	Ex $ 325	VG+ $ 220	Good $ 195	Poor $ 60

CZ 75 P-07 DUTY
NEW IN BOX

CZ 75 P-07 DUTY   REDBOOK CODE: RB-CZ-H-75P07D
Aluminum-alloy frame, compact design, improved trigger design, frame-mounted decocker safety, black finish.

Production: current	Caliber: .40 S&W, 9mm	Action: DA/SA, semi-auto		
Barrel Length: 3.8"	OA Length: 7.3"	Wt.: 28 oz.	Sights: fixed	
Capacity: 12, 15	Magazine: detachable box	Grips: synthetic		
D2C:	NIB $ 440	Ex $ 350	VG+ $ 245	Good $ 220	LMP $ 496
C2C:	Mint $ 430	Ex $ 325	VG+ $ 220	Good $ 200	Fair $ 105
Trade-In:	Mint $ 320	Ex $ 250	VG+ $ 180	Good $ 155	Poor $ 60

CZ 75 B STAINLESS
NEW IN BOX

CZ 75 B STAINLESS   REDBOOK CODE: RB-CZ-H-75BSTN
The second generation CZ 75. Features a steel and alloy frame, ergonomic design, modified internal firing pin safety, frame-mounted safety, and matte stainless finish.

Production: current	Caliber: 9mm	Action: DA/SA, semi-auto		
Barrel Length: 4.7"	OA Length: 8.1"	Wt.: 23 oz.	Sights: 3-dot	
Capacity: 10	Magazine: detachable box	Grips: synthetic		
D2C:	NIB $ 690	Ex $ 525	VG+ $ 380	Good $ 345	LMP $ 711
C2C:	Mint $ 670	Ex $ 500	VG+ $ 350	Good $ 315	Fair $ 160
Trade-In:	Mint $ 490	Ex $ 400	VG+ $ 270	Good $ 245	Poor $ 90

CZ 75 B SA
NEW IN BOX

CZ 75 BD
NEW IN BOX

CZ 75 COMPACT
NEW IN BOX

CZ 75 PO-6
NEW IN BOX

CZ 75 B SA   REDBOOK CODE: RB-CZ-H-75BSAA

The second generation CZ 75. Features a steel and alloy frame, ambi manual frame-mounted safety, and ergonomic design. Distinguished by its single-action firing mechanism.

Production: current Caliber: .40 S&W Action: SA, semi-auto

Barrel Length: 4.6" OA Length: 8.1" Wt.: 35 oz. Sights: fixed

Capacity: 10 Magazine: detachable box Grips: synthetic

D2C:	NIB	$ 520	Ex $ 400	VG+ $ 290	Good $ 260	LMP $ 654
C2C:	Mint	$ 500	Ex $ 375	VG+ $ 260	Good $ 235	Fair $ 120
Trade-In:	Mint	$ 370	Ex $ 300	VG+ $ 210	Good $ 185	Poor $ 60

CZ 75 BD   REDBOOK CODE: RB-CZ-H-75BDXX

Decocker version of the basic CZ 75B model. High-capacity, double-column magazine, ergonomic design, frame-mounted firing pin block and safety, safety stop on hammer.

Production: current Caliber: 9mm Action: DA/SA, semi-auto

Barrel Length: 4.61" OA Length: 8.1" Wt.: 35 oz. Sights: fixed

Capacity: 16 Magazine: detachable box Grips: synthetic

D2C:	NIB	$ 500	Ex $ 400	VG+ $ 275	Good $ 250	LMP $ 612
C2C:	Mint	$ 480	Ex $ 350	VG+ $ 250	Good $ 225	Fair $ 115
Trade-In:	Mint	$ 360	Ex $ 300	VG+ $ 200	Good $ 175	Poor $ 60

CZ 75 COMPACT   REDBOOK CODE: RB-CZ-H-75COMP

Steel frame, frame-mounted safety, firing pin block, shortened grip.

Production: 1994 - current Caliber: 9mm, .40 S&W Action: DA/SA, semi-auto

Barrel Length: 3.8" OA Length: 7.2" Wt.: 32 oz. Sights: fixed

Capacity: 14 Magazine: detachable box Grips: synthetic

D2C:	NIB	$ 470	Ex $ 375	VG+ $ 260	Good $ 235	LMP $ 581
C2C:	Mint	$ 460	Ex $ 325	VG+ $ 240	Good $ 215	Fair $ 110
Trade-In:	Mint	$ 340	Ex $ 275	VG+ $ 190	Good $ 165	Poor $ 60

CZ 75D PCR COMPACT   REDBOOK CODE: RB-CZ-H-75DPCR

Frame-mounted safety, firing pin block, black finish, light alloy frame, a compact version of the CZ 75 P-01.

Production: current Caliber: 9mm Action: DA/SA, semi-auto

Barrel Length: 3.7" OA Length: 7.21" Wt.: 27 oz.

Sights: fixed, snag free Capacity: 14 Magazine: detachable box Grips: rubber

D2C:	NIB	$ 545	Ex $ 425	VG+ $ 300	Good $ 275	LMP $ 599
C2C:	Mint	$ 530	Ex $ 400	VG+ $ 280	Good $ 250	Fair $ 130
Trade-In:	Mint	$ 390	Ex $ 325	VG+ $ 220	Good $ 195	Poor $ 60

CZ 75 P-06   REDBOOK CODE: RB-CZ-H-75P6X

The same as the P-01, but fires .40 S&W. Features a new Omega trigger system, polymer frame, and frame-mounted safety.

Production: current Caliber: .40 S&W Action: DA/SA, semi-auto

Barrel Length: 3.9" OA Length: 7.2" Wt.: 29 oz. Sights: fixed night sights

Capacity: 10 Magazine: detachable box Grips: synthetic

D2C:	NIB	$ 610	Ex $ 475	VG+ $ 340	Good $ 305	LMP $ 680
C2C:	Mint	$ 590	Ex $ 425	VG+ $ 310	Good $ 275	Fair $ 145
Trade-In:	Mint	$ 440	Ex $ 350	VG+ $ 240	Good $ 215	Poor $ 90

CZ-USA

CZ 75 CHAMPION   REDBOOK CODE: RB-CZ-H-75CHMP

Adjustable trigger, ambidextrous and frame-mounted safety, steel components, ergonomic design, blue-finish slide, nickel-finish frame. Comes with 3 port compensator.

Production: discontinued in 2010   Caliber: .40 S&W, 9mm, 9x21

Action: DA/SA, semi-auto   Barrel Length: 4.5"   OA Length: 9.45"   Wt.: 36 oz.

Sights: adjustable, 3-dot   Capacity: 10   Magazine: detachable box   Grips: rubber

D2C:	NIB	$1,380	Ex	$1,050	VG+	$ 760	Good	$ 690	
C2C:	Mint	$1,330	Ex	$ 975	VG+	$ 690	Good	$ 625	Fair $ 320
Trade-In:	Mint	$ 980	Ex	$ 775	VG+	$ 540	Good	$ 485	Poor $ 150

CZ 75 KADET PISTOL
NEW IN BOX

CZ 75 KADET PISTOL   REDBOOK CODE: RB-CZ-H-75KPS

Steel frame, frame-mounted safety, firing pin block, black finish.

Production: discontinued in 2012   Caliber: .22LR   Action: DA/SA, semi-auto

Barrel Length: 4.6"   OA Length: 8.1"   Wt.: 38 oz.

Sights: adjustable rear   Capacity: 10   Magazine: detachable box   Grips: synthetic

D2C:	NIB	$ 615	Ex	$ 475	VG+	$ 340	Good	$ 310	LMP $690
C2C:	Mint	$ 600	Ex	$ 425	VG+	$ 310	Good	$ 280	Fair $ 145
Trade-In:	Mint	$ 440	Ex	$ 350	VG+	$ 240	Good	$ 220	Poor $ 90

CZ 75 SP-01 PHANTOM
NEW IN BOX

CZ 75 SP-01 PHANTOM   REDBOOK CODE: RB-CZ-H-75SPPH

An updated variant of the SP-01. Designed for law enforcement and military use. Features a frame-mounted safety, black finish, polymer frame with accessory rail, and no firing pin block. 33% lighter than steel-framed models.

Production: discontinued in 2012   Caliber: 9mm   Action: DA/SA, semi-auto

Length: 4.7"   OA Length: 8.15"   Wt.: 41 oz.

Sights: fiber-optic front, Novak rear   Capacity: 19   Magazine: detachable box

Grips: synthetic

D2C:	NIB	$ 535	Ex	$ 425	VG+	$ 295	Good	$ 270	LMP $ 595
C2C:	Mint	$ 520	Ex	$ 375	VG+	$ 270	Good	$ 245	Fair $ 125
Trade-In:	Mint	$ 380	Ex	$ 300	VG+	$ 210	Good	$ 190	Poor $ 60

CZ 75 SP-01 TACTICAL
NEW IN BOX

CZ 75 SP-01 TACTICAL REDBOOK CODE: RB-CZ-H-75SPTC

Similar to the P-01. Features a steel frame, frame-mounted safety, firing pin block, accessory rail, and ambidextrous decocker. A variant of the CZ 75 Special-01. Pricing below reflects .9mm model. Models chambered for .40 caliber are about $80 more and hold 12 rounds.

Production: current   Caliber: .40 S&W, 9mmAction: DA/SA, semi-auto

Barrel Length: 4.6" OA Length: 8.15"   Wt.: 38 oz.   Sights: fixed night sights

Capacity: 18   Magazine: detachable box   Grips: rubber

D2C:	NIB	$ 620	Ex $ 475	VG+ $ 345	Good $ 310	LMP $ 680
C2C:	Mint	$ 600	Ex $ 450	VG+ $ 310	Good $ 280	Fair $ 145
Trade-In:	Mint	$ 450	Ex $ 350	VG+ $ 250	Good $ 220	Poor $ 90

CZ 75 SP-01
NEW IN BOX

CZ 75 SP-01 REDBOOK CODE: RB-CZ-H-75SP01

Similar to the P-01. Features a steel frame, frame-mounted safety, firing pin block, and accessory rail. Unlike the SP-01 Tactical, it doesn't feature an ambidextrous decocker device.

Production: current   Caliber: 9mm   Action: DA/SA, semi-auto

Barrel Length: 4.6" OA Length: 8.15"   Wt.: 38 oz.   Sights: fixed night sights

Capacity: 18   Magazine: detachable box   Grips: rubber

D2C:	NIB	$ 635	Ex $ 500	VG+ $ 350	Good $ 320	LMP $ 680
C2C:	Mint	$ 610	Ex $ 450	VG+ $ 320	Good $ 290	Fair $ 150
Trade-In:	Mint	$ 460	Ex $ 375	VG+ $ 250	Good $ 225	Poor $ 90

CZ 75 TACTICAL SPORT
NEW IN BOX

CZ 75 TACTICAL SPORT REDBOOK CODE: RB-CZ-H-75TACS

Ambidextrous design, extended magazine release, frame-mounted safety, matte stainless frame, blue slide, designed for competition. Pricing below reflects .40 model. 9mm models hold 20 rounds.

Production: 2005 - current   Caliber: .40 S&W, 9mmAction: SA, semi-auto

Barrel Length: 5.4"   OA Length: 8.9"   Wt: 45 oz.   Sights: fixed competition

Capacity: 16, 20   Magazine: detachable box   Grips: wood, checkered

D2C:	NIB	$1,195	Ex $ 925	VG+ $ 660	Good $ 600	LMP $1,310
C2C:	Mint	$1,150	Ex $ 825	VG+ $ 600	Good $ 540	Fair $ 275
Trade-In:	Mint	$ 850	Ex $ 675	VG+ $ 470	Good $ 420	Poor $ 120

CZ 75 SP-01 SHADOW REDBOOK CODE: RB-CZ-H-75SHAD

The first Shadow model, an updated variant of the SP-01. Features an extended magazine, drop free magazine break, black finish, custom weight springs, and frame-mounted safety. Designed for target and competition shooting.

Production: discontinued in 2013 Caliber: 9mm Action: DA/SA, semi-auto

Barrel Length: 4.7" Wt.: 41 oz. Sights: fiber-optic front Capacity: 18 Grips: synthetic

D2C:	NIB	$1,000	Ex $ 775	VG+ $ 550	Good $ 500	LMP $1,053
C2C:	Mint	$ 960	Ex $ 700	VG+ $ 500	Good $ 450	Fair $ 230
Trade-In:	Mint	$ 710	Ex $ 575	VG+ $ 390	Good $ 350	Poor $ 120

CZ 75 SHADOW T
NEW IN BOX

CZ 75 SHADOW T REDBOOK CODE: RB-CZ-H-75SHAT

Similar to the CZ 75 SP-01 Shadow, but features rubber grips and a fully-adjustable rear sight. Single-action only trigger.

Production: discontinued in 2013 Caliber: 9mm Action: DA/SA, semi-auto

Barrel Length: 4.7" Wt.: 41 oz. Sights: fiber-optic front, fully-adjustable rear

Capacity: 18   Magazine: detachable box   Grips: rubber

D2C:	NIB	$1,010	Ex $ 775	VG+ $ 560	Good $ 505	LMP $1,180
C2C:	Mint	$ 970	Ex $ 700	VG+ $ 510	Good $ 455	Fair $ 235
Trade-In:	Mint	$ 720	Ex $ 575	VG+ $ 400	Good $ 355	Poor $ 120

CZ 75 SHADOW T-SA REDBOOK CODE: RB-CZ-H-75STSA
Produced by the CZ Custom Shop for competition shooting. The same as the Shadow T, but features a single-action firing mechanism.

Production: discontinued in 2012		Caliber: 9mm		Action: SA, semi-auto		
Barrel Length: 4.7"	Sights: adjustable		Capacity: 18	Grips: rubber		
D2C:	NIB $1,010	Ex $ 775	VG+ $ 560	Good $ 505		
C2C:	Mint $ 970	Ex $ 700	VG+ $ 510	Good $ 455	Fair $ 235	
Trade-In:	Mint $ 720	Ex $ 575	VG+ $ 400	Good $ 355	Poor $ 120	

**CZ 75 SP-01
SHADOW TARGET**
NEW IN BOX

CZ 75 SP-01 SHADOW TARGET REDBOOK CODE: RB-CZ-H-SPCSHT
Similar to the SP-01 Shadow, but features an enhanced rear sight, aluminum grips, enhanced springs, and a modified guide rod.

Production: current		Caliber: 9mm	Action: DA/SA, semi-auto		Barrel Length: 4.6"
OA Length: 8.15"	Wt.: 38 oz.	Sights: competition style, fiber-optic front, adjustable rear			
Capacity: 18	Magazine: detachable box	Grips: aluminum			
D2C:	NIB $1,200	Ex $ 925	VG+ $ 660	Good $ 600	LMP $ 1,361
C2C:	Mint $1,160	Ex $ 850	VG+ $ 600	Good $ 540	Fair $ 280
Trade-In:	Mint $ 860	Ex $ 675	VG+ $ 470	Good $ 420	Poor $ 120

CZ 75 SHADOW SAO
NEW IN BOX

CZ 75 SHADOW SAO REDBOOK CODE: RB-CZ-H-75SHDW
Very similar to the SP-01 Shadow, but only operates as a single-action. Features a steel frame, frame-mounted safety, and enhanced sear.

Production: discontinued in 2013		Caliber: 9mm	Action: SA, semi-auto		
Barrel Length: 4.6"	OA Length: 8.1"Wt.: 35 oz.				
Sights: fiber-optic front	Capacity: 16	Magazine: detachable box	Grips: black synthetic		
D2C:	NIB $ 955	Ex $ 750	VG+ $ 530	Good $ 480	LMP $ 979
C2C:	Mint $ 920	Ex $ 675	VG+ $ 480	Good $ 430	Fair $ 220
Trade-In:	Mint $ 680	Ex $ 550	VG+ $ 380	Good $ 335	Poor $ 120

**CZ 75 SPECIAL-01
ACCU-SHADOW**
NEW IN BOX

CZ 75 SPECIAL-01 ACCU-SHADOW REDBOOK CODE: RB-CZ-H-75SPAC
Manufactured by the CZ Custom Shop. Features a steel frame, frame-mounted/ ambidextrous safety, modern design, and black finish. Barrel bushing threaded into slide and fitted to the barrel.

Production: current		Caliber: 9mm	Action: DA/SA, semi-auto		
Barrel Length: 4.6"	OA Length: 8.1"	Wt.: 38 oz.	Sights: fiber-optic front,		
HAJO rear	Capacity: 18	Magazine: detachable box	Grips: thin aluminum, black		
D2C:	NIB $1,560	Ex $1,200	VG+ $ 860	Good $ 780	LMP $1,665
C2C:	Mint $1,500	Ex $1,100	VG+ $ 780	Good $ 705	Fair $ 360
Trade-In:	Mint $ 1,110	Ex $ 875	VG+ $ 610	Good $ 550	Poor $ 180

**CZ 75 SHADOW
CTS LS-P**
NEW IN BOX

CZ 75 SHADOW CTS LS-P REDBOOK CODE: RB-CZ-H-75SHLS
Like the CZ 75 SP-01 Shadow, but features an elongated slide. Polymer guide rod.

Production: 2012 - current		Caliber: 9mm	Action: DA/SA, semi-auto		
Barrel Length: 5.4"	Sights: fiber-optic front, adjustable rear				
Capacity: 18	Magazine: detachable box	Grips: rubber			
D2C:	NIB $1,400	Ex $1,075	VG+ $ 770	Good $ 700	LMP $1,400
C2C:	Mint $1,350	Ex $ 975	VG+ $ 700	Good $ 630	Fair $ 325
Trade-In:	Mint $1,000	Ex $ 800	VG+ $ 550	Good $ 490	Poor $ 150

CZ 75 COMPACT SDP   REDBOOK CODE: RB-CZ-H-75CSDP

A smaller variant of the standard CZ 75. Built by the CZ Custom Shop and designed specifically for concealed carry.

Production: current Caliber: 9mm Action: DA/SA, semi-auto									
Barrel Length: 3.7" OA Length: 7.2" Wt.: 28.48 oz.									
Sights: tritium Capacity: 14 Magazine: detachable box Grips: aluminum									
D2C:	NIB	$1,230	Ex	$ 950	VG+	$ 680	Good	$ 615	LMP	$ 1,420
C2C:	Mint	$1,190	Ex	$ 850	VG+	$ 620	Good	$ 555	Fair	$ 285
Trade-In:	Mint	$ 880	Ex	$ 700	VG+	$ 480	Good	$ 435	Poor	$ 150

CZ 83   REDBOOK CODE: RB-CZ-H-83XXXX

A popular CZ pistol, the civilian version of the CZ 82. Features a compact design, blue or nickel finish, notable weight to decrease felt recoil, and frame-mounted safety.

CZ 83
NEW IN BOX

Production: 1986 - 2012 Caliber: .32 ACP, 9mm, .380 ACP Action: DA/SA, semi-auto									
Barrel Length: 3.8" OA Length: 6.8" Wt.: 29 oz. Sights: fixed, 3-dot									
Capacity: 12, 15 Magazine: detachable box Grips: black plastic									
D2C:	NIB	$ 415	Ex	$ 325	VG+	$ 230	Good	$ 210	LMP	$ 495
C2C:	Mint	$ 400	Ex	$ 300	VG+	$ 210	Good	$ 190	Fair	$ 100
Trade-In:	Mint	$ 300	Ex	$ 250	VG+	$ 170	Good	$ 150	Poor	$ 60

CZ 85 B   REDBOOK CODE: RB-CZ-H-85XXXX

An updated variant of the CZ 75. Features an ambidextrous design, firing pin block, frame-mounted safety, steel frame, black finish, and modified firing mechanisms, making it more reliable.

CZ 85 B
NEW IN BOX

Production: current Caliber: 9mm Action: DA/SA, semi-auto Barrel Length: 4.6"									
OA Length: 8.1" Wt.: 34 oz. Sights: fixed Capacity: 16									
Magazine: detachable box Grips: synthetic									
D2C:	NIB	$ 525	Ex	$ 400	VG+	$ 290	Good	$ 265	LMP	$ 628
C2C:	Mint	$ 510	Ex	$ 375	VG+	$ 270	Good	$ 240	Fair	$ 125
Trade-In:	Mint	$ 380	Ex	$ 300	VG+	$ 210	Good	$ 185	Poor	$ 60

CZ 85 COMBAT   REDBOOK CODE: RB-CZ-H-85CMBT

A variant of the CZ 85, but distinguished for its lack of a firing pin block. Steel frame, fully adjustable rear sight.

CZ 85 COMBAT
NEW IN BOX

Production: current Caliber: 9mm Action: DA/SA, semi-auto									
Barrel Length: 4.6" OA Length: 8.1" Wt.: 35 oz. Sights: notch front,									
adjustable rear Capacity: 16 Magazine: detachable box Grips: black synthetic									
D2C:	NIB	$ 545	Ex	$ 425	VG+	$ 300	Good	$ 275	LMP	$ 664
C2C:	Mint	$ 530	Ex	$ 400	VG+	$ 280	Good	$ 250	Fair	$ 130
Trade-In:	Mint	$ 390	Ex	$ 325	VG+	$ 220	Good	$ 195	Poor	$ 60

CZ 97 B   REDBOOK CODE: RB-CZ-H-97BXXX

A variant of the CZ 75, noted for firing .45 ACP rounds. Features a frame-mounted safety, chamber indicator, double-stack magazine, and improved ergonomics.

CZ 97 B
NEW IN BOX

Production: 1998 - current Caliber: .45 ACP Action: DA/SA, semi-auto									
Barrel Length: 4.8" Wt.: 41 oz. Sights: fiber-optic front Capacity: 10									
Magazine: detachable box Grips: thin aluminum or wood									
D2C:	NIB	$ 615	Ex	$ 475	VG+	$ 340	Good	$ 310	LMP	$ 707
C2C:	Mint	$ 600	Ex	$ 425	VG+	$ 310	Good	$ 280	Fair	$ 145
Trade-In:	Mint	$ 440	Ex	$ 350	VG+	$ 240	Good	$ 220	Poor	$ 90

CZ 97 BD   REDBOOK CODE: RB-CZ-H-97BDXX

Similar to the CZ 97 B, but distinguished by its decocker mechanism. First .45 caliber made by CZ.

CZ 97 BD
NEW IN BOX

Production: current   Caliber: .45 ACP   Action: DA/SA, semi-auto

Barrel Length: 4.8"   OA Length: 8.3"Wt.: 41 oz.   Sights: 3-dot tritium

Capacity: 10   Magazine: detachable box   Grips: black rubber

D2C:	NIB	$ 725	Ex $ 575	VG+ $ 400	Good $ 365	LMP $ 874
C2C:	Mint	$ 700	Ex $ 525	VG+ $ 370	Good $ 330	Fair $ 170
Trade-In:	Mint	$ 520	Ex $ 425	VG+ $ 290	Good $ 255	Poor $ 90

CZ P-09 DUTY   REDBOOK CODE: RB-CZ-H-75P09D

A full-size variant of the CZ 75 P-07 Duty. Also available in 9mm with 19 rounds. Flat Dark Earth frame available.

CZ P-09 DUTY
NEW IN BOX

Production: current   Caliber: .40 S&W   Action: DA/SA, semi-auto

Barrel Length: 4.5"   OA Length: 8.1"   Wt.: 27 oz.   Sights: fixed, 3-dot   Capacity: 15

Magazine: detachable box   Grips: polymer

D2C:	NIB	$ 490	Ex $ 375	VG+ $ 270	Good $ 245	LMP $ 514
C2C:	Mint	$ 480	Ex $ 350	VG+ $ 250	Good $ 225	Fair $ 115
Trade-In:	Mint	$ 350	Ex $ 275	VG+ $ 200	Good $ 175	Poor $ 60

CZ 100   REDBOOK CODE: RB-CZ-H-100CZX

Noted as one of the first CZ pistols to feature mostly polymer-based components. Includes a firing pin block, tactical rail, and matte black finish.

CZ 100
EXCELLENT

Production: 1995 - current   Caliber: 9mm, .40 S&W   Action: DAO, semi-auto

Barrel Length: 3.1"   Wt.: 23 oz.   Sights: 3-dot   Capacity: 12, 15

Magazine: detachable box   Grips: black synthetic

D2C:	NIB	$ 385	Ex $ 300	VG+ $ 215	Good $ 195	LMP $ 450
C2C:	Mint	$ 370	Ex $ 275	VG+ $ 200	Good $ 175	Fair $ 90
Trade-In:	Mint	$ 280	Ex $ 225	VG+ $ 160	Good $ 135	Poor $ 60

CZ 2075 RAMI   REDBOOK CODE: RB-CZ-H-2075RP

Similar to the CZ 75, but includes a polymer frame and more compact design. Features a frame-mounted safety, black finish, and aluminum-alloy frame.

Production: 2007 - current   Caliber: .40 S&W, 9mm   Action: DA/SA, semi-auto

Barrel Length: 3"   OA Length: 6.5"   Wt.: 25 oz.   Sights: fixed

Capacity: 7, 9, 10   Magazine: detachable box   Grips: rubber

D2C:	NIB	$ 545	Ex $ 425	VG+ $ 300	Good $ 275	LMP $ 633
C2C:	Mint	$ 530	Ex $ 400	VG+ $ 280	Good $ 250	Fair $ 130
Trade-In:	Mint	$ 390	Ex $ 325	VG+ $ 220	Good $ 195	Poor $ 60

CZ 2075 RAMI BD   REDBOOK CODE: RB-CZ-H-2075RB

Similar to the CZ 2075 Rami, but with decocker mechanism. Tritium 3-dot combat sights. Features aluminum-alloy frame, compact design, frame-mounted safety. Black finish.

CZ 2075 RAMI BD
NEW IN BOX

Production: 2009 - current   Caliber: 9mm   Action: DA/SA, semi-auto

Barrel Length: 3"   OA Length: 6.5"   Wt.: 24 oz.   Sights: 3-dot tritium

Capacity: 14   Magazine: detachable box   Grips: rubber

D2C:	NIB	$ 625	Ex $ 475	VG+ $ 345	Good $ 315	LMP $ 747
C2C:	Mint	$ 600	Ex $ 450	VG+ $ 320	Good $ 285	Fair $ 145
Trade-In:	Mint	$ 450	Ex $ 350	VG+ $ 250	Good $ 220	Poor $ 90

Dan Wesson

A subsidiary of C.Z.-U.S.A., Dan Wesson Firearms specializes in manufacturing M1911 pistols, which they offer in a wide variety. The company has existed in several incarnations over the years, most owned by D.B. Wesson or the Wesson family, but the company, as it currently stands, began operations in 2005 as a part of C.Z. Shooters have noted Dan Wesson M1911s for their accuracy and sturdy construction. The company is currently headquartered in Norwich, New York.

MODEL 715 SMALL FRAME REVOLVER
NEW IN BOX

MODEL 715 SMALL FRAME REVOLVER

REDBOOK CODE: RB-DW-H-M715SF

Interchangeable front sight, adjustable rear, interchangeable barrels, stainless steel frame and finish. New issue projected to be released in 2014.

Production: discontinued Caliber: .357 Magnum/.38 Special Action: DA/SA, revolver

Barrel Length: 2.5", 4", 5.6", 8", 10" Sights: adjustable, interchangeable

Capacity: 6 Grips: Hogue finger grip rubber

D2C:	NIB	$1,060	Ex	$ 825	VG+	$ 585	Good	$ 530		
C2C:	Mint	$1,020	Ex	$ 750	VG+	$ 530	Good	$ 480	Fair	$ 245
Trade-In:	Mint	$ 760	Ex	$ 600	VG+	$ 420	Good	$ 375	Poor	$ 120

CCO
NEW IN BOX

CCO REDBOOK CODE: RB-DW-H-CCOPIS

Compact, forged-aluminum frame, match barrel, undercut trigger guard, black ceramic coat.

Production: 2009 - current Caliber: .45 ACP Action: SA, semi-auto

Barrel Length: 4.25" OA Length: 8" Wt.: 26 oz. Sights: Novak-style night sights

Capacity: 7 Magazine: detachable box Grips: Stipple Shadow

D2C:	NIB	$ 1,385	Ex	$ 1,075	VG+	$ 765	Good	$ 695	LMP	$ 1,558
C2C:	Mint	$ 1,330	Ex	$ 975	VG+	$ 700	Good	$ 625	Fair	$ 320
Trade-In:	Mint	$ 990	Ex	$ 800	VG+	$ 550	Good	$ 485	Poor	$ 150

COMMANDER CLASSIC BOBTAIL
EXCELLENT

COMMANDER CLASSIC BOBTAIL REDBOOK CODE: RB-DW-H-CCBOBT

1911 pistol, stainless frame and forged stainless steel slide, grip safety, internal extractor, beveled magwell. Both calibers similar in price.

Production: 2005 - 2010 Caliber: .45 ACP, 10mm Action: SA, semi-auto

Barrel Length: 4.25" OA Length: 8" Wt.: 34 oz. Sights: fixed 3-dot tritium night sights Capacity: 8 Grips: diamond checkered cocobolo

D2C:	NIB	$1,560	Ex	$1,200	VG+	$ 860	Good	$ 780	LMP	$1,688
C2C:	Mint	$1,500	Ex	$1,100	VG+	$ 780	Good	$ 705	Fair	$ 360
Trade-In:	Mint	$ 1,110	Ex	$ 875	VG+	$ 610	Good	$ 550	Poor	$ 180

ECO
NEW IN BOX

ECO REDBOOK CODE: RB-DW-H-ECOPIS

Compact, black finish, high-ride grip safety. Anodized aluminum frame.

Production: 2012 - current Caliber: .45 ACP, 9mm Action: SA, semi-auto

Barrel Length: 3.5" OA Length: 7.25" Wt.: 25 oz. Sights: fixed night Capacity: 7 Grips: G10

D2C:	NIB	$1,560	Ex	$1,200	VG+	$ 860	Good	$ 780	LMP	$1,662
C2C:	Mint	$1,500	Ex	$1,100	VG+	$ 780	Good	$ 705	Fair	$ 360
Trade-In:	Mint	$ 1,110	Ex	$ 875	VG+	$ 610	Good	$ 550	Poor	$ 180

**ELITE SERIES
HAVOC**
NEW IN BOX

ELITE SERIES HAVOC REDBOOK CODE: RB-DW-H-ELIHAV

Competition pistol, black polycoat receiver finish, skeletonized trigger, ambidextrous thumb safety.

Production: 2011 - current Caliber: .38 Super, 9mm Action: SA, semi-auto

Barrel Length: 4.25" OA Length: 10.75" Wt.: 35 oz. Sights: C-More

Capacity: 21 Magazine: detachable box Grips: G10

D2C:	NIB	$ 3,990	Ex	$ 3,050	VG+	$ 2,195	Good	$ 1,995	LMP	$ 4,299
C2C:	Mint	$ 3,840	Ex	$ 2,775	VG+	$2,000	Good	$ 1,800	Fair	$ 920
Trade-In:	Mint	$ 2,840	Ex	$ 2,250	VG+	$ 1,560	Good	$ 1,400	Poor	$ 420

ELITE SERIES MAYHEM REDBOOK CODE: RB-DW-H-ELIMAY

Competition pistol, black polycoat receiver finish, skeletonized trigger, ambidextrous thumb safety, designed for USPSA Limited Division.

Production: 2011 - current Caliber: .40 S&W Action: SA, semi-auto

Barrel Length: 6" OA Length: 10" Wt.: 51 oz. Sights: adjustable

Capacity: 17 Magazine: detachable box Grips: G10

D2C:	NIB	$3,715	Ex	$2,825	VG+	$2,045	Good	$1,860	LMP	$3,899
C2C:	Mint	$3,570	Ex	$2,575	VG+	$1,860	Good	$1,675	Fair	$ 855
Trade-In:	Mint	$2,640	Ex	$2,100	VG+	$1,450	Good	$1,305	Poor	$ 390

ELITE SERIES TITAN REDBOOK CODE: RB-DW-H-ELITIT

Full-size, black polycoat receiver finish, skeletonized trigger, steel frame, bull barrel, tactical rail.

Production: 2011 - current Caliber: 10mm Action: SA, semi-auto

Barrel Length: 5" OA Length: 8.75" Sights: tritium adjustable

Capacity: 14 Magazine: detachable box Grips: G10

D2C:	NIB	$3,715	Ex	$2,825	VG+	$2,045	Good	$1,860	LMP	$3,829
C2C:	Mint	$3,570	Ex	$2,575	VG+	$1,860	Good	$1,675	Fair	$ 855
Trade-In:	Mint	$2,640	Ex	$2,100	VG+	$1,450	Good	$1,305	Poor	$ 390

GUARDIAN
NEW IN BOX

GUARDIAN REDBOOK CODE: RB-DW-H-GUARDN

Commander-size 1911, black anodized receiver finish, aluminum bobtail frame, manual thumb safety. For .45 caliber models, add $60.

Production: current Caliber: .45 ACP, .38 Super, 9mm Action: SA, semi-auto

Barrel Length: 4.25" OA Length: 8" Wt.: 29 oz. Sights: fixed 3-dot tritium night

Capacity: 8, 9 Magazine: detachable box Grips: wood

D2C:	NIB	$ 1,460	Ex	$ 1,125	VG+	$ 805	Good	$ 730	LMP	$ 1,530
C2C:	Mint	$ 1,410	Ex	$ 1,025	VG+	$ 730	Good	$ 660	Fair	$ 340
Trade-In:	Mint	$ 1,040	Ex	$ 825	VG+	$ 570	Good	$ 515	Poor	$ 150

CHAOS REDBOOK CODE: RB-DW-H-CHAOSX

Competition-style pistol, grip safety, 1911 frame, matte black finish.

Production: 2014 - current Caliber: 9mm Action: SA, semi-auto

Barrel Length: 5" OA Length: 8.75" Wt.: 51 oz.

Sights: adjustable fiber-optic Capacity: 21 Magazine: detachable box Grips: composite

D2C:	NIB	$ 3,829	Ex	$ 2,925	VG+	$ 2,110	Good	$ 1,915	LMP	$3,829
C2C:	Mint	$ 3,680	Ex	$ 2,650	VG+	$ 1,920	Good	$ 1,725	Fair	$ 885
Trade-In:	Mint	$ 2,720	Ex	$ 2,150	VG+	$ 1,500	Good	$ 1,345	Poor	$ 390

MARKSMAN REDBOOK CODE: RB-DW-H-MARKSM
Stainless frame, adjustable sights, grip safety.

Production: 2009 - discontinued Caliber: .45 ACP Action: SA, semi-auto

Barrel Length: 5" OA Length: 8.75" Wt.: 38 oz.

Sights: adjustable Capacity: 7 Magazine: detachable box Grips: cocobolo

D2C:	NIB	$1,315	Ex	$1,000	VG+	$ 725	Good	$ 660		
C2C:	Mint	$1,270	Ex	$ 925	VG+	$ 660	Good	$ 595	Fair	$ 305
Trade-In:	Mint	$ 940	Ex	$ 750	VG+	$ 520	Good	$ 465	Poor	$ 150

POINTMAN SEVEN REDBOOK CODE: RB-DW-H-PNT7XX
Stainless steel frame, forged stainless slide, internal extractor, match grade barrel, beveled magwell, adjustable sights, Commander-style match hammer.

Production: discontinued Caliber: .45 ACP Action: SA, semi-auto Barrel

Length: 5" OA Length: 8.75" Wt.: 38 oz. Sights: adjustable Capacity: 7

Magazine: detachable box Grips: diamond-checkered cocobolo

D2C:	NIB	$1,480	Ex	$1,125	VG+	$ 815	Good	$ 740	LMP	$1,597
C2C:	Mint	$1,430	Ex	$1,025	VG+	$ 740	Good	$ 670	Fair	$ 345
Trade-In:	Mint	$1,060	Ex	$ 850	VG+	$ 580	Good	$ 520	Poor	$ 150

POINTMAN NINE REDBOOK CODE: RB-DW-H-PNT9XX
Full-size, limited production, stainless finish, forged frame, match barrel, red fiber-optic front and adjustable rear sight, grip safety.

Production: current Caliber: 9mm Action: SA, semi-auto Barrel Length: 5"

OA Length: 8.75" Wt.: 39 oz. Sights: adjustable fiber-optic Capacity: 9

Magazine: detachable box Grips: double-diamond cocobolo

D2C:	NIB	$1,465	Ex	$1,125	VG+	$ 810	Good	$ 735	LMP	$1,558
C2C:	Mint	$1,410	Ex	$1,025	VG+	$ 740	Good	$ 660	Fair	$ 340
Trade-In:	Mint	$1,050	Ex	$ 825	VG+	$ 580	Good	$ 515	Poor	$ 150

POINTMAN MINOR REDBOOK CODE: RB-DW-H-PNTMIN
Blue finish slide and frame, interchangeable sights, beveled magwell.

Production: early 2000s - discontinued Caliber: .45 ACP Action: SA, semi-auto

Barrel Length: 5" Sights: adjustable, interchangeable Capacity: 8

Magazine: detachable box Grips: rosewood checkered

D2C:	NIB	$ 660	Ex	$ 525	VG+	$ 365	Good	$ 330		
C2C:	Mint	$ 640	Ex	$ 475	VG+	$ 330	Good	$ 300	Fair	$ 155
Trade-In:	Mint	$ 470	Ex	$ 375	VG+	$ 260	Good	$ 235	Poor	$ 90

POINTMAN MAJOR REDBOOK CODE: RB-DW-H-PNTMJR
Match grade barrel, beveled magwell, blue or stainless steel finish, front and rear slide serrations.

Production: 2000 - discontinued Caliber: .45 ACP Action: SA, semi-auto

Barrel Length: 5" Sights: interchangeable front, adjustable rear

Capacity: 8 Magazine: detachable box Grips: rosewood checkered

D2C:	NIB	$1,225	Ex	$ 950	VG+	$ 675	Good	$ 615		
C2C:	Mint	$1,180	Ex	$ 850	VG+	$ 620	Good	$ 555	Fair	$ 285
Trade-In:	Mint	$ 870	Ex	$ 700	VG+	$ 480	Good	$ 430	Poor	$ 150

POINTMAN HI-CAP REDBOOK CODE: RB-DW-H-PNTHIC
Blue finish, wide-frame design, extended slide release and thumb safety.

Production: 2001 - discontinued   Caliber: .45 ACP   Action: SA, semi-auto

Barrel Length: 5"   Sights: fixed rear target   Capacity: 10

Magazine: detachable box   Grips: black rubber

D2C:	NIB	$ 670	Ex $ 525	VG+ $ 370	Good $ 335			
C2C:	Mint	$ 650	Ex $ 475	VG+ $ 340	Good $ 305	Fair $ 155
Trade-In:	Mint	$ 480	Ex $ 400	VG+ $ 270	Good $ 235	Poor $ 90

POINTMAN GUARDIAN REDBOOK CODE: RB-DW-H-PNTGUA
Blue or stainless finish, choice of fixed or adjustable sights, and alloy frame. Designed for concealed carry.

Production: 2000 - discontinued   Caliber: .45 ACP   Action: SA, semi-auto

Barrel Length: 4.25"   Wt: 29 oz.   Sights: adjustable or fixed   Capacity: 8

Magazine: detachable box   Grips: black rubber

D2C:	NIB	$ 780	Ex $ 600	VG+ $ 430	Good $ 390			
C2C:	Mint	$ 750	Ex $ 550	VG+ $ 390	Good $ 355	Fair $ 180
Trade-In:	Mint	$ 560	Ex $ 450	VG+ $ 310	Good $ 275	Poor $ 90

RZ-10 RAZORBACK
NEW IN BOX
Courtesy of Bud's Gun Shop

RZ-10 REDBOOK CODE: RB-DW-H-RZ10RA
1911 full-size pistol, stainless frame, grip safety, fixed sights, undercut trigger guard.

Production: current   Caliber: 10mm   Action: SA, semi-auto   Barrel Length: 5"

OA Length: 8.75"   Wt.: 38 oz.   Sights: fixed   Capacity: 8

Magazine: detachable box   Grips: cocobolo

D2C:	NIB	$1,240	Ex $ 950	VG+ $ 685	Good $ 620	LMP $1,350
C2C:	Mint	$1,200	Ex $ 875	VG+ $ 620	Good $ 560	Fair $ 290
Trade-In:	Mint	$ 890	Ex $ 700	VG+ $ 490	Good $ 435	Poor $ 150

RZ-45 HERITAGE
EXCELLENT

RZ-45 HERITAGE REDBOOK CODE: RB-DW-H-RZ45HE
1911 full-size pistol, forged stainless steel frame, match barrel.

Production: 2009 - current   Caliber: .45 ACP   Action: SA, semi-auto

Barrel Length: 5"   OA Length: 8.75"   Wt.: 39 oz.   Sights: tactical fixed night sights

Capacity: 8   Magazine: detachable box   Grips: double diamond rubber

D2C:	NIB	$1,180	Ex $ 900	VG+ $ 650	Good $ 590	LMP $1,298
C2C:	Mint	$1,140	Ex $ 825	VG+ $ 590	Good $ 535	Fair $ 275
Trade-In:	Mint	$ 840	Ex $ 675	VG+ $ 470	Good $ 415	Poor $ 120

SPECIALIST
NEW IN BOX

SPECIALIST REDBOOK CODE: RB-DW-H-SPECIA
1911 full-size pistol, tactical rail frame, undercut trigger guard, detachable two-piece magwell, offered in matte black duty or stainless finish.

Production: 2012 - current   Caliber: .45 ACP   Action: SA, semi-auto

Barrel Length: 5"   Sights: white target ring front, tritium rear

Capacity: 8   Magazine: detachable box   Grips: composite

D2C:	NIB	$1,730	Ex $1,325	VG+ $ 955	Good $ 865	LMP $1,870
C2C:	Mint	$1,670	Ex $1,200	VG+ $ 870	Good $ 780	Fair $ 400
Trade-In:	Mint	$1,230	Ex $ 975	VG+ $ 680	Good $ 610	Poor $ 180

SS CUSTOM
NEW IN BOX

Courtesy of Bud's Gun Shop

VALOR
NEW IN BOX

VALOR
NEW IN BOX

SS CUSTOM   REDBOOK CODE: RB-DW-H-SSCUST
Stainless frame, ambidextrous thumb safety, magwell.

Production: 2008 - 2010   Caliber: .40 S&W   Action: SA, semi-auto

Barrel Length: 5"   OA Length: 8.75"   Wt.: 38 oz.

Sights: fiber-optic   Capacity: 9   Magazine: detachable box   Grips: Shark Skin

D2C:	NIB	$1,389	Ex	$1,075	VG+	$ 765	Good	$ 695	LMP	$ 1,389
C2C:	Mint	$1,340	Ex	$ 975	VG+	$ 700	Good	$ 630	Fair	$ 280
Trade-In:	Mint	$ 990	Ex	$ 800	VG+	$ 550	Good	$ 490	Poor	$ 150

VALOR   REDBOOK CODE: RB-DW-H-VALORX
1911 full-size pistol, forged frame, available in black or stainless finish, match barrel, grip safety, undercut trigger guard.

Production: 2008 - current   Caliber: .45 ACP   Action: SA, semi-auto

Barrel Length: 5"   OA Length: 8.75"   Wt.: 38 oz.   Sights: Heinie Ledge

Straight Eight night   Capacity: 8   Magazine: detachable box   Grips: Slim Line G10

D2C:	NIB	$ 1,495	Ex	$ 1,150	VG+	$ 825	Good	$ 750	LMP	$ 2,012
C2C:	Mint	$ 1,440	Ex	$ 1,050	VG+	$ 750	Good	$ 675	Fair	$ 345
Trade-In:	Mint	$ 1,070	Ex	$ 850	VG+	$ 590	Good	$ 525	Poor	$ 150

VALOR BOBTAIL COMMANDER "V-BOB"
REDBOOK CODE: RB-DW-H-VALORB
Compact design, black duty or matte stainless finish, forged stainless steel frame, grip safety, match barrel.

Production: 2010 - current   Caliber: .45 ACP   Action: SA, semi-auto

Barrel Length: 4.25"   OA Length: 8"   Wt.: 35 oz.   Sights: Heinie Ledge

Straight Eight night   Capacity: 8   Magazine: detachable box   Grips: Slim Line G10

D2C:	NIB	$ 1,610	Ex	$ 1,225	VG+	$ 890	Good	$ 805	LMP	$ 2,077
C2C:	Mint	$ 1,550	Ex	$ 1,125	VG+	$ 810	Good	$ 725	Fair	$ 375
Trade-In:	Mint	$ 1,150	Ex	$ 925	VG+	$ 630	Good	$ 565	Poor	$ 180

PATRIOT MARKSMAN   REDBOOK CODE: RB-DW-H-PAMARK
M1911 style, beveled magwell, flared ejection port, skeletonized trigger, grip and thumb safety.

Production: 2002 - discontinued   Caliber: .45 ACP   Action: SA, semi-auto

Barrel Length: 5"   Wt.: 38 oz.   Sights: Novak   Capacity: 8

Magazine: detachable box   Grips: checkered wood

D2C:	NIB	$ 1,300	Ex	$ 1,000	VG+	$ 715	Good	$ 650		
C2C:	Mint	$ 1,250	Ex	$ 900	VG+	$ 650	Good	$ 585	Fair	$ 300
Trade-In:	Mint	$ 930	Ex	$ 750	VG+	$ 510	Good	$ 455	Poor	$ 150

PATRIOT EXPERT   REDBOOK CODE: RB-DW-H-PAEXPT
Distinguished by its rear target sights. Flared ejection port, beveled magwell, skeletonized trigger and hammer, grip and thumb safety.

Production: 2002 - discontinued   Caliber: .45 ACP   Action: SA, semi-auto

Barrel Length: 5"   Wt.: 38 oz.   Sights: target-style adjustable

Capacity: 8   Magazine: detachable box   Grips: checkered cocobolo

D2C:	NIB	$ 930	Ex	$ 725	VG+	$ 515	Good	$ 465		
C2C:	Mint	$ 900	Ex	$ 650	VG+	$ 470	Good	$ 420	Fair	$ 215
Trade-In:	Mint	$ 670	Ex	$ 525	VG+	$ 370	Good	$ 330	Poor	$ 120

DAN WESSON

Ed Brown Products, Inc.

Ed Brown Products, Inc. traces its origins to 1968 when Ed Brown first began manufacturing guns under the name "Brown's Gun Shop." After several decades of development, Ed Brown Products, Inc. formed and introduced a full line of high-end M1911 pistols into the market. Since then, the company's pistols have gained acclaim for their superior quality and design. The company, located in Perry, Missouri, sells directly to consumers or through a host of distributors.

CLASSIC CUSTOM
NEW IN BOX

CLASSIC CUSTOM   REDBOOK CODE: RB-EB-H-CCXXXX

M1911-style design. Two-piece guide rod, mirror-finished slide, serrations on rear of the slide, glass-bead finish on frame, aluminum trigger, frame-mounted safety, stainless or blue finish.

Production: current Caliber: .45 ACP Action: SA, semi-auto Barrel Length: 5"

Wt.: 40 oz. Sights: cross dovetail front, adjustable rear Capacity: 7

Grips: double-diamond checkered laminate wood

	NIB	Ex	VG+	Good	LMP
D2C:	$ 3,415	$ 2,600	$1,880	Good $ 1,710	$ 3,495
C2C:	Mint $ 3,280	Ex $ 2,375	VG+ $ 1,710	Good $ 1,540	Fair $ 790
Trade-In:	Mint $ 2,430	Ex $ 1,925	VG+ $ 1,340	Good $ 1,200	Poor $ 360

CLASSIC CUSTOM ENHANCED
NEW IN BOX

CLASSIC CUSTOM ENHANCED   REDBOOK CODE: RB-EB-H-CCEXXX

M1911-style design. Like the Classic Custom, but noted for its unique engravings, square-cut serrations, and two-tone finish.

Production: current Caliber: .45 ACP Action: SA, semi-auto Barrel Length: 5"

Wt.: 40 oz. Sights: cross dovetail front, adjustable rear Capacity: 7

Grips: double-diamond checkered laminate

	NIB	Ex	VG+	Good	LMP
D2C:	$ 4,800	$ 3,650	$2,640	Good $ 2,400	$ 4,995
C2C:	Mint $ 4,610	Ex $ 3,325	VG+ $ 2,400	Good $ 2,160	Fair $ 1,105
Trade-In:	Mint $ 3,410	Ex $ 2,700	VG+ $ 1,880	Good $ 1,680	Poor $ 480

SIGNATURE EDITION
NEW IN BOX

SIGNATURE EDITION   REDBOOK CODE: RB-EB-H-SIGEDX

M1911-style design. Features serrations on the slide, a three-hole aluminum trigger, mirror-finished slide, two-piece guide rod, ornate engravings, and frame-mounted safety. Blue finish.

Production: current Caliber: .45 ACP Action: SA, semi-auto Barrel Length: 5"

Wt.: 40 oz. Sights: cross dovetail front, adjustable rear Capacity: 7

Grips: double-diamond checkered laminate

	NIB	Ex	VG+	Good	LMP
D2C:	$ 7,300	$ 5,550	$ 4,015	Good $ 3,650	$ 7,195
C2C:	Mint $ 7,010	Ex $ 5,050	VG+ $ 3,650	Good $ 3,285	Fair $ 1,680
Trade-In:	Mint $ 5,190	Ex $ 4,100	VG+ $ 2,850	Good $ 2,555	Poor $ 750

EXECUTIVE TARGET
NEW IN BOX

EXECUTIVE TARGET   REDBOOK CODE: RB-EB-H-ETXXXX

M1911-style design. Similar to the Executive Elite. Features a stainless or Gen 4 coating-finished frame and slide, three-hole aluminum trigger, recoil spring guide and plug, single-stack government frame, and frame-mounted safety.

Production: current Caliber: .45 ACP Action: SA, semi-auto Barrel Length: 5"

Wt.: 38 oz. Sights: cross dovetail front, adjustable rear Capacity: 7

Grips: double diamond checkered laminate wood

	NIB	Ex	VG+	Good	LMP
D2C:	$ 2,370	$ 1,825	$ 1,305	Good $ 1,185	$ 2,895
C2C:	Mint $ 2,280	Ex $ 1,650	VG+ $ 1,190	Good $ 1,070	Fair $ 550
Trade-In:	Mint $ 1,690	Ex $ 1,350	VG+ $ 930	Good $ 830	Poor $ 240

ED BROWN

ED BROWN

EXECUTIVE ELITE
NEW IN BOX

EXECUTIVE ELITE REDBOOK CODE: RB-EB-H-EEXXXX

M1911-style design. Features stainless or Gen 4 coating on frame and slide, square-cut cocking serrations on slide, three-hole aluminum trigger, recoil-spring guide and plug and frame-mounted safety.

Production: current   Caliber: .45 ACP   Action: SA, semi-auto   Barrel Length: 5"

Wt.: 38 oz.   Sights: fixed dovetail 3-dot night   Capacity: 7

Grips: double-diamond checkered laminate wood

D2C:	NIB	$ 2,250	Ex	$ 1,725	VG+	$ 1,240	Good	$ 1,125	LMP	$ 2,695
C2C:	Mint	$ 2,160	Ex	$ 1,575	VG+	$ 1,130	Good	$ 1,015	Fair	$ 520
Trade-In:	Mint	$ 1,600	Ex	$ 1,275	VG+	$ 880	Good	$ 790	Poor	$ 240

EXECUTIVE CARRY
NEW IN BOX

EXECUTIVE CARRY REDBOOK CODE: RB-EB-H-ECXXXX

M1911-style design. Features stainless or Gen III coating, a three-hole aluminum trigger, square-cut cocking serrations on the slide, recoil-spring guide and frame-mounted safety.

Production: current   Caliber: .45 ACP   Action: SA, semi-auto   Barrel Length: 4.25"

Wt.: 35 oz.   Sights: fixed, dovetail, 3-dot night   Capacity: 7

Grips: double-diamond checkered laminate wood

D2C:	NIB	$ 2,670	Ex	$ 2,050	VG+	$ 1,470	Good	$ 1,335	LMP	$ 2,945
C2C:	Mint	$ 2,570	Ex	$ 1,850	VG+	$ 1,340	Good	$ 1,205	Fair	$ 615
Trade-In:	Mint	$ 1,900	Ex	$ 1,500	VG+	$ 1,050	Good	$ 935	Poor	$ 270

KOBRA CARRY
NEW IN BOX

KOBRA CARRY REDBOOK CODE: RB-EB-H-KCXXXX

M1911-stye design. Features stainless finish, a three-hole aluminum trigger, compact design, snakeskin serrations on the slide and frame-mounted safety.

Production: current   Caliber: .45 ACP   Action: SA, semi-auto   Barrel Length: 4.25"

Wt.: 35 oz.   Sights: fixed dovetail 3-dot night sights   Capacity: 7

Grips: double-diamond checkered cocobolo

D2C:	NIB	$2,515	Ex	$1,925	VG+	$1,385	Good	$1,260	LMP	$2,745
C2C:	Mint	$2,420	Ex	$1,750	VG+	$1,260	Good	$1,135	Fair	$ 580
Trade-In:	Mint	$1,790	Ex	$1,425	VG+	$ 990	Good	$ 885	Poor	$ 270

SPECIAL FORCES
NEW IN BOX

SPECIAL FORCES REDBOOK CODE: RB-EB-H-SFXXXX

M1911-style design. Features a stainless steel or Gen 4 coating-finished frame, stainless or coated slide, available in stealth-grey with carbon fiber grips and tactical design.

Production: current   Caliber: .45 ACP   Action: SA, semi-auto

Barrel Length: 5"   Wt.: 38 oz.   Sights: fixed dovetail 3-dot night sights   Capacity: 7

Grips: black G10, black sand G10, carbon fiber

D2C:	NIB	$ 2,250	Ex	$ 1,725	VG+	$ 1,240	Good	$ 1,125	LMP	$ 2,495
C2C:	Mint	$ 2,160	Ex	$ 1,575	VG+	$ 1,130	Good	$ 1,015	Fair	$ 520
Trade-In:	Mint	$ 1,600	Ex	$ 1,275	VG+	$ 880	Good	$ 790	Poor	$ 240

JIM WILSON SPECIAL
NEW IN BOX

JIM WILSON SPECIAL REDBOOK CODE: RB-EB-H-JWSXXX

Sheriff Jim Wilson's signature engraved on the slide. Three-hole aluminum trigger, frame-mounted and grip safety, lowered and flared ejection port, matte finish on slide.

Production: 2007 - discontinued   Caliber: .45 ACP   Action: SA, semi-auto

Barrel Length: 5"   Wt.: 38 oz.   Sights: Novak front night sights, fixed dovetail rear

Capacity: 7   Grips: Tru-Ivory smooth grips

D2C:	NIB	$ 2,170	Ex	$ 1,650	VG+	$ 1,195	Good	$ 1,085	LMP	$ 2,295
C2C:	Mint	$2,090	Ex	$ 1,500	VG+	$ 1,090	Good	$ 980	Fair	$ 500
Trade-In:	Mint	$ 1,550	Ex	$ 1,225	VG+	$ 850	Good	$ 760	Poor	$ 240

EAA

European American Armory Corp. (EAA)

Though technically an importer and distributor, European American Armory Corp. has become heavily associated with weapons they handle, as most of these firearms bear their insignia. The Witness, manufactured by Tanfoglio of Italy, remains EAA's most beloved and noteworthy pistol, its design reminiscent of the C.Z. 75. Firearms critics and enthusiasts note the Witness for its accuracy and availability in .38 Super and 10mm. The company is located in Rockledge, Florida.

WINDICATOR-WEIHRAUCH
NEW IN BOX

WINDICATOR-WEIHRAUCH   REDBOOK CODE: RB-EA-H-WNDCTR

Though produced by Weihrauch, the gun is distributed, marked, and typically associated with EAA. Blue or nickel finish, hammer-block safety, steel or slightly lighter alloy frame. Pricing reflects .38 Special model with alloy frame and 2" barrel. Slight pricing increase for 4" barrel and steel frame.

Production: early 1990s - current	Caliber: .357 Magnum, .38 Special							
Action: DA/SA, revolver	Barrel Length: 2", 4"		Wt: 29 oz., 30 oz.					
Sights: fixed, blade	Capacity: 6		Grips: black rubber					
D2C:	NIB	$ 320	Ex	$ 250	VG+ $ 180	Good $ 160	LMP $ 325	
C2C:	Mint	$ 310	Ex	$ 225	VG+ $ 160	Good $ 145	Fair $ 75	
Trade-In:	Mint	$ 230	Ex	$ 200	VG+ $ 130	Good $ 115	Poor $ 60	

WITNESS FULL SIZE-TANFOGLIO
NEW IN BOX

WITNESS FULL SIZE-TANFOGLIO   REDBOOK CODE: RB-EA-H-WTFULL

While produced by Tanfoglio of Italy, the gun is distributed, marked and typically associated with EAA. Steel or polymer frame. Stainless "Wonder" or blue finish. Similar to the M1911. Slight premium for .45x22 caliber in blue finish.

Production: late 1990s - current	Caliber: 9mm, .38 Super, .40 S&W, 10mm, .45 ACP,							
.45x22	Action: DA/SA, semi-auto	Barrel Length: 4.5"	OA Length: 8.1"					
Sights: fixed front	Capacity: 10, 15, 17	Magazine: detachable box	Grips: synthetic					
D2C:	NIB	$ 515	Ex	$ 400	VG+ $ 285	Good $ 260	LMP $ 605	
C2C:	Mint	$ 500	Ex	$ 375	VG+ $ 260	Good $ 235	Fair $ 120	
Trade-In:	Mint	$ 370	Ex	$ 300	VG+ $ 210	Good $ 185	Poor $ 60	

WITNESS STEEL COMPACT-TANFOGLIO
NEW IN BOX

WITNESS STEEL COMPACT-TANFOGLIO

REDBOOK CODE: RB-E-H-WTSTCM
While actually produced by Tanfoglio of Italy, the gun is distributed, marked, and typically associated with EAA. Features a compact design and Stainless "Wonder" finish. Somewhat similar to the M1911 Commander.

Production: late 1990s - current	Caliber: 9mm, 10mm, .38 Super, .40 SW, .45 ACP							
Action: DA/SA, semi-auto	Barrel Length: 3.6"	OA Length: 7.3"	Sights: windage					
adjustable sight	Capacity: 8, 12, 14	Magazine: detachable box	Grips: synthetic					
D2C:	NIB	$ 515	Ex	$ 400	VG+ $ 285	Good $ 260	LMP $ 607	
C2C:	Mint	$ 500	Ex	$ 375	VG+ $ 260	Good $ 235	Fair $ 120	

WITNESS STOCK III-TANFOGLIO
NEW IN BOX

WITNESS STOCK III-TANFOGLIO   REDBOOK CODE: EA-H-WSIIIT

While actually produced by Tanfoglio of Italy, the gun is distributed, marked, and typically associated with EAA. Designed for competition shooting. Available in a host of different finishes. One of the largest Witness models.

Production: late 1990s - current   Caliber: 9mm, .40 S&W, .38 Super, .45 ACP, 10mm

Action: DA/SA, semi-auto   Barrel Length: 4.75"   OA Length: 8.9"   Wt: 37 oz.

Sights: fixed front, adjustable rear   Capacity: 10, 15, 17   Grips: wood, checkered

D2C:	NIB	$ 985	Ex $ 750	VG+ $ 545	Good $ 495	LMP $ 1,167
C2C:	Mint	$ 950	Ex $ 700	VG+ $ 500	Good $ 445	Fair $ 230
Trade-In:	Mint	$ 700	Ex $ 575	VG+ $ 390	Good $ 345	Poor $ 120

WITNESS CARRY-TANFOGLIO
NEW IN BOX

WITNESS POLYMER CARRY-TANFOGLIO

REDBOOK CODE: EA-H-WCARRY

While actually produced by Tanfoglio of Italy, the gun is distributed, marked, and typically associated with EAA. The smallest Witness variant, designed for concealed carry. Features a frame-mounted safety, accessory rail, and polymer frame.

Production: current   Caliber: 9mm, 10mm, .22 LR, .40 S&W, .45 ACP

Action: DA/SA, semi-auto   Barrel Length: 3.6"   OA Length: 7.5"   Wt: 29 oz.

Sights: fixed front, adjustable rear   Capacity: 10, 15, 17   Magazine: detachable box

Grips: black synthetic

D2C:	NIB	$ 610	Ex $ 475	VG+ $ 340	Good $ 305	LMP $ 691
C2C:	Mint	$ 590	Ex $ 425	VG+ $ 310	Good $ 275	Fair $ 145
Trade-In:	Mint	$ 440	Ex $ 350	VG+ $ 240	Good $ 215	Poor $ 90

WITNESS LIMITED-TANFOGLIO
NEW IN BOX

WITNESS ELITE LIMITED-TANFOGLIO

REDBOOK CODE: EA-H-WTLMTD

While actually produced by Tanfoglio of Italy, the gun is distributed, marked, and typically associated with EAA. A deluxe version of the Witness. Steel frame, frame-mounted safety, beveled magwell, enhanced rifling, extended magazine. Stainless "Wonder" finish.

Production: 2005 - current   Caliber: 9mm, .40 S&W, .38 Super, .45 ACP, 10mm

Action: SA, semi-auto   Barrel Length: 4.75"   OA Length: 9"   Sights: fully adjustable target   Capacity: 10, 15, 17   Magazine: detachable box   Grips: wood, checkered

D2C:	NIB	$1,245	Ex $ 950	VG+ $ 685	Good $ 625	LMP $1,462
C2C:	Mint	$1,200	Ex $ 875	VG+ $ 630	Good $ 565	Fair $ 290
Trade-In:	Mint	$ 890	Ex $ 700	VG+ $ 490	Good $ 440	Poor $ 150

WITNESS ELITE STOCK 1-TANFOGLIO
NEW IN BOX

WITNESS ELITE STOCK 1-TANFOGLIO

REDBOOK CODE: EA-H-WTELTS1

A modified variant of the Witness. Similar to the CZ 75. Beveled magwell, cone barrel, frame-mounted safety, blue or stainless "Wonder" finish.

Production: current   Caliber: 9mm, .40 S&W, 10mm, .45 ACP   Action: DA/SA, semi-auto

Barrel Length: 4.5"   OA Length: 8"   Wt: 39 oz.   Sights: adjustable

Capacity: 10, 15, 17   Magazine: detachable box   Grips: wood

D2C:	NIB	$ 975	Ex $ 750	VG+ $ 540	Good $ 490	LMP $ 1,102
C2C:	Mint	$ 940	Ex $ 675	VG+ $ 490	Good $ 440	Fair $ 225
Trade-In:	Mint	$ 700	Ex $ 550	VG+ $ 390	Good $ 345	Poor $ 120

**WITNESS ELITE GOLD
TEAM-TANFOGLIO**
NEW IN BOX

WITNESS ELITE GOLD TEAM-TANFOGLIO

REDBOOK CODE: EA-H-WTEGLD

A deluxe variant of the Witness. Steel frame, polished steel finish, modern tactical design. Frame-mounted safety, extended magazine, beveled magwell.

Production: early 1990s - current Caliber: 9mm, .40 S&W, .38 Super, .45 ACP, 10mm

Action: SA, semi-auto Barrel Length: 5.25" OA Length: 10.5" Sights: windage adjustable with scope mounts Capacity: 10, 15, 17 Grips: synthetic

D2C:	NIB	$ 2,025	Ex $ 1,550	VG+ $ 1,115	Good $ 1,015	LMP $ 2,336
C2C:	Mint $ 1,950	Ex $ 1,400	VG+ $ 1,020	Good $ 915	Fair $ 470
Trade-In:	Mint $ 1,440	Ex $ 1,150	VG+ $ 790	Good $ 710	Poor $ 210

**WITNESS HUNTER-
TANFOGLIO**
NEW IN BOX

WITNESS HUNTER-TANFOGLIO REDBOOK CODE: EA-H-WTHNTR

While actually produced by Tanfoglio of Italy, the gun is distributed, marked, and typically associated with EAA. Designed for hunting. Similar to the CZ 75. Steel frame, blue finish, frame-mounted safety, improved rifling.

Production: ~2005 - current Caliber: 10mm, .45 ACP Action: SA, semi-auto

Barrel Length: 6" OA Length: 10" Wt: 46 oz. Sights: 3-dot Capacity: 10, 15

Magazine: detachable box Grips: wood, checkered

D2C:	NIB $ 980	Ex $ 750	VG+ $ 540	Good $ 490	LMP $ 1,293
C2C:	Mint $ 950	Ex $ 700	VG+ $ 490	Good $ 445	Fair $ 230
Trade-In:	Mint $ 700	Ex $ 550	VG+ $ 390	Good $ 345	Poor $ 120

**WITNESS ELITE
MATCH-TANFOGLIO**
NEW IN BOX

WITNESS ELITE MATCH-TANFOGLIO

REDBOOK CODE: EA-H-WTELTMT

A variant of the Witness. Features two-tone finish, an extended magazine release, and frame-mounted safety.

Production: ~2005 - current Caliber: 9mm, .40 S&W, .45 ACP, 10MM, .38 Super

Action: SA, semi-auto Barrel Length: 4.75" OA Length: 8.75" Wt.: 44 oz.

Sights: fixed front, adjustable rear Capacity: 10, 15, 17 Magazine: detachable box

Grips: black synthetic

D2C:	NIB $ 810	Ex $ 625	VG+ $ 450	Good $ 405	LMP $ 778
C2C:	Mint $ 780	Ex $ 575	VG+ $ 410	Good $ 365	Fair $ 190
Trade-In:	Mint $ 580	Ex $ 475	VG+ $ 320	Good $ 285	Poor $ 90

F.A.B. 92- TANFOGLIO
GOOD

F.A.B. 92- TANFOGLIO REDBOOK CODE: EA-H-FAB92X

Manufactured by Tanfoglio, but marked and typically associated with EAA. A precursor to the Witness. Also available in compact model. Offered in blue, two-tone, or chrome.

Production: 1980s - mid-1990s Caliber: 9mm, .40 S&W Action: DA/SA, semi-auto

Barrel Length: 4.5" Sights: fixed front, adjustable rear Capacity: 15

Magazine: detachable box Grips: black synthetic or polished wood

D2C:	NIB $ 425	Ex $ 325	VG+ $ 235	Good $ 215	
C2C:	Mint $ 410	Ex $ 300	VG+ $ 220	Good $ 195	Fair $ 100
Trade-In:	Mint $ 310	Ex $ 175	VG+ $ 170	Good $ 150	Poor $ 60

ZASTAVA EZ VERY GOOD +
Courtesy of Bud's Gun Shop

ZASTAVA EZ REDBOOK CODE: EA-H-ZASTEZ
Manufactured by Zastava, but marked and typically associated with EAA. Compact design, frame-mounted safety, blue finish. Spurred hammer.

Production: 2007 - discontinued Caliber: .45 ACP, .40 S&W Action: DA/SA, semi-auto Barrel Length: 4", 4.25" OA Length: 7.75" Wt: 33 oz.

Sights: 3-dot Capacity: 10 Magazine: detachable box Grips: black synthetic

D2C:	NIB	$ 445	Ex $ 350	VG+ $ 245	Good $ 225		
C2C:	Mint	$ 430	Ex $ 325	VG+ $ 230	Good $ 205	Fair $ 105
Trade-In:	Mint	$ 320	Ex $ 250	VG+ $ 180	Good $ 160	Poor $ 60

ZASTAVA M88 REDBOOK CODE: EA-H-ZASM88
Manufactured by Zastava, but marked and typically associated with EAA. Similar to Tokarev pistols of the era. Steel frame. Blue finish.

Production: 2007 - discontinued Caliber: .40 S&W, 9mm Action: SA, semi-auto

Barrel Length: 3.8" OA Length: 6.9" Wt: 30 oz. Sights: fixed Capacity: 8

Magazine: detachable box Grips: black rubber

D2C:	NIB	$ 285	Ex $ 225	VG+ $ 160	Good $ 145	
C2C:	Mint	$ 280	Ex $ 200	VG+ $ 150	Good $ 130	Fair $ 70
Trade-In:	Mint	$ 210	Ex $ 175	VG+ $ 120	Good $ 100	Poor $ 30

Excel Arms (Accu-Tek)

Located in Ontario, California, Excel Arms manufactures a series of semi-automatic pistols and rifles designed for plinking, varmint hunting, and leisure shooting. Though not typical carry guns, their high-capacity magazines, affordability, tactical design, and durable construction have gained them strong following over the last several years. The company also owns Accu-Tek Firearms.

MP-22 REDBOOK CODE: EX-H-MP22XX
Polymer frame, blowback action, bull barrel, frame-mounted safety, firing pin block, scope mounts, stainless steel finish.

Production: ~2005 - current Caliber: .22 Magnum Action: SA, semi-auto

Barrel Length: 8.5" OA Length: 12.9" Wt: 54 oz. Sights: adjustable Capacity: 9

Magazine: detachable box Grips: black synthetic

D2C:	NIB	$ 455	Ex $ 350	VG+ $ 255	Good $ 230	LMP $ 480
C2C:	Mint	$ 440	Ex $ 325	VG+ $ 230	Good $ 205	Fair $ 105
Trade-In:	Mint	$ 330	Ex $ 275	VG+ $ 180	Good $ 160	Poor $ 60

MP-5.7 REDBOOK CODE: EX-H-MP57XX
Similar to the MP-22, but fires 5.7x28mm rounds. Polymer frame, blowback action, bull barrel, frame-mounted safety, firing pin block, scope mounts.

Production: ~2005 - current Caliber: 5.7x28mm, .22 WMR Action: SA, semi-auto

Barrel Length: 8.5" OA Length: 12.9" Wt: 54 oz. Sights: adjustable Capacity: 9

Magazine: detachable box Grips: black synthetic

D2C:	NIB	$ 570	Ex $ 450	VG+ $ 315	Good $ 285	LMP $ 615
C2C:	Mint	$ 550	Ex $ 400	VG+ $ 290	Good $ 260	Fair $ 135
Trade-In:	Mint	$ 410	Ex $ 325	VG+ $ 230	Good $ 200	Poor $ 60

MP-17 REDBOOK CODE: EX-H-MP17XX
Similar to the MP-22, but fires .17 HMR rounds. Polymer frame, blowback action, bull barrel, frame-mounted safety, firing pin block, scope mounts.

Production: 2007 - discontinued   Caliber: .17 HMR   Action: SA, semi-auto

Barrel Length: 8.5"   OA Length: 12.9"   Sights: adjustable   Capacity: 9

Magazine: detachable box   Grips: black synthetic

D2C:	NIB	$ 480	Ex	$ 375	VG+	$ 265	Good	$ 240		
C2C:	Mint	$ 470	Ex	$ 350	VG+	$ 240	Good	$ 220	Fair $	115
Trade-In:	Mint	$ 350	Ex	$ 275	VG+	$ 190	Good	$ 170	Poor $	60

ACCU-TEK MODEL L T-380 REDBOOK CODE: EX-H-MLT380
Sold and marked as an Accu-Tek, a subsidiary of Excel. Slide-mounted safety, aluminum frame, pocket-sized design.

Production: current   Caliber: .380 ACP   Action: SA, semi-auto

Barrel Length: 2.8"   OA Length: 6.1"   Sights: fixed   Capacity: 6

Magazine: detachable box   Grips: black synthetic

D2C:	NIB	$ 300	Ex	$ 250	VG+	$ 165	Good	$ 150		
C2C:	Mint	$ 290	Ex	$ 225	VG+	$ 150	Good	$ 135	Fair $	70
Trade-In:	Mint	$ 220	Ex	$ 175	VG+	$ 120	Good	$ 105	Poor $	30

ACCU-TEK MODEL XL-9SS REDBOOK CODE: EX-H-MXL9SS
Sold and marked as an Accu-Tek, a subsidiary of Excel. Pocket-sized design.

Production: late 1990s - 2003   Caliber: 9mm   Action: DAO, semi-auto

Barrel Length: 3"   Sights: fixed   Capacity: 5   Magazine: detachable box

Grips: black synthetic

D2C:	NIB	$ 245	Ex	$ 200	VG+	$ 135	Good	$ 125		
C2C:	Mint	$ 240	Ex	$ 175	VG+	$ 130	Good	$ 115	Fair $	60
Trade-In:	Mint	$ 180	Ex	$ 150	VG+	$ 100	Good	$ 90	Poor $	30

ACCU-TEK MODEL BL-9 REDBOOK CODE: EX-H-MBL9XX
Sold and marked as an Accu-Tek, a subsidiary of Excel. Steel frame, blowback design, pocket-sized design.

Production: late 1990s - 2002   Caliber: 9mm   Action: DAO, semi-auto

Barrel Length: 3"   OA Length: 5.6"   Sights: fixed   Capacity: 5

Magazine: detachable box   Grips: black synthetic

D2C:	NIB	$ 235	Ex	$ 200	VG+	$ 130	Good	$ 120		
C2C:	Mint	$ 230	Ex	$ 175	VG+	$ 120	Good	$ 110	Fair $	55
Trade-In:	Mint	$ 170	Ex	$ 150	VG+	$ 100	Good	$ 85	Poor $	30

ACCU-TEK MODEL HC-380 REDBOOK CODE: EX-H-MHC380
Sold and marked as an Accu-Tek, a subsidiary of Excel. Steel frame, blowback design, pocket-sized design, frame-mounted safety.

Production: early 1990s - current   Caliber: .380 ACP   Action: SA, semi-auto

Barrel Length: 2.8"   OA Length: 6"   Sights: adjustable rear   Capacity: 13

Magazine: detachable box   Grips: black synthetic

D2C:	NIB	$ 285	Ex	$ 225	VG+	$ 160	Good	$ 145		
C2C:	Mint	$ 280	Ex	$ 200	VG+	$ 150	Good	$ 130	Fair $	70
Trade-In:	Mint	$ 210	Ex	$ 175	VG+	$ 120	Good	$ 100	Poor $	30

FNH USA/FN Herstal

Belgium firearm manufacturer Fabrique Nationale d'Herstal, or FN Herstal, began operations in the late 1880s and has remained a significant force in the firearms industry ever since. Now a subsidiary of the Herstal Group, who also retains control of Browning and Winchester, FNH became a dominate force in the firearms industry after teaming with John Moses Browning in 1897 to develop and manufacture an array of significant weapons, such as the Browning Hi-Power and the M2 Machine Gun. It's well known within the firearms industry that the Herstal Group manufactures all Browning firearms, making the two companies essentially one and the same. In 1998, the Herstal Group formed FN USA, its American marketing and distribution subsidiary based in McLean, Virginia. The company also manufactures some weapons at their plant in Columbia, South Carolina, dubbed FN Manufacturing. FNH and Browning together stand as two of the most recognizable and influential names in the history of modern firearms.

FIVE-SEVEN   REDBOOK CODE: RB-FN-H-FNFIV7
Polymer grip frame, accessory rail, black finish, rounded trigger guard, rear cocking serrations, no slide release, replaced by a newer model with single-action trigger.

Production: late 1990s - discontinued Caliber: 5.7x28mm Action: DAO, semi-auto								
Barrel Length: 4.8" Sights: fixed front, dovetail rear Capacity: 10, 20								
Magazine: detachable box Grips: black synthetic								
D2C:	NIB	$ 985	Ex $ 750	VG+ $ 545	Good $ 495			
C2C:	Mint	$ 950	Ex $ 700	VG+ $ 500	Good $ 445	Fair $ 230		
Trade-In:	Mint	$ 700	Ex $ 575	VG+ $ 390	Good $ 345	Poor $ 120		

FIVE-SEVEN USG
NEW IN BOX

FIVE-SEVEN USG   REDBOOK CODE: RB-FN-H-FIVE7U
United States Government variant of the Five-SeveN. Features minor alterations including a squared trigger guard and a larger reversible magazine release switch. Similar model was available in Flat Dark Earth and OD Green.

Production: early 2000s - 2013 Caliber: 5.7x28mm Action: SA, semi-auto								
Barrel Length: 4.8" Sights: fixed or adjustable Capacity: 20								
Magazine: detachable box Grips: black synthetic								
D2C:	NIB	$1,150	Ex $ 875	VG+ $ 635	Good $ 575			
C2C:	Mint	$ 1,110	Ex $ 800	VG+ $ 580	Good $ 520	Fair $ 265		
Trade-In:	Mint	$ 820	Ex $ 650	VG+ $ 450	Good $ 405	Poor $ 120		

FIVE-SEVEN MK2
NEW IN BOX

FIVE-SEVEN MK2   REDBOOK CODE: RB-FN-H-FIVE7M
Current standard Five-seveN model, polymer grip frame, polymer covered steel slide, front and rear cocking serrations, cold hammer-forged stainless barrel, Picatinny rail, ambidextrous controls, serrated and squared trigger guard, black or Flat Dark Earth finish.

Production: 2013 - current Caliber: 5.7x28mm Action: SA, semi-auto								
Barrel Length: 4.8" Wt.: 23 oz. Sights: adjustable 3-dot Capacity: 20								
Magazine: detachable box Grips: black synthetic								
D2C:	NIB	$1,300	Ex $1,000	VG+ $ 715	Good $ 650			
C2C:	Mint	$1,250	Ex $ 900	VG+ $ 650	Good $ 585	Fair $ 300		
Trade-In:	Mint	$ 930	Ex $ 750	VG+ $ 510	Good $ 455	Poor $ 150		

HI-POWER DAC   REDBOOK CODE: RB-FN-H-HPSAXX

Double action compact. Blue finish, frame-mounted safety. Foreign production and sales by FN Herstal.

Production: early 2000s - discontinued   Caliber: 9mm   Action: DA/SA, semi-auto

Barrel Length: 3.75"   Sights: fixed   Capacity: 10   Magazine: detachable box

Grips: black synthetic

D2C:	NIB	$ 765	Ex	$ 600	VG+	$ 425	Good	$ 385	
C2C:	Mint	$ 740	Ex	$ 550	VG+	$ 390	Good	$ 345	Fair $ 180
Trade-In:	Mint	$ 550	Ex	$ 450	VG+	$ 300	Good	$ 270	Poor $ 90

HI-POWER-SA SFS   REDBOOK CODE: RB-FN-H-HPSASF

Features an improved cocking mechanism and fires as either SA or DA. Foreign production and sales by FN Herstal.

Production: 2003 - discontinued   Caliber: 9mm, .40 S&W   Action: DA/SA, semi-auto

Barrel Length: 4.6"   Sights: fixed front, dovetail rear   Capacity: 10, 13

Magazine: detachable box   Grips: black synthetic or rubber

D2C:	NIB	$ 880	Ex	$ 675	VG+	$ 485	Good	$ 440	
C2C:	Mint	$ 850	Ex	$ 625	VG+	$ 440	Good	$ 400	Fair $ 205
Trade-In:	Mint	$ 630	Ex	$ 500	VG+	$ 350	Good	$ 310	Poor $ 90

FNS-9
NEW IN BOX

FNS-9   REDBOOK CODE: RB-FN-H-FNS9XX

Full-size, black or silver matte finish, front and rear cocking serrations, accessory mounting rail, frame-mounted safety, ambidextrous design.

Production: early 2010s - current   Caliber: 9mm   Action: striker-fired, semi-auto

Barrel Length: 4"   Wt.: 25 oz.   Sights: 3-dot fixed   Capacity: 18

Magazine: detachable box   Grips: black synthetic

D2C:	NIB	$ 685	Ex	$ 525	VG+	$ 380	Good	$ 345	
C2C:	Mint	$ 660	Ex	$ 475	VG+	$ 350	Good	$ 310	Fair $ 160
Trade-In:	Mint	$ 490	Ex	$ 400	VG+	$ 270	Good	$ 240	Poor $ 90

FNS-40
NEW IN BOX

FNS-40   REDBOOK CODE: RB-FN-H-FNS40X

A .40 S&W variant of the FNS-9. Full-size, black or silver/matte black finish, front and rear cocking serrations, accessory mounting rail, frame-mounted safety, ambidextrous design.

Production: early 2010s - current   Caliber: .40 S&W   Action: striker-fired, semi-auto

Barrel Length: 4"   Sights: 3-dot fixed   Capacity: 15

Magazine: detachable box   Grips: black synthetic

D2C:	NIB	$ 655	Ex	$ 500	VG+	$ 365	Good	$ 330	
C2C:	Mint	$ 630	Ex	$ 475	VG+	$ 330	Good	$ 295	Fair $ 155
Trade-In:	Mint	$ 470	Ex	$ 375	VG+	$ 260	Good	$ 230	Poor $ 90

FNH USA

FNH USA

FNP-9/40
EXCELLENT

FNP-9/40   REDBOOK CODE: RB-FN-H-FNP9XX

Compact design, black finish, polymer frame, interchangeable backstraps, frame-mounted safety.

Production: 2003 - discontinued   Caliber: 9mm, .40 S&W   Action: DA/SA, semi-auto

Barrel Length: 4"   Sights: white dot or tritium   Capacity: 16

Magazine: detachable box   Grips: black synthetic

D2C:	NIB	$ 535	Ex	$ 425	VG+	$ 295	Good	$ 270		
C2C:	Mint	$ 520	Ex	$ 375	VG+	$ 270	Good	$ 245	Fair	$ 125
Trade-In:	Mint	$ 380	Ex	$ 300	VG+	$ 210	Good	$ 190	Poor	$ 60

FNP-357   REDBOOK CODE: RB-FN-H-FNP40X

A .357 Sig variant of FNP-9. Compact design, black finish, polymer frame, interchangeable backstraps, frame-mounted safety.

Production: ~2010 - discontinued   Caliber: .357 Sig   Action: DA/SA, semi-auto

Barrel Length: 4"   Sights: white dot or tritium   Capacity: 14

Magazine: detachable box   Grips: black synthetic

D2C:	NIB	$ 555	Ex	$ 425	VG+	$ 310	Good	$ 280		
C2C:	Mint	$ 540	Ex	$ 400	VG+	$ 280	Good	$ 250	Fair	$ 130
Trade-In:	Mint	$ 400	Ex	$ 325	VG+	$ 220	Good	$ 195	Poor	$ 60

FNP-45
EXCELLENT

FNP-45   REDBOOK CODE: RB-FN-H-FNP45X

A variant of FNP-9. Frame-mounted safety, extended barrel, black or gray matte finish, ambidextrous design. May feature accessory rail.

Production: discontinued   Caliber: .45 ACP   Action: DA/SA, semi-auto

Barrel Length: 4.5"   Sights: fixed, 3-dot   Capacity: 10, 15

Magazine: detachable box   Grips: black synthetic

D2C:	NIB	$ 685	Ex	$ 525	VG+	$ 380	Good	$ 345		
C2C:	Mint	$ 660	Ex	$ 475	VG+	$ 350	Good	$ 310	Fair	$ 160
Trade-In:	Mint	$ 490	Ex	$ 400	VG+	$ 270	Good	$ 240	Poor	$ 90

**FNP-45
COMPETITION**
EXCELLENT

FNP-45 COMPETITION   REDBOOK CODE: RB-FN-H-FNP45C

An enhanced variant of the FNP-45, designed for competition shooting. Matte black finish, frame-mounted safety, accessory rail.

Production: 2011 - discontinued   Caliber: .45 ACP   Action: DA/SA, semi-auto

Barrel Length: 4.5"   Sights: fiber-optic front   Capacity: 15

Magazine: detachable box   Grips: black synthetic

D2C:	NIB	$1,010	Ex	$ 775	VG+	$ 560	Good	$ 505		
C2C:	Mint	$ 970	Ex	$ 700	VG+	$ 510	Good	$ 455	Fair	$ 235
Trade-In:	Mint	$ 720	Ex	$ 575	VG+	$ 400	Good	$ 355	Poor	$ 120

**FNP-45
TACTICAL**
EXCELLENT

FNP-45 TACTICAL REDBOOK CODE: RB-FN-H-FNP45T

A variant of the FNP-45. Tactical design, threaded barrel, accessory rail, various finishes, frame-mounted safety.

Production: discontinued		Caliber: .45 ACP		Action: DA/SA, semi-auto				
Barrel Length: 4.5"		Wt.: 33 oz.		Sights: red and white dot, adjustable				
Capacity: 15		Magazine: detachable box		Grips: synthetic				
D2C:	NIB	$1,100	Ex $ 850	VG+ $ 605	Good $ 550			
C2C:	Mint	$1,060	Ex $ 775	VG+ $ 550	Good $ 495	Fair $	255	
Trade-In:	Mint	$ 790	Ex $ 625	VG+ $ 430	Good $ 385	Poor $	120	

MODEL FORTY-NINE
EXCELLENT

MODEL FORTY-NINE REDBOOK CODE: RB-FN-H-MODL49

Accessory rail, blowback action, steel slide, black or stainless finish, compact design.

Production: 2000 - ~2011		Caliber: 9mm, .40 S&W		Action: striker-fired, semi-auto				
Barrel Length: 4.3"		Sights: fixed, 3-dot		Capacity: 10, 14, 16				
Magazine: detachable box		Grips: synthetic						
D2C:	NIB	$ 530	Ex $ 425	VG+ $ 295	Good $ 265			
C2C:	Mint	$ 510	Ex $ 375	VG+ $ 270	Good $ 240	Fair $	125	
Trade-In:	Mint	$ 380	Ex $ 300	VG+ $ 210	Good $ 190	Poor $	60	

FNX-9
EXCELLENT

FNX-9 REDBOOK CODE: RB-FN-H-FNX9XX

An improved variant of the FN FNP series. Frame-mounted safety, decocker mechanism, ambidextrous design, black or silver matte finish, accessory rail, serrated trigger guard.

Production: early 2010s - current		Caliber: 9mm		Action: DA/SA, semi-auto				
Barrel Length: 4"		Sights: fixed, 3-dot		Capacity: 17		Magazine: detachable box		
Grips: black synthetic								
D2C:	NIB	$ 580	Ex $ 450	VG+ $ 320	Good $ 290			
C2C:	Mint	$ 560	Ex $ 425	VG+ $ 290	Good $ 265	Fair $	135	
Trade-In:	Mint	$ 420	Ex $ 325	VG+ $ 230	Good $ 205	Poor $	60	

FNX-40
EXCELLENT

FNX-40 REDBOOK CODE: RB-FN-H-FNX40X

An improved variant of the FN FNP series. Frame-mounted safety, decocker mechanism, ambidextrous design, polymer frame, black or silver/matte black finish, accessory rail.

Production: early 2010s - current		Caliber: .40 S&W		Action: DA/SA, semi-auto				
Barrel Length: 4"		Sights: fixed, 3-dot		Capacity: 14		Magazine: detachable box		
Grips: black synthetic								
D2C:	NIB	$ 620	Ex $ 475	VG+ $ 345	Good $ 310			
C2C:	Mint	$ 600	Ex $ 450	VG+ $ 310	Good $ 280	Fair $	145	
Trade-In:	Mint	$ 450	Ex $ 350	VG+ $ 250	Good $ 220	Poor $	90	

FNX-45
NEW IN BOX

FNX-45 TACTICAL
NEW IN BOX

FNX-45 REDBOOK CODE: RB-FN-H-FNX45X

Similar to the other FNX models, but chambered for .45 ACP and has an extended barrel. Various finishes.

Production: early 2010s - current	Caliber: .45 ACP	Action: DA/SA, semi-auto				
Barrel Length: 4.5"	Sights: fixed, 3-dot	Capacity: 10, 15				
Magazine: detachable box	Grips: synthetic					
D2C:	NIB $ 685	Ex $ 525	VG+ $ 380	Good $ 345		
C2C:	Mint $ 660	Ex $ 475	VG+ $ 350	Good $ 310	Fair $ 160	
Trade-In:	Mint $ 490	Ex $ 400	VG+ $ 270	Good $ 240	Poor $ 90	

FNX-45 TACTICAL REDBOOK CODE: RB-FN-H-FNX45T

An enhanced variant of the FNX-45. Distinguished by its extended barrel, threaded muzzle, accessory mounts, and updated design.

Production: early 2010s - current	Caliber: .45 ACP	Action: DA/SA, semi-auto				
Barrel Length: 5.3"	Sights: fixed, 3-dot	Capacity: 15				
Magazine: detachable box	Grips: synthetic					
D2C:	NIB $ 1,112	Ex $ 850	VG+ $ 615	Good $ 560		
C2C:	Mint $1,070	Ex $ 775	VG+ $ 560	Good $ 505	Fair $ 260	
Trade-In:	Mint $ 790	Ex $ 625	VG+ $ 440	Good $ 390	Poor $ 120	

BDA REDBOOK CODE: RB-FN-H-DBAXXX

Virtually identical to the Browning BDA, which FN also manufactures. Marked "Fabrique Nationale Herstal Belgique" with no Browning engravings.

Production: late 1990s - discontinued	Caliber: 9mm	Action: DA/SA, semi-auto				
Barrel Length: 4.6"	Sights: fixed	Capacity: 10, 14	Magazine: detachable box			
Grips: synthetic						
D2C:	NIB $ 715	Ex $ 550	VG+ $ 395	Good $ 360		
C2C:	Mint $ 690	Ex $ 500	VG+ $ 360	Good $ 325	Fair $ 165	
Trade-In:	Mint $ 510	Ex $ 425	VG+ $ 280	Good $ 255	Poor $ 90	

Federal Ordnance, Inc.

Now defunct, Federal Ordnance, Inc. of South El Monte, California manufactured pistols and rifles from the 1960s until the early 1990s. They produced a fine M1911A1, but the company gets mentioned more often for their Broomhandle, a replica of a somewhat obscure Mauser pistol of the same name. The original Broomhandle made appearances in a host of different conflicts in the early twentieth century, including the Spanish Civil War, World War I, and the Mexican Revolution. The Federal Ordnance replica is thus sought after by those interested in firearms of the era.

M714 BROOMHANDLE REDBOOK CODE: RB-FO-H-STNDBR

A reproduction of the classic C96 Mauser. Manufactured in China and briefly imported to the United States. Assembled from some surplus Mauser parts.

Production: late 1980s - ~1991	Caliber: 7.63mm, 9mm	Action: SA, semi-auto				
Barrel Length: 5.5"	Sights: fixed front, adjustable rear	Capacity: 10				
Magazine: detachable box	Grips: wood					
D2C:	NIB $ 855	Ex $ 650	VG+ $ 475	Good $ 430		
C2C:	Mint $ 830	Ex $ 600	VG+ $ 430	Good $ 385	Fair $ 200	
Trade-In:	Mint $ 610	Ex $ 500	VG+ $ 340	Good $ 300	Poor $ 90	

RANGER 1911A1   REDBOOK CODE: RB-FO-H-RANGA1

A reproduction of Colt 1911A1. Recoil operated action, grip and frame-mounted safety, likely manufactured in China.

Production: late 1980s - discontinued   Caliber: .45 ACP   Action: SA, semi-auto

Barrel Length: 5"   Sights: blade front, adjustable rear   Capacity: 7

Magazine: detachable box   Grips: checkered walnut

D2C:	NIB	$ 645	Ex $ 500	VG+ $ 355	Good $ 325					
C2C:	Mint	$ 620	Ex $ 450	VG+ $ 330	Good $ 295	Fair $	150			
Trade-In:	Mint	$ 460	Ex $ 375	VG+ $ 260	Good $ 230	Poor $	90			

Firestorm

Now defunct, Firestorm Inc. was imported by SGS Importers International, Inc. based in Wanamassa, New Jersey. Many of these pistols were assembled from parts manufactured by Bersa firearms in Argentina. Bersa now sells a "Firestorm" pistol, which varies slightly from the original Firestorm .380. Firestorm pistols remain a suitable option for carry use, given their compact size and affordability.

M1911 GOVERNMENT   REDBOOK CODE: RB-XF-H-1911GV

A reproduction of the popular Colt M1911 Government. Manufactured by Bersa. Blue finish, frame-mounted safety.

Production: early 2000s - discontinued   Caliber: .45 ACP   Action: SA, semi-auto

Barrel Length: 5"   Sights: 3-dot   Capacity: 7   Magazine: detachable box

Grips: wood or synthetic

D2C:	NIB	$ 325	Ex $ 250	VG+ $ 180	Good $ 165		
C2C:	Mint	$ 320	Ex $ 225	VG+ $ 170	Good $ 150	Fair $	75
Trade-In:	Mint	$ 240	Ex $ 200	VG+ $ 130	Good $ 115	Poor $	60

Freedom Arms

Freedom Arms, based in Freedom, Wyoming, manufactures a series of high-end revolvers, some of which fire large-caliber rounds, such as the .500 Wyoming Express or .475 Linebaugh. Collectors and shooting enthusiasts actively seek Freedom revolvers for their supreme quality and variety of caliber options, which help them retain much of their value aftermarket.

PATRIOT
NEW IN BOX

PATRIOT   REDBOOK CODE: RB-FA-H-PATRIO

Mini revolver, stainless steel finish, first gun manufactured by Freedom.

Production: 1978 - 1989   Caliber: .22 LR, .22 Short, .22 WMR

Action: SA, revolver   Barrel Length: 1.75"   Sights: blade front   Capacity: 5

Grips: rosewood, ivory, or synthetic

D2C:	NIB	$ 380	Ex $ 300	VG+ $ 210	Good $ 190		
C2C:	Mint	$ 370	Ex $ 275	VG+ $ 190	Good $ 175	Fair $	90
Trade-In:	Mint	$ 270	Ex $ 225	VG+ $ 150	Good $ 135	Poor $	60

BELT BUCKLE REVOLVER
NEW IN BOX
Courtesy of Rock Island Auction Company

BELT BUCKLE REVOLVER REDBOOK CODE: RB-FA-H-BELTXX

Sold with a belt buckle in which the gun can be snapped into place and worn. Somewhat of a novelty. Similar to the Patriot.

Production: late 1970s - discontinued   Caliber: .22 LR, .22 Short, .22 WMR

Action: SA, revolver   Barrel Length: 1.12"   Sights: blade front   Capacity: 5

Grips: rosewood, ivory, or synthetic

D2C:	NIB	$ 550	Ex $ 425	VG+ $ 305	Good $ 275		
C2C:	Mint	$ 530	Ex $ 400	VG+ $ 280	Good $ 250	Fair	$ 130
Trade-In:	Mint	$ 400	Ex $ 325	VG+ $ 220	Good $ 195	Poor	$ 60

MODEL 83 FIELD GRADE
NEW IN BOX
Courtesy of Rock Island Auction Company

MODEL 83 FIELD GRADE REDBOOK CODE: RB-FA-H-83FIEG

Similar to the Model 83. A very high quality gun. Matte stainless finish, smooth cylinder, partially-shrouded ejector rod.

Production: late 1980s - current   Caliber: .500 Wyoming Express, .475 Linebaugh, .454 Casull, .44 Magnum, .41 Magnum, .357 Magnum, .22 LR

Action: SA, revolver   Capacity: 5   Barrel Length: 4.75", 6", 7.5", 9", 10"

Sights: adjustable   Grips: rosewood

D2C:	NIB	$1,350	Ex $1,050	VG+ $ 745	Good $ 675		
C2C:	Mint	$1,300	Ex $ 950	VG+ $ 680	Good $ 610	Fair	$ 315
Trade-In:	Mint	$ 960	Ex $ 775	VG+ $ 530	Good $ 475	Poor	$ 150

MODEL 83 PREMIER GRADE
EXCELLENT

MODEL 83 PREMIER GRADE REDBOOK CODE: RB-FA-H-83PREG

High-grade bright brushed finish. Manual sliding bar safety. Stainless steel construction. Subtract $93 for magnum calibers. Subtract $98 for fixed sight model.

Production: current   Caliber: .500 Wyoming Express, .475 Linebaugh, .454 Casull, .44 Remington Magnum, .41 Remington Magnum, .357 Magnum   Action: SA, revolver

Barrel Length: 4.75," 6", 7.5", 9", 10"   Sights: adjustable   Capacity: 5   Grips: hardwood

D2C:	NIB	$2,534	Ex $1,950	VG+ $1,395	Good $1,270	LMP	$2,534
C2C:	Mint	$2,440	Ex $1,750	VG+ $1,270	Good $1,145	Fair	$ 585
Trade-In:	Mint	$1,800	Ex $1,425	VG+ $ 990	Good $ 890	Poor	$ 270

MODEL 97 PREMIER GRADE
EXCELLENT

MODEL 97 PREMIER GRADE REDBOOK CODE: RB-FA-H-97PREG

High-grade bright brushed finish. Transfer bar safety. Subtract $144 for fixed sights. Brushed stainless steel finish.

Production: 1997 - current   Caliber: .45 Colt, .44 Special, .41 Magnum, .357 Magnum, .327 Federal, .224-32 FA, .22 LR, .17 HMR

Action: SA, revolver   Barrel Length: 4.25", 5.5", 7.5", 10"

Sights: adjustable   Capacity: 5, 6   Grips: polished wood

D2C:	NIB	$ 1645	Ex $1275	VG+ $ 905	Good $ 825	LMP	$ 2,055
C2C:	Mint	$ 1580	Ex $1150	VG+ $ 830	Good $ 745	Fair	$ 380
Trade-In:	Mint	$ 1170	Ex $ 925	VG+ $ 650	Good $ 580	Poor	$ 180

Girsan

Based in Turkey, Girsan focuses on military and law enforcement specific firearms, though their pistols have gained popularity among general consumers. The company's MC 1911 G-2 has earned favorable reviews in the U.S. and the company seems primed to expand their presence in western markets. Currently, the company only distributes their pistols in the U.S. through American Tactical Imports of Summerville, South Carolina.

MC 27 S   REDBOOK CODE: RB-XG-H-MC27XX
Manufactured in Turkey. Slide-mounted safety, black finish, compact design.

Production: current Caliber: 9mm Action: DA/SA, semi-auto Barrel Length: 3.9"
Wt: 24 oz. Sights: dovetail front and rear Capacity: 15 Magazine: detachable box
Grips: black synthetic

D2C:	NIB	$ 415	Ex $ 325	VG+ $ 230	Good $ 210		
C2C:	Mint	$ 400	Ex $ 300	VG+ $ 210	Good $ 190	Fair $ 100
Trade-In:	Mint	$ 300	Ex $ 250	VG+ $ 170	Good $ 150	Poor $ 60

MC 17   REDBOOK CODE: RB-XG-H-MC17XX
Manufactured in Turkey. A larger variant of the MC 27.

Production: current Caliber: 9mm Action: DA/SA, semi-auto
Barrel Length: 4.3" Wt: 9.3 oz. Sights: dovetail front and rear Capacity: 15
Magazine: detachable box Grips: black synthetic

D2C:	NIB	$ 415	Ex $ 325	VG+ $ 230	Good $ 210		
C2C:	Mint	$ 400	Ex $ 300	VG+ $ 210	Good $ 190	Fair $ 100
Trade-In:	Mint	$ 300	Ex $ 250	VG+ $ 170	Good $ 150	Poor $ 60

MC 23   REDBOOK CODE: RB-XG-H-MC23XX
Manufactured in Turkey. A variant of the MC 27, but chambered for .40 S&W rounds.

Production: current Caliber: .40 S&W, 9mm, .45 ACP Action: DA/SA, semi-auto
Barrel Length: 4.19", 4.33" Wt: 32 oz. Sights: dovetail front and rear Capacity: 13
Magazine: detachable box Grips: black synthetic

D2C:	NIB	$ 430	Ex $ 350	VG+ $ 240	Good $ 215		
C2C:	Mint	$ 420	Ex $ 300	VG+ $ 220	Good $ 195	Fair $ 100
Trade-In:	Mint	$ 310	Ex $ 250	VG+ $ 170	Good $ 155	Poor $ 60

MC 1911 G-2
NEW IN BOX
Courtesy of Bud's Gun Shop

MC 1911 G-2   REDBOOK CODE: RB-XG-H-MC1911X
Manufactured in Turkey. A reasonably priced M1911, similar to Colt Government models. Frame mounted safety, blue or stainless finish. The second generation variant.

Production: current Caliber: .45 ACP Action: SA, semi-auto
Barrel Length: 5" Sights: dovetail foresight, notched sight Capacity: 8
Magazine: detachable box Grips: synthetic, wood

D2C:	NIB	$ 466	Ex $ 375	VG+ $ 260	Good $ 235		
C2C:	Mint	$ 450	Ex $ 325	VG+ $ 240	Good $ 210	Fair $ 110
Trade-In:	Mint	$ 340	Ex $ 275	VG+ $ 190	Good $ 165	Poor $ 60

GLOCK

Glock, Inc.

Based in Austria, Glock produced their first prototype in 1981, which would eventually lead to the development of the Glock 17, the company's first commercially available model. The Austrian military adopted the original Glock as their standard issue sidearm in 1982, as the pistol met and exceeded all military-outlined specs for a handgun. In 1985, Glock began commercial sales in the U.S. and quickly gained a strong following thanks to their ergonomic design, minimalist aesthetics, accuracy, and supreme reliability. After nearly four decades since their introduction, many shooters still consider Glock pistols the pinnacle of modern semi-automatic pistols. Given their simplistic design, Glocks are highly customizable, allowing shooters to add a host of aftermarket accessories.

17 (GEN. 1-3)
NEW IN BOX

17 (GEN. 1-3) REDBOOK CODE: RB-GK-H-17G123

Full-size reinforced polymer frame, "Safe Action" System, cold hammer-forged barrel with polygonal rifling, all metal parts treated with Nitration finish, Gen 3 model added accessory rail and finger grooves, less value for Gen 1 or 2, add roughly $25 for "C" model.

Production: 1985 - current Caliber: 9mm Action: striker-fired, semi-auto

Barrel Length: 4.48" OA Length: 8.03"

Wt.: 22.04 oz. Sights: fixed, adjustable, Glock NS, or Trijicon NS

Capacity: 17, optional 10,19,33 Magazine: detachable box Grips: polymer grip frame

D2C:	NIB	$ 545	Ex $ 425	VG+ $ 300	Good $ 275		
C2C:	Mint	$ 530	Ex $ 400	VG+ $ 280	Good $ 250	Fair $	130
Trade-In:	Mint	$ 390	Ex $ 325	VG+ $ 220	Good $ 195	Poor $	60

17 L
VERY GOOD +

17L REDBOOK CODE: RB-GK-H-17LXXX

Full-size reinforced polymer frame, extended competition slide and barrel, "Safe Action" System, cold-hammer-forged barrel with polygonal rifling, lightened trigger, early models had ported barrel that was soon dropped. Gen 3 models added an accessory rail and finger grooves. Glock produces this model in small runs. It has been replaced by the Glock 34.

Production: 1988 - limited runs Caliber: 9mm Action: striker-fired, semi-auto

Barrel Length: 6.02" Wt.: 26.5 oz. Sights: fixed, adjustable, Glock NS, or Trijicon NS

Capacity: 17, optional 10, 19, 33 Magazine: detachable box Grips: polymer grip frame

D2C:	NIB	$ 710	Ex $ 550	VG+ $ 395	Good $ 355		
C2C:	Mint	$ 690	Ex $ 500	VG+ $ 360	Good $ 320	Fair $	165
Trade-In:	Mint	$ 510	Ex $ 400	VG+ $ 280	Good $ 250	Poor $	90

17 GEN. 4
NEW IN BOX

17 GEN. 4 REDBOOK CODE: RB-GK-H-17GEN4

Full-size reinforced polymer frame with accessory rail, "Safe Action" System, cold hammer-forged barrel with polygonal rifling, all metal parts treated with Nitration finish, dual recoil spring assembly, reversible magazine catch for left or right-handed operation, Modular Back Strap, rough textured grip.

Production: 2010 - current Caliber: 9mm Action: striker-fired, semi-auto

Barrel Length: 4.48" OA Length: 7.95"

Wt.: 22.04 oz. Sights: fixed, adjustable, Glock NS, or Trijicon NS

Capacity: 17, optional 10, 19, 33 Magazine: detachable box

Grips: polymer, Modular Back Strap System

D2C:	NIB	$ 565	Ex $ 450	VG+ $ 315	Good $ 285		
C2C:	Mint	$ 550	Ex $ 400	VG+ $ 290	Good $ 255	Fair $	130
Trade-In:	Mint	$ 410	Ex $ 325	VG+ $ 230	Good $ 200	Poor $	60

19 (GEN. 1-3)
NEW IN BOX

19 (GEN. 1-3)   REDBOOK CODE: RB-GK-H-19XXXX

Compact reinforced polymer frame, "Safe Action" System, cold hammer-forged barrel with polygonal rifling, all metal parts treated with Nitration finish, Gen 3 model added accessory rail and finger grooves, less value for Gen 1 or 2, add roughly $25 for "C" model.

Production: 1988 - current   Caliber: 9mm   Action: striker-fired, semi-auto
Barrel Length: 4.02"   OA Length: 7.36"   Wt.: 21 oz.   Sights: fixed, adjustable, Glock NS, or Trijicon NS   Capacity: 15, optional 10, 17, 19, 33   Grips: polymer grip frame

D2C:	NIB	$ 520	Ex	$ 400	VG+	$ 290	Good	$ 260		
C2C:	Mint	$ 500	Ex	$ 375	VG+	$ 260	Good	$ 235	Fair	$ 120
Trade-In:	Mint	$ 370	Ex	$ 300	VG+	$ 210	Good	$ 185	Poor	$ 60

19 GEN. 4
NEW IN BOX

19 GEN. 4   REDBOOK CODE: RB-GK-H-19GEN4

Compact reinforced polymer frame with accessory rail, "Safe Action" System, cold hammer-forged barrel with polygonal rifling, all metal parts treated with Nitration finish, dual-recoil spring assembly, reversible magazine catch for left or right-handed operation, Modular Back Strap, rough textured grip.

Production: 2010 - current   Caliber: 9mm   Action: striker-fired, semi-auto
Barrel Length: 4.02"   OA Length: 7.28"   Wt.: 21 oz.   Sights: fixed, adjustable, Glock NS, or Trijicon NSCapacity: 15, optional 10, 17, 19, 33   Grips: polymer, Modular Back Strap System

D2C:	NIB	$ 535	Ex	$ 425	VG+	$ 295	Good	$ 270		
C2C:	Mint	$ 520	Ex	$ 375	VG+	$ 270	Good	$ 245	Fair	$ 125
Trade-In:	Mint	$ 380	Ex	$ 300	VG+	$ 210	Good	$ 190	Poor	$ 60

26
NEW IN BOX

26   REDBOOK CODE: RB-GK-H-26XXXX

Subcompact reinforced polymer frame, "Safe Action" System, cold hammer-forged barrel with polygonal rifling, all metal parts treated with Nitration finish, early models had no checkering on front-strap finger grooves.

Production: 1995 - current   Caliber: 9mm   Action: striker-fired, semi-auto
Barrel Length: 3.42"   OA Length: 6.49"   Wt.: 19.75 oz.   Sights: fixed, adjustable, Glock NS, or Trijicon NS   Capacity: 10, optional 12, 15, 17, 19, 33Grips: polymer grip frame

D2C:	NIB	$ 535	Ex	$ 425	VG+	$ 295	Good	$ 270		
C2C:	Mint	$ 520	Ex	$ 375	VG+	$ 270	Good	$ 245	Fair	$ 125
Trade-In:	Mint	$ 380	Ex	$ 300	VG+	$ 210	Good	$ 190	Poor	$ 60

26 GEN. 4
NEW IN BOX

26 GEN. 4   REDBOOK CODE: RB-GK-H-26GEN4

Subcompact reinforced polymer frame, "Safe Action" System, cold hammer-forged barrel with polygonal rifling, all metal parts treated with Nitration finish, dual recoil spring assembly, reversible magazine catch for left or right-handed operation, Modular Back Strap, rough textured grip.

Production: 2011 - current   Caliber: 9mm   Action: striker-fired, semi-auto
Barrel Length: 3.42"   OA Length: 6.41"   Wt.: 19.75 oz.   Sights: fixed, adjustable, Glock NS, or Trijicon NS   Capacity: 10, optional 12, 15, 17, 19, 33   Magazine: detachable box
Grips: polymer, Modular Back Strap System

D2C:	NIB	$ 535	Ex	$ 425	VG+	$ 295	Good	$ 270		
C2C:	Mint	$ 520	Ex	$ 375	VG+	$ 270	Good	$ 245	Fair	$ 125
Trade-In:	Mint	$ 380	Ex	$ 300	VG+	$ 210	Good	$ 190	Poor	$ 60

GLOCK

GLOCK

34
NEW IN BOX

34 REDBOOK CODE: RB-GK-H-34XXXX

Full-size reinforced polymer frame with accessory rail, extended slide and barrel, "Safe Action" System, cold-hammer-forged barrel with polygonal rifling, all metal parts treated with Nitration finish, ~4.5 lb. trigger, competition grade.

Production: 1998 - current Caliber: 9mm Action: striker-fired, semi-auto

Barrel Length: 5.31" OA Length: 8.81"

Wt.: 22.93 oz. Sights: fixed, adjustable, Glock NS, or Trijicon NS

Capacity: 17, optional 10, 19, 33 Magazine: detachable box Grips: polymer grip frame

D2C:	NIB	$ 590	Ex $ 450	VG+ $ 325	Good $ 295		
C2C:	Mint	$ 570	Ex $ 425	VG+ $ 300	Good $ 270	Fair $ 140	
Trade-In:	Mint	$ 420	Ex $ 350	VG+ $ 240	Good $ 210	Poor $ 60	

34 GEN. 4
NEW IN BOX

34 GEN. 4 REDBOOK CODE: RB-GK-H-34GEN4

Full-size reinforced polymer frame with accessory rail, extended slide and barrel, "Safe Action" System, cold-hammer-forged barrel with polygonal rifling, all metal parts treated with Nitration finish, dual recoil spring assembly, reversible magazine catch for left or right-handed operation, ~4.5 lb. trigger, Modular Back Strap, rough textured grip, competition grade.

Production: 2011 - current Caliber: 9mm Action: striker-fired, semi-auto Barrel Length: 5.31" OA Length: 8.74" Wt.: 23.10 oz. Sights: fixed, adjustable, Glock NS, or Trijicon NSCapacity: 17, optional 10,19,33 Magazine: detachable box

Grips: polymer, Modular Back Strap System

D2C:	NIB	$ 640	Ex $ 500	VG+ $ 355	Good $ 320		
C2C:	Mint	$ 620	Ex $ 450	VG+ $ 320	Good $ 290	Fair $ 150	
Trade-In:	Mint	$ 460	Ex $ 375	VG+ $ 250	Good $ 225	Poor $ 90	

22
NEW IN BOX

22 REDBOOK CODE: RB-GK-H-22XXXX

Full size reinforced polymer frame, "Safe Action" System, cold-hammer-forged barrel with polygonal rifling, all metal parts treated with Nitration finish, Gen 3 model added accessory rail and finger grooves, Gen 2 has slightly less value, add roughly $25 for "C" model.

Production: 1990 - current Caliber: .40 S&W Action: striker-fired, semi-auto

Barrel Length: 4.49" OA Length: 8.03" Wt.: 23 oz. Sights: fixed, adjustable, Glock NS, or Trijicon NS Capacity: 15, optional 10, 17, 22 Magazine: detachable box

Grips: polymer grip frame

D2C:	NIB	$ 555	Ex $ 425	VG+ $ 310	Good $ 280		
C2C:	Mint	$ 540	Ex $ 400	VG+ $ 280	Good $ 250	Fair $ 130	
Trade-In:	Mint	$ 400	Ex $ 325	VG+ $ 220	Good $ 195	Poor $ 60	

22 GEN. 4
NEW IN BOX

22 GEN. 4 REDBOOK CODE: RB-GK-H-22GEN4

Full-size reinforced polymer frame with accessory rail, "Safe Action" System, cold-hammer-forged barrel with polygonal rifling, all metal parts treated with Nitration finish, dual recoil spring assembly, reversible magazine catch for left or right-handed operation, Modular Back Strap, rough textured grip.

Production: 2010 - current Caliber: .40 S&W Action: striker-fired, semi-auto

Barrel Length: 4.48" OA Length: 7.95" Wt.: 23 oz.

Sights: fixed, adjustable, Glock NS, or Trijicon NS Capacity: 15

optional 10, 17, 22 Magazine: detachable box Grips: polymer, Modular Back Strap System

D2C:	NIB	$ 580	Ex $ 450	VG+ $ 320	Good $ 290		
C2C:	Mint	$ 560	Ex $ 425	VG+ $ 290	Good $ 265	Fair $ 135	
Trade-In:	Mint	$ 420	Ex $ 325	VG+ $ 230	Good $ 205	Poor $ 60	

23
NEW IN BOX

23 REDBOOK CODE: RB-GK-H-23XXXX

Compact reinforced polymer frame, "Safe Action" System, cold-hammer-forged barrel with polygonal rifling, all metal parts treated with Nitration finish, Gen 3 model added accessory rail and finger grooves, less value for Gen 2, add roughly $25 for "C" model.

Production: 1990 - current   Caliber: .40 S&W   Action: striker-fired, semi-auto

Barrel Length: 4.02"   OA Length: 7.36"   Wt.: 21.16 oz.

Sights: fixed, adjustable, Glock NS, or Trijicon NS   Capacity: 13,

optional 10,15,17,22   Magazine: detachable box   Grips: polymer grip frame

D2C:	NIB	$ 520	Ex	$ 400	VG+	$ 290	Good	$ 260	
C2C:	Mint	$ 500	Ex	$ 375	VG+	$ 260	Good	$ 235	Fair	$ 120
Trade-In:	Mint	$ 370	Ex	$ 300	VG+	$ 210	Good	$ 185	Poor	$ 60

23 GEN. 4
NEW IN BOX

23 GEN. 4 REDBOOK CODE: RB-GK-H-23GEN4

Compact reinforced polymer frame with accessory rail, "Safe Action" System, cold-hammer-forged barrel with polygonal rifling, all metal parts treated with Nitration finish, dual recoil spring assembly, reversible magazine catch for left or right-handed operation, Modular Back Strap, rough textured grip, compact.

Production: 2010 - current   Caliber: .40 S&W   Action: striker-fired, semi-auto

Barrel Length: 4.02"   OA Length: 7.28"   Wt.: 21.34 oz.

Sights: fixed, adjustable, Glock NS, or Trijicon NS   Capacity: 13, optional

10, 15, 17, 22   Magazine: detachable box   Grips: polymer, Modular Back Strap System

D2C:	NIB	$ 565	Ex	$ 450	VG+	$ 315	Good	$ 285	
C2C:	Mint	$ 550	Ex	$ 400	VG+	$ 290	Good	$ 255	Fair	$ 130
Trade-In:	Mint	$ 410	Ex	$ 325	VG+	$ 230	Good	$ 200	Poor	$ 60

24
NEW IN BOX

24 REDBOOK CODE: RB-GK-H-24XXXX

Similar to the Glock 17L but chambered for .40 S&W, full-size reinforced polymer frame, extended competition slide and barrel, "Safe Action" System, cold-hammer-forged barrel with polygonal rifling, lightened trigger, Glock produces this model in small runs.

Production: 1994 - discontinued   Caliber: .40 S&W   Action: striker-fired, semi-auto

Barrel Length: 6.02"   Wt.: 26.5 oz.   Sights: fixed, adjustable, Glock NS, or Trijicon NS

Capacity: 15, optional 10, 17, 22   Magazine: detachable box   Grips: polymer grip frame

D2C:	NIB	$ 660	Ex	$ 525	VG+	$ 365	Good	$ 330	
C2C:	Mint	$ 640	Ex	$ 475	VG+	$ 330	Good	$ 300	Fair	$ 155
Trade-In:	Mint	$ 470	Ex	$ 375	VG+	$ 260	Good	$ 235	Poor	$ 90

27
NEW IN BOX

27 REDBOOK CODE: RB-GK-H-27XXXX

Subcompact reinforced polymer frame, "Safe Action" System, cold-hammer-forged barrel with polygonal rifling, all metal parts treated with Nitration finish, early models had no checkering on front strap finger grooves.

Production: 1995 - current   Caliber: .40 S&W   Action: striker-fired, semi-auto

Barrel Length: 3.42"   OA Length: 6.49"Wt.: 19.75 oz.   Sights: fixed, adjustable, Glock NS, or Trijicon NS   Capacity: 9, optional 11, 13, 15, 17   Magazine: detachable box

Grips: polymer grip frame

D2C:	NIB	$ 530	Ex	$ 425	VG+	$ 295	Good	$ 265	
C2C:	Mint	$ 510	Ex	$ 375	VG+	$ 270	Good	$ 240	Fair	$ 125
Trade-In:	Mint	$ 380	Ex	$ 300	VG+	$ 210	Good	$ 190	Poor	$ 60

27 GEN. 4
NEW IN BOX

27 GEN. 4   REDBOOK CODE: RB-GK-H-27GEN4

Subcompact reinforced polymer frame, "Safe Action" System, cold hammer-forged barrel with polygonal rifling, all metal parts treated with Nitration finish, dual-recoil spring assembly, reversible magazine catch for left or right-handed operation, Modular Back Strap, rough textured grip.

Production: 2011 - current Caliber: .40 S&W Action: striker-fired, semi-auto

Barrel Length: 3.42" OA Length: 6.41"

Wt.: 19.75 oz. Sights: fixed, adjustable, Glock NS, or Trijicon NS Capacity: 9,

optional 11, 13, 15, 17 Magazine: detachable box Grips: polymer, Modular Back Strap System

D2C:	NIB	$ 570	Ex $ 450	VG+ $ 315	Good $ 285		
C2C:	Mint	$ 550	Ex $ 400	VG+ $ 290	Good $ 260	Fair $	135
Trade-In:	Mint	$ 410	Ex $ 325	VG+ $ 230	Good $ 200	Poor $	60

35   REDBOOK CODE: RB-GK-H-35XXXX

Full-size reinforced polymer frame with accessory rail, extended slide and barrel, "Safe Action" System, cold hammer-forged barrel with polygonal rifling, all metal parts treated with Nitration finish, ~4.5 lb. trigger, competition grade.

Production: 1998 - current Caliber: .40 S&W Action: striker-fired, semi-auto

Barrel Length: 5.31" OA Length: 8.81"

Wt.: 24.52 oz. Sights: fixed, adjustable, Glock NS, or Trijicon NS

Capacity: 15, optional 10, 17, 22 Grips: polymer grip frame

35
NEW IN BOX

D2C:	NIB	$ 620	Ex $ 475	VG+ $ 345	Good $ 310		
C2C:	Mint	$ 600	Ex $ 450	VG+ $ 310	Good $ 280	Fair $	145
Trade-In:	Mint	$ 450	Ex $ 350	VG+ $ 250	Good $ 220	Poor $	90

35 GEN. 4   REDBOOK CODE: RB-GK-H-35GEN4

Full-size reinforced polymer frame with accessory rail, extended slide and barrel, "Safe Action" System, cold hammer-forged barrel with polygonal rifling, all metal parts treated with Nitration finish, dual recoil spring assembly, reversible magazine catch for left or right-handed operation, ~4.5 lb. trigger, Modular Back Strap, rough-textured grip, competition grade.

Production: 2011 - current Caliber: .40 S&W Action: striker-fired, semi-auto

Barrel Length: 5.31" OA Length: 8.74" Wt.: 24.69 oz.

Sights: fixed, adjustable, Glock NS, or Trijicon NS Capacity: 15, optional

10, 17, 22 Magazine: detachable box Grips: polymer, Modular Back Strap System

35 GEN. 4
NEW IN BOX

D2C:	NIB	$ 640	Ex $ 500	VG+ $ 355	Good $ 320		
C2C:	Mint	$ 620	Ex $ 450	VG+ $ 320	Good $ 290	Fair $	150
Trade-In:	Mint	$ 460	Ex $ 375	VG+ $ 250	Good $ 225	Poor $	90

20   REDBOOK CODE: RB-GK-H-20XXXX

Full-size reinforced polymer frame, "Safe Action" System, cold hammer-forged barrel with polygonal rifling, all metal parts treated with Nitration finish, Gen 3 model added accessory rail and finger grooves, previous generation has slightly less value, add roughly $25 for "C" model.

Production: 1990 - 2014 Caliber: 10mm Action: striker-fired, semi-auto

Barrel Length: 4.60" OA Length: 8.22" Wt.: 27.69 oz.

Sights: fixed, adjustable, Glock NS, or Trijicon NS Capacity: 15

Magazine: detachable box Grips: polymer grip frame

20
NEW IN BOX

D2C:	NIB	$ 560	Ex $ 450	VG+ $ 310	Good $ 280		
C2C:	Mint	$ 540	Ex $ 400	VG+ $ 280	Good $ 255	Fair $	130
Trade-In:	Mint	$ 400	Ex $ 325	VG+ $ 220	Good $ 200	Poor $	60

20 SF
NEW IN BOX

20 SF   REDBOOK CODE: RB-GK-H-20SFXX

Full-size reinforced polymer frame with accessory rail, "Safe Action" System, cold hammer-forged barrel with polygonal rifling, all metal parts treated with Nitration finish, reduced backstrap frame area for smaller hands.

Production: 2009 - current   Caliber: 10mm   Action: striker-fired, semi-auto

Barrel Length: 4.60"   OA Length: 8.03"   Wt.: 27.51 oz.

Sights: fixed, adjustable, Glock NS, or Trijicon NS

Capacity: 15   Magazine: detachable box   Grips: polymer grip frame

D2C:	NIB	$ 565	Ex $ 450	VG+ $ 315	Good $ 285				
C2C:	Mint	$ 550	Ex $ 400	VG+ $ 290	Good $ 255	Fair $ 130			
Trade-In:	Mint	$ 410	Ex $ 325	VG+ $ 230	Good $ 200	Poor $ 60			

20 GEN. 4
NEW IN BOX

20 GEN. 4   REDBOOK CODE: RB-GK-H-20GEN4

Full-size reinforced polymer frame with accessory rail, "Safe Action" System, cold hammer-forged barrel with polygonal rifling, all metal parts treated with Nitration finish, dual recoil spring assembly, reversible magazine catch for left or right-handed operation, Modular Back Strap, rough-textured grip.

Production: 2012 - current   Caliber: 10mm   Action: striker-fired, semi-auto

Barrel Length: 4.61"   OA Length: 7.60"   Wt.: 27.51 oz.

Sights: fixed, adjustable, Glock NS, or Trijicon NS

Capacity: 15   Magazine: detachable box   Grips: polymer, Modular Back Strap System

D2C:	NIB	$ 620	Ex $ 475	VG+ $ 345	Good $ 310				
C2C:	Mint	$ 600	Ex $ 450	VG+ $ 310	Good $ 280	Fair $ 145			
Trade-In:	Mint	$ 450	Ex $ 350	VG+ $ 250	Good $ 220	Poor $ 90			

29
NEW IN BOX

29   REDBOOK CODE: RB-GK-H-29XXXX

Subcompact reinforced polymer frame, "Safe Action" System, cold hammer-forged barrel with polygonal rifling, all metal parts treated with Nitration finish.

Production: 1996 - 2014   Caliber: 10mm   Action: striker-fired, semi-auto

Barrel Length: 3.77"   OA Length: 6.96" Wt.: 24.69 oz.   Sights: fixed, adjustable, Glock NS, or Trijicon NS   Capacity: 10, optional 15   Grips: polymer grip frame

D2C:	NIB	$ 570	Ex $ 450	VG+ $ 315	Good $ 285				
C2C:	Mint	$ 550	Ex $ 400	VG+ $ 290	Good $ 260	Fair $ 135			
Trade-In:	Mint	$ 410	Ex $ 325	VG+ $ 230	Good $ 200	Poor $ 60			

29 SF
NEW IN BOX

29 SF   REDBOOK CODE: RB-GK-H-29SFXX

Subcompact reinforced polymer frame, "Safe Action" System, cold-hammer-forged barrel with polygonal rifling, all metal parts treated with Nitration finish, reduced backstrap frame area for smaller hands.

Production: 2009 - current   Caliber: 10mm   Action: striker-fired, semi-auto

Barrel Length: 3.77"   OA Length: 6.88"

Wt.: 24.52 oz.   Sights: fixed, adjustable, Glock NS, or Trijicon NS

Capacity: 10, optional 15   Magazine: detachable box   Grips: polymer grip frame

D2C:	NIB	$ 570	Ex $ 450	VG+ $ 315	Good $ 285				
C2C:	Mint	$ 550	Ex $ 400	VG+ $ 290	Good $ 260	Fair $ 135			
Trade-In:	Mint	$ 410	Ex $ 325	VG+ $ 230	Good $ 200	Poor $ 60			

GLOCK

29 GEN. 4 REDBOOK CODE: RB-GK-H-29GEN4

Subcompact reinforced polymer frame, "Safe Action" System, cold hammer-forged barrel with polygonal rifling, all metal parts treated with Nitration finish, dual-recoil spring assembly, reversible magazine catch for left or right-handed operation, Modular Back Strap, rough textured grip.

Production: 2012 - current Caliber: 10mm Action: striker-fired, semi-auto

Barrel Length: 3.78" OA Length: 6.77" Wt.: 24.34 oz.

Sights: fixed, adjustable, Glock NS, or Trijicon NS Capacity: 10

Magazine: detachable box Grips: polymer, Modular Back Strap System

D2C:	NIB	$ 590	Ex $ 450	VG+ $ 325	Good $ 295	
C2C:	Mint	$ 570	Ex $ 425	VG+ $ 300	Good $ 270	Fair $ 140
Trade-In:	Mint	$ 420	Ex $ 350	VG+ $ 240	Good $ 210	Poor $ 60

29 GEN. 4
NEW IN BOX

21 REDBOOK CODE: RB-GK-H-21XXXX

Full-size reinforced polymer frame, "Safe Action" System, cold hammer-forged barrel with polygonal rifling, all metal parts treated with Nitration finish, Gen 3 model added accessory rail and finger grooves, previous generation has slightly less value, add roughly $25 for "C" model.

Production: 1990 - 2014 Caliber: .45 ACP Action: striker-fired, semi-auto

Barrel Length: 4.60" OA Length: 8.22" Wt.: 26.46 oz.

Sights: fixed, adjustable, Glock NS, or Trijicon NS Capacity: 13, optional 10

Magazine: detachable box Grips: polymer grip frame

D2C:	NIB	$ 560	Ex $ 450	VG+ $ 310	Good $ 280	
C2C:	Mint	$ 540	Ex $ 400	VG+ $ 280	Good $ 255	Fair $ 130
Trade-In:	Mint	$ 400	Ex $ 325	VG+ $ 220	Good $ 200	Poor $ 60

21
NEW IN BOX

21 SF REDBOOK CODE: RB-GK-H-21SFXX

Full-size reinforced polymer frame with accessory rail, "Safe Action" System, cold hammer-forged barrel with polygonal rifling, all metal parts treated with Nitration finish, reduced backstrap frame area for smaller hands.

Production: 2009 - current Caliber: .45 ACP Action: striker-fired, semi-auto

Barrel Length: 4.60" OA Length: 8.03" Wt.: 26.28 oz.

Sights: fixed, adjustable, Glock NS, or Trijicon NS Capacity: 13, optional 10

Magazine: detachable box Grips: polymer grip frame

D2C:	NIB	$ 625	Ex $ 475	VG+ $ 345	Good $ 315	
C2C:	Mint	$ 600	Ex $ 450	VG+ $ 320	Good $ 285	Fair $ 145
Trade-In:	Mint	$ 450	Ex $ 350	VG+ $ 250	Good $ 220	Poor $ 90

21 SF
NEW IN BOX

21 GEN. 4 REDBOOK CODE: RB-GK-H-21GEN4

Full-size reinforced polymer frame with accessory rail, "Safe Action" System, cold hammer-forged barrel with polygonal rifling, all metal parts treated with Nitration finish, dual recoil spring assembly, reversible magazine catch for left or right-handed operation, Modular Back Strap, rough textured grip.

Production: 2011 - current Caliber: .45 ACP Action: striker-fired, semi-auto

Barrel Length: 4.61" OA Length: 8.23" Wt.: 26.46 oz.

Sights: fixed, adjustable, Glock NS, or Trijicon NS Capacity: 13, optional 10

Magazine: detachable box Grips: polymer, Modular Back Strap System

D2C:	NIB	$ 605	Ex $ 475	VG+ $ 335	Good $ 305	
C2C:	Mint	$ 590	Ex $ 425	VG+ $ 310	Good $ 275	Fair $ 140
Trade-In:	Mint	$ 430	Ex $ 350	VG+ $ 240	Good $ 215	Poor $ 90

21 GEN. 4
NEW IN BOX

30
NEW IN BOX

30 REDBOOK CODE: RB-GK-H-30XXXX

Subcompact reinforced polymer frame, "Safe Action" System, cold hammer-forged barrel with polygonal rifling, all metal parts treated with Nitration finish, accessory rail on later production.

Production: 1996 - 2014 Caliber: .45 ACP Action: striker-fired, semi-auto

Barrel Length: 3.77" OA Length: 6.96" Wt.: 23.99 oz.

Sights: fixed, adjustable, Glock NS, or Trijicon NS

Capacity: 10, optional 9, 13 Magazine: detachable box Grips: polymer grip frame

D2C:	NIB	$ 565	Ex $ 450	VG+ $ 315	Good $ 285		
C2C:	Mint	$ 550	Ex $ 400	VG+ $ 290	Good $ 255	Fair $ 130
Trade-In:	Mint	$ 410	Ex $ 325	VG+ $ 230	Good $ 200	Poor $ 60

30 SF
NEW IN BOX

30 SF REDBOOK CODE: RB-GK-H-30SFXX

Subcompact reinforced polymer frame with accessory rail, "Safe Action" System, cold hammer-forged barrel with polygonal rifling, all metal parts treated with Nitration finish, reduced backstrap frame area for smaller hands.

Production: 2008 - current Caliber: .45 ACP Action: striker-fired, semi-auto

Barrel Length: 3.77" OA Length: 6.88" Wt.: 20.28 oz.

Sights: fixed, adjustable, Glock NS, or Trijicon NS

Capacity: 10, optional 9, 13 Magazine: detachable box Grips: polymer grip frame

D2C:	NIB	$ 570	Ex $ 450	VG+ $ 315	Good $ 285		
C2C:	Mint	$ 550	Ex $ 400	VG+ $ 290	Good $ 260	Fair $ 135
Trade-In:	Mint	$ 410	Ex $ 325	VG+ $ 230	Good $ 200	Poor $ 60

30 SF
NEW IN BOX

30 GEN. 4 REDBOOK CODE: RB-GK-H-30GEN4

Subcompact reinforced polymer frame with accessory rail, "Safe Action" System, cold hammer-forged barrel with polygonal rifling, all metal parts treated with Nitration finish, dual recoil spring assembly, reversible magazine catch for left or right-handed operation, Modular Back Strap, rough-textured grip.

Production: 2012 - current Caliber: .45 ACP Action: striker-fired, semi-auto

Barrel Length: 3.78" OA Length: 6.77" Wt.: 23.81 oz.

Sights: fixed, adjustable, Glock NS, or Trijicon NS Capacity: 10, optional 13

Magazine: detachable box Grips: polymer, Modular Back Strap System

D2C:	NIB	$ 565	Ex $ 450	VG+ $ 315	Good $ 285		
C2C:	Mint	$ 550	Ex $ 400	VG+ $ 290	Good $ 255	Fair $ 130
Trade-In:	Mint	$ 410	Ex $ 325	VG+ $ 230	Good $ 200	Poor $ 60

30S
NEW IN BOX

30S REDBOOK CODE: RB-GK-H-30SXXX

Subcompact reinforced polymer frame with accessory rail, "Safe Action" System, cold hammer-forged barrel with polygonal rifling, all metal parts treated with Nitration finish, uses slimmer slide from Glock 36 with frame of Glock 30 SF.

Production: 2012 - current Caliber: .45 ACP Action: striker-fired, semi-auto

Barrel Length: 3.78" OA Length: 6.89" Wt.: 20.28 oz.

Sights: fixed, adjustable, Glock NS, or Trijicon NS

Capacity: 10, optional 13 Magazine: detachable box Grips: polymer grip frame

D2C:	NIB	$ 565	Ex $ 450	VG+ $ 315	Good $ 285		
C2C:	Mint	$ 550	Ex $ 400	VG+ $ 290	Good $ 255	Fair $ 130
Trade-In:	Mint	$ 410	Ex $ 325	VG+ $ 230	Good $ 200	Poor $ 60

GLOCK

36
NEW IN BOX

36 REDBOOK CODE: RB-GK-H-36XXXX

Subcompact slimline reinforced polymer frame, "Safe Action" System, cold hammer-forged barrel with polygonal rifling, all metal parts treated with Nitration finish.

Production: 1999 - current Caliber: .45 ACP Action: striker-fired, semi-auto

Barrel Length: 3.78" OA Length: 6.97"

Wt.: 20.11 oz. Sights: fixed, adjustable, Glock NS, or Trijicon NS Capacity: 6

Magazine: detachable box Grips: polymer grip frame

D2C:	NIB	$ 560	Ex $ 450	VG+ $ 310	Good $ 280		
C2C:	Mint	$ 540	Ex $ 400	VG+ $ 280	Good $ 255	Fair $ 130
Trade-In:	Mint	$ 400	Ex $ 325	VG+ $ 220	Good $ 200	Poor $ 60

41 REDBOOK CODE: RB-GK-H-41XXXX

Full-size reinforced polymer frame with accessory rail, extended slide and barrel, "Safe Action" System, cold hammer-forged barrel with polygonal rifling, all metal parts treated with Nitration finish, dual recoil spring assembly, reversible magazine catch for left or right-handed operation, lightened trigger, Modular Back Strap, rough textured grip, competition grade.

Production: 2014 - current Caliber: .45 ACP Action: striker-fired, semi-auto

Barrel Length: 5.31" OA Length: 8.9"

Wt.: 27 oz. Sights: fixed, adjustable, Glock NS, or Trijicon NS Capacity: 13, optional 10

Magazine: detachable box Grips: polymer, Modular Back Strap System

D2C:	NIB	$ 615	Ex $ 475	VG+ $ 340	Good $ 310		
C2C:	Mint	$ 600	Ex $ 425	VG+ $ 310	Good $ 280	Fair $ 145
Trade-In:	Mint	$ 440	Ex $ 350	VG+ $ 240	Good $ 220	Poor $ 90

37
NEW IN BOX

37 REDBOOK CODE: RB-GK-H-37XXXX

Full-size reinforced polymer frame with accessory rail, "Safe Action" System, cold hammer-forged barrel with polygonal rifling, all metal parts treated with Nitration finish.

Production: 2003 - current Caliber: .45 G.A.P. Action: striker-fired, semi-auto

Barrel Length: 4.48" OA Length: 8.03"

Wt.: 25.93 oz. Sights: fixed, adjustable, Glock NS, or Trijicon NS Capacity: 10, optional 11

Magazine: detachable box Grips: polymer grip frame

D2C:	NIB	$ 550	Ex $ 425	VG+ $ 305	Good $ 275		
C2C:	Mint	$ 530	Ex $ 400	VG+ $ 280	Good $ 250	Fair $ 130
Trade-In:	Mint	$ 400	Ex $ 325	VG+ $ 220	Good $ 195	Poor $ 60

37 GEN. 4
NEW IN BOX

37 GEN. 4 REDBOOK CODE: RB-GK-H-37GEN4

Full-size reinforced polymer frame with accessory rail, "Safe Action" System, cold hammer-forged barrel with polygonal rifling, all metal parts treated with Nitration finish, dual recoil spring assembly, reversible magazine catch for left or right-handed operation, Modular Back Strap, rough-textured grip.

Production: 2010 - current Caliber: .45 G.A.P. Action: striker-fired, semi-auto

Barrel Length: 4.48" OA Length: 7.95"

Wt.: 25.10 oz. Sights: fixed, adjustable, Glock NS, or Trijicon NS Capacity: 10, optional 11

Magazine: detachable box Grips: polymer, Modular Back Strap System

D2C:	NIB	$ 575	Ex $ 450	VG+ $ 320	Good $ 290		
C2C:	Mint	$ 560	Ex $ 400	VG+ $ 290	Good $ 260	Fair $ 135
Trade-In:	Mint	$ 410	Ex $ 325	VG+ $ 230	Good $ 205	Poor $ 60

38
NEW IN BOX

38   REDBOOK CODE: RB-GK-H-38XXXX

Compact reinforced polymer frame with accessory rail, "Safe Action" System, cold hammer-forged barrel with polygonal rifling, all metal parts treated with Nitration finish.

Production: 2005 - current   Caliber: .45 G.A.P.   Action: striker-fired, semi-auto

Barrel Length: 4.01"   OA Length: 7.36"

Wt.: 24.16 oz.   Sights: fixed, adjustable, Glock NS, or Trijicon NS   Capacity: 8, optional 10, 11

Magazine: detachable box   Grips: polymer grip frame

D2C:	NIB	$ 555	Ex $ 425	VG+ $ 310	Good $ 280			
C2C:	Mint	$ 540	Ex $ 400	VG+ $ 280	Good $ 250	Fair $ 130	
Trade-In:	Mint	$ 400	Ex $ 325	VG+ $ 220	Good $ 195	Poor $ 60	

39
NEW IN BOX

39   REDBOOK CODE: RB-GK-H-39XXXX

Subcompact reinforced polymer frame, "Safe Action" System, cold hammer-forged barrel with polygonal rifling, all metal parts treated with Nitration finish.

Production: 2005 - current   Caliber: .45 G.A.P.   Action: striker-fired, semi-auto

Barrel Length: 3.42"   OA Length: 6.49"

Wt.: 19.33 oz.   Sights: fixed, adjustable, Glock NS, or Trijicon NS

Capacity: 6, optional 8, 10, 11   Magazine: detachable box   Grips: polymer grip frame

D2C:	NIB	$ 555	Ex $ 425	VG+ $ 310	Good $ 280		
C2C:	Mint	$ 540	Ex $ 400	VG+ $ 280	Good $ 250	Fair $ 130
Trade-In:	Mint	$ 400	Ex $ 325	VG+ $ 220	Good $ 195	Poor $ 60

31
NEW IN BOX

31   REDBOOK CODE: RB-GK-H-31XXXX

Full-size reinforced polymer frame with accessory rail, "Safe Action" System, cold hammer-forged barrel with polygonal rifling, all metal parts treated with Nitration finish, add roughly $25 for "C" model.

Production: 1997 - current   Caliber: .357 SIG   Action: striker-fired, semi-auto

Barrel Length: 4.48"   OA Length: 8.03"

Wt.: 23.28 oz.   Sights: fixed, adjustable, Glock NS, or Trijicon NS   Capacity: 15, optional 17

Magazine: detachable box   Grips: polymer grip frame

D2C:	NIB	$ 550	Ex $ 425	VG+ $ 305	Good $ 275		
C2C:	Mint	$ 530	Ex $ 400	VG+ $ 280	Good $ 250	Fair $ 130
Trade-In:	Mint	$ 400	Ex $ 325	VG+ $ 220	Good $ 195	Poor $ 60

31 GEN. 4
NEW IN BOX

31 GEN. 4   REDBOOK CODE: RB-GK-H-31GEN4

Full-size reinforced polymer frame with accessory rail, "Safe Action" System, cold hammer-forged barrel with polygonal rifling, all metal parts treated with Nitration finish, dual recoil spring assembly, reversible magazine catch for left or right-handed operation, Modular Back Strap, rough-textured grip.

Production: 2010 - current   Caliber: .357 SIG   Action: striker-fired, semi-auto

Barrel Length: 4.48"   OA Length: 7.95"

Wt.: 23.28 oz.   Sights: fixed, adjustable, Glock NS, or Trijicon NS   Capacity: 15, optional 17

Magazine: detachable box   Grips: polymer, Modular Back Strap System

D2C:	NIB	$ 565	Ex $ 450	VG+ $ 315	Good $ 285		
C2C:	Mint	$ 550	Ex $ 400	VG+ $ 290	Good $ 255	Fair $ 130
Trade-In:	Mint	$ 410	Ex $ 325	VG+ $ 230	Good $ 200	Poor $ 60

32
NEW IN BOX

32 REDBOOK CODE: RB-GK-H-32XXXX

Compact reinforced polymer frame with accessory rail, "Safe Action" System, cold-hammer-forged barrel with polygonal rifling, all metal parts treated with Nitration finish, add roughly $25 for "C" model.

Production: 1998 - current   Caliber: .357 SIG   Action: striker-fired, semi-auto

Barrel Length: 4.01"   OA Length: 7.36"

Wt.: 21.52 oz.   Sights: fixed, adjustable, Glock NS, or Trijicon NS

Capacity: 13, optional 15, 17   Magazine: detachable box   Grips: polymer grip frame

D2C:	NIB	$ 545	Ex $ 425	VG+ $ 300	Good $ 275		
C2C:	Mint	$ 530	Ex $ 400	VG+ $ 280	Good $ 250	Fair $	130
Trade-In:	Mint	$ 390	Ex $ 325	VG+ $ 220	Good $ 195	Poor $	60

32 GEN. 4
NEW IN BOX

32 GEN. 4 REDBOOK CODE: RB-GK-H-32GEN4

Compact reinforced polymer frame with accessory rail, "Safe Action" System, cold-hammer-forged barrel with polygonal rifling, all metal parts treated with Nitration finish, dual recoil spring assembly, reversible magazine catch for left or right-handed operation, Modular Back Strap, rough-textured grip.

Production: 2012 - current   Caliber: .357 SIG   Action: striker-fired, semi-auto

Barrel Length: 4.01"   OA Length: 7.36"

Wt.: 21.52 oz.   Sights: fixed, adjustable, Glock NS, or Trijicon NS   Capacity: 13, optional 10, 14

Magazine: detachable box   Grips: polymer, Modular Back Strap System

D2C:	NIB	$ 540	Ex $ 425	VG+ $ 300	Good $ 270		
C2C:	Mint	$ 520	Ex $ 375	VG+ $ 270	Good $ 245	Fair $	125
Trade-In:	Mint	$ 390	Ex $ 325	VG+ $ 220	Good $ 190	Poor $	60

33
NEW IN BOX

33 REDBOOK CODE: RB-GK-H-33XXXX

Subcompact reinforced polymer frame, "Safe Action" System, cold hammer-forged barrel with polygonal rifling, all metal parts treated with Nitration finish.

Production: 1997 - current   Caliber: .357 SIG   Action: striker-fired, semi-auto

Barrel Length: 3.42"   OA Length: 6.49"

Wt.: 19.75 oz.   Sights: fixed, adjustable, Glock NS, or Trijicon NS

Capacity: 9, optional 11, 13, 15, 17 Grips: polymer grip frame

D2C:	NIB	$ 520	Ex $ 400	VG+ $ 290	Good $ 260		
C2C:	Mint	$ 500	Ex $ 375	VG+ $ 260	Good $ 235	Fair $	120
Trade-In:	Mint	$ 370	Ex $ 300	VG+ $ 210	Good $ 185	Poor $	60

33 GEN. 4
NEW IN BOX

33 GEN. 4 REDBOOK CODE: RB-GK-H-33GEN4

Subcompact reinforced polymer frame, "Safe Action" System, cold-hammer-forged barrel with polygonal rifling, all metal parts treated with Nitration finish, dual-recoil spring assembly, reversible magazine catch for left or right-handed operation, Modular Back Strap, rough textured grip.

Production: 2012 - current   Caliber: .357 SIG   Action: striker-fired, semi-auto

Barrel Length: 3.43"   OA Length: 6.3"

Wt.: 19.75 oz.   Sights: fixed, adjustable, Glock NS, or Trijicon NS   Capacity: 9, optional 11, 13, 15, 17   Magazine: detachable box   Grips: polymer, Modular Back Strap System

D2C:	NIB	$ 545	Ex $ 425	VG+ $ 300	Good $ 275		
C2C:	Mint	$ 530	Ex $ 400	VG+ $ 280	Good $ 250	Fair $	130
Trade-In:	Mint	$ 390	Ex $ 325	VG+ $ 220	Good $ 195	Poor $	60

42
NEW IN BOX

42 REDBOOK CODE: RB-GK-H-42XXXX

Subcompact slimline polymer frame, "Safe Action" System, cold hammer-forged barrel with polygonal rifling, all metal parts treated with Nitration finish.

Production: 2014 - current	Caliber: .380 ACP	Action: striker-fired, semi-auto					
Barrel Length: 3.25"	OA Length: 5.94"						
Wt.: 13.76 oz.	Sights: fixed, adjustable, Glock NS, or Trijicon NS	Capacity: 6					
Magazine: detachable box	Grips: polymer grip frame						
D2C:	NIB	$ 540	Ex $ 425	VG+ $ 300	Good $ 270		
C2C:	Mint	$ 520	Ex $ 375	VG+ $ 270	Good $ 245	Fair $ 125	
Trade-In:	Mint	$ 390	Ex $ 325	VG+ $ 220	Good $ 190	Poor $ 60	

Hammerli AG

Hammerli AG—or simply Hammerli—traces its origins back to 1863, though Walther, a subsidiary of Umarex, currently owns the brand. Over the past several decades, Hammerli has established itself as one of the premier target and sport shooting firearms manufacturers. The company's pistols have made appearances in the Olympics on multiple occasions, winning gold for several teams, and helped establish the brand's superiority in the world of competitive marksmanship. Though not designed for carry and self-defense purposes, the company's pistols have secured their place in the canon of modern target handguns. Hammerli is located in Ulm, Germany.

MODEL 206 REDBOOK CODE: RB-HG-H-206XXX

Designed for sport shooting. Made in Switzerland. The standard version of the Hammerli 200 series.

Production: early 1960s - 1969	Caliber: .22 LR	Action: SA, semi-auto					
Barrel Length: 7.5"	Sights: adjustable front and rear	Capacity: 8					
Magazine: detachable box	Grips: checkered walnut						
D2C:	NIB	$ 845	Ex $ 650	VG+ $ 465	Good $ 425		
C2C:	Mint	$ 820	Ex $ 600	VG+ $ 430	Good $ 385	Fair $ 195	
Trade-In:	Mint	$ 600	Ex $ 475	VG+ $ 330	Good $ 300	Poor $ 90	

MODEL 208 REDBOOK CODE: RB-HG-H-208XXX

Designed for competition and sport shooting. Though not a commercial success, the gun is considered one of the finest competition shooting pistols. Drilled and tapped barrel. Made in Switzerland.

Production: mid-1960s - late 1980s	Caliber: .22 LR	Action: SA, semi-auto					
Barrel Length: 6"	Sights: fixed front, adjustable rear	Capacity: 8					
Magazine: detachable box	Grips: molded walnut, thumb rests						
D2C:	NIB	$1,900	Ex $1,450	VG+ $1,045	Good $ 950		
C2C:	Mint	$1,830	Ex $1,325	VG+ $ 950	Good $ 855	Fair $ 440	
Trade-In:	Mint	$1,350	Ex $1,075	VG+ $ 750	Good $ 665	Poor $ 210	

HAMMERLI AG

MODEL 230 RAPID FIRE PISTOL REDBOOK CODE: RB-HG-H-230RFP

Designed for competition and sport shooting, though now it typically doesn't meet match requirements. Noted for firing .22 Short rounds.

Production: 1970 - 1980s Caliber: .22 Short Action: SA, semi-auto

Barrel Length: 6.3" Sights: adjustable front and rear Capacity: 5

Magazine: detachable box Grips: molded walnut

D2C:	NIB	$ 945	Ex $ 725	VG+ $ 520	Good $ 475		
C2C:	Mint	$ 910	Ex $ 675	VG+ $ 480	Good $ 430	Fair $ 220
Trade-In:	Mint	$ 680	Ex $ 550	VG+ $ 370	Good $ 335	Poor $ 120

MODEL 280
EXCELLENT

MODEL 280 REDBOOK CODE: RB-HG-H-280XXX

Perhaps the most well received and desired Hammerli pistol. Noted for its enhanced and modern design. Made in Switzerland.

Production: late 1980s - discontinued Caliber: .22 LR, .32 S&W Long

Action: SA, semi-auto Barrel Length: 4.6" Sights: adjustable front and rear

Capacity: 5, 6 Magazine: detachable box Grips: molded walnut thumb and palm rests

D2C:	NIB	$1,425	Ex $1,100	VG+ $ 785	Good $ 715		
C2C:	Mint	$1,370	Ex $1,000	VG+ $ 720	Good $ 645	Fair $ 330
Trade-In:	Mint	$1,020	Ex $ 800	VG+ $ 560	Good $ 500	Poor $ 150

Harrington & Richardson, Inc.

Established in 1871 by Gilbert H. Harrington, Harrington & Richardson 1871, LLC.—or simply H&R—produces an array of revolvers, shotguns, and single-shot pistols noted for their affordability. The company has traded hands several times, but Marlin Firearms purchased the company in 2000 and continues to oversee their operations. H&R is located in Madison, North Carolina.

HARRINGTON & RICHARDSON

**AMERICAN
DOUBLE ACTION**
POOR

AMERICAN DOUBLE ACTION REDBOOK CODE: RB-HR-H-AMRDUA

Roughly 1,000,000 produced. Some rare models marked "Bulldog." Blue or nickel finish. Solid frame.

Production: 1880s - 1940s Caliber: .32 S&W, .38 cal. , .44 cal.

Action: DA/SA, revolver Barrel Length: 2.5", 4.5", 6" Sights: blade front

Capacity: 5, 6 Grips: black rubber

D2C:	NIB	--	Ex $ 250	VG+ $ 175	Good $ 160		
C2C:	Mint	--	Ex $ 225	VG+ $ 160	Good $ 145	Fair $ 75
Trade-In:	Mint	--	Ex $ 200	VG+ $ 130	Good $ 115	Poor $ 60

YOUNG AMERICA DOUBLE ACTION

REDBOOK CODE: RB-HR-H-YAMERX

A slightly modified variant of the American Double Action. One of the most popular double-action revolvers of its era. Between approximately 1,000,000 and 2,000,000 produced. Blue or nickel finish. Early models fire black powder rounds, which increases their value.

Production: 1880s - 1940s Caliber: .22 LR, .32 S&W Action: DA/SA, revolver

Barrel Length: 2.5", 4.5", 6" Sights: blade front Capacity: 5, 7

Grips: black rubber

D2C:	NIB	--	Ex $ 225	VG+ $ 160	Good $ 145		
C2C:	Mint	--	Ex $ 225	VG+ $ 150	Good $ 135	Fair $ 70
Trade-In:	Mint	--	Ex $ 175	VG+ $ 120	Good $ 105	Poor $ 30

**YOUNG AMERICA
DOUBLE ACTION**
GOOD

HUNTER
VERY GOOD +

HUNTER   REDBOOK CODE: RB-HR-H-HUNTER
Designed for plinking and small game hunting. Noted for it's exaggerated barrel.

Production: 1930s - late 1930s   Caliber: .22 LR   Action: DA/SA, revolver

Barrel Length: 10"   Sights: blade front   Capacity: 9   Grips: checkered walnut

D2C:	NIB	--	Ex	$ 425	VG+ $ 295	Good	$ 265		
C2C:	Mint	--	Ex	$ 375	VG+ $ 270	Good	$ 240	Fair $	125
Trade-In:	Mint	--	Ex	$ 300	VG+ $ 210	Good	$ 190	Poor $	60

MODEL 32
EXCELLENT

MODEL 532   REDBOOK CODE: RB-HR-H-532XXX
Minimal design. Cylinder doesn't swing out and must be completely removed before reloading. Blue finish.

Production: mid-1980s   Caliber: .32 S&W Long   Action: DA/SA, revolver

Barrel Length: 4", 6"   Sights: adjustable   Capacity: 5

Grips: polished walnut, synthetic

D2C:	NIB	$ 385	Ex	$ 300	VG+ $ 215	Good	$ 195		
C2C:	Mint	$ 370	Ex	$ 275	VG+ $ 200	Good	$ 175	Fair $	90
Trade-In:	Mint	$ 280	Ex	$ 225	VG+ $ 160	Good	$ 135	Poor $	60

MODEL 586
GOOD

MODEL 586   REDBOOK CODE: RB-HR-H-586XXX
A somewhat improved variant of earlier H&R M500 series revolvers.

Production: 1980s   Caliber: .32 H&R Magnum   Action: DA/SA, revolver

Barrel Length: 4.5", 5.5", 7.5", 10"   Sights: adjustable   Capacity: 5

Grips: polished walnut, synthetic

D2C:	NIB	$ 430	Ex	$ 350	VG+ $ 240	Good	$ 215		
C2C:	Mint	$ 420	Ex	$ 300	VG+ $ 220	Good	$ 195	Fair $	100
Trade-In:	Mint	$ 310	Ex	$ 250	VG+ $ 170	Good	$ 155	Poor $	60

MODEL 603
GOOD

MODEL 603   REDBOOK CODE: RB-HR-H-603XXX
Blue finish, squared barrel, smooth rotating cylinder, steel frame.

Production: early 1990s - discontinued   Caliber: .22 Magnum   Action: DA/SA, revolver

Barrel Length: 6"   Sights: adjustable   Capacity: 6   Grips: smooth walnut

D2C:	NIB	$ 335	Ex	$ 275	VG+ $ 185	Good	$ 170		
C2C:	Mint	$ 330	Ex	$ 250	VG+ $ 170	Good	$ 155	Fair $	80
Trade-In:	Mint	$ 240	Ex	$ 200	VG+ $ 140	Good	$ 120	Poor $	60

MODEL 622
GOOD

MODEL 622   REDBOOK CODE: RB-HR-H-622XXX
Minimal design, blue finish, steel frame, noted for its reliability.

Production: 1960s - 1980s   Caliber: .22 Magnum   Action: DA/SA, revolver

Barrel Length: 2.5", 4"   Sights: blade front, adjustable rear   Capacity: 6

Grips: smooth walnut, synthetic

D2C:	NIB	$ 235	Ex	$ 200	VG+ $ 130	Good	$ 120		
C2C:	Mint	$ 230	Ex	$ 175	VG+ $ 120	Good	$ 110	Fair $	55
Trade-In:	Mint	$ 170	Ex	$ 150	VG+ $ 100	Good	$ 85	Poor $	30

MODEL 623   REDBOOK CODE: RB-HR-H-623XXX

Nickel-finish version of the M622. Minimal design, blue finish, steel frame, noted for its reliability.

Production: 1960s - 1980s   Caliber: .22 Magnum   Action: DA/SA, revolver

Barrel Length: 2.5", 4"   Sights: blade front, adjustable rear   Capacity: 6

Grips: smooth walnut, synthetic

D2C:	NIB	$ 245	Ex $ 200	VG+ $ 135	Good $ 125		
C2C:	Mint	$ 240	Ex $ 175	VG+ $ 130	Good $ 115	Fair $ 60	
Trade-In:	Mint	$ 180	Ex $ 150	VG+ $ 100	Good $ 90	Poor $ 30	

MODEL 632
GOOD

MODEL 632   REDBOOK CODE: RB-HR-H-632XXX

.32 S&W Long variant of the M622. Minimal design, blue finish, steel frame, noted for its reliability.

Production: 1960s - 1980s   Caliber: .32 S&W Long   Action: DA/SA, revolver

Barrel Length: 2.5", 4"   Sights: blade front, adjustable rear   Capacity: 6

Grips: smooth walnut, synthetic

D2C:	NIB	$ 215	Ex $ 175	VG+ $ 120	Good $ 110		
C2C:	Mint	$ 210	Ex $ 150	VG+ $ 110	Good $ 100	Fair $ 50	
Trade-In:	Mint	$ 160	Ex $ 125	VG+ $ 90	Good $ 80	Poor $ 30	

MODEL 649   REDBOOK CODE: RB-HR-H-649XXX

A variant of the M622, distinguished by its longer barrel length.

Production: 1960s - 1980s   Caliber: .32 S&W Long   Action: DA/SA, revolver

Barrel Length: 5.5", 7.5"   Sights: blade front, adjustable rear   Capacity: 6

Grips: smooth walnut, synthetic

D2C:	NIB	$ 215	Ex $ 175	VG+ $ 120	Good $ 110		
C2C:	Mint	$ 210	Ex $ 150	VG+ $ 110	Good $ 100	Fair $ 50	
Trade-In:	Mint	$ 160	Ex $ 125	VG+ $ 90	Good $ 80	Poor $ 30	

MODEL 660 GUNFIGHTER   REDBOOK CODE: RB-HR-H-660GUN

Similar to other H&R 600 revolvers of the era, blue finish, minimal design.

Production: 1960s   Caliber: .22 Magnum   Action: DA/SA, revolver

Barrel Length: 5.5"   Sights: blade front   Capacity: 6   Grips: smooth walnut, synthetic

D2C:	NIB	$ 330	Ex $ 275	VG+ $ 185	Good $ 165		
C2C:	Mint	$ 320	Ex $ 250	VG+ $ 170	Good $ 150	Fair $ 80	
Trade-In:	Mint	$ 240	Ex $ 200	VG+ $ 130	Good $ 120	Poor $ 60	

MODEL 666 CONVERTIBLE
GOOD

MODEL 666 CONVERTIBLE   REDBOOK CODE: RB-HR-H-666CON

One of the more popular H&R revolvers. Blue finish, minimal design, steel frame.

Production: late 1970s - early 1980s   Caliber: .22 Magnum, .22 LR

Action: DA/SA, revolver   Barrel Length: 6"   Sights: blade front

Capacity: 6   Grips: smooth walnut, synthetic

D2C:	NIB	$ 235	Ex $ 200	VG+ $ 130	Good $ 120		
C2C:	Mint	$ 300	Ex $ 175	VG+ $ 120	Good $ 110	Fair $ 55	
Trade-In:	Mint	$ 170	Ex $ 150	VG+ $ 100	Good $ 85	Poor $ 30	

MODEL 686 CONVERTIBLE REDBOOK CODE: RB-HR-H-686CON
A variant of the M666, distinguished by its longer barrel length.

Production: late 1970s - early 1980s   Caliber: .22 Magnum, .22 LR

Action: DA/SA, revolver   Barrel Length: 4.5", 5.5", 7.5", 10", 12"

Sights: blade front   Capacity: 6   Grips: smooth walnut, synthetic

D2C:	NIB	$ 325	Ex	$ 250	VG+	$ 180	Good	$ 165		
C2C:	Mint	$ 320	Ex	$ 225	VG+	$ 170	Good	$ 150	Fair $	75
Trade-In:	Mint	$ 240	Ex	$ 200	VG+	$ 130	Good	$ 115	Poor $	60

MODEL 826
EXCELLENT

MODEL 732 GUARDSMAN REDBOOK CODE: RB-HR-H-732GRD
Blue finish, compact design.

Production: late 1950s - early 1970s   Caliber: .32 S&W Long   Action: DA/SA, revolver

Barrel Length: 2.5", 4"   Sights: blade front   Capacity: 6   Grips: black plastic

D2C:	NIB	$ 240	Ex	$ 200	VG+	$ 135	Good	$ 120		
C2C:	Mint	$ 240	Ex	$ 175	VG+	$ 120	Good	$ 110	Fair $	60
Trade-In:	Mint	$ 180	Ex	$ 150	VG+	$ 100	Good	$ 85	Poor $	30

MODEL 733 GUARDSMAN REDBOOK CODE: RB-HR-H-733GRD
Nickel finish, compact design.

Production: early 1970s   Caliber: .32 S&W Long   Action: DA/SA, revolver

Barrel Length: 2.5"   Sights: blade front   Capacity: 6   Grips: black plastic

D2C:	NIB	$ 260	Ex	$ 200	VG+	$ 145	Good	$ 130		
C2C:	Mint	$ 250	Ex	$ 180	VG+	$ 130	Good	$ 120	Fair $	60
Trade-In:	Mint	$ 190	Ex	$ 150	VG+	$ 110	Good	$ 95	Poor $	30

MODEL 826
EXCELLENT

MODEL 826 REDBOOK CODE: RB-HR-H-826XXX
Blue finish, compact design.

Production: early 1980s - discontinued   Caliber: .22 Magnum, .22 LR

Action: DA/SA, revolver   Barrel Length: 3", 5"   Sights: adjustable

Capacity: 6   Grips: polished wood

D2C:	NIB	$ 205	Ex	$ 175	VG+	$ 115	Good	$ 105		
C2C:	Mint	$ 200	Ex	$ 150	VG+	$ 110	Good	$ 95	Fair $	50
Trade-In:	Mint	$ 150	Ex	$ 125	VG+	$ 80	Good	$ 75	Poor $	30

MODEL 829 REDBOOK CODE: RB-HR-H-829XXX
Blue finish, compact design, noted for its 9 round capacity.

Production: early 1980s - discontinued   Caliber: .22 LR   Action: DA/SA, revolver

Barrel Length: 3"   Sights: adjustable   Capacity: 9   Grips: polished wood

D2C:	NIB	$ 193	Ex	$ 150	VG+	$ 110	Good	$ 100		
C2C:	Mint	$ 190	Ex	$ 150	VG+	$ 100	Good	$ 90	Fair $	45
Trade-In:	Mint	$ 140	Ex	$ 125	VG+	$ 80	Good	$ 70	Poor $	30

MODEL 922
SECOND SERIES
GOOD

MODEL 922 SECOND ISSUE REDBOOK CODE: RB-HR-H-922SES
Solid steel frame, blue finish, minimal design.

Production: 1950s - 1980s   Caliber: .22 LR   Action: DA/SA, revolver

Barrel Length: 2.5", 4", 6"   Sights: fixed   Capacity: 6   Grips: wood or synthetic

D2C:	NIB	$ 220	Ex	$ 175	VG+	$ 125	Good	$ 110		
C2C:	Mint	$ 220	Ex	$ 175	VG+	$ 110	Good	$ 100	Fair $	55
Trade-In:	Mint	$ 160	Ex	$ 125	VG+	$ 90	Good	$ 80	Poor $	30

MODEL 922 CAMPER   REDBOOK CODE: RB-HR-H-922CMP

A variant of the M922, featuring nickel finish rather than blue.

Production: 1950s - 1980s   Caliber: .22 LR   Action: DA/SA, revolver

Barrel Length: 2.5", 4", 6"   Sights: fixed   Capacity: 6

Grips: wood, ivory, or synthetic

D2C:	NIB	$ 220	Ex $ 175	VG+ $ 125	Good $ 110		
C2C:	Mint	$ 220	Ex $ 175	VG+ $ 110	Good $ 100	Fair $	55
Trade-In:	Mint	$ 160	Ex $ 125	VG+ $ 90	Good $ 80	Poor $	30

MODEL 925   REDBOOK CODE: RB-HR-H-925XXX

Break-top action, blue finish, compact design, similar to other H&R revolvers of the era.

Production: mid-1960s - mid-1980s   Caliber: .38 S&W   Action: DA/SA, revolver

Barrel Length: 2.5"   Sights: adjustable   Capacity: 5   Grips: wood, one-piece

D2C:	NIB	$ 315	Ex $ 250	VG+ $ 175	Good $ 160		
C2C:	Mint	$ 310	Ex $ 225	VG+ $ 160	Good $ 145	Fair $	75
Trade-In:	Mint	$ 230	Ex $ 200	VG+ $ 130	Good $ 115	Poor $	60

MODEL 925
EXCELLENT

MODEL 929 NEW MODEL SIDEKICK

REDBOOK CODE: RB-HR-H-929XXX

Indicative of other H&R revolvers of the era. Most in blue finish, some squared barrels. All sold with storage case.

Production: mid-1990s - late 1990s   Caliber: .22 Short, .22 LC

Action: DA/SA, revolver   Barrel Length: 2.5", 4", 6"   Sights: fixed

Capacity: 9   Grips: wood or ivory

D2C:	NIB	$ 290	Ex $ 225	VG+ $ 160	Good $ 145		
C2C:	Mint	$ 280	Ex $ 225	VG+ $ 150	Good $ 135	Fair $	70
Trade-In:	Mint	$ 210	Ex $ 175	VG+ $ 120	Good $ 105	Poor $	30

**MODEL 929 SIDEKICK
TRAPPER EDITION**
VERY GOOD +

MODEL 929 SIDEKICK TRAPPER EDITION

REDBOOK CODE: RB-HR-H-929SDK

The same as the Model 929, but distinguished by its gray grips and "Trapper Edition" markings.

Production: late 1990s   Caliber: .22 Short, .22 LC   Action: DA/SA, revolver

Barrel Length: 2.5", 4", 6"   Sights: fixed   Capacity: 9   Grips: synthetic, gray

D2C:	NIB	$ 205	Ex $ 175	VG+ $ 115	Good $ 105		
C2C:	Mint	$ 200	Ex $ 150	VG+ $ 110	Good $ 95	Fair $	50
Trade-In:	Mint	$ 150	Ex $ 125	VG+ $ 80	Good $ 75	Poor $	30

MODEL 939 ULTRA SIDEKICK   REDBOOK CODE: RB-HR-H-939ULS

Vent rib barrel, squared barrel, high-polish nickel finish, key-lock safety device.

Production: late 1950s - early 1980s   Caliber: .22 Short, .22 LC   Action: DA/SA, revolver

Barrel Length: 6"   Sights: adjustable   Capacity: 9   Grips: wood

D2C:	NIB	$ 325	Ex $ 250	VG+ $ 180	Good $ 165		
C2C:	Mint	$ 320	Ex $ 225	VG+ $ 170	Good $ 150	Fair $	75
Trade-In:	Mint	$ 240	Ex $ 200	VG+ $ 130	Good $ 115	Poor $	60

**MODEL 939
ULTRA SIDEKICK**
GOOD

MODEL 939 PREMIER REDBOOK CODE: RB-HR-H-939USM
Similar to the Model 929 Ultra. Vent rib barrel, squared barrel, high-polish blue finish, key-lock safety device.

Production: mid-1990s - late 1990s Caliber: .22 Short, .22 LC Action: DA/SA, revolver
Barrel Length: 6" Sights: adjustable Capacity: 9 Grips: wood

D2C:	NIB	$ 330	Ex	$ 275	VG+	$ 185	Good	$ 165	
C2C:	Mint	$ 320	Ex	$ 250	VG+	$ 170	Good	$ 150	Fair $ 80
Trade-In:	Mint	$ 240	Ex	$ 200	VG+	$ 130	Good	$ 120	Poor $ 60

MODEL 949 "FORTY NINER"
EXCELLENT

MODEL 949 "FORTY NINER" REDBOOK CODE: RB-HR-H-949FRN
Marked "Forty Niner Model 949." Steel frame, blue finish, shrouded ejector rod.

Production: 1960 - mid-1980s Caliber: .22 LR Action: DA/SA, revolver
Barrel Length: 5.5" Sights: adjustable Capacity: 9 Grips: polished wood

D2C:	NIB	$ 325	Ex	$ 250	VG+	$ 180	Good	$ 165	
C2C:	Mint	$ 320	Ex	$ 225	VG+	$ 170	Good	$ 150	Fair $ 75
Trade-In:	Mint	$ 240	Ex	$ 200	VG+	$ 130	Good	$ 115	Poor $ 60

MODEL 949 "FORTY NINER" WESTERN
REDBOOK CODE: RB-HR-H-949FX2
Similar to the Model 949, but features a more western-style design and varying barrel lengths.

Production: 1960 - mid-1980s Caliber: .22 LR Action: DA/SA, revolver
Barrel Length: 5.5", 7.5" Sights: adjustable Capacity: 9 Grips: polished wood

D2C:	NIB	$ 345	Ex	$ 275	VG+	$ 190	Good	$ 175	
C2C:	Mint	$ 340	Ex	$ 250	VG+	$ 180	Good	$ 160	Fair $ 80
Trade-In:	Mint	$ 250	Ex	$ 200	VG+	$ 140	Good	$ 125	Poor $ 60

MODEL 950
EXCELLENT

MODEL 950 REDBOOK CODE: RB-HR-H-950XXX
Similar to the Model 949, but with nickel finish.

Production: 1960 - mid-1980s Caliber: .22 LR Action: DA/SA, revolver
Barrel Length: 5.5", 7.5" Sights: fixed Capacity: 9 Grips: polished wood

D2C:	NIB	$ 345	Ex	$ 275	VG+	$ 190	Good	$ 175	
C2C:	Mint	$ 340	Ex	$ 250	VG+	$ 180	Good	$ 160	Fair $ 80
Trade-In:	Mint	$ 250	Ex	$ 200	VG+	$ 140	Good	$ 125	Poor $ 60

MODEL 976
VERY GOOD +

MODEL 976 REDBOOK CODE: RB-HR-H-976XXX
Similar to the Model 949, but features a case-hardened frame.

Production: 1960 - mid-1980s Caliber: .22 LR Action: DA/SA, revolver
Barrel Length: 5.5", 7.5" Sights: adjustable Capacity: 9 Grips: polished wood

D2C:	NIB	$ 300	Ex	$ 250	VG+	$ 165	Good	$ 150	
C2C:	Mint	$ 290	Ex	$ 225	VG+	$ 150	Good	$ 135	Fair $ 70
Trade-In:	Mint	$ 220	Ex	$ 175	VG+	$ 120	Good	$ 105	Poor $ 30

MODEL 999 SPORTSMAN
EXCELLENT

MODEL 999 SPORTSMAN REDBOOK CODE: RB-HR-H-999SPR
Break-top action, vent rib barrel, blue finish.

Production: early 1990s - late 1990s Caliber: .22 LR Action: DA/SA, revolver
Barrel Length: 4", 6" Sights: adjustable Capacity: 9 Grips: polished wood

D2C:	NIB	$ 570	Ex	$ 450	VG+	$ 315	Good	$ 285	
C2C:	Mint	$ 550	Ex	$ 400	VG+	$ 290	Good	$ 260	Fair $ 135
Trade-In:	Mint	$ 410	Ex	$ 325	VG+	$ 230	Good	$ 200	Poor $ 60

HECKLER & KOCH

Heckler & Koch

Founded in 1949, Heckler & Koch, often referred to as H&K, is headquartered in Germany, but has a substantial presence in U.S. markets, given their sales offices in both Columbus, Georgia and Ashburn, Virginia. Their pistols typically feature tactical-design elements, a polymer grip frame, accessory mounts, and polygonal rifling and have gained a substantial following in civilian and military markets, given their exceptional reputation. Though not a handgun, the MP5 likely stands as the company's most recognizable model, as it often appears in movies and television.

HK4
VERY GOOD +

HK4 (FOUR CALIBER SET) REDBOOK CODE: RB-HK-H-HK4CAS
Interchangeable caliber conversions and magazines, blue finish, early models imported and sold as the Harrington & Richardson HK4, less value for individual caliber models without all four conversions.

Production: late 1960s - 1984 Caliber: .22 LR, .25 ACP, .32 ACP, .380 ACP

Action: DA/SA, semi-auto Barrel Length: 3.35" Wt.: 18 oz.

Sights: fixed front, windage adjustable rear Capacity: 7, 8 Magazine: detachable box

Grips: checkered black plastic

D2C:	NIB	$ 650	Ex	$ 500	VG+	$ 360	Good	$ 325	
C2C:	Mint	$ 630	Ex	$ 450	VG+	$ 330	Good	$ 295	Fair	$ 150
Trade-In:	Mint	$ 470	Ex	$ 375	VG+	$ 260	Good	$ 230	Poor	$ 90

PSP (P7)
GOOD

PSP (P7) REDBOOK CODE: RB-HK-H-PSPP7X
Blue finish, squeezing the front strap cocks the striker, a smaller run imported in 1990.

Production: late 1980s Caliber: 9mm Action: striker-fired, semi-auto

Barrel Length: 4.13" Wt.: 28 oz. Sights: fixed Capacity: 8

Magazine: detachable box Grips: checkered black plastic

D2C:	NIB	$ 890	Ex	$ 700	VG+	$ 490	Good	$ 445	
C2C:	Mint	$ 860	Ex	$ 625	VG+	$ 450	Good	$ 405	Fair	$ 205
Trade-In:	Mint	$ 640	Ex	$ 500	VG+	$ 350	Good	$ 315	Poor	$ 90

P7 M8
VERY GOOD +

P7 M8 REDBOOK CODE: RB-HK-H-P7M8XX
Matte blue or nickel finish, ambidextrous safety, heat shield, squeezing the front strap cocks the striker, premium for nickel finish and tritium sights.

Production: discontinued Caliber: 9mm Action: striker-fired, semi-auto Barrel Length: 4.13" Wt.: 28 oz. Sights: 3-dot with adjustable rear or tritium night sights

Capacity: 8 Magazine: detachable box Grips: black plastic with stippling

D2C:	NIB	$1,715	Ex	$1,325	VG+	$ 945	Good	$ 860	
C2C:	Mint	$1,650	Ex	$1,200	VG+	$ 860	Good	$ 775	Fair	$ 395
Trade-In:	Mint	$1,220	Ex	$ 975	VG+	$ 670	Good	$ 605	Poor	$ 180

P7 M13
EXCELLENT

P7 M13 REDBOOK CODE: RB-HK-H-P7M13X
Matte blue or nickel finish, double-stack magazine, ambidextrous safety, heat shield, squeezing the front strap cocks the striker, premium for nickel finish and tritium sights.

Production: 1994 Caliber: 9mm Action: striker-fired, semi-auto Barrel Length: 4.13"

Wt.: 30 oz. Sights: 3-dot with adjustable rear or tritium night sights

Capacity: 13 Magazine: detachable box Grips: black plastic with stippling

D2C:	NIB	$ 2,510	Ex	$ 1,925	VG+	$1,385	Good	$ 1,255	
C2C:	Mint	$ 2,410	Ex	$ 1,750	VG+	$1,260	Good	$ 1,130	Fair	$ 580
Trade-In:	Mint	$ 1,790	Ex	$ 1,425	VG+	$ 980	Good	$ 880	Poor	$ 270

P7 M10 REDBOOK CODE: RB-HK-H-P7M10X

Blue or nickel finish, heavier that the other P7s, ambidextrous safety, heat shield, squeezing the front strap cocks the striker, premium for nickel finish and tritium sights.

Production: early 1990s - discontinued Caliber: .40 S&W Action: striker-fired, semi-auto

Barrel Length: 4.13" Wt.: 43 oz. Sights: 3-dot with adjustable rear or tritium

night sights Capacity: 10 Magazine: detachable box Grips: black plastic with stippling

D2C:	NIB	$ 2,715	Ex	$ 2,075	VG+	$ 1,495	Good	$ 1,360		
C2C:	Mint	$ 2,610	Ex	$ 1,875	VG+	$ 1,360	Good	$ 1,225	Fair	$ 625
Trade-In:	Mint	$ 1,930	Ex	$ 1,525	VG+	$ 1,060	Good	$ 955	Poor	$ 300

P7 K3 REDBOOK CODE: RB-HK-H-P7K3XX

This shorter P7 uses an oil-filled hydraulic recoil buffer in a blowback action. Squeezing the front strap cocks the striker. Barrel is removable for use with .22 LR and .32 ACP conversion kits.

Production: 1988 - 1994 Caliber: .380 ACP, (.22 LR & .32 ACP conversions)

Action: striker-fired, semi-auto Barrel Length: 3.8" Wt.: 26 oz.

Sights: 3-dot with adjustable rear or tritium night sights Capacity: 8

Magazine: detachable box Grips: black plastic with stippling

D2C:	NIB	$ 3,065	Ex	$ 2,350	VG+	$ 1,690	Good	$ 1,535		
C2C:	Mint	$ 2,950	Ex	$ 2,125	VG+	$ 1,540	Good	$ 1,380	Fair	$ 705
Trade-In:	Mint	$ 2,180	Ex	$ 1,725	VG+	$ 1,200	Good	$ 1,075	Poor	$ 330

HK45
NEW IN BOX

HK45 REDBOOK CODE: RB-HK-H-HK45XX

Cold hammer-forged barrel with polygonal rifling, o-ring barrel, modified Browning linkless recoil system, ambidextrous controls, Picatinny rail, combination safety/decocking lever, can be converted to nine firing modes.

Production: current Caliber: .45 ACP Action: DA/SA, DAO, semi-auto

Barrel Length: 4.53" Wt.: 31 oz. Sights: 3-dot contrast or tritium night sights

Capacity: 10 Magazine: detachable box

Grips: polymer grip frame with interchangeable backstraps

D2C:	NIB	$1,230	Ex	$ 950	VG+	$ 680	Good	$ 615		
C2C:	Mint	$1,190	Ex	$ 850	VG+	$ 620	Good	$ 555	Fair	$ 285
Trade-In:	Mint	$ 880	Ex	$ 700	VG+	$ 480	Good	$ 435	Poor	$ 150

**HK45 COMPACT
TACTICAL**
NEW IN BOX

HK45 COMPACT TACTICAL REDBOOK CODE: RB-HK-H-HK45CT

Compact version of the HK45 Tactical, can be converted to 9 firing modes, cold hammer-forged barrel with polygonal rifling, o-ring barrel, modified Browning linkless recoil system, ambidextrous controls, Picatinny rail, combination safety/decocking lever.

Production: current Caliber: .45 ACP Action: DA/SA, DAO, semi-auto

Barrel Length: 4.57" Wt.: 29 oz. Sights: 3-dot contrast or tritium night sights

Capacity:8, 10 Magazine: detachable box

Grips: polymer grip frame with interchangeable backstraps

D2C:	NIB	$1,000	Ex	$ 775	VG+	$ 550	Good	$ 500		
C2C:	Mint	$ 960	Ex	$ 700	VG+	$ 500	Good	$ 450	Fair	$ 230
Trade-In:	Mint	$ 710	Ex	$ 575	VG+	$ 390	Good	$ 350	Poor	$ 120

HECKLER & KOCH

HK45 COMPACT
NEW IN BOX

HK45 COMPACT REDBOOK CODE: RB-HK-H-HK45CO

Cold hammer-forged barrel with polygonal rifling, o-ring barrel, modified Browning linkless recoil system, ambidextrous controls, Picatinny rail, combination safety/decocking lever.

Production: current Caliber: .45 ACP Action: DA/SA, DAO, semi-auto

Barrel Length: 3.94" Wt.: 28 oz. Sights: 3-dot contrast or tritium night sights

Capacity: 8, 10 Magazine: detachable box

Grips: polymer grip frame with interchangeable backstraps

D2C:	NIB	$1,270	Ex	$ 975	VG+	$ 700	Good	$ 635	
C2C:	Mint	$1,220	Ex	$ 900	VG+	$ 640	Good	$ 575	Fair $ 295
Trade-In:	Mint	$ 910	Ex	$ 725	VG+	$ 500	Good	$ 445	Poor $ 150

P30
NEW IN BOX

P30 REDBOOK CODE: RB-HK-H-P30XXX

Fiber-reinforced polymer frame, polygonal bore profile, HK recoil reduction system, Picatinny rail, nitro-carburized steel slide, ambidextrous controls, automatic hammer and firing pin safeties, P30S model has ambidextrous safety levers.

Production: 2007 - current Caliber: 9mm, .40 S&W Action: DA/SA, DAO, semi-auto

Barrel Length: 3.85" OA Length: 6.95" Wt.: 26 oz. Sights: fixed front luminous contrast, open notch luminous contrast rear Capacity: 15

Magazine: detachable box Grips: polymer grip frame with interchangeable backstraps and lateral plates

D2C:	NIB	$ 880	Ex	$ 675	VG+	$ 485	Good	$ 440	
C2C:	Mint	$ 850	Ex	$ 625	VG+	$ 440	Good	$ 400	Fair $ 205
Trade-In:	Mint	$ 630	Ex	$ 500	VG+	$ 350	Good	$ 310	Poor $ 90

P30L
NEW IN BOX

P30L REDBOOK CODE: RB-HK-H-P30LXX

Identical to the P30 but with a longer barrel length. P30LS model has ambidextrous safety levers.

Production: 2008-current Caliber: 9mm, .40 S&W Action: DA/SA, semi-auto

Barrel Length: 4.44" OA Length: 7.56" Wt.: 27.5 oz. Sights: fixed front luminous contrast, open notch luminous contrast rear

Capacity: 15 Magazine: detachable box

Grips: polymer grip frame with interchangeable backstraps and lateral plates

D2C:	NIB	$ 950	Ex	$ 725	VG+	$ 525	Good	$ 475	
C2C:	Mint	$ 920	Ex	$ 675	VG+	$ 480	Good	$ 430	Fair $ 220
Trade-In:	Mint	$ 680	Ex	$ 550	VG+	$ 380	Good	$ 335	Poor $ 120

P2000
NEW IN BOX

P2000 REDBOOK CODE: RB-HK-H-P2000X

Polymer frame, accessory rail, ambidextrous magazine release, dual slide release levers.

Production: current Caliber: 9mm, .40 S&W, .357 SIG Action: DA/SA, DAO, semi-auto

Capacity: 10, 12, 13 Magazine: detachable box

Grips: polymer grip frame with interchangeable backstraps

D2C:	NIB	$ 915	Ex	$ 700	VG+	$ 505	Good	$ 460	
C2C:	Mint	$ 880	Ex	$ 650	VG+	$ 460	Good	$ 415	Fair $ 215
Trade-In:	Mint	$ 860	Ex	$ 700	VG+	$ 480	Good	$ 425	Poor $ 150

USP
NEW IN BOX

USP   REDBOOK CODE: RB-HK-H-USPXXX

Fiber reinforced polymer frame with steel inserts, polygonal bore profile, can be converted to nine trigger firing modes, nitro-carburized steel slide, accessory rail, ambidextrous magazine release lever, extended slide release, modified Browning-type action with HK recoil reduction system.

Production: 1993 - current   Caliber: 9mm, .40 S&W, .45 ACP

Action: DA/SA, DAO, semi-auto   Barrel Length: 4.25" (9mm & .40), 4.41" (.45)

Wt.: 27 oz. - 31 oz.   Sights: 3-dot contrast   Capacity: 10, 12, 13, 15

Magazine: detachable box   Grips: polymer grip frame

D2C:	NIB	$ 865	Ex	$ 675	VG+	$ 480	Good	$ 435		
C2C:	Mint	$ 840	Ex	$ 600	VG+	$ 440	Good	$ 390	Fair	$ 200
Trade-In:	Mint	$ 620	Ex	$ 500	VG+	$ 340	Good	$ 305	Poor	$ 90

USP COMPACT
NEW IN BOX

USP COMPACT   REDBOOK CODE: RB-HK-H-USPCPT

Compact version of the USP, extended or flush magazine floor plate, modified linkless Browning-type action, specially designed captive recoil spring assembly with polymer absorber bushing, bobbed hammer, mounting grooves, extended slide release.

Production: 1994 - current   Caliber: 9mm, .40 S&W, .45 ACP   Action: DA/SA, DAO, semi-auto   Barrel Length: 3.58" (9mm & .40), 3.8" (.45)   Wt.: 26 oz. - 28 oz.

Sights: 3-dot contrast   Capacity: 8, 12, 13   Magazine: detachable box

Grips: polymer grip frame

D2C:	NIB	$ 910	Ex	$ 700	VG+	$ 505	Good	$ 455		
C2C:	Mint	$ 880	Ex	$ 650	VG+	$ 460	Good	$ 410	Fair	$ 210
Trade-In:	Mint	$ 650	Ex	$ 525	VG+	$ 360	Good	$ 320	Poor	$ 120

USP TACTICAL
NEW IN BOX

USP TACTICAL   REDBOOK CODE: RB-HK-H-USPTAC

Threaded barrel, match grade trigger, adjustable trigger stop, nitro-carburized steel slide, can be converted to nine firing modes, HK recoil reduction system, oversized trigger guard, universal mounting grooves, ambidextrous magazine release lever, extended slide release, polymer frame, 9mm variant known as the USP9 SD with threaded barrel.

Production: current   Caliber: 9mm, .40 S&W, .45 ACP   Action: DA/SA, DAO, semi-auto

Barrel Length: 4.86" (9mm), 4.9" (.40), 5.09" (.45)   OA Length: 7.94" - 8.64"

Sights: high-profile adjustable target sights   Capacity: 10, 12, 13, 15

Magazine: detachable box   Grips: polymer grip frame

D2C:	NIB	$ 1,170	Ex	$ 900	VG+	$ 645	Good	$ 585		
C2C:	Mint	$ 1,130	Ex	$ 825	VG+	$ 590	Good	$ 530	Fair	$ 270
Trade-In:	Mint	$ 840	Ex	$ 675	VG+	$ 460	Good	$ 410	Poor	$ 120

USP COMPACT TACTICAL
NEW IN BOX

USP COMPACT TACTICAL   REDBOOK CODE: RB-HK-H-USPCTA

Smaller variation of the full-size USP45 Tactical, extended and threaded o-ring barrel with polygonal bore profile, match grade trigger, adjustable trigger stop, nitro-carburized steel slide, can be converted to nine firing modes, HK recoil reduction system, oversized trigger guard, universal mounting grooves, ambidextrous magazine release lever, extended slide release, polymer frame.

Production: current   Caliber: .45 ACP   Action: DA/SA, DAO, semi-auto

Barrel Length: 4.46"   OA Length: 7.72"   Wt.: 27.5 oz.   Sights: 3-dot contrast

Capacity: 10   Magazine: detachable box   Grips: polymer grip frame

D2C:	NIB	$ 1,210	Ex	$ 925	VG+	$ 670	Good	$ 605		
C2C:	Mint	$ 1,170	Ex	$ 850	VG+	$ 610	Good	$ 545	Fair	$ 280
Trade-In:	Mint	$ 860	Ex	$ 700	VG+	$ 480	Good	$ 425	Poor	$ 150

USP CUSTOM COMBAT
NEW IN BOX

USP EXPERT
NEW IN BOX
Courtesy of Bud's Gun Shop

USP ELITE
NEW IN BOX

MARK 23
EXCELLENT

USP CUSTOM COMBAT   REDBOOK CODE: RB-HK-H-USPCUC

HK recoil reduction system, ambidextrous magazine release, extended slide release, jet funnel magwell attachment, extended magazines.

Production: discontinued Caliber: 9mm, .40 S&W Action: DA/SA, semi-auto
Barrel Length: 4.25" Sights: Novak fiber-optic Capacity: 18 (9mm), 16 (.40)
Magazine: detachable box Grips: polymer grip frame

D2C:	NIB	$1,200	Ex	$ 925	VG+	$ 660	Good	$ 600	LMP	$ 1,199
C2C:	Mint	$1,160	Ex	$ 850	VG+	$ 600	Good	$ 540	Fair	$ 280
Trade-In:	Mint	$ 860	Ex	$ 675	VG+	$ 470	Good	$ 420	Poor	$ 120

USP EXPERT   REDBOOK CODE: RB-HK-H-USPEXP

O-ring barrel with polygonal bore profile, match grade trigger, ambidextrous control levers, adjustable trigger stop, HK recoil reduction system, universal mounting grooves, ambidextrous magazine release, available jet funnel magazine well attachment.

Production: late 1990s - discontinued Caliber: 9mm, .40 S&W, .45 ACP
Action: DA/SA, DAO, semi-auto Barrel Length: 5.2" OA Length: 8.7"
Sights: adjustable target Capacity: 12, 16, 18 Magazine: detachable box
Grips: polymer grip frame

D2C:	NIB	$1,270	Ex	$ 975	VG+	$ 700	Good	$ 635	LMP	$ 1,339
C2C:	Mint	$1,220	Ex	$ 900	VG+	$ 640	Good	$ 575	Fair	$ 295
Trade-In:	Mint	$ 910	Ex	$ 725	VG+	$ 500	Good	$ 445	Poor	$ 150

USP ELITE   REDBOOK CODE: RB-HK-H-USPELT

O-ring barrel, elongated milled target slide, ambidextrous control levers, match grade trigger, adjustable trigger stop, HK recoil reduction system, polymer frame, extended slide release.

Production: early 2000s - discontinued Caliber: 9mm, .45 ACP
Action: DA/SA, DAO, semi-auto Barrel Length: 6.02" Sights: adjustable target
Capacity: 12, 18 Magazine: detachable box Grips: polymer grip frame

D2C:	NIB	$1,345	Ex	$1,025	VG+	$ 740	Good	$ 675	LMP	$ 1,339
C2C:	Mint	$1,300	Ex	$ 950	VG+	$ 680	Good	$ 610	Fair	$ 310
Trade-In:	Mint	$ 960	Ex	$ 775	VG+	$ 530	Good	$ 475	Poor	$ 150

MARK 23   REDBOOK CODE: RB-HK-H-MARK23

Polymer frame, machined steel slide, threaded o-ring barrel with polygonal bore profile, available in 10 firing variants, match grade trigger, decocking lever and separate ambidextrous safety lever, HK recoil reduction system, universal mounting grooves, ambidextrous magazine release, extended slide release.

Production: 1996 - current Caliber: .45 ACP Action: DA/SA, DAO, semi-auto
Barrel Length: 5.9" OA Length: 9.7" Wt.: 39 oz. Sights: 3-dot contrast
Capacity: 10, 12 Magazine: detachable box Grips: polymer grip frame

D2C:	NIB	$ 2,125	Ex	$ 1,625	VG+	$ 1,170	Good	$ 1,065		
C2C:	Mint	$ 2,040	Ex	$ 1,475	VG+	$ 1,070	Good	$ 960	Fair	$ 490
Trade-In:	Mint	$ 1,510	Ex	$ 1,200	VG+	$ 830	Good	$ 745	Poor	$ 240

Heritage Manufacturing, Inc.

Established in the early 1990s, Heritage Manufacturing, Inc. produces a series of revolvers and pistols, the Rough Rider being the most noted among these. As of 2012, Taurus purchased Heritage and now manufactures all of the company's firearms. The company is based in Miami, Florida.

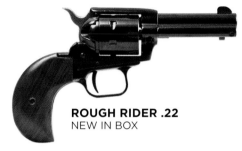

ROUGH RIDER .22
NEW IN BOX

ROUGH RIDER .22 (MB3BH)   REDBOOK CODE: RB-HM-H-RRXXX1

Small bore revolver similar to the Colt Single Action Army. Blue finish, aluminum-alloy frame, hammer block, also available in silver satin finish and other grip materials.

Production: current Caliber: .22 LR, .22 Magnum Action: SA, revolver

Barrel Length: 3.5" Wt.: 33 oz. Sights: fixed, notch at rear

Capacity: 6 Grips: polished cocobolo

D2C:	NIB	$ 215	Ex	$ 175	VG+ $ 120	Good	$ 110	LMP	$ 225
C2C:	Mint	$ 210	Ex	$ 150	VG+ $ 110	Good	$ 100	Fair	$ 50
Trade-In:	Mint	$ 160	Ex	$ 125	VG+ $ 90	Good	$ 80	Poor	$ 30

ROUGH RIDER .22
NEW IN BOX

ROUGH RIDER .22 (MB6PRL)   REDBOOK CODE: RB-HM-H-RRXXX2

Small bore revolver similar to the Colt Single Action Army. Blue finish, aluminum-alloy frame, hammer block, optional silver satin finish.

Production: current Caliber: .22 LR, .22 Magnum Action: SA, revolver

Barrel Length: 6.5" Wt.: 34 oz. Sights: fixed, notch at rear

Capacity: 6 Grips: white mother-of-pearl

D2C:	NIB	$ 225	Ex	$ 175	VG+ $ 125	Good	$ 115	LMP	$ 250
C2C:	Mint	$ 220	Ex	$ 175	VG+ $ 120	Good	$ 105	Fair	$ 55
Trade-In:	Mint	$ 160	Ex	$ 150	VG+ $ 90	Good	$ 80	Poor	$ 30

ROUGH RIDER .22 (MBS4)   REDBOOK CODE: RB-HM-H-RRXXX3

Small bore revolver similar to other Rough Rider models, but includes an interchangeable cylinder, allowing the gun to fire .22 LR and .22 Magnum. Black satin finish, aluminum alloy frame.

Production: current Caliber: .22 LR, .22 Magnum (interchangeable cylinder included)

Action: SA, revolver Barrel Length: 4.75" Wt: 33 oz. Sights: blade Capacity: 6

Grips: camo green laminate

D2C:	NIB	$ 240	Ex	$ 200	VG+ $ 135	Good	$ 120	LMP	$ 258
C2C:	Mint	$ 240	Ex	$ 175	VG+ $ 120	Good	$ 110	Fair	$ 60
Trade-In:	Mint	$ 180	Ex	$ 150	VG+ $ 100	Good	$ 85	Poor	$ 30

ROUGH RIDER .45
NEW IN BOX

ROUGH RIDER .45   REDBOOK CODE: RB-HM-H-RRXXX4

Big bore revolver similar to the Colt Single Action Army. Blue finish, steel components, partially shrouded ejector rod, various finishes.

Production: current Caliber: .45 LC Action: SA, revolver Barrel Length: 4.75"

Wt.: 36 oz. Sights: open fixed, notch at rear Capacity: 6 Grips: polished cocobolo

D2C:	NIB	$ 440	Ex	$ 350	VG+ $ 245	Good	$ 220	LMP	$ 467
C2C:	Mint	$ 430	Ex	$ 325	VG+ $ 220	Good	$ 200	Fair	$ 105
Trade-In:	Mint	$ 320	Ex	$ 250	VG+ $ 180	Good	$ 155	Poor	$ 60

HERITAGE

ROUGH RIDER .357 (B4)   REDBOOK CODE: RB-HM-H-RRXXX5

A big bore .357 Magnum variant of the Rough Rider. Blue or nickel finish, steel components, various finishes.

Production: current Caliber: .357 Magnum Action: SA, revolver

Barrel Length: 4.75" Wt.: 36 oz. Sights: open fixed, notch at rear

Capacity: 6 Grips: polished cocobolo

D2C:	NIB	$ 430	Ex $ 350	VG+ $ 240	Good $ 215	LMP $ 467
C2C:	Mint	$ 420	Ex $ 300	VG+ $ 220	Good $ 195	Fair $ 100
Trade-In:	Mint	$ 310	Ex $ 250	VG+ $ 170	Good $ 155	Poor $ 60

ROUGH RIDER .357 (B5)
NEW IN BOX

ROUGH RIDER .357 (B5)   REDBOOK CODE: RB-HM-H-RRXXX6

A big bore .357 Magnum revolver variant of the Rough Rider, distinguished by its extended barrel. Blue finish, steel components, various finishes.

Production: current Caliber: .357 Magnum Action: SA, revolver Barrel Length: 5.5"

Wt.: 36 oz. Sights: fixed, notch at rear Capacity: 6 Grips: polished cocobolo

D2C:	NIB	$ 435	Ex $ 350	VG+ $ 240	Good $ 220	
C2C:	Mint	$ 420	Ex $ 325	VG+ $ 220	Good $ 200	Fair $ 105
Trade-In:	Mint	$ 310	Ex $ 250	VG+ $ 170	Good $ 155	Poor $ 60

SENTRY   REDBOOK CODE: RB-HM-H-SENTRY

Compact snub-nose design, blue or nickel finish, cylinder doesn't swing out from frame.

Production: late 1990s - early 2000s   Caliber: .38 Special, .32 Mag, 9mm,

.22 LR, .22 WMR   Action: DA/SA, revolver   Barrel Length: 2", 4"

Sights: fixed   Capacity: 6, 8   Grips: wood or synthetic

D2C:	NIB	$ 130	Ex $ 100	VG+ $ 75	Good $ 65	
C2C:	Mint	$ 130	Ex $ 95	VG+ $ 70	Good $ 60	Fair $ 30
Trade-In:	Mint	$ 100	Ex $ 75	VG+ $ 60	Good $ 55	Poor $ 25

H25   REDBOOK CODE: RB-HM-H-H25XXX

Pocket-sized design, blue or nickel finish, briefly produced.

Production: mid-1990s - late 1990s Caliber: .25 ACP Action: SA, semi-auto

Barrel Length: 2.25" Wt.: 14 oz. Sights: fixed Capacity: 6

Magazine: detachable box Grips: wood or synthetic

D2C:	NIB	$ 165	Ex $ 150	VG+ $ 95	Good $ 85	
C2C:	Mint	$ 160	Ex $ 125	VG+ $ 90	Good $ 75	Fair $ 40
Trade-In:	Mint	$ 120	Ex $ 100	VG+ $ 70	Good $ 60	Poor $ 30

STEALTH   REDBOOK CODE: RB-HM-STELTH

Heritage's most successful striker-fired pistol. Modern design, polymer frame, blue or stainless steel slide, frame-mounted safety. Offered in black, two-tone black chrome/stainless, or black chrome finishes.

Production: mid-1990s - early 2000s Caliber: 9mm, .40 S&W

Action: striker-fired, semi-auto Barrel Length: 4" Wt.: 20 oz. Sights: fixed

Capacity: 10 Magazine: detachable box Grips: black synthetic

D2C:	NIB	$ 335	Ex $ 275	VG+ $ 185	Good $ 170	
C2C:	Mint	$ 330	Ex $ 250	VG+ $ 170	Good $ 155	Fair $ 80
Trade-In:	Mint	$ 240	Ex $ 200	VG+ $ 140	Good $ 120	Poor $ 60

Hi-Point Firearms

Established in the early 1990s, Hi-Point Firearms manufactures a series of affordable striker-fired pistols, most of which feature minimalist, unique designs. Firearms critics and enthusiasts have noted Hi-Point pistols for their reliability, heavy frames, and sturdy construction, especially considering their low retail price. Hi-Point is owned by Beemiller, Inc. and based in Mansfield, Ohio.

CF-380
NEW IN BOX

CF-380   REDBOOK CODE: RB-HP-H-CF380X
Option of 10-shot magazine, black powder coat, chrome rail, polymer frame, extra rear peep sight and trigger lock.

Production: current Caliber: .380 ACP Action: striker-fired, semi-auto
Barrel Length: 3.5" OA Length: 6.75" Wt.: 29 oz. Sights: 3-dot, adjustable
Capacity: 8, 10 Grips: polymer

D2C:	NIB	$ 145	Ex $ 125	VG+ $ 80	Good $ 75	LMP $ 158
C2C:	Mint	$ 140	Ex $ 125	VG+ $ 80	Good $ 70	Fair $ 35
Trade-In:	Mint	$ 110	Ex $ 100	VG+ $ 60	Good $ 55	Poor $ 30

C-9
NEW IN BOX

C-9   REDBOOK CODE: RB-HP-H-MC9XXX
Option of 10-shot magazine, black powder coat, polymer frame, extra rear peep sight and trigger lock.

Production: current Caliber: 9mm Action: striker-fired, semi-auto
Barrel Length: 3.5" OA Length: 6.75" Wt.: 29 oz.
Sights: 3-dot, adjustable rear Capacity: 8, 10 Grips: polymer

D2C:	NIB	$ 170	Ex $ 150	VG+ $ 95	Good $ 85	LMP $ 190
C2C:	Mint	$ 170	Ex $ 125	VG+ $ 90	Good $ 80	Fair $ 40
Trade-In:	Mint	$ 130	Ex $ 100	VG+ $ 70	Good $ 60	Poor $ 30

JCP-40
NEW IN BOX

JCP-40   REDBOOK CODE: RB-HP-H-JCP40X
Black powder coat, polymer frame, extra rear peep sight and trigger lock. Optional laser, green slide, and hard case each for an additional cost.

Production: current Caliber: .40 S&W Action: striker-fired, semi-auto
Barrel Length: 4.5" OA Length: 7.75" Wt.: 35 oz. Sights: 3-dot, adjustable rear
Capacity: 10 Grips: polymer

D2C:	NIB	$ 185	Ex $ 150	VG+ $ 105	Good $ 95	LMP $ 210
C2C:	Mint	$ 180	Ex $ 150	VG+ $ 100	Good $ 85	Fair $ 45
Trade-In:	Mint	$ 140	Ex $ 125	VG+ $ 80	Good $ 65	Poor $ 30

JCP-45
NEW IN BOX

JHP-45   REDBOOK CODE: RB-HP-H-JHP45X
Black powder coat, polymer frame, trigger lock. Optional laser, green slide, and hard case each for an additional price.

Production: current Caliber: .45 ACP Action: striker-fired, semi-auto
Barrel Length: 4.5" OA Length: 7.75" Wt.: 35 oz. Sights: 3-dot, adjustable rear
Capacity: 9 Grips: polymer

D2C:	NIB	$ 180	Ex $ 150	VG+ $ 100	Good $ 90	LMP $ 200
C2C:	Mint	$ 180	Ex $ 125	VG+ $ 90	Good $ 85	Fair $ 45
Trade-In:	Mint	$ 130	Ex $ 125	VG+ $ 80	Good $ 65	Poor $ 30

High Standard

The High Standard Manufacturing Company, Inc. of Houston, Texas traces its origins to High Standard of Connecticut, but the two companies operated as separate entities and had different ownership. A 15-year break in production existed from the time High Standard ceased operations and the time High Standard Manufacturing of Houston, Texas purchased the rights, trademarks, and assets of the previous company. High Standard Manufacturing, like High Standard, produces several target-style .22 LR pistols, but also manufactures a host of different semi-automatic and M1911-style pistols.

DERRINGER (GEN 1) REDBOOK CODE: RB-HS-H-DRRXX1
Marked "D-100" or "DM-100." Over/under action, pocket-size design, stainless or blue finish.

Production: 1962 - 1967 Caliber: .22 LR, .22 Magnum Action: DAO, derringer
Barrel Length: 3.5" Sights: fixed Capacity: 2
Grips: synthetic

D2C:	NIB	$ 385	Ex $ 300	VG+ $ 215	Good $ 195		
C2C:	Mint	$ 370	Ex $ 275	VG+ $ 200	Good $ 175	Fair $ 90
Trade-In:	Mint	$ 280	Ex $ 225	VG+ $ 160	Good $ 135	Poor $ 60

DERRINGER (GEN 2) REDBOOK CODE: RB-HS-H-DDRXX2
Not marked with an eagle emblem, as seen on the Gen 1 Derringer. Similar design otherwise.

Production: 1967 - 1970 Caliber: .22 LR, .22 Magnum Action: DAO, derringer
Barrel Length: 3.5" Sights: fixed Capacity: 2 Grips: synthetic

D2C:	NIB	$ 315	Ex $ 250	VG+ $ 175	Good $ 160		
C2C:	Mint	$ 310	Ex $ 225	VG+ $ 160	Good $ 145	Fair $ 75
Trade-In:	Mint	$ 230	Ex $ 200	VG+ $ 130	Good $ 115	Poor $ 60

MODEL C
VERY GOOD +

MODEL C REDBOOK CODE: RB-HS-H-MCXXXX
Serial numbers begin around approximately 18,700. Fires .22 short rounds. Less than 5,000 produced.

Production: mid-1930s - discontinued Caliber: .22 Short Action: SA, semi-auto
Barrel Length: 4.5", 6.75" Sights: fixed Capacity: 10 Magazine: detachable box
Grips: checkered synthetic

D2C:	NIB	--	Ex $ 1,000	VG+ $ 715	Good $ 650		
C2C:	Mint	--	Ex $ 900	VG+ $ 650	Good $ 585	Fair $ 300
Trade-In:	Mint	--	Ex $ 750	VG+ $ 510	Good $ 455	Poor $ 150

MODEL A
VERY GOOD +

MODEL A REDBOOK CODE: RB-HS-H-MAXXXX
Similar to the Model A, distinguished by its light-weight barrel, walnut grips, and modified sights. Less than 7,500 manufactured.

Production: late 1930s - discontinued Caliber: .22 LR Action: SA, semi-auto
Barrel Length: 4.5", 6.75" Sights: adjustable Capacity: 10
Magazine: detachable box Grips: checkered walnut

D2C:	NIB	--	Ex $ 1,075	VG+ $ 770	Good $ 700		
C2C:	Mint	--	Ex $ 975	VG+ $ 700	Good $ 630	Fair $ 325
Trade-In:	Mint	--	Ex $ 800	VG+ $ 550	Good $ 490	Poor $ 150

MODEL D REDBOOK CODE: RB-HS-H-MDXXXX
Serial numbers begin around approximately 39,400. A bull-barrel variant of the Model A. Less than 3,000 manufactured.

Production: late 1930s - discontinued Caliber: .22 LR Action: SA, semi-auto

Barrel Length: 4.5", 6.75" Sights: adjustable Capacity: 10 Grips: walnut or synthetic

D2C:	NIB	--	Ex	$1,050	VG+	$ 745	Good	$ 675	
C2C:	Mint	--	Ex	$ 950	VG+	$ 680	Good	$ 610	Fair	$ 315
Trade-In:	Mint	--	Ex	$ 775	VG+	$ 530	Good	$ 475	Poor	$ 150

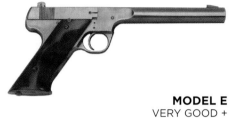

MODEL E
VERY GOOD +

MODEL E REDBOOK CODE: RB-HS-H-MEXXXX
Serial numbers begin around approximately 39,400. A variant of the Model D, but features an even thicker bull-barrel. Less than 3,000 manufactured.

Production: late 1930s - discontinued Caliber: .22 LR Action: SA, semi-auto

Barrel Length: 4.5", 6.75" Sights: adjustable Capacity: 10 Magazine: detachable box

Grips: walnut or synthetic

D2C:	NIB	--	Ex	$ 1,275	VG+	$ 905	Good	$ 825	
C2C:	Mint	--	Ex	$ 1,150	VG+	$ 830	Good	$ 745	Fair	$ 380
Trade-In:	Mint	--	Ex	$ 925	VG+	$ 650	Good	$ 580	Poor	$ 180

MODEL H-D
VERY GOOD +

MODEL H-D REDBOOK CODE: RB-HS-H-MHDXXX
Serial numbers begin around approximately 70,700. An exposed-hammer variant of the Model D. Less than 7,000 manufactured.

Production: 1940 - discontinued Caliber: .22 LR Action: SA, semi-auto

Barrel Length: 4.5", 6.75" Sights: adjustable Capacity: 10 Magazine: detachable box

Grips: walnut or synthetic

D2C:	NIB	--	Ex	$ 750	VG+	$ 540	Good	$ 490	
C2C:	Mint	--	Ex	$ 700	VG+	$ 490	Good	$ 445	Fair	$ 230
Trade-In:	Mint	--	Ex	$ 550	VG+	$ 390	Good	$ 345	Poor	$ 120

MODEL G-E REDBOOK CODE: RB-HS-H-MGEXXX
Serial numbers begin around approximately 326,000. A variant of the Model E. Less than 3,000 manufactured.

Production: late 1940s - 1950 Caliber: .22 LR Action: striker-fired, semi-auto

Barrel Length: 4.5", 6.75" Sights: walnut, molded Capacity: 10

Magazine: detachable box Grips: walnut

D2C:	NIB	--	Ex	$ 825	VG+	$ 595	Good	$ 540	
C2C:	Mint	--	Ex	$ 750	VG+	$ 540	Good	$ 485	Fair	$ 250
Trade-In:	Mint	--	Ex	$ 625	VG+	$ 420	Good	$ 380	Poor	$ 120

OLYMPIC (G-O) REDBOOK CODE: RB-HS-H-OLYXXX
Modified magazine design. Less than 1,500 produced. Includes a frame-mounted safety.

Production: 1949 - 1950 Caliber: .22 Short Action: striker-fired, semi-auto

Barrel Length: 4.5", 6.75" Sights: adjustable Capacity: 10 Magazine: detachable box

Grips: walnut, molded

D2C:	NIB	--	Ex	$1,350	VG+	$ 690	Good	$ 625	
C2C:	Mint	--	Ex	$ 875	VG+	$ 630	Good	$ 565	Fair	$ 290
Trade-In:	Mint	--	Ex	$ 700	VG+	$ 490	Good	$ 440	Poor	$ 150

HIGH STANDARD

SPORT KING
NEW IN BOX

SPORT KING   REDBOOK CODE: RB-HS-H-SPTKNG
Blue finish and lightweight barrel.

Production: 1950 - mid-1950s   Caliber: .22 Short   Action: striker-fired, semi-auto
Barrel Length: 4.5", 6.75"   Sights: fixed   Capacity: 10   Magazine: detachable box
Grips: synthetic, molded

D2C:	NIB	--	Ex	$ 500	VG+	$ 365	Good	$ 330	
C2C:	Mint	--	Ex	$ 475	VG+	$ 330	Good	$ 295	Fair	$ 155
Trade-In:	Mint	--	Ex	$ 375	VG+	$ 260	Good	$ 230	Poor	$ 90

FIELD KING
NEW IN BOX

FIELD KING   REDBOOK CODE: RB-HS-H-FLDKNG
Bull or standard barrel. Frame-mounted safety.

Production: early 1950s - 1953   Caliber: .22 LR   Action: striker-fired, semi-auto
Barrel Length: 4.5", 6.75"   Sights: adjustable   Capacity: 10   Magazine: detachable box
Grips: synthetic, molded

D2C:	NIB	--	Ex	$ 600	VG+	$ 435	Good	$ 395	
C2C:	Mint	--	Ex	$ 550	VG+	$ 400	Good	$ 355	Fair	$ 185
Trade-In:	Mint	--	Ex	$ 450	VG+	$ 310	Good	$ 275	Poor	$ 90

SUPERMATIC
VERY GOOD +

SUPERMATIC   REDBOOK CODE: RB-HS-H-SPRMTC
Updated design. Ribbed barrel, frame-mounted safety.

Production: 1949 - 1957   Caliber: .22 LR   Action: striker-fired, semi-auto
Barrel Length: 4.5", 6.75"   Sights: adjustable   Capacity: 10   Magazine: detachable box
Grips: synthetic, molded

D2C:	NIB	--	Ex	$ 700	VG+	$ 505	Good	$ 460	
C2C:	Mint	--	Ex	$ 650	VG+	$ 460	Good	$ 415	Fair	$ 215
Trade-In:	Mint	--	Ex	$ 525	VG+	$ 360	Good	$ 325	Poor	$ 120

OLYMPIC
GOOD

OLYMPIC   REDBOOK CODE: RB-HS-H-OLYXX1
Bull barrel, blue finish, ribbed barrel, scope mounts.

Production: late 1940s - late 1950s   Caliber: .22 Short   Action: striker-fired, semi-auto
Barrel Length: 4.5", 6.75"   Sights: adjustable   Capacity: 10   Magazine: detachable box
Grips: synthetic, molded

D2C:	NIB	--	Ex	$1,100	VG+	$790	Good	$ 720	
C2C:	Mint	--	Ex	$1,000	VG+	$720	Good	$ 650	Fair	$ 335
Trade-In:	Mint	--	Ex	$ 825	VG+	$560	Good	$ 505	Poor	$ 150

FLITE KING LW-100
VERY GOOD +

FLITE KING LW-100   REDBOOK CODE: RB-HS-H-FLTKNG
Serial numbers range between approximately 475,000 and 778,000. The Flite King is noted for its aluminum frame. Series 100 models typically feature detachable barrels and barrel weights.

Production: 1954 - 1957   Caliber: .22 Short   Action: striker-fired, semi-auto
Barrel Length: 4.5", 6.75"   Sights: fixed   Capacity: 10   Magazine: detachable box
Grips: synthetic, molded, checkered

D2C:	NIB	--	Ex	$ 450	VG+	$ 315	Good	$ 285	
C2C:	Mint	--	Ex	$ 400	VG+	$ 290	Good	$ 260	Fair	$ 135
Trade-In:	Mint	--	Ex	$ 325	VG+	$ 230	Good	$ 200	Poor	$ 60

SUPERMATIC S-100 REDBOOK CODE: RB-HS-H-1FLTXX
Included barrel weights. Back strap, frame-mounted safety, limited production model.

Production: 1954 Caliber: .22 LR Action: striker-fired, semi-auto

Barrel Length: 4.5", 6.75" Sights: adjustable Capacity: 10 Magazine: detachable box

Grips: synthetic, molded

D2C:	NIB	--	Ex	$ 800	VG+ $ 580	Good $ 525			
C2C:	Mint	--	Ex	$ 725	VG+ $ 530	Good $ 475	Fair	$ 245	
Trade-In:	Mint	--	Ex	$ 600	VG+ $ 410	Good $ 370	Poor	$ 120	

OLYMPIC O-100 REDBOOK CODE: RB-HS-H-1OLYXX
Interchangeable barrel capabilities. Weighted barrel options, frame-mounted safety.

Production: 1954 Caliber: .22 LR Action: striker-fired, semi-auto

Barrel Length: 4.5", 6.75" Sights: adjustable Capacity: 10 Magazine: detachable box

Grips: synthetic, molded

D2C:	NIB	--	Ex	$ 850	VG+ $ 605	Good $ 550			
C2C:	Mint	--	Ex	$ 775	VG+ $ 550	Good $ 495	Fair	$ 255	
Trade-In:	Mint	--	Ex	$ 625	VG+ $ 430	Good $ 385	Poor	$ 120	

FIELD KING FK-100 REDBOOK CODE: RB-HS-H-1FLDKG
Similar to the Field King model, limited production model.

Production: 1954 Caliber: .22 LR Action: striker-fired, semi-auto

Barrel Length: 4.5", 6.75" Sights: adjustable Capacity: 10 Magazine: detachable box

Grips: synthetic, molded

D2C:	NIB	--	Ex	$ 725	VG+ $ 525	Good $ 475			
C2C:	Mint	--	Ex	$ 675	VG+ $ 480	Good $ 430	Fair	$ 220	
Trade-In:	Mint	--	Ex	$ 550	VG+ $ 380	Good $ 335	Poor	$ 120	

SPORT KING SK-100
VERY GOOD +

SPORT KING SK-100 REDBOOK CODE: RB-HS-H-1SPRTX
Serial numbers range between approximately 475,000 and 778,000. Similar to other 100 Series models of the era, noted for their detachable barrels.

Production: 1954 - late 1950s Caliber: .22 LR Action: striker-fired, semi-auto

Barrel Length: 4.5", 6.75" Sights: blade front, dovetail rear Capacity: 10

Magazine: detachable box Grips: synthetic

D2C:	NIB	--	Ex	$ 500	VG+ $ 350	Good $ 320			
C2C:	Mint	--	Ex	$ 450	VG+ $ 320	Good $ 290	Fair	$ 150	
Trade-In:	Mint	--	Ex	$ 375	VG+ $ 250	Good $ 225	Poor	$ 90	

DURA-MATIC M-100 REDBOOK CODE: RB-HS-H-1DURXX
Serial numbers begin around approximately 450,000. Similar to the Series 100.

Production: 1954 Caliber: .22 LR Action: striker-fired, semi-auto

Barrel Length: 4.5", 6.75" Sights: blade front, fixed rear Capacity: 10

Magazine: detachable box Grips: synthetic, molded

D2C:	NIB	--	Ex	$ 425	VG+ $ 300	Good $ 275			
C2C:	Mint	--	Ex	$ 400	VG+ $ 280	Good $ 250	Fair	$ 130	
Trade-In:	Mint	--	Ex	$ 325	VG+ $ 220	Good $ 195	Poor	$ 60	

HIGH STANDARD

OLYMPIC O-101
VERY GOOD +

DURA-MATIC M-101
GOOD

SPORT KING 102 SERIES
VERY GOOD +

OLYMPIC O-101 REDBOOK CODE: RB-HS-H-101XXX

The 101 Series models are an improved variant of the earlier 100 Series and noted for their push-button detachable barrels and modified design.

Production: 1954 - 1957 Caliber: .22 Short Action: striker-fired, semi-auto

Barrel Length: 4.5", 6.75" Sights: blade front, dovetail rear Capacity: 10

Magazine: detachable box Grips: synthetic, molded

D2C:	NIB	--	Ex	$ 950	VG+ $ 680	Good	$ 620		
C2C:	Mint	--	Ex	$ 875	VG+ $ 620	Good	$ 560	Fair	$ 285
Trade-In:	Mint	--	Ex	$ 700	VG+ $ 490	Good	$ 435	Poor	$ 150

DURA-MATIC M-101 REDBOOK CODE: RB-HS-H-101XX1

Similar to the Dura-Mastic M-100, but features a modified barrel takedown.

Production: 1954 - early 1970s Caliber: .22 LR Action: striker-fired, semi-auto

Barrel Length: 4.5", 6.75" Sights: blade front, fixed rear Capacity: 10

Magazine: detachable box Grips: synthetic, molded

D2C:	NIB	--	Ex	$ 350	VG+ $ 240	Good	$ 220		
C2C:	Mint	--	Ex	$ 325	VG+ $ 220	Good	$ 200	Fair	$ 105
Trade-In:	Mint	--	Ex	$ 250	VG+ $ 170	Good	$ 155	Poor	$ 60

SPORT KING 102 SERIES REDBOOK CODE: RB-HS-H-102XXX

Serial numbers range between approximately 778,000 and 1,120,000. Series 102 pistols feature an enhanced takedown button and updated frame. Lightweight barrel, enhanced design, frame-mounted safety.

Production: 1958 - 1960 Caliber: .22 LR Action: striker-fired, semi-auto

Barrel Length: 4.5", 6.75" Sights: fixed Capacity: 10 Magazine: detachable box

Grips: synthetic, molded

D2C:	NIB	--	Ex	$ 325	VG+ $ 230	Good	$ 205		
C2C:	Mint	--	Ex	$ 300	VG+ $ 210	Good	$ 185	Fair	$ 95
Trade-In:	Mint	--	Ex	$ 250	VG+ $ 160	Good	$ 145	Poor	$ 60

FLITE KING 102 SERIES REDBOOK CODE: RB-HS-H-102XX1

Distinguished by its steel frame.

Production: 1958 - 1960 Caliber: .22 LR Action: striker-fired, semi-auto

Barrel Length: 4.5", 6.75" Sights: fixed Capacity: 10 Magazine: detachable box

Grips: synthetic

D2C:	NIB	--	Ex	$ 500	VG+ $ 360	Good	$ 325		
C2C:	Mint	--	Ex	$ 450	VG+ $ 330	Good	$ 295	Fair	$ 150
Trade-In:	Mint	--	Ex	$ 375	VG+ $ 260	Good	$ 230	Poor	$ 90

SUPERMATIC TOURNAMENT 102 SERIES

REDBOOK CODE: RB-HS-H-102XX2

Blue finish, was also issued to United States military. Military models brings slightly higher premium.

Production: 1958 - 1960 Caliber: .22 LR Action: striker-fired, semi-auto

Barrel Length: 4.5", 6.75" Sights: adjustable Capacity: 10 Magazine: detachable box

Grips: synthetic

D2C:	NIB	--	Ex	$ 700	VG+ $ 495	Good	$ 450		
C2C:	Mint	--	Ex	$ 625	VG+ $ 450	Good	$ 405	Fair	$ 210
Trade-In:	Mint	--	Ex	$ 525	VG+ $ 360	Good	$ 315	Poor	$ 90

SUPERMATIC CITATION 102 SERIES
EXCELLENT

SUPERMATIC CITATION 102 SERIES REDBOOK CODE: RB-HS-H-102XX3
High-polish finish, grooved straps. Some marked "U.S."

Production: 1958 - 1960 Caliber: .22 LR Action: striker-fired, semi-auto

Barrel Length: 6.75", 8", 10" Sights: synthetic Capacity: 10

Magazine: detachable box Grips: synthetic

D2C:	NIB	--	Ex $ 900	VG+ $ 650	Good $ 590		
C2C:	Mint	--	Ex $ 825	VG+ $ 590	Good $ 530	Fair $ 275
Trade-In:	Mint	--	Ex $ 675	VG+ $ 460	Good $ 415	Poor $ 120

SUPERMATIC TROPHY 102 SERIES
EXCELLENT

SUPERMATIC TROPHY 102 SERIES REDBOOK CODE: RB-HS-H-102XX4
Polished finish, tapered barrel, fitted for a muzzle break, frame-mounted safety.

Production: 1958 - 1960 Caliber: .22 LR Action: striker-fired, semi-auto

Barrel Length: 6.75", 8", 10" Sights: adjustable Capacity: 10

Magazine: detachable box Grips: walnut, molded thumb rest

D2C:	NIB	--	Ex $1,175	VG+ $ 845	Good $ 765		
C2C:	Mint	--	Ex $1,075	VG+ $ 770	Good $ 690	Fair $ 355
Trade-In:	Mint	--	Ex $ 875	VG+ $ 600	Good $ 540	Poor $ 180

OLYMPIC 102 SERIES
VERY GOOD +

OLYMPIC 102 SERIES REDBOOK CODE: RB-HS-H-102XX5
Grooved straps, blue finish, frame-mounted safety.

Production: 1958 - 1960 Caliber: .22 Short Action: striker-fired, semi-auto

Barrel Length: 6.75", 8", 10" Sights: adjustable Capacity: 10

Magazine: detachable box Grips: synthetic, molded

D2C:	NIB	--	Ex $1,125	VG+ $ 805	Good $ 730		
C2C:	Mint	--	Ex $1,025	VG+ $ 730	Good $ 655	Fair $ 335
Trade-In:	Mint	--	Ex $ 825	VG+ $ 570	Good $ 510	Poor $ 150

SPORT KING 103 SERIES
EXCELLENT

SPORT KING 103 SERIES REDBOOK CODE: RB-HS-H-103XXX
Lightweight and tapered barrel. Series 103 models are similar to the Series 102 and feature a modified frame and enlarged takedown button.

Production: 1960 - 1977 Caliber: .22 LR Action: striker-fired, semi-auto

Barrel Length: 4.5", 6.75" Sights: fixed Capacity: 10 Magazine: detachable box

Grips: synthetic, checkered

D2C:	NIB	--	Ex $ 325	VG+ $ 220	Good $ 200		
C2C:	Mint	--	Ex $ 300	VG+ $ 200	Good $ 180	Fair $ 95
Trade-In:	Mint	--	Ex $ 225	VG+ $ 160	Good $ 140	Poor $ 60

FLITE KING 103 SERIES
VERY GOOD +

FLITE KING 103 SERIES REDBOOK CODE: RB-HS-H-103XX1
Similar to the Series 103. Noted for its steel frame.

Production: 1960 - 1966 Caliber: .22 Short Action: striker-fired, semi-auto

Barrel Length: 4.5", 6.75" Sights: fixed Capacity: 10 Magazine: detachable box

Grips: synthetic, checkered

D2C:	NIB	--	Ex $ 450	VG+ $ 320	Good $ 290		
C2C:	Mint	--	Ex $ 400	VG+ $ 290	Good $ 260	Fair $ 135
Trade-In:	Mint	--	Ex $ 325	VG+ $ 230	Good $ 205	Poor $ 60

HIGH STANDARD

SHARPSHOOTER 103 SERIES
VERY GOOD +

SUPERMATIC TOURNAMENT 103 SERIES
EXCELLENT

SUPERMATIC CITATION 103 SERIES
GOOD

SHARPSHOOTER 103 SERIES   REDBOOK CODE: RB-HS-H-103XX3
Noted for its bull barrel. Frame-mounted safety.

Production: 1969 - 1978   Caliber: .22 LR   Action: striker-fired, semi-auto

Barrel Length: 5.5"   Sights: adjustable   Capacity: 10   Magazine: detachable box

Grips: walnut, molded

D2C:	NIB	--	Ex	$ 475	VG+ $ 330	Good	$ 300		
C2C:	Mint	--	Ex	$ 425	VG+ $ 300	Good	$ 270	Fair	$ 140
Trade-In:	Mint	--	Ex	$ 350	VG+ $ 240	Good	$ 210	Poor	$ 60

SUPERMATIC TOURNAMENT 103 SERIES
REDBOOK CODE: RB-HS-H-103XX2
Some models purchased and marked by the U.S. Military.

Production: 1960 - 1965   Caliber: .22 LR   Action: striker-fired, semi-auto

Barrel Length: 4.5", 5.5", 6.75"   Sights: adjustable   Capacity: 10

Magazine: detachable box   Grips: synthetic, checkered

D2C:	NIB	--	Ex	$ 575	VG+ $ 410	Good	$ 370		
C2C:	Mint	--	Ex	$ 525	VG+ $ 370	Good	$ 335	Fair	$ 175
Trade-In:	Mint	--	Ex	$ 425	VG+ $ 290	Good	$ 260	Poor	$ 90

SUPERMATIC CITATION 103 SERIES   REDBOOK CODE: RB-HS-H-103XX4
Similar to Supermatic Trophy, but features synthetic grips and grooved straps. High-polish finish.

Production: 1960 - 1962   Caliber: .22 LR   Action: striker-fired, semi-auto

Barrel Length: 5.5", 6.75", 8", 10"   Sights: adjustable   Capacity: 10

Magazine: detachable box   Grips: synthetic, checkered

D2C:	NIB	--	Ex	$ 750	VG+ $ 540	Good	$ 490		
C2C:	Mint	--	Ex	$ 675	VG+ $ 490	Good	$ 440	Fair	$ 225
Trade-In:	Mint	--	Ex	$ 550	VG+ $ 390	Good	$ 345	Poor	$ 120

SUPERMATIC TROPHY 103 SERIES   REDBOOK CODE: RB-HS-H-103XX5
High-polish blue finish, barrel weights, grooved straps, fitted for muzzle brake, tapered barrel, frame-mounted safety.

Production: 1960 - 1962   Caliber: .22 LR   Action: striker-fired, semi-auto

Barrel Length: 5", 6.75", 8", 10"   Sights: adjustable   Capacity: 10

Magazine: detachable box   Grips: walnut, molded

D2C:	NIB	--	Ex	$1,000	VG+ $ 715	Good	$ 650		
C2C:	Mint	--	Ex	$ 900	VG+ $ 650	Good	$ 585	Fair	$ 300
Trade-In:	Mint	--	Ex	$ 750	VG+ $ 510	Good	$ 455	Poor	$ 150

OLYMPIC 103 SERIES   REDBOOK CODE: RB-HS-H-103XX6
A .22 variant of the Supermatic Citation 103 Series.

Production: 1960 - 1963   Caliber: .22 Short   Action: striker-fired, semi-auto

Barrel Length: 6.75", 8", 10"   Sights: adjustable   Capacity: 10

Magazine: detachable box   Grips: synthetic, checkered

D2C:	NIB	--	Ex	$1,100	VG+ $ 785	Good	$ 715		
C2C:	Mint	--	Ex	$1,000	VG+ $ 720	Good	$ 645	Fair	$ 330
Trade-In:	Mint	--	Ex	$ 800	VG+ $ 560	Good	$ 500	Poor	$ 150

**SUPERMATIC CITATION
104 SERIES**
VERY GOOD +

SUPERMATIC CITATION 104 SERIES

REDBOOK CODE: RB-HS-H-104XX1

Serial numbers range between approximately 1,300,000 and G20,150. High-polish blue finish, grooved straps, minimal design. The Series 104 is very similar to the Series 103, but some models are unmarked.

Production: 1964 - late 1970s Caliber: .22 LR Action: striker-fired, semi-auto

Barrel Length: 5.5", 6.75", 8" Sights: adjustable Capacity: 10

Magazine: detachable box Grips: walnut, molded thumb rest

D2C:	NIB	--	Ex	$ 725	VG+ $ 525	Good $ 475			
C2C:	Mint	--	Ex	$ 675	VG+ $ 480	Good $ 430	Fair	$ 220	
Trade-In:	Mint	--	Ex	$ 550	VG+ $ 380	Good $ 335	Poor	$ 120	

**SUPERMATIC TROPHY
104 SERIES**
VERY GOOD +

SUPERMATIC TROPHY 104 SERIES REDBOOK CODE: RB-HS-H-104XX2

Fluted barrel, high-polish blue finish, grooved straps.

Production: 1964 - 1965 Caliber: .22 LR Action: striker-fired, semi-auto

Barrel Length: 5.5", 7.25" Sights: adjustable Capacity: 10

Magazine: detachable box Grips: walnut, molded thumb rest, checkered

D2C:	NIB	--	Ex	$ 1,150	VG+ $ 825	Good $ 750			
C2C:	Mint	--	Ex	$ 1,050	VG+ $ 750	Good $ 675	Fair	$ 345	
Trade-In:	Mint	--	Ex	$ 850	VG+ $ 590	Good $ 525	Poor	$ 150	

OLYMPIC ISU 104 SERIES REDBOOK CODE: RB-HS-H-104XX3

Similar to the previous model Olympic ISU, but may not include accessories.

Production: 1964 - late 1970s Caliber: .22 Short Action: striker-fired, semi-auto

Barrel Length: 5.5", 6.75" Sights: adjustable Capacity: 10

Magazine: detachable box Grips: synthetic, checkered

D2C:	NIB	--	Ex	$1,075	VG+ $ 780	Good $ 705			
C2C:	Mint	--	Ex	$ 975	VG+ $ 710	Good $ 635	Fair	$ 325	
Trade-In:	Mint	--	Ex	$ 800	VG+ $ 550	Good $ 495	Poor	$ 150	

**SUPERMATIC TOURNAMENT
MILITARY 106 SERIES**
VERY GOOD +

SUPERMATIC TOURNAMENT MILITARY 106 SERIES

REDBOOK CODE: RB-HS-H-106XXX

Serial numbers range between approximately 1,484,750 and 1,940,000. Designed to fire like a M1911, but resemble previous High Standard pistols.

Production: mid-1960s - 1968 Caliber: .22 LR Action: striker-fired, semi-auto

Barrel Length: 5.5", 6.75" Sights: adjustable Capacity: 10

Magazine: detachable box Grips: walnut, molded thumb rest, checkered

D2C:	NIB	$ 730	Ex	$ 575	VG+ $ 405	Good $ 365			
C2C:	Mint	$ 710	Ex	$ 525	VG+ $ 370	Good $ 330	Fair	$ 170	
Trade-In:	Mint	$ 520	Ex	$ 425	VG+ $ 290	Good $ 260	Poor	$ 90	

SUPERMATIC CITATION MILITARY 106 SERIES
GOOD

SUPERMATIC TROPHY MILITARY 106 SERIES
EXCELLENT

SPORT KING 107 SERIES
EXCELLENT

SUPERMATIC CITATION MILITARY 106 SERIES

REDBOOK CODE: RB-HS-H-106XX1

Updated military-style frame, frame-mounted safety, blue finish.

Production: mid-1960s - 1968   Caliber: .22 LR   Action: striker-fired, semi-auto

Barrel Length: 5.5", 7.25"   Sights: adjustable   Capacity: 10

Magazine: detachable box   Grips: walnut, molded thumb rest, checkered

D2C:	NIB	--	Ex	$ 575	VG+	$ 410	Good	$ 370		
C2C:	Mint	--	Ex	$ 525	VG+	$ 370	Good	$ 335	Fair	$ 175
Trade-In:	Mint	--	Ex	$ 425	VG+	$ 290	Good	$ 260	Poor	$ 90

SUPERMATIC TROPHY MILITARY 106 SERIES

REDBOOK CODE: RB-HS-H-106XX2

Updated military-style frame, stippled straps, frame-mounted safety, minimal design.

Production: mid-1960s - 1968   Caliber: .22 LR   Action: striker-fired, semi-auto

Barrel Length: 5.5", 7.25"   Sights: adjustable   Capacity: 10

Magazine: detachable box   Grips: walnut, molded thumb rest, checkered

D2C:	NIB	--	Ex	$ 950	VG+	$ 690	Good	$ 625		
C2C:	Mint	--	Ex	$ 875	VG+	$ 630	Good	$ 565	Fair	$ 290
Trade-In:	Mint	--	Ex	$ 700	VG+	$ 490	Good	$ 440	Poor	$ 150

OLYMPIC ISU MILITARY 106 SERIES   REDBOOK CODE: RB-HS-H-106XX3

Updated military-style frame, similar to the ISU Series 104.

Production: mid-1960s - 1968   Caliber: .22 Short   Action: striker-fired, semi-auto

Barrel Length: 6.75"   Sights: adjustable   Capacity: 10   Magazine: detachable box

Grips: walnut, molded thumb rest, checkered

D2C:	NIB	--	Ex	$ 950	VG+	$ 685	Good	$ 625		
C2C:	Mint	--	Ex	$ 875	VG+	$ 630	Good	$ 565	Fair	$ 290
Trade-In:	Mint	--	Ex	$ 700	VG+	$ 490	Good	$ 440	Poor	$ 150

SPORT KING 107 SERIES   REDBOOK CODE: RB-HS-H-107XXX

Military-style frame, frame-mounted safety. Series 107 noted for its modified spring hole.

Production: 1978 - early 1980s   Caliber: .22 LR   Action: striker-fired, semi-auto

Barrel Length: 4.5", 6.75"   Sights: fixed   Capacity: 10   Magazine: detachable box

Grips: walnut

D2C:	NIB	$ 330	Ex	$ 275	VG+	$ 185	Good	$ 165		
C2C:	Mint	$ 320	Ex	$ 250	VG+	$ 170	Good	$ 150	Fair	$ 80
Trade-In:	Mint	$ 240	Ex	$ 200	VG+	$ 130	Good	$ 120	Poor	$ 60

SUPERMATIC TOURNAMENT MILITARY 107 SERIES

REDBOOK CODE: RB-HS-H-107XX1

Military-style frame, minimal design, frame-mounted safety.

Production: 1968 - 1971   Caliber: .22 LR   Action: striker-fired, semi-auto

Barrel Length: 5.5", 6.75"   Sights: adjustable   Capacity: 10

Magazine: detachable box   Grips: walnut, molded thumb rest, checkered

D2C:	NIB	$ 590	Ex	$ 450	VG+	$ 325	Good	$ 295		
C2C:	Mint	$ 570	Ex	$ 425	VG+	$ 300	Good	$ 270	Fair	$ 140
Trade-In:	Mint	$ 420	Ex	$ 350	VG+	$ 240	Good	$ 210	Poor	$ 60

**SUPERMATIC TROPHY
MILITARY 107 SERIES**
EXCELLENT

SENTINEL
EXCELLENT

SENTINEL DELUXE
EXCELLENT

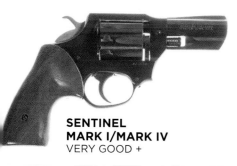

**SENTINEL
MARK I/MARK IV**
VERY GOOD +

SENTINEL MARK II/MARK III
EXCELLENT

HIGH STANDARD

SUPERMATIC TROPHY MILITARY 107 SERIES
REDBOOK CODE: RB-HS-H-107XX2
Almost identical to the Series 106 model. Updated military-style frame, minimal design.

Production: 1968 - early 1980s   Caliber: .22 LR   Action: striker-fired, semi-auto

Barrel Length: 5.5", 7.25"   Sights: adjustable   Capacity: 10

Magazine: detachable box   Grips: walnut, molded thumb rest, checkered

D2C:	NIB	--	Ex	$ 875	VG+ $ 635	Good	$ 575		
C2C:	Mint	--	Ex	$ 800	VG+ $ 580	Good	$ 520	Fair	$ 265
Trade-In:	Mint	--	Ex	$ 650	VG+ $ 450	Good	$ 405	Poor	$ 120

SENTINEL (R-1/R-6)   REDBOOK CODE: RB-H2-H-SNTXXX
Many produced for and distributed by Sears. Similar to the S&W Model 10. Minor variations between models. Aluminum frame, bobbed or spurred hammers, blue or nickel finish.

Production: mid-1950s - early 1960s   Caliber: .22 LR, .22 Short

Action: DA/SA, revolver   Barrel Length: 2.4", 4", 6"   Sights: fixed   Capacity: 9

Grips: synthetic

D2C:	NIB	--	Ex	$ 325	VG+ $ 225	Good	$ 205		
C2C:	Mint	--	Ex	$ 300	VG+ $ 210	Good	$ 185	Fair	$ 95
Trade-In:	Mint	--	Ex	$ 250	VG+ $ 160	Good	$ 145	Poor	$ 60

SENTINEL DELUXE (R-7/R-9)   REDBOOK CODE: RB-H2-H-SNTXX1
Later production, minor variations between models. Similar to previous models, some models have a steel frame and adjustable sights.

Production: late 1960s - 1973   Caliber: .22 LR, .22 Magnum   Action: DA/SA, revolver

Barrel Length: 2.4", 4", 6"   Sights: fixed or adjustable   Capacity: 9

Grips: synthetic or wood

D2C:	NIB	--	Ex	$ 300	VG+ $ 215	Good	$ 195		
C2C:	Mint	--	Ex	$ 275	VG+ $ 200	Good	$ 175	Fair	$ 90
Trade-In:	Mint	--	Ex	$ 225	VG+ $ 160	Good	$ 135	Poor	$ 60

SENTINEL MARK I/MARK IV   REDBOOK CODE: RB-H2-H-SNTXX2
MK I chambered for .22 LR, but the guns are almost identical. MK IV chambered for .22 Magnum. Shrouded ejector rod.

Production: 1974 - 1975   Caliber: .22 LR, .22 Magnum   Action: DA/SA, revolver

Barrel Length: 2", 3", 4"   Sights: fixed or adjustable   Capacity: 9   Grips: wood

D2C:	NIB	--	Ex	$ 300	VG+ $ 205	Good	$ 185		
C2C:	Mint	--	Ex	$ 275	VG+ $ 190	Good	$ 165	Fair	$ 85
Trade-In:	Mint	--	Ex	$ 225	VG+ $ 150	Good	$ 130	Poor	$ 60

SENTINEL MARK II/MARK III   REDBOOK CODE: RB-H2-H-SNTXX3
Steel frame, blue finish, shrouded ejector rod, swing-out barrel.

Production: 1974 - 1975   Caliber: .357 Magnum   Action: DA/SA, revolver

Barrel Length: 4"   Sights: fixed or adjustable   Capacity: 6   Grips: wood

D2C:	NIB	--	Ex	$ 325	VG+ $ 235	Good	$ 215		
C2C:	Mint	--	Ex	$ 300	VG+ $ 220	Good	$ 195	Fair	$ 100
Trade-In:	Mint	--	Ex	$ 250	VG+ $ 170	Good	$ 150	Poor	$ 60

CRUSADER
EXCELLENT

DOUBLE-NINE
GOOD

MARSHALL
VERY GOOD +

CRUSADER REDBOOK CODE: RB-H2-H-SNTXX4
Most include a gold knight emblem on the frame. Blue finish, swing-out cylinder, high-polish blue finish, shrouded-ejector rod. Approximately 500 manufactured.

Production: late 1970s - 1983 Caliber: .44 Magnum, .45 LC Action: DA, revolver

Barrel Length: 6.5" Sights: adjustable Capacity: 6 Grips: polished wood

D2C:	NIB	--	Ex $ 950	VG+ $ 690	Good $ 625		
C2C:	Mint	--	Ex $ 875	VG+ $ 630	Good $ 565	Fair $ 290
Trade-In:	Mint	--	Ex $ 700	VG+ $ 490	Good $ 440	Poor $ 150

DOUBLE-NINE (EARLY PRODUCTION) REDBOOK CODE: RB-H2-H-DDNXXX
Aluminum frame, high-polish blue or nickel frame, some with interchangeable cylinders.

Production: late 1950s - 1971 Caliber: .22 LR, .22 Magnum Action: DA/SA, revolver

Barrel Length: 5.5" Sights: fixed or adjustable Capacity: 9

Grips: synthetic, some walnut

D2C:	NIB	--	Ex $ 325	VG+ $ 230	Good $210	
C2C:	Mint	--	Ex $ 300	VG+ $ 210	Good $ 190	Fair $ 100
Trade-In:	Mint	--	Ex $ 250	VG+ $ 170	Good $ 150	Poor $ 60

DOUBLE-NINE (LATER PRODUCTION) REDBOOK CODE: RB-H2-H-DDBXX1
Serial numbers range between approximately M001000 and M090050. Steel frame, high-polish blue or nickel frame, some with interchangeable cylinders.

Production: 1972 - 1984 Caliber: .22 LR, .22 Magnum Action: DA/SA, revolver

Barrel Length: 5.5" Sights: fixed or adjustable Capacity: 9

Grips: synthetic or walnut

D2C:	NIB	--	Ex $ 300	VG+ $ 210	Good $ 190	
C2C:	Mint	--	Ex $ 275	VG+ $ 190	Good $ 170	Fair $ 90
Trade-In:	Mint	--	Ex $ 225	VG+ $ 150	Good $ 135	Poor $ 60

LONGHORN REDBOOK CODE: RB-H2-H-LNGHRN
Aluminum or steel frame, blue finish, modified trigger guard.

Production: 1973 - discontinued Caliber: .22 Short, .22 LR, .22 Magnum

Action: DA/SA, revolver Barrel Length: 9", 9.5" Sights: fixed Capacity: 9

Grips: polished wood

D2C:	NIB	--	Ex $ 375	VG+ $ 270	Good $ 245	
C2C:	Mint	--	Ex $ 350	VG+ $ 250	Good $ 220	Fair $ 115
Trade-In:	Mint	--	Ex $ 275	VG+ $ 190	Good $ 170	Poor $ 60

MARSHALL REDBOOK CODE: RB-H2-H-MRSHL
Aluminum frame, blue finish, indicative of other High Standard revolvers of the era.

Production: early 1970s Caliber: .22 Short, .22 LR, .22 Magnum

Action: DA/SA, revolver Barrel Length: 5.5" Sights: fixed Capacity: 9

Grips: white plastic, unique markings

D2C:	NIB	--	Ex $ 350	VG+ $ 240	Good $ 220	
C2C:	Mint	--	Ex $ 325	VG+ $ 220	Good $ 200	Fair $ 105
Trade-In:	Mint	--	Ex $ 250	VG+ $ 170	Good $ 155	Poor $ 60

POSSE   REDBOOK CODE: RB-H2-H-POSSEX
Aluminum frame, blue finish, similar to the Double-Nine revolver.

Production: early 1960s - mid-1960s Caliber: .22 Short, .22 LR, .22 Magnum

Action: DA/SA, revolver   Barrel Length: 3.5" Sights: fixed   Capacity: 9

Grips: walnut or synthetic

D2C:	NIB	--	Ex	$ 425	VG+	$ 305	Good	$ 275		
C2C:	Mint	--	Ex	$ 400	VG+	$ 280	Good	$ 250	Fair	$ 130
Trade-In:	Mint	--	Ex	$ 325	VG+	$ 220	Good	$ 195	Poor	$ 60

NATCHEZ   REDBOOK CODE: RB-H2-H-NTCHEZ
Aluminum frame, blue finish, similar to the Double-Nine revolver.

Production: early 1960s - mid-1960s Caliber: .22 Short, .22 LR, .22 Magnum

Action: DA/SA, revolver   Barrel Length: 4.5" Sights: fixed   Capacity: 9

Grips: plastic ivory, bird's-head style

D2C:	NIB	--	Ex	$ 450	VG+	$ 315	Good	$ 285		
C2C:	Mint	--	Ex	$ 400	VG+	$ 290	Good	$ 260	Fair	$ 135
Trade-In:	Mint	--	Ex	$ 325	VG+	$ 230	Good	$ 200	Poor	$ 60

HOMBRE   REDBOOK CODE: RB-H2-H-HOMBRE
Noted for its 4" barrel. Aluminum frame, blue or nickel finish (which commands slightly higher premium).

Production: early 1970s - discontinued Caliber: .22 Short, .22 LR Action: DA/SA, revolver

Barrel Length: 4" Sights: fixed   Capacity: 9   Grips: walnut

D2C:	NIB	$ 440	Ex	$ 350	VG+	$ 245	Good	$ 220		
C2C:	Mint	$ 430	Ex	$ 325	VG+	$ 220	Good	$ 200	Fair	$ 105
Trade-In:	Mint	$ 320	Ex	$ 250	VG+	$ 180	Good	$ 155	Poor	$ 60

DURANGO
VERY GOOD +

DURANGO   REDBOOK CODE: RB-H2-H-DURNGO
Aluminum or steel frame, blue or nickel finish.

Production: mid-1970s   Caliber: .22 Short, .22 LR Action: DA/SA, revolver

Barrel Length: 4.5", 5.5" Sights: fixed or adjustable   Capacity: 9

Grips: walnut or plastic ivory

D2C:	NIB	--	Ex	$ 400	VG+	$ 280	Good	$ 255		
C2C:	Mint	--	Ex	$ 350	VG+	$ 260	Good	$ 230	Fair	$ 120
Trade-In:	Mint	--	Ex	$ 300	VG+	$ 200	Good	$ 180	Poor	$ 60

HIGH SIERRA
EXCELLENT

Courtesy of Bud's Gun Shop

HIGH SIERRA (THE GUN)   REDBOOK CODE: RB-H2-H-SIERRA
Noted for its octagonal barrel. Blue finish, gold accents, steel frame. Some models marked "The Gun."

Production: early 1970s   Caliber: .22 Short, .22 LR, .22 Magnum

Action: DA/SA, revolver   Barrel Length: 7" Sights: fixed or adjustable   Capacity: 9

Grips: smooth wood

D2C:	NIB	--	Ex	$ 400	VG+	$ 290	Good	$ 260		
C2C:	Mint	--	Ex	$ 375	VG+	$ 260	Good	$ 235	Fair	$ 120
Trade-In:	Mint	--	Ex	$ 300	VG+	$ 210	Good	$ 185	Poor	$ 60

M1911A1 GOVERNMENT REDBOOK CODE: RB-H2-H-1911XX

A classic reproduction Colt M1911A1. Matte blue finish, frame-mounted and grip safety.

Production: current Caliber: .45 ACP, .38 Super +P Action: SA, semi-auto

Barrel Length: 5" Sights: GI-style fixed, blade front Capacity: 7

Magazine: detachable box Grips: checkered walnut

D2C:	NIB	$ 580	Ex $ 450	VG+ $ 320	Good $ 290				
C2C:	Mint	$ 560	Ex $ 425	VG+ $ 290	Good $ 265	Fair $	135		
Trade-In:	Mint	$ 420	Ex $ 325	VG+ $ 230	Good $ 205	Poor $	60		

M1911A1 GOVERNMENT CUSTOM REDBOOK CODE: RB-H2-H-1911X1

A modern reproduction Colt M1911A1. Parkerized finish, skeletonized trigger, tactical design, frame-mounted and grip safety. Noted for its Novak-style sights.

Production: current Caliber: .45 ACP Action: SA, semi-auto Barrel Length: 5"

Sights: Novak Capacity: 7 Magazine: detachable box Grips: checkered walnut

D2C:	NIB	$ 630	Ex $ 500	VG+ $ 350	Good $ 315		
C2C:	Mint	$ 610	Ex $ 450	VG+ $ 320	Good $ 285	Fair $ 145
Trade-In:	Mint	$ 450	Ex $ 375	VG+ $ 250	Good $ 225	Poor $ 90

M1911 CAMP PERRY MODEL REDBOOK CODE: RB-H2-H-1911X2

Designed for competitive shooting. Flared ejection port, skeletonized trigger, frame-mounted and grip safety, steel frame and slide, parkerized finish.

Production: current Caliber: .45 ACP Action: SA, semi-auto Barrel Length: 5"

Sights: fixed front, adjustable rear Capacity: 7 Magazine: detachable box

Grips: checkered walnut

D2C:	NIB	$1,025	Ex $ 800	VG+ $ 565	Good $ 515		
C2C:	Mint	$ 990	Ex $ 725	VG+ $ 520	Good $ 465	Fair $ 240
Trade-In:	Mint	$ 730	Ex $ 575	VG+ $ 400	Good $ 360	Poor $ 120

M1911 SUPERMATIC CUSTOM REDBOOK CODE: RB-H2-H-1911X3

Designed for competitive shooting. Noted for its extended barrel. Skeletonized trigger, frame-mounted and grip safety, steel frame and slide, parkerized finish.

Production: current Caliber: .45 ACP Action: SA, semi-auto Barrel Length: 6"

Sights: fixed front, adjustable rear Capacity: 7 Magazine: detachable box

Grips: checkered walnut

D2C:	NIB	$1,200	Ex $ 925	VG+ $ 660	Good $ 600		
C2C:	Mint	$1,160	Ex $ 850	VG+ $ 600	Good $ 540	Fair $ 280
Trade-In:	Mint	$ 860	Ex $ 675	VG+ $ 470	Good $ 420	Poor $ 120

M1911 COMPACT REDBOOK CODE: RB-H2-H-1911X4

Compact design, parkerized finish, frame-mounted and grip safety, skeletonized trigger and hammer.

Production: current Caliber: .45 ACP Action: SA, semi-auto Barrel Length: 3.6"

Sights: Novak, adjustable Capacity: 6 Magazine: detachable box Grips: walnut

D2C:	NIB	$ 560	Ex $ 450	VG+ $ 310	Good $ 280		
C2C:	Mint	$ 540	Ex $ 400	VG+ $ 280	Good $ 255	Fair $ 130
Trade-In:	Mint	$ 400	Ex $ 325	VG+ $ 220	Good $ 200	Poor $ 60

Iver Johnson

The Iver Johnson name traces its origins back to the 1880s, during which time the company manufactured a host of different derringers, single-shot revolvers, double-action revolvers, and shotguns. This incarnation of Iver Johnson—Iver Johnson Arms & Cycle Works—folded in the early 1990s. Established around 2006, Iver Johnson Arms, Inc., located in Rockledge, Florida, bears no connection to the original Iver Johnson aside from its name. The current namesake, which some suggest began as a subsidiary of Squires Bingham International, Inc., manufactures and imports M1911s and several other semi-automatic pistols.

SAFETY AUTOMATIC DOUBLE-ACTION
GOOD

SAFETY AUTOMATIC DOUBLE-ACTION
REDBOOK CODE: RB-IJ-H-SAFXXX
The first Iver Johnson revolver to enter production, many variants exist given the model's extensive production. Break-top action, transfer bar safety, most with nickel finish, compact design, steel frame.

Production: late 1890s - 1950s Caliber: .22 LR, .32 S&W, .38 S&W

Action: DA/SA, revolver Barrel Length: 3", 5" Sights: blade front, fixed notch rear

Capacity: 3, 5, 6 Grips: synthetic

D2C:	NIB	--	Ex $ 300	VG+ $ 205	Good $ 185		
C2C:	Mint	$ 360	Ex $ 275	VG+ $ 190	Good $ 165	Fair $ 85
Trade-In:	Mint	$ 260	Ex $ 225	VG+ $ 150	Good $ 130	Poor $ 60

MODEL 1900
EXCELLENT

MODEL 1900 REDBOOK CODE: RB-IJ-H-1900XX
The gun's designs vary widely, given its extensive time of production. Blue or nickel finish, some with octagonal barrel, compact design.

Production: 1900 - 1947 Caliber: .22 LR, .32 S&W, .38 S&W Action: SA, revolver

Barrel Length: 2.5", 4.5", 6" Sights: blade front Capacity: 6, 7 Grips: rubber

D2C:	NIB	--	Ex $ 200	VG+ $ 130	Good $ 120		
C2C:	Mint	--	Ex $ 175	VG+ $ 120	Good $ 110	Fair $ 55
Trade-In:	Mint	--	Ex $ 150	VG+ $ 100	Good $ 85	Poor $ 30

PETITE
POOR

PETITE REDBOOK CODE: RB-IJ-H-PETITE
Little is known about this gun, less than 1,000 believed to be produced. Partially shrouded hammer, nickel finish, compact design, no trigger guard.

Production: ~1909 Caliber: .22 LR Action: DA/SA, revolver

Barrel Length: 1" Sights: blade front Capacity: 5, 7 Grips: rubber

D2C:	NIB	--	Ex $ 300	VG+ $ 210	Good $ 190		
C2C:	Mint	--	Ex $ 275	VG+ $ 190	Good $ 170	Fair $ 90
Trade-In:	Mint	--	Ex $ 225	VG+ $ 150	Good $ 135	Poor $ 60

SUPERSHOT SEALED 8
VERY GOOD +

SUPERSHOT SEALED 8 REDBOOK CODE: RB-IJ-H-SUP8XX
Similar to the Safety Automatic, but features an extended barrel. Blue or nickel finish, break-top action.

Production: 1930s - late 1950s Caliber: .22 LR Action: DA/SA, revolver

Barrel Length: 6" Sights: fixed front, adjustable rear Capacity: 8 Grips: rubber

D2C:	NIB	--	Ex $ 250	VG+ $ 165	Good $ 150		
C2C:	Mint	$ 290	Ex $ 225	VG+ $ 150	Good $ 135	Fair $ 70
Trade-In:	Mint	$ 220	Ex $ 175	VG+ $ 120	Good $ 105	Poor $ 30

PROTECTOR SEALED 8
VERY GOOD +

PROTECTOR SEALED 8   REDBOOK CODE: RB-IJ-H-PRO8XX

The same as the Supershot Sealed 8, but features a 2.5" barrel.

Production: 1930s - late 1950s Caliber: .22 LR Action: DA/SA, revolver									
Barrel Length: 2.5" Sights: fixed front, adjustable rear Capacity: 8 Grips: wood									
D2C:	NIB	$ 375	Ex $ 300	VG+ $ 210	Good $ 190				
C2C:	Mint	$ 360	Ex $ 275	VG+ $ 190	Good $ 170	Fair $	90		
Trade-In:	Mint	$ 270	Ex $ 225	VG+ $ 150	Good $ 135	Poor $	60		

SUPERSHOT 9
EXCELLENT

SUPERSHOT 9   REDBOOK CODE: RB-IJ-H-SUPER9

A variant of the Supershot Sealed 8, but chambers 9 rounds and features a modified cylinder.

Production: 1930s - late 1940s Caliber: .22 LR Action: DA/SA, revolver								
Barrel Length: 6" Sights: fixed front, adjustable rear Capacity: 9 Grips: wood								
D2C:	NIB	$ 355	Ex $ 275	VG+ $ 200	Good $ 180			
C2C:	Mint	$ 350	Ex $ 250	VG+ $ 180	Good $ 160	Fair $	85	
Trade-In:	Mint	$ 260	Ex $ 200	VG+ $ 140	Good $ 125	Poor $	60	

TRIGGER COCKER
VERY GOOD +

TRIGGER COCKER   REDBOOK CODE: RB-IJ-H-TRGGER

Similar to Supershot 9 and Supershot Sealed 8, but operates as a single-action.

Production: 1940 - late 1940s Caliber: .22 LR Action: SA, revolver								
Barrel Length: 6" Sights: fixed front, adjustable rear Capacity: 8 Grips: wood								
D2C:	NIB	$ 310	Ex $ 250	VG+ $ 175	Good $ 155			
C2C:	Mint	$ 300	Ex $ 225	VG+ $ 160	Good $ 140	Fair $	75	
Trade-In:	Mint	$ 230	Ex $ 175	VG+ $ 130	Good $ 110	Poor $	60	

MODEL 844
VERY GOOD +

MODEL 844   REDBOOK CODE: RB-IJ-H-M844XX

Break-top action, blue finish, triangular barrel.

Production: 1950s Caliber: .22 LR Action: DA/SA, revolver								
Barrel Length: 4.5", 6" Sights: adjustable Capacity: 8 Grips: wood								
D2C:	NIB	$ 340	Ex $ 275	VG+ $ 190	Good $ 170			
C2C:	Mint	$ 330	Ex $ 250	VG+ $ 170	Good $ 155	Fair $	80	
Trade-In:	Mint	$ 250	Ex $ 200	VG+ $ 140	Good $ 120	Poor $	60	

MODEL 855
EXCELLENT

MODEL 855   REDBOOK CODE: RB-IJ-H-M855XX

A single-action variant of the Model 844, break-top action, blue finish, triangular barrel.

Production: mid-1950s - late 1950s Caliber: .22 LR Action: SA, revolver								
Barrel Length: 4.5", 6" Sights: adjustable Capacity: 8 Grips: wood								
D2C:	NIB	$ 370	Ex $ 300	VG+ $ 205	Good $ 185			
C2C:	Mint	$ 360	Ex $ 275	VG+ $ 190	Good $ 170	Fair $	90	
Trade-In:	Mint	$ 270	Ex $ 225	VG+ $ 150	Good $ 130	Poor $	60	

MODEL 55 TARGET   REDBOOK CODE: RB-IJ-H-M55XXX

The first solid frame, double action I.J. revolver of the era, swing-out cylinder, steel frame, blue finish, minimal design.

Production: mid-1950s - 1960 Caliber: .22 LR Action: DA/SA, revolver								
Barrel Length: 4.5", 6" Sights: fixed Capacity: 8 Grips: synthetic								
D2C:	NIB	$ 215	Ex $ 175	VG+ $ 120	Good $ 110			
C2C:	Mint	$ 210	Ex $ 150	VG+ $ 110	Good $ 100	Fair $	50	
Trade-In:	Mint	$ 160	Ex $ 125	VG+ $ 90	Good $ 80	Poor $	30	

MODEL 55A SPORTSMAN TARGET REDBOOK CODE: RB-IJ-H-M55AXX
A variant of the Model 55, marked "Sportsman" or "Target," though the same gun, distinguished by its slim barrel and modified loading gate.

Production: early 1960s - late 1970s Caliber: .22 LR Action: DA/SA, revolver

Barrel Length: 4.5", 6" Sights: fixed, blade front Capacity: 8

Grips: checkered wood

D2C:	NIB	$ 285	Ex $ 225	VG+ $ 160	Good $ 145				
C2C:	Mint	$ 280	Ex $ 200	VG+ $ 150	Good $ 130	Fair $	70		
Trade-In:	Mint	$ 210	Ex $ 175	VG+ $ 120	Good $ 100	Poor $	30		

MODEL 55S-A CADET
EXCELLENT

MODEL 55S-A CADET REDBOOK CODE: RB-IJ-H-M55SAC
A variant of the Model 55A, distinguished by its shortened barrel and modified loading gate, steel frame.

Production: mid-1950s - discontinued Caliber: .22 LR, .38 S&W, .32 S&W

Action: DA/SA, revolver Barrel Length: 2.5" Sights: fixed, blade front

Capacity: 8 Grips: wood

D2C:	NIB	$ 265	Ex $ 225	VG+ $ 150	Good $ 135		
C2C:	Mint	$ 260	Ex $ 200	VG+ $ 140	Good $ 120	Fair $ 65
Trade-In:	Mint	$ 190	Ex $ 150	VG+ $ 110	Good $ 95	Poor $ 30

MODEL 57A TARGET
EXCELLENT

MODEL 57A TARGET REDBOOK CODE: RB-IJ-H-M57AXX
A variant of the Model 55, distinguished by its adjustable sights and weighted loading gate.

Production: mid-1950s - late 1970s Caliber: .22 LR Action: DA/SA, revolver

Barrel Length: 4.5", 6" Sights: adjustable Capacity: 8 Grips: wood or synthetic

D2C:	NIB	$ 225	Ex $ 175	VG+ $ 125	Good $ 115		
C2C:	Mint	$ 220	Ex $ 175	VG+ $ 120	Good $ 105	Fair $ 55
Trade-In:	Mint	$ 160	Ex $ 150	VG+ $ 90	Good $ 80	Poor $ 30

MODEL 66 TRAILSMAN
EXCELLENT

MODEL 66 TRAILSMAN REDBOOK CODE: RB-IJ-H-M66TRL
Break-top action, blue finish, large frame.

Production: late 1950s - mid-1970s Caliber: .22 LR Action: DA/SA, revolver

Barrel Length: 4.5", 6" Sights: adjustable Capacity: 8

Grips: wood or synthetic, thumb molded

D2C:	NIB	$ 270	Ex $ 225	VG+ $ 150	Good $ 135		
C2C:	Mint	$ 260	Ex $ 200	VG+ $ 140	Good $ 125	Fair $ 65
Trade-In:	Mint	$ 200	Ex $ 175	VG+ $ 110	Good $ 95	Poor $ 30

MODEL 67 VIKING
VERY GOOD +

MODEL 67 VIKING REDBOOK CODE: RB-IJ-H-M67VKG
A variant of the Model 66 Trailsman, includes a hammer safety device, break-top action, blue finish.

Production: mid-1960s - late 1970s Caliber: .22 LR Action: DA/SA, revolver

Barrel Length: 4.5", 6" Sights: adjustable Capacity: 8

Grips: wood or synthetic, thumb molded

D2C:	NIB	$ 230	Ex $ 175	VG+ $ 130	Good $ 115		
C2C:	Mint	$ 230	Ex $ 175	VG+ $ 120	Good $ 105	Fair $ 55
Trade-In:	Mint	$ 170	Ex $ 150	VG+ $ 90	Good $ 85	Poor $ 30

MODEL 67S VIKING SNUB
EXCELLENT

MODEL 67S VIKING SNUB   REDBOOK CODE: RB-IJ-H-67VKGS

A variant of the Model 67 Viking, though it's perhaps more popular, distinguished by its 2" barrel and varying calibers.

Production: mid-1960s - late 1970s   Caliber: .22 LR, .38 S&W, .32 S&W

Action: DA/SA, revolver   Barrel Length: 2"   Sights: adjustable   Capacity: 5, 8

Grips: wood or synthetic, thumb molded

D2C:	NIB	$ 220	Ex $ 175	VG+ $ 125	Good $ 110				
C2C:	Mint	$ 220	Ex $ 175	VG+ $ 110	Good $ 100	Fair $	55		
Trade-In:	Mint	$ 160	Ex $ 125	VG+ $ 90	Good $ 80	Poor $	30		

MODEL 50   REDBOOK CODE: RB-IJ-H-M50XXX

Western-style design, blue or stainless finish, steel frame, sometimes referred to as the "Sidewinder."

Production: early 1960s - late 1970s   Caliber: .22 LR, .22 Magnum   Action: DA/SA, revolver

Barrel Length: 4.5", 6"   Sights: fixed, adjustable   Capacity: 8   Grips: wood or synthetic

D2C:	NIB	$ 200	Ex $ 175	VG+ $ 110	Good $ 100		
C2C:	Mint	$ 200	Ex $ 150	VG+ $ 100	Good $ 90	Fair $	50
Trade-In:	Mint	$ 150	Ex $ 125	VG+ $ 80	Good $ 70	Poor $	30

AMERICAN BULLDOG
EXCELLENT

AMERICAN BULLDOG   REDBOOK CODE: RB-IJ-H-AMRBLL

Not to be confused with the original I.J. Bulldog series circa 1882-1889. Blue or nickel finish, compact design, comparable to S&W and Colt DA/SA revolvers of the era.

Production: ~1974 - 1978   Caliber: .22 LR, .22 Magnum, .38 Special

Action: DA/SA, revolver   Barrel Length: 2.5", 4"   Sights: adjustable

Capacity: 5, 6   Grips: black synthetic

D2C:	NIB	$ 250	Ex $ 200	VG+ $ 140	Good $ 125		
C2C:	Mint	$ 240	Ex $ 175	VG+ $ 130	Good $ 115	Fair $	60
Trade-In:	Mint	$ 180	Ex $ 150	VG+ $ 100	Good $ 90	Poor $	30

ROOKIE   REDBOOK CODE: RB-IJ-H-ROOKIE

A variant of the Model 55A, blue or nickel finish, pin loading gate, compact design.

Production: mid-1950s   Caliber: .38 Special   Action: DA/SA, revolver

Barrel Length: 4"   Sights: fixed   Capacity: 5   Grips: synthetic

D2C:	NIB	$ 275	Ex $ 225	VG+ $ 155	Good $ 140		
C2C:	Mint	$ 270	Ex $ 200	VG+ $ 140	Good $ 125	Fair $	65
Trade-In:	Mint	$ 200	Ex $ 175	VG+ $ 110	Good $ 100	Poor $	30

CATTLEMAN
VERY GOOD +

CATTLEMAN   REDBOOK CODE: RB-IJ-H-CATXXX

Produced by Uberti, marked as an Iver Johnson, similar to the Colt Single Action Army, most popular with 7.5" barrel, blue finish, solid frame.

Production: early 1970s - late 1970s   Caliber: .45 LC, .44 Magnum, .357 Magnum

Action: SA, revolver   Barrel Length: 4.75", 5.5", 7.5", 12" (rare)

Sights: fixed or adjustable   Capacity: 6   Grips: smooth wood

D2C:	NIB	$ 315	Ex $ 250	VG+ $ 175	Good $ 160		
C2C:	Mint	$ 310	Ex $ 225	VG+ $ 160	Good $ 145	Fair $	75
Trade-In:	Mint	$ 230	Ex $ 200	VG+ $ 130	Good $ 115	Poor $	60

CATTLEMAN TRAILBLAZER EXCELLENT

CATTLEMAN TRAILBLAZER REDBOOK CODE: RB-IJ-H-CATXX2
A variant of the Cattleman, Similar to the Colt SAA. Produced by Uberti. Only fires .22 caliber rounds, case-hardened frame, blue barrel, brass accents.

Production: 1970s   Caliber: .22 LR, .22 Magnum   Action: SA, revolver

Barrel Length: 4.75", 6"   Sights: fixed   Capacity: 6   Grips: smooth wood

D2C:	NIB	$ 305	Ex $ 250	VG+ $ 170	Good $ 155		
C2C:	Mint	$ 300	Ex $ 225	VG+ $ 160	Good $ 140	Fair $	75
Trade-In:	Mint	$ 220	Ex $ 175	VG+ $ 120	Good $ 110	Poor $	60

PONY (MODEL X300) REDBOOK CODE: RB-IJ-H-MPONYX
Similar to the Colt M1911 Compact, blue or nickel finish, compact design, frame-mounted safety.

Production: late 1970s - early 1990s Caliber: .380 ACP Action: SA, semi-auto

Barrel Length: 3" Sights: fixed front, adjustable rear Capacity: 6

Magazine: detachable box Grips: wood

D2C:	NIB	$ 395	Ex $ 325	VG+ $ 220	Good $ 200		
C2C:	Mint	$ 380	Ex $ 275	VG+ $ 200	Good $ 180	Fair $	95
Trade-In:	Mint	$ 290	Ex $ 225	VG+ $ 160	Good $ 140	Poor $	60

U.S. BORDER PATROL REDBOOK CODE: RB-IJ-H-BORDER
A variant of the Model X300 Pony, marked "1924-U.S. Border Patrol-1984," sold in collector's case.

Production: 1984 - only Caliber: .380 ACP Action: SA, semi-auto

Barrel Length: 3" Sights: blade front Capacity: 6 Magazine: detachable box

Grips: polished wood

D2C:	NIB	$ 835	Ex $ 650	VG+ $ 460	Good $ 420		
C2C:	Mint	$ 810	Ex $ 600	VG+ $ 420	Good $ 380	Fair $	195
Trade-In:	Mint	$ 600	Ex $ 475	VG+ $ 330	Good $ 295	Poor $	90

TRAILSMAN EXCELLENT

TRAILSMAN REDBOOK CODE: RB-IJ-H-TRAILS
Similar to the Colt Woodsman, blue finish, frame-mounted safety, modern design.

Production: mid-1980s - early 1990s Caliber: .22 LR Action: SA, semi-auto

Barrel Length: 4.5", 6" Sights: fixed Capacity: 10 Magazine: detachable box

Grips: wood or plastic, thumb-molded

D2C:	NIB	$ 395	Ex $ 325	VG+ $ 220	Good $ 200		
C2C:	Mint	$ 380	Ex $ 275	VG+ $ 200	Good $ 180	Fair $	95
Trade-In:	Mint	$ 290	Ex $ 225	VG+ $ 160	Good $ 140	Poor $	60

TP22 EXCELLENT

TP22 REDBOOK CODE: RB-IJ-H-TP22XX
Similar to the Walther PP7, blue or nickel finish, pocket-size design, frame-mounted safety.

Production: early 1980s - late 1980s Caliber: .22 LR Action: DA/SA, semi-auto

Barrel Length: 2.8" Sights: fixed Capacity: 7 Magazine: detachable box

Grips: black synthetic

D2C:	NIB	$ 335	Ex $ 275	VG+ $ 185	Good $ 170		
C2C:	Mint	$ 330	Ex $ 250	VG+ $ 170	Good $ 155	Fair $	80
Trade-In:	Mint	$ 240	Ex $ 200	VG+ $ 140	Good $ 120	Poor $	60

IVER JOHNSON

TP25
EXCELLENT

TP25 REDBOOK CODE: RB-IJ-H-TP25XX
A variant of the TP22, limited production, similar to the Walther PP7.

Production: early 1980s		Caliber: .25 ACP		Action: DA/SA, semi-auto			
Barrel Length: 2.8"		Sights: fixed		Capacity: 7		Magazine: detachable box	
Grips: black synthetic							
D2C:	NIB	$ 340	Ex $ 275	VG+ $ 190	Good $ 170		
C2C:	Mint	$ 330	Ex $ 250	VG+ $ 170	Good $ 155	Fair $	80
Trade-In:	Mint	$ 250	Ex $ 200	VG+ $ 140	Good $ 120	Poor $	60

COMPACT 25 REDBOOK CODE: RB-IJ-H-COMPCT
Less than 500 believed to be manufactured, pocket-size design, blue finish, frame-mounted safety.

Production: early 1990s		Caliber: .25 ACP		Action: SA, semi-auto			
Barrel Length: 2"		Sights: fixed		Capacity: 6		Magazine: detachable box	
Grips: black synthetic							
D2C:	NIB	$ 470	Ex $ 375	VG+ $ 260	Good $ 235		
C2C:	Mint	$ 460	Ex $ 325	VG+ $ 240	Good $ 215	Fair $	110
Trade-In:	Mint	$ 340	Ex $ 275	VG+ $ 190	Good $ 165	Poor $	60

**M1911A1 EAGLE
POLISHED & MATTE**
NEW IN BOX
Courtesy of Bud's Gun Shop

M1911A1 EAGLE (POLISHED/MATTE) REDBOOK CODE: RB-IJ-H-1911XX
Similar to the Colt M1911A1 Government, offered in high-polish or matte black finish, skeletonized trigger, modern design, steel frame, frame-mounted and grip safety.

Production: current		Caliber: .45 ACP		Action: SA, semi-auto		Barrel Length: 5"	
Sights: front fixed, adjustable rear			Capacity: 7		Magazine: detachable box		
Grips: rosewood or walnut							
D2C:	NIB	$ 665	Ex $ 525	VG+ $ 370	Good $ 335	LMP $	770
C2C:	Mint	$ 640	Ex $ 475	VG+ $ 340	Good $ 300	Fair $	155
Trade-In:	Mint	$ 480	Ex $ 375	VG+ $ 260	Good $ 235	Poor $	90

**M1911A1 EAGLE
LIGHTRAIL**
NEW IN BOX
Courtesy of Bud's Gun Shop

M1911A1 EAGLE LIGHTRAIL REDBOOK CODE: RB-IJ-H-1911X1
A tactical variant of the M1911A1 Eagle, Picatinny accessory rail and tactical design, frame-mounted and grip safety, skeletonized trigger, matte blue finish.

Production: current		Caliber: .45 ACP		Action: SA, semi-auto		Barrel Length: 5"	
Sights: front fixed, adjustable rear			Capacity: 7		Magazine: detachable box		
Grips: walnut							
D2C:	NIB	$ 845	Ex $ 650	VG+ $ 465	Good $ 425	LMP $	960
C2C:	Mint	$ 820	Ex $ 600	VG+ $ 430	Good $ 385	Fair $	195
Trade-In:	Mint	$ 600	Ex $ 475	VG+ $ 330	Good $ 300	Poor $	90

M1911A1 PINK REDBOOK CODE: RB-IJ-H-1911X2
A variant of the M1911A1 Eagle, noted for its pink finish.

Production: current		Caliber: .45 ACP		Action: SA, semi-auto		Barrel Length: 5"	
Sights: fixed		Capacity: 7	Magazine: detachable box		Grips: synthetic, multi-textured		
D2C:	NIB	$ 610	Ex $ 475	VG+ $ 340	Good $ 305	LMP $	700
C2C:	Mint	$ 590	Ex $ 425	VG+ $ 310	Good $ 275	Fair $	145
Trade-In:	Mint	$ 440	Ex $ 350	VG+ $ 240	Good $ 215	Poor $	90

M1911 FALCON   REDBOOK CODE: RB-IJ-H-1911X3
Similar to the Colt M1911 Commander, blue finish, frame-mounted and grip safety, compact design.

Production: post-2006 - discontinued   Caliber: .45 ACP   Action: SA, semi-auto

Barrel Length: 4.25"   Sights: fixed   Capacity: 8   Magazine: detachable box

Grips: checkered wood

D2C:	NIB	$ 515	Ex $ 400	VG+ $ 285	Good $ 260	LMP $ 617
C2C:	Mint	$ 500	Ex $ 375	VG+ $ 260	Good $ 235	Fair $ 120
Trade-In:	Mint	$ 370	Ex $ 300	VG+ $ 210	Good $ 185	Poor $ 60

M1911 HAWK   REDBOOK CODE: RB-IJ-H-1911X4
An updated variant of M1911 Falcon, beveled magwell, compact design, frame-mounted and grip safety, ambidextrous design.

Production: current   Caliber: .45 ACP   Action: SA, semi-auto   Barrel Length: 4.25"

Sights: Novak   Capacity: 8   Magazine: detachable box   Grips: checkered wood

D2C:	NIB	$ 680	Ex $ 525	VG+ $ 375	Good $ 340	LMP $ 780
C2C:	Mint	$ 660	Ex $ 475	VG+ $ 340	Good $ 310	Fair $ 160
Trade-In:	Mint	$ 490	Ex $ 400	VG+ $ 270	Good $ 240	Poor $ 90

M1911 THRASHER
EXCELLENT

M1911 THRASHER   REDBOOK CODE: RB-IJ-H-TRASHR
Similar to the Colt M1911 Officer, blue or stainless finish, beveled magwell, frame-mounted and grip safety, skeletonized trigger.

Production: current   Caliber: .45 ACP, 9mm   Action: SA, semi-auto

Barrel Length: 3", 3.5"   Sights: fixed or adjustable   Capacity: 7

Magazine: detachable box   Grips: checkered wood

D2C:	NIB	$ 585	Ex $ 450	VG+ $ 325	Good $ 295	LMP $ 636
C2C:	Mint	$ 570	Ex $ 425	VG+ $ 300	Good $ 265	Fair $ 135
Trade-In:	Mint	$ 420	Ex $ 350	VG+ $ 230	Good $ 205	Poor $ 60

JIMENEZ

Jimenez Arms

Founded in 2004, Jimenez Arms, Inc. manufactures a series of compact, semi-automatic pistols, most of which are incredibly inexpensive. Sources state former employees of Bryco Arms formed Jimenez after the former filed for bankruptcy. Jimenez is based in Las Vegas, Nevada.

JA-NINE
NEW IN BOX

JA-NINE   REDBOOK CODE: RB-JA-H-JANINE
Loaded chamber indicator, frame-mounted safety, black or stainless finish, aluminum frame.

Production: current   Caliber: 9mm   Action: striker-fired, semi-auto
Barrel Length: 3.75"   Wt.: 30 oz.   Sights: adjustable   Capacity: 10, 12
Magazine: detachable box   Grips: black plastic

D2C:	NIB	$ 180	Ex $ 150	VG+ $ 100	Good $ 90					
C2C:	Mint	$ 180	Ex $ 125	VG+ $ 90	Good $ 85	Fair $	45			
Trade-In:	Mint	$ 130	Ex $ 125	VG+ $ 80	Good $ 65	Poor $	30			

JA-22
NEW IN BOX

JA-22   REDBOOK CODE: RB-JA-H-22LRXX
Compact design, frame-mounted safety, loaded chamber indicator, black or stainless finish, aluminum frame.

Production: early 2000s - current   Caliber: .22 LR   Action: striker-fired, semi-auto
Barrel Length: 2.5"   Wt.: 14 oz.   Sights: fixed   Capacity: 6   Magazine: detachable box
Grips: black plastic

D2C:	NIB	$ 155	Ex $ 125	VG+ $ 90	Good $ 80		
C2C:	Mint	$ 150	Ex $ 125	VG+ $ 80	Good $ 70	Fair $ 40
Trade-In:	Mint	$ 120	Ex $ 100	VG+ $ 70	Good $ 55	Poor $ 30

JA-25
NEW IN BOX

JA-25   REDBOOK CODE: RB-JA-H-JA25XX
Compact design, frame-mounted safety, loaded chamber indicator, black or stainless finish, aluminum frame.

Production: early 2000s - current   Caliber: .25 ACP   Action: striker-fired, semi-auto
Barrel Length: 2.5"   OA Length: 5"   Wt.: 14 oz.   Sights: fixed   Capacity: 6
Magazine: detachable box   Grips: black plastic

D2C:	NIB	$ 165	Ex $ 150	VG+ $ 95	Good $ 85		
C2C:	Mint	$ 160	Ex $ 125	VG+ $ 90	Good $ 75	Fair $ 40
Trade-In:	Mint	$ 120	Ex $ 100	VG+ $ 70	Good $ 60	Poor $ 30

JA-32
NEW IN BOX

JA-32   REDBOOK CODE: RB-JA-H-JA32XX
Compact design, loaded chamber indicator, frame-mounted safety, aluminum frame, black or stainless finish.

Production: current   Caliber: .32 ACP   Action: striker-fired, semi-auto
Barrel Length: 2.75"Wt.: 20 oz.   Sights: fixed   Capacity: 6
Magazine: detachable box   Grips: black plastic

D2C:	NIB	$ 170	Ex $ 150	VG+ $ 95	Good $ 85		
C2C:	Mint	$ 170	Ex $ 125	VG+ $ 90	Good $ 80	Fair $ 40
Trade-In:	Mint	$ 130	Ex $ 100	VG+ $ 70	Good $ 60	Poor $ 30

JA-380
EXCELLENT

JA-380 REDBOOK CODE: RB-JA-H-JA380X
Loaded chamber indicator, compact size, sold with an extra magazine, black or stainless finish, frame-mounted safety, aluminum frame.

Production: current	Caliber: .380 ACP	Action: striker-fired, semi-auto						
Barrel Length: 2.75"	OA Length: 5.3"	Wt.: 20 oz.	Sights: fixed					
Capacity: 6	Magazine: detachable box	Grips: black plastic						
D2C:	NIB	$ 145	Ex $ 125	VG+ $ 80	Good $ 75			
C2C:	Mint	$ 140	Ex $ 125	VG+ $ 80	Good $ 70	Fair $ 35		
Trade-In:	Mint	$ 110	Ex $ 100	VG+ $ 60	Good $ 55	Poor $ 30		

Kahr Arms

Kahr Arms CEO and President Justin Moon formed the company in 1994 and the company is currently based in Blauvelt, New York with various manufacturing facilities. The company produces a wide array of compact, semi-automatic pistols, most of which operate on striker-fired actions, the CM and PM series standing as the most recognized among these. In the late 1990s, the company purchased Magnum Research and Auto Ordnance, which they still operate. Kahr has developed the reputation of manufacturing reliable, moderately-priced handguns.

P380
NEW IN BOX

P380 REDBOOK CODE: RB-KA-H-P380LC
Matte stainless slide, black polymer frame, loaded chamber indicator, premium Lothar Walther match grade barrel.

Production: current	Caliber: .380 ACP	Action: striker-fired, semi-auto				
Barrel Length: 2.5"	Sights: white bar-dot combat					
Capacity: 6	Magazine: detachable box	Grips: polymer				
D2C:	NIB	$ 695	Ex $ 550	VG+ $ 385	Good $ 350	LMP $ 733
C2C:	Mint	$ 670	Ex $ 500	VG+ $ 350	Good $ 315	Fair $ 160
Trade-In:	Mint	$ 500	Ex $ 400	VG+ $ 280	Good $ 245	Poor $ 90

P380 BLACK ROSE
NEW IN BOX

P380 BLACK ROSE REDBOOK CODE: RB-KA-H-P380BR
High-polish custom engraved stainless slide, black polymer frame, premium Lothar Walther match grade barrel.

Production: current	Caliber: .380 ACP	Action: striker-fired, semi-auto				
Barrel Length: 2.5"	OA Length: 5"	Sights: white bar-dot combat				
Capacity: 6	Magazine: detachable box	Grips: polymer				
D2C:	NIB	$ 890	Ex $ 700	VG+ $ 490	Good $ 445	LMP $ 949
C2C:	Mint	$ 860	Ex $ 625	VG+ $ 450	Good $ 405	Fair $ 205
Trade-In:	Mint	$ 640	Ex $ 500	VG+ $ 350	Good $ 315	Poor $ 90

P9
NEW IN BOX

P9 REDBOOK CODE: RB-KA-H-P9XXXX
P Series pistols feature match grade barrel rifling and an enhanced slide stop, slightly larger than the PM series. Matte stainless slide, black polymer frame, textured polymer grips.

Production: current	Caliber: 9mm	Action: striker-fired, semi-auto				
Barrel Length: 3.5"	OA Length: 5.8"	Sights: white bar-dot combat				
Capacity: 7	Magazine: detachable box	Grips: polymer				
D2C:	NIB	$ 630	Ex $ 500	VG+ $ 350	Good $ 315	LMP $ 739
C2C:	Mint	$ 610	Ex $ 450	VG+ $ 320	Good $ 285	Fair $ 145
Trade-In:	Mint	$ 450	Ex $ 375	VG+ $ 250	Good $ 225	Poor $ 90

KAHR

P9 COVERT
NEW IN BOX

P40
NEW IN BOX

P40 COVERT
NEW IN BOX

P40
NEW IN BOX

P45
NEW IN BOX

P9 COVERT   REDBOOK CODE: RB-KA-H-P9COVE

Black polymer frame, textured polymer grips, internal safety.

Production: 2002 - 2006   Caliber: 9mm   Action: striker-fired, semi-auto

Barrel Length: 3.5"   OA Length: 6"   Sights: white bar-dot combat

Capacity: 6   Magazine: detachable box   Grips: polymer

D2C:	NIB	$ 640	Ex $ 500	VG+ $ 355	Good $ 320	LMP $ 697
C2C:	Mint	$ 620	Ex $ 450	VG+ $ 320	Good $ 290	Fair $ 150
Trade-In:	Mint	$ 460	Ex $ 375	VG+ $ 250	Good $ 225	Poor $ 90

P40   REDBOOK CODE: RB-KA-H-P40XXX

Black polymer frame, matte stainless steel slide, also available with night sights.

Production: current   Caliber: .40 S&W   Action: striker-fired, semi-auto

Barrel Length: 3.6"   OA Length: 6"   Wt.: 17 oz.   Sights: white bar-dot combat Capacity: 6   Magazine: detachable box   Grips: polymer

D2C:	NIB	$ 580	Ex $ 450	VG+ $ 320	Good $ 290	LMP $ 739
C2C:	Mint	$ 560	Ex $ 425	VG+ $ 290	Good $ 265	Fair $ 135
Trade-In:	Mint	$ 420	Ex $ 325	VG+ $ 230	Good $ 205	Poor $ 60

P40 COVERT   REDBOOK CODE: RB-KA-H-P40COV

Black polymer frame, stainless steel slide, extended magazine.

Production: early 2000s - discontinued   Caliber: .40 S&W   Action: striker-fired, semi-auto

Barrel Length: 3.5"   Wt.: 18 oz.   Sights: white bar-dot combat

Capacity: 5, 6   Magazine: detachable box   Grips: polymer

D2C:	NIB	$ 605	Ex $ 475	VG+ $ 335	Good $ 305	LMP $ 695
C2C:	Mint	$ 590	Ex $ 425	VG+ $ 310	Good $ 275	Fair $ 140
Trade-In:	Mint	$ 430	Ex $ 350	VG+ $ 240	Good $ 215	Poor $ 90

P40 (NIGHT SIGHTS, LCI, EXTERNAL SAFETY)   REDBOOK CODE: RB-KA-H-P40NSX

Black polymer frame, matte stainless slide, loaded chamber indicator, enhanced trigger, external safety.

Production: 2013 - current   Caliber: .40 S&W   Action: striker-fired, semi-auto

Barrel Length: 3.6"   OA Length: 6"   Wt.: 17 oz.   Sights: tritium night

Capacity: 6   Magazine: detachable box   Grips: polymer

D2C:	NIB	$ 745	Ex $ 575	VG+ $ 410	Good $ 375	LMP $ 857
C2C:	Mint	$ 720	Ex $ 525	VG+ $ 380	Good $ 340	Fair $ 175
Trade-In:	Mint	$ 530	Ex $ 425	VG+ $ 300	Good $ 265	Poor $ 90

P45   REDBOOK CODE: RB-KA-H-P45XXX

Black polymer frame, matte stainless-steel slide, also available with night sights.

Production: current   Caliber: .45 ACP   Action: striker-fired, semi-auto

Barrel Length: 3.5"   OA Length: 6"   Sights: white bar-dot combat

Capacity: 6   Magazine: detachable box   Grips: polymer

D2C:	NIB	$ 725	Ex $ 575	VG+ $ 400	Good $ 365	LMP $ 805
C2C:	Mint	$ 700	Ex $ 525	VG+ $ 370	Good $ 330	Fair $ 170
Trade-In:	Mint	$ 520	Ex $ 425	VG+ $ 290	Good $ 255	Poor $ 90

CM9
NEW IN BOX

CM9 REDBOOK CODE: RB-KA-H-CM9XXX

CM Series pistols feature conventional barrel rifling and minimal design and markings. Matte stainless slide, black polymer frame, optional extended magazine.

Production: current Caliber: 9mm Action: striker-fired, semi-auto

Barrel Length: 3" OA Length: 5.5" Wt.: 14 oz. Sights: white

bar-dot combat Capacity: 6 Magazine: detachable box Grips: polymer

D2C:	NIB	$ 425	Ex $ 325	VG+ $ 235	Good $ 215	LMP $ 517
C2C:	Mint	$ 410	Ex $ 300	VG+ $ 220	Good $ 195	Fair $ 100
Trade-In:	Mint	$ 310	Ex $ 250	VG+ $ 170	Good $ 150	Poor $ 60

CM40
NEW IN BOX

CM40 REDBOOK CODE: RB-KA-H-CM40XX

Similar to the CM Series. Matte stainless steel slide, black polymer frame, optional extended magazine.

Production: current Caliber: .40 S&W Action: striker-fired, semi-auto

Barrel Length: 3" OA Length: 5.5" Wt.: 16 oz. Sights: polymer

front, white bar-dot combat rear Capacity: 5 Grips: polymer

D2C:	NIB	$ 425	Ex $ 325	VG+ $ 235	Good $ 215	LMP $ 517
C2C:	Mint	$ 410	Ex $ 300	VG+ $ 220	Good $ 195	Fair $ 100
Trade-In:	Mint	$ 310	Ex $ 250	VG+ $ 170	Good $ 150	Poor $ 60

CM45
NEW IN BOX

CM45 REDBOOK CODE: RB-KA-H-CM45XX

Similar to CM Series. Matte stainless slide, black polymer frame, textured polymer grips.

Production: 2014 - current Caliber: .45 ACP Action: striker-fired, semi-auto

Barrel Length: 3.24" OA Length: 5.8" Wt.: 17 oz. Sights: polymer

front, white bar-dot combat rear Capacity: 5 Magazine: detachable box

Grips: polymer

D2C:	NIB	$ 450	Ex $ 350	VG+ $ 250	Good $ 225	LMP $ 517
C2C:	Mint	$ 440	Ex $ 325	VG+ $ 230	Good $ 205	Fair $ 105
Trade-In:	Mint	$ 320	Ex $ 275	VG+ $ 180	Good $ 160	Poor $ 60

CW380
NEW IN BOX

CW380 REDBOOK CODE: RB-KA-H-CW380X

CW Series pistols features conventional barrel rifling and a minimal design. Black polymer frame, matte stainless steel slide, compact design.

Production: current Caliber: .380 ACP Action: striker-fired, semi-auto

Barrel Length: 2.58" OA Length: 5" Sights: polymer front,

white bar-dot combat rear Capacity: 6 Magazine: detachable box

Grips: polymer

D2C:	NIB	$ 355	Ex $ 275	VG+ $ 200	Good $ 180	LMP $ 419
C2C:	Mint	$ 350	Ex $ 250	VG+ $ 180	Good $ 160	Fair $ 85
Trade-In:	Mint	$ 260	Ex $ 200	VG+ $ 140	Good $ 125	Poor $ 60

CW9
NEW IN BOX

CW9   REDBOOK CODE: RB-KA-H-CW9XXX

Matte stainless slide, black polymer frame.

Production: current   Caliber: 9mm   Action: striker-fired, semi-auto

Barrel Length: 3.6"   OA Length: 6"   Wt.: 16 oz.   Sights: polymer

front, white bar-dot combat rear   Capacity: 7

Magazine: detachable box   Grips: polymer

D2C:	NIB	$ 385	Ex $ 300	VG+ $ 215	Good $ 195	LMP $ 485
C2C:	Mint	$ 370	Ex $ 275	VG+ $ 200	Good $ 175	Fair $ 90
Trade-In:	Mint	$ 280	Ex $ 225	VG+ $ 160	Good $ 135	Poor $ 60

CW40
NEW IN BOX

CW40   REDBOOK CODE: RB-KA-H-CW40XX

Black polymer frame, matte stainless steel slide.

Production: current   Caliber: .40 S&W   Action: striker-fired, semi-auto

Barrel Length: 3.6"   OA Length: 6.4"   Wt.: 17 oz.   Sights: polymer front,

white bar-dot combat rear   Capacity: 6   Magazine: detachable box   Grips: polymer

D2C:	NIB	$ 375	Ex $ 300	VG+ $ 210	Good $ 190	LMP $ 485
C2C:	Mint	$ 360	Ex $ 275	VG+ $ 190	Good $ 170	Fair $ 90
Trade-In:	Mint	$ 270	Ex $ 225	VG+ $ 150	Good $ 135	Poor $ 60

CW45
NEW IN BOX

CW45   REDBOOK CODE: RB-KA-H-CW45XX

Black polymer frame, matte stainless slide.

Production: current   Caliber: .45 ACP   Action: striker-fired, semi-auto

Barrel Length: 3.6"   OA Length: 6.3"   Wt.: 20 oz.   Sights: polymer

front, white bar-dot combat rear   Capacity: 6   Grips: polymer

D2C:	NIB	$ 395	Ex $ 325	VG+ $ 220	Good $ 200	LMP $ 485
C2C:	Mint	$ 380	Ex $ 275	VG+ $ 200	Good $ 180	Fair $ 95
Trade-In:	Mint	$ 290	Ex $ 225	VG+ $ 160	Good $ 140	Poor $ 60

CT45
NEW IN BOX

CT45   REDBOOK CODE: RB-KA-H-CT45XX

Matte stainless steel slide, black polymer frame, full-size design, some models with accessory mounts.

Production: 2014 - current   Caliber: .45 ACP   Action: striker-fired, semi-auto

Barrel Length: 4"   OA Length: 6.6"   Wt.: 24 oz.   Sights: polymer

front, white bar-dot combat rear   Capacity: 7   Magazine: detachable box

Grips: polymer

D2C:	NIB	$ 415	Ex $ 325	VG+ $ 230	Good $ 210	LMP $ 449
C2C:	Mint	$ 400	Ex $ 300	VG+ $ 210	Good $ 190	Fair $ 100
Trade-In:	Mint	$ 300	Ex $ 250	VG+ $ 170	Good $ 150	Poor $ 60

CT40
NEW IN BOX

CT40   REDBOOK CODE: RB-KA-H-CT40XX

Matte stainless slide, black polymer frame, full-size design.

Production: 2014 - current   Caliber: .40 S&W   Action: striker-fired, semi-auto

Barrel Length: 4"   OA Length: 6.5"Wt.: 22 oz.   Sights: polymer

front, white bar-dot combat rear   Capacity: 7   Magazine: detachable box

Grips: polymer

D2C:	NIB	$ 395	Ex $ 325	VG+ $ 220	Good $ 200	LMP $ 449
C2C:	Mint	$ 380	Ex $ 275	VG+ $ 200	Good $ 180	Fair $ 95
Trade-In:	Mint	$ 290	Ex $ 225	VG+ $ 160	Good $ 140	Poor $ 60

E9
NEW IN BOX

E9   REDBOOK CODE: RB-KA-H-E9XXXX

Duo-tone finish, matte black frame, matte stainless steel slide, full-size design.

Production: discontinued   Caliber: 9mm   Action: striker-fired, semi-auto
Barrel Length: 3.5"   OA Length: 6"   Wt.: 23 oz.   Sights: polymer
front, drift adjustable rear   Capacity: 7   Magazine: detachable box   Grips: soft polymer

D2C:	NIB	$ 385	Ex $ 300	VG+ $ 215	Good $ 195		
C2C:	Mint	$ 370	Ex $ 275	VG+ $ 200	Good $ 175	Fair $	90
Trade-In:	Mint	$ 280	Ex $ 225	VG+ $ 160	Good $ 135	Poor $	60

K9
NEW IN BOX

K9   REDBOOK CODE: RB-KA-H-K9XXXX

Matte stainless finish, also available in all black finish, night sights at an additional price.

Production: current   Caliber: 9mm   Action: striker-fired, semi-auto
Barrel Length: 3.5"   OA Length: 6"   Wt.: 23 oz.   Sights: white bar-dot combat
Capacity: 7   Magazine: detachable box   Grips: soft polymer

D2C:	NIB	$ 660	Ex $ 525	VG+ $ 365	Good $ 330	LMP $	855
C2C:	Mint	$ 640	Ex $ 475	VG+ $ 330	Good $ 300	Fair $	155
Trade-In:	Mint	$ 470	Ex $ 375	VG+ $ 260	Good $ 235	Poor $	90

K9 ELITE
NEW IN BOX

K9 ELITE   REDBOOK CODE: RB-KA-H-K9ELIT

Distinguished by its smoother trigger pull and shortened frame. Polished stainless-steel finish, laser-etched markings, also available with night sights at an additional price.

Production: current   Caliber: 9mm   Action: striker-fired, semi-auto
Barrel Length: 3.5"   OA Length: 6"   Wt.: 23 oz.   Sights: white bar-dot combat
Capacity: 7   Magazine: detachable box   Grips: polymer

D2C:	NIB	$ 685	Ex $ 525	VG+ $ 380	Good $ 345	LMP $	932
C2C:	Mint	$ 660	Ex $ 475	VG+ $ 350	Good $ 310	Fair $	160
Trade-In:	Mint	$ 490	Ex $ 400	VG+ $ 270	Good $ 240	Poor $	90

MK9
NEW IN BOX

MK9   REDBOOK CODE: RB-KA-H-MK9XXX

A shortened grip and sight radius variant of the K9, matte stainless or blue finish, also available with night sights at an additional price.

Production: current   Caliber: 9mm   Action: striker-fired, semi-auto
Barrel Length: 3"   OA Length: 5.3"   Wt.: 22 oz.   Sights: white bar-dot combat
Capacity: 6, 7   Magazine: detachable box   Grips: hard nylon

D2C:	NIB	$ 750	Ex $ 575	VG+ $ 415	Good $ 375	LMP $	855
C2C:	Mint	$ 720	Ex $ 525	VG+ $ 380	Good $ 340	Fair $	175
Trade-In:	Mint	$ 540	Ex $ 425	VG+ $ 300	Good $ 265	Poor $	90

MK9 ELITE
NEW IN BOX

MK9 ELITE   REDBOOK CODE: RB-KA-H-MK9ELI

Distinguished by its polished finish and components, beveled magwell, and laser-etched markings. Hard nylon frame, magazine grip extension, also available with night sights at an additional price.

Production: current   Caliber: 9mm   Action: striker-fired, semi-auto
Barrel Length: 3"   OA Length: 5.3"   Wt.: 22 oz.   Sights: white bar-dot combat
Capacity: 6, 7   Magazine: detachable box   Grips: hard nylon

D2C:	NIB	$ 815	Ex $ 625	VG+ $ 450	Good $ 410	LMP $	932
C2C:	Mint	$ 790	Ex $ 575	VG+ $ 410	Good $ 370	Fair $	190
Trade-In:	Mint	$ 580	Ex $ 475	VG+ $ 320	Good $ 290	Poor $	90

KAHR

MK40
NEW IN BOX

MK40   REDBOOK CODE: RB-KA-H-MK40XX

Matte stainless steel finish, compact design, polygonal barrel rifling, magazine grip extension, also available with night sights at an additional price.

Production: current   Caliber: .40 S&W   Action: striker-fired, semi-auto

Barrel Length: 3"   OA Length: 5.35"   Wt.: 23 oz.   Sights: white bar-dot combat

Capacity: 5, 6   Magazine: detachable box   Grips: hard nylon

D2C:	NIB	$ 750	Ex $ 575	VG+ $ 415	Good $ 375	LMP $ 855
C2C:	Mint	$ 720	Ex $ 525	VG+ $ 380	Good $ 340	Fair $ 175
Trade-In:	Mint	$ 540	Ex $ 425	VG+ $ 300	Good $ 265	Poor $ 90

MK40 ELITE
NEW IN BOX

MK40 ELITE   REDBOOK CODE: RB-KA-H-MK40EL

Polished stainless steel finish, compact design, magazine grip extension, laser-etched markings, also available with night sights at an additional price.

Production: current   Caliber: .40 S&W   Action: striker-fired, semi-auto

Barrel Length: 3"   OA Length: 5.35"   Wt.: 23 oz.   Sights: white bar-dot combat

Capacity: 5, 6   Magazine: detachable box   Grips: hard nylon

D2C:	NIB	$ 810	Ex $ 625	VG+ $ 450	Good $ 405	LMP $ 932
C2C:	Mint	$ 780	Ex $ 575	VG+ $ 410	Good $ 365	Fair $ 190
Trade-In:	Mint	$ 580	Ex $ 475	VG+ $ 320	Good $ 285	Poor $ 90

PM9
NEW IN BOX

PM9   REDBOOK CODE: RB-KA-H-PM9XXX

PM Series pistols feature match grade rifling, enhanced slide stops, compact frames, and rounded features, though similar to the P series. Black polymer frame, stainless slide, magazine grip extension, some with Crimson Trace laser sights.

Production: current   Caliber: 9mm   Action: striker-fired, semi-auto

Barrel Length: 3"   OA Length: 5.4"   Sights: white bar-dot combat

Capacity: 6, 7   Magazine: detachable box   Grips: polymer

D2C:	NIB	$ 680	Ex $ 525	VG+ $ 375	Good $ 340	LMP $ 786
C2C:	Mint	$ 660	Ex $ 475	VG+ $ 340	Good $ 310	Fair $ 160
Trade-In:	Mint	$ 490	Ex $ 400	VG+ $ 270	Good $ 240	Poor $ 90

PM9 BLACK ROSE
NEW IN BOX

PM9 BLACK ROSE   REDBOOK CODE: RB-KA-H-PM9BRX

Black polymer frame, high polish engraved stainless slide, magazine with grip extension, textured polymer grips.

Production: current   Caliber: 9mm   Action: striker-fired, semi-auto

Barrel Length: 3"   OA Length: 5.4"   Wt.: 14 oz.   Sights: white bar-dot combat

Capacity: 6, 7   Magazine: detachable box   Grips: polymer

D2C:	NIB	$ 935	Ex $ 725	VG+ $ 515	Good $ 470	LMP $ 1049
C2C:	Mint	$ 900	Ex $ 650	VG+ $ 470	Good $ 425	Fair $ 220
Trade-In:	Mint	$ 670	Ex $ 525	VG+ $ 370	Good $ 330	Poor $ 120

PM40
NEW IN BOX

PM40   REDBOOK CODE: RB-KA-H-PM40XX

Black polymer frame, matte stainless slide, also available with night sights.

Production: current   Caliber: .40 S&W   Action: striker-fired, semi-auto

Barrel Length: 3"   OA Length: 5.5"   Wt.: 16 oz.   Sights: white bar-dot combat

Capacity: 5, 6   Magazine: detachable box   Grips: polymer

D2C:	NIB	$ 675	Ex $ 525	VG+ $ 375	Good $ 340	LMP $ 786
C2C:	Mint	$ 650	Ex $ 475	VG+ $ 340	Good $ 305	Fair $ 160
Trade-In:	Mint	$ 480	Ex $ 400	VG+ $ 270	Good $ 240	Poor $ 90

PM45

PM45   REDBOOK CODE: RB-KA-H-PM45XX

Black polymer frame, matte stainless slide.

Production: current Caliber: .45 ACP Action: striker-fired, semi-auto									
Barrel Length: 3.24" OA Length: 5.8"Sights: white bar-dot combat Capacity: 5									
Magazine: detachable box Grips: polymer									
D2C:	NIB	$ 690	Ex $ 525	VG+ $ 380	Good $ 345	LMP $ 855
C2C:	Mint	$ 670	Ex $ 500	VG+ $ 350	Good $ 315	Fair $ 160
Trade-In:	Mint	$ 490	Ex $ 400	VG+ $ 270	Good $ 245	Poor $ 90

T9

T9   REDBOOK CODE: RB-KA-H-T9XXXX

Full-size design, matte stainless finish, recoil-operated design, polygonal barrel rifling, also available with night sights.

Production: current Caliber: 9mm Action: striker-fired, semi-auto									
Barrel Length: 4" OA Length: 6.5"Wt.: 26 oz. Sights: Novak or bar-dot combat									
Capacity: 8 Magazine: detachable box Grips: checkered Hogue Pau Ferro wood									
D2C:	NIB	$ 710	Ex $ 550	VG+ $ 395	Good $ 355	LMP $ 831
C2C:	Mint	$ 690	Ex $ 500	VG+ $ 360	Good $ 320	Fair $ 165
Trade-In:	Mint	$ 510	Ex $ 400	VG+ $ 280	Good $ 250	Poor $ 90

T40

T40   REDBOOK CODE: RB-KA-H-T40XXX

Full-size design, matte stainless finish, recoil-operated design, polygonal barrel rifling, also available with Novak sights at an additional price.

Production: current Caliber: .40 S&W Action: striker-fired, semi-auto									
Barrel Length: 4" OA Length: 6.6" Wt.: 27 oz. Sights: white bar-dot combat									
Capacity: 7 Magazine: detachable box Grips: checkered Hogue Pau Ferro wood									
D2C:	NIB	$ 700	Ex $ 550	VG+ $ 385	Good $ 350	LMP $ 831
C2C:	Mint	$ 680	Ex $ 500	VG+ $ 350	Good $ 315	Fair $ 165
Trade-In:	Mint	$ 500	Ex $ 400	VG+ $ 280	Good $ 245	Poor $ 90

TP9

TP9   REDBOOK CODE: RB-KA-H-TP9XXX

TP Series is smaller and lighter than Kahr's full-sized pistols. Matte stainless slide, black polymer frame, also available with Novak night sights at an additional price.

Production: current Caliber: 9mm Action: striker-fired, semi-auto									
Barrel Length: 4" OA Length: 6.5"Wt.: 18 oz. Sights: white bar-dot combat									
Capacity: 8 Magazine: detachable box Grips: polymer									
D2C:	NIB	$ 610	Ex $ 475	VG+ $ 340	Good $ 305	LMP $ 697
C2C:	Mint	$ 590	Ex $ 425	VG+ $ 310	Good $ 275	Fair $ 145
Trade-In:	Mint	$ 440	Ex $ 350	VG+ $ 240	Good $ 215	Poor $ 90

TP40

TP40   REDBOOK CODE: RB-KA-H-TP40XX

Black polymer frame, matte stainless slide, textured polymer grips, also available with Novak night sights at an additional price.

Production: current Caliber: .40 S&W Action: striker-fired, semi-auto									
Barrel Length: 4" OA Length: 6.5"Wt.: 20 oz. Sights: white bar-dot combat									
Capacity: 7 Magazine: detachable box Grips: polymer									
D2C:	NIB	$ 615	Ex $ 475	VG+ $ 340	Good $ 310	LMP $ 697
C2C:	Mint	$ 600	Ex $ 425	VG+ $ 310	Good $ 280	Fair $ 145
Trade-In:	Mint	$ 440	Ex $ 350	VG+ $ 240	Good $ 220	Poor $ 90

KAHR

TP45
NEW IN BOX

TP45   REDBOOK CODE: RB-KA-H-TP45XX

Black polymer frame, matte stainless slide.

Production: current   Caliber: .45 ACP   Action: striker-fired, semi-auto

Barrel Length: 4"   OA Length: 6.6"   Wt.: 21 oz.   Sights: white bar-dot combat

Capacity: 7   Magazine: detachable box   Grips: polymer

D2C:	NIB	$ 600	Ex $ 475	VG+ $ 330	Good $ 300	LMP $ 697
C2C:	Mint	$ 580	Ex $ 425	VG+ $ 300	Good $ 270	Fair $ 140
Trade-In:	Mint	$ 430	Ex $ 350	VG+ $ 240	Good $ 210	Poor $ 60

K40
NEW IN BOX

K40   REDBOOK CODE: RB-KA-H-K40XXX

Matte stainless steel finish, polygonal barrel rifling, available with night sights for an additional price.

Production: current   Caliber: .40 S&W   Action: striker-fired, semi-auto

Barrel Length: 3.5"   OA Length: 6"   Wt.: 24 oz.   Sights: white bar-dot combat

Capacity: 6   Magazine: detachable box   Grips: soft polymer

D2C:	NIB	$ 725	Ex $ 575	VG+ $ 400	Good $ 365	LMP $ 855
C2C:	Mint	$ 700	Ex $ 525	VG+ $ 370	Good $ 330	Fair $ 170
Trade-In:	Mint	$ 520	Ex $ 425	VG+ $ 290	Good $ 255	Poor $ 90

K40 ELITE
NEW IN BOX

K40 ELITE   REDBOOK CODE: RB-KA-H-K40ELI

Polished stainless finish, laser-etched markings, beveled magwell, enhanced tactical design.

Production: current   Caliber: .40 S&W   Action: striker-fired, semi-auto

Barrel Length: 3.5"   OA Length: 6"   Wt.: 24 oz.   Sights: white bar-dot combat

Capacity: 6   Magazine: detachable box   Grips: soft polymer

D2C:	NIB	$ 775	Ex $ 600	VG+ $ 430	Good $ 390	LMP $ 932
C2C:	Mint	$ 750	Ex $ 550	VG+ $ 390	Good $ 350	Fair $ 180
Trade-In:	Mint	$ 560	Ex $ 450	VG+ $ 310	Good $ 275	Poor $ 90

Kel-Tec CNC.

Kel-Tec CNC Industrues, Inc.—or simply Kel-Tec—began operations in 1991 as a CNC shop and didn't produce their first firearm until around 1995. Since then, the company has established itself as one of the leaders in manufacturing quality, affordable compact pistols. The P-11 stands as one of the company's most recognizable and favored pistol, as it fires 9mm rounds—popular for self-defense use—and its compact size allows shooters to easily conceal it on their person. In 2006, the company released the PF-9, which the company claimed was the thinnest, lightest 9mm semi-automatic pistol to ever enter production. Kel-Tec is currently based in Cocoa, Florida.

PF-9
NEW IN BOX

PF-9   REDBOOK CODE: RB-KT-H-PF9XXX

Transfer bar safety, polymer frame, accessory rails, bobbed hammer, and various finishes and grip colors, very similar to both the P-11 and P-3AT.

Production: 2006 - current   Caliber: 9mm   Action: DAO, semi-auto

Barrel Length: 3"   OA Length: 5.85"   Sights: 3-dot

Capacity: 7   Magazine: detachable box   Grips: polymer

D2C:	NIB	$ 280	Ex $ 225	VG+ $ 155	Good $ 140	LMP $ 333
C2C:	Mint	$ 270	Ex $ 200	VG+ $ 140	Good $ 130	Fair $ 65
Trade-In:	Mint	$ 200	Ex $ 175	VG+ $ 110	Good $ 100	Poor $ 30

P-11
NEW IN BOX

Courtesy of Bud's Gun Shop

P-11 REDBOOK CODE: RB-KT-H-P11XXX

Transfer bar safety, 4140 steel slide, 7075-T6 aluminum frame, ultra high impact polymer grip, blue or (rare) stainless finish on slide.

Production: 1995 - current Caliber: 9mm Action: DAO, semi-auto

Barrel Length: 3.1" OA Length: 5.6" Wt.: 14 oz. Sights: fixed

Capacity: 10 Magazine: detachable box Grips: polymer

D2C:	NIB	$ 315	Ex $ 250	VG+ $ 175	Good $ 160	LMP $ 400
C2C:	Mint	$ 310	Ex $ 225	VG+ $ 160	Good $ 145	Fair $ 75
Trade-In:	Mint	$ 230	Ex $ 200	VG+ $ 130	Good $ 115	Poor $ 60

P-40
NEW IN BOX

P-40 REDBOOK CODE: RB-KT-H-P40XXX

A .40 S&W variant of the P-11, polymer frame, transfer bar safety, various finishes, compact design.

Production: late 1990s - discontinued Caliber: .40 S&W Action: DAO, semi-auto

Barrel Length: 3.3" Wt.: 16 oz. Sights: fixed Capacity: 9 Magazine: detachable box

Grips: polymer

D2C:	NIB	$ 345	Ex $ 275	VG+ $ 190	Good $ 175	
C2C:	Mint	$ 340	Ex $ 250	VG+ $ 180	Good $ 160	Fair $ 80
Trade-In:	Mint	$ 250	Ex $ 200	VG+ $ 140	Good $ 125	Poor $ 60

P-32
VERY GOOD +

P-32 REDBOOK CODE: RB-KT-H-P32XXX

First generation models were produced between approximately 1999 and 2005 with second generation models in production since then. Finish and grip colors vary, polymer frame, transfer bar safety, compact design, lightweight parts.

Production: 1999 - current Caliber: .32 ACP Action: DAO, semi-auto

Barrel Length: 2.6" Sights: fixed Capacity: 7 Magazine: detachable box

Grips: polymer

D2C:	NIB	$ 255	Ex $ 200	VG+ $ 145	Good $ 130	LMP $ 385
C2C:	Mint	$ 250	Ex $ 200	VG+ $ 130	Good $ 115	Fair $ 60
Trade-In:	Mint	$ 190	Ex $ 150	VG+ $ 100	Good $ 90	Poor $ 30

P-3AT
NEW IN BOX

Courtesy of Bud's Gun Shop

P-3AT REDBOOK CODE: RB-KT-H-P3ATXX

Polymer frame, transfer bar safety, various finishes and grip colors, compact design, one of the smallest pocket-size .380s to enter production.

Production: early 2000s - current Caliber: .380 ACP Action: DAO, semi-auto

Barrel Length: 2.7" Sights: fixed Capacity: 6 Magazine: detachable box

Grips: polymer

D2C:	NIB	$ 275	Ex $ 225	VG+ $ 155	Good $ 140	
C2C:	Mint	$ 270	Ex $ 200	VG+ $ 140	Good $ 125	Fair $ 65
Trade-In:	Mint	$ 200	Ex $ 175	VG+ $ 110	Good $ 100	Poor $ 30

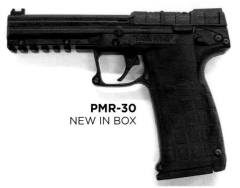

PMR-30
NEW IN BOX

PMR-30 REDBOOK CODE: RB-KT-H-PMR30X

Modern full-size design, frame-mounted safety, nylon slide cover, Picatinny rails, recoil buffer, matte blue finish, steel slide, aluminum frame.

Production: 2010 - current Caliber: .22 Magnum Action: SA, semi-auto

Barrel Length: 4.3" OA Length: 7.9" Wt.: 14 oz. Sights: fiber-optic front

Capacity: 30 Magazine: detachable box Grips: Zytel, synthetic

D2C:	NIB	$ 450	Ex $ 350	VG+ $ 250	Good $ 225	LMP $ 415
C2C:	Mint	$ 440	Ex $ 325	VG+ $ 230	Good $ 205	Fair $ 105
Trade-In:	Mint	$ 320	Ex $ 275	VG+ $ 180	Good $ 160	Poor $ 60

Kimber

Kimber Manufacturing traces its origins to Kimber of Oregon, founded in the late 1970s, but Kimber, in its current incarnation, was formed in the mid-1990s, though they'd already begun importing and distributing some weapons under various names. Over the years, the company has imported and manufactured a host of different firearms for themselves and other companies, but their line of top-tier M1911 pistols remains their most noteworthy product. The company is currently located in Elmsford, New York, where they manufacture a variety of premier pistols and rifles.

GOLD GUARDIAN REDBOOK CODE: RB-KI-H-GOLDGU
Hand-fit match barrel and bushing, high polished stainless steel slide and frame, ambidextrous safety, extended magwell, only 300 produced from the Kimber Custom Shop.

Production: 1998   Caliber: .45 ACP   Action: SA, semi-auto

Barrel Length: 5"   Wt.: 38 oz.   Sights: tritium night   Capacity: 7

Magazine: detachable box   Grips: checkered rosewood

D2C:	NIB $ 1,445	Ex $ 1,100	VG+ $ 795	Good $ 725		
C2C:	Mint $ 1,390	Ex $ 1,000	VG+ $ 730	Good $ 655	Fair $ 335	
Trade-In:	Mint $ 1,030	Ex $ 825	VG+ $ 570	Good $ 510	Poor $ 150	

ELITE CARRY
EXCELLENT
Courtesy of Bud's Gun Shop

ELITE CARRY REDBOOK CODE: RB-KI-H-ELICAR
Stainless finish slide, black aluminum frame, beavertail grip safety, ambidextrous thumb safety, only 1,200 produced.

Production: 1998 - late 1990s   Caliber: .45 ACP   Action: SA, semi-auto

Barrel Length: 4"   Sights: tritium night   Capacity: 7   Magazine: detachable box

Grips: checkered rosewood

D2C:	NIB $1,015	Ex $ 775	VG+ $ 560	Good $ 510		
C2C:	Mint $ 980	Ex $ 725	VG+ $ 510	Good $ 460	Fair $ 235	
Trade-In:	Mint $ 730	Ex $ 575	VG+ $ 400	Good $ 360	Poor $ 120	

STAINLESS COVERT REDBOOK CODE: RB-KI-H-STNCOV
Carry Melt treatment, satin silver KimPro finish, premium aluminum trigger, ambidextrous thumb safety, roughly 1,000 produced by the Kimber Custom Shop.

Production: 1999   Caliber: .45 ACP   Action: SA, semi-auto

Barrel Length: 4"   Wt.: 34 oz.   Sights: tritium night   Capacity: 7

Magazine: detachable box   Grips: checkered rosewood

D2C:	NIB $1,230	Ex $ 950	VG+ $ 680	Good $ 615		
C2C:	Mint $1,190	Ex $ 850	VG+ $ 620	Good $ 555	Fair $ 285	
Trade-In:	Mint $ 880	Ex $ 700	VG+ $ 480	Good $ 435	Poor $ 150	

PRO ELITE REDBOOK CODE: RB-KI-H-PROELI
Similar to the Pro Carry model, aluminum frame, match barrel and chamber, beveled slide serrations, Carry Melt treatment, KimPro finished black slide and silver finish frame, only 2,500 produced by Kimber Custom Shop.

Production: 1999   Caliber: .45 ACP   Action: SA, semi-auto   Barrel Length: 4"

Sights: 3-dot tritium night sights   Capacity: 7   Magazine: detachable box

Grips: checkered rosewood

D2C:	NIB $1,635	Ex $1,250	VG+ $ 900	Good $ 820		
C2C:	Mint $1,570	Ex $1,150	VG+ $ 820	Good $ 740	Fair $ 380	
Trade-In:	Mint $1,170	Ex $ 925	VG+ $ 640	Good $ 575	Poor $ 180	

ULTRA ELITE   REDBOOK CODE: RB-KI-H-ULTELI
Custom Shop produced, Carry Melt treatment, KimPro silver finish slide, black finish aluminum frame, ambidextrous thumb safety, beveled magwell, limited production.

Production: 1999   Caliber: .45 ACP   Action: SA, semi-auto   Barrel Length: 3"
Sights: 3-dot tritium night sights   Capacity: 6   Magazine: detachable box
Grips: checkered rosewood

D2C:	NIB	$1,265	Ex	$ 975	VG+	$ 700	Good	$ 635		
C2C:	Mint	$1,220	Ex	$ 875	VG+	$ 640	Good	$ 570	Fair	$ 295
Trade-In:	Mint	$ 900	Ex	$ 725	VG+	$ 500	Good	$ 445	Poor	$ 150

CUSTOM HERITAGE EDITION   REDBOOK CODE: RB-KI-H-HEREDI
Black oxide finished steel frame and slide, beavertail grip safety, extended ambidextrous thumb safeties, adjustable target trigger, checkered front strap and mainspring housing, limited production.

Production: 2000 - discontinued   Caliber: .45 ACP   Action: SA, semi-auto
Barrel Length: 5"   Sights: low-profile fixed   Capacity: 7   Magazine: detachable box
Grips: checkered rosewood with gold medallions

D2C:	NIB	$ 860	Ex	$ 675	VG+	$ 475	Good	$ 430		
C2C:	Mint	$ 830	Ex	$ 600	VG+	$ 430	Good	$ 390	Fair	$ 200
Trade-In:	Mint	$ 620	Ex	$ 500	VG+	$ 340	Good	$ 305	Poor	$ 90

ULTRA CDP ELITE II   REDBOOK CODE: RB-KI-H-ULCDPE
Black anodized aluminum frame, black oxide steel slide, ramped match barrel, Carry Melt treatment, 30 lpi checkering, ambidextrous safety, very limited production from Custom Shop.

Production: early 2000s   Caliber: .45 ACP   Action: SA, semi-auto
Barrel Length: 3"   Wt.: 25 oz.   Sights: 3-dot tritium night sights
Capacity: 7   Magazine: detachable box   Grips: ruby/charcoal laminated

D2C:	NIB	$ 1,180	Ex	$ 900	VG+	$ 650	Good	$ 590		
C2C:	Mint	$ 1,140	Ex	$ 825	VG+	$ 590	Good	$ 535	Fair	$ 275
Trade-In:	Mint	$ 840	Ex	$ 675	VG+	$ 470	Good	$ 415	Poor	$ 120

ULTRA SP II   REDBOOK CODE: RB-KI-H-ULSP2X
Bushingless barrel, Carry Melt treatment, black oxide slide, anodized frame colors available in black/blue, black/red, or black/silver.

Production: 2003 - discontinued   Caliber: .45 ACP   Action: SA, semi-auto
Barrel Length: 3"   Wt.: 25 oz.   Sights: fixed   Capacity: 7
Magazine: detachable box   Grips: Micarta

D2C:	NIB	$1,105	Ex	$ 850	VG+	$ 610	Good	$ 555		
C2C:	Mint	$1,070	Ex	$ 775	VG+	$ 560	Good	$ 500	Fair	$ 255
Trade-In:	Mint	$ 790	Ex	$ 625	VG+	$ 440	Good	$ 390	Poor	$ 120

PRO CARRY TEN II
NEW IN BOX

PRO CARRY TEN II   REDBOOK CODE: RB-KI-H-PCT2XX
Black polymer frame, beavertail grip safety, bushingless barrel, stainless slide, double stack frame, thumb safety.

Production: early 2000s   Caliber: .45 ACP   Action: SA, semi-auto
Barrel Length: 4"   Wt.: 30 oz.   Sights: fixed   Capacity: 13   Magazine: detachable box
Grips: polymer grip frame

D2C:	NIB	$ 900	Ex	$ 700	VG+	$ 495	Good	$ 450		
C2C:	Mint	$ 870	Ex	$ 625	VG+	$ 450	Good	$ 405	Fair	$ 210
Trade-In:	Mint	$ 640	Ex	$ 525	VG+	$ 360	Good	$ 315	Poor	$ 90

PRO BP TEN II
NEW IN BOX

PRO BP TEN II   REDBOOK CODE: RB-KI-H-PROBPT

Black matte finish, polymer double stack frame, steel slide, beavertail grip safety, thumb safety.

Production: 2003 - 2007	Caliber: .45 ACP	Action: SA, semi-auto			
Barrel Length: 4" Wt.: 31 oz.	Sights: fixed	Capacity: 10, 13			
Magazine: detachable box	Grips: polymer				
D2C:	NIB $ 765	Ex $ 600	VG+ $ 425	Good $ 385	
C2C:	Mint $ 740	Ex $ 550	VG+ $ 390	Good $ 345	Fair $ 180
Trade-In:	Mint $ 550	Ex $ 450	VG+ $ 300	Good $ 270	Poor $ 90

LTP II
NEW IN BOX

LTP II   REDBOOK CODE: RB-KI-H-LTP2XX

Beveled magazine well, external extractor, match grade barrel, steel slide and frame, ambidextrous thumb safety, tungsten guide rod, adjustable trigger, extended 10-round magazine.

Production: 2002 - 2006	Caliber: .45 ACP	Action: SA, semi-auto			
Barrel Length: 5" Wt.: 38 oz.	Sights: adjustable	Capacity: 10			
Magazine: detachable box	Grips: double diamond checkered rosewood				
D2C:	NIB $ 1,885	Ex $ 1,450	VG+ $ 1,040	Good $ 945	
C2C:	Mint $ 1,810	Ex $ 1,325	VG+ $ 950	Good $ 850	Fair $ 435
Trade-In:	Mint $ 1,340	Ex $ 1,075	VG+ $ 740	Good $ 660	Poor $ 210

CUSTOM II
NEW IN BOX

CUSTOM II   REDBOOK CODE: RB-KI-H-CUSTWO

Stainless match barrel, lowered and flared ejection port, rounded and blended edges, Kimber Firing Pin Safety, loaded chamber indicator, high-ride beavertail grip safety, extended thumb safety, checkered slide release, beveled magazine well, polished breech face.

Production: late 1990s - current	Caliber: .45 ACP	Action: SA, semi-auto			
Barrel Length: 5" OA Length: 8.7" Wt.: 38 oz.					
Sights: fixed low-profile, optional night sights	Capacity: 7				
Magazine: detachable box	Grips: black synthetic with double diamond, optional walnut				
D2C:	NIB $ 855	Ex $ 650	VG+ $ 475	Good $ 430	LMP $ 871
C2C:	Mint $ 830	Ex $ 600	VG+ $ 430	Good $ 385	Fair $ 200
Trade-In:	Mint $ 610	Ex $ 500	VG+ $ 340	Good $ 300	Poor $ 90

CUSTOM TARGET II
NEW IN BOX

CUSTOM TARGET II   REDBOOK CODE: RB-KI-H-CUSTA2

Stainless match barrel, lowered and flared ejection port, rounded and blended edges, Kimber Firing Pin Safety, loaded chamber indicator, high-ride beavertail grip safety, extended thumb safety, checkered slide release, beveled magazine well, polished breech face.

Production: 1998 - current	Caliber: .45 ACP	Action: SA, semi-auto			
Barrel Length: 5" OA Length: 8.7" Wt.: 38 oz.					
Sights: Kimber adjustable target	Capacity: 7				
Magazine: detachable box	Grips: black synthetic with double diamond				
D2C:	NIB $ 920	Ex $ 700	VG+ $ 510	Good $ 460	LMP $ 974
C2C:	Mint $ 890	Ex $ 650	VG+ $ 460	Good $ 415	Fair $ 215
Trade-In:	Mint $ 660	Ex $ 525	VG+ $ 360	Good $ 325	Poor $ 120

ROYAL II
NEW IN BOX

ROYAL II   REDBOOK CODE: RB-KI-H-ROYAL2

Deep charcoal blue from Turnbull Restoration, high polished flats, beveled magwell, polished breech face, match grade components, wide front and rear slide serrations, extended thumb safety, beavertail grip safety.

Production: mid-1990s - current Caliber: .45 ACP Action: SA, semi-auto					
Barrel Length: 5" OA Length: 8.7" Wt.: 38 oz. Sights: fixed low-profile					
Capacity: 7 Magazine: detachable box Grips: smooth solid bone					
D2C:	NIB $ 1,350	Ex $ 1,050	VG+ $ 745	Good $ 675	LMP $2,020
C2C:	Mint $ 1,300	Ex $ 950	VG+ $ 680	Good $ 610	Fair $ 315
Trade-In: Mint $ 960	Ex $ 775	VG+ $ 530	Good $ 475	Poor $ 150	

STAINLESS II
NEW IN BOX

STAINLESS II   REDBOOK CODE: RB-KI-H-ST2XXX

Stainless slide and frame, polished breech face, stainless match barrel, match chamber and trigger, lowered and flared ejection port, rounded and blended edges, Kimber Firing Pin Safety, loaded chamber indicator, high-ride beavertail grip safety, extended thumb safety, beveled magwell.

Production: late 1990s Caliber: .45 ACP, 9mm Action: SA, semi-auto					
Barrel Length: 5" OA Length: 8.7" Wt.: 38 oz.					
Sights: fixed low-profile, optional night sights Capacity: 7					
Magazine: detachable box Grips: checkered black double diamond					
D2C:	NIB $ 940	Ex $ 725	VG+ $ 520	Good $ 470	LMP $ 998
C2C:	Mint $ 910	Ex $ 650	VG+ $ 470	Good $ 425	Fair $ 220
Trade-In: Mint $ 670	Ex $ 550	VG+ $ 370	Good $ 330	Poor $ 120	

STAINLESS TARGET II
NEW IN BOX

STAINLESS TARGET II   REDBOOK CODE: RB-KI-H-STTAR2

Stainless slide and frame, polished breech face, stainless match barrel, match chamber and trigger, lowered and flared ejection port, rounded and blended edges, Kimber Firing Pin Safety, loaded chamber indicator, high-ride beavertail grip safety, extended thumb safety, beveled magwell.

Production: late 1990s Caliber: .45 ACP, 9mm, .38 Super, 10mm Action: SA, semi-auto					
Barrel Length: 5" OA Length: 8.7" Wt.: 38 oz. Sights: Kimber adjustable target					
Capacity: 7, 8, 9 Magazine: detachable box Grips: checkered black double diamond					
D2C:	NIB $1,050	Ex $ 800	VG+ $ 580	Good $ 525	LMP $ 1,108
C2C:	Mint $1,010	Ex $ 725	VG+ $ 530	Good $ 475	Fair $ 245
Trade-In: Mint $ 750	Ex $ 600	VG+ $ 410	Good $ 370	Poor $ 120	

CUSTOM TLE II
NEW IN BOX

CUSTOM TLE II   REDBOOK CODE: RB-KI-H-TLE2CU

Matte black finish, stainless match barrel, match chamber and trigger, lowered and flared ejection port, rounded and blended edges, fining pin safety, loaded chamber indicator, beavertail grip safety, extended thumb safety, beveled magwell, polished breech face, front strap checkering.

Production: early 2000s - current Caliber: .45 ACP, 10mm Action: SA, semi-auto					
Barrel Length: 5" OA Length: 8.7"					
Wt.: 38 oz. Sights: Meprolight tritium 3-dot night sights Capacity: 7, 9					
Magazine: detachable box Grips: checkered black double diamond					
D2C:	NIB $1,020	Ex $ 800	VG+ $ 565	Good $ 510	LMP $1,080
C2C:	Mint $ 980	Ex $ 725	VG+ $ 510	Good $ 460	Fair $ 235
Trade-In: Mint $ 730	Ex $ 575	VG+ $ 400	Good $ 360	Poor $ 120	

KIMBER

CUSTOM TLE/RL II
NEW IN BOX

CUSTOM TLE/RL II   REDBOOK CODE: RB-KI-H-TLRL2C

Similar to the Custom TLE II but with added tactical rail.

Production: 2003 - discontinued   Caliber: .45 ACP, 10mm   Action: SA, semi-auto

Barrel Length: 5" OA Length: 8.7"Wt.: 39 oz.

Sights: Low Profile Combat Meprolight tritium 3-dot night sights

Capacity: 7, 9   Magazine: detachable box   Grips: checkered black double diamond

D2C:	NIB	$ 1,115	Ex $ 850	VG+ $ 615	Good $ 560	LMP $1,178
C2C:	Mint	$1,080	Ex $ 775	VG+ $ 560	Good $ 505	Fair $ 260
Trade-In:	Mint	$ 800	Ex $ 625	VG+ $ 440	Good $ 395	Poor $ 120

STAINLESS TLE II
NEW IN BOX

STAINLESS TLE II   REDBOOK CODE: RB-KI-H-TLE2SS

Slide and frame machined from solid stainless steel, extended and beveled magwell, polished breech face, beavertail grip safety, wide front and rear slide serrations, front strap checkering.

Production: 2004 - discontinued   Caliber: .45 ACP   Action: SA, semi-auto

Barrel Length: 5" OA Length: 8.7"   Wt.: 38 oz.   Sights: Meprolight tritium 3-dot night sights   Capacity: 7Grips: checkered black double diamond

D2C:	NIB	$ 1,155	Ex $ 900	VG+ $ 640	Good $ 580	LMP $ 1,211
C2C:	Mint	$ 1,110	Ex $ 800	VG+ $ 580	Good $ 520	Fair $ 270
Trade-In:	Mint	$ 830	Ex $ 650	VG+ $ 460	Good $ 405	Poor $ 120

**STAINLESS
TLE/RL II**
NEW IN BOX

STAINLESS TLE/RL II   REDBOOK CODE: RB-KI-H-TLERLS

Similar to the Stainless TLE II but with tactical rail.

Production: 2003   Caliber: .45 ACP   Action: SA, semi-auto

Barrel Length: 5"   OA Length: 8.7"   Wt.: 39 oz.   Sights: fixed Low Profile Combat Meprolight tritium 3-dot night sights   Capacity: 7   Magazine: detachable box

Grips: checkered black double diamond

D2C:	NIB	$1,255	Ex $ 975	VG+ $ 695	Good $ 630	LMP $1,323
C2C:	Mint	$1,210	Ex $ 875	VG+ $ 630	Good $ 565	Fair $ 290
Trade-In:	Mint	$ 900	Ex $ 725	VG+ $ 490	Good $ 440	Poor $ 150

WARRIOR
NEW IN BOX

WARRIOR   REDBOOK CODE: RB-KI-H-WARRIO

Matte black KimPro II finish, tactical rail, ambidextrous thumb safety, lanyard ring, beavertail grip safety, military-length guide rod.

Production: 2004 - current   Caliber: .45 ACP   Action: SA, semi-auto

Barrel Length: 5"   OA Length: 8.7" Wt.: 40 oz.

Sights: Tactical Wedge 3-dot night sights   Capacity: 7

Magazine: detachable box   Grips: G-10 textured

D2C:	NIB	$ 1,435	Ex $ 1,100	VG+ $ 790	Good $ 720	LMP $ 1,512
C2C:	Mint	$ 1,380	Ex $1,000	VG+ $ 720	Good $ 650	Fair $ 335
Trade-In:	Mint	$ 1,020	Ex $ 825	VG+ $ 560	Good $ 505	Poor $ 150

KIMBER

DESERT WARRIOR
NEW IN BOX

DESERT WARRIOR REDBOOK CODE: RB-KI-H-WARRDE

Desert Tan KimPro II finish, tactical rail, ambidextrous thumb safety, lanyard ring, beavertail grip safety, military-length guide rod.

Production: 2005 - current   Caliber: .45 ACP   Action: SA, semi-auto

Barrel Length: 5"   OA Length: 8.7"

Wt.: 40 oz.   Sights: Tactical Wedge 3-dot night sights   Capacity: 7

Magazine: detachable box   Grips: G-10 textured

D2C:	NIB $ 1,405	Ex $ 1,075	VG+ $ 775	Good $ 705	LMP $ 1,512
C2C:	Mint $ 1,350	Ex $ 975	VG+ $ 710	Good $ 635	Fair $ 325
Trade-In:	Mint $ 1,000	Ex $ 800	VG+ $ 550	Good $ 495	Poor $ 150

WARRIOR SOC
NEW IN BOX

WARRIOR SOC REDBOOK CODE: RB-KI-H-WARRIS

Tan/green KimPro II finish, tactical rail, Crimson Trace Rail Master laser sight, ambidextrous thumb safety, lanyard ring, beavertail grip safety, military-length guide rod.

Production: current   Caliber: .45 ACP   Action: SA, semi-auto   Barrel Length: 5"

OA Length: 8.7"   Wt.: 40 oz.   Sights: Tactical Wedge 3-dot night sights   Capacity: 7

Magazine: detachable box   Grips: G-10 textured

D2C:	NIB $ 1,505	Ex $ 1,150	VG+ $ 830	Good $ 755	LMP $ 1,665
C2C:	Mint $ 1,450	Ex $ 1,050	VG+ $ 760	Good $ 680	Fair $ 350
Trade-In:	Mint $ 1,070	Ex $ 850	VG+ $ 590	Good $ 530	Poor $ 180

GOLD MATCH II
NEW IN BOX

GOLD MATCH II REDBOOK CODE: RB-KI-H-GM2XXX

Hand-fitted stainless match barrel, Premium Aluminum Trigger, loaded chamber indicator, high ride beavertail grip safety, polished breech face, extended ambidextrous thumb safety.

Production: early 2000s - current   Caliber: .45 ACP   Action: SA, semi-auto

Barrel Length: 5"   OA Length: 8.7"   Wt.: 38 oz.

Sights: Kimber adjustable   Capacity: 8   Magazine: detachable box

Grips: double diamond checkered rosewood

D2C:	NIB $ 1,315	Ex $ 1,000	VG+ $ 725	Good $ 660	LMP $ 1,393
C2C:	Mint $ 1,270	Ex $ 925	VG+ $ 660	Good $ 595	Fair $ 305
Trade-In:	Mint $ 940	Ex $ 750	VG+ $ 520	Good $ 465	Poor $ 150

STAINLESS GOLD MATCH II
NEW IN BOX

STAINLESS GOLD MATCH II REDBOOK CODE: RB-KI-H-GM2SSX

Similar to the Gold Match II but with stainless steel slide and frame.

Production: early 2000s - current   Caliber: .45 ACP, 9mm   Action: SA, semi-auto

Barrel Length: 5"   OA Length: 8.7"   Wt.: 38 oz.   Sights: Kimber adjustable

Capacity: 8 Grips: double diamond checkered rosewood

D2C:	NIB $ 1,500	Ex $ 1,150	VG+ $ 825	Good $ 750	LMP $ 1,574
C2C:	Mint $ 1,440	Ex $ 1,050	VG+ $ 750	Good $ 675	Fair $ 345
Trade-In:	Mint $ 1,070	Ex $ 850	VG+ $ 590	Good $ 525	Poor $ 150

KIMBER

TEAM MATCH II
NEW IN BOX

TEAM MATCH II   REDBOOK CODE: RB-KI-H-TM2XXX

Black DLC slide finish, extended magwell, hand-fitted stainless match barrel, Premium Aluminum Trigger, high ride beavertail grip safety, polished breech face, extended ambidextrous thumb safety.

Production: early 2000s - 2012 Caliber: .45 ACP, 9mm Action: SA, semi-auto

Barrel Length: 5" OA Length: 8.7" Wt.: 39 oz. Sights: Kimber adjustable

Capacity: 8, 9Magazine: detachable box Grips: red/blue G-10

D2C:	NIB	$ 1,570	Ex	$ 1,200	VG+	$ 865	Good	$ 785	LMP $ 1,868
C2C:	Mint	$ 1,510	Ex	$ 1,100	VG+	$ 790	Good	$ 710	Fair $ 365
Trade-In:	Mint	$ 1,120	Ex	$ 900	VG+	$ 620	Good	$ 550	Poor $ 180

COMPACT STAINLESS II
NEW IN BOX

COMPACT STAINLESS II   REDBOOK CODE: RB-KI-H-CSS2XX

Short grip, stainless steel slide and aluminum frame, stainless steel bushingless match bull barrel, rounded and blended edges, Kimber Firing Pin Safety, loaded chamber indicator, high ride beavertail grip safety, extended thumb safety, polished breech face.

Production: 1998 - current Caliber: .45 ACP Action: SA, semi-auto

Barrel Length: 4" OA Length: 7.7" Wt.: 27 oz.

Sights: fixed low-profile Capacity: 7 Magazine: detachable box

Grips: checkered black double diamond

D2C:	NIB	$ 965	Ex	$ 750	VG+	$ 535	Good	$ 485	LMP $ 1,052
C2C:	Mint	$ 930	Ex	$ 675	VG+	$ 490	Good	$ 435	Fair $ 225
Trade-In:	Mint	$ 690	Ex	$ 550	VG+	$ 380	Good	$ 340	Poor $ 120

PRO CARRY II
NEW IN BOX

PRO CARRY II   REDBOOK CODE: RB-KI-H-PRC2XX

Full-length grip, steel slide and aluminum frame, matte black finish, stainless bushingless match bull barrel, rounded and blended edges, Kimber Firing Pin Safety, loaded chamber indicator, high-ride beavertail grip safety, extended thumb safety, polished breech face.

Production: current Caliber: .45 ACP, 9mm Action: SA, semi-auto

Barrel Length: 4" OA Length: 7.7" Wt.: 28 oz.

Sights: fixed low-profile, optional night sights Capacity: 7, 9

Magazine: detachable box Grips: checkered black double diamond

D2C:	NIB	$ 860	Ex	$ 675	VG+	$ 475	Good	$ 430	LMP $ 919
C2C:	Mint	$ 830	Ex	$ 600	VG+	$ 430	Good	$ 390	Fair $ 200
Trade-In:	Mint	$ 620	Ex	$ 500	VG+	$ 340	Good	$ 305	Poor $ 90

STAINLESS PRO CARRY II
NEW IN BOX

STAINLESS PRO CARRY II   REDBOOK CODE: RB-KI-H-PRC2SS

Similar to the Pro Carry II but with stainless steel slide and satin silver finish.

Production: current Caliber: .45 ACP, 9mm Action: SA, semi-auto

Barrel Length: 4" OA Length: 7.7" Wt.: 28 oz.

Sights: fixed low-profile, optional night sights Capacity: 7, 9

Magazine: detachable box Grips: checkered black double diamond

D2C:	NIB	$ 990	Ex	$ 775	VG+	$ 545	Good	$ 495	LMP $ 1,016
C2C:	Mint	$ 960	Ex	$ 700	VG+	$ 500	Good	$ 450	Fair $ 230
Trade-In:	Mint	$ 710	Ex	$ 575	VG+	$ 390	Good	$ 350	Poor $ 120

PRO CARRY HD II
NEW IN BOX

PRO CARRY HD II REDBOOK CODE: RB-KI-H-PRCHD2

Full length grip, stainless steel slide and frame, stainless bushingless match bull barrel, firing pin safety, loaded chamber indicator, beavertail grip safety, extended thumb safety, polished breech face.

Production: 2001 - current   Caliber: .45 ACP, .38 Super   Action: SA, semi-auto

Barrel Length: 4"   OA Length: 7.7" Wt.: 35 oz.

Sights: fixed low-profile   Capacity: 7, 9   Magazine: detachable box

Grips: checkered black double diamond

D2C:	NIB	$1,000	Ex $ 775	VG+ $ 550	Good $ 500	LMP $1,046
C2C:	Mint	$ 960	Ex $ 700	VG+ $ 500	Good $ 450	Fair $ 230
Trade-In:	Mint	$ 710	Ex $ 575	VG+ $ 390	Good $ 350	Poor $ 120

PRO TLE II
NEW IN BOX

PRO TLE II REDBOOK CODE: RB-KI-H-PRTLE2

Full length grip, steel slide and frame, matte black finish, stainless bushingless match bull barrel, loaded chamber indicator, beavertail grip safety, extended thumb safety, polished breech face.

Production: 2006 - current   Caliber: .45 ACP   Action: SA, semi-auto

Barrel Length: 4"   OA Length: 7.7"   Wt.: 35 oz.

Sights: Meprolight tritium 3-dot night sights   Capacity: 7   Magazine: detachable box

Grips: checkered black double diamond

D2C:	NIB	$1,010	Ex $ 775	VG+ $ 560	Good $ 505	LMP $ 1,150
C2C:	Mint	$ 970	Ex $ 700	VG+ $ 510	Good $ 455	Fair $ 235
Trade-In:	Mint	$ 720	Ex $ 575	VG+ $ 400	Good $ 355	Poor $ 120

PRO TLE/RL II
NEW IN BOX

PRO TLE/RL II REDBOOK CODE: RB-KI-H-PRTLEX

Similar to the Pro TLE II but with tactical rail.

Production: 2004 - current   Caliber: .45 ACP   Action: SA, semi-auto

Barrel Length: 4"   OA Length: 7.7"   Wt.: 36 oz.

Sights: Meprolight tritium 3-dot night sights   Capacity: 7   Magazine: detachable box

Grips: checkered black double diamond

D2C:	NIB	$ 1,175	Ex $ 900	VG+ $ 650	Good $ 590	LMP $1,248
C2C:	Mint	$1,130	Ex $ 825	VG+ $ 590	Good $ 530	Fair $ 275
Trade-In:	Mint	$ 840	Ex $ 675	VG+ $ 460	Good $ 415	Poor $ 120

STAINLESS PRO TLE II
NEW IN BOX

STAINLESS PRO TLE II REDBOOK CODE: RB-KI-H-PRTLES

Features of the Pro TLE II but with a stainless steel slide and frame with satin silver finish.

Production: current   Caliber: .45 ACP   Action: SA, semi-auto   Barrel Length: 4"

OA Length: 7.7"   Wt.: 35 oz.   Sights: Meprolight tritium 3-dot night sights

Capacity: 7   Magazine: detachable box   Grips: checkered black double diamond

D2C:	NIB	$1,210	Ex $ 925	VG+ $ 670	Good $ 605	LMP $1,253
C2C:	Mint	$1,170	Ex $ 850	VG+ $ 610	Good $ 545	Fair $ 280
Trade-In:	Mint	$ 860	Ex $ 700	VG+ $ 480	Good $ 425	Poor $ 150

STAINLESS PRO TLE/RL II
NEW IN BOX

STAINLESS PRO TLE/RL II   REDBOOK CODE: RB-KI-H-PRTLER
Features of the Stainless Pro TLE II but with added tactical rail.

Production: current   Caliber: .45 ACP   Action: SA, semi-auto   Barrel Length: 4"
OA Length: 7.7"   Wt.: 36 oz.
Sights: Meprolight tritium 3-dot night sights   Capacity: 7   Magazine: detachable box
Grips: checkered black double diamond

D2C:	NIB	$1,285	Ex	$1,000	VG+ $ 710	Good	$ 645	LMP	$1,379
C2C:	Mint	$1,240	Ex	$ 900	VG+ $ 650	Good	$ 580	Fair	$ 300
Trade-In:	Mint	$ 920	Ex	$ 725	VG+ $ 510	Good	$ 450	Poor	$ 150

ULTRA CARRY II
NEW IN BOX

ULTRA CARRY II   REDBOOK CODE: RB-KI-H-UC2XXX
Short grip, aluminum frame, steel slide, matte black finish, 30 lpi front strap checkering, beavertail grip safety, ramped bushingless bull barrel.

Production: late 1990s - current   Caliber: .45 ACP   Action: SA, semi-auto
Barrel Length: 3"   OA Length: 6.8"   Wt.: 25 oz.
Sights: fixed low-profile   Capacity: 7   Magazine: detachable box
Grips: checkered black double diamond

D2C:	NIB	$ 885	Ex	$ 675	VG+ $ 490	Good	$ 445	LMP	$ 919
C2C:	Mint	$ 850	Ex	$ 625	VG+ $ 450	Good	$ 400	Fair	$ 205
Trade-In:	Mint	$ 630	Ex	$ 500	VG+ $ 350	Good	$ 310	Poor	$ 90

SUPER AMERICA
NEW IN BOX

SUPER AMERICA (M1911)   REDBOOK CODE: RB-KI-H-SUPAME
Limited production of 200, wood presentation case, includes knife with matching ivory handle, high polished flats, fine English scroll gold engraving.

Caliber: .45 ACP   Action: SA, semi-auto   Barrel Length: 5"
Sights: Kimber adjustable   Capacity: 7   Magazine: detachable box   Grips: mammoth ivory

D2C:	NIB	$ 3,700	Ex	$ 2,825	VG+ $2,035	Good	$ 1,850		
C2C:	Mint	$ 3,560	Ex	$ 2,575	VG+ $ 1,850	Good	$ 1,665	Fair	$ 855
Trade-In:	Mint	$ 2,630	Ex	$ 2,075	VG+ $ 1,450	Good	$ 1,295	Poor	$ 390

ULTRA TLE II
NEW IN BOX

ULTRA TLE II   REDBOOK CODE: RB-KI-H-ULTLE2
Similar features of the Ultra Carry II but with night sights.

Production: current   Caliber: .45 ACP   Action: SA, semi-auto   Barrel Length: 3"
OA Length: 6.8"   Wt.: 25 oz.
Sights: Meprolight tritium 3-dot night sights   Capacity: 7   Magazine: detachable box
Grips: checkered black double diamond

D2C:	NIB	$ 990	Ex	$ 775	VG+ $ 545	Good	$ 495	LMP	$1,150
C2C:	Mint	$ 960	Ex	$ 700	VG+ $ 500	Good	$ 450	Fair	$ 230
Trade-In:	Mint	$ 710	Ex	$ 575	VG+ $ 390	Good	$ 350	Poor	$ 120

ULTRA TLE II (LG)
NEW IN BOX

ULTRA TLE II (LG)   REDBOOK CODE: RB-KI-H-ULTLEL
Added Crimson Trace Lasergrips to the Ultra TLE II.

Production: current   Caliber: .45 ACP   Action: SA, semi-auto   Barrel Length: 3"
OA Length: 6.8"   Wt.: 25 oz.
Sights: Meprolight tritium 3-dot night sights   Capacity: 7   Magazine: detachable box
Grips: tactical gray with Crimson Trace Lasergrips

D2C:	NIB	$1,100	Ex	$ 850	VG+ $ 605	Good	$ 550	LMP	$1,393
C2C:	Mint	$1,060	Ex	$ 775	VG+ $ 550	Good	$ 495	Fair	$ 255
Trade-In:	Mint	$ 790	Ex	$ 625	VG+ $ 430	Good	$ 385	Poor	$ 120

STAINLESS ULTRA TLE II
NEW IN BOX

STAINLESS ULTRA TLE II   REDBOOK CODE: RB-KI-H-ULTLES

Stainless steel slide, aluminum frame, satin silver finish, similar to Stainless Ultra Carry II with added night sights.

Production: current   Caliber: .45 ACP   Action: SA, semi-auto   Barrel Length: 3"

OA Length: 6.8"   Wt.: 25 oz.   Sights: Meprolight tritium 3-dot night sights

Capacity: 7   Magazine: detachable box   Grips: checkered black double diamond

D2C:	NIB	$1,195	Ex	$ 925	VG+	$ 660	Good	$ 600	LMP	$1,253
C2C:	Mint	$1,150	Ex	$ 825	VG+	$ 600	Good	$ 540	Fair	$ 275
Trade-In:	Mint	$ 850	Ex	$ 675	VG+	$ 470	Good	$ 420	Poor	$ 120

STAINLESS ULTRA TLE II (LG)
NEW IN BOX

STAINLESS ULTRA TLE II (LG)   REDBOOK CODE: RB-KI-H-TLE2LG

Added Crimson Trace Lasergrips to the Stainless Ultra TLE II.

Production: current   Caliber: .45 ACP   Action: SA, semi-auto   Barrel Length: 3"

OA Length: 6.8"   Wt.: 25 oz.   Sights: Meprolight tritium 3-dot night sights

Capacity: 7   Magazine: detachable box

Grips: tactical gray with Crimson Trace Lasergrips

D2C:	NIB	$1,185	Ex	$ 925	VG+	$ 655	Good	$ 595	LMP	$1,518
C2C:	Mint	$1,140	Ex	$ 825	VG+	$ 600	Good	$ 535	Fair	$ 275
Trade-In:	Mint	$ 850	Ex	$ 675	VG+	$ 470	Good	$ 415	Poor	$ 120

ULTRA CRIMSON CARRY II
NEW IN BOX

ULTRA CRIMSON CARRY II   REDBOOK CODE: RB-KI-H-UCC2XX

Short grip, matte black steel slide and satin silver aluminum frame, Crimson Trace Lasergrips, stainless steel bushingless match grade bull barrel.

Production: current   Caliber: .45 ACP   Action: SA, semi-auto   Barrel Length: 3"

OA Length: 6.8"   Wt.: 25 oz.   Sights: fixed low-profile

Capacity: 7   Magazine: detachable box

Grips: rosewood with Crimson Trace Lasergrips

D2C:	NIB	$1,070	Ex	$ 825	VG+	$ 590	Good	$ 535	LMP	$1,206
C2C:	Mint	$1,030	Ex	$ 750	VG+	$ 540	Good	$ 485	Fair	$ 250
Trade-In:	Mint	$ 760	Ex	$ 600	VG+	$ 420	Good	$ 375	Poor	$ 120

PRO CRIMSON CARRY II
NEW IN BOX

PRO CRIMSON CARRY II   REDBOOK CODE: RB-KI-H-PCC2XX

Full-length grip, matte black steel slide and satin silver aluminum frame, Crimson Trace Lasergrips, stainless steel bushingless match grade bull barrel.

Production: current   Caliber: .45 ACP   Action: SA, semi-auto   Barrel Length: 4"

OA Length: 7.7"   Wt.: 28 oz.   Sights: fixed low-profile

Capacity: 8   Magazine: detachable box

Grips: rosewood with Crimson Trace Lasergrips

D2C:	NIB	$1,070	Ex	$ 825	VG+	$ 590	Good	$ 535	LMP	$1,206
C2C:	Mint	$1,030	Ex	$ 750	VG+	$ 540	Good	$ 485	Fair	$ 250
Trade-In:	Mint	$ 760	Ex	$ 600	VG+	$ 420	Good	$ 375	Poor	$ 120

KIMBER

CUSTOM CRIMSON CARRY II
NEW IN BOX

CUSTOM CRIMSON CARRY II REDBOOK CODE: RB-KI-H-CCC2XX
Full-length grip, matte black steel slide and satin silver aluminum frame, Crimson Trace Lasergrips, stainless steel match grade bull barrel with stainless bushing.

Production: current Caliber: .45 ACP Action: SA, semi-auto Barrel Length: 5"

OA Length: 8.7" Wt.: 31 oz.

Sights: fixed low-profile Capacity: 8 Magazine: detachable box

Grips: rosewood with Crimson Trace Lasergrips

D2C:	NIB $1,105	Ex $850	VG+ $610	Good $555	LMP $1,206
C2C:	Mint $1,070	Ex $775	VG+ $560	Good $500	Fair $255
Trade-In:	Mint $790	Ex $625	VG+ $440	Good $390	Poor $120

MASTER CARRY ULTRA
NEW IN BOX

MASTER CARRY ULTRA REDBOOK CODE: RB-KI-H-MCUXXX
Short grip, black KimPro II finished slide, satin silver finished aluminum Round Heel Frame, Crimson Trace Master Series Lasergrips.

Production: current Caliber: .45 ACP Action: SA, semi-auto Barrel Length: 3"

OA Length: 6.8" Wt.: 25 oz.

Sights: fixed low-profile Capacity: 7 Magazine: detachable box

Grips: Crimson Trace Master Series Lasergrips

D2C:	NIB $1,300	Ex $1,000	VG+ $715	Good $650	LMP $1,568
C2C:	Mint $1,250	Ex $900	VG+ $650	Good $585	Fair $300
Trade-In:	Mint $930	Ex $750	VG+ $510	Good $455	Poor $150

MASTER CARRY PRO
NEW IN BOX

MASTER CARRY PRO REDBOOK CODE: RB-KI-H-MCPXXX
Full-length grip, black KimPro II finished slide, satin silver finished aluminum Round Heel Frame, Crimson Trace Master Series Lasergrips.

Production: current Caliber: .45 ACP Action: SA, semi-auto Barrel Length: 4"

OA Length: 7.7" Wt.: 28 oz.

Sights: fixed low-profile Capacity: 8 Magazine: detachable box

Grips: Crimson Trace Master Series Lasergrips

D2C:	NIB $1,300	Ex $1,000	VG+ $715	Good $650	LMP $1,568
C2C:	Mint $1,250	Ex $900	VG+ $650	Good $585	Fair $300
Trade-In:	Mint $930	Ex $750	VG+ $510	Good $455	Poor $150

MASTER CARRY CUSTOM
NEW IN BOX

MASTER CARRY CUSTOM REDBOOK CODE: RB-KI-H-MCCXXX
Full-length grip, black KimPro II finished slide, satin silver finished aluminum Round Heel Frame, Crimson Trace Master Series Lasergrips.

Production: current Caliber: .45 ACP Action: SA, semi-auto Barrel Length: 5"

OA Length: 8.7" Wt.: 38 oz.

Sights: fixed low-profile Capacity: 8 Magazine: detachable box

Grips: Crimson Trace Master Series Lasergrips

D2C:	NIB $1,325	Ex $1,025	VG+ $730	Good $665	LMP $1,568
C2C:	Mint $1,280	Ex $925	VG+ $670	Good $600	Fair $305
Trade-In:	Mint $950	Ex $750	VG+ $520	Good $465	Poor $150

**TACTICAL
ULTRA II**
NEW IN BOX

TACTICAL ULTRA II REDBOOK CODE: RB-KI-H-TU2XXX

Short grip, charcoal gray KimPro II finished aluminum frame, ramped stainless bushingless barrel, extended ambidextrous thumb safety.

Production: 2003 - current   Caliber: .45 ACP   Action: SA, semi-auto

Barrel Length: 3"   OA Length: 6.8"

Wt.: 26 oz.   Sights: Meprolight tritium 3-dot night sights   Capacity: 7

Magazine: detachable box   Grips: laminated with double diamond and Kimber logo

D2C:	NIB	$ 1,115	Ex $ 850	VG+ $ 615	Good $ 560	LMP $ 1,317
C2C:	Mint	$1,080	Ex $ 775	VG+ $ 560	Good $ 505	Fair $ 260
Trade-In:	Mint	$ 800	Ex $ 625	VG+ $ 440	Good $ 395	Poor $ 120

**TACTICAL
PRO II**
NEW IN BOX

TACTICAL PRO II REDBOOK CODE: RB-KI-H-TP2XXX

Full-length grip, charcoal gray KimPro II finished aluminum frame, ramped stainless bushingless barrel, extended ambidextrous thumb safety.

Production: 2003 - current   Caliber: .45 ACP, 9mm   Action: SA, semi-auto

Barrel Length: 4"   OA Length: 7.7"   Wt.: 29 oz.

Sights: Meprolight tritium 3-dot night sights   Capacity: 7, 9   Magazine: detachable box

Grips: laminated with double diamond and Kimber logo

D2C:	NIB	$1,205	Ex $ 925	VG+ $ 665	Good $ 605	LMP $ 1,317
C2C:	Mint	$1,160	Ex $ 850	VG+ $ 610	Good $ 545	Fair $ 280
Trade-In:	Mint	$ 860	Ex $ 675	VG+ $ 470	Good $ 425	Poor $ 150

**TACTICAL
CUSTOM II**
NEW IN BOX

TACTICAL CUSTOM II REDBOOK CODE: RB-KI-H-TC2XXX

Full-length grip, charcoal gray KimPro II finished aluminum frame, match stainless barrel and bushing, extended ambidextrous thumb safety.

Production: 2003 - current   Caliber: .45 ACP   Action: SA, semi-auto

Barrel Length: 5"   OA Length: 8.7"   Wt.: 32 oz.

Sights: Meprolight tritium 3-dot night sights   Capacity: 7

Magazine: detachable box   Grips: laminated with double diamond and Kimber logo

D2C:	NIB	$1,205	Ex $ 925	VG+ $ 665	Good $ 605	LMP $ 1,317
C2C:	Mint	$1,160	Ex $ 850	VG+ $ 610	Good $ 545	Fair $ 280
Trade-In:	Mint	$ 860	Ex $ 675	VG+ $ 470	Good $ 425	Poor $ 150

**TACTICAL
CUSTOM HD II**
NEW IN BOX

TACTICAL CUSTOM HD II REDBOOK CODE: RB-KI-H-TCHD2X

Features of Tactical Custom II but with stainless steel frame.

Production: current   Caliber: .45 ACP   Action: SA, semi-auto   Barrel Length: 5"

OA Length: 8.7"   Wt.: 39 oz.

Sights: Meprolight tritium 3-dot night sights   Capacity: 7   Magazine: detachable box

Grips: laminated with double diamond and Kimber logo

D2C:	NIB	$1,245	Ex $ 950	VG+ $ 685	Good $ 625	LMP $ 1,387
C2C:	Mint	$1,200	Ex $ 875	VG+ $ 630	Good $ 565	Fair $ 290
Trade-In:	Mint	$ 890	Ex $ 700	VG+ $ 490	Good $ 440	Poor $ 150

TACTICAL ENTRY II
NEW IN BOX

TACTICAL ENTRY II   REDBOOK CODE: RB-KI-H-TE2XXX

Features of Tactical Custom HD II but with added tactical rail.

Production: current   Caliber: .45 ACP   Action: SA, semi-auto   Barrel Length: 5"

OA Length: 8.7"   Wt.: 40 oz.

Sights: Meprolight tritium 3-dot night sights   Capacity: 7

Magazine: detachable box   Grips: laminated with double diamond and Kimber logo

D2C:	NIB	$1,360	Ex	$1,050	VG+	$ 750	Good	$ 680	LMP	$1,490
C2C:	Mint	$1,310	Ex	$ 950	VG+	$ 680	Good	$ 615	Fair	$ 315
Trade-In:	Mint	$ 970	Ex	$ 775	VG+	$ 540	Good	$ 480	Poor	$ 150

ECLIPSE ULTRA II
NEW IN BOX

ECLIPSE ULTRA II   REDBOOK CODE: RB-KI-H-EU2XXX

Short grip, brush-polished stainless steel slide and frame, front strap checkering, ramped match stainless bushingless barrel.

Production: 2002 - current   Caliber: .45 ACP   Action: SA, semi-auto

Barrel Length: 3"   OA Length: 6.8"   Wt.: 31 oz.

Sights: Meprolight tritium 3-dot night sights   Capacity: 7   Magazine: detachable box

Grips: laminated with double diamond

D2C:	NIB	$ 1,125	Ex	$ 875	VG+	$ 620	Good	$ 565	LMP	$1,289
C2C:	Mint	$1,080	Ex	$ 800	VG+	$ 570	Good	$ 510	Fair	$ 260
Trade-In:	Mint	$ 800	Ex	$ 650	VG+	$ 440	Good	$ 395	Poor	$ 120

ECLIPSE PRO II
NEW IN BOX

ECLIPSE PRO II   REDBOOK CODE: RB-KI-H-EP2XXX

Full-length grip, brush-polished stainless steel slide and frame, front strap checkering, ramped match stainless bushingless barrel.

Production: early 2000s - current   Caliber: .45 ACP   Action: SA, semi-auto

Barrel Length: 4"   OA Length: 7.7"   Wt.: 35 oz.

Sights: Meprolight tritium 3-dot night sights   Capacity: 8

Magazine: detachable box   Grips: laminated with double diamond

D2C:	NIB	$ 1,125	Ex	$ 875	VG+	$ 620	Good	$ 565	LMP	$1,289
C2C:	Mint	$1,080	Ex	$ 800	VG+	$ 570	Good	$ 510	Fair	$ 260
Trade-In:	Mint	$ 800	Ex	$ 650	VG+	$ 440	Good	$ 395	Poor	$ 120

ECLIPSE PRO TARGET II
NEW IN BOX

ECLIPSE PRO TARGET II   REDBOOK CODE: RB-KI-H-EPT2XX

Full-length grip, brush-polished stainless steel slide and frame, front strap checkering, ramped match stainless bushingless barrel, premium aluminum trigger.

Production: 2002 - current   Caliber: .45 ACP   Action: SA, semi-auto

Barrel Length: 4"   OA Length: 7.7"   Wt.: 35 oz.

Sights: Meprolight tritium bar-dot adjustable night sight   Capacity: 8

Magazine: detachable box   Grips: laminated with double diamond

D2C:	NIB	$1,200	Ex	$ 925	VG+	$ 660	Good	$ 600	LMP	$1,359
C2C:	Mint	$1,160	Ex	$ 850	VG+	$ 600	Good	$ 540	Fair	$ 280
Trade-In:	Mint	$ 860	Ex	$ 675	VG+	$ 470	Good	$ 420	Poor	$ 120

ECLIPSE CUSTOM II NEW IN BOX

ECLIPSE CUSTOM II REDBOOK CODE: RB-KI-H-EC2XXX
Full-length grip, brush-polished stainless steel slide and frame, front strap checkering, ramped match stainless barrel with bushing, premium aluminum trigger.

Production: ~2004 - current Caliber: .45 ACP, 10mm Action: SA, semi-auto

Barrel Length: 5" OA Length: 8.7" Wt.: 38 oz.

Sights: Meprolight tritium 3-dot night sights Capacity: 8 Magazine: detachable box

Grips: laminated with double diamond

D2C:	NIB	$1,130	Ex	$ 875	VG+	$ 625	Good	$ 565	LMP $1,289
C2C:	Mint	$1,090	Ex	$ 800	VG+	$ 570	Good	$ 510	Fair $ 260
Trade-In:	Mint	$ 810	Ex	$ 650	VG+	$ 450	Good	$ 400	Poor $ 120

ECLIPSE TARGET II NEW IN BOX

ECLIPSE TARGET II REDBOOK CODE: RB-KI-H-ET2XXX
Full-length grip, brush-polished stainless steel slide and frame, front strap checkering, ramped match stainless barrel with bushing, premium aluminum trigger.

Production: 2002 - current Caliber: .45 ACP Action: SA, semi-auto

Barrel Length: 5" OA Length: 8.7" Wt.: 38 oz.

Sights: Meprolight tritium bar-dot adjustable night sight Capacity: 8

Magazine: detachable box Grips: laminated with double diamond

D2C:	NIB	$1,245	Ex	$ 950	VG+	$ 685	Good	$ 625	LMP $1,393
C2C:	Mint	$1,200	Ex	$ 875	VG+	$ 630	Good	$ 565	Fair $ 290
Trade-In:	Mint	$ 890	Ex	$ 700	VG+	$ 490	Good	$ 440	Poor $ 150

ULTRA COVERT II NEW IN BOX

ULTRA COVERT II REDBOOK CODE: RB-KI-H-ULC2XX
Short grip, Dark Earth KimPro II finished steel slide and matte black aluminum frame, front strap checkering, Carry Melt treatment, ramped stainless match bushingless barrel.

Production: 2007 - current Caliber: .45 ACP Action: SA, semi-auto

Barrel Length: 3" OA Length: 6.8" Wt.: 25 oz.

Sights: Tactical Wedge tritium night sights Capacity: 7 Magazine: detachable box

Grips: digital camo Crimson Trace Lasergrips

D2C:	NIB	$1,475	Ex	$1,125	VG+	$ 815	Good	$ 740	LMP $ 1,657
C2C:	Mint	$1,420	Ex	$1,025	VG+	$ 740	Good	$ 665	Fair $ 340
Trade-In:	Mint	$1,050	Ex	$ 850	VG+	$ 580	Good	$ 520	Poor $ 150

PRO COVERT II NEW IN BOX

PRO COVERT II REDBOOK CODE: RB-KI-H-PRCO2X
Full-length grip, Dark Earth KimPro II finished steel slide and matte black aluminum frame, front strap checkering, Carry Melt treatment, ramped stainless match bushingless barrel.

Production: 2007 - current Caliber: .45 ACP Action: SA, semi-auto

Barrel Length: 4" OA Length: 7.7" Wt.: 28 oz.

Sights: Tactical Wedge tritium night sights Capacity: 7

Magazine: detachable box Grips: digital camo Crimson Trace Lasergrips

D2C:	NIB	$1,475	Ex	$1,125	VG+	$ 815	Good	$ 740	LMP $ 1,657
C2C:	Mint	$1,420	Ex	$1,025	VG+	$ 740	Good	$ 665	Fair $ 340
Trade-In:	Mint	$1,050	Ex	$ 850	VG+	$ 580	Good	$ 520	Poor $ 150

KIMBER

COVERT II CUSTOM
NEW IN BOX

CUSTOM COVERT II REDBOOK CODE: RB-KI-H-CO2CUS

Full-length grip, Dark Earth KimPro II finished steel slide and matte black aluminum frame, front strap checkering, Carry Melt treatment, stainless match barrel and bushing.

Production: 2007 - current Caliber: .45 ACP Action: SA, semi-auto

Barrel Length: 5" OA Length: 8.7" Wt.: 31 oz.

Sights: Tactical Wedge tritium night sights Capacity: 7 Magazine: detachable box

Grips: digital camo Crimson Trace Lasergrips

D2C:	NIB $ 1,415	Ex $ 1,100	VG+ $ 780	Good $ 710	LMP $ 1,657	
C2C:	Mint $ 1,360	Ex $ 1,000	VG+ $ 710	Good $ 640	Fair $ 330	
Trade-In:	Mint $ 1,010	Ex $ 800	VG+ $ 560	Good $ 500	Poor $ 150	

ULTRA AEGIS II
NEW IN BOX

ULTRA AEGIS II REDBOOK CODE: RB-KI-H-AEG2UL

Short grip, matte black KimPro II finished steel slide and satin silver aluminum frame, ramped stainless and bushingless barrel, front strap checkering, Service Melt treatment, bobbed thumb safety.

Production: 2006 - current Caliber: 9mm Action: SA, semi-auto

Barrel Length: 3" OA Length: 6.8" Wt.: 25 oz.

Sights: Tactical Wedge tritium night sights Capacity: 8 Magazine: detachable box

Grips: thin fluted rosewood

D2C:	NIB $1,230	Ex $ 950	VG+ $ 680	Good $ 615	LMP $ 1,331	
C2C:	Mint $1,190	Ex $ 850	VG+ $ 620	Good $ 555	Fair $ 285	
Trade-In:	Mint $ 880	Ex $ 700	VG+ $ 480	Good $ 435	Poor $ 150	

PRO AEGIS II
NEW IN BOX

PRO AEGIS II REDBOOK CODE: RB-KI-H-AEG2PR

Full-length grip, matte black KimPro II finished steel slide and satin silver aluminum frame, ramped stainless and bushingless barrel, front strap checkering, Service Melt treatment, bobbed thumb safety.

Production: 2006 - current Caliber: 9mm Action: SA, semi-auto

Barrel Length: 4" OA Length: 7.7" Wt.: 28 oz.

Sights: Tactical Wedge tritium night sights Capacity: 9 Magazine: detachable box

Grips: thin fluted rosewood

D2C:	NIB $ 1,230	Ex $ 950	VG+ $ 680	Good $ 615	LMP $ 1,331	
C2C:	Mint $ 1,190	Ex $ 850	VG+ $ 620	Good $ 555	Fair $ 285	
Trade-In:	Mint $ 880	Ex $ 700	VG+ $ 480	Good $ 435	Poor $ 150	

CUSTOM AEGIS II
NEW IN BOX

CUSTOM AEGIS II REDBOOK CODE: RB-KI-H-AEG2CU

Full-length grip, matte black KimPro II finished steel slide and satin silver aluminum frame, ramped stainless barrel with bushing, front strap checkering, Service Melt treatment, bobbed thumb safety.

Production: 2006 - current Caliber: 9mm Action: SA, semi-auto

Barrel Length: 5" OA Length: 8.7" Wt.: 31 oz.

Sights: Tactical Wedge tritium night sights Capacity: 9 Magazine: detachable box

Grips: thin fluted rosewood

D2C:	NIB $1,160	Ex $ 900	VG+ $ 640	Good $ 580	LMP $ 1,331	
C2C:	Mint $1,120	Ex $ 825	VG+ $ 580	Good $ 525	Fair $ 270	
Trade-In:	Mint $ 830	Ex $ 650	VG+ $ 460	Good $ 410	Poor $ 120	

ULTRA CDP II
NEW IN BOX

ULTRA CDP II REDBOOK CODE: RB-KI-H-ULCDP2

Short grip, matte black KimPro II finished aluminum frame, satin silver finished stainless steel slide, Carry Melt treatment, ramped stainless and bushingless barrel, ambidextrous thumb safety.

Production: 2000 - current Caliber: .45 ACP, 9mm Action: SA, semi-auto

Barrel Length: 3" OA Length: 6.8" Wt.: 25 oz.

Sights: Meprolight tritium 3-dot night sights Capacity: 7, 8 Magazine: detachable box

Grips: rosewood with double diamond

D2C:	NIB	$ 1,300	Ex	$ 1,000	VG+ $	715	Good $	650	LMP	$ 1,331
C2C:	Mint	$ 1,250	Ex	$ 900	VG+ $	650	Good $	585	Fair	$ 300
Trade-In:	Mint $	930	Ex $	750	VG+ $	510	Good $	455	Poor	$ 150

ULTRA CDP II (LG)
NEW IN BOX

ULTRA CDP II (LG) REDBOOK CODE: RB-KI-H-ULLG2X

Added Crimson Trace Lasergrips to the Ultra CDP II.

Production: current Caliber: .45 ACP Action: SA, semi-auto Barrel Length: 3"

OA Length: 6.8" Wt.: 25 oz.

Sights: Meprolight tritium 3-dot night sights Capacity: 7 Magazine: detachable box

Grips: rosewood with Crimson Trace Lasergrips

D2C:	NIB	$ 1,415	Ex	$ 1,100	VG+ $	780	Good $	710	LMP	$ 1,631
C2C:	Mint	$ 1,360	Ex	$ 1,000	VG+ $	710	Good $	640	Fair	$ 330
Trade-In:	Mint $	1,010	Ex $	800	VG+ $	560	Good $	500	Poor	$ 150

COMPACT CDP II
NEW IN BOX

COMPACT CDP II REDBOOK CODE: RB-KI-H-CDP2CO

Short grip with 4" barrel, matte black KimPro II finished aluminum frame, satin silver finished stainless steel slide, Carry Melt treatment, ramped stainless and bushingless barrel, ambidextrous thumb safety.

Production: 2000 - current Caliber: .45 ACP Action: SA, semi-auto

Barrel Length: 4" OA Length: 7.7" Wt.: 27 oz.

Sights: Meprolight tritium 3-dot night sights Capacity: 7 Magazine: detachable box

Grips: rosewood with double diamond

D2C:	NIB	$ 1,150	Ex	$ 875	VG+ $	635	Good $	575	LMP	$ 1,331
C2C:	Mint	$ 1,110	Ex	$ 800	VG+ $	580	Good $	520	Fair	$ 265
Trade-In:	Mint $	820	Ex $	650	VG+ $	450	Good $	405	Poor	$ 120

PRO CDP II
NEW IN BOX

PRO CDP II REDBOOK CODE: RB-KI-H-CDP2PR

Full-length grip, matte black KimPro II finished aluminum frame, satin silver finished stainless steel slide, Carry Melt treatment, ramped stainless and bushingless barrel, ambidextrous thumb safety.

Production: 2000 - current Caliber: .45 ACP Action: SA, semi-auto

Barrel Length: 4" OA Length: 7.7" Wt.: 28 oz.

Sights: Meprolight tritium 3-dot night sights Capacity: 7 Magazine: detachable box

Grips: rosewood with double diamond

D2C:	NIB	$ 1,135	Ex	$ 875	VG+ $	625	Good $	570	LMP	$ 1,331
C2C:	Mint	$1,090	Ex	$ 800	VG+ $	570	Good $	515	Fair	$ 265
Trade-In:	Mint $	810	Ex $	650	VG+ $	450	Good $	400	Poor	$ 120

CUSTOM CDP II
NEW IN BOX

CUSTOM CDP II   REDBOOK CODE: RB-KI-H-CDP2CU

Full-length grip, matte black KimPro II finished aluminum frame, satin silver finished stainless steel slide, Carry Melt treatment, ramped stainless barrel with bushing, ambidextrous thumb safety.

Production: 2000 - current   Caliber: .45 ACP   Action: SA, semi-auto

Barrel Length: 5"   OA Length: 8.7"   Wt.: 31 oz.

Sights: Meprolight tritium 3-dot night sights   Capacity: 7   Magazine: detachable box

Grips: rosewood with double diamond

D2C:	NIB	$1,120	Ex $ 875	VG+ $ 620	Good $ 560	LMP $ 1,331
C2C:	Mint	$1,080	Ex $ 775	VG+ $ 560	Good $ 505	Fair $ 260
Trade-In:	Mint	$ 800	Ex $ 650	VG+ $ 440	Good $ 395	Poor $ 120

ULTRA RAPTOR II
NEW IN BOX

ULTRA RAPTOR II   REDBOOK CODE: RB-KI-H-ULRAP2

Short grip, matte black KimPro II finished aluminum frame, matte black steel slide, scaled serrations on front strap and slide, ramped stainless and bushingless barrel.

Production: 2005 - current   Caliber: .45 ACP   Action: SA, semi-auto

Barrel Length: 3"   OA Length: 6.8"   Wt.: 25 oz.

Sights: Tactical Wedge tritium night sights   Capacity: 7   Magazine: detachable box

Grips: scale patterned zebra wood

D2C:	NIB	$1,125	Ex $ 875	VG+ $ 620	Good $ 565	LMP $1,295
C2C:	Mint	$1,080	Ex $ 800	VG+ $ 570	Good $ 510	Fair $ 260
Trade-In:	Mint	$ 800	Ex $ 650	VG+ $ 440	Good $ 395	Poor $ 120

**STAINLESS
ULTRA RAPTOR II**
NEW IN BOX

STAINLESS ULTRA RAPTOR II   REDBOOK CODE: RB-KI-H-STULR2

Short grip, satin silver KimPro II finished aluminum frame, satin silver stainless steel slide, scaled serrations on front strap and slide, ramped stainless and bushingless barrel.

Production: current   Caliber: .45 ACP   Action: SA, semi-auto   Barrel Length: 3"

OA Length: 6.8"   Wt.: 25 oz.

Sights: Tactical Wedge tritium night sights   Capacity: 7   Magazine: detachable box

Grips: scale patterned zebra wood

D2C:	NIB	$1,210	Ex $ 925	VG+ $ 670	Good $ 605	LMP $ 1,415
C2C:	Mint	$1,170	Ex $ 850	VG+ $ 610	Good $ 545	Fair $ 280
Trade-In:	Mint	$ 860	Ex $ 700	VG+ $ 480	Good $ 425	Poor $ 150

PRO RAPTOR II
NEW IN BOX

PRO RAPTOR II   REDBOOK CODE: RB-KI-H-PRRAP2

Full-length grip, matte black KimPro II finished aluminum frame, matte black steel slide, scaled serrations on front strap and slide, ramped stainless and bushingless barrel.

Production: 2004 - current   Caliber: .45 ACP   Action: SA, semi-auto

Barrel Length: 4"   OA Length: 7.7"   Wt.: 35 oz.

Sights: Tactical Wedge tritium night sights   Capacity: 8   Magazine: detachable box

Grips: scale patterned zebra wood

D2C:	NIB	$1,120	Ex $ 875	VG+ $ 620	Good $ 560	LMP $ 1,295
C2C:	Mint	$1,080	Ex $ 775	VG+ $ 560	Good $ 505	Fair $ 260
Trade-In:	Mint	$ 800	Ex $ 650	VG+ $ 440	Good $ 395	Poor $ 120

STAINLESS PRO RAPTOR II
NEW IN BOX

STAINLESS PRO RAPTOR II   REDBOOK CODE: RB-KI-H-STPRRA2

Full-length grip, satin silver KimPro II finished stainless steel frame and slide, scaled serrations on front strap and slide, ramped stainless and bushingless barrel.

Production: current   Caliber: .45 ACP   Action: SA, semi-auto   Barrel Length: 4"
OA Length: 7.7"   Wt.: 35 oz.
Sights: Tactical Wedge tritium night sights   Capacity: 8   Magazine: detachable box
Grips: scale patterned zebra wood

D2C:	NIB	$1,250	Ex	$ 950	VG+	$ 690	Good	$ 625	LMP	$ 1,415
C2C:	Mint	$1,200	Ex	$ 875	VG+	$ 630	Good	$ 565	Fair	$ 290
Trade-In:	Mint	$ 890	Ex	$ 700	VG+	$ 490	Good	$ 440	Poor	$ 150

RAPTOR II
NEW IN BOX

RAPTOR II   REDBOOK CODE: RB-KI-H-RAPT2X

Full-length grip, matte black KimPro II finished steel slide and frame, scaled serrations on front strap and slide, ramped stainless barrel and bushing.

Production: current   Caliber: .45 ACP   Action: SA, semi-auto
Barrel Length: 5"   OA Length: 8.7"   Wt.: 38 oz.
Sights: Tactical Wedge tritium night sights   Capacity: 8   Magazine: detachable box
Grips: scale patterned zebra wood

D2C:	NIB	$1,265	Ex	$ 975	VG+	$ 700	Good	$ 635	LMP	$ 1,434
C2C:	Mint	$1,220	Ex	$ 875	VG+	$ 640	Good	$ 570	Fair	$ 295
Trade-In:	Mint	$ 900	Ex	$ 725	VG+	$ 500	Good	$ 445	Poor	$ 150

STAINLESS RAPTOR II
NEW IN BOX

STAINLESS RAPTOR II   REDBOOK CODE: RB-KI-H-STRAP2

Full-length grip, satin silver KimPro II finished aluminum frame, satin silver finished stainless steel slide, scaled serrations on front strap and slide, ramped stainless barrel and bushing.

Production: current   Caliber: .45 ACP   Action: SA, semi-auto
Barrel Length: 5"   OA Length: 8.7"   Wt.: 38 oz.
Sights: Tactical Wedge tritium night sights   Capacity: 8   Magazine: detachable box
Grips: scale patterned zebra wood

D2C:	NIB	$ 1,445	Ex	$ 1,100	VG+	$ 795	Good	$ 725	LMP	$ 1,568
C2C:	Mint	$ 1,390	Ex	$ 1,000	VG+	$ 730	Good	$ 655	Fair	$ 335
Trade-In:	Mint	$ 1,030	Ex	$ 825	VG+	$ 570	Good	$ 510	Poor	$ 150

GRAND RAPTOR II
NEW IN BOX

GRAND RAPTOR II   REDBOOK CODE: RB-KI-H-GRRAP2

Full-length grip, stainless steel frame and steel slide, scaled serrations on front strap and slide, highly-polished flats, stainless match grade barrel and bushing.

Production: 2006 - current   Caliber: .45 ACP   Action: SA, semi-auto
Barrel Length: 5"   OA Length: 8.7"   Wt.: 38 oz.
Sights: Meprolight tritium bar-dot adjustable night sight   Capacity: 8
Magazine: detachable box   Grips: rosewood scale pattern with Kimber logo

D2C:	NIB	$ 1,515	Ex	$ 1,175	VG+	$ 835	Good	$ 760	LMP	$ 1,657
C2C:	Mint	$ 1,460	Ex	$ 1,050	VG+	$ 760	Good	$ 685	Fair	$ 350
Trade-In:	Mint	$ 1,080	Ex	$ 850	VG+	$ 600	Good	$ 535	Poor	$ 180

SUPER CARRY ULTRA
NEW IN BOX

SUPER CARRY ULTRA   REDBOOK CODE: RB-KI-H-SUCAUL

Short grip, KimPro II finished satin silver aluminum frame and matte black stainless slide, Carry Melt treatment, ramped stainless and bushingless barrel, high-cut under trigger guard, front strap serrations, Round Heel Frame.

Production: current   Caliber: .45 ACP   Action: SA, semi-auto   Barrel Length: 3"

OA Length: 6.8"   Wt.: 25 oz.

Sights: tritium night sights with cocking shoulder   Capacity: 7

Magazine: detachable box   Grips: Micarta/laminated wood

D2C:	NIB $	1,415	Ex $	1,100	VG+ $	780	Good $	710	LMP $	1,596
C2C:	Mint $	1,360	Ex $	1,000	VG+ $	710	Good $	640	Fair $	330
Trade-In:	Mint $	1,010	Ex $	800	VG+ $	560	Good $	500	Poor $	150

SUPER CARRY ULTRA HD
NEW IN BOX

SUPER CARRY ULTRA HD   REDBOOK CODE: RB-KI-H-SUCAHD

Short grip, KimPro II finished matte black stainless slide and frame, Carry Melt treatment, ramped stainless and bushingless barrel, high-cut under trigger guard, front strap serrations, Round Heel Frame.

Production: current   Caliber: .45 ACP   Action: SA, semi-auto

Barrel Length: 3"   OA Length: 6.8"   Wt.: 32 oz.

Sights: tritium night sights with cocking shoulder   Capacity: 8

Magazine: detachable box   Grips: checkered G10

D2C:	NIB $	1,480	Ex $	1,125	VG+ $	815	Good $	740	LMP $	1,699
C2C:	Mint $	1,430	Ex $	1,025	VG+ $	740	Good $	670	Fair $	345
Trade-In:	Mint $	1,060	Ex $	850	VG+ $	580	Good $	520	Poor $	150

SUPER CARRY ULTRA +
NEW IN BOX

SUPER CARRY ULTRA+   REDBOOK CODE: RB-KI-H-SUCAUP

Full-length grip, KimPro II finished satin silver aluminum frame and matte black stainless slide, Carry Melt treatment, ramped stainless and bushingless barrel, high-cut under trigger guard, front strap serrations, Round Heel Frame.

Production: current   Caliber: .45 ACP   Action: SA, semi-auto

Barrel Length: 3"   OA Length: 6.8"   Wt.: 27 oz.

Sights: tritium night sights with cocking shoulder   Capacity: 8

Magazine: detachable box   Grips: Micarta/laminated wood checkered with border

D2C:	NIB $	1,430	Ex $	1,100	VG+ $	790	Good $	715	LMP $	1,596
C2C:	Mint $	1,380	Ex $	1,000	VG+ $	720	Good $	645	Fair $	330
Trade-In:	Mint $	1,020	Ex $	825	VG+ $	560	Good $	505	Poor $	150

SUPER CARRY PRO
NEW IN BOX

SUPER CARRY PRO   REDBOOK CODE: RB-KI-H-SUCAPR

Full-length grip with 4" barrel, KimPro II finished satin silver aluminum frame and matte black stainless slide, Carry Melt treatment, ramped stainless and bushingless barrel, high-cut under trigger guard, front strap serrations, Round Heel Frame.

Production: current   Caliber: .45 ACP   Action: SA, semi-auto

Barrel Length: 4"   OA Length: 7.7"   Wt.: 28 oz.

Sights: tritium night sights with cocking shoulder   Capacity: 8

Magazine: detachable box   Grips: Micarta/laminated wood

D2C:	NIB $	1,445	Ex $	1,100	VG+ $	795	Good $	725	LMP $	1,596
C2C:	Mint $	1,390	Ex $	1,000	VG+ $	730	Good $	655	Fair $	335
Trade-In:	Mint $	1,030	Ex $	825	VG+ $	570	Good $	510	Poor $	150

SUPER CARRY PRO HD
NEW IN BOX

SUPER CARRY PRO HD   REDBOOK CODE: RB-KI-H-SUCPHD

Full-length grip, KimPro II finished matte black stainless slide and frame, Carry Melt treatment, ramped stainless and bushingless barrel, high-cut under trigger guard, front strap serrations, Round Heel Frame.

Production: current   Caliber: .45 ACP   Action: SA, semi-auto   Barrel Length: 4"

OA Length: 7.7"   Wt.: 35 oz.

Sights: tritium night sights with cocking shoulder   Capacity: 8

Magazine: detachable box   Grips: checkered G10

D2C:	NIB	$ 1,510	Ex	$ 1,150	VG+	$ 835	Good	$ 755	LMP	$ 1,699
C2C:	Mint	$ 1,450	Ex	$ 1,050	VG+	$ 760	Good	$ 680	Fair	$ 350
Trade-In:	Mint $ 1,080		Ex	$ 850	VG+	$ 590	Good	$ 530	Poor	$ 180

SUPER CARRY CUSTOM
NEW IN BOX

SUPER CARRY CUSTOM   REDBOOK CODE: RB-KI-H-SUCACU

Full-length grip, KimPro II finished satin silver aluminum frame and matte black stainless slide, Carry Melt treatment, ramped stainless barrel with bushing, high-cut under trigger guard, front strap serrations, Round Heel Frame.

Production: current   Caliber: .45 ACP   Action: SA, semi-auto

Barrel Length: 5"   OA Length: 8.7"   Wt.: 38 oz.

Sights: tritium night sights with cocking shoulder   Capacity: 8

Magazine: detachable box   Grips: Micarta/laminated wood

D2C:	NIB	$ 1,485	Ex	$ 1,150	VG+	$ 820	Good	$ 745	LMP	$ 1,596
C2C:	Mint	$ 1,430	Ex	$ 1,025	VG+	$ 750	Good	$ 670	Fair	$ 345
Trade-In:	Mint $ 1,060		Ex	$ 850	VG+	$ 580	Good	$ 520	Poor	$ 150

SUPER CARRY CUSTOM HD
NEW IN BOX

SUPER CARRY CUSTOM HD   REDBOOK CODE: RB-KI-H-SUCACH

Full-length grip, KimPro II finished matte black stainless slide and frame, Carry Melt treatment, ramped stainless barrel with bushing, high-cut under trigger guard, front strap serrations, Round Heel Frame.

Production: current   Caliber: .45 ACP   Action: SA, semi-auto

Barrel Length: 5"   OA Length: 8.7"   Wt.: 38 oz.

Sights: tritium night sights with cocking shoulder   Capacity: 8

Magazine: detachable box   Grips: checkered G10

D2C:	NIB	$ 1,580	Ex	$ 1,225	VG+	$ 870	Good	$ 790	LMP	$ 1,699
C2C:	Mint	$ 1,520	Ex	$ 1,100	VG+	$ 790	Good	$ 715	Fair	$ 365
Trade-In:	Mint $ 1,130		Ex	$ 900	VG+	$ 620	Good	$ 555	Poor	$ 180

SAPPHIRE ULTRA II
NEW IN BOX

SAPPHIRE ULTRA II   REDBOOK CODE: RB-KI-H-SAULT2

Aluminum frame with PVD coating on small parts, high-polish bright blue PVD stainless steel slide with cut scroll engraving and border, front strap checkering with ball-milled grooves, grooved aluminum mainspring housing, round heel.

Production: current   Caliber: 9mm   Action: SA, semi-auto   Barrel Length: 3"

OA Length: 6.8"   Wt.: 25 oz.

Sights: Tactical Wedge tritium night sights   Capacity: 8   Magazine: detachable box

Grips: blue/black ball-milled G-10

D2C:	NIB	$ 1,550	Ex	$ 1,200	VG+	$ 855	Good	$ 775	LMP	$ 1,652
C2C:	Mint	$ 1,490	Ex	$ 1,075	VG+	$ 780	Good	$ 700	Fair	$ 360
Trade-In:	Mint $ 1,110		Ex	$ 875	VG+	$ 610	Good	$ 545	Poor	$ 180

KIMBER

ULTRA RCP II NEW IN BOX

ULTRA RCP II   REDBOOK CODE: RB-KI-H-ULRCP2

Ramped bushingless barrel, KimPro II finished steel slide and aluminum frame, rounded mainspring housing and frame, Carry Melt treatment, bobbed controls.

Production: 2003 - current   Caliber: .45 ACP   Action: SA, semi-auto   Barrel Length: 3"
OA Length: 6.8"   Wt.: 25 oz.

Sights: sighting trough   Capacity: 7   Magazine: detachable box   Grips: fluted rosewood

D2C:	NIB $1,280	Ex $ 975	VG+ $ 705	Good $ 640	LMP $ 1,351	
C2C:	Mint $1,230	Ex $ 900	VG+ $ 640	Good $ 580	Fair $ 295	
Trade-In:	Mint $ 910	Ex $ 725	VG+ $ 500	Good $ 450	Poor $ 150	

CLASSIC CARRY PRO NEW IN BOX

CLASSIC CARRY PRO   REDBOOK CODE: RB-KI-H-CLCAPR

Charcoal blue finished frame and slide, brush polished flats, Round Heel Frame, ambidextrous safety, aluminum match trigger.

Production: current   Caliber: .45 ACP   Action: SA, semi-auto   Barrel Length: 4"
OA Length: 7.7"   Wt.: 35 oz.

Sights: fixed low-profile night sights, 3-dot, cocking shoulder   Capacity: 8
Magazine: detachable box   Grips: bone

D2C:	NIB $ 1,890	Ex $ 1,450	VG+ $1,040	Good $ 945	LMP $2,056	
C2C:	Mint $ 1,820	Ex $ 1,325	VG+ $ 950	Good $ 855	Fair $ 435	
Trade-In:	Mint $ 1,350	Ex $ 1,075	VG+ $ 740	Good $ 665	Poor $ 210	

GOLD COMBAT II NEW IN BOX

GOLD COMBAT II   REDBOOK CODE: RB-KI-H-GLDCM2

Stainless steel bushingless match grade bull barrel with deep crown, matte black KimPro II finished stainless steel slide and frame, flat top slide with front serrations, ambidextrous thumb safety.

Production: late 1990s - current   Caliber: .45 ACP   Action: SA, semi-auto
Barrel Length: 5"   OA Length: 8.7"   Wt.: 39 oz.

Sights: tritium night sights with cocking shoulder   Capacity: 8
Magazine: detachable box   Grips: Micarta

D2C:	NIB $ 1,975	Ex $ 1,525	VG+ $1,090	Good $ 990	LMP $2,307	
C2C:	Mint $ 1,900	Ex $ 1,375	VG+ $ 990	Good $ 890	Fair $ 455	
Trade-In:	Mint $ 1,410	Ex $ 1,125	VG+ $ 780	Good $ 695	Poor $ 210	

GOLD COMBAT RL II NEW IN BOX

GOLD COMBAT RL II   REDBOOK CODE: RB-KI-H-GLDCMR

Tactical rail, stainless steel bushingless match grade bull barrel with deep crown, matte black KimPro II finished stainless steel slide and frame, flat top slide with front serrations, ambidextrous thumb safety.

Production: early 2000s - current   Caliber: .45 ACP   Action: SA, semi-auto
Barrel Length: 5"   OA Length: 8.7"   Wt.: 40 oz.

Sights: tritium night sights with cocking shoulder   Capacity: 8
Magazine: detachable box   Grips: Micarta

D2C:	NIB $ 2,135	Ex $ 1,625	VG+ $ 1,175	Good $ 1,070	LMP $2,405	
C2C:	Mint $ 2,050	Ex $ 1,475	VG+ $ 1,070	Good $ 965	Fair $ 495	
Trade-In:	Mint $ 1,520	Ex $ 1,200	VG+ $ 840	Good $ 750	Poor $ 240	

GOLD COMBAT STAINLESS II
NEW IN BOX

GOLD COMBAT STAINLESS II   REDBOOK CODE: RB-KI-H-GLDCMS

Stainless steel bushingless match grade bull barrel with deep crown, satin silver KimPro II finished stainless steel slide and frame, flat-top slide with front serrations, ambidextrous thumb safety.

Production: current   Caliber: .45 ACP   Action: SA, semi-auto   Barrel Length: 5"
OA Length: 8.7"   Wt.: 39 oz.   Sights: tritium night sights with cocking shoulder
Capacity: 8   Magazine: detachable box   Grips: Micarta

D2C:	NIB	$1,895	Ex	$1,450	VG+	$1,045	Good	$ 950	LMP $ 2,251
C2C:	Mint	$1,820	Ex	$1,325	VG+	$ 950	Good	$ 855	Fair $ 440
Trade-In:	Mint	$1,350	Ex	$1,075	VG+	$ 740	Good	$ 665	Poor $ 210

SUPER MATCH II
NEW IN BOX

SUPER MATCH II   REDBOOK CODE: RB-KI-H-SUPMT2

Matte black KimPro II finished stainless steel slide, satin silver stainless steel frame, stainless match barrel and bushing, extended magwell, ambidextrous thumb safety, premium aluminum trigger.

Production: 1999 - current   Caliber: .45 ACP   Action: SA, semi-auto
Barrel Length: 5"   OA Length: 8.7"   Wt.: 39 oz.
Sights: Kimber adjustable   Capacity: 8   Magazine: detachable box
Grips: rosewood with double diamond

D2C:	NIB	$ 2,115	Ex	$ 1,625	VG+	$ 1,165	Good	$ 1,060	LMP $ 2,313
C2C:	Mint	$ 2,040	Ex	$ 1,475	VG+	$ 1,060	Good	$ 955	Fair $ 490
Trade-In:	Mint	$ 1,510	Ex	$ 1,200	VG+	$ 830	Good	$ 745	Poor $ 240

RIMFIRE TARGET - BLACK
NEW IN BOX

RIMFIRE TARGET - BLACK   REDBOOK CODE: RB-KI-H-RTBLAK

Ramped match grade barrel and chamber, aluminum match grade trigger, aluminum frame and slide, matte black finish.

Production: 2003 - current   Caliber: .22 LR   Action: SA, semi-auto
Barrel Length: 5"   OA Length: 8.7"   Wt.: 23 oz.
Sights: Kimber adjustable   Capacity: 10   Magazine: detachable box
Grips: checkered black double diamond

D2C:	NIB	$ 735	Ex	$ 575	VG+	$ 405	Good	$ 370	LMP $ 871
C2C:	Mint	$ 710	Ex	$ 525	VG+	$ 370	Good	$ 335	Fair $ 170
Trade-In:	Mint	$ 530	Ex	$ 425	VG+	$ 290	Good	$ 260	Poor $ 90

RIMFIRE SUPER
NEW IN BOX

RIMFIRE SUPER   REDBOOK CODE: RB-KI-H-RTSUPR

Satin silver frame finish, match grade barrel and trigger, matte black slide finish, ramped barrel, full-length guide rod, front slide serrations.

Production: current   Caliber: .22 LR   Action: SA, semi-auto   Barrel Length: 5"
OA Length: 8.7"   Wt.: 23 oz.
Sights: Kimber adjustable   Capacity: 10   Magazine: detachable box
Grips: rosewood with double diamond

D2C:	NIB	$1,125	Ex	$ 675	VG+	$ 480	Good	$ 435	LMP $1,220
C2C:	Mint	$ 840	Ex	$ 600	VG+	$ 440	Good	$ 390	Fair $ 200
Trade-In:	Mint	$ 620	Ex	$ 500	VG+	$ 340	Good	$ 305	Poor $ 90

KIMBER

MICRO CARRY
NEW IN BOX

MICRO CARRY   REDBOOK CODE: RB-KI-H-MICCAR

Matte black finished aluminum frame and steel slide, ramped stainless steel barrel, Carry Melt treatment.

Production: current   Caliber: .380 ACP   Action: SA, semi-auto

Barrel Length: 2.75"   OA Length: 5.6"   Wt.: 13.4 oz.

Sights: fixed low-profile   Capacity: 6   Magazine: detachable box

Grips: checkered black double diamond

D2C:	NIB	$ 545	Ex $ 425	VG+ $ 300	Good $ 275	LMP $ 651
C2C:	Mint	$ 530	Ex $ 400	VG+ $ 280	Good $ 250	Fair $ 130
Trade-In:	Mint	$ 390	Ex $ 325	VG+ $ 220	Good $ 195	Poor $ 60

**MICRO CARRY
STAINLESS**
NEW IN BOX

MICRO CARRY STAINLESS   REDBOOK CODE: RB-KI-H-MCSSXX

Satin silver finished aluminum frame and stainless steel slide, ramped stainless steel barrel, Carry Melt treatment.

Production: current   Caliber: .380 ACP   Action: SA, semi-auto

Barrel Length: 2.75"   OA Length: 5.6"   Wt.: 13.4 oz.

Sights: fixed low-profile   Capacity: 6   Magazine: detachable box

Grips: checkered black double diamond

D2C:	NIB	$ 570	Ex $ 450	VG+ $ 315	Good $ 285	LMP $ 679
C2C:	Mint	$ 550	Ex $ 400	VG+ $ 290	Good $ 260	Fair $ 135
Trade-In:	Mint	$ 410	Ex $ 325	VG+ $ 230	Good $ 200	Poor $ 60

MICRO CDP
NEW IN BOX

MICRO CDP   REDBOOK CODE: RB-KI-H-MICCDP

Matte black aluminum frame and satin silver stainless steel slide, ramped stainless steel barrel, Carry Melt treatment, ambidextrous thumb safety, front strap checkering, high cut under trigger guard.

Production: current   Caliber: .380 ACP   Action: SA, semi-auto

Barrel Length: 2.75"   OA Length: 5.6"

Wt.: 13.4 oz.   Sights: Meprolight tritium 3-dot night sights   Capacity: 6

Magazine: detachable box   Grips: rosewood with double diamond

D2C:	NIB	$1,005	Ex $ 775	VG+ $ 555	Good $ 505	LMP $ 1,121
C2C:	Mint	$ 970	Ex $ 700	VG+ $ 510	Good $ 455	Fair $ 235
Trade-In:	Mint	$ 720	Ex $ 575	VG+ $ 400	Good $ 355	Poor $ 120

MICRO CDP (LG)
NEW IN BOX

MICRO CDP (LG)   REDBOOK CODE: RB-KI-H-MCDPLG

Matte black aluminum frame and satin silver stainless steel slide, ramped stainless steel barrel, Carry Melt treatment, ambidextrous thumb safety, front strap checkering, high cut under trigger guard, CT Lasergrips.

Production: current   Caliber: .380 ACP   Action: SA, semi-auto

Barrel Length: 2.75"   OA Length: 5.6"   Wt.: 13.4 oz.   Sights: Meprolight tritium 3-dot night sights   Capacity: 6   Grips: rosewood with Crimson Trace Lasergrips

D2C:	NIB	$ 1,290	Ex $ 1,000	VG+ $ 710	Good $ 645	LMP $ 1,406
C2C:	Mint	$ 1,240	Ex $ 900	VG+ $ 650	Good $ 585	Fair $ 300
Trade-In:	Mint	$ 920	Ex $ 725	VG+ $ 510	Good $ 455	Poor $ 150

SOLO CARRY
NEW IN BOX

SOLO CARRY   REDBOOK CODE: RB-KI-H-SOLOCX

Stainless steel barrel, ambidextrous thumb safety and magazine release, matte black KimPro II finished aluminum frame, satin silver finished stainless steel slide.

Production: 2011 - current   Caliber: 9mm   Action: striker-fired, semi-auto

Barrel Length: 2.7"   OA Length: 5.5"   Wt.: 17 oz.

Sights: fixed low-profile   Capacity: 6   Magazine: detachable box

Grips: black synthetic, checkered/smooth

D2C:	NIB	$ 680	Ex $ 525	VG+ $ 375	Good $ 340	LMP $ 815
C2C:	Mint	$ 660	Ex $ 475	VG+ $ 340	Good $ 310	Fair $ 160
Trade-In:	Mint	$ 490	Ex $ 400	VG+ $ 270	Good $ 240	Poor $ 90

**SOLO CARRY
STAINLESS**
NEW IN BOX

SOLO CARRY STAINLESS   REDBOOK CODE: RB-KI-H-SOLOCS

Stainless steel barrel, ambidextrous thumb safety and magazine release, KimPro II finished aluminum frame, stainless steel slide, satin silver finish.

Production: current   Caliber: 9mm   Action: striker-fired, semi-auto

Barrel Length: 2.7"   OA Length: 5.5"   Wt.: 17 oz.

Sights: fixed low-profile   Capacity: 6   Magazine: detachable box

Grips: black synthetic, checkered/smooth

D2C:	NIB	$ 685	Ex $ 525	VG+ $ 380	Good $ 345	LMP $ 815
C2C:	Mint	$ 660	Ex $ 475	VG+ $ 350	Good $ 310	Fair $ 160
Trade-In:	Mint	$ 490	Ex $ 400	VG+ $ 270	Good $ 240	Poor $ 90

SOLO CARRY DC
NEW IN BOX

SOLO CARRY DC   REDBOOK CODE: RB-KI-H-SOLODC

Stainless steel barrel and slide, aluminum frame, DLC (diamond-like coating) finish, Carry Melt treatment, ambidextrous thumb safety and magazine release.

Production: current   Caliber: 9mm   Action: striker-fired, semi-auto

Barrel Length: 2.7"   OA Length: 5.5"   Wt.: 17 oz.

Sights: Meprolight tritium 3-dot night sights   Capacity: 6   Magazine: detachable box

Grips: Micarta checkered

D2C:	NIB	$ 815	Ex $ 625	VG+ $ 450	Good $ 410	LMP $ 904
C2C:	Mint	$ 790	Ex $ 575	VG+ $ 410	Good $ 370	Fair $ 190
Trade-In:	Mint	$ 580	Ex $ 475	VG+ $ 320	Good $ 290	Poor $ 90

**SOLO CARRY
DC (LG)**
NEW IN BOX

SOLO CARRY DC (LG)   REDBOOK CODE: RB-KI-H-SOLOLG

Stainless steel barrel and slide, aluminum frame, DLC (diamond-like coating) finish, Carry Melt treatment, ambidextrous thumb safety and magazine release, CT Lasergrips.

Production: current   Caliber: 9mm   Action: striker-fired, semi-auto

Barrel Length: 2.7"   OA Length: 5.5"   Wt.: 17 oz.

Sights: Meprolight tritium 3-dot night sights   Capacity: 6   Magazine: detachable box

Grips: Crimson Trace Lasergrips

D2C:	NIB	$1,065	Ex $ 825	VG+ $ 590	Good $ 535	LMP $1,204
C2C:	Mint	$1,030	Ex $ 750	VG+ $ 540	Good $ 480	Fair $ 245
Trade-In:	Mint	$ 760	Ex $ 600	VG+ $ 420	Good $ 375	Poor $ 120

SOLO CDP (LG)
NEW IN BOX

SOLO CDP (LG) REDBOOK CODE: RB-KI-H-SOLOCD

Duo-tone, matte black KimPro II finished aluminum frame, satin silver stainless steel slide, ambidextrous thumb safety and magazine release.

Production: current Caliber: 9mm Action: striker-fired, semi-auto Barrel Length: 2.7"
OA Length: 5.5" Wt.: 17 oz. Sights: 3-dot tritium night sights Capacity: 6
Grips: rosewood with Crimson Trace Lasergrips

D2C:	NIB	$1,080	Ex $	825	VG+ $	595	Good $	540	LMP	$1,223
C2C:	Mint	$1,040	Ex $	750	VG+ $	540	Good $	490	Fair $	250
Trade-In:	Mint $	770	Ex $	625	VG+ $	430	Good $	380	Poor $	120

Korth

Established in the early 1950s, the German-based Korth specializes in high-quality revolvers and pistols, most of which retail at over $4,000. The company annually produces a very limited amount of handguns, but they've become some of the most renowned high-end, luxury revolvers and pistols commercially available. While based in Germany, the company imports firearms through its American subsidiary, located in Tewksbury, Massachusetts.

COMBAT REVOLVER
EXCELLENT

COMBAT REVOLVER REDBOOK CODE: RB-KK-H-COMBRV

Custom polished or plasma coated finishes available, interchangeable cylinders and barrels, made entirely of forged, hardened tool steel, adjustable trigger, +$1,500 for hardened finish (blue price listed, +$300-$1500 for stainless or plasma).

Production: current Caliber: .22 WMR, .22 LR, .357 Magnum, .38 Special
Action: DA/SA, revolver Barrel Length: 3", 4", 5.25", 6" Wt.: 40 oz.
Sights: Baughman ramp front, adjustable rear Capacity: 6 Grips: walnut round butt

D2C:	NIB	$4,000	Ex	$3,050	VG+	$2,200	Good	$2,000
C2C:	Mint	$3,840	Ex	$2,775	VG+	$2,000	Good	$1,800	Fair
Trade-In:	Mint	$2,840	Ex	$2,250	VG+	$1,560	Good	$1,400	Poor

SPORT REVOLVER
EXCELLENT

SPORT REVOLVER REDBOOK CODE: RB-KK-H-SPRTRV

Custom polished or plasma coated finishes available, auxiliary interchangeable cylinder available, forged tool steel construction (blue price listed, +$300-$1500 for stainless or plasma).

Production: current Caliber: .22 LR, .22 WMR, .32 S&W Long, .38 Special,
.357 Magnum, 9mm Action: DA/SA, revolver Barrel Length: 3", 4", 5.25", 6"
Sights: patridge front, adjustable rear Capacity: 6 Grips: walnut square butt

D2C:	NIB	$4,000	Ex	$3,050	VG+	$2,200	Good	$2,000
C2C:	Mint	$3,840	Ex	$2,775	VG+	$2,000	Good	$1,800	Fair
Trade-In:	Mint	$2,840	Ex	$2,250	VG+	$1,560	Good	$1,400	Poor

TARGET MODEL REDBOOK CODE: RB-KK-H-TRGTML

Available with auxiliary interchangeable cylinder, custom polished or plasma coated finishes available, forged tool steel construction, removable wide-milled trigger shoe.

Production: current Caliber: .22 LR, .22 WMR, .32 S&W Long, .38 Special,
.357 Magnum, 9mm Action: DA/SA, revolver Barrel Length: 5.25", 6"
Sights: patridge front, interchangeable adjustable rear Capacity: 6
Grips: walnut adjustable orthopedic

D2C:	NIB	$5,400	Ex	$4,125	VG+	$2,970	Good	$2,700
C2C:	Mint	$5,190	Ex	$3,750	VG+	$2,700	Good	$2,430	Fair
Trade-In:	Mint	$3,840	Ex	$3,025	VG+	$2,110	Good	$1,890	Poor

SEMI-AUTOMATIC PISTOL REDBOOK CODE: RB-KK-H-SMATPS

Extremely sought after, the model represented is the basic silver or blue model, blue finish listed, +$200-$600 for varying finishes.

Production: mid-1980s - current	Caliber: 9mm, 9x21, .45 ACP, .40 S&W				
Action: DA/SA, semi-auto	Barrel Length: 4", 5" Sights: adjustable Capacity: 10				
Magazine: detachable box Grips: walnut					
D2C:	NIB $5,400	Ex $4,125	VG+ $2,970	Good $2,700	
C2C:	Mint $5,190	Ex $3,750	VG+ $2,700	Good $2,430	Fair
Trade-In:	Mint $3,840	Ex $3,025	VG+ $2,110	Good $1,890	Poor

L.A.R. Manufacturing, Inc.

Now defunct, L.A.R. Manufacturing, Inc. specialized in their Grizzly series of M1911-style pistols, which were available in .45 Winchester Magnum and 10mm, along with the conventional .45 ACP. The Grizzly Bear pistols have retained much of their value over the years, given their ability to chamber large-caliber rounds. The company also produced AR-15-style rifles. The Freedom Group, who also owns Remington Arms, among others, purchased the company in 2012 and has since shut it down.

GRIZZLY MARK I
EXCELLENT

GRIZZLY MARK I REDBOOK CODE: RB-LM-H-GRIZM1

Blue, hard-chrome, or parkerized finish, ambidextrous safety, option of compensator and caliber conversion kits.

Production: mid-1980s Caliber: .45 ACP, .45 Win. Mag., 10mm Action: SA, semi-					
auto Barrel Length: 5.5", 6.5", 8", 10" Wt.: 48 oz. Sights: ramped blade front,					
adjustable rear Capacity: 7 Magazine: detachable Grips: rubber					
D2C:	NIB $1,445	Ex $1,100	VG+ $ 795	Good $ 725	
C2C:	Mint $1,390	Ex $1,000	VG+ $ 730	Good $ 655	Fair $ 335
Trade-In:	Mint $1,030	Ex $ 825	VG+ $ 570	Good $ 510	Poor $ 150

GRIZZLY MARK II REDBOOK CODE: RB-LM-H-GRIZM2

Blue, hard-chrome, or parkerized finish, option of compensator and caliber conversion kits.

Production: mid-1980s Caliber: .45 ACP, .45 Win. Mag., 10mm Action: SA, semi-					
auto Barrel Length: 5.5", 6.5", 8", 10" Sights: fixed Capacity: 7 Grips: rubber					
D2C:	NIB $1,375	Ex $1,050	VG+ $ 760	Good $ 690	
C2C:	Mint $1,320	Ex $ 950	VG+ $ 690	Good $ 620	Fair $ 320
Trade-In:	Mint $ 980	Ex $ 775	VG+ $ 540	Good $ 485	Poor $ 150

GRIZZLY MARK IV REDBOOK CODE: RB-LM-H-GRIZM3

Parkerized or blue finish, ambidextrous safety, beavertail grip safety.

Production: early 1990s Caliber: .44 Magnum Action: SA, semi-auto					
Barrel Length: 5.4", 6.5" Wt.: 50 oz. Sights: ramp front, adjustable rear					
Capacity: 7 Magazine: detachable Grips: rubber					
D2C:	NIB $1,635	Ex $1,250	VG+ $ 900	Good $ 820	
C2C:	Mint $1,570	Ex $1,150	VG+ $ 820	Good $ 740	Fair $ 380
Trade-In:	Mint $ 1,170	Ex $ 925	VG+ $ 640	Good $ 575	Poor $ 180

KORTH

L.A.R. MANUFACTURING

GRIZZLY MARK V
EXCELLENT

GRIZZLY MARK V REDBOOK CODE: RB-LM-H-GRIZM4

Ambidextrous safety, caliber conversion kits available, beavertail grip safety, blue or nickel finish, serrated slide release.

Production: 1993 Caliber: .50 AE Action: SA, semi-auto Barrel Length: 5.5", 6.5"

Sights: ramped blade front, adjustable rear Capacity: 6 Grips: rubber

D2C:	NIB	$1,420	Ex $1,100	VG+ $ 785	Good $ 710		
C2C:	Mint	$1,370	Ex $1,000	VG+ $ 710	Good $ 640	Fair $ 330	
Trade-In:	Mint	$1,010	Ex $ 800	VG+ $ 560	Good $ 500	Poor $ 150	

Les Baer Custom, Inc.

Formed in the early 1990s, Les Baer Custom, Inc. produces a wide variety of custom M1911 pistols, most of which feature adjustable sights, a skeletonized trigger, and tasteful design. The company somewhat famously moved their headquarters from Illinois after the state began imposing harsher restrictions on firearm carry and ownership. Thus, the company is now based in LeClaire, Iowa.

1911 .38 SUPER STINGER
NEW IN BOX

1911 .38 SUPER STINGER REDBOOK CODE: RB-LB-H-38SUPS

Beveled magwell, extended ejector, deluxe hammer and sear. Blue, chrome, or DuPont S finish.

Production: current Caliber: .38 Super Action: SA, semi-auto

Barrel Length: 4.25" Wt.: 34 oz. Sights: deluxe fixed combat rear, night sights

Capacity: 8 Magazine: detachable box Grips: premium checkered cocobolo

D2C:	NIB $2,160	Ex $1,650	VG+ $1,190	Good $1,080	LMP $2,495	
C2C:	Mint $2,080	Ex $1,500	VG+ $1,080	Good $ 975	Fair $ 500	
Trade-In:	Mint $1,540	Ex $1,225	VG+ $ 850	Good $ 760	Poor $ 240	

1911 BOSS
NEW IN BOX

1911 BOSS REDBOOK CODE: RB-LB-H-19BOSS

Rear cocking serrations, blue slide, grip safety, special tactical package.

Production: current Caliber: .45 ACP Action: SA, semi-auto

Barrel Length: 5" Wt.: 37 oz. Sights: Baer red fiber-optic front Capacity: 8

Magazine: detachable box Grips: checkered

D2C:	NIB $2,260	Ex $1,725	VG+ $1,245	Good $1,130	LMP $2,260	
C2C:	Mint $2,170	Ex $1,575	VG+ $1,130	Good $1,020	Fair $ 520	
Trade-In:	Mint $1,610	Ex $1,275	VG+ $890	Good $ 795	Poor $ 240	

1911 PREMIER II
NEW IN BOX

1911 PREMIER II REDBOOK CODE: RB-LB-H-PRE2XX

Blue finish, grip safety, extended ejector, extended ambidextrous safety, beveled magwell, add $320 for .400 Corbon, add $500 for .38 Super.

Production: current Caliber: .45 ACP, .38 Super, .400 Corbon Action: SA, semi-auto

Barrel Length: 5", 6" Sights: dovetail front, adjustable rear Capacity: 8

Magazine: detachable box Grips: premium checkered cocobolo

D2C:	NIB $1,820	Ex $1,400	VG+ $1,005	Good $910	LMP $1,905	
C2C:	Mint $1,750	Ex $1,275	VG+ $910	Good $820	Fair $420	
Trade-In:	Mint $1,300	Ex $1,025	VG+ $710	Good $640	Poor $210	

**1911 CMP-LEGAL
NATIONAL MATCH
HARDBALL PISTOL**
NEW IN BOX

1911 CMP-LEGAL NATIONAL MATCH HARDBALL PISTOL

REDBOOK CODE: RB-LB-H-CMPNMH

Blue finish, beveled magwell, extended ejector, rear serrated slide.

Production: current   Caliber: .45 ACP   Action: SA, semi-auto

Barrel Length: 5" Wt.: 36 oz. Sights: dovetail front, LBC adjustable rear   Capacity: 7

Magazine: detachable box   Grips: premium checkered cocobolo

D2C:	NIB $1,880	Ex $1,450	VG+ $1,035	Good $ 940	
C2C:	Mint $1,810	Ex $1,300	VG+ $ 940	Good $ 850	Fair $ 435
Trade-In:	Mint $1,340	Ex $1,075	VG+ $ 740	Good $ 660	Poor $ 210

**1911 ULTIMATE
MASTER 5**
NEW IN BOX

1911 ULTIMATE MASTER 5   REDBOOK CODE: RB-LB-H-UM5XXX

Blue finish, beveled magwell, deluxe Commander hammer and sear.

Production: current   Caliber: .45 ACP, .38 Super, .400 Corbon   Action: SA, semi-auto

Barrel Length: 5" Sights: dovetail front, LBC adjustable rear   Capacity: 8

Magazine: detachable box   Grips: premium checkered cocobolo

D2C:	NIB $2,630	Ex $2,000	VG+ $1,450	Good $1,315	
C2C:	Mint $2,530	Ex $1,825	VG+ $1,320	Good $1,185	Fair $ 605
Trade-In:	Mint $1,870	Ex $1,475	VG+ $1,030	Good $ 925	Poor $ 270

**1911 ULTIMATE
MASTER 6**
NEW IN BOX

1911 ULTIMATE MASTER 6   REDBOOK CODE: RB-LB-H-UM6XXX

Blue finish, deluxe Commander hammer and sear, extended ejector, Beavertail grip safety.

Production: current   Caliber: .45 ACP, .38 Super, .400 Corbon   Action: SA, semi-auto

Barrel Length: 6" Sights: dovetail front, LBC adjustable rear

Capacity: 8   Magazine: detachable box   Grips: premium checkered cocobolo

D2C:	NIB $2,630	Ex $2,000	VG+ $1,450	Good $1,315	
C2C:	Mint $2,530	Ex $1,825	VG+ $1,320	Good $1,185	Fair $ 605
Trade-In:	Mint $1,870	Ex $1,475	VG+ $1,030	Good $ 925	Poor $ 270

**1911 ULTIMATE
MASTER COMBAT
COMPENSATED**
NEW IN BOX

1911 ULTIMATE MASTER COMBAT COMPENSATED

REDBOOK CODE: RB-LB-H-UMCCXX

Blue finish, extended ejector, beveled magwell, deluxe Commander hammer and sear, compensator.

Production: current   Caliber: .45 ACP, .38 Super   Action: SA, semi-auto

Barrel Length: 5" Sights: dovetail front, LBC adjustable rear

Capacity: 8   Magazine: detachable box   Grips: premium checkered cocobolo

D2C:	NIB $2,700	Ex $2,075	VG+ $,1485	Good $1,350	
C2C:	Mint $2,600	Ex $1,875	VG+ $1,350	Good $1,215	Fair $ 625
Trade-In:	Mint $1,920	Ex $1,525	VG+ $1,060	Good $ 945	Poor $ 270

**1911 BULLSEYE
WADCUTTER PISTOL**
NEW IN BOX

1911 BULLSEYE WADCUTTER PISTOL

REDBOOK CODE: RB-LB-H-BWPXXX

Blue finish, beveled and extended magwell, deluxe Commander hammer and sear, Beavertail grip safety.

Production: current   Caliber: .45 ACP   Action: SA, semi-auto   Barrel Length: 5"

Sights: LBC adjustable or slide mount   Capacity: 7   Magazine: detachable box

Grips: premium checkered cocobolo

D2C:	NIB $2,140	Ex $1,650	VG+ $1,180	Good $1,070	LMP $2,140
C2C:	Mint $2,060	Ex $1,500	VG+ $1,070	Good $ 965	Fair $ 495
Trade-In:	Mint $1,520	Ex $1,200	VG+ $ 840	Good $ 750	Poor $ 240

LES BAER CUSTOM, INC.

1911 PPC DISTINGUISHED MATCH
NEW IN BOX

1911 PPC OPEN CLASS
NEW IN BOX

1911 PREMIER II SUPER-TAC
NEW IN BOX

1911 CUSTOM CARRY
NEW IN BOX

1911 SUPER COMANCHE
NEW IN BOX

1911 PPC DISTINGUISHED MATCH   REDBOOK CODE: RB-LB-H-PPCDMX

Blue finish, beveled and extended magwell, deluxe Commander hammer and sear, Beavertail grip safety, add $395 for 9mm model.

Production: late 1990s - current   Caliber: .45 ACP, 9mm   Action: SA, semi-auto

Barrel Length: 5" Wt.: 36 oz.   Sights: dovetail front   Capacity: 8

Magazine: detachable box   Grips: checkered cocobolo

D2C:	NIB	$2,240	Ex	$1,725	VG+	$1,235	Good	$1,120	LMP	$2,240
C2C:	Mint	$2,160	Ex	$1,550	VG+	$1,120	Good	$1,010	Fair	$520
Trade-In:	Mint	$1,600	Ex	$1,275	VG+	$880	Good	$785	Poor	$240

1911 PPC OPEN CLASS   REDBOOK CODE: RB-LB-H-PPCOCD

Blue finish, extended ejector, deluxe Commander hammer and sear, beveled magwell, add $395 for 9mm model.

Production: late 1990s - current   Caliber: .45 ACP, 9mm   Action: SA, semi-auto

Barrel Length: 6" Wt.: 37 oz.   Sights: dovetail front   Capacity: 8

Magazine: detachable box   Grips: premium checkered cocobolo

D2C:	NIB	$2,350	Ex	$1,800	VG+	$1,295	Good	$1,175	LMP	$2,350
C2C:	Mint	$2,260	Ex	$1,625	VG+	$1,180	Good	$1,060	Fair	$545
Trade-In:	Mint	$1,670	Ex	$1,325	VG+	$920	Good	$825	Poor	$240

1911 PREMIER II SUPER-TAC   REDBOOK CODE: RB-LB-H-PRE2ST

DuPont S finish, Baer deluxe Commander hammer and sear, beavertail grip safety.

Production: current   Caliber: .45 ACP, .38 Super, .400 Corbon   Action: SA, semi-auto

Barrel Length: 5" Wt.: 37 oz.   Sights: dovetail front, adjustable rear

Capacity: 8   Magazine: detachable box   Grips: checkered cocobolo

D2C:	NIB	$2,420	Ex	$1,850	VG+	$1,335	Good	$1,210		
C2C:	Mint	$2,330	Ex	$1,675	VG+	$1,210	Good	$1,090	Fair	$60
Trade-In:	Mint	$1,720	Ex	$1,375	VG+	$950	Good	$850	Poor	$270

1911 CUSTOM CARRY   REDBOOK CODE: RB-LB-H-CUSTCA

Beveled magwell, Baer deluxe hammer and sear, extended ejector, blue finish, add $300 for .38 Super model with 4.25" barrel.

Production: current   Caliber: .45 ACP, .38 Super   Action: SA, semi-auto

Barrel Length: 5", 4.25" Wt.: 37 oz.   Sights: deluxe fixed combat, ramp style night

Capacity: 8   Magazine: detachable box   Grips: premium checkered cocobolo

D2C:	NIB	$1,920	Ex	$1,475	VG+	$1,060	Good	$960	LMP	$1,920
C2C:	Mint	$1,850	Ex	$1,325	VG+	$960	Good	$865	Fair	$445
Trade-In:	Mint	$1,370	Ex	$1,100	VG+	$750	Good	$675	Poor	$210

1911 SUPER COMANCHE   REDBOOK CODE: RB-LB-H-SUCOMA

Blue finish, chrome, or DuPont S finish, Baer deluxe hammer and sear, beveled magwell, extended ejector.

Production: current   Caliber: .38 Super   Action: SA, semi-auto   Barrel Length: 4.3"

Sights: dovetail front, deluxe fixed combat rear, night sights   Capacity: 9

Magazine: detachable box   Grips: premium checkered cocobolo

D2C:	NIB	$2,060	Ex	$1,575	VG+	$1,135	Good	$1,030		
C2C:	Mint	$1,980	Ex	$1,425	VG+	$1,030	Good	$930	Fair	$475
Trade-In:	Mint	$1,470	Ex	$1,175	VG+	$810	Good	$725	Poor	$210

1911 ULTIMATE RECON PISTOL
NEW IN BOX

1911 ULTIMATE RECON PISTOL REDBOOK CODE: RB-LB-H-ULRECP

Beveled magwell, Baer deluxe hammer and sear, extended ejector, integral Picatinny rail, comes with Streamlight TLR-1, SureFire X-300 is available for additional price, add $240 for chrome finish model.

Production: 2006 - current Caliber: .45 ACP Action: SA, semi-auto

Barrel Length: 5" Wt.: 37 oz. Sights: dovetail front, deluxe fixed combat rear,

night sights Capacity: 8 Magazine: detachable box Grips: premium checkered cocobolo

D2C:	NIB	$2,370	Ex	$1,825	VG+	$1,305	Good	$1,185	LMP	$2,370
C2C:	Mint	$2,280	Ex	$1,650	VG+	$1,190	Good	$1,070	Fair	$550
Trade-In:	Mint	$1,690	Ex	$1,350	VG+	$930	Good	$830	Poor	$240

1911 ULTIMATE TACTICAL CARRY
NEW IN BOX

1911 ULTIMATE TACTICAL CARRY REDBOOK CODE: RB-LB-H-ULTACA

Blue finish, beveled magwell, Beavertail grip safety, deluxe Commander hammer and sear.

Production: 2013 - current Caliber: .45 ACP Action: SA, semi-auto

Barrel Length: 5" Sights: dovetail front, LBC deluxe fixed combat rear

Capacity: 8 Magazine: detachable box Grips: deluxe

D2C:	NIB	$1,930	Ex	$1,475	VG+	$1,065	Good	$965	LMP	$1,930
C2C:	Mint	$1,860	Ex	$1,350	VG+	$970	Good	$870	Fair	$445
Trade-In:	Mint	$1,380	Ex	$1,100	VG+	$760	Good	$680	Poor	$210

1911 THUNDER RANCH SPECIAL
NEW IN BOX

1911 THUNDER RANCH SPECIAL REDBOOK CODE: RB-LB-H-THUNRS

Thunder Ranch logo on deluxe grips, available with night sights, Beavertail grip safety, blue finish, also comes in special engraved model.

Production: current Caliber: .45 ACP Action: SA, semi-auto

Barrel Length: 5" Sights: fixed combat rear Capacity: 7

Magazine: detachable box Grips: deluxe

D2C:	NIB	$1,820	Ex	$1,400	VG+	$1,005	Good	$910	LMP	$1,980
C2C:	Mint	$1,750	Ex	$1,275	VG+	$910	Good	$820	Fair	$420
Trade-In:	Mint	$1,300	Ex	$1,025	VG+	$710	Good	$640	Poor	$210

1911 SHOOTING USA CUSTOM PISTOL
NEW IN BOX

1911 SHOOTING USA CUSTOM PISTOL

REDBOOK CODE: RB-LB-H-USACUS

Beveled magwell, deluxe Commander hammer and sear, DuPont S finish, presentation box included, extended ejector, speed trigger, comes with DVD by Shooting USA.

Production: current Caliber: .45 ACP Action: SA, semi-auto

Barrel Length: 5" Sights: tritium night, dovetail front Capacity: 8

Magazine: detachable box Grips: checkered cocobolo

D2C:	NIB	$1,905	Ex	$1,450	VG+	$1,050	Good	$955	LMP	$1,905
C2C:	Mint	$1,830	Ex	$1,325	VG+	$960	Good	$860	Fair	$440
Trade-In:	Mint	$1,360	Ex	$1,075	VG+	$750	Good	$670	Poor	$210

1911 CUSTOM CENTENNIAL MODEL
NEW IN BOX

1911 CUSTOM CENTENNIAL MODEL REDBOOK CODE: RB-LB-H-CUSCEN

Presentation box included, deluxe charcoal blue finish, model name engraved on slide, rear serrated slide.

Production: current Caliber: .45 ACP Action: SA, semi-auto Barrel Length: 5"

Sights: fixed Capacity: 8 Magazine: detachable box Grips: ivory

D2C:	NIB	$4,050	Ex	$3,100	VG+	$2,230	Good	$2,025	LMP	$4,050
C2C:	Mint	$3,890	Ex	$2,800	VG+	$2,030	Good	$1,825	Fair	$935
Trade-In:	Mint	$2,880	Ex	$2,275	VG+	$1,580	Good	$1,420	Poor	$420

1911 S.R.P.
NEW IN BOX

1911 S.R.P. REDBOOK CODE: RB-LB-H-SRPXXX

Swift Response Pistol, DuPont S finish, beveled magwell, stainless bushing, Commander hammer and sear, speed trigger, front and rear slide serrations, presentation box included.

Production: mid-1990s - current Caliber: .45 ACP Action: SA, semi-auto

Barrel Length: 5" Wt.: 37 oz. Sights: tritium night sights Capacity: 8

Magazine: detachable box Grips: checkered cocobolo

D2C:	NIB	$2,310	Ex	$1,775	VG+	$1,275	Good	$1,155	LMP	$2,490
C2C:	Mint	$2,220	Ex	$1,600	VG+	$1,160	Good	$1,040	Fair	$ 535
Trade-In:	Mint	$1,650	Ex	$1,300	VG+	$ 910	Good	$810	Poor	$ 240

1911 MONOLITH
NEW IN BOX

1911 MONOLITH REDBOOK CODE: RB-LB-H-MONOLH

Blue finish, beavertail grip safety, deluxe Commander hammer and sear, extended ejector, beveled magwell, add $400 for .38 Super model.

Production: late 1990s - current Caliber: .45 ACP Action: SA, semi-auto

Barrel Length: 5" Wt.: 37 oz. Sights: adjustable target Capacity: 8

Magazine: detachable box Grips: premium checkered cocobolo

D2C:	NIB	$1,990	Ex	$1,525	VG+	$1,095	Good	$ 995	LMP	$1,990
C2C:	Mint	$1,920	Ex	$1,375	VG+	$1,000	Good	$ 900	Fair	$ 460
Trade-In:	Mint	$1,420	Ex	$1,125	VG+	$ 780	Good	$ 700	Poor	$ 210

1911 MONOLITH COMANCHE
NEW IN BOX

1911 MONOLITH COMANCHE REDBOOK CODE: RB-LB-H-MONOCO

Beavertail grip safety, deluxe Commander hammer and sear, blue finish, extended ejector, beveled magwell.

Production: early 2000s - current Caliber: .45 ACP Action: SA, semi-auto

Barrel Length: 4.25" Sights: deluxe fixed tritium night Capacity: 8

Magazine: detachable box Grips: premium checkered cocobolo

D2C:	NIB	$1,920	Ex	$1,475	VG+	$1,060	Good	$ 960
C2C:	Mint	$1,850	Ex	$1,325	VG+	$ 960	Good	$ 865	Fair	$ 445
Trade-In:	Mint	$1,370	Ex	$1,100	VG+	$ 750	Good	$ 675	Poor	$ 210

1911 MONOLITH COMANCHE HEAVYWEIGHT
NEW IN BOX

1911 MONOLITH COMANCHE HEAVYWEIGHT

REDBOOK CODE: RB-LB-H-MONOCH

Beavertail grip safety, deluxe Commander hammer and sear, blue finish, extended ejector, beveled magwell.

Production: early 2000s - current Caliber: .45 ACP Action: SA, semi-auto

Barrel Length: 4.25" Sights: deluxe fixed tritium night Capacity: 8

Magazine: detachable box Grips: premium checkered cocobolo

D2C:	NIB	$2,060	Ex	$1,575	VG+	$1,135	Good	$1,030	LMP	$2,060
C2C:	Mint	$1,980	Ex	$1,425	VG+	$1,030	Good	$ 930	Fair	$ 475
Trade-In:	Mint	$1,470	Ex	$1,175	VG+	$ 810	Good	$ 725	Poor	$ 210

1911 MONOLITH TACTICAL ILLUMINATOR
NEW IN BOX

1911 MONOLITH TACTICAL ILLUMINATOR

REDBOOK CODE: RB-LB-H-MONOTI

Blue finish, Streamlight TLR-1 mounted light, Commander-style hammer, speed trigger.

Production: late 1990s - current Caliber: .45 ACP, 9mm, .38 Super

Action: SA, semi-auto Barrel Length: 5" Wt.: 40 oz. Sights: adjustable target

Capacity: 8 Magazine: detachable box Grips: checkered

D2C:	NIB	$2,290	Ex	$1,750	VG+	$1,260	Good	$1,145
C2C:	Mint	$2,200	Ex	$1,600	VG+	$1,150	Good	$1,035	Fair	$ 530
Trade-In:	Mint	$1,630	Ex	$1,300	VG+	$ 900	Good	$ 805	Poor	$ 240

1911 STINGER
NEW IN BOX

1911 STINGER   REDBOOK CODE: RB-LB-H-STINGX

Beavertail grip safety, deluxe Commander hammer and sear, blue finish, extended ejector, beveled magwell, available in a stainless steel model.

Production: late 1990s - current   Caliber: .45 ACP   Action: SA, semi-auto

Barrel Length: 4.25"   Wt.: 34 oz.   Sights: dovetail front, combat fixed rear

Capacity: 7   Magazine: detachable box   Grips: premium checkered cocobolo

D2C:	NIB	$1,930	Ex	$1,475	VG+	$1,065	Good	$965	LMP	$1,930
C2C:	Mint	$1,860	Ex	$1,350	VG+	$970	Good	$870	Fair	$ 445
Trade-In:	Mint	$1,380	Ex	$1,100	VG+	$760	Good	$680	Poor	$ 210

1911 CONCEPT I
NEW IN BOX

1911 CONCEPT I   REDBOOK CODE: RB-LB-H-CON1XX

Blue finish, beveled magwell, extended ejector, speed trigger.

Production: current   Caliber: .45 ACP   Action: SA, semi-auto   Barrel Length: 5"

Wt.: 37 oz.   Sights: dovetail front, LBC adjustable deluxe rear   Capacity: 8

Magazine: detachable box   Grips: premium checkered cocobolo

D2C:	NIB	$1,770	Ex	$1,350	VG+	$ 975	Good	$ 885	LMP	$1,770
C2C:	Mint	$1,700	Ex	$1,225	VG+	$ 890	Good	$ 800	Fair	$ 410
Trade-In:	Mint	$1,260	Ex	$1,000	VG+	$ 700	Good	$ 620	Poor	$ 180

1911 CONCEPT II
NEW IN BOX

1911 CONCEPT II   REDBOOK CODE: RB-LB-H-CON2XX

Blue finish, extended ejector, beveled magwell, speed trigger.

Production: current   Caliber: .45 ACP   Action: SA, semi-auto

Barrel Length: 5"   Wt.: 37 oz.   Sights: adjustable   Capacity: 8

Magazine: detachable box   Grips: premium checkered cocobolo

D2C:	NIB	$1,770	Ex	$1,350	VG+	$ 975	Good	$ 885	LMP	$1,770
C2C:	Mint	$1,700	Ex	$1,225	VG+	$ 890	Good	$ 800	Fair	$ 410
Trade-In:	Mint	$1,260	Ex	$1,000	VG+	$ 700	Good	$ 620	Poor	$ 180

1911 CONCEPT III
NEW IN BOX

1911 CONCEPT III   REDBOOK CODE: RB-LB-H-CON3XX

Stainless steel with blue steel slide, beveled magwell, extended ejector.

Production: current   Caliber: .45 ACP   Action: SA, semi-auto

Barrel Length: 5"   Wt.: 37 oz.   Sights: adjustable rear   Capacity: 8

Magazine: detachable box   Grips: premium checkered cocobolo

D2C:	NIB	$1,910	Ex	$1,475	VG+	$1,055	Good	$ 955	LMP	$1,910
C2C:	Mint	$1,840	Ex	$1,325	VG+	$960	Good	$ 860	Fair	$ 440
Trade-In:	Mint	$1,360	Ex	$1,075	VG+	$750	Good	$ 670	Poor	$ 210

1911 CONCEPT IV
NEW IN BOX

1911 CONCEPT IV   REDBOOK CODE: RB-LB-H-CON4XX

Stainless steel frame with blue steel slide, beveled magwell, extended ejector.

Production: current   Caliber: .45 ACP   Action: SA, semi-auto   Barrel Length: 5"

Wt.: 37 oz.   Sights: deluxe fixed combat rear   Capacity: 8   Magazine: detachable box

Grips: premium checkered cocobolo

D2C:	NIB	$1,910	Ex	$1,475	VG+	$1,055	Good	$ 955	LMP	$1,910
C2C:	Mint	$1,840	Ex	$1,325	VG+	$960	Good	$ 860	Fair	$ 440
Trade-In:	Mint	$1,360	Ex	$1,075	VG+	$750	Good	$ 670	Poor	$ 210

1911 CONCEPT V
NEW IN BOX

1911 CONCEPT VI
NEW IN BOX

1911 CONCEPT VII
NEW IN BOX

1911 CONCEPT VIII
NEW IN BOX

1911 HEMI 572
NEW IN BOX

1911 CONCEPT V   REDBOOK CODE: RB-LB-H-CON5XX

Stainless steel, beveled magwell, extended ejector, also available with a 6" barrel.

Production: current   Caliber: .45 ACP   Action: SA, semi-auto   Barrel Length: 5"

Wt.: 37 oz.   Sights: adjustable rear   Capacity: 8   Magazine: detachable box

Grips: premium checkered cocobolo

D2C:	NIB $1,940	Ex $1,475	VG+ $1,070	Good $ 970	LMP $1,940
C2C:	Mint $1,870	Ex $1,350	VG+ $970	Good $ 875	Fair $ 450
Trade-In:	Mint $1,380	Ex $1,100	VG+ $760	Good $ 680	Poor $ 210

1911 CONCEPT VI   REDBOOK CODE: RB-LB-H-CON7XX

Stainless steel, beveled magwell, extended ejector.

Production: current   Caliber: .45 ACP   Action: SA, semi-auto   Barrel Length: 5"

Wt.: 37 oz.   Sights: deluxe fixed combat   Capacity: 8   Magazine: detachable box

Grips: premium checkered cocobolo

D2C:	NIB $1,910	Ex $1,475	VG+ $1,055	Good $ 955	LMP $1,910
C2C:	Mint $1,840	Ex $1,325	VG+ $960	Good $ 860	Fair $ 440
Trade-In:	Mint $1,360	Ex $1,075	VG+ $750	Good $ 670	Poor $ 210

1911 CONCEPT VII   REDBOOK CODE: RB-LB-H-CON8XX

Blue steel, beveled magwell, extended ejector.

Production: current   Caliber: .45 ACP   Action: SA, semi-auto   Barrel Length: 5"

Wt.: 37 oz.   Sights: deluxe fixed combat   Capacity: 8   Magazine: detachable box

Grips: premium checkered cocobolo

D2C:	NIB $1,920	Ex $1,475	VG+ $1,060	Good $ 960	LMP $1,920
C2C:	Mint $1,850	Ex $1,325	VG+ $960	Good $ 865	Fair $ 445
Trade-In:	Mint $1,370	Ex $1,100	VG+ $750	Good $ 675	Poor $ 210

1911 CONCEPT VIII   REDBOOK CODE: RB-LB-H-CON9XX

Stainless steel, beveled magwell, extended ejector.

Production: current   Caliber: .45 ACP   Action: SA, semi-auto   Barrel Length: 5"

Wt.: 37 oz.   Sights: deluxe fixed combat   Capacity: 8   Magazine: detachable box

Grips: premium checkered cocobolo

D2C:	NIB $1,955	Ex $1,500	VG+ $1,080	Good $ 980	LMP $1,955
C2C:	Mint $1,880	Ex $1,350	VG+ $980	Good $ 880	Fair $ 450
Trade-In:	Mint $1,390	Ex $1,100	VG+ $770	Good $ 685	Poor $ 210

1911 HEMI 572   REDBOOK CODE: RB-LB-H-HEM572

Hard chrome finish, DuPont S coating, special tactical package with ambidextrous safety.

Production: 2013 - current   Caliber: .45 ACP   Action: SA, semi-auto

Barrel Length: 5"   Sights: fiber-optic front with green insert

Capacity: 8   Magazine: detachable box   Grips: VZ black recon

D2C:	NIB $2,395	Ex $1,825	VG+ $1,320	Good $1,200	LMP $2,395
C2C:	Mint $2,300	Ex $1,675	VG+ $1,200	Good $1,080	Fair $ 555
Trade-In:	Mint $ 1,710	Ex $1,350	VG+ $940	Good $840	Poor $ 240

CUSTOM 25TH ANNIVERSARY MODEL 1911 PISTOL SPECIAL COLLECTORS' MODEL
NEW IN BOX

CUSTOM 25TH ANNIVERSARY 1911 PISTOL SPECIAL COLLECTORS' MODEL REDBOOK CODE: RB-LB-H-C25AMP

Blue finish, Les Baer signature and "25th Anniversary" in white gold engraving on the slide, presentation box included.

Production: current Caliber: .45 ACP Action: SA, semi-auto Barrel Length: 5"					
Sights: fixed Capacity: 8 Magazine: detachable box Grips: ivory					
D2C:	NIB $6,995	Ex $5,325	VG+ $3,850	Good	
C2C:	Mint $6,720	Ex $4,850	VG+ $3,500	Good	Fair
Trade-In:	Mint $4,970	Ex $3,925	VG+ $2,730	Good	Poor

Luger

Luger is an umbrella term applied to a host of manufacturers that produced the Luger pistol for various German, Nazi, and Swiss military campaigns and commercial use. Some of these manufacturers include Deutch Waffen und Munitions (often abbreviated DWM), Heinrich Krieghoff, and Mauser. While most of these pistols look very similar and adhere to Georg Luger's original design, minor variations distinguish the models and thus their values. The history of the Luger pistol is convoluted, given the sheer number produced and the involvement of numerous manufacturers, so the editors of this book would encourage extreme discretion before purchase. The Luger made appearances in both the First and Second World Wars, the German Revolution, the Spanish Civil War, and the Chinese Civil War, among others, making it one of the most widely used semi-automatic pistols of its era. Lugers still command substantial collector interest. If no pricing is listed for a model, too few sales examples exist on the secondary market to accurately assess values at this time.

1900 COMMERCIAL
EXCELLENT

Courtesy of Rock Island Auction Company

1900 COMMERCIAL REDBOOK CODE: RB-LU-H-1900XX

Serial numbers range between approximately 1 and 20,000, though less than 6,000 produced. These models don't feature any Swiss engravings and are thus somewhat less valuable. Imported to North America, Germany, and several other countries.

Production: ~1900 Caliber: 7.65mm Action: SA, semi-auto						
Barrel Length: 4.75" Sights: fixed Capacity: 8 Magazine: detachable box Grips: wood						
D2C:	NIB	Ex $5,400	VG+ $3,175	Good $2,885		
C2C:	Mint	Ex $5,000	VG+ $2,890	Good $2,600	Fair $1,330	
Trade-In:	Mint	Ex $4,250	VG+ $2,260	Good $2,020	Poor $ 600	

1900 SWISS COMMERCIAL
GOOD

Courtesy of Rock Island Auction Company

1900 SWISS COMMERCIAL REDBOOK CODE: RB-LU-H-1900X1

Serial numbers range between approximately 1 and 2,130. Less than 2,500 produced. Features a Swiss Cross engraved near the barrel, above the chamber.

Production: ~1900 Caliber: 7.65mm Action: SA, semi-auto						
Barrel Length: 4.75" Sights: fixed Capacity: 8 Magazine: detachable box Grips: wood						
D2C:	NIB	Ex $5,825	VG+ $3,490	Good $3,175		
C2C:	Mint	Ex $5,400	VG+ $3,180	Good $2,860	Fair $1,460	
Trade-In:	Mint	Ex $4,575	VG+ $2,480	Good $2,225	Poor $ 660	

LUGER

**1900 SWISS
MILITARY CONTRACT**
EXCELLENT

Courtesy of Rock Island Auction Company

1900 SWISS MILITARY CONTRACT
REDBOOK CODE: RB-LU-H-1900X2

Serial numbers range between approximately 2,001 and 5,001. Less than 3,000 units produced for the Swiss military use. Marked with the Swiss Cross.

Production: ~1900	Caliber: 7.65mm	Action: SA, semi-auto				
Barrel Length: 4.75"	Sights: fixed	Capacity: 8	Magazine: detachable box	Grips: wood		
D2C:	NIB	Ex	$5,300	VG+ $3,100	Good $2,815	
C2C:	Mint	Ex	$4,900	VG+ $2,820	Good $2,535	Fair $1,295
Trade-In:	Mint	Ex	$4,175	VG+ $2,200	Good $1,975	Poor $ 570

**1900 AMERICAN
EAGLE**
EXCELLENT

Courtesy of Rock Island Auction Company

1900 AMERICAN EAGLE REDBOOK CODE: RB-LU-H-1900X3

Serial numbers range anywhere between approximately 1,950 and 200,050. 11,500 or so engraved to denote German military use, though some were unmarked.

Production: ~1900	Caliber: 7.65mm	Action: SA, semi-auto				
Barrel Length: 4.75"	Sights: fixed	Capacity: 8	Magazine: detachable box	Grips: wood		
D2C:	NIB	-- Ex	$5,400	VG+ $3,175	Good $2,885	
C2C:	Mint	-- Ex	$5,000	VG+ $2,890	Good $2,595	Fair $1,330
Trade-In:	Mint	-- Ex	$4,250	VG+ $2,250	Good $2,020	Poor $ 600

1902 COMMERCIAL
GOOD

Courtesy of Rock Island Auction Company

1902 COMMERCIAL REDBOOK CODE: RB-LU-H-1902X1

Less than 650 units produced. Serial numbers range between approximately 22,250 and 23,550. Thick, or "fat," barrel.

Production: ~1902	Caliber: 9mm	Action: SA, semi-auto	Barrel Length: 4"			
Sights: fixed	Capacity: 8	Magazine: detachable box	Grips: wood			
D2C:	NIB	Ex $11,500	VG+ $9,000	Good $3,800		
C2C:	Mint	Ex $10,500	VG+ $8,200	Good $3,450	Fair $1,400	
Trade-In:	Mint	Ex $8,000	VG+ $6,300	Good $2,700	Poor $750	

**1902 AMERICAN
EAGLE**
VERY GOOD +

Courtesy of Rock Island Auction Company

1902 AMERICAN EAGLE REDBOOK CODE: RB-LU-H-1902X2

Less than 750 produced with serial numbers ranging between approximately 22,010 and 23,000. Blue finish, frame-mounted safety.

Production: 1902	Caliber: 9mm	Action: SA, semi-auto	Barrel Length: 4"			
Sights: fixed	Capacity: 8	Magazine: detachable box	Grips: wood			
D2C:	NIB	Ex $14,000	VG+ $10,900	Good $7,300		
C2C:	Mint	Ex $12,740	VG+ $9,900	Good $7,000	Fair $1,100	
Trade-In:	Mint	Ex $9,800	VG+ $7,700	Good $5,100	Poor $600	

1904 NAVY
GOOD

Courtesy of Rock Island Auction Company

1904 NAVY REDBOOK CODE: RB-LU-H-1904XX

Less than 1,600 produced with varying serial numbers. Marked both "Geladen" and "Gesichert," but many fakes exist so an expert should be consulted before purchase. Frame-mounted safety. According to auction sources, only 21 known to exist. Accurate pricing is questionable, so pricing is theoretical.

Production: 1904	Caliber: 9mm	Action: SA, semi-auto				
Barrel Length: 6"	Sights: fixed	Capacity: 8	Magazine: detachable box	Grips: wood		
D2C:	NIB	Ex	$60,000	VG+ $43,000	Good	
C2C:	Mint	Ex	$55,000	VG+ $39,000	Good	Fair --
Trade-In:	Mint	Ex	$42,000	VG+ $30,000	Good	Poor --

1906 COMMERCIAL
EXCELLENT

Courtesy of Rock Island Auction Company

1906 COMMERCIAL (STANDARD) REDBOOK CODE: RB-LU-H-1906X1

Serial numbers range between approximately 26,490 and 68,000, less than 5,000 produced. Manufactured for civilian use. Marked "Geladen," frame and chamber-area not marked. Frame-mounted safety.

Production: 1906 - ~1908	Caliber: 9mm	Action: SA, semi-auto			
Barrel Length: 4" Sights: fixed Capacity: 8 Magazine: detachable box Grips: wood					
D2C:	NIB $13,525	Ex $10,300	VG+ $7,440	Good $6,765	
C2C:	Mint $12,990	Ex $9,350	VG+ $6,770	Good $6,090	Fair $3,115
Trade-In:	Mint $9,610	Ex $7,575	VG+ $5,280	Good $4,735	Poor $1,380

1906 AMERICAN EAGLE
(STANDARD 9MM)
EXCELLENT

Courtesy of Rock Island Auction Company

1906 AMERICAN EAGLE (STANDARD 9MM)
REDBOOK CODE: RB-LU-H-1906X3

Serial numbers range between approximately 25,500 and 70,000 with 3,000 or so units produced. "Loaded" engraved on the extractor. No lug in the grip, frame-mounted safety not marked.

Production: 1906 - discontinued	Caliber: 9mm	Action: SA, semi-auto			
Barrel Length: 4" Sights: fixed Capacity: 8 Grips: wood					
D2C:	NIB $4,125	Ex $3,150	VG+ $2,270	Good $2,065	
C2C:	Mint $3,960	Ex $2,850	VG+ $2,070	Good $1,860	Fair $ 950
Trade-In:	Mint $2,930	Ex $2,325	VG+ $1,610	Good $1,445	Poor $ 420

**1906 NAVY MILITARY
FIRST MODEL**
EXCELLENT

Courtesy of Rock Island Auction Company

1906 NAVY MILITARY FIRST MODEL REDBOOK CODE: RB-LU-H-1906X5

Manufactured for the German Navy, less than 12,500 units produced. Marked "Gesichert" behind the safety, DWM logo on the toggle. Doesn't feature a hammer-securing safety mechanism like the 1906 Navy Second Model.

Production: 1906	Caliber: 9mm	Action: SA, semi-auto			
Barrel Length: 6" Sights: fixed Capacity: 8 Magazine: detachable box Grips: wood					
D2C:	NIB $7,215	Ex $5,500	VG+ $3,970	Good $3,610	
C2C:	Mint $6,930	Ex $5,000	VG+ $3,610	Good $3,250	Fair $1,660
Trade-In:	Mint $5,130	Ex $4,050	VG+ $2,820	Good $2,530	Poor $ 750

1906 NAVY MILITARY SECOND MODEL
REDBOOK CODE: RB-LU-H-1906X6

Similar to the 1906 Navy First Model, but features an altered safety device, securing the weapon, albeit a minor detail. An expert should be consulted before purchasing the weapon. Marked "Gesichert" behind the safety, DWM logo on the frame.

**1906 NAVY MILITARY
SECOND MODEL**
GOOD

Courtesy of Rock Island Auction Company

Production: 1906	Caliber: 9mm	Action: SA, semi-auto			
Barrel Length: 6" Sights: fixed Capacity: 8 Magazine: detachable box Grips: wood					
D2C:	NIB $6,915	Ex $5,275	VG+ $3,805	Good $3,460	
C2C:	Mint $6,640	Ex $4,775	VG+ $3,460	Good $3,115	Fair $1,595
Trade-In:	Mint $4,910	Ex $3,875	VG+ $2,700	Good $2,425	Poor $ 720

1906 SWISS COMMERCIAL REDBOOK CODE: RB-LU-H-1906X7

Serial numbers range between approximately 34,950 and 55,050, less than 1,500 manufactured. Marked with the Swiss Cross on the chamber and "Geladen" on the extractor. No lug in the grip (sometimes called stock lug).

**1906 SWISS
COMMERCIAL**
EXCELLENT

Courtesy of Rock Island Auction Company

Production: 1906	Caliber: 7.65mm	Action: SA, semi-auto			
Barrel Length: 4.75" Sights: fixed Capacity: 8 Magazine: detachable box Grips: wood					
D2C:	NIB $4,015	Ex $3,075	VG+ $2,210	Good $2,010	
C2C:	Mint $3,860	Ex $2,775	VG+ $2,010	Good $1,810	Fair $ 925
Trade-In:	Mint $2,860	Ex $2,250	VG+ $1,570	Good $1,410	Poor $ 420

LUGER

1908 NAVY
VERY GOOD +

Courtesy of Rock Island Auction Company

1908 NAVY REDBOOK CODE: RB-LU-H-1908X1

Marked "Crown M" and with an "a" or "b" suffix in the serial number. Less than 45,000 units produced.

Production: 1908	Caliber: 9mm	Action: SA, semi-auto		
Barrel Length: 6" Sights: adjustable Capacity: 8 Magazine: detachable box Grips: wood				
D2C:	NIB $3,665	Ex $2,800	VG+ $2,020 Good $1,835	
C2C:	Mint $3,520	Ex $2,550	VG+ $1,840 Good $1,650	Fair $ 845
Trade-In:	Mint $2,610	Ex $2,075	VG+ $1,430 Good $1,285	Poor $ 390

1908 COMMERCIAL
EXCELLENT

Courtesy of Rock Island Auction Company

1908 COMMERCIAL REDBOOK CODE: RB-LU-H-1908X2

Serial numbers range between approximately 38,500 and 72,000. Manufactured for civilian sales, less than 10,000 initially produced. All markings in German.

Production: 1908	Caliber: 9mm	Action: SA, semi-auto		
Barrel Length: 4" Sights: fixed Capacity: 8 Magazine: detachable box Grips: wood				
D2C:	NIB $2,815	Ex $2,150	VG+ $1,550 Good $1,410	
C2C:	Mint $2,710	Ex $1,950	VG+ $1,410 Good $1,270	Fair $ 650
Trade-In:	Mint $2,000	Ex $1,600	VG+ $1,100 Good $ 990	Poor $ 300

1908 MILITARY
EXCELLENT

Courtesy of Rock Island Auction Company

1908 MILITARY REDBOOK CODE: RB-LU-H-1908X3

Marked "Gesichert" behind the frame-mounted safety. Serial numbers have an "a" suffix. No grip lug and no markings on the chamber.

Production: 1908	Caliber: 9mm	Action: SA, semi-auto		
Barrel Length: 4" Sights: fixed Capacity: 8 Magazine: detachable box Grips: wood				
D2C:	NIB $2,840	Ex $2,175	VG+ $1,565 Good $1,420	
C2C:	Mint $2,730	Ex $1,975	VG+ $1,420 Good $1,280	Fair $ 655
Trade-In:	Mint $2,020	Ex $1,600	VG+ $1,110 Good $ 995	Poor $ 300

1913 COMMERCIAL
EXCELLENT

Courtesy of Rock Island Auction Company

1913 COMMERCIAL REDBOOK CODE: RB-LU-H-1913CM

Serial numbers range between approximately 70,000 and 72,000. These models feature a manual thumb safety. Similar to 1908 Commercial.

Production: ~1913	Caliber: 9mm	Action: SA, semi-auto	Barrel Length: 4"	
Capacity: 8 Magazine: detachable box				
D2C:	NIB $3,215	Ex $2,450	VG+ $1,770 Good $1,610	
C2C:	Mint $3,090	Ex $2,225	VG+ $1,610 Good $1,450	Fair $ 740
Trade-In:	Mint $2,290	Ex $1,825	VG+ $1,260 Good $1,130	Poor $ 330

1920 COMMERCIAL
VERY GOOD +

Courtesy of Rock Island Auction Company

1920 COMMERCIAL REDBOOK CODE: RB-LU-H-1920XX

Serial numbers vary, but most are marked "Germany" or "Made in Germany." Frame-mounted safety, DWM logo on the toggle.

Production: 1920	Caliber: 7.65mm, 9mm	Action: SA, semi-auto		
Barrel Length: 3.5", 4", 4.75", 6" Sights: fixed Capacity: 8 Magazine: detachable box				
Grips: wood				
D2C:	NIB $2,425	Ex $1,850	VG+ $1,335 Good $1,215	
C2C:	Mint $2,330	Ex $1,675	VG+ $1,220 Good $1,095	Fair $ 560
Trade-In:	Mint $1,730	Ex $1,375	VG+ $ 950 Good $ 850	Poor $ 270

1921 DWM/ KRIEGHOFF COMMERCIAL
EXCELLENT

Courtesy of Rock Island Auction Company

1921 DWM/KRIEGHOFF COMMERCIAL

REDBOOK CODE: RB-LU-H-1921XX

Marked "Krieghoff Suhl" on the rear of the frame, DWM logo on the toggle, "i" suffix in the serial number.

Production: 1921	Caliber: 7.65mm, 9mm	Action: SA, semi-auto				
Barrel Length: 4"	Sights: fixed front	Capacity: 8	Grips: wood			
D2C:	NIB $4,235	Ex $3,225	VG+ $2,330	Good $2,120		
C2C:	Mint $4,070	Ex $2,925	VG+ $2,120	Good $1,910	Fair $ 975	
Trade-In:	Mint $3,010	Ex $2,375	VG+ $1,660	Good $1,485	Poor $ 450	

1923 COMMERCIAL
EXCELLENT

Courtesy of Rock Island Auction Company

1923 COMMERCIAL   REDBOOK CODE: RB-LU-H-1923XX

Serial numbers range between approximately 73,000 and 96,500, less than 18,500 units produced. Marked "DWM."

Production: 1923	Caliber: 7.65mm	Action: SA, semi-auto			
Barrel Length: 7.5"	Sights: fixed front	Capacity: 8	Grips: wood		
D2C:	NIB Ex	$ 2,100	VG+ $1,100	Good $ 775	
C2C:	MintEx	$ 1,900	VG+ $1,060	Good $ 705	Fair $ 600
Trade-In:	MintEx	$ 1,475	VG+ $750	Good $ 550	Poor $ 420

1923 STOEGER AMERICAN EAGLE
EXCELLENT

Courtesy of Rock Island Auction Company

1923 STOEGER AMERICAN EAGLE

REDBOOK CODE: RB-LU-H-1923SA

Marked "A.F. Stoeger Inc New York" on the frame, "Gesichert" on the safety, and "Geladen" on the extractor. Manufactured by DWM for A.F. Stoeger Co. Longer barrel variations sell at a premium. In 2013, an 8" version in pristine condition sold for $51,750 at Rock Island Auction.

Production: 1923	Caliber: 7.65mm, 9mm	Action: SA, semi-auto			
Barrel Length: 3.875", 8", 11"	Sights: fixed front, "V" notch rear	Capacity: 8			
Magazine: detachable box	Grips: wood or plastic				
D2C:	NIB --	Ex --	VG+ --	Good --	
C2C:	Mint $18,400	Ex --	VG+ --	Good --	Fair --
Trade-In:	Mint $13,800	Ex --	VG+ --	Good --	Poor --

KRIEGHOFF S CODE
GOOD

Courtesy of Rock Island Auction Company

KRIEGHOFF S CODE   REDBOOK CODE: RB-LU-H-KRGSXX

Marked with an "S" on the chamber and with the Krieghoff logo on the front toggle. Early models have wood grips, while later have plastic.

Production: mid-1930s	Caliber: 9mm	Action: SA, semi-auto			
Barrel Length: 4"	Sights: fixed	Capacity: 8	Grips: wood or plastic		
D2C:	NIB $8,755	Ex $6,675	VG+ $4,820	Good $4,380	
C2C:	Mint $8,410	Ex $6,050	VG+ $4,380	Good $3,940	Fair $2,015
Trade-In:	Mint $6,220	Ex $4,925	VG+ $3,420	Good $3,065	Poor $ 900

KRIEGHOFF 36 DATE   REDBOOK CODE: RB-LU-H-KRG36D

Serial numbers between approximately 3,750 and 4,600. Less than 750 produced. Marked with the Krieghoff logo on the toggle and "36" on the frame.

Production: early 1920s	Caliber: 9mm	Action: SA, semi-auto			
Barrel Length: 4"	Sights: fixed	Capacity: 8	Grips: wood or plastic		
D2C:	NIB $13,450	Ex $10,225	VG+ $7,400	Good $6,725	
C2C:	Mint $12,920	Ex $9,300	VG+ $6,730	Good $6,055	Fair $3,095
Trade-In:	Mint $9,550	Ex $7,550	VG+ $5,250	Good $4,710	Poor $1,350

KRIEGHOFF 36 DATE
EXCELLENT

Courtesy of Rock Island Auction Company

KRIEGHOFF POST-WAR REDBOOK CODE: RB-LU-H-KRGPST

Less than 320 units produced for military forces occupying Germany at the time of production. Most made from salvaged components of older models, denoted by their fairly large-type serial numbers.

Production: mid-1940s Caliber: 9mm Action: SA, semi-auto

Barrel Length: 4" Sights: fixed front Capacity: 8 Magazine: detachable box

Grips: walnut or plastic

D2C:	NIB	$8,125	Ex	$6,175	VG+	$4,470	Good	$4,065	
C2C:	Mint	$7,800	Ex	$5,625	VG+	$4,070	Good	$3,660	Fair $1,870
Trade-In:	Mint	$5,770	Ex	$4,550	VG+	$3,170	Good	$2,845	Poor $ 840

1934 COMMERCIAL MAUSER REDBOOK CODE: RB-LU-H-1934X1

Serial numbers begin around approximately 500,000. Mauser logo on toggle. A couple thousand produced and sold throughout Germany.

Production: 1934 Caliber: 7.65mm, 9mm Action: SA, semi-auto

Barrel Length: 4" Sights: fixed front Capacity: 8 Magazine: detachable box

Grips: walnut or plastic

D2C:	NIB	$4,745	Ex	$3,625	VG+	$2,610	Good	$2,375	
C2C:	Mint	$4,560	Ex	$3,275	VG+	$2,380	Good	$2,140	Fair $1,095
Trade-In:	Mint	$3,370	Ex	$2,675	VG+	$1,860	Good	$1,665	Poor $ 480

1934 GERMAN CONTRACT MAUSER REDBOOK CODE: RB-LU-H-1934X2

Serial numbers begin around approximately 500,000. Several thousand produced for German police and paramilitary forces. Marked with Mauser logo.

Production: late 1930s - 1942 Caliber: 9mm Action: SA, semi-auto Barrel Length: 4"

Sights: fixed front Capacity: 8 Magazine: detachable box Grips: walnut or plastic

D2C:	NIB	$ 4,115	Ex	$3,150	VG+	$2,265	Good	$2,060	
C2C:	Mint	$3,960	Ex	$2,850	VG+	$2,060	Good	$1,855	Fair $ 950
Trade-In:	Mint	$2,930	Ex	$2,325	VG+	$1,610	Good	$1,445	Poor $ 420

S/42 K DATE
EXCELLENT

Courtesy of Rock Island Auction Company

S/42 K DATE REDBOOK CODE: RB-LU-H-43KXXX

Less than 11,000 produced. Serial numbers vary. "K" indicates 1934 as its year of production. Gothic-style type rather than script. Indicativ t Capacity: 8 Grips: walnut

D2C:	NIB	$3,665	Ex	$2,800	VG+	$2,020	Good	$1,835	
C2C:	Mint	$3,520	Ex	$2,550	VG+	$1,840	Good	$1,650	Fair $ 845
Trade-In:	Mint	$2,610	Ex	$2,075	VG+	$1,430	Good	$1,285	Poor $ 390

S/42 G DATE
VERY GOOD +

Courtesy of Rock Island Auction Company

S/42 G DATE REDBOOK CODE: RB-LU-H-42GXXX

Serial numbers begin around approximately 506,500. Widely produced, features a "G" to denote 1935 as its year of production. Script type, rather than gothic. Simplified design.

Production: 1935 Caliber: 9mm Action: SA, semi-auto

Barrel Length: 4" Sights: fixed front Capacity: 8 Grips: walnut

D2C:	NIB	$3,360	Ex	$2,575	VG+	$1,850	Good	$1,680	
C2C:	Mint	$3,230	Ex	$2,325	VG+	$1,680	Good	$1,515	Fair $ 775
Trade-In:	Mint	$2,390	Ex	$1,900	VG+	$1,320	Good	$1,180	Poor $ 360

Magnum Research

Israel Military Industries formed Magnum Research, Inc. in the 1980s, which led to the development of the Desert Eagle pistol, one of the most recognizable high-power, semi-automatic pistols to ever enter production. Kahr purchased Magnum Research in 2010 and the company continues to manufacture Desert Eagles, variants of the Desert Eagle, and a series of large-caliber revolvers.

BFR .30/30 WIN
NEW IN BOX

BFR .30/30 WIN   REDBOOK CODE: RB-MR-H-BFR30W
Stainless steel frame and finish, long cylinder, cut rifled barrel, transfer-bar safety.

Production: 2014 - current   Caliber: .30-30 Win.   Action: SA, revolver
Barrel Length: 7.5"   Sights: fixed front, rear adjustable   Capacity: 5   Grips: rubber

D2C:	NIB	$ 865	Ex $ 675	VG+ $ 480	Good $ 435	LMP $1,050
C2C:	Mint $ 840	Ex $ 600	VG+ $ 440	Good $ 390	Fair $ 200
Trade-In:	Mint $ 620	Ex $ 500	VG+ $ 340	Good $ 305	Poor $ 90

BFR .44 MAGNUM REVOLVER
NEW IN BOX

BFR .44 MAGNUM REVOLVER   REDBOOK CODE: RB-MR-H-BFR44M
Stainless steel frame and slide, cut rifled barrel, short cylinder.

Production: current   Caliber: .44 Magnum   Action: SA, revolver   Barrel Length: 5"
OA Length: 11.25"   Sights: fixed front, rear adjustable   Capacity: 5   Grips: rubber

D2C:	NIB	$ 865	Ex $ 675	VG+ $ 480	Good $ 435	LMP $1,050
C2C:	Mint $ 840	Ex $ 600	VG+ $ 440	Good $ 390	Fair $ 200
Trade-In:	Mint $ 620	Ex $ 500	VG+ $ 340	Good $ 305	Poor $ 90

BFR .454 CASULL
NEW IN BOX

BFR .454 CASULL   REDBOOK CODE: RB-MR-H-BFR454
Stainless steel frame and finish, cut rifled barrel.

Production: 1999 - current   Caliber: .454 Casull   Action: SA, revolver
Barrel Length: 6.5", 7.5"   OA Length: 15"   Sights: fixed front, adjustable rear
Capacity: 5   Grips: rubber

D2C:	NIB	$ 890	Ex $ 700	VG+ $ 490	Good $ 445	LMP $1,050
C2C:	Mint $ 860	Ex $ 625	VG+ $ 450	Good $ 405	Fair $ 205
Trade-In:	Mint $ 640	Ex $ 500	VG+ $ 350	Good $ 315	Poor $ 90

**BFR .480 RUGER/
.475 LINEBAUGH**
NEW IN BOX

BFR .480 RUGER/.475 LINEBAUGH   REDBOOK CODE: RB-MR-H-BFR480
Stainless steel frame and finish, short cylinder, cut rifled barrel.

Production: 2002 - current   Caliber: .480 Ruger/.475 Linebaugh
Action: SA, revolver   Barrel Length: 6.5", 7.5"   Sights: fixed front, adjustable rear
Capacity: 5   Grips: rubber

D2C:	NIB	$ 890	Ex $ 700	VG+ $ 490	Good $ 445	LMP $1,050
C2C:	Mint $ 860	Ex $ 625	VG+ $ 450	Good $ 405	Fair $ 205
Trade-In:	Mint $ 640	Ex $ 500	VG+ $ 350	Good $ 315	Poor $ 90

BFR .460 S&W
NEW IN BOX

BFR .460 S&W   REDBOOK CODE: RB-MR-H-BFR460
Stainless steel frame and finish, long cylinder, cut rifled barrel.

Production: 2006 - current   Caliber: .460 S&W Magnum
Action: SA, revolver   Barrel Length: 7.5", 10"   Sights: fixed front, adjustable rear
Capacity: 5   Grips: rubber

D2C:	NIB	$ 880	Ex $ 675	VG+ $ 485	Good $ 440	LMP $1,050
C2C:	Mint $ 850	Ex $ 625	VG+ $ 440	Good $ 400	Fair $ 205
Trade-In:	Mint $ 630	Ex $ 500	VG+ $ 350	Good $ 310	Poor $ 90

MAGNUM RESEARCH

BFR .500 JRH
NEW IN BOX

BFR .500 JRH   REDBOOK CODE: RB-MR-H-BFR500

Stainless steel frame and slide, short cylinder, cut rifled barrel.

Production: current	Caliber: .500 JRH	Action: SA, revolver		
Barrel Length: 5.5"	Sights: fixed front, adjustable rear	Capacity: 5	Grips: rubber	
D2C:	NIB $ 880	Ex $ 675	VG+ $ 485	Good $ 440	LMP $1,050
C2C:	Mint $ 850	Ex $ 625	VG+ $ 440	Good $ 400	Fair $ 205
Trade-In:	Mint $ 630	Ex $ 500	VG+ $ 350	Good $ 310	Poor $ 90

**BABY DESERT
EAGLE II .40 S&W**
NEW IN BOX

BABY DESERT EAGLE II .40 S&W

REDBOOK CODE: RB-MR-H-BDE40S

Black oxide finish, tactical rail, slide-mounted decocking safety, steel or polymer frame, also available in a semi-compact model.

Production: 2012 - current	Caliber: .40 S&W	Action: DA/SA, semi-auto		
Barrel Length: 4.52", 3.93"	Sights: combat, fixed white 3-dot			
Capacity: 12	Magazine: detachable box	Grips: black polymer		
D2C:	NIB $ 570	Ex $ 450	VG+ $ 315	Good $ 285	LMP $ 629
C2C:	Mint $ 550	Ex $ 400	VG+ $ 290	Good $ 260	Fair $ 135
Trade-In:	Mint $ 410	Ex $ 325	VG+ $ 230	Good $ 200	Poor $ 60

**BABY DESERT
EAGLE II 9MM**
NEW IN BOX

BABY DESERT EAGLE II 9MM   REDBOOK CODE: RB-MR-H-BDE9MM

Black oxide finish, tactical rail, either steel or polymer frame, slide-mounted decocking safety, available in full-size, semi-compact, and compact models.

Production: 2012 - current	Caliber: 9mm	Action: DA/SA, semi-auto		
Barrel Length: 4.52", 3.93", 3.64"	Sights: combat, fixed white 3-dot			
Capacity: 15	Magazine: detachable box	Grips: black polymer		
D2C:	NIB $ 570	Ex $ 450	VG+ $ 315	Good $ 285	LMP $ 656
C2C:	Mint $ 550	Ex $ 400	VG+ $ 290	Good $ 260	Fair $ 135
Trade-In:	Mint $ 410	Ex $ 325	VG+ $ 230	Good $ 200	Poor $ 60

**BABY DESERT
EAGLE II .45 ACP**
NEW IN BOX

BABY DESERT EAGLE II .45 ACP

REDBOOK CODE: RB-MR-H-BDE45A

Black oxide finish, steel frame, semi-compact, tactical rail, slide-mounted decocking safety.

Production: 2012 - current	Caliber: .45 ACP	Action: DA/SA, semi-auto		
Barrel Length: 3.93"	Sights: combat, fixed white 3-dot	Capacity: 10		
Magazine: detachable box	Grips: black polymer			
D2C:	NIB $ 570	Ex $ 450	VG+ $ 315	Good $ 285	LMP $ 656
C2C:	Mint $ 550	Ex $ 400	VG+ $ 290	Good $ 260	Fair $ 135
Trade-In:	Mint $ 410	Ex $ 325	VG+ $ 230	Good $ 200	Poor $ 60

**MICRO
DESERT EAGLE**
NEW IN BOX

MICRO DESERT EAGLE   REDBOOK CODE: RB-MR-H-MICDEA

Nickel finish, gas-assisted blowback system, also available in blue and nickel finish with a blue slide and barrel.

Production: 2009 - current	Caliber: .380 ACP	Action: DAO, semi-auto		
Barrel Length: 2.22"	OA Length: 4.52"	Wt.: 14 oz.	Sights: fixed	
Capacity: 6	Magazine: detachable box	Grips: polymer		
D2C:	NIB $ 430	Ex $ 350	VG+ $ 240	Good $ 215	LMP $ 479
C2C:	Mint $ 420	Ex $ 300	VG+ $ 220	Good $ 195	Fair $ 100
Trade-In:	Mint $ 310	Ex $ 250	VG+ $ 170	Good $ 155	Poor $ 60

MOUNTAIN EAGLE
EXCELLENT

MOUNTAIN EAGLE REDBOOK CODE: RB-MR-H-MEEAGL

Blue/polymer finish, alloy receiver.

Production: early 1990s Caliber: .22 LR Action: SA, semi-auto

Barrel Length: 6.5", 8" Wt.: 20 oz. Sights: ramp front, adjustable rear Capacity: 15

Magazine: detachable box Grips: checkered polymer

D2C:	NIB $	205	Ex $	175	VG+ $	115	Good $	105	
C2C:	Mint $	200	Ex $	150	VG+ $	110	Good $	95	Fair $ 50
Trade-In:	Mint $	150	Ex $	125	VG+ $	80	Good $	75	Poor $ 30

MOUNTAIN EAGLE COMPACT EDITION

REDBOOK CODE: RB-MR-H-MECOME

Black oxide finish, features a shorter grip, and adjustable rear sight.

Production: mid-1990s Caliber: .22 LR Action: SA, semi-auto

Barrel Length: 4.5" Wt.: 19 oz. Sights: adjustable rear Capacity: 15

Magazine: detachable box Grips: checkered polymer

D2C:	NIB $	189	Ex $	150	VG+ $	105	Good $	95	
C2C:	Mint $	190	Ex $	150	VG+ $	100	Good $	90	Fair $ 45
Trade-In:	Mint $	140	Ex $	125	VG+ $	80	Good $	70	Poor $ 30

MOUNTAIN EAGLE TARGET EDITION

REDBOOK CODE: RB-MR-H-MEAGTE

Interchangeable blades, two-stage target trigger, black finish.

Production: early 1990s Caliber: .22 LR Action: SA, semi-auto

Barrel Length: 8" Wt.: 20 oz. Sights: adjustable rear Capacity: 15

Magazine: detachable box Grips: checkered polymer

D2C:	NIB $	220	Ex $	175	VG+ $	125	Good $	110	
C2C:	Mint $	220	Ex $	175	VG+ $	110	Good $	100	Fair $ 55
Trade-In:	Mint $	160	Ex $	125	VG+ $	90	Good $	80	Poor $ 30

IMI SP-21 REDBOOK CODE: RB-MR-H-IMISP2

Black finish, polymer frame, steel slide and barrel, tactical rail, made in Israel.

Production: discontinued Caliber: 9mm, .40 S&W, .45 ACP Action: DA/SA, semi-auto

Barrel Length: 4" Wt.: 26 oz. Sights: 3-dot Capacity: 10

Magazine: detachable box Grips: polymer

D2C:	NIB $	575	Ex $	450	VG+ $	320	Good $	290	
C2C:	Mint $	560	Ex $	400	VG+ $	290	Good $	260	Fair $ 135
Trade-In:	Mint $	410	Ex $	325	VG+ $	230	Good $	205	Poor $ 60

**DESERT EAGLE
MARK XIX .50 AE**
NEW IN BOX

DESERT EAGLE MARK XIX .50 AE REDBOOK CODE: RB-MR-H-DEMX50

Interchangeable barrels, carbon steel barrel, Picatinny rail, rotating bolt, black oxide finish, many custom finishes available, optional 10" barrel and muzzle brake.

Production: 1996 - current Caliber: .50 AE Action: SA, semi-auto

Barrel Length: 6" Sights: fixed, combat Capacity: 7 Magazine: detachable box

Grips: polymer

D2C:	NIB $1,360	Ex $1,050	VG+ $ 750	Good $ 680	LMP $1,594
C2C:	Mint $1,310	Ex $ 950	VG+ $ 680	Good $ 615	Fair $ 315
Trade-In:	Mint $ 970	Ex $ 775	VG+ $ 540	Good $ 480	Poor $ 150

DESERT EAGLE MARK XIX .44 MAG.
NEW IN BOX

DESERT EAGLE MARK XIX .357 MAG.
NEW IN BOX

Courtesy of Bud's Gun Shop

DESERT EAGLE MARK XIX 25TH ANNIVERSARY
NEW IN BOX

DESERT EAGLE 1911 U MODEL
NEW IN BOX

DESERT EAGLE 1911 G MODEL
NEW IN BOX

DESERT EAGLE MARK XIX .44 MAG.

REDBOOK CODE: RB-MR-H-DEMX44

Interchangeable carbon steel barrels, Picatinny rail, rotating bolt, black oxide finish, many custom finishes, optional 10" barrel and muzzle brake.

Production: current   Caliber: .44 Magnum   Action: SA, semi-auto   Barrel Length: 6"
Sights: fixed, combat   Capacity: 8   Magazine: detachable box   Grips: polymer

D2C:	NIB	$1,410	Ex	$1,075	VG+	$780	Good	$705	LMP $1,563
C2C:	Mint	$1,360	Ex	$975	VG+	$710	Good	$635	Fair $325
Trade-In:	Mint	$1,010	Ex	$800	VG+	$550	Good	$495	Poor $150

DESERT EAGLE MARK XIX .357 MAG.

REDBOOK CODE: RB-MR-H-DEM357

Interchangeable carbon steel barrels, Picatinny rail, rotating bolt, black oxide finish, many custom finishes available.

Production: current   Caliber: .357 Magnum   Action: SA, semi-auto   Barrel Length: 6"
Sights: fixed, combat   Capacity: 9   Magazine: detachable box   Grips: polymer

D2C:	NIB	$1,360	Ex	$1,050	VG+	$750	Good	$680	
C2C:	Mint	$1,310	Ex	$950	VG+	$680	Good	$615	Fair $315
Trade-In:	Mint	$970	Ex	$775	VG+	$540	Good	$480	Poor $150

DESERT EAGLE MARK XIX 25TH ANNIVERSARY

REDBOOK CODE: RB-MR-H-DEM25A

Made in Israel, matte stainless finish with steel frame, two-stage dry-firing trigger.

Production: 2010   Caliber: .50 AE, .44 Magnum, .357 Magnum   Action: SA, semi-auto
Barrel Length: 6"   Wt.: 70 oz.   Sights: fixed   Capacity: 7 (.50 AE), 8 (.44 Magnum),
9 (.357 Magnum)   Magazine: detachable box   Grips: checkered wood

D2C:	NIB	$2,300	Ex	$1,750	VG+	$1,265	Good	$1,150	
C2C:	Mint	$2,210	Ex	$1,600	VG+	$1,150	Good	$1,035	Fair $530
Trade-In:	Mint	$1,640	Ex	$1,300	VG+	$900	Good	$805	Poor $240

DESERT EAGLE 1911 U MODEL   REDBOOK CODE: RB-MR-H-DEUMDL

Aluminum alloy frame, carbon steel slide, black finish, skeletonized hammer, aluminum trigger, bull barrel, beavertail safety, extended magazine release, extended thumb safety, stainless steel guide rod.

Production: 2014 - current   Caliber: .45 ACP   Action: SA, semi-auto
Barrel Length: 3"   OA Length: 6.85"   Wt.: 26 oz.   Sights: adjustable rear
Capacity: 6   Magazine: detachable box   Grips: double-diamond checkered wood

D2C:	NIB	$765	Ex	$600	VG+	$425	Good	$385	LMP $946
C2C:	Mint	$740	Ex	$550	VG+	$390	Good	$345	Fair $180
Trade-In:	Mint	$550	Ex	$450	VG+	$300	Good	$270	Poor $90

DESERT EAGLE 1911 G MODEL   REDBOOK CODE: RB-MR-H-DEGMDL

Black finish, steel frame, skeletonized trigger, extended magazine release, extended thumb safety, stainless steel guide rod, beavertail grip safety, aluminum trigger, beveled magazine well.

Production: 2010 - current   Caliber: .45 ACP   Action: SA, semi-auto
Barrel Length: 5", 4.33"   OA Length: 8.6"   Wt.: 36 oz.
Sights: pinned-in front, adjustable rear   Capacity: 8   Magazine: detachable box
Grips: double-diamond checkered wood

D2C:	NIB	$670	Ex	$525	VG+	$370	Good	$335	LMP $800
C2C:	Mint	$650	Ex	$475	VG+	$340	Good	$305	Fair $155
Trade-In:	Mint	$480	Ex	$400	VG+	$270	Good	$235	Poor $90

**DESERT EAGLE
1911 C MODEL**
NEW IN BOX

DESERT EAGLE 1911 C MODEL   REDBOOK CODE: RB-MR-H-DECMDL

Black finish, steel frame, skeletonized trigger, extended magazine release, extended thumb safety, stainless steel guide rod.

Production: 2010 - current   Caliber: .45 ACP   Action: SA, semi-auto   Barrel Length: 4.33"   OA Length: 7.8"   Wt.: 36 oz.   Sights: pinned-in front, adjustable rear   Capacity: 8   Magazine: detachable box   Grips: wood

		NIB		Ex		VG+		Good		LMP/Fair/Poor	
D2C:	NIB $	690	Ex $	525	VG+ $	380	Good $	345	LMP $	874
C2C:	Mint $	670	Ex $	500	VG+ $	350	Good $	315	Fair $	160
Trade-In:	Mint $	490	Ex $	400	VG+ $	270	Good $	245	Poor $	90

MR9 EAGLE
NEW IN BOX

MR9 EAGLE   REDBOOK CODE: RB-MR-H-MR9EAX

Polymer frame, Picatinny rail, matte stainless slide with engraved markings, manual decocker, replaceable front sight blades, replaceable palm swells.

Production: current   Caliber: 9mm   Action: striker-fired, semi-auto
Barrel Length: 4.5"   Wt.: 25 oz.   Sights: white 3-dot, adjustable rear
Capacity: 10, 15   Magazine: detachable box   Grips: polymer grip frame

D2C:	NIB $	499	Ex $	400	VG+ $	275	Good $	250	LMP $	559
C2C:	Mint $	480	Ex $	350	VG+ $	250	Good $	225	Fair $	115
Trade-In:	Mint $	360	Ex $	300	VG+ $	200	Good $	175	Poor $	60

MR40 EAGLE
NEW IN BOX

MR40 EAGLE   REDBOOK CODE: RB-MR-H-MR40EA

Polymer frame, matte stainless slide with engraved markings, manual striker decocker, replaceable front sight blades, replaceable palm swells.

Production: current   Caliber: .40 S&W   Action: striker-fired, semi-auto
Barrel Length: 4.5"   Sights: white 3-dot, adjustable rear
Capacity: 10, 11   Magazine: detachable box   Grips: polymer grip frame

D2C:	NIB $	499	Ex $	400	VG+ $	275	Good $	250	LMP $	559
C2C:	Mint $	480	Ex $	350	VG+ $	250	Good $	225	Fair $	115
Trade-In:	Mint $	360	Ex $	300	VG+ $	200	Good $	175	Poor $	60

Mauser

Now somewhat defunct, the German-based Mauser produced a series of derringers, pistols, revolvers, and rifles from the 1870s until 2000 or so. The company built many firearms for the German military, most notably the Model 1898 rifle, the Pocket Model pistol, and the C96 pistol. In 2000, the original Mauser imprint was sold to SIG Sauer, though a division of the company still manufactures hunting rifles. Original Mauser handguns have retained moderate-to-substantial value over the years.

M2   REDBOOK CODE: RB-MA-H-M2XXXX

Short recoil, blue finish, alloy frame and steel slide, manual safety at rear of the frame, rotating barrel, loaded chamber indicator, imported by SIG Sauer.

Production: 2000   Caliber: .40 S&W, .357 SIG, .45 ACP
Action: striker-fired, semi-auto   Barrel Length: 3.5"   OA Length: detachable box
Wt.: 32.7 oz.   Sights: combat   Capacity: 8, 10   Grips: black plastic

D2C:	NIB $	525	Ex $	400	VG+ $	290	Good $	265		
C2C:	Mint $	510	Ex $	375	VG+ $	270	Good $	240	Fair $	125
Trade-In:	Mint $	380	Ex $	300	VG+ $	210	Good $	185	Poor $	60

MODEL 80 REDBOOK CODE: RB-MA-H-MDL80X
Blue finish, similar to Browning Hi-Power, manufactured by FEG of Hungary.

Production: early 1990s	Caliber: 9mm	Action: SA, semi-auto			
Barrel Length: 4.7"	OA Length: detachable box	Wt.: 35 oz.	Sights: fixed		
Capacity: 14	Grips: walnut				
D2C:	NIB $ 530	Ex $ 425	VG+ $ 295	Good $ 265	
C2C:	Mint $ 510	Ex $ 375	VG+ $ 270	Good $ 240	Fair $ 125
Trade-In:	Mint $ 380	Ex $ 300	VG+ $ 210	Good $ 190	Poor $ 60

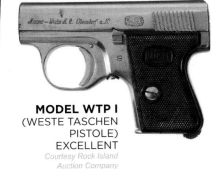

MODEL WTP I
(WESTE TASCHEN
PISTOLE)
EXCELLENT
*Courtesy Rock Island
Auction Company*

MODEL WTP I (WESTE TASCHEN PISTOLE) REDBOOK CODE: RB-MA-H-WTPXXX
Vest pocket pistol, blue finish, steel frame, wraparound grips, made in Germany, "W.T.P. - 6.35 - D.R.G." marking on slide.

Production: 1921 - 1938	Caliber: .25 ACP	Action: SA, semi-auto			
Barrel Length: 2.5"	OA Length: detachable box	Wt.: 12 oz.	Sights: notch groove		
Capacity: 6	Grips: checkered, hard plastic				
D2C:	NIB $ 710	Ex $ 550	VG+ $ 395	Good $ 355	
C2C:	Mint $ 690	Ex $ 500	VG+ $ 360	Good $ 320	Fair $ 165
Trade-In:	Mint $ 510	Ex $ 400	VG+ $ 280	Good $ 250	Poor $ 90

MODEL WTP II (WESTE TASCHEN PISTOLE) REDBOOK CODE: RB-MA-H-WTPIIX
Blue finish, compact design, steel frame, made in Germany, signal pin, separate grip panels, smaller grips than the WTP, "T. - 6.35" marking on the slide.

Production: late 1930s	Caliber: .25 ACP	Action: SA, semi-auto	Barrel Length: 2"		
Magazine: detachable box	Wt.: 10.5 oz.	Sights: notch groove	Capacity: 6		
Grips: checkered plastic					
D2C:	NIB $1,620	Ex $1,250	VG+ $ 895	Good $ 810	
C2C:	Mint $1,560	Ex $1,125	VG+ $ 810	Good $ 730	Fair $ 375
Trade-In:	Mint $1,160	Ex $ 925	VG+ $ 640	Good $ 570	Poor $ 180

**MODEL 1878
ZIG-ZAG REVOLVER**
EXCELLENT

MODEL 1878 ZIG-ZAG REVOLVER REDBOOK CODE: RB-MA-H-1878ZR
Zig-zag pattern on the cylinder, ring cylinder locking lever, original model featured a loading gate and solid frame, "GEBR.MAUSER & CIE OBERNDORF A/N. WURTTEMBERG 1878" on top of barrel, premium for 10.6mm models.

Production: 1878	Caliber: 7.6mm, 9mm, 10.6mm	Action: SA, revolver			
Barrel Length: 5.5"	Magazine: detachable box	Sights: fixed	Capacity: 6		
Grips: hard rubber or walnut checkered					
D2C:	NIB	Ex $7,000	VG+ $4,100	Good $2,900	
C2C:	Mint	Ex $6,400	VG+ $3,700	Good $2,600	Fair $ 900
Trade-In:	Mint	Ex $5,250	VG+ $3,100	Good $2,200	Poor $ 800

POCKET MODEL 1910
VERY GOOD +

POCKET MODEL 1910 REDBOOK CODE: RB-MA-H-PM1910
Blue or factory nickel finish, blowback operated, add 80% for early "sidelatch" model.

Production: 1910 - 1934	Caliber: .25 ACP	Action: SA, semi-auto			
Barrel Length: 3.14"	Magazine: detachable box	Wt.: 15 oz.	Sights: blade front,		
notch rear	Capacity: 9	Grips: checkered wood, black plastic, or hard rubber			
D2C:	NIB $ 720	Ex $ 550	VG+ $ 400	Good $ 360	
C2C:	Mint $ 700	Ex $ 500	VG+ $ 360	Good $ 325	Fair $ 170
Trade-In:	Mint $ 520	Ex $ 425	VG+ $ 290	Good $ 255	Poor $ 90

POCKET MODEL 1914
VERY GOOD +

POCKET MODEL 1934
EXCELLENT

Courtesy Rock Island Auction Company

POCKET MODEL 1914 REDBOOK CODE: RB-MA-H-PM1914

Blowback operated, factory nickel or blue finish, early models called "Humpbacks," distinguished by the hump on top of the slide, worth approx. 10X in value.

Production: 1914 - 1934 Caliber: .32 ACP (7.65mm) Action: SA, semi-auto
Barrel Length: 3.5", 4.5" Magazine: detachable box Wt.: 21 oz.
Sights: blade front, notch rear Capacity: 8 Grips: walnut

D2C:	NIB $ 770	Ex $ 600	VG+ $ 425	Good $ 385		
C2C:	Mint $ 740	Ex $ 550	VG+ $ 390	Good $ 350	Fair $ 180	
Trade-In:	Mint $ 550	Ex $ 450	VG+ $ 310	Good $ 270	Poor $ 90	

POCKET MODEL 1934 REDBOOK CODE: RB-MA-H-PM1934

Distinguished by its curved ergonomic grip, military and police-marked models have increased value.

Production: 1934 Caliber: .32 ACP (7.65mm) Action: SA, semi-auto
Barrel Length: 3.5" Magazine: detachable box Wt.: 21 oz.
Sights: blade front, notch rear Capacity: 8 Grips: walnut

D2C:	NIB $ 885	Ex $ 675	VG+ $ 490	Good $ 445		
C2C:	Mint $ 850	Ex $ 625	VG+ $ 450	Good $ 400	Fair $ 205	
Trade-In:	Mint $ 630	Ex $ 500	VG+ $ 350	Good $ 310	Poor $ 90	

Nighthawk Custom

Based in Berryville, Arkansas, Nighthawk Custom specializes in high-end, premier M1911 pistols, which many handgun experts and critics note as some of the best available. The company formed in 2003 when several Wilson Combat designers and gunsmiths left to start their own venture. Since then, Nighthawk has become a favorite among M1911 enthusiasts, given the detail and quality of each of their guns. The company manufactures each of their weapons in-house and can make custom upgrades per customer specs. In 2012, the company began machining all of their parts from solid-billet metals, further improving their firearms' quality.

AAC REDBOOK CODE: RB-NC-H-AACXXX

Proprietary frame and slide cuts, forged frame and slide, blacked-out threaded barrel with thread protector, match grade blacked-out trigger, AAC logos, forged slide stop-cut flush with chamfered frame, complete de-horn treatment.

Production: current Caliber: .45 ACP, 9mm Action: SA, semi-auto
Barrel Length: 5" Wt.: 38.9 oz. Sights: Heinie Ledge Straight-Eight Suppressor
Capacity: 7, 8, 9, 10 Magazine: detachable box
Grips: ultra-thin Aluma grips with Nighthawk logo

	NIB	Ex	VG+	Good	LMP/Fair/Poor
D2C:	NIB $3,295	Ex $2,525	VG+ $1,815	Good $1,650	LMP $3,295
C2C:	Mint $3,170	Ex $2,275	VG+ $1,650	Good $1,485	Fair $ 760
Trade-In:	Mint $2,340	Ex $1,850	VG+ $1,290	Good $1,155	Poor $ 330

AAC RECON REDBOOK CODE: RB-NC-H-AACREC

Similar to AAC model but with added tactical rail.

Production: current Caliber: .45 ACP, 9mm Action: SA, semi-auto
Barrel Length: 5" Wt.: 40.4 oz. Sights: Heinie Ledge Straight-Eight Suppressor
Capacity: 7, 8, 9, 10 Magazine: detachable box
Grips: ultra-thin Aluma grips with Nighthawk logo

	NIB	Ex	VG+	Good	LMP/Fair/Poor
D2C:	NIB $3,395	Ex $2,600	VG+ $1,870	Good $1,700	LMP $3,395
C2C:	Mint $3,260	Ex $2,350	VG+ $1,700	Good $1,530	Fair $ 785
Trade-In:	Mint $2,420	Ex $1,925	VG+ $1,330	Good $1,190	Poor $ 360

AAC RECON
NEW IN BOX

BOB MARVEL CUSTOM
NEW IN BOX

BOB MARVEL CUSTOM REDBOOK CODE: RB-NC-H-BOBMAC

Hand stippled top of slide, bull nose front taper, proprietary barrel, French border on top of slide, heavy bevel on bottom of slide, one-piece mainspring housing/magwell, high cut front strap, aluminum match trigger, Nighthawk Custom/Marvel EVERLAST Recoil System, tactical single-side safety, black melonite finish, all stainless upgrade available.

Production: current Caliber: .45 ACP Action: SA, semi-auto Barrel Length: 4.25"

Wt.: 37.6 oz. Sights: fully adjustable recessed Bob Marvel Capacity: 7,8

Magazine: detachable box Grips: G10

D2C:	NIB $3,995	Ex $3,050	VG+ $2,200	Good $2,000	LMP $3,995
C2C:	Mint $3,840	Ex $2,775	VG+ $2,000	Good $1,800	Fair $ 920
Trade-In:	Mint $2,840	Ex $2,250	VG+ $1,560	Good $1,400	Poor $ 420

COSTA COMPACT
NEW IN BOX

COSTA COMPACT REDBOOK CODE: RB-NC-H-COSCOM

Tri-cut slide, serrated arrow-style slide top, heavy bevel on bottom of slide, hand serrated rear of slide, crowned barrel is beveled flush with bushing, magwell, high cut front strap, aluminum match trigger, extended tactical mag catch, Nighthawk Custom/Marvel EVERLAST Recoil System, COSTA logo engraved in silver.

Production: current Caliber: .45 ACP Action: SA, semi-auto Barrel Length: 4.25"

Wt.: 34.7 oz. Sights: red fiber-optic front, Heinie Slant Pro rear Capacity: 7,8

Magazine: detachable box Grips: black 10-8 Performance 5 lpi grips

D2C:	NIB $3,695	Ex $2,825	VG+ $2,035	Good $1,850	LMP $3,695
C2C:	Mint $3,550	Ex $2,550	VG+ $1,850	Good $1,665	Fair $ 850
Trade-In:	Mint $2,630	Ex $2,075	VG+ $1,450	Good $1,295	Poor $ 390

COSTA RECON
NEW IN BOX

COSTA RECON REDBOOK CODE: RB-NC-H-COSREC

Integrated Recon Light Rail, multi-faceted slide, serrated arrow-style slide top, heavy bevel on bottom of slide, hand serrated rear of slide, crowned barrel is beveled flush with bushing, magwell, high cut front strap, aluminum match trigger, extended tactical mag catch, Nighthawk Custom/Marvel EVERLAST Recoil System, COSTA logo engraved in silver.

Production: current Caliber: .45 ACP, 9mm Action: SA, semi-auto

Barrel Length: 5" Wt.: 40.6 oz. Sights: red fiber-optic front, Jardine rear

Capacity: 7, 8, 9, 10 Magazine: detachable box

Grips: 10-8 Performance Hyena Brown grips

D2C:	NIB $3,695	Ex $2,825	VG+ $2,035	Good $1,850	LMP $3,695
C2C:	Mint $3,550	Ex $2,550	VG+ $1,850	Good $1,665	Fair $ 850
Trade-In:	Mint $2,630	Ex $2,075	VG+ $1,450	Good $1,295	Poor $ 390

DOMINATOR REDBOOK CODE: RB-NC-H-DOMINA

Crowned barrel is beveled flush with the bushing, stainless steel frame with black nitride slide, serrated slide top, 25 lpi checkering on front strap and mainspring housing, serrated rear of slide, complete de-horn, front and rear cocking serrations, tactical single-side safety.

Production: current Caliber: .45 ACP Action: SA, semi-auto Barrel Length: 5"

Wt.: 37.6 oz. Sights: fully-adjustable night Capacity: 7, 8 Magazine: detachable box

Grips: cocobolo grips with Nighthawk logo

D2C:	NIB $3,250	Ex $2,475	VG+ $1,790	Good $1,625	LMP $3,250
C2C:	Mint $3,120	Ex $2,250	VG+ $1,630	Good $1,465	Fair $ 750
Trade-In:	Mint $2,310	Ex $1,825	VG+ $1,270	Good $1,140	Poor $ 330

ENFORCER
NEW IN BOX

ENFORCER   REDBOOK CODE: RB-NC-H-ENFORC

Fully machined one-piece mainspring housing and magwell, lanyard loop cutout, ball cut radius under cocking serrations, serrated slide top cut flush to frame, chamfered slide stop hole, crowned barrel, hand serrated rear slide, tactical single-side safety.

Production: current Caliber: .45 ACP Action: SA, semi-auto Barrel Length: 5"
Wt.: 39 oz. Sights: Heinie Slant Pro Straight-Eight Night Capacity: 7, 8
Magazine: detachable box Grips: golf-ball dimple pattern G10 or Mil-Tac G10 Spiral
Logo in either black and gray or black and green

D2C:	NIB	$3,395	Ex	$2,600	VG+	$1,870	Good	$1,700	LMP $3,395
C2C:	Mint	$3,260	Ex	$2,350	VG+	$1,700	Good	$1,530	Fair $ 785
Trade-In:	Mint	$2,420	Ex	$1,925	VG+	$1,330	Good	$1,190	Poor $ 360

FALCON
NEW IN BOX

FALCON   REDBOOK CODE: RB-NC-H-FALCON

One-piece fully machined mainspring/magwell, three ball-radius cuts on top of slide, rear cocking serrations, chamfered extra-thick barrel bushing, recessed front slide, crowned barrel, custom barrel options, custom grips available.

Production: current Caliber: .45 ACP, 9mm, 10mm Action: SA, semi-auto
Barrel Length: 5" Wt.: 38.8 oz. Sights: Heinie Ledge Capacity: 7, 8, 9, 10
Magazine: detachable box Grips: golf-ball dimple pattern G10

D2C:	NIB	$3,295	Ex	$2,525	VG+	$1,815	Good	$1,650	LMP $3,295
C2C:	Mint	$3,170	Ex	$2,275	VG+	$1,650	Good	$1,485	Fair $ 760
Trade-In:	Mint	$2,340	Ex	$1,850	VG+	$1,290	Good	$1,155	Poor $ 330

FALCON COMMANDER   REDBOOK CODE: RB-NC-H-FALCOM

Commander size, one-piece fully machined mainspring/magwell, three ball-radius cuts on top of slide, rear cocking serrations, chamfered extra thick barrel bushing, recessed front slide, crowned barrel, custom barrel options, custom grips available.

Production: current Caliber: .45 ACP Action: SA, semi-auto Barrel Length: 4.25"
Wt.: 36.7 oz. Sights: Heinie Ledge Capacity: 7, 8 Magazine: detachable box
Grips: golf-ball dimple pattern G10

D2C:	NIB	$3,295	Ex	$2,525	VG+	$1,815	Good	$1,650	LMP $3,295
C2C:	Mint	$3,170	Ex	$2,275	VG+	$1,650	Good	$1,485	Fair $ 760
Trade-In:	Mint	$2,340	Ex	$1,850	VG+	$1,290	Good	$1,155	Poor $ 330

GRP
NEW IN BOX

GRP   REDBOOK CODE: RB-NC-H-GRPXXX

Lanyard loop, forged-slide stop axle cut flush with frame, chamfered around frame, tactical single-side safety, custom Perma Kote colors available, aluminum frame available.

Production: current Caliber: .45 ACP, 9mm, 10mm Action: SA, semi-auto
Barrel Length: 5" Wt.: 36.9 oz. Sights: Heinie Slant Pro Straight-Eight night
sights or 3-dot night Capacity: 7, 8, 9, 10 Grips: Micarta Gator Grips

D2C:	NIB	$2,895	Ex	$2,225	VG+	$1,595	Good	$1,450	LMP $2,895
C2C:	Mint	$2,780	Ex	$2,000	VG+	$1,450	Good	$1,305	Fair $ 670
Trade-In:	Mint	$2,060	Ex	$1,625	VG+	$1,130	Good	$1,015	Poor $ 300

GRP RECON
NEW IN BOX

GRP RECON REDBOOK CODE: RB-NC-H-GRPREC

Lanyard loop, forged slide stop axle cut flush with frame, chamfered around frame, tactical single-side or ambidextrous safety, available in any color or combination of Perma Kote finishes, integrated rail, includes a Surefire X 300 Tactical Light, Crimson Trace Lasergrips available with Nighthawk logo with optional hand-stippling.

Production: current Caliber: .45 ACP, 9mm, 10mm Action: SA, semi-auto

Barrel Length: 5" Wt.: 39 oz. Sights: Heinie Slant Pro Straight-Eight night

sights or 3-dot night Capacity: 7, 8, 9, 10 Magazine: detachable box

Grips: Micarta Gator Grips, Crimson Trace available

D2C:	NIB	$3,099	Ex	$2,375	VG+	$1,705	Good	$1,550	LMP $3,099
C2C:	Mint	$2,980	Ex	$2,150	VG+	$1,550	Good	$1,395	Fair $ 715
Trade-In:	Mint	$2,210	Ex	$1,750	VG+	$1,210	Good	$1,085	Poor $ 330

HEINIE LADY HAWK REDBOOK CODE: RB-NC-H-HEINLH

Crowned and recessed barrel, Heinie Signature scalloped front strap and mainspring housing, reduced frame circumference, forged carbon steel frame, Heinie magazine release, Heinie stainless steel hammer and sear, Titanium Blue finish with stainless controls, magazine well, tactical magazine release, hand serrated rear of slide, tactical single-side or ambidextrous safety.

Production: current Caliber: .45 ACP, 9mm Action: SA, semi-auto

Barrel Length: 4.25" Wt.: 36 oz. Sights: Heinie Slant-Pro Straight Eight night

Capacity: 7, 8, 9, 10 Magazine: detachable box Grips: Nighthawk ultra-thin Alumagrips

D2C:	NIB	$3,450	Ex	$2,625	VG+	$1,900	Good	$1,725	LMP $3,450
C2C:	Mint	$3,320	Ex	$2,400	VG+	$1,730	Good	$1,555	Fair $ 795
Trade-In:	Mint	$2,450	Ex	$1,950	VG+	$1,350	Good	$1,210	Poor $ 360

HEINIE LADY HAWK
NEW IN BOX

HEINIE SIGNATURE COMPETITION 5"

REDBOOK CODE: RB-NC-H-HEINSC

Forged frame and slide, match grade barrel, crowned barrel and chamfered bushing, scalloped cut front strap and mainspring housing, 40 lpi serrated rear of slide, one-piece magwell with rounded comfort heel, Heinie logo.

Production: current Caliber: .45 ACP, 9mm Action: SA, semi-auto

Barrel Length: 5" Wt.: 40.3 oz. Sights: Heinie Slant-Pro rear and red fiber-optic front

Capacity: 7, 8, 9, 10 Magazine: detachable box Grips: thin G10 proprietary grips

D2C:	NIB	$3,450	Ex	$2,625	VG+	$1,900	Good	$1,725	LMP $3,450
C2C:	Mint	$3,320	Ex	$2,400	VG+	$1,730	Good	$1,555	Fair $ 795
Trade-In:	Mint	$2,450	Ex	$1,950	VG+	$1,350	Good	$1,210	Poor $ 360

HEINIE SIGNATURE COMPETITION 5"
NEW IN BOX

HEINIE SIGNATURE RECON
NEW IN BOX

HEINIE SIGNATURE RECON REDBOOK CODE: RB-NC-H-HEINSR

Forged frame and slide, crowned barrel and chamfered bushing, scalloped cut front strap and mainspring housing, 40 lpi serrated rear of slide, one-piece magwell with rounded comfort heel, Heinie logo, tactical rail.

Production: current Caliber: .45 ACP, 9mm Action: SA, semi-auto

Barrel Length: 5" Wt.: 38.8 oz. Sights: Heinie Slant-Pro Straight Eight night

Capacity: 7, 8, 9, 10 Magazine: detachable box Grips: thin G10 proprietary grips

D2C:	NIB	$3,550	Ex	$2,700	VG+	$1,955	Good	$1,775	LMP $3,550
C2C:	Mint	$3,410	Ex	$2,450	VG+	$1,780	Good	$1,600	Fair $ 820
Trade-In:	Mint	$2,530	Ex	$2,000	VG+	$1,390	Good	$1,245	Poor $ 360

HEINIE PDP
NEW IN BOX

HEINIE PDP   REDBOOK CODE: RB-NC-H-HEINPD

Forged frame, Government or Commander size, rear of slide hand serrated, serrated slide top, Heinie scalloped front strap and mainspring housing, magwell, extended safety, Heinie match hammer and sear, Heinie match grade barrel, Heinie thick bushing and plug, Heinie aluminum trigger, highly polished feed ramp, firing pin and extractor made of tool steel, Heinie logo, extended tactical mag catch, tactical single-side or ambidextrous safety.

Production: current   Caliber: .45 ACP   Action: SA, semi-auto

Barrel Length: 4.25", 5"   Wt.: 34.6 oz.   Sights: Heinie Straight-Eight

Capacity: 7, 8   Magazine: detachable box   Grips: cocobolo with Heinie logo

D2C:	NIB	$3,395	Ex	$2,600	VG+	$1,870	Good	$1,700	LMP	$3,395
C2C:	Mint	$3,260	Ex	$2,350	VG+	$1,700	Good	$1,530	Fair	$ 785
Trade-In:	Mint	$2,420	Ex	$1,925	VG+	$1,330	Good	$1,190	Poor	$ 360

PREDATOR
NEW IN BOX

PREDATOR   REDBOOK CODE: RB-NC-H-PREDAT

All specifications of the Talon family but with upgraded 416d stainless barrel, relieved slide and fitted barrel, tactical single-side or ambidextrous safety, aluminum frame option, Perma Kote finishes available.

Production: current   Caliber: .45 ACP, 9mm, 10mm   Action: SA, semi-auto

Barrel Length: 5"   Wt.: 40.3 oz.   Sights: 3-dot night sights or Heinie Slant Pro

Straight Eight   Capacity: 7, 8, 9, 10   Magazine: detachable box

Grips: golf-ball dimple pattern G10 with Nighthawk logo

D2C:	NIB	$3,450	Ex	$2,625	VG+	$1,900	Good	$1,725	LMP	$3,450
C2C:	Mint	$3,320	Ex	$2,400	VG+	$1,730	Good	$1,555	Fair	$ 795
Trade-In:	Mint	$2,450	Ex	$1,950	VG+	$1,350	Good	$1,210	Poor	$ 360

PREDATOR II
NEW IN BOX

PREDATOR II   REDBOOK CODE: RB-NC-H-PREDA2

All specifications of the Talon family but with upgraded 416d stainless barrel, relieved slide and fitted barrel, tactical single-side or ambidextrous safety, finish options include Black Nitride or Perma Kote finishes.

Production: current   Caliber: .45 ACP, 9mm, 10mm   Action: SA, semi-auto

Barrel Length: 4.25"   Wt.: 35.8 oz.   Sights: 3-dot night sights or Heinie Slant Pro

Straight Eight   Capacity: 7, 8, 9, 10   Magazine: detachable box   Grips: cocobolo

D2C:	NIB	$3,450	Ex	$2,625	VG+	$1,900	Good	$1,725	LMP	$3,450
C2C:	Mint	$3,320	Ex	$2,400	VG+	$1,730	Good	$1,555	Fair	$ 795
Trade-In:	Mint	$2,450	Ex	$1,950	VG+	$1,350	Good	$1,210	Poor	$ 360

PREDATOR III   REDBOOK CODE: RB-NC-H-PREDA3

All specifications of the Talon family but with upgraded 416d stainless barrel, relieved slide and fitted barrel, tactical single-side or ambidextrous safety, available in stainless steel, Perma Kote finishes available.

Production: current   Caliber: .45 ACP   Action: SA, semi-auto

Barrel Length: 4.25"   Wt.: 35.8 oz.   Sights: 3-dot night sights or Heinie Slant Pro

Straight Eight   Capacity: 7, 8   Magazine: detachable box   Grips: cocobolo

D2C:	NIB	$3,450	Ex	$2,625	VG+	$1,900	Good	$1,725	LMP	$3,450
C2C:	Mint	$3,320	Ex	$2,400	VG+	$1,730	Good	$1,555	Fair	$ 795
Trade-In:	Mint	$2,450	Ex	$1,950	VG+	$1,350	Good	$1,210	Poor	$ 360

T3
NEW IN BOX

T3 REDBOOK CODE: RB-NC-H-T3XXXX

Commander length slide, crowned barrel, frame based on officer model but with lightening cut extended magazine well, horizontally serrated mainspring housing and rear of slide, serrated slide top, tactical magazine release, 40 lpi serrated rear of slide, Nighthawk aluminum trigger, extended tactical mag catch, complete de-horn, tactical single-side or ambidextrous safety, Perma Kote finishes available.

Production: current Caliber: .45 ACP, 9mm, .40 S&W Action: SA, semi-auto

Barrel Length: 4.25" Wt.: 34.8 oz. Sights: Heinie Slant-Pro Straight Eight night

Capacity: 7, 8, 9, 10 Magazine: detachable box Grips: G-10 Alien grips

D2C:	NIB $3,250	Ex $2,475	VG+ $1,790	Good $1,625	LMP $3,250
C2C:	Mint $3,120	Ex $2,250	VG+ $1,630	Good $1,465	Fair $ 750
Trade-In:	Mint $2,310	Ex $1,825	VG+ $1,270	Good $1,140	Poor $ 330

T4
NEW IN BOX

T4 REDBOOK CODE: RB-NC-H-T4XXXX

Proprietary-crowned barrel cut and flush with slide, tapered slide cuts, Bob Marvel Everlast Recoil System, thinned frame, shorter 9mm platform, other features similar to T3, available in steel or aluminum frame upgrade.

Production: current Caliber: 9mm Action: SA, semi-auto Barrel Length: 3.8"

Wt.: 28.2 oz., 34.3 oz. Sights: Heinie Straight-Eight Capacity: 9, 10

Magazine: detachable box Grips: thin G-10 Alien grips

D2C:	NIB $3,350	Ex $2,550	VG+ $1,845	Good $1,675	LMP $3,350
C2C:	Mint $3,220	Ex $2,325	VG+ $1,680	Good $1,510	Fair $ 775
Trade-In:	Mint $2,380	Ex $1,900	VG+ $1,310	Good $1,175	Poor $ 360

TALON
NEW IN BOX

TALON REDBOOK CODE: RB-NC-H-TALONX

Forged steel frame, checkered rear of slide, serrated slide top, complete de-horn package, lightweight aluminum match trigger with bar stock sear, tool steel hammer and hammer strut, extended magwell fully blended into frame, 25 lpi checkering on front strap and mainspring housing, tactical single-side or ambidextrous safety, tactical mag release, aluminum frame available, Perma Kote finishes available.

Production: current Caliber: .45 ACP Action: SA, semi-auto Barrel Length: 5"

Wt.: 37.4 oz. Sights: Heinie Slant-Pro 2-dot tritium Capacity: 7, 8

Magazine: detachable box Grips: double-diamond checkered wood grips

D2C:	NIB $3,095	Ex $2,375	VG+ $1,705	Good $1,550	LMP $3,095
C2C:	Mint $2,980	Ex $2,150	VG+ $1,550	Good $1,395	Fair $ 715
Trade-In:	Mint $2,200	Ex $1,750	VG+ $1,210	Good $1,085	Poor $ 330

TALON 2
NEW IN BOX

TALON 2 REDBOOK CODE: RB-NC-H-TALON2

Commander-size, checkered rear of slide, serrated slide top, forged steel frame, complete de-horn package, aluminum match trigger with bar stock sear, tool steel hammer and hammer strut, 25 lpi checkering on front strap and mainspring housing, tactical single-side or ambidextrous safety, tactical magazine release, aluminum frame available, Perma Kote finishes available.

Production: current Caliber: .45 ACP Action: SA, semi-auto Barrel Length: 4.25"

Wt.: 34.7 oz. Sights: low mount tritium night sights or Heinie Slant-Pro Straight Eight

Capacity: 7, 8 Magazine: detachable box Grips: double-diamond checkered wood grips

D2C:	NIB $3,095	Ex $2,375	VG+ $1,705	Good $1,550	LMP $3,095
C2C:	Mint $2,980	Ex $2,150	VG+ $1,550	Good $1,395	Fair $ 715
Trade-In:	Mint $2,200	Ex $1,750	VG+ $1,210	Good $1,085	Poor $ 330

TALON 2 BOBTAIL REDBOOK CODE: RB-NC-H-TAL2BO

Commander-size with Ed Brown Bobtail, checkered rear of slide, serrated slide top, forged steel frame, complete de-horn package, aluminum match trigger with bar stock sear, tool steel hammer and hammer strut, 25 lpi checkering on front strap and mainspring housing, tactical single-side or ambidextrous safety, tactical magazine release, hex head grip screws, aluminum frame available, Perma Kote finishes.

Production: current Caliber: .45 ACP Action: SA, semi-auto Barrel Length: 4.25"

Wt.: 34.1 oz. Sights: low mount tritium night sights or Heinie Slant-Pro Straight Eight

Capacity: 7, 8 Magazine: detachable box Grips: double-diamond checkered wood grips

D2C:	NIB	$3,245	Ex	$2,475	VG+	$1,785	Good	$1,625	LMP	$3,245
C2C:	Mint	$3,120	Ex	$2,250	VG+	$1,630	Good	$1,465	Fair	$ 750
Trade-In:	Mint	$2,310	Ex	$1,825	VG+	$1,270	Good	$1,140	Poor	$ 330

North American Arms, Inc.

Founded in 1971 as Rocky Mountain Arms, North American Arms is best known for their line of pocket-sized revolvers, which have become a popular concealed-carry option. The company acquired the rights for the mini-revolvers from Freedom Arms in 1990, and those are now their most popular models. The company is based in Provo, Utah.

GUARDIAN .25 NAA
NEW IN BOX

GUARDIAN .25 NAA REDBOOK CODE: RB-NA-H-GUAR25

Proprietary cartridge that is a necked-down .32 caliber, grip options available, stainless finish.

Production: current Caliber: .25 NAA Action: DAO, semi-auto Barrel Length: 2.2"

OA Length: 4.4" Wt.: 13.57 oz. Sights: fixed Capacity: 6 Magazine: detachable box

Grips: black polymer

D2C:	NIB	$ 365	Ex	$ 300	VG+	$ 205	Good	$ 185	LMP	$ 402
C2C:	Mint	$ 360	Ex	$ 275	VG+	$ 190	Good	$ 165	Fair	$ 85
Trade-In:	Mint	$ 260	Ex	$ 225	VG+	$ 150	Good	$ 130	Poor	$ 60

GUARDIAN .32 ACP
NEW IN BOX

GUARDIAN .32 ACP REDBOOK CODE: RB-NA-H-GUAR32

Grip options available, stainless finish, available with an integral locking system safety.

Production: 1997 - current Caliber: .32 ACP Action: DAO, semi-auto

Barrel Length: 2.49" OA Length: 4.75" Wt.: 13.5 oz. Sights: fixed

Capacity: 6 Magazine: detachable box Grips: black polymer

D2C:	NIB	$ 365	Ex	$ 300	VG+	$ 205	Good	$ 185	LMP	$ 402
C2C:	Mint	$ 360	Ex	$ 275	VG+	$ 190	Good	$ 165	Fair	$ 85
Trade-In:	Mint	$ 260	Ex	$ 225	VG+	$ 150	Good	$ 130	Poor	$ 60

**GUARDIAN
.380 ACP**
NEW IN BOX

GUARDIAN .380 ACP REDBOOK CODE: RB-NA-H-GUAR38

Grip options available, stainless finish, available with an integral locking system safety.

Production: 2001 - current Caliber: .380 ACP Action: DAO, semi-auto

Barrel Length: 2.49" OA Length: 4.75" Wt.: 18.72 oz.

Sights: fixed Capacity: 6 Magazine: detachable box Grips: black polymer

D2C:	NIB	$ 400	Ex	$ 325	VG+	$ 220	Good	$ 200	LMP	$ 449
C2C:	Mint	$ 390	Ex	$ 300	VG+	$ 200	Good	$ 180	Fair	$ 95
Trade-In:	Mint	$ 290	Ex	$ 225	VG+	$ 160	Good	$ 140	Poor	$ 60

NAA PUG
NEW IN BOX

NAA PUG REDBOOK CODE: RB-NA-H-NAAPUG
Stainless finish, also available with a tritium sight at an additional cost.

Production:	current	Caliber: .22 Magnum		Action: SA, revolver		Barrel Length: 1"	
OA Length: 4.5"	Wt.: 6.4 oz.	Sights: white dot		Capacity: 5	Grips: black rubber		
D2C:	NIB $ 299	Ex $ 250	VG+ $ 165	Good $ 150	LMP $ 314		
C2C:	Mint $ 290	Ex $ 225	VG+ $ 150	Good $ 135	Fair $ 70		
Trade-In:	Mint $ 220	Ex $ 175	VG+ $ 120	Good $ 105	Poor $ 30		

MINI-MASTER
NEW IN BOX

MINI-MASTER REDBOOK CODE: RB-NA-H-MINIMA
Stainless steel finish, heavy vent barrel, bull cylinder.

Production: early 2000s - current		Caliber: .22 LR, .22 Magnum		Action: SA, revolver	
Barrel Length: 4"	OA Length: 7.875"	Wt.: 10.7 oz.			
Sights: fixed or adjustable	Capacity: 5	Grips: black rubber			
D2C:	NIB $ 299	Ex $ 250	VG+ $ 165	Good $ 150	LMP $ 314
C2C:	Mint $ 290	Ex $ 225	VG+ $ 150	Good $ 135	Fair $ 70
Trade-In:	Mint $ 220	Ex $ 175	VG+ $ 120	Good $ 105	Poor $ 30

THE EARL
NEW IN BOX

THE EARL REDBOOK CODE: RB-NA-H-EARLXX
Mini-revolver, 1860 replica, stainless finish, octagonal barrel, top strap channel.

Production:	current	Caliber: .22 Magnum		Action: SA, revolver		Barrel Length: 4"	
OA Length: 7.75"	Wt.: 6.8 oz.	Sights: bead front		Capacity: 5	Grips: wood		
D2C:	NIB $ 284	Ex $ 225	VG+ $ 160	Good $ 145	LMP $ 284		
C2C:	Mint $ 280	Ex $ 200	VG+ $ 150	Good $ 130	Fair $ 70		
Trade-In:	Mint $ 210	Ex $ 175	VG+ $ 120	Good $ 100	Poor $ 30		

BLACK WIDOW
NEW IN BOX

BLACK WIDOW REDBOOK CODE: RB-NA-H-BLKWID
Stainless steel finish, adjustable sights and conversion model available at an additional cost, heavy vent barrel, bull cylinder.

Production: early 1990s - current		Caliber: .22 LR, .22 Magnum		Action: SA, revolver	
Barrel Length: 2"	OA Length: 5.875"	Wt.: 8.8 oz.	Sights: fixed	Capacity: 5	
Grips: black rubber					
D2C:	NIB $ 274	Ex $ 225	VG+ $ 155	Good $ 140	LMP $ 274
C2C:	Mint $ 270	Ex $ 200	VG+ $ 140	Good $ 125	Fair $ 65
Trade-In:	Mint $ 200	Ex $ 175	VG+ $ 110	Good $ 100	Poor $ 30

NAA COMPANION
NEW IN BOX

NAA COMPANION REDBOOK CODE: RB-NA-H-COMPAN
Stainless steel, percussion cap and ball mini revolver, also available in a Super Companion model at an additional cost.

Production:	current	Caliber: .22 LR	Action: SA, revolver		Barrel Length: 1.125"	
OA Length: 4"	Wt.: 5.1 oz.	Sights: blade front		Capacity: 5	Grips: wood	
D2C:	NIB $ 220	Ex $ 175	VG+ $ 125	Good $ 110	LMP $ 234	
C2C:	Mint $ 220	Ex $ 175	VG+ $ 110	Good $ 100	Fair $ 55	
Trade-In:	Mint $ 160	Ex $ 125	VG+ $ 90	Good $ 80	Poor $ 30	

SIDEWINDER
NEW IN BOX

SIDEWINDER REDBOOK CODE: RB-NA-H-SIDEWI
Stainless finish, LR conversion cylinder also available at an additional cost, swing-out cylinder.

Production:	current	Caliber: .22 Magnum		Action: SA, revolver	
Barrel Length: 1"	OA Length: 5"	Wt.: 6.7 oz.			
Sights: stainless steel post	Capacity: 5	Grips: wood			
D2C:	NIB $ 330	Ex $ 275	VG+ $ 185	Good $ 165	LMP $ 349
C2C:	Mint $ 320	Ex $ 250	VG+ $ 170	Good $ 150	Fair $ 80
Trade-In:	Mint $ 240	Ex $ 200	VG+ $ 130	Good $ 120	Poor $ 60

Olympic Arms, Inc.

Based in Olympia, Washington, Olympic Arms produces an array of AR-15 rifles and M1911 pistols. Founded in 1975 by Robert Schuetz, the company manufactures all of their firearms in-house, which have gained a reputation for their relative affordability. The company's M1911-style Big Deuce pistol stands as the company's most acclaimed and recognized handgun.

BIG DEUCE
NEW IN BOX

BIG DEUCE REDBOOK CODE: RB-OA-H-BIGDEU

1911-style pistol, parkerized slide, satin bead-blast full-size frame, lowered ejection port, beavertail grip safety, beveled magwell.

Production: 1995 - current   Caliber: .45 ACP   Action: SA, semi-auto

Barrel Length: 6"   Wt.: 44 oz.   Sights: adjustable rear   Capacity: 7

Magazine: detachable box   Grips: double-diamond checkered cocobolo

D2C:	NIB	$ 940	Ex	$ 725	VG+	$ 520	Good	$ 470	LMP	$ 1,163
C2C:	Mint	$ 910	Ex	$ 650	VG+	$ 470	Good	$ 425	Fair	$ 220
Trade-In:	Mint	$ 670	Ex	$ 550	VG+	$ 370	Good	$ 330	Poor	$ 120

MATCHMASTER 5 REDBOOK CODE: RB-OA-H-MTMAS5

1911-style pistol, clear finish, available as a custom order with a different trigger guard and front strap, also offered with a 6" barrel at an additional cost.

Production: early 2000s - current   Caliber: .45 ACP   Action: SA, semi-auto

Barrel Length: 5"   Wt.: 40 oz.   Sights: adjustable rear   Capacity: 7

Magazine: detachable box   Grips: walnut

D2C:	NIB	$ 780	Ex	$ 600	VG+	$ 430	Good	$ 390	LMP	$ 1,033
C2C:	Mint	$ 750	Ex	$ 550	VG+	$ 390	Good	$ 355	Fair	$ 180
Trade-In:	Mint	$ 560	Ex	$ 450	VG+	$ 310	Good	$ 275	Poor	$ 90

STREET DEUCE REDBOOK CODE: RB-OA-H-STDEUC

1911-style custom pistol, two-tone finish, bull barrel.

Production: 2006 - current   Caliber: .45 ACP   Action: SA, semi-auto

Barrel Length: 5.2"   Wt.: 38 oz.   Sights: LPA adjustable or Heinie Slant-pro fixed rear

Capacity: 7   Magazine: detachable box   Grips: double-diamond checkered wood

D2C:	NIB	$1,520	Ex	$1,175	VG+	$840	Good	$ 760	LMP	$1,683
C2C:	Mint	$1,460	Ex	$1,050	VG+	$760	Good	$ 685	Fair	$ 350
Trade-In:	Mint	$1,080	Ex	$875	VG+	$600	Good	$ 535	Poor	$ 180

WESTERNER
NEW IN BOX

WESTERNER REDBOOK CODE: RB-OA-H-WESTRN

1911-style pistol, color case-hardened frame, stainless steel barrel, ivory grip.

Production: early 2000s - current   Caliber: .45 ACP   Action: SA, semi-auto

Barrel Length: 5"   Wt.: 39 oz.   Sights: adjustable rear   Capacity: 7

Magazine: detachable box   Grips: custom laser etched

D2C:	NIB	$ 940	Ex	$ 725	VG+	$ 520	Good	$ 470	LMP	$ 1,163
C2C:	Mint	$ 910	Ex	$ 650	VG+	$ 470	Good	$ 425	Fair	$ 220
Trade-In:	Mint	$ 670	Ex	$ 550	VG+	$ 370	Good	$ 330	Poor	$ 120

Para USA

Para, USA, LLC., founded in the mid-1980s, is based in Pineville, North Carolina and almost exclusively produces M1911 pistols. The company is noted for producing some of the first high-capacity M1911s, which quickly gained a following among shooting enthusiasts, law enforcement agencies, and military units. The Freedom Group, which also owns Bushmaster, DPMS, and Remington, purchased Para in 2012.

EXECUTIVE CARRY
(AGENT)
NEW IN BOX

EXECUTIVE CARRY (AGENT) REDBOOK CODE: RB-PA-H-EXECRY
Ramped stainless bull barrel, stainless slide, lightweight aluminum frame, Ed Brown Bobtail mainspring housing, skeletonized match grade trigger, IonBond/anodized finish.

Production: 2013 - current   Caliber: .45 ACP   Action: SA, semi-auto

Barrel Length: 3"   Wt.: 26 oz.   Sights: Trijicon tritium night sights

Capacity: 8   Magazine: detachable box   Grips: VZ machined G10

D2C:	NIB	$ 1,167	Ex $	900	VG+ $	645	Good $	585	LMP $1,400
C2C:	Mint	$ 1,130	Ex $	825	VG+ $	590	Good $	530	Fair $ 270
Trade-In:	Mint	$ 830	Ex $	675	VG+ $	460	Good $	410	Poor $ 120

PRO COMP 9/.40
NEW IN BOX

PRO COMP 9/.40 REDBOOK CODE: RB-PA-H-PROC94
Stainless match grade ramped barrel, EGW HD extractor, full-length guide rod, stainless steel slide and frame, skeletonized match grade trigger, beavertail grip safety, IonBond PVD finish.

Production: current   Caliber: 9mm, .40 S&W   Action: SA, semi-auto

Barrel Length: 5"   Wt.: 40 oz.   Sights: adjustable target rear, fiber-optic front

Capacity: 8, 9   Magazine: detachable box   Grips: G10

D2C:	NIB	$ 1,190	Ex $	925	VG+ $	655	Good $	595	LMP $1,299
C2C:	Mint	$ 1,150	Ex $	825	VG+ $	600	Good $	540	Fair $ 275
Trade-In:	Mint	$ 850	Ex $	675	VG+ $	470	Good $	420	Poor $ 120

PRO CUSTOM 14.45
NEW IN BOX

Courtesy of Bud's Gun Shop

PRO CUSTOM 14.45 REDBOOK CODE: RB-PA-H-PROC14
Stainless ramped barrel, EGW HD extractor, full-length guide rod, stainless slide and frame, IonBond PVD finish, skeletonized match grade trigger, ambidextrous thumb safety, beavertail grip safety.

Production: current   Caliber: .45 ACP   Action: SA, semi-auto   Barrel Length: 5"

Wt.: 40 oz.   Sights: adjustable target rear, fiber-optic front   Capacity: 14

Magazine: detachable box   Grips: G10

D2C:	NIB	$1,258	Ex $	975	VG+ $	695	Good $	630	LMP $1,449
C2C:	Mint	$1,210	Ex $	875	VG+ $	630	Good $	570	Fair $ 290
Trade-In:	Mint	$ 900	Ex $	725	VG+ $	500	Good $	445	Poor $ 150

PRO CUSTOM 16.40
NEW IN BOX

Courtesy of Bud's Gun Shop

PRO CUSTOM 16.40 REDBOOK CODE: RB-PA-H-PROC16
Stainless ramped barrel, EGW HD extractor, full-length guide rod, stainless slide and frame, IonBond PVD finish, skeletonized match grade trigger, ambidextrous thumb safety, beavertail grip safety.

Production: current   Caliber: .40 S&W   Action: SA, semi-auto   Barrel Length: 5"

Wt.: 40 oz.   Sights: adjustable target rear, fiber-optic front   Capacity: 16

Magazine: detachable box   Grips: G10

D2C:	NIB	$1,258	Ex $	975	VG+ $	695	Good $	630	LMP $1,449
C2C:	Mint	$1,210	Ex $	875	VG+ $	630	Good $	570	Fair $ 290
Trade-In:	Mint	$ 900	Ex $	725	VG+ $	500	Good $	445	Poor $ 150

PRO CUSTOM 18.9
NEW IN BOX

PARA USA

PRO CUSTOM 18.9 REDBOOK CODE: RB-PA-H-PROC18

Stainless ramped barrel, EGW HD extractor, full-length guide rod, stainless slide and frame, IonBond PVD finish, skeletonized match grade trigger, ambidextrous thumb safety, beavertail grip safety.

Production: current Caliber: 9mm Action: SA, semi-auto Barrel Length: 5"

Wt.: 40 oz. Sights: adjustable target rear, fiber-optic front Capacity: 18

Magazine: detachable box Grips: G10

D2C:	NIB $1,258	Ex $ 975	VG+ $ 695	Good $ 630	LMP $1,449
C2C:	Mint $1,210	Ex $ 875	VG+ $ 630	Good $ 570	Fair $ 290
Trade-In:	Mint $ 900	Ex $ 725	VG+ $ 500	Good $ 445	Poor $ 150

BLACK OPS 1911
NEW IN BOX

BLACK OPS 1911 REDBOOK CODE: RB-PA-H-BKOP11

Ramped barrel, stainless steel frame and slide, EGW HD extractor, IonBond PVD finish, checkered front strap, integral accessory rail, beavertail grip and ambidextrous thumb safety, skeletonized match grade trigger.

Production: current Caliber: .45 ACP Action: SA, semi-auto Barrel Length: 5"

Wt.: 40 oz. Sights: Trijicon tritium night sights Capacity: 8

Magazine: detachable box Grips: G10

D2C:	NIB $1,098	Ex $ 850	VG+ $ 605	Good $ 550	LMP $1,257
C2C:	Mint $1,060	Ex $ 775	VG+ $ 550	Good $ 495	Fair $ 255
Trade-In:	Mint $ 780	Ex $ 625	VG+ $ 430	Good $ 385	Poor $ 120

BLACK OPS 14.45
NEW IN BOX
Courtesy of Bud's Gun Shop

BLACK OPS 14.45 REDBOOK CODE: RB-PA-H-BKOP14

Ramped barrel, stainless steel frame and slide, EGW HD extractor, IonBond PVD finish, checkered front strap, integral accessory rail, beavertail grip and ambidextrous thumb safety, skeletonized match grade trigger.

Production: current Caliber: .45 ACP Action: SA, semi-auto Barrel Length: 5"

Wt.: 41 oz. Sights: Trijicon tritium night sights Capacity: 14

Magazine: detachable box Grips: G10

D2C:	NIB $ 1,141	Ex $ 875	VG+ $ 630	Good $ 575	LMP $1,299
C2C:	Mint $1,100	Ex $ 800	VG+ $ 580	Good $ 515	Fair $ 265
Trade-In:	Mint $ 820	Ex $ 650	VG+ $ 450	Good $ 400	Poor $ 120

BLACK OPS COMBAT
NEW IN BOX

BLACK OPS COMBAT REDBOOK CODE: RB-PA-H-BKOPCO

Threaded stainless ramped barrel, stainless frame and slide, EGW HD extractor, IonBond PVD finish, checkered front strap, integral accessory rail, beavertail grip and ambidextrous thumb safety, skeletonized match grade trigger.

Production: current Caliber: .45 ACP Action: SA, semi-auto Barrel Length: 5.5"

Wt.: 41 oz. Sights: high-profile tritium combat night sights Capacity: 14

Magazine: detachable box Grips: G10

D2C:	NIB $1,180	Ex $ 900	VG+ $ 650	Good $ 590	LMP $1,325
C2C:	Mint $1,140	Ex $ 825	VG+ $ 590	Good $ 535	Fair $ 275
Trade-In:	Mint $ 840	Ex $ 675	VG+ $ 470	Good $ 415	Poor $ 120

PARA USA

BLACK OPS RECON
NEW IN BOX

BLACK OPS RECON   REDBOOK CODE: RB-PA-H-BKOPRE
Ramped barrel, stainless frame and slide, EGW HD extractor, IonBond PVD finish, checkered front strap, integral accessory rail, beavertail grip and ambidextrous thumb safety, skeletonized match grade trigger.

Production: 2013 - current   Caliber: 9mm, .45 ACP   Action: SA, semi-auto

Barrel Length: 4.25"   Wt.: 41 oz.   Sights: Trijicon tritium night sights

Capacity: 14, 18 (9mm)   Magazine: detachable box   Grips: G10

D2C:	NIB	$ 1,140	Ex $ 875	VG+ $ 630	Good $ 570	LMP $1,299
C2C:	Mint	$1,100	Ex $ 800	VG+ $ 570	Good $ 515	Fair $ 265
Trade-In:	Mint	$ 810	Ex $ 650	VG+ $ 450	Good $ 400	Poor $ 120

ELITE
NEW IN BOX

ELITE   REDBOOK CODE: RB-PA-H-ELITEX
Match grade ramped barrel, stainless frame and slide, EGW HD extractor, IonBond PVD finish, oversized and flared ejection port, beavertail grip safety, skeletonized match grade trigger.

Production: 2013 - current   Caliber: .45 ACP   Action: SA, semi-auto

Barrel Length: 5"   Wt.: 39 oz.   Sights: 2-dot rear, green fiber-optic front

Capacity: 8   Magazine: detachable box   Grips: double-diamond walnut

D2C:	NIB	$ 815	Ex $ 625	VG+ $ 450	Good $ 410	LMP $ 949
C2C:	Mint	$ 790	Ex $ 575	VG+ $ 410	Good $ 370	Fair $ 190
Trade-In:	Mint	$ 580	Ex $ 475	VG+ $ 320	Good $ 290	Poor $ 90

ELITE STAINLESS
NEW IN BOX

ELITE STAINLESS   REDBOOK CODE: RB-PA-H-ELITES
Match grade ramped barrel, stainless frame and slide, satin stainless finish, EGW HD extractor, oversized and flared ejection port, beavertail grip safety, skeletonized match grade trigger.

Production: 2013 - current   Caliber: .45 ACP   Action: SA, semi-auto

Barrel Length: 5"   Wt.: 39 oz.   Sights: 2-dot rear, green fiber-optic front

Capacity: 8   Magazine: detachable box   Grips: double-diamond walnut

D2C:	NIB	$ 849	Ex $ 650	VG+ $ 470	Good $ 425	LMP $ 999
C2C:	Mint	$ 820	Ex $ 600	VG+ $ 430	Good $ 385	Fair $ 200
Trade-In:	Mint	$ 610	Ex $ 500	VG+ $ 340	Good $ 300	Poor $ 90

ELITE TARGET
NEW IN BOX

ELITE TARGET   REDBOOK CODE: RB-PA-H-ELITET
Satin stainless finished slide, black IonBond PVD frame, match grade ramped barrel, stainless frame and slide, EGW HD extractor, oversized and flared ejection port, beavertail grip safety, skeletonized match grade trigger.

Production: 2013 - current   Caliber: .45 ACP   Action: SA, semi-auto

Barrel Length: 5"   Wt.: 39 oz.   Sights: adjustable target rear, green fiber-optic front

Capacity: 8   Magazine: detachable box   Grips: double-diamond walnut

D2C:	NIB	$ 849	Ex $ 650	VG+ $ 470	Good $ 425	LMP $ 999
C2C:	Mint	$ 820	Ex $ 600	VG+ $ 430	Good $ 385	Fair $ 200
Trade-In:	Mint	$ 610	Ex $ 500	VG+ $ 340	Good $ 300	Poor $ 90

ELITE PRO
NEW IN BOX

ELITE PRO   REDBOOK CODE: RB-PA-H-ELITEP
Match grade ramped barrel, stainless frame and slide, EGW HD extractor, IonBond PVD finish, oversized and flared ejection port, beavertail grip safety, skeletonized match grade trigger.

Production: 2013 - current   Caliber: .45 ACP   Action: SA, semi-auto   Capacity: 8

Barrel Length: 5"   Wt.: 39 oz.   Sights: Trijicon tritium night sights   Grips: G10

D2C:	NIB	$1,090	Ex $ 850	VG+ $ 600	Good $ 545	LMP $1,249
C2C:	Mint	$1,050	Ex $ 775	VG+ $ 550	Good $ 495	Fair $ 255
Trade-In:	Mint	$ 780	Ex $ 625	VG+ $ 430	Good $ 385	Poor $ 120

ELITE COMMANDER
NEW IN BOX

ELITE COMMANDER REDBOOK CODE: RB-PA-H-ELTCOM

Stainless match grade ramped barrel, frame, and slide, EGW HD extractor, IonBond PVD finish, oversized and flared ejection port, beavertail grip safety, skeletonized match grade trigger.

Production: 2013 - current Caliber: .45 ACP Action: SA, semi-auto

Barrel Length: 4.25" Wt.: 35 oz. Sights: 2-dot rear, green fiber-optic front

Capacity: 8 Magazine: detachable box Grips: double-diamond walnut

D2C:	NIB $ 840	Ex $ 650	VG+ $ 465	Good $ 420	LMP $ 949
C2C:	Mint $ 810	Ex $ 600	VG+ $ 420	Good $ 380	Fair $ 195
Trade-In:	Mint $ 600	Ex $ 475	VG+ $ 330	Good $ 295	Poor $ 90

ELITE OFFICER
NEW IN BOX

ELITE OFFICER REDBOOK CODE: RB-PA-H-ELTOFF

Match grade ramped barrel, stainless frame and slide, EGW HD extractor, IonBond PVD finish, oversized and flared ejection port, beavertail grip safety, skeletonized match grade trigger.

Production: 2013 - current Caliber: .45 ACP Action: SA, semi-auto

Barrel Length: 3.5" Wt.: 32 oz. Sights: 2-dot rear, green fiber-optic front

Capacity: 7 Magazine: detachable box Grips: double-diamond walnut

D2C:	NIB $ 840	Ex $ 650	VG+ $ 465	Good $ 420	LMP $ 949
C2C:	Mint $ 810	Ex $ 600	VG+ $ 420	Good $ 380	Fair $ 195
Trade-In:	Mint $ 600	Ex $ 475	VG+ $ 330	Good $ 295	Poor $ 90

ELITE CARRY
NEW IN BOX

ELITE CARRY REDBOOK CODE: RB-PA-H-ELTCRY

Match grade ramped barrel, stainless slide and aluminum frame, EGW HD extractor, IonBond PVD finish slide with anodized frame, oversized and flared ejection port, beavertail grip safety, skeletonized match grade trigger.

Production: 2013 - current Caliber: .45 ACP Action: SA, semi-auto

Barrel Length: 3" Wt.: 24 oz. Sights: 2-dot rear, green fiber-optic front

Capacity: 6 Magazine: detachable box Grips: double-diamond walnut

D2C:	NIB $ 840	Ex $ 650	VG+ $ 465	Good $ 420	LMP $ 949
C2C:	Mint $ 810	Ex $ 600	VG+ $ 420	Good $ 380	Fair $ 195
Trade-In:	Mint $ 600	Ex $ 475	VG+ $ 330	Good $ 295	Poor $ 90

EXPERT
NEW IN BOX

EXPERT REDBOOK CODE: RB-PA-H-EXPERT

Match grade barrel, stainless frame and steel slide, black nitride finish, oversized and flared ejection port, beavertail grip safety, skeletonized match grade trigger.

Production: current Caliber: 9mm, .45 ACP Action: SA, semi-auto

Barrel Length: 5" Wt.: 39 oz. Sights: 2-dot rear, green fiber-optic front

Capacity: 8, 9 Magazine: detachable box Grips: polymer

D2C:	NIB $ 599	Ex $ 475	VG+ $ 330	Good $ 300	LMP $ 663
C2C:	Mint $ 580	Ex $ 425	VG+ $ 300	Good $ 270	Fair $ 140
Trade-In:	Mint $ 430	Ex $ 350	VG+ $ 240	Good $ 210	Poor $ 60

EXPERT COMMANDER
NEW IN BOX

EXPERT COMMANDER REDBOOK CODE: RB-PA-H-EXPCOM

Match grade barrel, stainless slide and aluminum frame, beavertail grip safety, skeletonized match grade trigger, black nitride/anodized finish.

Production: current Caliber: .45 ACP Action: SA, semi-auto Barrel Length: 4.25"

Wt.: 28 oz. Sights: 2-dot rear, green fiber-optic front Capacity: 8 Grips: polymer

D2C:	NIB $ 710	Ex $ 550	VG+ $395	Good $ 355	LMP $ 799
C2C:	Mint $ 690	Ex $ 500	VG+ $360	Good $ 320	Fair $ 165
Trade-In:	Mint $ 510	Ex $ 400	VG+ $280	Good $ 250	Poor $ 90

PARA USA

EXPERT CARRY
NEW IN BOX

EXPERT CARRY   REDBOOK CODE: RB-PA-H-EXPCRY

Match grade barrel, stainless slide & aluminum frame, oversized and flared ejection port, beavertail grip safety, black nitride/anodize finish, skeletonized match grade trigger.

Production: current   Caliber: .45 ACP   Action: SA, semi-auto   Barrel Length: 3"

Wt.: 26 oz.   Sights: 2-dot rear, green fiber-optic front   Capacity: 8   Grips: polymer

D2C:	NIB $ 710	Ex $ 550	VG+ $ 395	Good $ 355	LMP $ 799
C2C:	Mint $ 690	Ex $ 500	VG+ $ 360	Good $ 320	Fair $ 165
Trade-In:	Mint $ 510	Ex $ 400	VG+ $ 280	Good $ 250	Poor $ 90

EXPERT 14.45
NEW IN BOX

EXPERT 14.45   REDBOOK CODE: RB-PA-H-EXPT14

Match grade barrel, double-stack stainless frame, stainless slide, oversized and flared ejection port, black nitride finish, beavertail grip safety, skeletonized match grade trigger.

Production: current   Caliber: .45 ACP   Action: SA, semi-auto   Barrel Length: 5"

Wt.: 40 oz.   Sights: 2-dot rear, green fiber-optic front   Capacity: 14

Magazine: detachable box   Grips: polymer

D2C:	NIB $ 775	Ex $ 600	VG+ $ 430	Good $ 390	LMP $ 884
C2C:	Mint $ 750	Ex $ 550	VG+ $ 390	Good $ 350	Fair $ 180
Trade-In:	Mint $ 560	Ex $ 450	VG+ $ 310	Good $ 275	Poor $ 90

WARTHOG (P10)
NEW IN BOX
Courtesy of Bud's Gun Shop

WARTHOG (P10)   REDBOOK CODE: RB-PA-H-WARTHG

Stainless match barrel, stainless slide & aluminum frame, EGW HD extractor, beavertail grip safety, black nitride/anodized finish, previously named the P10.

Production: 2004 - current   Caliber: .45 ACP, 9mm   Action: SA, semi-auto

Barrel Length: 3"   Wt.: 24 oz.   Sights: 2-dot rear, green fiber-optic front

Capacity: 10, 12 (9mm)   Magazine: detachable box   Grips: polymer

D2C:	NIB $ 795	Ex $ 625	VG+ $ 440	Good $ 400	LMP $ 884
C2C:	Mint $ 770	Ex $ 550	VG+ $ 400	Good $ 360	Fair $ 185
Trade-In:	Mint $ 570	Ex $ 450	VG+ $ 320	Good $ 280	Poor $ 90

STEALTH
NEW IN BOX
Courtesy of Bud's Gun Shop

STEALTH   REDBOOK CODE: RB-PA-H-STELTH

IonBond/anodized finish, alloy frame, Light Double Action trigger, HD extractor, ambidextrous thumb safety, grip safety, checkered front strap. This gun is a reintroduction and is extremely rare.

Production: 2012   Caliber: .45 ACP   Action: LDA (DAO), semi-auto

Barrel Length: 3"   Wt.: 24 oz.   Sights: tritium night sights   Capacity: 6

Magazine: detachable box   Grips: G10

D2C:	NIB $ 1,150	Ex $ 875	VG+ $ 635	Good $ 575	
C2C:	Mint $ 1,110	Ex $ 800	VG+ $ 580	Good $ 520	Fair $ 265
Trade-In:	Mint $ 820	Ex $ 650	VG+ $ 450	Good $ 405	Poor $ 120

CARRY C6.45 LDA   REDBOOK CODE: RB-PA-H-CRY645

1911-style with Light Double Action trigger, stainless steel frame and slide, Griptor grooves on front strap and slide.

Production: 2001 - discontinued   Caliber: .45 ACP   Action: LDA (DAO), semi-auto

Barrel Length: 3"   Wt.: 30 oz.   Sights: fixed   Capacity: 6   Magazine: detachable box

D2C:	NIB $ 700	Ex $ 550	VG+ $ 385	Good $ 350	
C2C:	Mint $ 680	Ex $ 500	VG+ $ 350	Good $ 315	Fair $ 165
Trade-In:	Mint $ 500	Ex $ 400	VG+ $ 280	Good $ 245	Poor $ 90

P10   REDBOOK CODE: RB-PA-H-P10XXX

Sub-compact, alloy frame with stainless steel option, duo-tone finish available, double-stack frame, name later changed to Warthog.

Production: 1996 - 2004   Caliber: 9mm, .40 S&W, .45 ACP   Action: SA, semi-auto
Barrel Length: 3"   Wt.: 21 oz. (alloy), 31 oz. (stainless)   Sights: 3-dot   Capacity: 10
Magazine: detachable box   Grips: black polymer

D2C:	NIB $ 700	Ex $ 550	VG+ $ 385	Good $ 350	
C2C:	Mint $ 680	Ex $ 500	VG+ $ 350	Good $ 315	Fair $ 165
Trade-In:	Mint $ 500	Ex $ 400	VG+ $ 280	Good $ 245	Poor $ 90

P12
EXCELLENT

Courtesy of Bud's Gun Shop

P12   REDBOOK CODE: RB-PA-H-P12XXX

Compact frame, alloy or steel double stack frame, grip safety, thumb safety, slight premium for steel or duo-tone finish, Power Extractor added in 2004.

Production: 1993 - discontinued   Caliber: .40 S&W, .45 ACP   Action: SA, semi-auto
Barrel Length: 3.5"   Wt.: 24 oz. (alloy), 33 oz. (steel)   Sights: 3-dot   Capacity: 10, 11, 12
Magazine: detachable box   Grips: black polymer

D2C:	NIB $ 780	Ex $ 600	VG+ $ 430	Good $ 390	
C2C:	Mint $ 750	Ex $ 550	VG+ $ 390	Good $ 355	Fair $ 180
Trade-In:	Mint $ 560	Ex $ 450	VG+ $ 310	Good $ 275	Poor $ 90

P12.45 LDA
EXCELLENT

Courtesy of Bud's Gun Shop

P12.45 LDA   REDBOOK CODE: RB-PA-H-P12LDA

Similar to P12 with Light Double Action trigger, compact alloy or steel frame, stainless or black finish, slight premium for stainless finish.

Production: 1999   Caliber: .45 ACP   Action: LDA (DAO), semi-auto
Barrel Length: 3.5"   Wt.: 34 oz.   Sights: fixed   Capacity: 12   Magazine: detachable box
Grips: black polymer

D2C:	NIB $ 785	Ex $ 600	VG+ $ 435	Good $ 395	
C2C:	Mint $ 760	Ex $ 550	VG+ $ 400	Good $ 355	Fair $ 185
Trade-In:	Mint $ 560	Ex $ 450	VG+ $ 310	Good $ 275	Poor $ 90

P13.45
EXCELLENT

P13.45   REDBOOK CODE: RB-PA-H-P1345X

Slightly longer frame than the P12, alloy or steel frame, slight premium for steel or duo-tone model, Power Extractor added in 2004.

Production: 1995 - discontinued   Caliber: .45 ACP   Action: SA, semi-auto
Barrel Length: 4.25"   Wt.: 36 oz. (steel), 28 oz. (alloy)   Sights: 3-dot low-mount
Capacity: 13   Magazine: detachable box   Grips: black polymer

D2C:	NIB $ 780	Ex $ 600	VG+ $ 430	Good $ 390	
C2C:	Mint $ 750	Ex $ 550	VG+ $ 390	Good $ 355	Fair $ 180
Trade-In:	Mint $ 560	Ex $ 450	VG+ $ 310	Good $ 275	Poor $ 90

P14.45
EXCELLENT

P14.45   REDBOOK CODE: RB-PA-H-P1445X

Stainless frame, black finish, PCT extractor on newer models, flared ejection port, beavertail grip safety, thumb safety, double stack frame.

Production: 1993   Caliber: .45 ACP   Action: SA, semi-auto   Barrel Length: 5"
Wt.: 40 oz.   Sights: 3-dot, fiber-optic front   Capacity: 14   Grips: black polymer

D2C:	NIB $ 735	Ex $ 575	VG+ $ 405	Good $ 370	
C2C:	Mint $ 710	Ex $ 525	VG+ $ 370	Good $ 335	Fair $ 170
Trade-In:	Mint $ 530	Ex $ 425	VG+ $ 290	Good $ 260	Poor $ 90

P14.45 STAINLESS REDBOOK CODE: RB-PA-H-P1445S

Stainless slide and frame, stainless finish, competition trigger, beavertail grip safety, double stack frame.

Production: discontinued Caliber: .45 ACP Action: SA, semi-auto							
Barrel Length: 5" Wt.: 40 oz. Sights: 3-dot, fiber-optic front							
Capacity: 14 Magazine: detachable box Grips: black polymer							
D2C:	NIB $ 788	Ex $ 600	VG+ $ 435	Good $ 395			
C2C:	Mint $ 760	Ex $ 550	VG+ $ 400	Good $ 355	Fair $ 185		
Trade-In:	Mint $ 560	Ex $ 450	VG+ $ 310	Good $ 280	Poor $ 90		

P14.45 LIMITED REDBOOK CODE: RB-PA-H-P1445L

Black or stainless finish, front and rear slide serrations, Ed Brown barrel bushing, competition hammer, competition trigger.

Production: discontinued Caliber: .45 ACP Action: SA, semi-auto					
Barrel Length: 5" Wt.: 40 oz. Sights: adjustable Capacity: 14					
Magazine: detachable box Grips: black polymer					
D2C:	NIB $1,080	Ex $ 825	VG+ $ 595	Good $ 540	
C2C:	Mint $1,040	Ex $ 750	VG+ $ 540	Good $ 490	Fair $ 250
Trade-In:	Mint $ 770	Ex $ 625	VG+ $ 430	Good $ 380	Poor $ 120

P14.45 LDA REDBOOK CODE: RB-PA-H-P14LDA

Black finish, Light Double Action trigger, similar to P14.45, stainless model available for slight premium.

Production: 1999-discontinued Caliber: .45 ACP Action: LDA (DAO), semi-auto					
Barrel Length: 5" Wt.: 40 oz. Sights: 3-dot, fiber-optic front					
Capacity: 14 Magazine: detachable box Grips: black polymer					
D2C:	NIB $ 950	Ex $ 725	VG+ $ 525	Good $ 475	
C2C:	Mint $ 920	Ex $ 675	VG+ $ 480	Good $ 430	Fair $ 220
Trade-In:	Mint $ 680	Ex $ 550	VG+ $ 380	Good $ 335	Poor $ 120

P16.40 REDBOOK CODE: RB-PA-H-P1640X

Black or stainless finish, beavertail grip safety, thumb safety, double-stack frame.

P16.40
EXCELLENT

Caliber: .40 S&W Action: SA, semi-auto					
Sights: fiber-optic front, adjustable rear Capacity: 16 Magazine: detachable box					
D2C:	NIB $ 750	Ex $ 575	VG+ $ 415	Good $ 375	
C2C:	Mint $ 720	Ex $ 525	VG+ $ 380	Good $ 340	Fair $ 175
Trade-In:	Mint $ 540	Ex $ 425	VG+ $ 300	Good $ 265	Poor $ 90

P16.40 LDA REDBOOK CODE: RB-PA-H-P16LDA

Similar to P16.40 with Light Double Action trigger.

Caliber: .40 S&W Action: LDA (DAO), semi-auto					
Sights: fiber-optic front, adjustable rear Capacity: 16 Magazine: detachable box					
D2C:	NIB $ 790	Ex $ 625	VG+ $ 435	Good $ 395	
C2C:	Mint $ 760	Ex $ 550	VG+ $ 400	Good $ 360	Fair $ 185
Trade-In:	Mint $ 570	Ex $ 450	VG+ $ 310	Good $ 280	Poor $ 90

P18.9 REDBOOK CODE: RB-PA-H-P189XX
Stainless steel finish, double-stack frame, grip safety, thumb safety.

Caliber: 9mm Action: SA, semi-auto Barrel Length: 5" Wt.: 40 oz.

Sights: adjustable rear Capacity: 18 Magazine: detachable box

D2C:	NIB $ 930	Ex $ 725	VG+ $ 515	Good $ 465	
C2C:	Mint $ 900	Ex $ 650	VG+ $ 470	Good $ 420	Fair $ 215
Trade-In:	Mint $ 670	Ex $ 525	VG+ $ 370	Good $ 330	Poor $ 120

14.45 TACTICAL
EXCELLENT
Courtesy of Bud's Gun Shop

14.45 TACTICAL REDBOOK CODE: RB-PA-H-1445TA
Stainless frame, Stealth IonBond coating, Dawson magwell, double-stack magazines, tactical rail, flat-top slide.

Caliber: .45 ACP Action: SA, semi-auto Barrel Length: 5" Wt.: 42 oz.

Sights: fiber-optic front, adjustable rear Capacity: 14 Magazine: detachable box

Grips: VZ Grips

D2C:	NIB $1,285	Ex $1,000	VG+ $ 710	Good $ 645	
C2C:	Mint $1,240	Ex $900	VG+ $ 650	Good $ 580	Fair $ 300
Trade-In:	Mint $ 920	Ex $ 725	VG+ $ 510	Good $ 450	Poor $ 150

18.9 LIMITED
NEW IN BOX
Courtesy of Bud's Gun Shop

18.9 LIMITED REDBOOK CODE: RB-PA-H-189LIM
Stainless steel frame and slide, stainless finish, ambidextrous thumb safeties, beavertail grip safety, skeletonized competition trigger.

Production: discontinued Caliber: 9mm Action: SA, semi-auto Barrel Length: 5"

Wt.: 40 oz. Sights: fiber-optic front, adjustable rear Capacity: 18 Magazine: detachable box

Grips: black polymer

D2C:	NIB $1,205	Ex $ 925	VG+ $ 665	Good $ 605	
C2C:	Mint $1,160	Ex $ 850	VG+ $ 610	Good $ 545	Fair $ 280
Trade-In:	Mint $ 860	Ex $ 675	VG+ $ 470	Good $ 425	Poor $ 150

S14.45 LIMITED
NEW IN BOX
Courtesy of Bud's Gun Shop

S14.45 LIMITED REDBOOK CODE: RB-PA-H-S1445L
Sterling finish, beavertail grip safety, ambidextrous thumb safeties, stainless frame, Power Extractor.

Caliber: .45 ACP Action: SA, semi-auto Barrel Length: 5" Wt.: 40 oz.

Sights: adjustable Capacity: 14 Magazine: detachable box Grips: black polymer

D2C:	NIB $ 930	Ex $ 725	VG+ $ 515	Good $ 465	
C2C:	Mint $ 900	Ex $ 650	VG+ $ 470	Good $ 420	Fair $ 215
Trade-In:	Mint $ 670	Ex $ 525	VG+ $ 370	Good $ 330	Poor $ 120

OPS REDBOOK CODE: RB-PA-H-OPSXXX
Stainless steel slide and frame, stainless finish, single-stack frame, Griptor grooves on front strap and slide.

Production: 2005-discontinued Caliber: .45 ACP Action: SA, semi-auto

Barrel Length: 3.5" Wt.: 32 oz. Sights: 3-dot Capacity: 7

Magazine: detachable box Grips: cocobolo

D2C:	NIB $1,030	Ex $ 800	VG+ $ 570	Good $ 515	
C2C:	Mint $ 990	Ex $ 725	VG+ $ 520	Good $ 465	Fair $ 240
Trade-In:	Mint $ 740	Ex $ 600	VG+ $ 410	Good $ 365	Poor $ 120

SLIM HAWG
NEW IN BOX

SLIM HAWG   REDBOOK CODE: RB-PA-H-SLMHWG

Stainless frame and slide, stainless finish, Griptor grooves in front strap and slide, beavertail grip safety, thumb safety, single-stack frame.

Production: 2005-discontinued   Caliber: .45 ACP   Action: SA, semi-auto

Barrel Length: 3"   Wt.: 30 oz.   Sights: 3-dot   Capacity: 6   Grips: checkered wood

D2C:	NIB $ 760	Ex $ 600	VG+ $ 420	Good $ 380		
C2C:	Mint $ 730	Ex $ 525	VG+ $ 380	Good $ 345	Fair $ 175
Trade-In:	Mint $ 540	Ex $ 450	VG+ $ 300	Good $ 270	Poor $ 90

HAWG 7
NEW IN BOX
Courtesy of Bud's Gun Shop

HAWG 7   REDBOOK CODE: RB-PA-H-HAWG7X

Black IonBond finish, stainless steel frame, Griptor grooves, integral ramp match barrel.

Production: 2011 - discontinued   Caliber: .45 ACP   Action: SA, semi-auto

Barrel Length: 3.5"   Wt.: 32 oz.   Sights: fiber-optic front, 2-dot rear   Capacity: 7

Grips: black polymer

D2C:	NIB $ 768	Ex $ 600	VG+ $ 425	Good $ 385		
C2C:	Mint $ 740	Ex $ 550	VG+ $ 390	Good $ 350	Fair $ 180
Trade-In:	Mint $ 550	Ex $ 450	VG+ $ 300	Good $ 270	Poor $ 90

HAWG 9
EXCELLENT

Courtesy of Bud's Gun Shop

HAWG 9   REDBOOK CODE: RB-PA-H-HAWG9X

Regal finish, steel frame, bushingless recoil system, Power Extractor, beavertail grip safety.

Production: 2005 - 2010   Caliber: 9mm   Action: SA, semi-auto   Barrel Length: 3"

Wt.: 24 oz.   Sights: 3-dot, fixed   Capacity: 12   Magazine: detachable box   Grips: black polymer

D2C:	NIB $ 770	Ex $ 600	VG+ $ 425	Good $ 385		
C2C:	Mint $ 740	Ex $ 550	VG+ $ 390	Good $ 350	Fair $ 180
Trade-In:	Mint $ 550	Ex $ 450	VG+ $ 310	Good $ 270	Poor $ 90

NITE HAWG
NEW IN BOX
Courtesy of Bud's Gun Shop

NITE HAWG   REDBOOK CODE: RB-PA-H-NITEHG

Alloy frame, Covert Black finish, PXT extractor, beavertail grip safety, double-stack frame.

Production: discontinued   Caliber: .45 ACP   Action: SA, semi-auto

Barrel Length: 3"   Wt.: 24 oz.   Sights: tritium night sights

Capacity: 10   Magazine: detachable box   Grips: black polymer

D2C:	NIB $ 930	Ex $ 725	VG+ $ 515	Good $ 465		
C2C:	Mint $ 900	Ex $ 650	VG+ $ 470	Good $ 420	Fair $ 215
Trade-In:	Mint $ 670	Ex $ 525	VG+ $ 370	Good $ 330	Poor $ 120

BIG HAWG
EXCELLENT

BIG HAWG   REDBOOK CODE: RB-PA-H-BIGHWG

Regal finish, beavertail grip safety, spurred hammer, PXT extractor, double-stack frame.

Production: 2007-discontinued   Caliber: .45 ACP   Action: SA, semi-auto

Barrel Length: 5"   Wt.: 28 oz.   Sights: 3-dot   Capacity: 14   Grips: black polymer

D2C:	NIB $ 895	Ex $ 700	VG+ $ 495	Good $ 450		
C2C:	Mint $ 860	Ex $ 625	VG+ $ 450	Good $ 405	Fair $ 210
Trade-In:	Mint $ 640	Ex $ 525	VG+ $ 350	Good $ 315	Poor $ 90

1911 LTC   REDBOOK CODE: RB-PA-H-LTCXXX

Regal finish (.45) or Covert Black (9mm) finish, alloy frame, skeletonized hammer and trigger, beavertail grip safety, steel or stainless frame available for slight premium.

Production: 2005-discontinued		Caliber: .45 ACP, 9mm		Action: SA, semi-auto						
Barrel Length: 4.25"		Wt.: 37 oz., 28 oz. (alloy)		Sights: 3-dot, fixed						
Capacity: 8, 9		Magazine: detachable box		Grips: cocobolo wood						
D2C:	NIB $ 720	Ex $ 550	VG+ $ 400	Good $ 360	
C2C:	Mint $ 700	Ex $ 500	VG+ $ 360	Good $ 325	Fair $ 170
Trade-In:	Mint $ 520	Ex $ 425	VG+ $ 290	Good $ 255	Poor $ 90

1911 LTC
NEW IN BOX
Courtesy of Bud's Gun Shop

HI-CAP LTC   REDBOOK CODE: RB-PA-H-HICLTC

Stainless frame, green Spec-Ops slide and frame finish, beavertail grip safety, thumb safety, PXT extractor.

Production: 2005-discontinued		Caliber: .45 ACP		Action: SA, semi-auto	
Barrel Length: 4.25"	Wt.: 37 oz.	Sights: fixed	Capacity: 14	Grips: black polymer	
D2C:	NIB $ 700	Ex $ 550	VG+ $ 385	Good $ 350	
C2C:	Mint $ 680	Ex $ 500	VG+ $ 350	Good $ 315	Fair $ 165
Trade-In:	Mint $ 500	Ex $ 400	VG+ $ 280	Good $ 245	Poor $ 90

STEALTH CARRY   REDBOOK CODE: RB-PA-H-STLCRY

Compact 1911-style with LDA trigger, Stealth Black IonBond finish, stainless slide, stainless frame, PXT extractor, Griptor front strap.

Production: 2004 - discontinued		Caliber: .45 ACP		Action: LDA (DAO), semi-auto	
Barrel Length: 3"	OA Length: 6.5"	Wt.: 30 oz. or 24 oz. (newer model)			
Sights: Novak adjustable night sights		Capacity: 6	Magazine: detachable box		
Grips: black checkered polymer or G10 (newer model)					
D2C:	NIB $ 875	Ex $ 675	VG+ $ 485	Good $ 440	
C2C:	Mint $ 840	Ex $ 625	VG+ $ 440	Good $ 395	Fair $ 205
Trade-In:	Mint $ 630	Ex $ 500	VG+ $ 350	Good $ 310	Poor $ 90

STEALTH CARRY
NEW IN BOX
Courtesy of Bud's Gun Shop

CCO-COMPANION CARRY OPTION   REDBOOK CODE: RB-PA-H-CCOCCY

1911-style with LDA trigger, stainless frame, stainless finish, thumb and grip safety, Griptor grooves in front strap and slide.

Production: 2003-discontinued		Caliber: .45 ACP		Action: LDA (DAO), semi-auto	
Barrel Length: 3.5"	Wt.: 32 oz.	Sights: 3-dot, night	Capacity: 7		
D2C:	NIB $ 760	Ex $ 600	VG+ $ 420	Good $ 380	
C2C:	Mint $ 730	Ex $ 525	VG+ $ 380	Good $ 345	Fair $ 175
Trade-In:	Mint $ 540	Ex $ 450	VG+ $ 300	Good $ 270	Poor $ 90

CCO-COMPANION CARRY OPTION
EXCELLENT
Courtesy of Bud's Gun Shop

TAC-S   REDBOOK CODE: RB-PA-H-TACSXX

1911-style with Light Double Action trigger, Spec Ops OD Green finish, beavertail grip safety, steel frame.

Production: 2008-discontinued		Caliber: .45 ACP		Action: LDA (DAO), semi-auto	
Barrel Length: 4.25"	Wt.: 35 oz.	Sights: 3-dot	Capacity: 7	Grips: black polymer	
D2C:	NIB $ 890	Ex $ 700	VG+ $ 490	Good $ 445	
C2C:	Mint $ 860	Ex $ 625	VG+ $ 450	Good $ 405	Fair $ 205
Trade-In:	Mint $ 640	Ex $ 500	VG+ $ 350	Good $ 315	Poor $ 90

NITE-TAC .45
EXCELLENT
Courtesy of Bud's Gun Shop

NITE-TAC .45 REDBOOK CODE: RB-PA-H-NTAC45

Stainless steel frame and slide, LDA trigger, beavertail grip safety, thumb safety, accessory rail.

Production: 2005 - discontinued   Caliber: .45 ACP   Action: LDA (DAO), semi-auto
Barrel Length: 5" Wt.: 40 oz. Sights: 3-dot Capacity: 14 Grips: black polymer

D2C:	NIB $ 930	Ex $ 725	VG+ $ 515	Good $ 465	
C2C:	Mint $ 900	Ex $ 650	VG+ $ 470	Good $ 420	Fair $ 215
Trade-In:	Mint $ 670	Ex $ 525	VG+ $ 370	Good $ 330	Poor $ 120

Remington Arms Co.

In 1816, Eliphalet Remington II founded E. Remington and Sons in Ilion, New York. The company manufactured some of the most well-known and best-selling firearms in history. After some restructuring and acquisitions, the company changed its name to Remington Arms Company. Currently, Cerberus Capital Management owns Remington Arms as part of the Freedom Group, a conglomerate that owns a host of various firearm manufacturers. Through its long history, Remington has produced nearly every type of firearm and continues to produce a staggering array of weapons.

REMINGTON-BEALS 1ST MODEL POCKET REVOLVER
REDBOOK CODE: RB-RE-H-BE1MPR

Remington's first revolver. Came in a three-compartment pasteboard box with powder flask and bullet mold. Octagonal barrel. Marked with "F. Beal's" patent markings along with dates. Distinct differences in markings between issues. Roughly 5,000 produced. Cased models will bring a high premium.

Production: 1857 - 1858 Caliber: .31 cal. Action: SA, revolver   Barrel Length: 3"
Sights: fixed front Capacity: 5 Grips: gutta-percha wood grip, smooth

D2C:	NIB --	Ex $2,650	VG+ $1,915	Good $1,740	
C2C:	Mint --	Ex $2,425	VG+ $1,740	Good $1,570	Fair $ 805
Trade-In:	Mint --	Ex $1,950	VG+ $1,360	Good $1,220	Poor $ 360

REMINGTON-BEALS 2ND MODEL POCKET REVOLVER
VERY GOOD +
Courtesy of Rock Island Auction Company

REMINGTON-BEALS 2ND MODEL POCKET REVOLVER
REDBOOK CODE: RB-RE-H-BE2MPR

Marked "Beals Patent 1856 & 57, Manufactured by Remingtons Ilion, N.Y." on octagonal barrel. Distinct date differences in markings between issues. Blue finish. Less than 1,200 produced.

Production: 1858 - 1859 Caliber: .31 cal. Action: SA, revolver   Barrel Length: 3"
Sights: bead front Capacity: 5 Grips: gutta-percha wood, checkered or smooth

D2C:	NIB --	Ex $5,325	VG+ $2,900	Good $2,590	
C2C:	Mint --	Ex $4,850	VG+ $2,700	Good $2,330	Fair $1,200
Trade-In:	Mint --	Ex $3,920	VG+ $2,000	Good $1,815	Poor $ 540

BEALS' 3RD MODEL POCKET REVOLVER
GOOD
Courtesy of Rock Island Auction Company

REMINGTON-BEALS 3RD MODEL POCKET REVOLVER
REDBOOK CODE: RB-RE-H-BE3MPR

Marked "Beals Pat. 1856, 57, 58" and "Manufactured by Remingtons, Ilion, N.Y." Blue finish, loading lever. Roughly 1,500 produced.

Production: 1859 - 1860 Caliber: .31 cal. Action: SA, revolver   Barrel Length: 4"
Sights: fixed front Capacity: 5 Grips: gutta-percha wood grip or rubber

D2C:	NIB --	Ex $4,000	VG+ $2,900	Good $1,485	
C2C:	Mint --	Ex $3,600	VG+ $2,700	Good $1,335	Fair $ 685
Trade-In:	Mint --	Ex $3,000	VG+ $2,000	Good $1,040	Poor $ 300

**1861 ARMY
REVOLVER**
(OLD MODEL)
VERY GOOD +

Courtesy of Rock Island Auction Company

1861 ARMY REVOLVER (OLD MODEL)

REDBOOK CODE: RB-RE-H-1861AR

Marked with 1858 patent dates, though not produced until 1861. Octagonal barrel. No Beals patent markings. Higher value is placed on .46 caliber conversion.

Production: 1861 - 1862 Caliber: .44 cal., .46 cal. Action: SA, revolver

Barrel Length: 8" Sights: bead front Capacity: 6 Grips: walnut

D2C:	NIB	--	Ex	$3,275	VG+	$2,365	Good	$1,410		
C2C:	Mint	--	Ex	$2,975	VG+	$2,150	Good	$1,270	Fair	$ 650
Trade-In:	Mint	--	Ex	$2,425	VG+	$1,680	Good	$990	Poor	$ 300

**1861 NAVY
REVOLVER**
(OLD MODEL)
EXCELLENT

Courtesy of Rock Island Auction Company

1861 NAVY REVOLVER (OLD MODEL) REDBOOK CODE: RB-RE-H-1861NR

A short-barrel variant of the 1861 Army. Blue finish. Manufactured for U.S. government contracts. Less than 7,500 produced. No Beals patent markings.

Production: 1861 - 1862 Caliber: .36 cal. Action: SA, revolver Barrel Length: 7.3"

Sights: bead front Capacity: 6 Grips: walnut

D2C:	NIB	--	Ex	$2,950	VG+	$2,125	Good	$1,930		
C2C:	Mint	--	Ex	$2,675	VG+	$1,930	Good	$1,735	Fair	$ 890
Trade-In:	Mint	--	Ex	$2,175	VG+	$1,510	Good	$1,350	Poor	$ 390

REMINGTON-RIDER MAGAZINE PISTOL

REDBOOK CODE: BREH-RRMPXX An unusual tubular magazine pistol. Features hammer with breechblock, octagon barrel, most are engraved with nickel plating. Models with case hardened frames and blue barrel worth high premium.

Production: 1871 - 1888 Caliber: .32 RF cal. Action: SA Barrel Length: 3"

Sights: fixed front Capacity: 5 Grips: walnut, rosewood, pearl, or ivory

D2C:	NIB	--	Ex	$4,275	VG+	$3,080	Good	$845		
C2C:	Mint	$12,650	Ex	$3,875	VG+	$2,800	Good	$765	Fair	$390
Trade-In:	Mint	$9,360	Ex	$3,150	VG+	$2,190	Good	$595	Poor	$180

**REMINGTON-RIDER
POCKET REVOLVER**
EXCELLENT

REMINGTON-RIDER POCKET REVOLVER

REDBOOK CODE: RB-RE-H-REMRPR

Octagonal barrel, unique "mushroom" style cylinder, 2-line Remington address and Rider's patent markings. Factory engraved and cased models may bring high premium.

Production: 1860 - early 1870s Caliber: 31. cal. Action: DA/SA, revolver

Barrel Length: 2", 3" Sights: bead front Capacity: 5 Grips: gutta-percha wood

D2C:	NIB	--	Ex	$2,350	VG+	$1,050	Good	$955		
C2C:	Mint	--	Ex	$2,150	VG+	$960	Good	$860	Fair	$440
Trade-In:	Mint	--	Ex	$1,725	VG+	$750	Good	$670	Poor	$210

**REMINGTON-BEALS
NAVY REVOLVER**
VERY GOOD +

Courtesy of Rock Island Auction Company

REMINGTON-BEALS NAVY REVOLVER

REDBOOK CODE: RB-RE-H-RBNRXX

Marked with Beals 1858 patent dates, octagon barrel, martially marked models worth a 25% premium.

Production: ~1860 - ~1862 Caliber: .36 cal. Action: SA, revolver

Barrel Length: 7.5" Sights: bead front Capacity: 6 Grips: walnut

D2C:	NIB	--	Ex	$6,850	VG+	$4,950	Good	$4,500		
C2C:	Mint	--	Ex	$6,225	VG+	$4,500	Good	$4,050	Fair	$2,070
Trade-In:	Mint	--	Ex	$5,050	VG+	$3,510	Good	$3,150	Poor	$900

REMINGTON-BEALS ARMY REVOLVER
VERY GOOD +

REMINGTON-BEALS ARMY REVOLVER

REDBOOK CODE: RB-RE-H-RBARXX

Marked with Beals 1858 patent dates. Similar to the 1861 Navy Revolver. Blue finish, octagonal barrel. Less than 3,000 produced. Maritally marked models will bring a high premium of at least 50%.

Production: ~1860 - ~1862   Caliber: .44 cal.   Action: SA, revolver

Barrel Length: 8"   Sights: bead front   Capacity: 6   Grips: walnut

D2C:	NIB	--	Ex	$8,900	VG+	$6,435	Good	$2,900		
C2C:	Mint	--	Ex	$8,075	VG+	$5,850	Good	$2,610	Fair	$1,335
Trade-In:	Mint	--	Ex	$6,575	VG+	$4,570	Good	$2,030	Poor	$ 600

NEW MODEL ARMY REVOLVER
EXCELLENT

Courtesy of Rock Island Auction Company

NEW MODEL ARMY REVOLVER   REDBOOK CODE: RB-RE-H-NMARXX

Marked "New Model" after the 1858 patent date and address. Slightly less value for civilian model with no inspector markings. Roughly 20% less value for conversion model.

Production: 1863 - mid-1870s   Caliber: .44 cal.   .Action: SA, revolver

Barrel Length: 8"   Sights: fixed front   Capacity: 6   Grips: walnut

D2C:	NIB	--	Ex	$6,325	VG+	$3,165	Good	$1,495		
C2C:	Mint	--	Ex	$5,750	VG+	$2,880	Good	$1,350	Fair	$690
Trade-In:	Mint	--	Ex	$4,650	VG+	$2,250	Good	$1,050	Poor	$300

NEW MODEL NAVY REVOLVER
GOOD

Courtesy of Rock Island Auction Company

NEW MODEL NAVY REVOLVER   REDBOOK CODE: RB-RE-H-NMNRXX

A slightly shorter barrel variant of the New Model Army. Blue finish, octagonal barrel. .38 cal. conversions exist and are traditionally less valuable. Less than 28,000 produced. Less value for caliber conversions.

Production: 1863 - mid-1870s   Caliber: .36 cal.   Action: SA, revolver

Barrel Length: 7.2"   Sights: fixed front   Capacity: 6   Grips: walnut

D2C:	NIB	--	Ex	$6,100	VG+	$2,915	Good	$2,250		
C2C:	Mint	--	Ex	$5,525	VG+	$2,650	Good	$2,025	Fair	$1,035
Trade-In:	Mint	--	Ex	$4,500	VG+	$2,070	Good	$1,575	Poor	$ 450

NEW MODEL SINGLE-ACTION BELT REVOLVER

REDBOOK CODE: RB-RE-H-NMSABR

A smaller variant of the New Model Navy, may feature fluted cylinders to decrease weight. Octagonal barrel, brass trigger guard. Less than roughly 3,000 or so produced. Value traditionally decreases for caliber conversions.

Production: 1863 - mid-1870s   Caliber: .36 cal.   Action: SA, revolver

Barrel Length: 6.5"   Sights: fixed front   Capacity: 6   Grips: walnut

D2C:	NIB	--	Ex	$5,775	VG+	$3,440	Good	$2,105		
C2C:	Mint	--	Ex	$5,250	VG+	$3,130	Good	$1,895	Fair	$970
Trade-In:	Mint	--	Ex	$4,275	VG+	$2,440	Good	$1,475	Poor	$450

NEW MODEL SINGLE-ACTION BELT REVOLVER
VERY GOOD +

Courtesy of Rock Island Auction Company

REMINGTON-RIDER DOUBLE-ACTION BELT REVOLVER
GOOD

Courtesy of Rock Island Auction Company

REMINGTON-RIDER DOUBLE-ACTION BELT REVOLVER

REDBOOK CODE: RB-RE-H-RRDABR

Marked to indicate Rider's patent date of 1858, 1859. Blue or nickel finish. Octagonal barrel. Caliber conversions slightly reduce value. Fluted cylinder models are worth a high premium of roughly 30%. Roughly 5,000 manufactured.

Production: 1863 - early 1870s   Caliber: .36 cal.   Action: DA/SA, revolver

Barrel Length: 6.5"   Sights: fixed front   Capacity: 6   Grips: walnut

D2C:	NIB	--	Ex	$5,750	VG+	$2,435	Good	$2,210		
C2C:	Mint	--	Ex	$5,225	VG+	$2,210	Good	$1,990	Fair	$1,020
Trade-In:	Mint	--	Ex	$4,250	VG+	$1,730	Good	$1,550	Poor	$ 450

**NEW MODEL
POLICE REVOLVER**
VERY GOOD +

Courtesy of Rock Island Auction Company

NEW MODEL POLICE REVOLVER   REDBOOK CODE: RB-RE-H-NMPRXX

Patent dates and company markings on top of barrel. Roughhly 30% less value for rimfire conversion. High premium for cased models. Blue or nickel finish with higher value for blue model.

Production: mid-1860s - 1873   Caliber: .36 cal.   Action: SA, revolver

Barrel Length: 3.5", 4.5", 5.5", 6.5"   Sights: blade front   Capacity: 5   Grips: walnut

D2C:	NIB	--	Ex	$2,950	VG+	$2,135	Good	$1,940	
C2C:	Mint	--	Ex	$2,700	VG+	$1,940	Good	$1,750	Fair	$ 895
Trade-In:	Mint	--	Ex	$2,175	VG+	$1,520	Good	$1,360	Poor	$ 390

NEW MODEL POCKET REVOLVER FIRST SERIES

REDBOOK CODE: RB-RE-H-NMPRXX

Marked "Patented Sept. 14, 1858, March 17, 1863 E. Remington & Sons, Ilion, New York U.S.A. New Model." Very Rare. Nickel or blue finish, spurred trigger, compact design, octagonal barrel. Distinguished by its brass frame and trigger. Less value for .32 RF conversions.

Production: mid-1860s - 1873   Caliber: .31 cal.   Action: SA, revolver

Barrel Length: 3", 3.5", 4", 4.5"   Sights: blade front   Capacity: 5   Grips: walnut

D2C:	NIB	--	Ex	$2,650	VG+	$1,905	Good	$1,730	
C2C:	Mint	--	Ex	$2,400	VG+	$1,730	Good	$1,560	Fair	$ 800
Trade-In:	Mint	--	Ex	$1,950	VG+	$1,350	Good	$1,215	Poor	$ 360

NEW MODEL POCKET REVOLVER SECOND SERIES

REDBOOK CODE: RB-RE-H-NMPRX1

Marked "Patented Sept. 14, 1858, March 17, 1863 E. Remington & Sons, Ilion, New York U.S.A. New Model." Similar to first series but distinguished by its iron frame with brass trigger. Less value for .32 rimfire conversions.

Production: mid-1860s - 1873   Caliber: .31 cal.   Action: SA, revolver

Barrel Length: 3", 3.5", 4", 4.5"   Sights: blade front   Capacity: 5   Grips: walnut

D2C:	NIB	--	Ex	$2,375	VG+	$1,720	Good	$1,560	
C2C:	Mint	--	Ex	$2,175	VG+	$1,560	Good	$1,405	Fair	$ 720
Trade-In:	Mint	--	Ex	$1,750	VG+	$1,220	Good	$1,095	Poor	$ 330

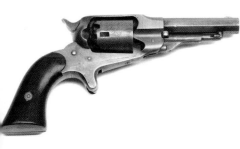

NEW MODEL POCKET REVOLVER THIRD SERIES

REDBOOK CODE: RB-RE-H-NMPRX2

Marked "Patented Sept. 14, 1858, March 17, 1863 E. Remington & Sons, Ilion, New York U.S.A. New Model." An all-iron variant of the New Model Pocket Revolver. No brass components. Less value for .32 rimfire conversions.

Production: mid-1860s - 1873   Caliber: .31 cal.   Action: SA, revolver

Barrel Length: 3", 3.5", 4", 4.5"   Sights: blade front   Capacity: 5   Grips: walnut

D2C:	NIB	--	Ex	$2,000	VG+	$1,430	Good	$1,050	
C2C:	Mint	--	Ex	$1,800	VG+	$1,300	Good	$ 945	Fair	$ 485
Trade-In:	Mint	--	Ex	$1,475	VG+	$1,020	Good	$ 735	Poor	$ 210

ZIG-ZAG DERRINGER
VERY GOOD +

ZIG-ZAG DERRINGER   REDBOOK CODE: RB-RE-H-ZZDXXX

Marked with "Elliots' Patents/Aug.17, 1858/May 29, 1860." Less than 1,000 produced. Noted for its six revolving barrels. Trigger-cocking action.

Production: 1861 - 1862   Caliber: .22 Short   Action: SA, revolver   Barrel Length: 3.25"

Sights: none   Capacity: 6   Grips: gutta percha (rubber)

D2C:	NIB	--	Ex	$7,475	VG+	$5,405	Good	$4,915	
C2C:	Mint	--	Ex	$6,800	VG+	$4,920	Good	$4,425	Fair	$2,260
Trade-In:	Mint	--	Ex	$5,525	VG+	$3,840	Good	$3,440	Poor	$990

**MODEL 1875
SINGLE-ACTION ARMY**
VERY GOOD +

Courtesy of Rock Island Auction Company

MODEL 1875 SINGLE-ACTION ARMY   REDBOOK CODE: RB-RE-H-1875SA

Marked "E. Remington & Sons. Ilion, N.Y. U.S.A." Built to compete with the Colt SAA. Nickel or blue finish. Some variations exist in markings.

Production: 1875 - late 1880s Caliber: .44 WCF, .45 Colt, .44 Rem. Action: SA, revolver									
Barrel Length: 7.5" Sights: blade front Capacity: 6 Grips: walnut or ivory									
D2C:	NIB	--	Ex	$12,550	VG+	$6,960	Good	$6,325	
C2C:	Mint	--	Ex	$11,400	VG+	$6,330	Good	$5,695	Fair $2,910
Trade-In:	Mint	--	Ex	$9,250	VG+	$4,940	Good	$4,430	Poor $1,290

**MODEL 1890
SINGLE-ACTION ARMY**
VERY GOOD +

Courtesy of Rock Island Auction Company

MODEL 1890 SINGLE-ACTION ARMY

REDBOOK CODE: RB-RE-H-1890SA
Marked "Remington Arms Co. Ilion, N.Y." on barrel. Frame marked "44C.F.W."

Production: 1891 - 1894 Caliber: .44 WCF (.44-40) Action: SA, revolver									
Barrel Length: 5.75", 7.5" Sights: blade front Capacity: 6 Grips: wood, ivory, or synthetic									
D2C:	NIB	--	Ex	$17,200	VG+	$12,430	Good	$11,300	
C2C:	Mint	--	Ex	$15,600	VG+	$11,300	Good	$10,170	Fair $5,200
Trade-In:	Mint	--	Ex	$12,675	VG+	$8,820	Good	$7,910	Poor $2,280

**REMINGTON-RIDER
DERRINGER
PARLOR PISTOL**
EXCELLENT

Courtesy of Rock Island Auction Company

REMINGTON-RIDER DERRINGER PARLOR PISTOL

REDBOOK CODE: RB-RE-H-REMRID
Marked with Rider's patent dates. Less than 700 believed to exist. Single-shot. Brass frame, barrel, and grip. Many fakes in circulation. Fires the .17 percussion caliber.

Production: 1860 - 1863 Caliber: .17 cal. Action: SA, derringer									
Barrel Length: 3" Sights: fixed front Capacity: 1 Grips: brass									
D2C:	NIB	--	Ex	$19,550	VG+	$14,135	Good	$12,850	
C2C:	Mint	--	Ex	$17,750	VG+	$12,850	Good	$11,565	Fair $5,915
Trade-In:	Mint	--	Ex	$14,400	VG+	$10,030	Good	$8,995	Poor $2,580

**REMINGTON-
ELLIOT PEPPERBOX
DERRINGER**
GOOD

Courtesy of Rock Island Auction Company

REMINGTON-ELLIOT PEPPERBOX DERRINGER

REDBOOK CODE: RB-RE-H-RERXXX
Marked with Elliot patent dates, 1860/1861. Blue or nickel finish. Features four barrels on the .32 caliber and five barrels on the .22 caliber. Also known as the "Ring-Trigger." High premium with original box.

Production: 1863 - 1888 Caliber: .22 cal., .32 cal. Action: SA, derringer Barrel									
Length: 3", 3.375" Sights: fixed front Capacity: 4, 5" Grips: wood, ivory, or synthetic									
D2C:	NIB	--	Ex	$5,075	VG+	$3,660	Good	$1,600	
C2C:	Mint	--	Ex	$4,600	VG+	$3,300	Good	$1,440	Fair $ 740
Trade-In:	Mint	--	Ex	$3,725	VG+	$2,600	Good	$ 1,120	Poor $ 330

VEST POCKET NO. 1
GOOD

Courtesy of Rock Island Auction Company

VEST POCKET NO. 1   REDBOOK CODE: RB-RE-H-VPXXX1

No marking on early production. Later marked with address and patent date. Nickel or blued finish. Over 20,000 produced in .22 caliber. High premium for blued model. Some higher priced exceptions exist for factory engraved or silver inlayed models with box.

Production: 1865 - 1888 Caliber: .22 cal. Action: SA, derringer									
Barrel Length: 3", 3.25" Sights: fixed front Capacity: 1 Grips: smooth wood or ivory									
D2C:	NIB	--	Ex	$2,825	VG+	$2,035	Good	$1,850	
C2C:	Mint	--	Ex	$2,575	VG+	$1,850	Good	$1,665	Fair $ 855
Trade-In:	Mint	--	Ex	$2,075	VG+	$1,450	Good	$1,295	Poor $ 390

VEST POCKET NO. 2
EXCELLENT

Courtesy of Rock Island Auction Company

VEST POCKET NO. 2   REDBOOK CODE: RB-RE-H-VPXXX2

A .32 cal. and .30 cal. variant of the original Vest Pocket .22. Marked Ilion production and 1861 patent date. Nickel or blue finish. High premium for engraved or silver inlay models with box.

Production: 1865 - 1888 Caliber: .32 cal., .30 cal. Action: SA, derringer									
Barrel Length: 3.5", 4" Sights: fixed front Capacity: 1 Grips: smooth wood or ivory									
D2C:	NIB	--	Ex	$1,900	VG+	$1,365	Good	$1,240	
C2C:	Mint	--	Ex	$1,725	VG+	$1,240	Good	$1,120	Fair	$575
Trade-In:	Mint	--	Ex	$1,400	VG+	$970	Good	$870	Poor	$270

VEST POCKET NO. 3
EXCELLENT

Courtesy of Rock Island Auction Company

VEST POCKET NO. 3   REDBOOK CODE: RB-RE-H-VPXXX3

A .41 cal. variant of the original Vest Pocket .22. Noted for its extended barrel and slightly larger, octagonal frame. Round barrel, blue or nickel finish.

Production: 1865 - 1888 Caliber: .41 cal. Action: SA, derringer									
Barrel Length: 4" Sights: fixed front Capacity: 1 Grips: smooth wood or ivory									
D2C:	NIB	--	Ex	$1,800	VG+	$1,295	Good	$1,175	
C2C:	Mint	--	Ex	$1,625	VG+	$1,180	Good	$1,060	Fair	$545
Trade-In:	Mint	--	Ex	$1,325	VG+	$920	Good	$825	Poor	$240

**REMINGTON OVER/
UNDER DERRINGER
TYPE I**
EXCELLENT

Courtesy of Rock Island Auction Company

REMINGTON OVER/UNDER DERRINGER MODEL 95 (TYPE I)

REDBOOK CODE: RB-RE-H-ROUDT1

Marked "Elliot's Patent Dec. 12, 1865" on one side between barrels. Marked "E. Remington & Sons, Ilion, N.Y." on other side between barrels. Rare early models with no extractor have higher value. Nickel or blue finish. Higher value for blue model.

Production: 1865 - 1888 Caliber: .41 cal. Action: SA, derringer									
Barrel Length: 3" Sights: blade front Capacity: 2 Grips: wood, ivory, or synthetic									
D2C:	NIB	--	Ex	$2,250	VG+	$1,625	Good	$1,475	
C2C:	Mint	--	Ex	$2,050	VG+	$1,480	Good	$1,330	Fair	$680
Trade-In:	Mint	--	Ex	$1,675	VG+	$1,160	Good	$1,035	Poor	$300

**REMINGTON OVER/
UNDER DERRINGER
TYPE II**
EXCELLENT

Courtesy of Rock Island Auction Company

REMINGTON OVER/UNDER DERRINGER MODEL 95

(TYPE II)   REDBOOK CODE: RB-RE-H-ROUDMP

Similar to Type I, but marked "Remington Arms Co., Ilion, N.Y." on the top of the barrel rib. Several variations exist in the address. Nickel or blue finish. Higher value for blue model.

Production: 1888 - early 1910 Caliber: .41 cal. Action: SA, derringer									
Barrel Length: 3" Sights: blade front Capacity: 2 Grips: wood, ivory, or synthetic									
D2C:	NIB	--	Ex	$2,075	VG+	$1,485	Good	$1,350	
C2C:	Mint	--	Ex	$1,875	VG+	$1,350	Good	$1,315	Fair	$625
Trade-In:	Mint	--	Ex	$1,525	VG+	$1,060	Good	$945	Poor	$270

**REMINGTON OVER/UNDER
DERRINGER TYPE III**
EXCELLENT

Courtesy of Rock Island Auction Company

REMINGTON OVER/UNDER DERRINGER MODEL 95

(TYPE III)   REDBOOK CODE: RB-RE-H-ROUDT3

Similar to previous models but marked "Remington Arms - U.M.C. Co. Ilion, N.Y." Blue or nickel finish. Higher value for blue model.

Production: 1910 - 1935 Caliber: .41 cal. Action: SA, derringer Barrel Length: 3"									
Sights: blade front Capacity: 2 Grips: wood, ivory, or synthetic									
D2C:	NIB	--	Ex	$2,450	VG+	$1,760	Good	$1,600	
C2C:	Mint	--	Ex	$2,225	VG+	$1,600	Good	$1,440	Fair	$740
Trade-In:	Mint	--	Ex	$1,800	VG+	$1,250	Good	$1,120	Poor	$330

**MODEL 1911A1
REMINGTON-RAND**
GOOD

Courtesy of Rock Island Auction Company

MODEL 1911A1 REMINGTON-RAND   REDBOOK CODE: RB-RE-H-1911RR

Produced by the Remington Rand Company and not the Remington Arms Company, although Remington Rand was originally owned by E. Remington & Sons. Most marked "REMINGTON RAND, INC. SYRACUSE, NEW YORK." Parkerized finish. Markings include a proof mark, ordnance inspector mark, and ordnance department inspector's stamp.

Production: 1942 - ~1945 Caliber: .45 ACP Action: SA, semi-auto

Barrel Length: 5" Sights: fixed Capacity: 7 Grips: walnut or synthetic

D2C:	NIB	--	Ex	$2,600	VG+	$1,870	Good $1,700	
C2C:	Mint	--	Ex	$2,350	VG+	$1,700	Good $1,530	Fair $ 785
Trade-In:	Mint	--	Ex	$1,925	VG+	$1,330	Good $1,190	Poor $ 360

MODEL 51
EXCELLENT

Courtesy of Rock Island Auction Company

MODEL 51   REDBOOK CODE: RB-RE-H-MOD51

Marked "Remington UMC" on the grips and Pedersen patent. Less than 70,000 produced. Parkerized finish, frame-mounted and grip safety. Engraved models with original box will bring a high premium.

Production: 1918 - early 1930s Caliber: .32 ACP, .380 ACP Action: SA, semi-auto

Barrel Length: 3.5" Sights: fixed Capacity: 7, 8 Grips: black synthetic

D2C:	NIB	--	Ex	$875	VG+	$635	Good $575	
C2C:	Mint	--	Ex	$800	VG+	$580	Good $520	Fair $265
Trade-In:	Mint	--	Ex	$650	VG+	$450	Good $405	Poor $120

MODEL R51
NEW IN BOX

MODEL R51   REDBOOK CODE: RB-RE-H-MR51XX

An updated reissue of the Model 51. Aluminum frame, steel barrel, frame-mounted and grip safety.

Production: 2014 - current Caliber: 9mm Action: SA, semi-auto

Barrel Length: 3.4" Sights: 3-dot or adjustable Capacity: 7 Magazine: detachable box

Grips: black synthetic

D2C:	NIB	$ 399	Ex $ 325	VG+ $ 220	Good $ 200	
C2C:	Mint	$ 390	Ex $ 300	VG+ $ 200	Good $ 180	Fair $ 95
Trade-In:	Mint	$ 290	Ex $ 225	VG+ $ 160	Good $ 140	Poor $ 60

M1911 R1
NEW IN BOX

M1911 R1   REDBOOK CODE: RB-RE-H-1911R1

Flared ejection port, satin-black finish, frame-mounted safety, also available in stainless for a slight premium.

Production: current Caliber: .45 ACP Action: SA, semi-auto   Barrel Length: 5"

Wt.: 38 oz. Sights: dovetail, 3-dot Capacity: 7 Grips: double-diamond walnut wood

D2C:	NIB $ 698	Ex $ 550	VG+ $385	Good $ 350	LMP $729
C2C:	Mint $ 680	Ex $ 500	VG+ $350	Good $ 315	Fair $165
Trade-In:	Mint $ 500	Ex $ 400	VG+ $280	Good $ 245	Poor $90

**M1911 R1
ENHANCED**
NEW IN BOX

M1911 R1 ENHANCED   REDBOOK CODE: RB-RE-H-1911EN

Updated tactical design, black oxide finish, enhanced hammer, front and rear slide serrations, beavertail grip safety, aluminum-match trigger, also available in stainless at an additional price.

Production: current Caliber: .45 ACP Action: SA, semi-auto   Barrel Length: 5"

Wt.: 40 oz Sights: red fiber-optic front, adjustable rear Capacity: 8

Magazine: detachable box Grips: enhanced wood laminate

D2C:	NIB $ 885	Ex $ 675	VG+ $490	Good $ 445	LMP $940
C2C:	Mint $ 850	Ex $ 625	VG+ $450	Good $ 400	Fair $ 205
Trade-In:	Mint $ 630	Ex $ 500	VG+ $350	Good $ 310	Poor $90

M1911 R1 CARRY
NEW IN BOX

M1911 R1 CARRY   REDBOOK CODE: RB-RE-H-1911CA

Noted for being more lightweight than the M1911 R1. Grip and frame-mounted safety, enhanced hammer, stainless steel barrel with target crown, black oxide finish, skeletonized match trigger.

Production: current   Caliber: .45 ACP   Action: SA, semi-auto   Barrel Length: 5"

Wt.: 39 oz.   Sights: Novak, tritium night front   Capacity: 7, 8   Grips: cocobolo

D2C:	NIB	$ 1,110	Ex $ 850	VG+ $ 615	Good $ 555	LMP	$1,299
C2C:	Mint	$1,070	Ex $ 775	VG+ $ 560	Good $ 500	Fair	$260
Trade-In:	Mint	$ 790	Ex $ 625	VG+ $ 440	Good $ 390	Poor	$120

M1911 R1 CARRY COMMANDER
NEW IN BOX

M1911 R1 CARRY COMMANDER   REDBOOK CODE: RB-RE-H-1911CC

Indicative of the Colt M1911 Commander. Enhanced safety, flared ejection port, black oxide finish, de-horned frame and slide, match grade barrel, also available with Crimson Trace grips for an additional price.

Production: current   Caliber: .45 ACP   Action: SA, semi-auto   Barrel Length: 4.25"

Wt.: 39 oz.   Sights: Novak, tritium front   Capacity: 7   Grips: cocobolo

D2C:	NIB	$1,098	Ex $ 850	VG+ $ 605	Good $ 550	LMP	$1299
C2C:	Mint	$1,060	Ex $ 775	VG+ $ 550	Good $ 495	Fair	$255
Trade-In:	Mint	$ 780	Ex $ 625	VG+ $ 430	Good $ 385	Poor	$120

M1911 R1 COMMANDER
NEW IN BOX

M1911 R1 COMMANDER   REDBOOK CODE: RB-RE-H-1911CO

Indicative to the Colt M1911 Commander. Enhanced safety, flared ejection port, black oxide finish, frame-mounted and grip safety, match grade barrel.

Production: current   Caliber: .45 ACP   Action: SA, semi-auto   Barrel Length: 4.25"

Wt.: 39 oz.   Sights: Novak, tritium front   Capacity: 7   Magazine: detachable box

Grips: walnut

D2C:	NIB	$ 660	Ex $ 525	VG+ $ 365	Good $ 330	LMP	$ 729
C2C:	Mint	$ 640	Ex $ 475	VG+ $ 330	Good $ 300	Fair	$ 155
Trade-In:	Mint	$ 470	Ex $ 375	VG+ $ 260	Good $ 235	Poor	$ 90

M1911 R1 CENTENNIAL
NEW IN BOX

M1911 R1 CENTENNIAL   REDBOOK CODE: RB-RE-H-1911CE

Celebrates the 100th year of the M1911's production. Unique engravings on the slide, black oxide finish. Centennial edition certificate included.

Production: current   Caliber: .45 ACP   Action: SA, semi-auto   Barrel Length: 5"

Wt.: 39 oz.   Sights: dovetail   Capacity: 7   Magazine: detachable box

Grips: rosewood laminate

D2C:	NIB	$1,030	Ex $ 800	VG+ $ 570	Good $ 515	LMP	$1,250
C2C:	Mint	$ 990	Ex $ 725	VG+ $ 520	Good $ 465	Fair	$240
Trade-In:	Mint	$ 740	Ex $ 600	VG+ $ 410	Good $ 365	Poor	$120

M1911 R1 CENTENNIAL LIMITED
NEW IN BOX

M1911 R1 CENTENNIAL LIMITED   REDBOOK CODE: RB-RE-H-1911CL

Celebrates the 100th year of the M1911's production. Features 24-kt. gold engravings on the slide. Blue finish, frame-mounted and grip safety, includes custom carrying case.

Production: current   Caliber: .45 ACP   Action: SA, semi-auto   Barrel Length: 5"

Wt.: 39 oz.   Sights: gold bead front   Capacity: 7   Magazine: detachable box

Grips: exhibition-grade smooth walnut

D2C:	NIB	$1,760	Ex $1,350	VG+ $ 970	Good $ 880	LMP	$2,250
C2C:	Mint	$1,690	Ex $1,225	VG+ $ 880	Good $ 795	Fair	$ 405
Trade-In:	Mint	$1,250	Ex $1,000	VG+ $ 690	Good $ 620	Poor	$ 180

Republic Forge *As featured on the front cover*

Located in Perryville, Texas, Republic Forge produces custom, hand-crafted 1911 pistols. The company prides itself on employing all American-made machinery and parts in production of all of their firearms. The company's website allows buyers to fully customize and build their own 1911 with an array of options for sights, grips, Cerakote colors, frame modifications, and other custom features.

REPUBLIC
NEW IN BOX

REPUBLIC   REDBOOK CODE: RB-RI-H-1911A2
Full size Government frame and slide, Cerakote finish, standard or ambi safety, add $200 for double stack frame, add $575 for elephant ivory, add $850 for mammoth ivory, add $1,200 for mammoth tooth ivory. Add $125 for calibers other than .45.

Production: 2014 - current   Caliber: .45 ACP, 10mm, .40 S&W, .38 Super, 9mm
Action: SA, semi-auto Barrel Length: 5" Sights: Novak night sights or Bomar adjustable
Capacity: 8 Magazine: detachable box   Grips: color coordinated VZ grips

D2C:	NIB	$2,795	Ex	$2,125	VG+	--	Good	--	
C2C:	Mint	$2,690	Ex	$1,950	VG+	--	Good	--	Fair --
Trade-In:	Mint	$1,990	Ex	$1,575	VG+	--	Good	--	Poor --

Rock Island Armory (Armscor)

Rock Island Armory is a brand owned by the Philippine company Armscor. Armscor opened a U.S.-based manufacturing plant in 1985 before purchasing the Rock Island Armory brand. Now located in Pharump, Nevada, Rock Island Armory produces a full line of M1911 pistols, polymer pistols, revolvers, rifles, and shotguns. They are best known for their quality, reasonably priced M1911 pistols.

M1911-A2   REDBOOK CODE: RB-RI-H-1911A2
Blue finish, additional price for two-tone or chrome finish, magwell, combat hammer, skeletonized trigger, high capacity.

Production: discontinued Caliber: .45 ACP, .40 S&W, .22 TCM Action: SA, semi-auto
Barrel Length: 5"   Sights: fiber-optic front, adjustable rear Capacity: 14, 16, 17

D2C:	NIB	$ 485	Ex	$ 375	VG+	$ 270	Good	$ 245	
C2C:	Mint	$ 470	Ex	$ 350	VG+	$ 250	Good	$ 220	Fair $ 115
Trade-In:	Mint	$ 350	Ex	$ 275	VG+	$ 190	Good	$ 170	Poor $ 60

1911 COMPACT GI
NEW IN BOX
Courtesy of Bud's Gun Shop

1911 COMPACT GI   REDBOOK CODE: RB-RI-H-1911CG
Indicative of the Colt M1911 Compact. Parkerized finish, frame-mounted and grip safety, compact design.

Production: current Caliber: .45 ACP Action: SA, semi-auto   Barrel Length: 3.5"
Sights: fixed Capacity: 7 Magazine: detachable box Grips: smoothed wood or black synthetic

D2C:	NIB	$ 470	Ex	$ 375	VG+	$ 260	Good	$ 235	
C2C:	Mint	$ 460	Ex	$ 325	VG+	$ 240	Good	$ 215	Fair $ 110
Trade-In:	Mint	$ 340	Ex	$ 275	VG+	$ 190	Good	$ 165	Poor $ 60

1911 MIDSIZE GI   REDBOOK CODE: RB-RI-H-1911MG
Indicative of the Colt M1911 Commander. Parkerized finish, frame-mounted and grip safety, classic design.

Production: current Caliber: .45 ACP Action: SA, semi-auto   Barrel Length: 4.25"
Sights: fixed Capacity: 8 Magazine: detachable box Grips: smoothed wood

D2C:	NIB	$ 470	Ex	$ 375	VG+	$ 260	Good	$ 235	
C2C:	Mint	$ 460	Ex	$ 325	VG+	$ 240	Good	$ 215	Fair $ 110
Trade-In:	Mint	$ 340	Ex	$ 275	VG+	$ 190	Good	$ 165	Poor $ 60

1911 STANDARD GI
NEW IN BOX

Courtesy of Bud's Gun Shop

1911 STANDARD GI REDBOOK CODE: RB-RI-H-1911SG

Indicative of the Colt M1911 Government. Full-sized design, parkerized finish, frame-mounted and grip safety.

Production: current Caliber: 9mm, .38 Super, .45 ACP Action: SA, semi-auto

Barrel Length: 5" Sights: fixed Capacity: 8, 9 Magazine: detachable box

Grips: smoothed wood

D2C:	NIB $ 470	Ex $ 375	VG+ $ 260	Good $ 235	
C2C:	Mint $ 460	Ex $ 325	VG+ $ 240	Good $ 215	Fair $ 110
Trade-In:	Mint $ 340	Ex $ 275	VG+ $ 190	Good $ 165	Poor $ 60

TAC 1911 FS
NEW IN BOX

Courtesy of Bud's Gun Shop

TAC 1911 FS REDBOOK CODE: RB-RI-H-T19112

Indicative of the Colt M1911 Government. Tactical design, parkerized finish, skeletonized trigger and hammer, frame-mounted and grip safety.

Production: current Caliber: 9mm, .45 ACP Action: SA, semi-auto

Barrel Length: 5" Sights: combat-style Capacity: 8, 9 Magazine: detachable box

Grips: smoothed wood or black synthetic

D2C:	NIB $ 499	Ex $ 400	VG+ $ 275	Good $ 250	
C2C:	Mint $ 480	Ex $ 350	VG+ $ 250	Good $ 225	Fair $ 115
Trade-In:	Mint $ 360	Ex $ 300	VG+ $ 200	Good $ 175	Poor $ 60

TAC 1911 COMPACT
NEW IN BOX

Courtesy of Bud's Gun Shop

TAC 1911 COMPACT REDBOOK CODE: RB-RI-H-T1911C

Indicative of M1911 Compact. Tactical design, parkerized finish, full guide rod, frame-mounted and grip safety, compact design.

Production: current Caliber: 9mm, .45 ACP Action: SA, semi-auto

Barrel Length: 3.5" Sights: low profile dovetailed front, low profile dovetailed rear

Capacity: 7, 8 Magazine: detachable box Grips: smoothed wood

D2C:	NIB $ 520	Ex $ 400	VG+ $ 290	Good $ 260	
C2C:	Mint $ 500	Ex $ 375	VG+ $ 260	Good $ 235	Fair $ 120
Trade-In:	Mint $ 370	Ex $ 300	VG+ $ 210	Good $ 185	Poor $ 60

TAC 2011 VZ
NEW IN BOX

Courtesy of Bud's Gun Shop

TAC 2011 VZ REDBOOK CODE: RB-RI-H-2011V1

Indicative of the Colt M1911 Government. Tactical design, tactical rail, parkerized finish, frame-mounted and grip safety, distinguished by its VZ-brand grips.

Production: current Caliber: 10mm, .45 ACP Action: SA, semi-auto

Barrel Length: 5" Sights: fiber-optic front, 2-dot tactical adjustable rear

Capacity: 8 Magazine: detachable box Grips: VZ tactical, synthetic

D2C:	NIB $ 678	Ex $ 525	VG+ $ 375	Good $ 340	
C2C:	Mint $ 660	Ex $ 475	VG+ $ 340	Good $ 310	Fair $ 160
Trade-In:	Mint $ 490	Ex $ 400	VG+ $ 270	Good $ 240	Poor $ 90

TAC 2011 HI CAP REDBOOK CODE: RB-RI-H-201145

Modern design, tactical rail, parkerized finish, ambidextrous/skeletonized hammer and trigger, frame-mounted and grip safety.

Production: current Caliber: .45 ACP, 9mm Action: SA, semi-auto

Barrel Length: 5" Sights: orange fiber-optic front, white dot rear

Capacity: 13, 17 Magazine: detachable box Grips: VZ Operator II or synthetic

D2C:	NIB $ 725	Ex $ 575	VG+ $ 400	Good $ 365	
C2C:	Mint $ 700	Ex $ 525	VG+ $ 370	Good $ 330	Fair $ 170
Trade-In:	Mint $ 520	Ex $ 425	VG+ $ 290	Good $ 255	Poor $ 90

MAPP FULL SIZE
NEW IN BOX
Courtesy of Bud's Gun Shop

MAPP FULL SIZE   REDBOOK CODE: RB-RI-H-MAPPFS

Matte black frame, parkerized slide, frame-mounted safety, accessory rail, also available with full parkerized finish.

Production: current   Caliber: 9mm   Action: DA/SA, semi-auto

Sights: integrated front, snag-free rear   Capacity: 16   Magazine: detachable box

Grips: black synthetic

D2C:	NIB	$ 450	Ex $ 350	VG+ $ 250	Good $ 225				
C2C:	Mint $ 440	Ex $ 325	VG+ $ 230	Good $ 205	Fair $ 105				
Trade-In:	Mint $ 320	Ex $ 275	VG+ $ 180	Good $ 160	Poor $ 60				

Rock River Arms, Inc.

Brothers Mark and Chuck Larson founded Rock River Arms in 1996 after working for Springfield Armory. Originally, the brothers focused on custom M1911 pistols, but soon ventured into the AR-15 market, eventually dropping their pistol line to focus on rifles. After winning several government contracts, the company quickly rose to fame in the AR world. Currently, Rock River Arms produces many variants of their LAR series rifles, as well as a host of parts and accessories. The company recently announced they anticipate reintroducing a M1911 pistol.

LIMITED MATCH   REDBOOK CODE: RB-RR-H-LMTCHX

Match Commander hammer, beavertail grip safety, flared ejection port. Available in hard chrome, blue, or duo-tone finishes.

Production: discontinued   Caliber: .45 ACP   Action: SA, semi-auto

Barrel Length: 5"   Sights: dovetail front, Bomar rear   Capacity: 7   Grips: checkered rosewood

D2C:	NIB $2,495	Ex $1,900	VG+ $1,375	Good $1,250	
C2C:	Mint $2,400	Ex $1,725	VG+ $1,250	Good $1,125	Fair $ 575
Trade-In:	Mint $1,780	Ex $1,400	VG+ $ 980	Good $ 875	Poor $ 270

NATIONAL MATCH HARDBALL   REDBOOK CODE: RB-RR-H-NMTCHH

Indicative of the Colt M1911 Commander. Blue or chrome finish, beveled magwell, modified ejector, steel slide, frame-mounted safety.

Production: early 2000s   Caliber: .45 ACP   Action: SA, semi-auto

Barrel Length: 4"   Sights: dovetail front, Bomar adjustable rear   Capacity: 7

Magazine: detachable box   Grips: cocobolo or walnut, checkered

D2C:	NIB $1,530	Ex $1,175	VG+ $ 845	Good $ 765	
C2C:	Mint $1,470	Ex $1,075	VG+ $ 770	Good $ 690	Fair $ 355
Trade-In:	Mint $1,090	Ex $ 875	VG+ $ 600	Good $ 540	Poor $ 180

BULLSEYE WADCUTTER   REDBOOK CODE: RB-RR-H-BULWAD

Slide scope mount, blue finish, beavertail grip and frame-mounted safety, modified ejector, skeletonized trigger.

Production: early 2000s   Caliber: .45 ACP   Action: SA, semi-auto

Barrel Length: 5"   Sights: scope rails   Capacity: 7   Magazine: detachable box

Grips: checkered rosewood

D2C:	NIB $1,730	Ex $1,325	VG+ $ 955	Good $ 865	
C2C:	Mint $1,670	Ex $1,200	VG+ $ 870	Good $ 780	Fair $ 400
Trade-In:	Mint $1,230	Ex $ 975	VG+ $ 680	Good $ 610	Poor $ 180

ROCK ISLAND ARMORY

ROCK RIVER ARMS

BASIC CARRY   REDBOOK CODE: RB-RR-H-BASCAR
Flared ejection port, de-horned/match barrel, blue finish.

Production: 2005-discontinued Caliber: .45 ACP Action: SA, semi-auto   Barrel Length: 5"

Sights: Heinie sights   Capacity: 7   Grips: rosewood

D2C:	NIB	$1,699	Ex	$1,300	VG+	$ 935	Good	$ 850
C2C:	Mint	$1,640	Ex	$1,175	VG+	$ 850	Good	$ 765	Fair	$ 395
Trade-In:	Mint	$1,210	Ex	$ 975	VG+	$ 670	Good	$ 595	Poor	$ 180

PRO CARRY   REDBOOK CODE: RB-RR-H-PROCAR
Blue finish, extended extractor, optional Black T finish, lowered ejection port.

Production: discontinued   Caliber: .45 ACP   Action: SA, semi-auto

Barrel Length: 4.25", 5", 6"   Sights: Novak tritium or Heinie   Capacity: 7   Grips: rosewood

D2C:	NIB	$1,730	Ex	$1,325	VG+	$ 955	Good	$ 865
C2C:	Mint	$1,670	Ex	$1,200	VG+	$ 870	Good	$ 780	Fair	$ 400
Trade-In:	Mint	$1,230	Ex	$ 975	VG+	$ 680	Good	$ 610	Poor	$ 180

Rossi

Amadeo Rossi founded the company in 1889 and members of his family still run the company today. Importation to the United States began with Interarms, but now BrazTech is the exclusive importer of the Rossi brand. While Taurus manufactures several revolver models under contract from the company, Rossi produces most of their revolvers, rifles, and shotguns in Sao Leupoldo, Brazil.

R97206
EXCELLENT

R97206   REDBOOK CODE: RB-RO-H-97206X
Stainless steel finish, contoured grip, Taurus Security System.

Production: current Caliber: .357 Magnum Action: DA/SA, revolver

Barrel Length: 6" OA Length: 10.5" Wt.: 35 oz. Sights: red ramp front,

adjustable rear Capacity: 6 Grips: rubber

D2C:	NIB	$ 460	Ex	$ 350	VG+	$ 255	Good	$ 230	LMP	$511.37
C2C:	Mint	$ 450	Ex	$ 325	VG+	$ 230	Good	$ 210	Fair	$ 110
Trade-In:	Mint	$ 330	Ex	$ 275	VG+	$ 180	Good	$ 165	Poor	$ 60

R97206FC   REDBOOK CODE: RB-RO-H-97206F
Stainless steel finish, contoured finger groove black rubber grip.

Production: discontinued   Caliber: .357 Magnum   Action: DA/SA, revolver

Barrel Length: 6"   Sights: red ramp front, adjustable rear   Capacity: 6   Grips: rubber

D2C:	NIB	$ 450	Ex	$ 350	VG+	$ 250	Good	$ 225	LMP	$ 537
C2C:	Mint	$ 440	Ex	$ 325	VG+	$ 230	Good	$ 205	Fair	$ 105
Trade-In:	Mint	$ 320	Ex	$ 275	VG+	$ 180	Good	$ 160	Poor	$ 60

R97104FC   REDBOOK CODE: RB-RO-H-97104F
Blue finish, contoured black rubber grip.

Production: discontinued   Caliber: .357 Magnum   Action: DA/SA, revolver

Barrel Length: 4"   Sights: adjustable rear   Capacity: 6   Grips: rubber

D2C:	NIB	$ 375	Ex	$ 300	VG+	$ 210	Good	$ 190	LMP	$ 479
C2C:	Mint	$ 360	Ex	$ 275	VG+	$ 190	Good	$ 170	Fair	$ 90
Trade-In:	Mint	$ 270	Ex	$ 225	VG+	$ 150	Good	$ 135	Poor	$ 60

ROSSI

R97104
NEW IN BOX

R97104 REDBOOK CODE: RB-RO-H-97104X
Blue finish, Taurus Security System, contoured black rubber grips.

Production: current Caliber: .357 Magnum Action: DA/SA, revolver

Barrel Length: 4" OA Length: 8.5" Wt.: 32 oz. Sights: adjustable

Capacity: 6 Grips: rubber

D2C:	NIB $ 390	Ex $300	VG+ $ 215	Good $ 195	LMP $ 454
C2C:	Mint $ 380	Ex $ 275	VG+ $ 200	Good $ 180	Fair $ 90
Trade-In:	Mint $ 280	Ex $ 225	VG+ $ 160	Good $ 140	Poor $ 60

R85104
NEW IN BOX

R85104 REDBOOK CODE: RB-RO-H-85104X
Blue finish, contoured grip, vent rib, Taurus Security System.

Production: current Caliber: .38 Special +P Action: DA/SA, revolver

Barrel Length: 4" Wt.: 32 oz. Sights: adjustable Capacity: 6 Grips: rubber

D2C:	NIB $ 354	Ex $ 275	VG+ $ 195	Good $ 180	LMP $ 390
C2C:	Mint $ 340	Ex $ 250	VG+ $ 180	Good $ 160	Fair $ 85
Trade-In:	Mint $ 260	Ex $ 200	VG+ $ 140	Good $ 125	Poor $ 60

R85104FC
NEW IN BOX

R85104FC REDBOOK CODE: RB-RO-H-85104F
Blue finish, vent rib barrel.

Production: discontinued Caliber: .38 Special +P Action: DA/SA, revolver

Barrel Length: 4" Sights: adjustable Capacity: 6 Grips: rubber

D2C:	NIB $ 350	Ex $ 275	VG+ $ 195	Good $ 175	LMP $ 414
C2C:	Mint $ 340	Ex $ 250	VG+ $ 180	Good $ 160	Fair $ 85
Trade-In:	Mint $ 250	Ex $ 200	VG+ $ 140	Good $ 125	Poor $ 60

R46202FC
NEW IN BOX

R46202FC REDBOOK CODE: RB-RO-H-46202F
Stainless steel finish.

Production: discontinued Caliber: .357 Magnum Action: DA/SA, revolver

Barrel Length: 2" Sights: fixed Capacity: 6 Grips: rubber

D2C:	NIB $ 380	Ex $ 300	VG+ $ 210	Good $ 190	LMP $ 479
C2C:	Mint $ 370	Ex $ 275	VG+ $ 190	Good $ 175	Fair $ 90
Trade-In:	Mint $ 270	Ex $ 225	VG+ $ 150	Good $ 135	Poor $ 60

R46202
NEW IN BOX

R46202 REDBOOK CODE: RB-RO-H-46202X
Stainless steel finish, Taurus Security System.

Production: current Caliber: .357 Magnum Action: DA/SA, revolver

Barrel Length: 2" Wt.: 26 oz. Sights: fixed Capacity: 6 Grips: rubber

D2C:	NIB $ 380	Ex $ 300	VG+ $ 210	Good $ 190	LMP $ 454
C2C:	Mint $ 370	Ex $ 275	VG+ $ 190	Good $ 175	Fair $ 90
Trade-In:	Mint $ 270	Ex $ 225	VG+ $ 150	Good $ 135	Poor $ 60

R46102
NEW IN BOX

R46102 REDBOOK CODE: RB-RO-H-46102X
Blue finish, Taurus Security System, contoured grips.

Production: current Caliber: .357 Magnum Action: DA/SA, revolver

Barrel Length: 2" Wt.: 26 oz. Sights: fixed Capacity: 6 Grips: rubber

D2C:	NIB $ 360	Ex $ 275	VG+ $ 200	Good $ 180	LMP $ 390
C2C:	Mint $ 350	Ex $ 250	VG+ $ 180	Good $ 165	Fair $ 85
Trade-In:	Mint $ 260	Ex $ 225	VG+ $ 150	Good $ 130	Poor $ 60

whitman.com

R46102FC
NEW IN BOX

ROSSI

R46102FC REDBOOK CODE: RB-RO-H-46102F
Blue finish.

Production: discontinued Caliber: .357 Magnum Action: DA/SA, revolver											
Barrel Length: 2" Sights: fixed Capacity: 6 Grips: rubber											
D2C:	NIB $	365	Ex $	300	VG+ $	205	Good $	185	LMP $	414	
C2C:	Mint $	360	Ex $	275	VG+ $	190	Good $	165	Fair $	85	
Trade-In:	Mint $	260	Ex $	225	VG+ $	150	Good $	130	Poor $	60	

R35102
NEW IN BOX

R35102 REDBOOK CODE: RB-RO-H-35102X
Blue finish, transfer-bar safety, steel frame, integral locking system.

Production: current Caliber: .38 Special +P Action: DA/SA, revolver											
Barrel Length: 2" Wt.: 24 oz. Sights: fixed Capacity: 5 Grips: rubber											
D2C:	NIB $	365	Ex $	300	VG+ $	205	Good $	185	LMP $	414	
C2C:	Mint $	360	Ex $	275	VG+ $	190	Good $	165	Fair $	85	
Trade-In:	Mint $	260	Ex $	225	VG+ $	150	Good $	130	Poor $	60	

R35102FC
NEW IN BOX

R35102FC REDBOOK CODE: RB-RO-H-35102F
Blue finish, contoured grip, vent rib.

Production: discontinued Caliber: .38 Special +P Action: DA/SA, revolver											
Barrel Length: 2" Sights: fixed Capacity: 5 Grips: rubber											
D2C:	NIB $	365	Ex $	300	VG+ $	205	Good $	185	LMP $	414	
C2C:	Mint $	360	Ex $	275	VG+ $	190	Good $	165	Fair $	85	
Trade-In:	Mint $	260	Ex $	225	VG+ $	150	Good $	130	Poor $	60	

R35202
NEW IN BOX

R35202 REDBOOK CODE: RB-RO-H-35202X
Stainless steel finish, Taurus Security System, contoured black rubber grip.

Production: current Caliber: .38 Special +P Action: DA/SA, revolver											
Barrel Length: 2" OA Length: 6.5" Wt.: 24 oz. Sights: fixed											
Capacity: 5 Grips: rubber											
D2C:	NIB $	410	Ex $	325	VG+ $	230	Good $	205	LMP $	454	
C2C:	Mint $	400	Ex $	300	VG+ $	210	Good $	185	Fair $	95	
Trade-In:	Mint $	300	Ex $	250	VG+ $	160	Good $	145	Poor $	60	

R35202FC
NEW IN BOX

R35202FC REDBOOK CODE: RB-RO-H-35202F
Stainless steel finish.

Production: discontinued Caliber: .38 Special +P Action: DA/SA, revolver											
Barrel Length: 2" Sights: fixed Capacity: 5 Grips: rubber											
D2C:	NIB $	410	Ex $	325	VG+ $	230	Good $	205	LMP $	479	
C2C:	Mint $	400	Ex $	300	VG+ $	210	Good $	185	Fair $	95	
Trade-In:	Mint $	300	Ex $	250	VG+ $	160	Good $	145	Poor $	60	

R98106 PLINKER
NEW IN BOX

R98106 PLINKER REDBOOK CODE: RB-RO-H-98106P
Blue finish, hammer block, Taurus Security System, alloy steel cylinder.

Production: current Caliber: .22 LR Action: DA/SA, revolver Barrel Length: 6"											
Wt.: 29.5 oz. Sights: fixed red fiber-optic front, adjustable rear Capacity: 8 Grips: rubber											
D2C:	NIB $	370	Ex $	300	VG+ $	205	Good $	185	LMP $	407	
C2C:	Mint $	360	Ex $	275	VG+ $	190	Good $	170	Fair $	90	
Trade-In:	Mint $	270	Ex $	225	VG+ $	150	Good $	130	Poor $	60	

R98104 PLINKER
NEW IN BOX

R98104 PLINKER   REDBOOK CODE: RB-RO-H-98104P
Blue finish, hammer block, Taurus Security System, alloy steel cylinder.

Production: current	Caliber: .22 LR	Action: DA/SA, revolver		Barrel Length: 4"	
Wt.: 29 oz.	Sights: fixed red fiber-optic, adjustable rear		Capacity: 8	Grips: rubber	
D2C:	NIB $ 370	Ex $ 300	VG+ $ 205	Good $ 185	LMP $407
C2C:	Mint $ 360	Ex $ 275	VG+ $ 190	Good $ 170	Fair $90
Trade-In:	Mint $ 270	Ex $ 225	VG+ $ 150	Good $ 130	Poor $60

MODEL 31   REDBOOK CODE: RB-RO-H-MDL31X
Nickel or blue finish, steel frame.

Production: late 1970s - mid-1980s	Caliber: .38 Special	Action: DA/SA, revolver			
Barrel Length: 4"	Wt.: 22 oz.	Sights: ramp front, adjustable rear			
Capacity: 5	Grips: walnut				
D2C:	NIB $ 210	Ex $ 175	VG+ $ 120	Good $ 105	
C2C:	Mint $ 210	Ex $ 150	VG+ $ 110	Good $ 95	Fair $ 50
Trade-In:	Mint $ 150	Ex $ 125	VG+ $ 90	Good $ 75	Poor $ 30

MODEL 51   REDBOOK CODE: RB-RO-H-MDL51X
Blue finish, wood grips.

Production: discontinued in 1908	Caliber: .22 LR	Action: DA/SA, revolver			
Barrel Length: 6"	Sights: adjustable	Capacity: 6	Grips: walnut		
D2C:	NIB $ 190	Ex $ 150	VG+ $ 105	Good $ 95	
C2C:	Mint $ 190	Ex $ 150	VG+ $ 100	Good $ 90	Fair $ 45
Trade-In:	Mint $ 140	Ex $ 125	VG+ $ 80	Good $ 70	Poor $ 30

MODEL 511 SPORTSMAN   REDBOOK CODE: RB-RO-H-MDL511
Stainless steel, shrouded ejector rod, noted for its ribbed barrel, made in Brazil.

Production: 1986 - discontinued	Caliber: .22 LR	Action: DA/SA, revolver			
Barrel Length: 4"	Wt.: 30 oz.	Sights: ramp front, adjustable rear			
Capacity: 6	Grips: walnut				
D2C:	NIB $ 300	Ex $ 250	VG+ $ 165	Good $ 150	
C2C:	Mint $ 290	Ex $ 225	VG+ $ 150	Good $ 135	Fair $ 70
Trade-In:	Mint $ 220	Ex $ 175	VG+ $ 120	Good $ 105	Poor $ 30

MODEL 68   REDBOOK CODE: RB-RO-H-MDL68X
Blue or nickel finish with steel frame, swing-out cylinder, made in Brazil.

Production: discontinued	Caliber: .38 Special	Action: DA/SA, revolver			
Barrel Length: 3"	Sights: ramp front, adjustable rear	Capacity: 5			
Grips: wood or rubber					
D2C:	NIB $ 310	Ex $ 250	VG+ $ 175	Good $ 155	
C2C:	Mint $ 300	Ex $ 225	VG+ $ 160	Good $ 140	Fair $ 75
Trade-In:	Mint $ 230	Ex $ 175	VG+ $ 130	Good $ 110	Poor $ 60

MODEL 68S   REDBOOK CODE: RB-RO-H-MDL68S
Blue or nickel finish, shrouded ejector rod, steel frame.

Production: discontinued	Caliber: .38 Special	Action: DA/SA, revolver			
Barrel Length: 2", 3"	Sights: fixed	Capacity: 5	Grips: wood or rubber		
D2C:	NIB $ 300	Ex $ 250	VG+ $ 165	Good $ 150	
C2C:	Mint $ 290	Ex $ 225	VG+ $ 150	Good $ 135	Fair $ 70
Trade-In:	Mint $ 220	Ex $ 175	VG+ $ 120	Good $ 105	Poor $ 30

MODEL 69   REDBOOK CODE: RB-RO-H-MDL69X
Blue or nickel finish with steel frame, swing-out cylinder, made in Brazil.

Production: discontinued Caliber: .32 S&W Action: DA/SA, revolver						
Barrel Length: 3" Wt.: 23 oz. Sights: fixed Capacity: 6 Grips: wood or rubber						
D2C:	NIB $ 180	Ex $ 150	VG+ $ 100	Good $ 90		
C2C:	Mint $ 180	Ex $ 125	VG+ $ 90	Good $ 85	Fair $ 45	
Trade-In:	Mint $ 130	Ex $ 125	VG+ $ 80	Good $ 65	Poor $ 30	

MODEL 70   REDBOOK CODE: RB-RO-H-MDL70X
Nickel or blue finish with steel frame, swing-out cylinder, made in Brazil.

Production: late 1970s - discontinued Caliber: .22 LR Action: DA/SA, revolver						
Barrel Length: 3" Sights: ramp front, adjustable rear Capacity: 6 Grips: wood						
D2C:	NIB $ 320	Ex $ 250	VG+ $ 180	Good $ 160		
C2C:	Mint $ 310	Ex $ 225	VG+ $ 160	Good $ 145	Fair $ 75	
Trade-In:	Mint $ 230	Ex $ 200	VG+ $ 130	Good $ 115	Poor $ 60	

MODEL 84 STAINLESS   REDBOOK CODE: RB-RO-H-MDL84X
Stainless steel finish, fixed sights, ribbed barrel.

Production: early 1980s - discontinued Caliber: .38 Special Action: DA/SA, revolver						
Barrel Length: 3", 4" Sights: fixed Capacity: 6 Grips: wood						
D2C:	NIB $ 325	Ex $ 250	VG+ $ 180	Good $ 165		
C2C:	Mint $ 320	Ex $ 225	VG+ $ 170	Good $ 150	Fair $ 75	
Trade-In:	Mint $ 240	Ex $ 200	VG+ $ 130	Good $ 115	Poor $ 60	

MODEL 85   REDBOOK CODE: RB-RO-H-MDL85X
Stainless steel, vent rib barrel.

Production: discontinued Caliber: .38 Special Action: DA/SA, revolver						
Barrel Length: 3" Sights: adjustable Capacity: 6 Grips: walnut						
D2C:	NIB $ 250	Ex $ 200	VG+ $ 140	Good $ 125		
C2C:	Mint $ 240	Ex $ 175	VG+ $ 130	Good $ 115	Fair $ 60	
Trade-In:	Mint $ 180	Ex $ 150	VG+ $ 100	Good $ 90	Poor $ 30	

MODEL 85
VERY GOOD +

MODEL 88 STAINLESS   REDBOOK CODE: RB-RO-H-MDL88S
Small frame revolver, blue, nickel, or stainless finish.

Production: discontinued Caliber: .38 Special Action: DA/SA, revolver						
Barrel Length: 2", 3" Wt.: 22 oz. Sights: adjustable Capacity: 5						
Grips: wood or rubber						
D2C:	NIB $ 310	Ex $ 250	VG+ $ 175	Good $ 155		
C2C:	Mint $ 300	Ex $ 225	VG+ $ 160	Good $ 140	Fair $ 75	
Trade-In:	Mint $ 230	Ex $ 175	VG+ $ 130	Good $ 110	Poor $ 60	

MODEL 88 LADY ROSSI   REDBOOK CODE: RB-RO-H-MDL88L
Stainless finish, high-polish, swing-out cylinder, steel frame, compact design.

Production: mid-1990s - discontinued Caliber: .38 Special Action: DA/SA, revolver						
Barrel Length: 2" Sights: fixed Capacity: 5 Grips: wood						
D2C:	NIB $ 260	Ex $ 200	VG+ $ 145	Good $ 130		
C2C:	Mint $ 250	Ex $ 200	VG+ $ 130	Good $ 120	Fair $ 60	
Trade-In:	Mint $ 190	Ex $ 150	VG+ $ 110	Good $ 95	Poor $ 30	

ROSSI

MODEL 89 STAINLESS   REDBOOK CODE: RB-RO-H-MDL89X
Matte stainless finish, compact design.

Production: mid-1980s - discontinued Caliber: .32 S&W Action: DA/SA, revolver									
Barrel Length: 3" Wt.: 20 oz. Sights: ramp front, adjustable rear Capacity: 6									
Grips: wood or rubber									
D2C:	NIB $	225	Ex $	175	VG+ $	125	Good $	115	
C2C:	Mint $	220	Ex $	175	VG+ $	120	Good $	105	Fair $ 55
Trade-In:	Mint $	160	Ex $	150	VG+ $	90	Good $	80	Poor $ 30

MODEL 515   REDBOOK CODE: RB-RO-H-MDL515
Small frame revolver, stainless finish, shrouded ejector rod.

Production: discontinued Caliber: .22 LR Action: DA/SA, revolver									
Barrel Length: 4" Wt.: 30 oz. Sights: blade or fiber-optic front, adjustable rear									
Capacity: 6 Grips: wood or rubber									
D2C:	NIB $	220	Ex $	175	VG+ $	125	Good $	110	
C2C:	Mint $	220	Ex $	175	VG+ $	110	Good $	100	Fair $ 55
Trade-In:	Mint $	160	Ex $	125	VG+ $	90	Good $	80	Poor $ 30

MODEL 518   REDBOOK CODE: RB-RO-H-MDL518
Stainless steel finish, ribbed barrel, shrouded ejector rod.

Production: discontinued Caliber: .22 LR Action: DA/SA, revolver									
Barrel Length: 4" Wt.: 30 oz. Sights: blade front, adjustable rear									
Capacity: 6 Grips: rubber									
D2C:	NIB $	295	Ex $	225	VG+ $	165	Good $	150	
C2C:	Mint $	290	Ex $	225	VG+ $	150	Good $	135	Fair $ 70
Trade-In:	Mint $	210	Ex $	175	VG+ $	120	Good $	105	Poor $ 30

MODEL 677   REDBOOK CODE: RB-RO-H-MDL677
Blue finish, fluted cylinder, shrouded ejector rod.

Production: late 1990s-discontinued Caliber: .357 Magnum Action: DA/SA, revolver									
Barrel Length: 2" Wt.: 26 oz. Sights: fixed Capacity: 6 Grips: rubber									
D2C:	NIB $	275	Ex $	225	VG+ $	155	Good $	140	
C2C:	Mint $	270	Ex $	200	VG+ $	140	Good $	125	Fair $ 65
Trade-In:	Mint $	200	Ex $	175	VG+ $	110	Good $	100	Poor $ 30

MODEL 720
EXCELLENT

MODEL 720   REDBOOK CODE: RB-RO-H-MDL720
Stainless steel finish, shrouded ejector rod, snub-nose design, heavy barrel.

Production: early 1990s-discontinued Caliber: .44 Special Action: DA/SA, revolver									
Barrel Length: 3" Wt.: 28 oz. Sights: ramp or fiber-optic front, adjustable rear									
Capacity: 5 Grips: combat									
D2C:	NIB $	355	Ex $	275	VG+ $	200	Good $	180	
C2C:	Mint $	350	Ex $	250	VG+ $	180	Good $	160	Fair $ 85
Trade-In:	Mint $	260	Ex $	200	VG+ $	140	Good $	125	Poor $ 60

MODEL 877   REDBOOK CODE: RB-RO-H-MDL877
Stainless steel finish, heavy barrel, shrouded ejector rod.

Production: late 1990s - discontinued Caliber: .357 Magnum Action: DA/SA, revolver									
Barrel Length: 2" Sights: blade front, fixed rear Capacity: 6 Grips: rubber									
D2C:	NIB $	380	Ex $	300	VG+ $	210	Good $	190	
C2C:	Mint $	370	Ex $	275	VG+ $	190	Good $	175	Fair $ 90
Trade-In:	Mint $	270	Ex $	225	VG+ $	150	Good $	135	Poor $ 60

MODEL 877, VERY GOOD +

MODEL 951   REDBOOK CODE: RB-RO-H-MDL951
Blue finish, vent rib barrel, colored insert with front sight, made in Brazil.

Production: mid-1980s - discontinued		Caliber: .38 Special		Action: DA/SA, revolver		
Barrel Length: 3", 4" Sights: blade front, adjustable rear Capacity: 6						
Grips: walnut						
D2C:	NIB $ 380	Ex $ 300	VG+ $ 210	Good $ 190		
C2C:	Mint $ 370	Ex $ 275	VG+ $ 190	Good $ 175	Fair $ 90	
Trade-In:	Mint $ 270	Ex $ 225	VG+ $ 150	Good $ 135	Poor $ 60	

MODEL 971   REDBOOK CODE: RB-RO-H-971XXX
Stainless steel or blue finish, ribbed barrel, made in Brazil, some with non-fluted cylinders.

Production: late 1980s - discontinued		Caliber: .357 Magnum		Action: DA/SA, revolver		
Barrel Length: 2.5", 4", 6" Sights: blade or fiber-optic front, adjustable rear						
Capacity: 6 Grips: walnut						
D2C:	NIB $ 520	Ex $ 400	VG+ $ 290	Good $ 260		
C2C:	Mint $ 500	Ex $ 375	VG+ $ 260	Good $ 235	Fair $ 120	
Trade-In:	Mint $ 370	Ex $ 300	VG+ $ 210	Good $ 185	Poor $ 60	

MODEL 971 COMPENSATED
VERY GOOD +

MODEL 971 COMPENSATED   REDBOOK CODE: RB-RO-H-971COM
Compensator on barrel, stainless finish, fluted cylinder, full shroud.

Production: early 1990s - discontinued		Caliber: .357 Magnum		Action: DA/SA, revolver		
Barrel Length: 3.25" Sights: adjustable rear Capacity: 6 Grips: rubber						
D2C:	NIB $ 520	Ex $ 400	VG+ $ 290	Good $ 260		
C2C:	Mint $ 500	Ex $ 375	VG+ $ 260	Good $ 235	Fair $ 120	
Trade-In:	Mint $ 370	Ex $ 300	VG+ $ 210	Good $ 185	Poor $ 60	

MODEL 971 VRC   REDBOOK CODE: RB-RO-H-971VRC
Stainless steel finish, compensator, vent rib barrel, full-length ejector rod shroud.

Production: mid-1990s - discontinued		Caliber: .357 Magnum		Action: DA/SA, revolver		
Barrel Length: 2.5", 4", 6" Sights: adjustable, some with fiber-optic front						
Capacity: 6 Grips: black rubber						
D2C:	NIB $ 485	Ex $ 375	VG+ $ 270	Good $ 245		
C2C:	Mint $ 470	Ex $ 350	VG+ $ 250	Good $ 220	Fair $ 115	
Trade-In:	Mint $ 350	Ex $ 275	VG+ $ 190	Good $ 170	Poor $ 60	

MODEL 988 CYCLOPS   REDBOOK CODE: RB-RO-H-988CYC
Stainless steel finish, fitted with ported barrel, scope mounts, flat-sided barrel.

Production: late 1990s		Caliber: .357 Magnum		Action: DA/SA, revolver		
Barrel Length: 6", 8" Wt.: 45 oz., 50 oz. Sights: blade front, adjustable rear						
Capacity: 6 Grips: rubber						
D2C:	NIB $ 530	Ex $ 425	VG+ $ 295	Good $ 265		
C2C:	Mint $ 510	Ex $ 375	VG+ $ 270	Good $ 240	Fair $ 125	
Trade-In:	Mint $ 380	Ex $ 300	VG+ $ 210	Good $ 190	Poor $ 60	

Savage Arms

Arthur Savage founded Savage Arms in 1894. The company would develop the first hammerless lever-action rifle, which eventually developed into the renowned Model 1899, best known as the Model 99. Savage quickly began producing rifles, handguns, and ammunition. By World War I, the company acquired and merged with several others to become the world's largest firearm manufacturer at the time. In recent years, Savage has become well known for their AccuTrigger and AccuStock rifle accessories. Through the years, the company has changed hands several times and is now a part of the ATK Sporting Group.

MODEL 1907 AUTO
EXCELLENT

MODEL 1907 AUTO   REDBOOK CODE: RB-SV-H-1907XX

An early model striker-fired pistol, but features a rear cocking lever. Blue finish, frame-mounted safety, compact design, bull barrel. Sometimes called the Model 1905 because "Nov. 21, 1905" is stamped on the slide.

Production: 1907 - 1920   Caliber: .32 ACP, .380 ACP   Action: striker-fired, semi-auto

Barrel Length: 3.75"   Sights: fixed   Capacity: 9, 10   Magazine: detachable box

Grips: synthetic or wood

D2C:	NIB	$1,020	Ex $ 800	VG+ $ 565	Good $ 510		
C2C:	Mint	$ 980	Ex $ 725	VG+ $ 510	Good $ 460	Fair	$ 235
Trade-In:	Mint	$ 730	Ex $ 575	VG+ $ 400	Good $ 360	Poor	$ 120

MODEL 1917 AUTO
EXCELLENT

MODEL 1917 AUTO   REDBOOK CODE: RB-SV-H-1917XX

An updated variant of the Model 1907 Auto that features a larger grip and modified cocking lever. Blue finish, frame-mounted safety, compact design, bull barrel.

Production: 1920 - 1928   Caliber: .32 ACP, .380 ACP   Action: striker-fired, semi-auto

Barrel Length: 3.75"   Sights: fixed   Capacity: 9, 10   Magazine: detachable box

Grips: synthetic or wood

D2C:	NIB	$ 990	Ex $ 775	VG+ $ 545	Good $ 495		
C2C:	Mint	$ 960	Ex $ 700	VG+ $ 500	Good $ 450	Fair	$ 230
Trade-In:	Mint	$ 710	Ex $ 575	VG+ $ 390	Good $ 350	Poor	$ 120

Sig Sauer

SIG Sauer is part of a larger group of manufacturers that includes J.P Sauer & Sohn, Blaser, Gmbh. of Germany and Swiss Arms AG of Switzerland. Sig began in 1853 as the Swiss Wagon Factory and later developed a rifle for which they won a contract for Switzerland's Federal Ministry of Defense. By 1985, the company created SIGARMS, Inc. as their U.S. subsidiary and manufacturing plant, located in Herndon, Virginia and later moved to Exeter, New Hampshire. SIGARMS changed its name to SIG Sauer in 2007 and remains one of the top suppliers to law enforcement and military units, manufacturing an array of pistols and rifles.

P210 LEGEND
NEW IN BOX

P210 LEGEND   REDBOOK CODE: RB-SS-H-P210L1

Nitron stainless steel slide and frame, improved manual safety, American magazine release.

Production: 2011 - current   Caliber: 9mm   Action: SA, semi-auto

Barrel Length: 4.7"   OA Length: 8.5"   Wt.: 37 oz.

Sights: drift adjustable post & notch   Capacity: 8   Magazine: detachable box

Grips: custom wood

D2C:	NIB	$2,275	Ex $1,750	VG+ $1,265	Good $1,140	LMP	$2,199
C2C:	Mint	$2,190	Ex $1,575	VG+ $1,140	Good $1,025	Fair	$ 525
Trade-In:	Mint	$1,620	Ex $1,275	VG+ $ 890	Good $ 800	Poor	$ 240

P210 LEGEND TARGET
NEW IN BOX

P210 LEGEND TARGET REDBOOK CODE: RB-SS-H-P210LT

Nitron coated stainless steel slide and frame, manual safety, American magazine release.

Production: 2011 - current   Caliber: 9mm   Action: SA, semi-auto
Barrel Length: 4.7"   OA Length: 8.5"   Wt.: 37 oz.
Sights: adjustable target   Capacity: 8   Magazine: detachable box   Grips: custom wood

D2C:	NIB	$2,650	Ex	$2,025	VG+	$1,475	Good	$1,325	LMP $2,399
C2C:	Mint	$2,550	Ex	$1,850	VG+	$1,330	Good	$1,195	Fair $ 610
Trade-In:	Mint	$1,890	Ex	$1,500	VG+	$1,040	Good	$ 930	Poor $ 270

P210 SUPER LEGEND
NEW IN BOX

P210 SUPER LEGEND REDBOOK CODE: RB-SS-H-P210SL

Nitron stainless steel slide and frame, 1911-style safety lever, extended slide catch lever, black controls, American magazine release.

Production: 2013 - current   Caliber: 9mm   Action: SA, semi-auto
Barrel Length: 5"   Sights: drift adjustable post & notch   Capacity: 8
Magazine: detachable box   Grips: custom wood

D2C:	NIB	$2,279	Ex	$1,750	VG+	$1,265	Good	$1,140	LMP $2,279
C2C:	Mint	$2,190	Ex	$1,575	VG+	$1,140	Good	$1,030	Fair $ 525
Trade-In:	Mint	$1,620	Ex	$1,300	VG+	$ 890	Good	$ 800	Poor $ 240

P210 SUPERTARGET
NEW IN BOX

P210 SUPERTARGET REDBOOK CODE: RB-SS-H-P210ST

Nitron stainless steel slide and frame, 1911-style safety lever, extended slide catch lever.

Production: 2012 - current   Caliber: 9mm   Action: SA, semi-auto
Barrel Length: 6"   OA Length: 10"   Wt.: 43 oz.
Sights: adjustable target   Capacity: 8   Magazine: detachable box
Grips: custom wood with integral magwell

D2C:	NIB	$3,650	Ex	$2,775	VG+	$2,030	Good	$1,825	LMP $3,626
C2C:	Mint	$3,510	Ex	$2,525	VG+	$1,830	Good	$1,645	Fair $840
Trade-In:	Mint	$2,600	Ex	$2,050	VG+	$1,430	Good	$1,280	Poor $390

P220
NEW IN BOX

P220 REDBOOK CODE: RB-SS-H-P22076

Black hard anodized lightweight alloy frame with accessory rail, stainless steel slide coated in Nitron finish, four-point safety system, SIGLITE night sights available.

Production: 1976 - current   Caliber: .45 ACP   Action: DA/SA, semi-auto
Barrel Length: 4.4"   OA Length: 7.7"
Wt.: 30 oz.   Sights: post and dot contrast or SIGLITE night sights
Capacity: 8   Magazine: detachable box   Grips: black polymer factory

D2C:	NIB	$ 795	Ex	$ 625	VG+	$ 445	Good	$ 400	LMP $993
C2C:	Mint	$ 770	Ex	$ 550	VG+	$ 400	Good	$ 360	Fair $185
Trade-In:	Mint	$ 570	Ex	$ 450	VG+	$ 320	Good	$ 280	Poor $90

P220 DAK
NEW IN BOX

P220 DAK REDBOOK CODE: RB-SS-H-220DAK

Black hard anodized frame, Nitron slide finish, SIG's exclusive DAK trigger system, accessory rail.

Production: 2006 - current   Caliber: .45 ACP   Action: DAK (DAO), semi-auto
Barrel Length: 4.4"   OA Length: 7.7"   Wt.: 30 oz.
Sights: post and dot contrast   Capacity: 8   Magazine: detachable box
Grips: black polymer factory

D2C:	NIB	$1,085	Ex	$ 825	VG+	$ 605	Good	$ 545	LMP $1,085
C2C:	Mint	$1,050	Ex	$ 750	VG+	$ 550	Good	$ 490	Fair $250
Trade-In:	Mint	$ 780	Ex	$ 625	VG+	$ 430	Good	$ 380	Poor $120

SIG SAUER

**P220
TWO-TONE DAK**
EXCELLENT

P220 TWO-TONE DAK REDBOOK CODE: RB-SS-H-220TTD

SIG's exclusive DAK trigger, natural stainless slide finish, Nitron frame finish.

Caliber: .45 ACP   Action: DAK (DAO), semi-auto   Barrel Length: 4.4"
OA Length: 7.7"   Wt.: 31 oz.
Sights: SIGLITE night sights   Capacity: 8   Magazine: detachable box
Grips: black polymer factory

D2C:	NIB	$1,040	Ex	$ 800	VG+	$ 580	Good	$ 520	LMP	$1,085
C2C:	Mint	$1,000	Ex	$ 725	VG+	$ 520	Good	$ 470	Fair	$ 240
Trade-In:	Mint	$740	Ex	$ 600	VG+	$ 410	Good	$ 365	Poor	$ 120

P220 CARRY
NEW IN BOX

P220 CARRY REDBOOK CODE: RB-SS-H-P220CC

Full-size frame with short slide, accessory rail, black hard anodized frame finish, Nitron slide finish, optional night sights.

Production: 2006 - current   Caliber: .45 ACP   Action: DA/SA, semi-auto
Barrel Length: 3.9"   OA Length: 7.1"   Wt.: 30 oz.
Sights: SIGLITE or post and dot contrast   Capacity: 8   Magazine: detachable box
Grips: black polymer factory

D2C:	NIB	$1,030	Ex	$800	VG+	$ 575	Good	$ 515	LMP	$1,068
C2C:	Mint	$ 990	Ex	$ 725	VG+	$ 520	Good	$ 465	Fair	$240
Trade-In:	Mint	$ 740	Ex	$600	VG+	$ 410	Good	$ 365	Poor	$120

P220 CARRY SAS
(TWO-TONE)
EXCELLENT

P220 CARRY SAS (TWO-TONE) REDBOOK CODE: RB-SS-H-220CST

SIG Anti-Snag (SAS) treatment, stainless steel slide, black hard anodized alloy frame, more compact version of P220 SAS.

Production: discontinued   Caliber: .45 ACP   Action: DAK (DAO), semi-auto
Barrel Length: 3.9"   OA Length: 7.1"   Wt.: 30 oz.
Sights: SIGLITE night sights   Capacity: 8   Magazine: detachable box
Grips: custom shop wood

D2C:	NIB	$ 810	Ex	$ 625	VG+	$ 450	Good	$ 405	LMP	$1,020
C2C:	Mint	$ 780	Ex	$ 575	VG+	$ 410	Good	$ 365	Fair	$190
Trade-In:	Mint	$ 580	Ex	$ 475	VG+	$ 320	Good	$ 285	Poor	$90

**P220 CARRY SAS
GEN 2 NITRON**
NEW IN BOX

P220 CARRY SAS GEN 2 NITRON REDBOOK CODE: RB-SS-H-220SG2

Short Reset Trigger (SRT), SIG Anti-Snag (SAS) treatment, black hard anodized frame finish, Nitron slide finish, two-tone model available.

Caliber: .45 ACP   Action: DA/SA, semi-auto   Barrel Length: 3.9"
OA Length: 7.1"   Wt.: 30 oz.   Sights: SIGLITE night sights   Capacity: 8
Magazine: detachable box   Grips: black polymer factory

D2C:	NIB	$ 930	Ex	$ 725	VG+	$ 520	Good	$ 465	LMP	$1,126
C2C:	Mint	$ 900	Ex	$ 650	VG+	$ 470	Good	$ 420	Fair	$ 215
Trade-In:	Mint	$ 670	Ex	$ 525	VG+	$ 370	Good	$ 330	Poor	$ 120

P220 CARRY DAK
NEW IN BOX

P220 CARRY DAK REDBOOK CODE: RB-SS-H-220CDK

SIG's exclusive DAK trigger, black hard anodized frame, Nitron slide finish, two-tone variation available.

Caliber: .45 ACP   Action: DAK (DAO), semi-auto   Barrel Length: 3.9"
OA Length: 7.1"   Wt.: 30 oz.   Sights: SIGLITE night sights
Capacity: 8   Magazine: detachable box   Grips: black polymer factory

D2C:	NIB	$1,065	Ex	$ 825	VG+	$ 595	Good	$ 535	LMP	$1,085
C2C:	Mint	$1,030	Ex	$ 750	VG+	$ 540	Good	$ 480	Fair	$ 245
Trade-In:	Mint	$ 760	Ex	$ 600	VG+	$ 420	Good	$ 375	Poor	$ 120

P220 CARRY SAO
NEW IN BOX

P220 CARRY SAO REDBOOK CODE: RB-SS-H-220CSA

Black hard anodized finish, 5-pound SAO trigger, accessory rail, Nitron slide finish, two-tone variation available.

Caliber: .45 ACP   Action: SA, semi-auto   Barrel Length: 3.9"
OA Length: 7.1"   Wt.: 30 oz.
Sights: SIGLITE night sights   Capacity: 8   Magazine: detachable box
Grips: black polymer

D2C:	NIB	$ 920	Ex	$ 700	VG+	$ 515	Good	$ 460	LMP	$1,106
C2C:	Mint	$ 890	Ex	$ 650	VG+	$ 460	Good	$ 415	Fair	$ 215
Trade-In:	Mint	$ 660	Ex	$ 525	VG+	$ 360	Good	$ 325	Poor	$ 120

P220 EQUINOX
NEW IN BOX

P220 EQUINOX REDBOOK CODE: RB-SS-H-220EQX

Lightweight black hard anodized alloy frame, two-tone Nitron finished stainless steel slide, nickel accents, accessory rail.

Caliber: .45 ACP   Action: DA/SA, semi-auto   Barrel Length: 4.4"
OA Length: 7.7"   Wt.: 30 oz.
Sights: TRUGLO Front, SIGLITE Rear/Front   Capacity: 8   Magazine: detachable box
Grips: gray laminated wood

D2C:	NIB	$ 955	Ex	$ 750	VG+	$ 535	Good	$ 480	LMP	$1,218
C2C:	Mint	$ 920	Ex	$ 675	VG+	$ 480	Good	$ 430	Fair	$ 220
Trade-In:	Mint	$ 680	Ex	$ 550	VG+	$ 380	Good	$ 335	Poor	$ 120

P220 CARRY EQUINOX
NEW IN BOX

P220 CARRY EQUINOX REDBOOK CODE: RB-SS-H-220CEQ

Lightweight black hard anodized alloy frame, two-tone Nitron finished stainless steel slide, nickel accents, accessory rail.

Caliber: .45 ACP   Action: DA/SA, semi-auto   Barrel Length: 3.9"
OA Length: 7.1"   Wt.: 30 oz.
Sights: TRUGLO TFO Night Sights   Capacity: 8   Magazine: detachable box
Grips: custom wood

D2C:	NIB	$ 830	Ex	$ 650	VG+	$ 465	Good	$ 415	LMP	$1,218
C2C:	Mint	$ 800	Ex	$ 575	VG+	$ 420	Good	$ 375	Fair	$ 195
Trade-In:	Mint	$ 590	Ex	$ 475	VG+	$ 330	Good	$ 295	Poor	$ 90

P220 SAO REDBOOK CODE: RB-SS-H-220SAO

SAO trigger, cold hammer-forged barrel, Nitron-coated stainless steel slide, hard-coat anodized lightweight alloy frame, accessory rail.

Caliber: .45 ACP   Action: SA, semi-auto   Barrel Length: 4.4"
OA Length: 7.7"   Wt.: 30 oz.
Sights: SIGLITE night sights   Capacity: 8   Magazine: detachable box
Grips: black polymer factory

D2C:	NIB	$1,060	Ex	$ 825	VG+	$ 590	Good	$ 530	LMP	$1,085
C2C:	Mint	$1,020	Ex	$ 750	VG+	$ 530	Good	$ 480	Fair	$ 245
Trade-In:	Mint	$ 760	Ex	$ 600	VG+	$ 420	Good	$ 375	Poor	$ 120

P220 COMBAT
NEW IN BOX

P220 COMBAT REDBOOK CODE: RB-SS-H-220COM

Hard-chrome lined barrel finished in Nitron, Flat Dark Earth alloy frame, Nitron coated stainless slide, accessory rail, model available with threaded barrel.

Production: current   Caliber: .45 ACP   Action: DA/SA, semi-auto
Barrel Length: 4.4"   OA Length: 7.7"   Wt.: 30 oz.
Sights: SIGLITE night sights   Capacity: 8, 10   Magazine: detachable box
Grips: Flat Dark Earth polymer

D2C:	NIB	$ 815	Ex	$ 625	VG+	$ 455	Good	$ 410	LMP	$1,228
C2C:	Mint	$ 790	Ex	$ 575	VG+	$ 410	Good	$ 370	Fair	$ 190
Trade-In:	Mint	$ 580	Ex	$ 475	VG+	$ 320	Good	$ 290	Poor	$ 90

P220 CLASSIC 22
NEW IN BOX

P220 CLASSIC 22 REDBOOK CODE: RB-SS-H-220C22

Four-point safety system, black anodized frame and slide finish, DA/SA or SAO version with manual safety available.

Caliber: .22 LR Action: DA/SA, SA, semi-auto Barrel Length: 4.5"
OA Length: 7.7" Wt.: 30 oz. Sights: adjustable
Capacity: 10 Magazine: detachable box Grips: black polymer factory

D2C:	NIB	$ 580	Ex	$ 450	VG+	$ 325	Good	$ 290	LMP	$ 626
C2C:	Mint	$ 560	Ex	$ 425	VG+	$ 290	Good	$ 265	Fair	$ 135
Trade-In:	Mint	$ 420	Ex	$ 325	VG+	$ 230	Good	$ 205	Poor	$ 60

P220 ELITE
NEW IN BOX

P220 ELITE REDBOOK CODE: RB-SS-H-220ELT

Nitron-finished stainless slide, black hard anodized frame with beavertail, (SRT) Short Reset Trigger, accessory rail, front cocking serrations, front-strap checkering, decocking lever.

Production: discontinued Caliber: .45 ACP Action: DA/SA, semi-auto
Barrel Length: 4.4" OA Length: 8.32" Wt.: 30 oz.
Sights: SIGLITE night sights Capacity: 8 Magazine: detachable box
Grips: custom rosewood

D2C:	NIB	$ 845	Ex	$ 650	VG+	$ 470	Good	$ 425	LMP	$1,200
C2C:	Mint	$ 820	Ex	$ 600	VG+	$ 430	Good	$ 385	Fair	$ 195
Trade-In:	Mint	$ 600	Ex	$ 475	VG+	$ 330	Good	$ 300	Poor	$ 90

P220 ELITE STAINLESS
NEW IN BOX

P220 ELITE STAINLESS REDBOOK CODE: RB-SS-H-220ELS

Natural stainless steel slide, elite stainless frame, accessory rail.

Caliber: .45 ACP Action: DA/SA, semi-auto Barrel Length: 4.4"
OA Length: 8.32" Wt.: 40 oz.
Sights: SIGLITE night sights Capacity: 8 Magazine: detachable box
Grips: custom rosewood

D2C:	NIB	$1,180	Ex	$ 900	VG+	$ 655	Good	$ 590	LMP	$1,368
C2C:	Mint	$1,140	Ex	$ 825	VG+	$ 590	Good	$ 535	Fair	$ 275
Trade-In:	Mint	$ 840	Ex	$ 675	VG+	$ 470	Good	$ 415	Poor	$ 120

P220 CARRY ELITE
NEW IN BOX

P220 CARRY ELITE REDBOOK CODE: RB-SS-H-220CEL

Nitron-finished stainless slide, black hard anodized frame with beavertail, (SRT) Short Reset Trigger, compact length with full-size frame, accessory rail, front cocking serrations, front strap checkering, decocking lever.

Production: discontinued Caliber: .45 ACP Action: DA/SA, semi-auto
Barrel Length: 3.9" OA Length: 7.6" Wt.: 30 oz.
Sights: SIGLITE night sights Capacity: 8 Magazine: detachable box
Grips: custom rosewood

D2C:	NIB	$ 1,105	Ex	$850	VG+	$ 615	Good	$ 555	LMP	$1,200
C2C:	Mint	$1,070	Ex	$ 775	VG+	$ 560	Good	$ 500	Fair	$ 255
Trade-In:	Mint	$ 790	Ex	$ 625	VG+	$ 440	Good	$ 390	Poor	$ 120

P220 CARRY ELITE STAINLESS
NEW IN BOX

P220 CARRY ELITE STAINLESS REDBOOK CODE: RB-SS-H-220CES

Short Reset Trigger (SRT), beavertail grip, front strap checkering, front cocking serrations, stainless slide and frame.

Caliber: .45 ACP Action: DA/SA, semi-auto Barrel Length: 3.9"
OA Length: 7.6" Wt.: 39 oz.
Sights: SIGLITE night sights Capacity: 8 Magazine: detachable box
Grips: custom rosewood

D2C:	NIB	$1,205	Ex	$ 925	VG+	$670	Good	$ 605	LMP	$1,368
C2C:	Mint	$1,160	Ex	$ 850	VG+	$ 610	Good	$ 545	Fair	$ 280
Trade-In:	Mint	$ 860	Ex	$ 675	VG+	$ 470	Good	$ 425	Poor	$ 150

**P220
PLATINUM ELITE**
NEW IN BOX

P220 PLATINUM ELITE REDBOOK CODE: RB-SS-H-220PPLE

Stainless finish slide with ground side flats, black hard anodized alloy frame with beavertail, accessory rail, front cocking serrations, front strap checkering, decocking lever.

Production: discontinued Caliber: .45 ACP Action: DA/SA, semi-auto

Barrel Length: 4.4" OA Length: 8.2" Wt.: 30 oz. Sights: adjustable combat night sights

Capacity: 8 Magazine: detachable box Grips: aluminum

D2C:	NIB	$1,200	Ex	$ 925	VG+	$670	Good	$600	LMP	$1,289
C2C:	Mint	$1,160	Ex	$ 850	VG+	$600	Good	$ 540	Fair	$ 280
Trade-In:	Mint	$ 860	Ex	$ 675	VG+	$470	Good	$ 420	Poor	$ 120

P220 ELITE DARK
NEW IN BOX

P220 ELITE DARK REDBOOK CODE: RB-SS-H-220ELD

Short Reset Trigger (SRT), accessory rail, black hard anodized frame, Nitron slide finish, model with threaded barrel available

Caliber: .45 ACP Action: DA/SA, semi-auto Barrel Length: 4.4" OA Length: 8.3"

Wt.: 30 oz. Sights: adjustable combat night sights Capacity: 8 Magazine: detachable box

Grips: custom aluminum

D2C:	NIB	$ 850	Ex	$650	VG+	$ 475	Good	$425	LMP	$1,228
C2C:	Mint	$ 820	Ex	$600	VG+	$430	Good	$385	Fair	$ 200
Trade-In:	Mint	$ 610	Ex	$500	VG+	$340	Good	$300	Poor	$ 90

**P220 CARRY
ELITE DARK**
NEW IN BOX

P220 CARRY ELITE DARK REDBOOK CODE: RB-SS-H-220CED

Short Reset Trigger (SRT), accessory rail, black hard-coat anodized frame finish, Nitron slide finish.

Caliber: .45 ACP Action: DA/SA, semi-auto Barrel Length: 4"

OA Length: 7.6" Wt.: 30 oz.

Sights: adjustable combat night sights Capacity: 8 Magazine: detachable box

Grips: custom aluminum

D2C:	NIB	$1,140	Ex	$ 875	VG+	$ 635	Good	$ 570	LMP	$1,218
C2C:	Mint	$1,100	Ex	$ 800	VG+	$ 570	Good	$ 515	Fair	$ 265
Trade-In:	Mint	$ 810	Ex	$ 650	VG+	$ 450	Good	$ 400	Poor	$ 120

P220 SCORPION
NEW IN BOX

P220 SCORPION REDBOOK CODE: RB-SS-H-220SCO

Custom Flat Dark Earth finished slide and frame, beavertail grip frame, front cocking serrations, model with threaded barrel available.

Caliber: .45 ACP Action: DA/SA, semi-auto Barrel Length: 4.4"

OA Length: 7.7" Wt.: 30 oz.

Sights: SIGLITE night sights Capacity: 8 Grips: Hogue G-10 Parana

D2C:	NIB	$ 845	Ex	$650	VG+	$ 470	Good	$425	LMP	$1,285
C2C:	Mint	$ 820	Ex	$600	VG+	$ 430	Good	$ 385	Fair	$ 195
Trade-In:	Mint	$ 600	Ex	$ 475	VG+	$ 330	Good	$300	Poor	$ 90

P220 MATCH REDBOOK CODE: RB-SS-H-220MTC

Black hard-coat anodized frame, stainless slide, accessory rail, cold hammer-forged barrel.

Production: discontinued Caliber: .45 ACP Action: DA/SA, semi-auto

Barrel Length: 5" OA Length: 8.9" Wt.: 33.6 oz.

Sights: adjustable target Capacity: 8 Grips: black polymer factory

D2C:	NIB	$1,105	Ex	$ 850	VG+	$ 615	Good	$ 555	LMP	$1,170
C2C:	Mint	$1,070	Ex	$ 775	VG+	$ 560	Good	$ 500	Fair	$ 255
Trade-In:	Mint	$ 790	Ex	$ 625	VG+	$ 440	Good	$ 390	Poor	$ 120

P220 MATCH SAO REDBOOK CODE: RB-SS-H-220MSA

Single action only trigger, black hard-coat anodized frame, stainless slide, accessory rail, cold hammer-forged barrel.

Production: discontinued   Caliber: .45 ACP   Action: SA, semi-auto

Barrel Length: 5"   OA Length: 8.9"   Wt.: 33.6 oz.

Sights: adjustable target   Capacity: 8   Magazine: detachable box

Grips: black polymer factory

D2C:	NIB	$1,075	Ex	$ 825	VG+	$600	Good	$ 540	
C2C:	Mint	$1,040	Ex	$ 750	VG+	$540	Good	$ 485	Fair $ 250
Trade-In:	Mint	$ 770	Ex	$ 625	VG+	$ 420	Good	$ 380	Poor $ 120

P220 MATCH ELITE REDBOOK CODE: RB-SS-H-220MEL

Short Reset Trigger (SRT), accessory rail, stainless slide and frame finish.

Caliber: .45 ACP   Action: DA/SA, semi-auto   Barrel Length: 5"

OA Length: 8.9"   Wt.: 39 oz.

Sights: adjustable target   Capacity: 8   Magazine: detachable box   Grips: aluminum

D2C:	NIB	$ 860	Ex	$ 675	VG+	$ 480	Good	$ 430	LMP $1,368
C2C:	Mint	$ 830	Ex	$600	VG+	$ 430	Good	$ 390	Fair $200
Trade-In:	Mint	$ 620	Ex	$500	VG+	$ 340	Good	$ 305	Poor $ 90

P220 SUPER MATCH REDBOOK CODE: RB-SS-H-220SUM

Lightweight-alloy beavertail frame with hard-coat anodized finish, stainless slide, front strap checkering, front cocking serrations, adjustable target sights, SIG Custom Shop logo engraved on slide.

Caliber: .45 ACP   Action: SA, semi-auto   Barrel Length: 5"

OA Length: 8.92"   Wt.: 34 oz.

Sights: adjustable   Capacity: 8   Magazine: detachable box   Grips: custom wood

D2C:	NIB	$1,525	Ex	$1,175	VG+	$ 850	Good	$ 765	LMP $1,375
C2C:	Mint	$1,470	Ex	$1,075	VG+	$ 770	Good	$690	Fair $ 355
Trade-In:	Mint	$1,090	Ex	$ 875	VG+	$ 600	Good	$ 535	Poor $ 180

P220 SUPER MATCH
NEW IN BOX

P220 STAINLESS REDBOOK CODE: RB-SS-H-220STA

Stainless steel slide and frame, natural stainless finish, accessory rail, decocking lever.

Production: discontinued   Caliber: .45 ACP   Action: DA/SA, semi-auto

Barrel Length: 4.4"   OA Length: 7.8"   Wt.: 40 oz.

Sights: contrast or night sights   Capacity: 8   Magazine: detachable box

Grips: black polymer factory

D2C:	NIB	$820	Ex	$ 625	VG+	$460	Good	$410	LMP $ 935
C2C:	Mint	$790	Ex	$ 575	VG+	$ 410	Good	$ 370	Fair $ 190
Trade-In:	Mint	$ 590	Ex	$ 475	VG+	$320	Good	$290	Poor $ 90

P220 STAINLESS
NEW IN BOX

P220 STAINLESS NITRON REDBOOK CODE: RB-SS-H-220STN

Stainless steel slide and frame with Nitron finish, accessory rail.

Caliber: .45 ACP   Action: DA/SA, semi-auto   Barrel Length: 4.4"   OA Length: 7.7"

Wt.: 40 oz.   Sights: SIGLITE night sights   Capacity: 8   Magazine: detachable box

Grips: Hogue rubber

D2C:	NIB	$ 830	Ex	$650	VG+	$465	Good	$ 415	LMP $1,129
C2C:	Mint	$ 800	Ex	$ 575	VG+	$420	Good	$ 375	Fair $ 195
Trade-In:	Mint	$ 590	Ex	$ 475	VG+	$330	Good	$ 295	Poor $ 90

P220 STAINLESS NITRON
NEW IN BOX

**P220 STAINLESS
REVERSE TWO-TONE**
NEW IN BOX

P220 STAINLESS REVERSE TWO-TONE
REDBOOK CODE: RB-SS-H-220SRT
Natural stainless frame finish, accessory rail, Nitron-finished stainless steel slide.

Caliber: .45 ACP	Action: DA/SA, semi-auto	Barrel Length: 4.4"	OA Length: 7.7"		
Wt.: 40 oz.	Sights: SIGLITE night sights	Capacity: 8	Magazine: detachable box		
Grips: Hogue rubber					
D2C:	NIB $1,200	Ex $925	VG+ $670	Good $600	LMP $1,115
C2C:	Mint $1,160	Ex $850	VG+ $600	Good $540	Fair $280
Trade-In:	Mint $860	Ex $675	VG+ $470	Good $420	Poor $120

P220 EXTREME
NEW IN BOX

P220 EXTREME REDBOOK CODE: RB-SS-H-220EXT
Black hard-coat anodized alloy frame, accessory rail, stainless slide with Nitron finish, front cocking serrations, Short Reset Trigger (SRT).

Caliber: .45 ACP	Action: DA/SA, semi-auto	Barrel Length: 4.4"		
OA Length: 8.3"	Wt.: 30 oz.			
Sights: SIGLITE night sights	Capacity: 8	Magazine: detachable box		
Grips: Hogue Extreme G-10				
D2C:	NIB $785	Ex $600	VG+ $440	Good $395	LMP $1,146
C2C:	Mint $760	Ex $550	VG+ $400	Good $355	Fair $185
Trade-In:	Mint $560	Ex $450	VG+ $310	Good $275	Poor $90

P220 X-SIX
NEW IN BOX

P220 X-SIX REDBOOK CODE: RB-SS-H-220XSX
Stainless steel frame and slide, beavertail grip frame, adjustable trigger, front cocking serrations.

Production: current	Caliber: .45 ACP	Action: SA, semi-auto	Barrel Length: 6"		
OA Length: 9"	Wt.: 43 oz.	Sights: low profile integrated adjustable competition			
Capacity: 8	Magazine: detachable box	Grips: ergonomic custom extended wood			
D2C:	NIB $2,600	Ex $2,000	VG+ $1,445	Good $1,300	LMP $2,599
C2C:	Mint $2,500	Ex $1,800	VG+ $1,300	Good $1,170	Fair $600
Trade-In:	Mint $1,850	Ex $1,475	VG+ $1,020	Good $910	Poor $270

P220R COMPACT
NEW IN BOX

P220R COMPACT REDBOOK CODE: RB-SS-H-220RCO
Compact frame with short slide, accessory rail, black hard-coat anodized frame, Nitron slide finish, decocking lever, two-tone finish available.

Caliber: .45 ACP	Action: DA/SA, semi-auto	Barrel Length: 3.9"	OA Length: 7"		
Wt.: 30 oz.	Sights: SIGLITE night sights	Capacity: 6	Magazine: detachable box		
Grips: black polymer factory					
D2C:	NIB $800	Ex $625	VG+ $445	Good $400	LMP $1,068
C2C:	Mint $770	Ex $575	VG+ $400	Good $360	Fair $185
Trade-In:	Mint $570	Ex $450	VG+ $320	Good $280	Poor $90

P220 COMPACT REDBOOK CODE: RB-SS-H-220CPT
Nitron-finished stainless slide, compact black hard anodized alloy frame with beavertail, extended magazine base, decocking lever.

Production: discontinued	Caliber: .45 ACP	Action: DA/SA, semi-auto		
Barrel Length: 3.9"	OA Length: 7.6"	Wt.: 30 oz.	Sights: SIGLITE night sights	
Capacity: 6	Magazine: detachable box	Grips: black polymer factory		
D2C:	NIB $760	Ex $600	VG+ $425	Good $380	LMP $1,085
C2C:	Mint $730	Ex $525	VG+ $380	Good $345	Fair $175
Trade-In:	Mint $540	Ex $450	VG+ $300	Good $270	Poor $90

P220 COMPACT, NEW IN BOX

P220 COMPACT SAS REDBOOK CODE: RB-SS-H-220COS
Stainless-finished slide, black alloy beavertail frame, SIG Anti-Snag (SAS) treatment.

Production: discontinued Caliber: .45 ACP Action: DAK (DAO), semi-auto					
Barrel Length: 3.9" OA Length: 7.79" Wt.: 30 oz. Sights: SIGLITE night sights					
Capacity: 6 Magazine: detachable box Grips: custom rosewood					
D2C:	NIB $ 920	Ex $ 700	VG+ $ 515	Good $ 460	LMP $1,093
C2C:	Mint $ 890	Ex $ 650	VG+ $ 460	Good $ 415	Fair $ 215
Trade-In:	Mint $ 660	Ex $ 525	VG+ $ 360	Good $ 325	Poor $ 120

P220 COMPACT SAS GEN 2
NEW IN BOX

P220 COMPACT SAS GEN 2 REDBOOK CODE: RB-SS-H-220CS2
Compact frame with short slide, SIG Anti-Snag (SAS) treatment, black hard anodized frame finish, Nitron slide finish, accessory rail, two-tone model available.

Production: 2009 - current Caliber: .45 ACP Action: DA/SA, semi-auto					
Barrel Length: 3.9" OA Length: 7.1" Wt.: 30 oz. Sights: SIGLITE night sights					
Capacity: 6 Magazine: detachable box Grips: black polymer factory					
D2C:	NIB $ 915	Ex $700	VG+ $ 510	Good $ 460	LMP $1,125
C2C:	Mint $ 880	Ex $650	VG+ $460	Good $ 415	Fair $ 215
Trade-In:	Mint $ 650	Ex $ 525	VG+ $360	Good $ 325	Poor $ 120

P224
NEW IN BOX

P224 REDBOOK CODE: RB-SS-H-224XXX
Black Nitron-finish slide, black hard anodized alloy frame, subcompact size, optional night sights.

Production: 2012 - current Caliber: 9mm, .40 S&W, .357 SIG Action: DA/SA,					
DAK (DAO), semi-auto Barrel Length: 3.5" Wt.: 29 oz. Sights: contrast or SIGLITE					
night sights Capacity: 10, 12, 15 Magazine: detachable box Grips: black polymer					
D2C:	NIB $ 890	Ex $700	VG+ $495	Good $ 445	LMP $ 993
C2C:	Mint $ 860	Ex $ 625	VG+ $450	Good $ 405	Fair $ 205
Trade-In:	Mint $ 640	Ex $500	VG+ $350	Good $ 315	Poor $ 90

P224 NICKEL
NEW IN BOX

P224 NICKEL REDBOOK CODE: RB-SS-H-224NIC
Nickel-plated slide and controls, black hard anodized frame finish, subcompact frame, optional DA/SA or SIG's exclusive DAK trigger.

Production: 2012 - current Caliber: 9mm, .40 S&W, .357 SIG Action: DA/SA,					
DAK (DAO), semi-auto Barrel Length: 3.5" OA Length: 6.7" Wt.: 29 oz. Sights:					
SIGLITE night sights Capacity: 10, 12, 15					
Magazine: detachable box Grips: Hogue custom black G-10					
D2C:	NIB $ 900	Ex $700	VG+ $ 500	Good $ 450	LMP $ 1,125
C2C:	Mint $ 870	Ex $ 625	VG+ $ 450	Good $ 405	Fair $ 210
Trade-In:	Mint $ 640	Ex $ 525	VG+ $ 360	Good $ 315	Poor $ 90

P224 EQUINOX
NEW IN BOX

P224 EQUINOX REDBOOK CODE: RB-SS-H-224EQI
Two-tone slide with nickel controls, black hard-coat anodized subcompact frame.

Production: 2012 - current Caliber: .40 S&W Action: DA/SA, DAK (DAO), semi-auto					
Barrel Length: 3.5" OA Length: 6.7" Wt.: 29 oz.					
Sights: TRUGLO Tritium Fiber-optic front, SIGLITE night sights rear					
Capacity: 10, 12, 15 Magazine: detachable box Grips: Hogue black G-10					
D2C:	NIB $ 950	Ex $ 725	VG+ $530	Good $475	LMP $1,218
C2C:	Mint $ 920	Ex $ 675	VG+ $480	Good $430	Fair $ 220
Trade-In:	Mint $ 680	Ex $ 550	VG+ $380	Good $335	Poor $ 120

P224 EXTREME
NEW IN BOX

P224 EXTREME REDBOOK CODE: RB-SS-H-224EXT

Black hard-coat anodized frame finish, Nitron stainless steel slide, DA/SA model features a Short Reset Trigger (SRT).

Production: 2012 - current   Caliber: 9mm, .40 S&W, .357 SIG   Action: DA/SA, DAK (DAO), semi-auto   Barrel Length: 3.5"   OA Length: 6.7"   Wt.: 29 oz.   Sights: SIGLITE night sights   Capacity: 10, 12, 15

Magazine: detachable box   Grips: Hogue black and gray Extreme Series G-10

D2C:	NIB	$ 880	Ex $ 675	VG+ $ 490	Good $ 440	LMP $1,146
C2C:	Mint $ 850	Ex $ 625	VG+ $ 440	Good $ 400	Fair $ 205	
Trade-In:	Mint $ 630	Ex $ 500	VG+ $ 350	Good $ 310	Poor $ 90	

P224 SAS
NEW IN BOX

P224 SAS REDBOOK CODE: RB-SS-H-224SAS

Black hard anodized lightweight alloy frame, Nitron-finished slide, subcompact size, SIG Anti-Snag (SAS) treatment, optional DA/SA or SIG's exclusive DAK trigger, SRT trigger system on DA/SA model.

Production: 2012 - current   Caliber: 9mm, .40 S&W, .357 SIG   Action: DA/SA, DAK (DAO), semi-auto   Barrel Length: 3.5"   OA Length: 6.7"   Wt.: 29 oz.   Sights: SIGLITE night sights   Capacity: 10, 12, 15

Magazine: detachable box   Grips: one-piece ergonomic grip

D2C:	NIB	$1,100	Ex $850	VG+ $ 615	Good $ 550	LMP $ 1,125
C2C:	Mint $1,060	Ex $ 775	VG+ $550	Good $ 495	Fair $ 255	
Trade-In:	Mint $ 790	Ex $ 625	VG+ $430	Good $ 385	Poor $ 120	

P225
EXCELLENT

P225 REDBOOK CODE: RB-SS-H-225XXX

Black phosphate alloy frame, black phosphate slide, de-cocker.

Production: 1987 - 1998   Caliber: 9mm   Action: DA/SA, DAO, semi-auto

Barrel Length: 3.9"   OA Length: 7.1"   Wt.: 26 oz.

Sights: fixed contrast, SIGLITE night sights   Capacity: 8

Magazine: detachable box   Grips: black checkered polymer

D2C:	NIB	$ 930	Ex $725	VG+ $520	Good $465	LMP $ 725
C2C:	Mint $ 900	Ex $650	VG+ $470	Good $420	Fair $ 215	
Trade-In:	Mint $ 670	Ex $525	VG+ $370	Good $330	Poor $ 120	

P226
NEW IN BOX

P226 REDBOOK CODE: RB-SS-H-226XXX

Nitron-finished slide, accessory rail, black hard anodized frame finish, optional night sights.

Production: ~1987 - current   Caliber: 9mm, .357 SIG, .40 S&W   Action: DA/SA, semi-auto   Barrel Length: 4.4"   OA Length: 7.7"

Wt.: 34 oz.   Sights: contrast or SIGLITE night sights   Capacity: 10, 12, 15

Magazine: detachable box   Grips: one-piece ergo

D2C:	NIB	$ 780	Ex $600	VG+ $435	Good $ 390	LMP $1,085
C2C:	Mint $ 750	Ex $550	VG+ $390	Good $ 355	Fair $ 180	
Trade-In:	Mint $ 560	Ex $450	VG+ $310	Good $ 275	Poor $ 90	

P226 TWO-TONE
NEW IN BOX

P226 TWO-TONE REDBOOK CODE: RB-SS-H-226TWT

Stainless slide finish, black hard-coat anodized frame, reversible magazine release, accessory rail.

Caliber: 9mm, .40 S&W Action: DA/SA, semi-auto Barrel Length: 4.4"										
OA Length: 7.7" Wt.: 34 oz. Sights: SIGLITE night sights										
Capacity: 10, 15, 20 Magazine: detachable box Grips: black polymer										
D2C:	NIB	$	775	Ex	$600	VG+	$435	Good	$ 390	LMP	$1,096
C2C:	Mint	$	750	Ex	$ 550	VG+	$390	Good	$ 350	Fair	$ 180
Trade-In:	Mint	$	560	Ex	$ 450	VG+	$ 310	Good	$ 275	Poor	$ 90

P226 ELITE STAINLESS
NEW IN BOX

P226 ELITE STAINLESS REDBOOK CODE: RB-SS-H-226STA

All stainless frame and slide, natural stainless finish, accessory rail, four-point safety system of decocking lever, patented automatic firing pin safety block, safety intercept notch and trigger bar disconnector.

Production: discontinued Caliber: 9mm, .357 SIG, .40 S&W Action: DA/SA, semi-auto										
Barrel Length: 4.4" OA Length: 7.7" Wt.: 42.2 oz. Sights: SIGLITE night sights										
Capacity: 10, 12, 15 Magazine: detachable box Grips: black polymer										
D2C:	NIB	$1,925	Ex	$1,475	VG+	$1,070	Good	$ 965	LMP	$1,396
C2C:	Mint	$1,850	Ex	$1,350	VG+	$ 970	Good	$ 870	Fair	$ 445
Trade-In:	Mint	$1,370	Ex	$1,100	VG+	$ 760	Good	$ 675	Poor	$ 210

P226 EQUINOX
NEW IN BOX

P226 EQUINOX REDBOOK CODE: RB-SS-H-226EQI

Accessory rail, gray laminated wood grips, black hard anodized frame finish.

Caliber: .40 S&W Action: DA/SA, semi-auto Barrel Length: 4.4"										
OA Length: 7.7" Wt.: 34 oz. Sights: TRUGLO tritium fiber-optic front, SIGLITE rear										
Capacity: 12 Magazine: detachable box Grips: custom-shop wood										
D2C:	NIB	$ 840	Ex	$ 650	VG+	$ 470	Good	$ 420	LMP	$1,218
C2C:	Mint	$ 810	Ex	$ 600	VG+	$ 420	Good	$ 380	Fair	$ 195
Trade-In:	Mint	$ 600	Ex	$ 475	VG+	$ 330	Good	$ 295	Poor	$ 90

P226 DAK
NEW IN BOX

P226 DAK REDBOOK CODE: RB-SS-H-226DAK

Accessory rail, DAK trigger features a consistent DAO trigger pull and double-strike capability, black hard anodized frame finish.

Production: 2005 - current Caliber: 9mm, .357 SIG, .40 S&W Action: DAK (DAO),										
semi-auto Barrel Length: 4.4" OA Length: 7.7"										
Wt.: 34 oz. Sights: post and dot contrast, SIGLITE night sights Capacity: 10, 12, 15										
Magazine: detachable box Grips: black polymer factory										
D2C:	NIB	$ 875	Ex	$ 675	VG+	$490	Good	$ 440	LMP	$1,108
C2C:	Mint	$ 840	Ex	$ 625	VG+	$440	Good	$ 395	Fair	$ 205
Trade-In:	Mint	$ 630	Ex	$ 500	VG+	$350	Good	$ 310	Poor	$ 90

P226 X-FIVE
NEW IN BOX

P226 X-FIVE REDBOOK CODE: RB-SS-H-226XFE

Natural stainless steel frame and slide, adjustable trigger, beavertail grip frame, jet funnel magwell, front cocking serrations, ambidextrous thumb safety.

Production: current Caliber: 9mm, .40 S&W Action: SA, semi-auto Barrel Length: 5"										
OA Length: 8.8" Wt.: 47 oz. Sights: low profile integrated adjustable competition										
sights Capacity: 19, 14 Grips: custom Nill wood grips										
D2C:	NIB	$2,380	Ex	$1,825	VG+	$1,325	Good	$1,190	LMP	$2,747
C2C:	Mint	$2,290	Ex	$1,650	VG+	$1,190	Good	$1,075	Fair	$ 550
Trade-In:	Mint	$1,690	Ex	$1,350	VG+	$ 930	Good	$ 835	Poor	$ 240

P226 X-FIVE COMPETITION
NEW IN BOX

P226 X-FIVE COMPETITION REDBOOK CODE: RB-SS-H-226XFC

Natural stainless steel slide, stainless beavertail frame, cold hammer-forged barrel, jet funnel magwell, front cocking serrations, ambidextrous thumb safety.

Production: current Caliber: 9mm, .40 S&W Action: SA, semi-auto

Barrel Length: 5" OA Length: 8.8" Wt.: 47 oz. Sights: adjustable target

Capacity: 19, 14 Magazine: detachable box Grips: black polymer

D2C:	NIB	$1,855	Ex	$1,425	VG+	$1,030	Good	$ 930	LMP	$1,976
C2C:	Mint	$1,790	Ex	$1,300	VG+	$ 930	Good	$ 835	Fair	$ 430
Trade-In:	Mint	$1,320	Ex	$1,050	VG+	$ 730	Good	$ 650	Poor	$ 210

P226 X-FIVE TACTICAL
NEW IN BOX

P226 X-FIVE TACTICAL REDBOOK CODE: RB-SS-H-226XFT

Nitron stainless steel slide and frame, front cocking serrations, ambidextrous thumb safety, accessory rail.

Production: current Caliber: 9mm Action: SA, semi-auto Barrel Length: 5"

OA Length: 8.8" Wt.: 35.5 oz. Sights: SIGLITE night sights Capacity: 15

Magazine: detachable box Grips: black polymer

D2C:	NIB	$1,455	Ex	$1,125	VG+	$ 810	Good	$ 730	LMP	$1,696
C2C:	Mint	$1,400	Ex	$1,025	VG+	$730	Good	$ 655	Fair	$ 335
Trade-In:	Mint	$1,040	Ex	$ 825	VG+	$570	Good	$ 510	Poor	$ 150

P226 X-FIVE ALLROUND
NEW IN BOX

P226 X-FIVE ALLROUND REDBOOK CODE: RB-SS-H-226XFA

Natural stainless steel slide, stainless beavertail frame, rubber-coated magazine floor, cross-grooves in front of frame and on trigger, grip grooves on front of frame, undercut front sight.

Production: current Caliber: 9mm, .40 S&W Action: DA/SA, semi-auto

Barrel Length: 5" OA Length: 8.8" Wt.: 45 oz. Sights: adjustable target

Capacity: 17, 12 Magazine: detachable box Grips: black polymer

D2C:	NIB	$1,455	Ex	$1,125	VG+	$ 810	Good	$ 730	LMP	$1,696
C2C:	Mint	$1,400	Ex	$1,025	VG+	$ 730	Good	$ 655	Fair	$ 335
Trade-In:	Mint	$1,040	Ex	$ 825	VG+	$ 570	Good	$ 510	Poor	$ 150

P226 X-FIVE LIGHTWEIGHT
NEW IN BOX

P226 X-FIVE LIGHTWEIGHT REDBOOK CODE: RB-SS-H-226XFL

Natural stainless steel slide, lightweight alloy beavertail frame, adjustable trigger, front-cocking serrations, jet-funnel magwell, accessory rail.

Production: current Caliber: 9mm Action: SA, semi-auto Barrel Length: 5"

OA Length: 8.8" Wt.: 36 oz. Sights: adjustable target

Capacity: 19 Magazine: detachable box Grips: custom wood

D2C:	NIB	$1,900	Ex	$1,450	VG+	$1,055	Good	$ 950	LMP	$2,149
C2C:	Mint	$1,830	Ex	$1,325	VG+	$950	Good	$ 855	Fair	$ 440
Trade-In:	Mint	$1,350	Ex	$1,075	VG+	$ 750	Good	$ 665	Poor	$ 210

P226 X-FIVE SHORT REDBOOK CODE: RB-SS-H-226XFS

Shortened X5 stainless frame, stainless slide, ported barrel, skeletonized hammer and trigger, extended magwell.

Production: current Caliber: 9mm Action: DA/SA, semi-auto

Barrel Length: 4.4" OA Length: 8.2" Wt.: 43 oz. Sights: micrometer target

Capacity: 19 Magazine: detachable box Grips: custom wood

D2C:	NIB	$2,600	Ex	$2,000	VG+	$1,445	Good	$1,300	LMP	$2,749
C2C:	Mint	$2,500	Ex	$1,800	VG+	$1,300	Good	$1,170	Fair	$ 600
Trade-In:	Mint	$1,850	Ex	$1,475	VG+	$1,020	Good	$ 910	Poor	$ 270

SIG SAUER

P226 X-FIVE SHORT & SMART
NEW IN BOX

P226 X-FIVE GOLDEN DRAGON
NEW IN BOX

P226 X-SIX (9MM)
NEW IN BOX

P226 X-SIX
(.45 ACP)
NEW IN BOX

P226 X-SIX LIGHTWEIGHT
NEW IN BOX

P226 X-FIVE SHORT & SMART   REDBOOK CODE: RB-SS-H-226X5S

Shortened X5 stainless frame, stainless slide, adjustable trigger, jet-funnel magwell.

Production: current   Caliber: 9mm   Action: SA, semi-auto   Barrel Length: 4.4"

OA Length: 8.2"   Wt.: 43 oz.   Sights: adjustable target

Capacity: 19   Magazine: detachable box   Grips: custom laminate hardwood

D2C:	NIB	$2,400	Ex	$1,825	VG+	$1,335	Good	$1,200	LMP $2,599
C2C:	Mint	$2,310	Ex	$1,675	VG+	$1,200	Good	$1,080	Fair $ 555
Trade-In:	Mint	$ 1,710	Ex	$1,350	VG+	$940	Good	$ 840	Poor $ 240

P226 X-FIVE GOLDEN DRAGON   REDBOOK CODE: RB-SS-H-226X5G

Stainless frame and slide, finished in Copper Pearl Ilaflon coating, gold accents, Golden Dragon slide-top engraving, adjustable trigger, extended magwell.

Production: current   Caliber: 9mm   Action: SA, semi-auto   Barrel Length: 5"

OA Length: 8.8"   Wt.: 45 oz.   Sights: adjustable target

Capacity: 19   Magazine: detachable box   Grips: custom laminate wood

D2C:	NIB	$3,100	Ex	$2,375	VG+	$1,725	Good	$1,550	LMP $3,199
C2C:	Mint	$2,980	Ex	$2,150	VG+	$1,550	Good	$1,395	Fair $ 715
Trade-In:	Mint	$2,210	Ex	$1,750	VG+	$1,210	Good	$1,085	Poor $ 330

P226 X-SIX (9MM)   REDBOOK CODE: RB-SS-H-226X69

Natural stainless steel slide and frame, beavertail grip, adjustable trigger, jet-funnel magwell, front cocking serrations.

Caliber: 9mm   Action: SA, semi-auto   Barrel Length: 6"

OA Length: 9.8"   Wt.: 43 oz.

Sights: low-profile integrated adjustable competition sights

Capacity: 19   Magazine: detachable box   Grips: custom wood grip plates

D2C:	NIB	$2,500	Ex	$1,900	VG+	$1,390	Good	$1,250	LMP $2,950
C2C:	Mint	$2,400	Ex	$1,725	VG+	$1,250	Good	$1,125	Fair $ 575
Trade-In:	Mint	$1,780	Ex	$1,400	VG+	$980	Good	$ 875	Poor $ 270

P226 X-SIX (.45 ACP)   REDBOOK CODE: RB-SS-H-226X65

Natural stainless steel slide and frame, beavertail grip, adjustable trigger, integral magwell, front cocking serrations.

Caliber: .45 ACP   Action: SA, semi-auto   Barrel Length: 6"

OA Length: 9.9"   Wt.: 43 oz.

Sights: low-profile integrated adjustable competition sights   Capacity: 8

Magazine: detachable box   Grips: custom extended wood grip plates

D2C:	NIB	$2,400	Ex	$1,825	VG+	$1,335	Good	$1,200	LMP $2,599
C2C:	Mint	$2,310	Ex	$1,675	VG+	$1,200	Good	$1,080	Fair $ 555
Trade-In:	Mint	$ 1,710	Ex	$1,350	VG+	$940	Good	$ 840	Poor $ 240

P226 X-SIX LIGHTWEIGHT   REDBOOK CODE: RB-SS-H-226X6L

Natural stainless steel slide, lightweight alloy beavertail frame, adjustable trigger, front cocking serrations, jet-funnel magwell, accessory rail.

Caliber: 9mm   Action: SA, semi-auto   Barrel Length: 6"

OA Length: 9.8"   Wt.: 38 oz.   Sights: adjustable target

Capacity: 19   Magazine: detachable box   Grips: custom wood

D2C:	NIB	$2,100	Ex	$1,600	VG+	$1,170	Good	$1,050	LMP $2,249
C2C:	Mint	$2,020	Ex	$1,450	VG+	$1,050	Good	$ 945	Fair $ 485
Trade-In:	Mint	$1,500	Ex	$1,200	VG+	$820	Good	$ 735	Poor $ 210

P226 X-SIX SCANDIC
NEW IN BOX

P226 X-SIX SCANDIC BLUE
NEW IN BOX

P226 X-SIX SCANDIC   REDBOOK CODE: RB-SS-H-226X6S

Stainless beavertail frame, stainless slide with hand-polished slide flats, adjustable trigger, gold-plated accents, extended magwell.

Caliber: 9mm   Action: SA, semi-auto   Barrel Length: 6"

OA Length: 9.8"   Wt.: 46 oz.   Sights: adjustable target

Capacity: 19   Magazine: detachable box   Grips: custom Scandinavian birch wood

D2C:	NIB	$3,100	Ex	$2,375	VG+	$1,725	Good	$1,550	LMP	$3,199
C2C:	Mint	$2,980	Ex	$2,150	VG+	$1,550	Good	$1,395	Fair	$ 715
Trade-In:	Mint	$2,210	Ex	$1,750	VG+	$1,210	Good	$1,085	Poor	$ 330

P226 X-SIX SCANDIC BLUE   REDBOOK CODE: RB-SS-H-2266SB

Stainless steel frame and slide, Blue Pearl Ilaflon coating, tribal engraving on slide flats, gold plated hammer and controls, adjustable trigger, extended magwell.

Caliber: 9mm   Action: SA, semi-auto   Barrel Length: 6"

OA Length: 8.8"   Wt.: 45 oz.   Sights: adjustable target

Capacity: 19   Magazine: detachable box   Grips: custom Scandinavian birch wood

D2C:	NIB	$3,100	Ex	$2,375	VG+	$1,725	Good	$1,550	LMP	$3,199
C2C:	Mint	$2,980	Ex	$2,150	VG+	$1,550	Good	$1,395	Fair	$ 715
Trade-In:	Mint	$2,210	Ex	$1,750	VG+	$1,210	Good	$1,085	Poor	$ 330

P226 X-SIX PPC   REDBOOK CODE: RB-SS-H-226X6P

Stainless steel frame and slide, beavertail grip frame, adjustable trigger, front cocking serrations, accessory rail.

Caliber: 9mm   Action: SA, semi-auto   Barrel Length: 6"

OA Length: 9.8"   Wt.: 57 oz.

Sights: Aristocrat adjustable PPC sight set   Capacity: 19

Magazine: detachable box   Grips: ergonomic custom magwell wood grip plates

D2C:	NIB	$2,500	Ex	$1,900	VG+	$1,390	Good	$1,250	LMP	$3,199
C2C:	Mint	$2,400	Ex	$1,725	VG+	$1,250	Good	$1,125	Fair	$ 575
Trade-In:	Mint	$1,780	Ex	$1,400	VG+	$ 980	Good	$ 875	Poor	$ 270

P226 COMBAT
NEW IN BOX

P226 COMBAT   REDBOOK CODE: RB-SS-H-226COM

M1913 Picatinny rail, Flat Dark Earth frame finish, Nitron-finished slide, front-strap serrations, model with threaded barrel available.

Caliber: 9mm   Action: DA/SA, semi-auto   Barrel Length: 4.4"

OA Length: 7.7"   Wt.: 34 oz.

Sights: SIGLITE night sights   Capacity: 10, 15   Magazine: detachable box

Grips: Flat Dark Earth polymer

D2C:	NIB	$1,105	Ex	$850	VG+	$ 615	Good	$ 555	LMP	$1,218
C2C:	Mint	$1,070	Ex	$ 775	VG+	$560	Good	$ 500	Fair	$ 255
Trade-In:	Mint	$ 790	Ex	$ 625	VG+	$440	Good	$ 390	Poor	$ 120

P226 CLASSIC 22   REDBOOK CODE: RB-SS-H-226C22

Four-point safety system, accessory rail, black anodized-finished slide and frame.

Caliber: .22 LR   Action: DA/SA, semi-auto   Barrel Length: 4.5"

OA Length: 7.7"   Wt.: 24 oz.   Sights: adjustable

Capacity: 10   Magazine: detachable box   Grips: black polymer factory

D2C:	NIB	$ 540	Ex	$ 425	VG+	$300	Good	$270	LMP	$ 626
C2C:	Mint	$ 520	Ex	$ 375	VG+	$270	Good	$245	Fair	$ 125
Trade-In:	Mint	$ 390	Ex	$ 325	VG+	$220	Good	$190	Poor	$ 60

P226 CLASSIC 22 BEAVERTAIL
NEW IN BOX

P226 CLASSIC 22 BEAVERTAIL REDBOOK CODE: RB-SS-H-22622B

Four-point safety system, accessory rail, black anodized-finished slide and frame, beavertail frame.

Caliber: .22 LR Action: DA/SA, semi-auto Barrel Length: 4.5"					
OA Length: 7.7" Wt.: 24 oz. Sights: adjustable					
Capacity: 10 Magazine: detachable box Grips: standard polymer or E2 grip					
D2C:	NIB $ 540	Ex $ 425	VG+ $ 300	Good $ 270	LMP $ 626
C2C:	Mint $ 520	Ex $ 375	VG+ $ 270	Good $ 245	Fair $ 125
Trade-In:	Mint $ 390	Ex $ 325	VG+ $ 220	Good $ 190	Poor $ 60

P226 TACTICAL OPERATIONS
(TACOPS)
NEW IN BOX

P226 TACTICAL OPERATIONS (TACOPS)

REDBOOK CODE: RB-SS-H-226TAC

Short Reset Trigger (SRT), Nitron-finished slide, black hard anodized frame with beavertail, accessory rail, front cocking serrations, 4 super-capacity magazines, available with threaded barrel.

Caliber: 9mm, .40 S&W, .357 SIG Action: DA/SA, semi-auto Barrel Length: 4.4"					
OA Length: 8.2" Wt.: 34 oz. Sights: SIGLITE night sights, TRUGLO tritium fiber-optic					
front Capacity: 20, 15, 10 Magazine: detachable box Grips: polymer magwell grips					
D2C:	NIB $ 875	Ex $ 675	VG+ $490	Good $ 440	LMP $1,302
C2C:	Mint $ 840	Ex $ 625	VG+ $440	Good $ 395	Fair $ 205
Trade-In:	Mint $ 630	Ex $500	VG+ $350	Good $ 310	Poor $ 90

P226 EXTREME
NEW IN BOX

P226 EXTREME REDBOOK CODE: RB-SS-H-226EXT

Short Reset Trigger (SRT), black hard anodized frame, Nitron-finished slide, front cocking serrations, accessory rail.

Caliber: 9mm, .40 S&W Action: DA/SA, semi-auto Barrel Length: 4.4"					
OA Length: 8.2" Wt.: 34 oz. Sights: SIGLITE night sights Capacity: 15, 12					
Magazine: detachable box Grips: Hogue custom G10					
D2C:	NIB $ 825	Ex $ 650	VG+ $ 460	Good $ 415	LMP $1,213
C2C:	Mint $ 800	Ex $ 575	VG+ $ 420	Good $ 375	Fair $ 190
Trade-In:	Mint $ 590	Ex $ 475	VG+ $ 330	Good $ 290	Poor $ 90

P226 BLACK STAINLESS
NEW IN BOX

P226 BLACK STAINLESS REDBOOK CODE: RB-SS-H-226BST

All-stainless frame and slide with Nitron finish, accessory rail.

Caliber: 9mm Action: DA/SA, semi-auto Barrel Length: 4.4"					
OA Length: 7.7" Wt.: 42 oz. Sights: SIGLITE night sights Capacity: 15					
Magazine: detachable box Grips: Hogue rubber					
D2C:	NIB $1,010	Ex $ 775	VG+ $565	Good $505	LMP $1,129
C2C:	Mint $ 970	Ex $700	VG+ $510	Good $455	Fair $235
Trade-In:	Mint $ 720	Ex $ 575	VG+ $400	Good $355	Poor $120

P226 NAVY
EXCELLENT

P226 NAVY REDBOOK CODE: RB-SS-H-226NAV

Nitron-finished stainless slide with engraved anchor on left side, black hard anodized alloy frame, accessory rail, some early versions with no rail, this model continued as the MK25.

Production: discontinued Caliber: 9mm Action: DA/SA, semi-auto					
Barrel Length: 4.4" OA Length: 7.7" Wt.: 34 oz.					
Sights: contrast Capacity: 15 Magazine: detachable box Grips: black polymer					
D2C:	NIB $ 915	Ex $700	VG+ $510	Good $ 460	LMP $1,038
C2C:	Mint $ 880	Ex $650	VG+ $460	Good $ 415	Fair $ 215
Trade-In:	Mint $ 650	Ex $ 525	VG+ $360	Good $ 325	Poor $ 120

P226 MK25
NEW IN BOX

P226 MK25 REDBOOK CODE: RB-SS-H-26MK25

Black hard anodized frame, Nitron-finished slide, anti-corrosion coatings on controls and internal components, M1913 accessory rail, UID identification label, anchor engraving.

Production: 2011 - current Caliber: 9mm Action: DA/SA, semi-auto Barrel Length: 4.4"
OA Length: 7.7" Wt.: 34 oz. Sights: SIGLITE night sights Capacity: 10, 15
Magazine: detachable box Grips: black polymer factory

D2C:	NIB	$ 805	Ex $ 625	VG+ $450	Good $ 405	LMP	$1,142
C2C:	Mint	$ 780	Ex $ 575	VG+ $410	Good $ 365	Fair	$ 190
Trade-In:	Mint	$ 580	Ex $ 475	VG+ $320	Good $ 285	Poor	$ 90

P226 MK25 DESERT
NEW IN BOX

P226 MK25 DESERT REDBOOK CODE: RB-SS-H-MK25DE

Black hard anodized frame, Nitron-finished slide, anti-corrosion coatings on controls and internal components, M1913-accessory rail, UID-identification label, anchor engraving.

Caliber: 9mm Action: DA/SA, semi-auto Barrel Length: 4.4" OA Length: 7.7"
Wt.: 34 oz. Sights: SIGLITE night sights Capacity: 10, 15 Magazine: detachable box
Grips: black polymer factory

D2C:	NIB	$ 885	Ex $ 675	VG+ $495	Good $ 445	LMP	$1,236
C2C:	Mint	$ 850	Ex $ 625	VG+ $450	Good $400	Fair	$ 205
Trade-In:	Mint	$ 630	Ex $500	VG+ $350	Good $ 310	Poor	$ 90

P226 SCORPION
NEW IN BOX

P226 SCORPION REDBOOK CODE: RB-SS-H-226SCO

Short Reset Trigger (SRT), Flat Dark Earth finish on frame and slide, beavertail, accessory rail, model with threaded barrel available.

Caliber: 9mm Action: DA/SA, semi-auto Barrel Length: 4.4"
OA Length: 8.2" Wt.: 34 oz. Sights: SIGLITE night sights Capacity: 10, 15
Magazine: detachable box Grips: Hogue Extreme G10

D2C:	NIB	$ 885	Ex $ 675	VG+ $495	Good $ 445	LMP	$1,285
C2C:	Mint	$ 850	Ex $ 625	VG+ $450	Good $ 400	Fair	$ 205
Trade-In:	Mint	$ 630	Ex $ 500	VG+ $350	Good $ 310	Poor	$ 90

P226 ELITE
NEW IN BOX

P226 ELITE REDBOOK CODE: RB-SS-H-226ELT

Nitron stainless slide, black hard anodized frame with beavertail, (SRT) Short Reset Trigger, decocking lever, accessory rail.

Production: discontinued Caliber: 9mm, .40 S&W Action: DA/SA, semi-auto
Barrel Length: 4.4" OA Length: 8.2" Wt.: 34 oz.
Sights: SIGLITE night sights Capacity: 10, 12, 15 Magazine: detachable box
Grips: custom rosewood

D2C:	NIB	$ 815	Ex $625	VG+ $455	Good $ 410	LMP	$1,200
C2C:	Mint	$ 790	Ex $575	VG+ $410	Good $ 370	Fair	$ 190
Trade-In:	Mint	$ 580	Ex $475	VG+ $320	Good $ 290	Poor	$ 90

P226 ELITE STAINLESS
NEW IN BOX

P226 ELITE STAINLESS REDBOOK CODE: RB-SS-H-226ELS

Short Reset Trigger (SRT), accessory rail, stainless frame and slide.

Caliber: 9mm, .40 S&W Action: DA/SA, semi-auto Barrel Length: 4.4"
OA Length: 8.2" Wt.: 42 oz.
Sights: SIGLITE night sights Capacity: 10, 12, 15 Magazine: detachable box
Grips: custom rosewood

D2C:	NIB	$1,330	Ex $1,025	VG+ $740	Good $ 665	LMP	$1,368
C2C:	Mint	$1,280	Ex $ 925	VG+ $670	Good $ 600	Fair	$ 310
Trade-In:	Mint	$ 950	Ex $ 750	VG+ $520	Good $ 470	Poor	$ 150

P226 ELITE SAO
NEW IN BOX

P226 ELITE SAO REDBOOK CODE: RB-SS-H-226ESA
Black hard anodized frame, Nitron slide finish, front cocking serrations, beavertail Elite frame, accessory rail.

Caliber: 9mm	Action: SA, semi-auto	Barrel Length: 4.4"	OA Length: 8.2"						
Wt.: 34 oz.	Sights: SIGLITE night sights	Capacity: 15	Magazine: detachable box						
Grips: 2-piece polymer									
D2C:	NIB	$1,280	Ex $ 975	VG+ $ 715	Good $ 640	LMP	$1,218		
C2C:	Mint	$1,230	Ex $ 900	VG+ $640	Good $ 580	Fair	$ 295		
Trade-In:	Mint	$ 910	Ex $ 725	VG+ $500	Good $ 450	Poor	$ 150		

P226 ELITE DARK
NEW IN BOX

P226 ELITE DARK REDBOOK CODE: RB-SS-H-226ELD
Short Reset Trigger (SRT), black hard anodized frame, Nitron slide finish, beavertail, accessory rail, front strap checkering, front cocking serrations, model with threaded barrel available.

Caliber: 9mm, .40 S&W, .357 SIG	Action: DA/SA, semi-auto	Barrel Length: 4.4					
OA Length: 8.2"	Wt.: 34 oz.	Sights: adjustable combat night sights					
Capacity: 15, 12	Magazine: detachable box	Grips: aluminum					
D2C:	NIB	$1,280	Ex $ 975	VG+ $ 715	Good $ 640	LMP	$1,218
C2C:	Mint	$1,230	Ex $ 900	VG+ $640	Good $ 580	Fair	$ 295
Trade-In:	Mint	$ 910	Ex $ 725	VG+ $500	Good $ 450	Poor	$ 150

P226 ENHANCED ELITE
NEW IN BOX

P226 ENHANCED ELITE REDBOOK CODE: RB-SS-H-226ENE
Short Reset Trigger (SRT), black hard-coat anodized frame, Nitron slide finish, front cocking serrations, beavertail frame, front strap checkering, accessory rail.

Caliber: 9mm, .357 SIG, .40 S&W	Action: DA/SA, semi-auto						
Barrel Length: 4.4	OA Length: 8.2"	Wt.: 34 oz.					
Sights: SIGLITE night sights	Capacity: 15, 12	Magazine: detachable box					
Grips: one-piece reduced reach ergo grip							
D2C:	NIB	$1,200	Ex $925	VG+ $670	Good $ 600	LMP	$1,175
C2C:	Mint	$1,160	Ex $850	VG+ $600	Good $ 540	Fair	$ 280
Trade-In:	Mint	$ 860	Ex $ 675	VG+ $470	Good $ 420	Poor	$ 120

P226 PLATINUM ELITE
NEW IN BOX

P226 PLATINUM ELITE REDBOOK CODE: RB-SS-H-226PLE
Stainless slide, black hard anodized alloy frame with beavertail, Short Reset Trigger (SRT), front cocking serrations, front-strap checkering, decocking lever.

Production: discontinued	Caliber: 9mm, .40 S&W	Action: DA/SA, semi-auto					
Barrel Length: 4.4	OA Length: 8.2"	Wt.: 34 oz.					
Sights: adjustable SIGLITE combat night sights	Capacity: 12, 15						
Magazine: detachable box	Grips: custom aluminum						
D2C:	NIB	$1,290	Ex $1,000	VG+ $720	Good $ 645	LMP	$1,289
C2C:	Mint	$1,240	Ex $900	VG+ $650	Good $ 585	Fair	$ 300
Trade-In:	Mint	$ 920	Ex $ 725	VG+ $ 510	Good $ 455	Poor	$ 150

P226 ENGRAVED
NEW IN BOX

P226 ENGRAVED REDBOOK CODE: RB-SS-H-226ENG
Polished-Nitron stainless steel slide with custom engraving, black hard anodized frame.

Caliber: 9mm	Action: DA/SA, semi-auto	Barrel Length: 4.4					
OA Length: 7.7"	Wt.: 34 oz.	Sights: SIGLITE night sights	Capacity: 15	Magazine:			
detachable box	Grips: custom wood with SIG medallion						
D2C:	NIB	$1,285	Ex $1,000	VG+ $ 715	Good $ 645	LMP	$1,289
C2C:	Mint	$1,240	Ex $ 900	VG+ $650	Good $ 580	Fair	$ 300
Trade-In:	Mint	$ 920	Ex $ 725	VG+ $ 510	Good $ 450	Poor	$ 150

P226 ENGRAVED STAINLESS
NEW IN BOX

P226 ENGRAVED STAINLESS REDBOOK CODE: RB-SS-H-226ENS

Custom-engraved stainless steel slide, natural-stainless frame finish, accessory rail.

Caliber: 9mm Action: DA/SA, semi-auto Barrel Length: 4.4

OA Length: 7.7" Wt.: 34 oz. Sights: SIGLITE night sights

Capacity: 15 Magazine: detachable box Grips: custom wood, SIG medallion

D2C:	NIB	$1,360	Ex	$1,050	VG+	$ 755	Good	$ 680	LMP	$1,402
C2C:	Mint	$1,310	Ex	$ 950	VG+	$ 680	Good	$ 615	Fair	$ 315
Trade-In:	Mint	$ 970	Ex	$ 775	VG+	$ 540	Good	$ 480	Poor	$ 150

P226 TRIBAL
NEW IN BOX

P226 TRIBAL REDBOOK CODE: RB-SS-H-226TRI

Custom engraved and polished Nitron stainless steel slide, black hard-coat anodized frame, front cocking serrations.

Caliber: 9mm Action: DA/SA, semi-auto OA Length: 7.7"

Sights: SIGLITE night sights Capacity: 15 Magazine: detachable box

Grips: custom engraved aluminum

D2C:	NIB	$1,280	Ex	$ 975	VG+	$ 715	Good	$ 640	LMP	$1,279
C2C:	Mint	$1,230	Ex	$ 900	VG+	$ 640	Good	$ 580	Fair	$ 295
Trade-In:	Mint	$910	Ex	$ 725	VG+	$ 500	Good	$ 450	Poor	$ 150

P226 TRIBAL TWO-TONE
NEW IN BOX

P226 TRIBAL TWO-TONE REDBOOK CODE: RB-SS-H-226TTT

Custom-engraved natural-stainless steel slide, black hard-coat anodized frame.

Caliber: 9mm Action: DA/SA, semi-auto OA Length: 7.7

Sights: SIGLITE night sights Capacity: 15

Magazine: detachable box Grips: custom engraved aluminum

D2C:	NIB	$ 1,130	Ex	$ 875	VG+	$630	Good	$ 565	LMP	$1,145
C2C:	Mint	$1,090	Ex	$ 800	VG+	$570	Good	$ 510	Fair	$ 260
Trade-In:	Mint	$810	Ex	$ 650	VG+	$450	Good	$400	Poor	$ 120

P226 USPSA REDBOOK CODE: RB-SS-H-226USP

Available in either a black anodized alloy frame with Nitron slide and nickel-accent controls or a natural-stainless finish frame with Nitron slide and black controls, USPSA laser engraved logo, Short Reset Trigger (SRT), beavertail style frame, accessory rail.

Production: discontinued Caliber: 9mm Action: DA/SA, semi-auto Barrel Length: 4.4"

OA Length: 8.2" Wt.: 34 oz., 42.2 oz. Sights: Dawson fiber-optic front and Warren

tactical rear Capacity: 15 Magazine: detachable box Grips: black polymer

D2C:	NIB	$ 1,110	Ex	$850	VG+	$ 620	Good	$ 555	LMP	$1,246
C2C:	Mint	$1,070	Ex	$ 775	VG+	$ 560	Good	$500	Fair	$ 260
Trade-In:	Mint	$ 790	Ex	$ 625	VG+	$ 440	Good	$ 390	Poor	$ 120

P226 LIMITED EDITION GADSDEN
NEW IN BOX
Courtesy of Bud's Gun Shop

P226 LIMITED EDITION GADSDEN
REDBOOK CODE: RB-SS-H-226LEG

Nitron-finished stainless slide with engraving, nickel accents, black hard anodized frame, accessory rail, only 1,776 manufactured.

Production: discontinued Caliber: 9mm Action: DA/SA, semi-auto

Barrel Length: 4.4" OA Length: 7.7" Wt.: 34 oz. Sights: SIGLITE night sights

Capacity: 15 Magazine: detachable box Grips: engraved wood

D2C:	NIB	$1,210	Ex	$ 925	VG+	$ 675	Good	$605	LMP	$1,200
C2C:	Mint	$1,170	Ex	$ 850	VG+	$ 610	Good	$ 545	Fair	$ 280
Trade-In:	Mint	$ 860	Ex	$ 700	VG+	$ 480	Good	$ 425	Poor	$ 150

SIG SAUER

P226 SAS REDBOOK CODE: RB-SS-H-226SAS

SIG Anti-Snag treatment (SAS), limited production from custom shop, black hard-coat anodized alloy frame, stainless slide.

Production: discontinued Caliber: .40 S&W Action: DAK (DAO), semi-auto				
Barrel Length: 4.4" OA Length: 7.7" Wt.: 34 oz. Sights: SIGLITE front, contrast rear				
Capacity: 12 Magazine: detachable box Grips: custom shop wood				
D2C:	NIB $1,025	Ex $ 800	VG+ $ 570	Good $ 515	LMP $1,067
C2C:	Mint $ 990	Ex $ 725	VG+ $ 520	Good $ 465	Fair $ 240
Trade-In:	Mint $ 730	Ex $ 575	VG+ $ 400	Good $ 360	Poor $ 120

P227 NITRON
NEW IN BOX

P227 NITRON REDBOOK CODE: RB-SS-H-227NIT

Nitron stainless steel frame, black hard-coat anodized frame finish, accessory rail, optional night sights.

Production: current Caliber: .45 ACP Action: DA/SA, semi-auto				
Barrel Length: 4.4" Wt.: 32 oz. Sights: contrast, SIGLITE night sights Capacity: 10				
Magazine: detachable box Grips: one-piece polymer				
D2C:	NIB $ 880	Ex $ 675	VG+ $ 490	Good $ 440	LMP $ 993
C2C:	Mint $ 850	Ex $ 625	VG+ $ 440	Good $ 400	Fair $ 205
Trade-In:	Mint $ 630	Ex $ 500	VG+ $ 350	Good $ 310	Poor $ 90

P227 CARRY NITRON
NEW IN BOX

P227 CARRY NITRON REDBOOK CODE: RB-SS-H-227CAN

Nitron stainless steel or black hard-coat anodized frame, accessory rail, optional night sights.

Production: current Caliber: .45 ACP Action: DA/SA, semi-auto Barrel Length: 4"				
OA Length: 7.1" Wt.: 31 oz. Sights: SIGLITE night sights Capacity: 10				
Magazine: detachable box Grips: one-piece polymer				
D2C:	NIB $ 980	Ex $750	VG+ $545	Good $490	LMP $1,085
C2C:	Mint $ 950	Ex $700	VG+ $490	Good $445	Fair $ 230
Trade-In:	Mint $ 700	Ex $550	VG+ $390	Good $345	Poor $ 120

P227 CARRY SAS GEN 2
NEW IN BOX

P227 CARRY SAS GEN 2 REDBOOK CODE: RB-SS-H-227CS2

Short Reset Trigger (SRT), Nitron stainless steel slide, black hard-coat anodized alloy frame, SIG Anti-Snag Treatment on slide and frame.

Production: current Caliber: .45 ACP Action: DA/SA, semi-auto				
Barrel Length: 4" OA Length: 7.1" Wt.: 31 oz.				
Sights: SIGLITE night sights Capacity: 10 Grips: one-piece polymer				
D2C:	NIB $1,025	Ex $ 800	VG+ $ 570	Good $ 515	LMP $ 1,125
C2C:	Mint $ 990	Ex $ 725	VG+ $ 520	Good $ 465	Fair $ 240
Trade-In:	Mint $ 730	Ex $ 575	VG+ $ 400	Good $ 360	Poor $ 120

P228 REDBOOK CODE: RB-SS-H-228XXX

Forged carbon steel slide, blue finish, K-Kote or nickel finish available for slight premium, night sights optional, decocking lever, the U.S. military selected this model as the M11.

Production: 1990 - 1997 Caliber: 9mm Action: DA/SA, semi-auto or DAO				
Barrel Length: 4" OA Length: 7.1" Wt.: 30 oz.				
Sights: contrast, SIGLITE night sights Capacity: 10, 13				
Magazine: detachable box Grips: black polymer factory				
D2C:	NIB $1,450	Ex $1,125	VG+ $ 805	Good $ 725	LMP $ 850
C2C:	Mint $1,400	Ex $1,025	VG+ $ 730	Good $ 655	Fair $ 335
Trade-In:	Mint $1,030	Ex $ 825	VG+ $ 570	Good $ 510	Poor $ 150

M11-A1
NEW IN BOX

M11-A1 REDBOOK CODE: RB-SS-H-M11A1X

Short Reset Trigger (SRT), black hard-coat anodized frame finish, Nitron-coated stainless steel slide, corrosion-resistant internal parts.

Production: 2012 - current   Caliber: 9mm   Action: DA/SA, semi-auto

Barrel Length: 4"   OA Length: 7.1"   Wt.: 32 oz.   Sights: SIGLITE night sights

Capacity: 15   Magazine: detachable box   Grips: black polymer

D2C:	NIB	$1,005	Ex	$ 775	VG+	$560	Good	$ 505	LMP	$1,125
C2C:	Mint	$ 970	Ex	$ 700	VG+	$ 510	Good	$ 455	Fair	$ 235
Trade-In:	Mint	$ 720	Ex	$ 575	VG+	$400	Good	$ 355	Poor	$ 120

M11-A1 DESERT
NEW IN BOX

M11-A1 DESERT REDBOOK CODE: RB-SS-H-M11A1D

Short Reset Trigger (SRT), Flat Dark Earth coated stainless slide and frame, corrosion-resistant internal parts.

Production: 2012 - current   Caliber: 9mm   Action: DA/SA, semi-auto

Barrel Length: 4"   OA Length: 7.1"   Wt.: 32 oz.   Sights: SIGLITE night sights

Capacity: 15   Magazine: detachable box   Grips: black polymer factory

D2C:	NIB	$1,095	Ex	$850	VG+	$ 610	Good	$ 550	LMP	$1,220
C2C:	Mint	$1,060	Ex	$ 775	VG+	$550	Good	$ 495	Fair	$ 255
Trade-In:	Mint	$ 780	Ex	$ 625	VG+	$430	Good	$ 385	Poor	$ 120

P229
NEW IN BOX

P229 REDBOOK CODE: RB-SS-H-229XXX

Nitron-finished slide, black hard-coat anodized frame, accessory rail, optional night sights.

Production: 1992 - current   Caliber: 9mm, .40 S&W, .357 SIG   Action: DA/SA, semi-auto

Barrel Length: 4"   OA Length: 7.1"   Wt.: 32 oz.   Sights: contrast, SIGLITE night sights

Capacity: 10, 12, 13   Magazine: detachable box   Grips: black polymer factory

D2C:	NIB	$ 950	Ex	$ 725	VG+	$ 530	Good	$ 475	LMP	$ 993
C2C:	Mint	$ 920	Ex	$ 675	VG+	$ 480	Good	$ 430	Fair	$ 220
Trade-In:	Mint	$ 680	Ex	$ 550	VG+	$ 380	Good	$ 335	Poor	$ 120

P229 EQUINOX
NEW IN BOX

P229 EQUINOX REDBOOK CODE: RB-SS-H-229EQI

Two-tone accented Nitron stainless steel slide, lightweight black hard-coat anodized alloy frame, brush polished flats, nickel accents, accessory rail.

Caliber: .40 S&W   Action: DA/SA, semi-auto   Barrel Length: 4

OA Length: 7.1"   Wt.: 32 oz.   Sights: TRUGLO front, SIGLITE rear

Capacity: 12, 10   Magazine: detachable box

Grips: custom shop wood

D2C:	NIB	$1,200	Ex	$ 925	VG+	$ 670	Good	$ 600	LMP	$1,218
C2C:	Mint	$1,160	Ex	$ 850	VG+	$ 600	Good	$ 540	Fair	$280
Trade-In:	Mint	$ 860	Ex	$ 675	VG+	$ 470	Good	$ 420	Poor	$ 120

P229 TWO-TONE REDBOOK CODE: RB-SS-H-229TWT

Black hard-coat anodized frame, stainless slide finish, accessory rail.

Caliber: 9mm   Action: DA/SA, semi-auto   Barrel Length: 4

OA Length: 7.1"   Wt.: 32 oz.   Sights: SIGLITE night sights

Capacity: 15   Magazine: detachable box

Grips: black polymer factory

D2C:	NIB	$1,050	Ex	$ 800	VG+	$ 585	Good	$ 525	LMP	$1,096
C2C:	Mint	$1,010	Ex	$ 725	VG+	$ 530	Good	$ 475	Fair	$ 245
Trade-In:	Mint	$ 750	Ex	$ 600	VG+	$ 410	Good	$ 370	Poor	$ 120

P229 DAK
NEW IN BOX

P229 DAK REDBOOK CODE: RB-SS-H-229DAK

Black hard-coat anodized frame, Nitron-finished slide, accessory rail, Sig's exclusive DAK trigger, optional night sights.

Caliber: 9mm, .357 SIG, .40 S&W Action: DAK (DAO), semi-auto
Barrel Length: 4" OA Length: 7.1" Wt.: 32 oz. Sights: SIGLITE night sights
Capacity: 13, 12 Magazine: detachable box Grips: black polymer factory

D2C:	NIB $ 880	Ex $ 675	VG+ $ 490	Good $ 440	LMP $ 920
C2C:	Mint $ 850	Ex $ 625	VG+ $ 440	Good $ 400	Fair $ 205
Trade-In:	Mint $ 630	Ex $ 500	VG+ $ 350	Good $ 310	Poor $ 90

P229 SAS
NEW IN BOX
Courtesy of Bud's Gun Shop

P229 SAS REDBOOK CODE: RB-SS-H-229SAS

SIG Anti-Snag (SAS) dehorn treatment, stainless steel slide, black hard-coat anodized alloy frame, limited production.

Production: discontinued Caliber: .40 S&W Action: DAK (DAO), semi-auto
Barrel Length: 4" OA Length: 7.1" Wt.: 32 oz. Sights: SIGLITE front, contrast rear
Capacity: 10, 12 Magazine: detachable box Grips: custom shop wood

D2C:	NIB $ 980	Ex $ 750	VG+ $ 545	Good $ 490	LMP $1,093
C2C:	Mint $ 950	Ex $ 700	VG+ $ 490	Good $ 445	Fair $ 230
Trade-In:	Mint $ 700	Ex $ 550	VG+ $ 390	Good $ 345	Poor $ 120

P229 SAS GEN 2 REDBOOK CODE: RB-SS-H-229SG2

Short Reset Trigger (SRT), SIG Anti-Snag (SAS) dehorn treatment, Nitron-finished stainless steel slide, black hard-coat anodized frame, two-tone model available.

Production: current Caliber: 9mm, .357 SIG, .40 S&W Action: DA/SA, semi-auto
Barrel Length: 4" OA Length: 7.1" Wt.: 32 oz. Sights: SIGLITE night sights
Capacity: 13, 12 Magazine: detachable box Grips: black polymer factory

D2C:	NIB $ 975	Ex $ 750	VG+ $ 545	Good $ 490	LMP $1,125
C2C:	Mint $ 940	Ex $ 675	VG+ $ 490	Good $ 440	Fair $ 225
Trade-In:	Mint $ 700	Ex $ 550	VG+ $ 390	Good $ 345	Poor $ 120

P229 SAS GEN 2
NEW IN BOX

P229 EXTREME
NEW IN BOX

P229 EXTREME REDBOOK CODE: RB-SS-H-229EXT

Short Reset Trigger (SRT), black hard-coat anodized frame, Nitron finish stainless steel slide, accessory rail.

Caliber: 9mm Action: DA/SA, semi-auto Barrel Length: 4"
OA Length: 7.6" Wt.: 32 oz. Sights: SIGLITE night sights Capacity: 15, 10
Magazine: detachable box Grips: Hogue Extreme G10

D2C:	NIB $ 865	Ex $ 675	VG+ $ 485	Good $ 435	LMP $1,213
C2C:	Mint $ 840	Ex $ 600	VG+ $ 440	Good $ 390	Fair $ 200
Trade-In:	Mint $ 620	Ex $ 500	VG+ $ 340	Good $ 305	Poor $ 90

P229R HSP REDBOOK CODE: RB-SS-H-229RHS

Nitron-finished stainless slide, alloy frame, integral rail, limited-edition Homeland Security Pistol, engraved with the American flag and "Homeland Security # of 1000" with the # being the serial number of that particular gun.

Production: discontinued Caliber: .40 S&W Action: DAK (DAO), semi-auto
Barrel Length: 4" OA Length: 7.1" Wt.: 32 oz. Sights: SIGLITE night sights
Capacity: 10, 12 Magazine: detachable box Grips: black polymer factory

D2C:	NIB $ 850	Ex $ 650	VG+ $ 475	Good $ 425	LMP $ 900
C2C:	Mint $ 820	Ex $ 600	VG+ $ 430	Good $ 385	Fair $ 200
Trade-In:	Mint $ 610	Ex $ 500	VG+ $ 340	Good $ 300	Poor $ 90

P229 ELITE
NEW IN BOX

P229 ELITE REDBOOK CODE: RB-SS-H-229ELT

Nitron-finished stainless slide, black hard-coat anodized frame with beavertail, (SRT) Short Reset Trigger, accessory rail, decocking lever.

Production: discontinued   Caliber: 9mm, .40 S&W   Action: DA/SA, semi-auto

Barrel Length: 4"   OA Length: 7.6"   Wt.: 32 oz.

Sights: SIGLITE night sights   Capacity: 12, 13   Magazine: detachable box

Grips: custom rosewood

D2C:	NIB	$ 990	Ex	$ 775	VG+	$550	Good	$495	LMP	$1,200
C2C:	Mint	$ 960	Ex	$700	VG+	$500	Good	$450	Fair	$ 230
Trade-In:	Mint	$ 710	Ex	$ 575	VG+	$390	Good	$350	Poor	$ 120

P229 ELITE STAINLESS
NEW IN BOX

P229 ELITE STAINLESS REDBOOK CODE: RB-SS-H-229ELS

Short Reset Trigger (SRT), beavertail grip, front cocking serrations, front-strap checkering, stainless steel slide and frame in natural-stainless finish, accessory rail.

Caliber: 9mm, .40 S&W   Action: DA/SA, semi-auto   Barrel Length: 4"

OA Length: 7.6"   Wt.: 40 oz.

Sights: SIGLITE night sights   Capacity: 10, 12   Magazine: detachable box

Grips: custom rosewood

D2C:	NIB	$1,160	Ex	$ 900	VG+	$ 645	Good	$ 580	LMP	$1,368
C2C:	Mint	$1,120	Ex	$ 825	VG+	$ 580	Good	$ 525	Fair	$ 270
Trade-In:	Mint	$ 830	Ex	$ 650	VG+	$ 460	Good	$ 410	Poor	$ 120

P229 ELITE DARK
NEW IN BOX

P229 ELITE DARK REDBOOK CODE: RB-SS-H-229ELD

Short Reset Trigger (SRT), beavertail grip, front cocking serrations, front-strap checkering, black hard-coat anodized frame, Nitron-finished slide, accessory rail, threaded barrel available for 9mm.

Caliber: 9mm, .357 SIG, .40 S&W   Action: DA/SA, semi-auto

Barrel Length: 4"   OA Length: 7.6"   Wt.: 32 oz.

Sights: adjustable combat night sights   Capacity: 13, 12   Magazine: detachable box

Grips: aluminum

D2C:	NIB	$1,150	Ex	$ 875	VG+	$ 640	Good	$ 575	LMP	$1,218
C2C:	Mint	$1,110	Ex	$ 800	VG+	$ 580	Good	$ 520	Fair	$ 265
Trade-In:	Mint	$ 820	Ex	$ 650	VG+	$ 450	Good	$ 405	Poor	$ 120

P229 ENHANCED ELITE
NEW IN BOX

P229 ENHANCED ELITE REDBOOK CODE: RB-SS-H-229ENE

Short Reset Trigger (SRT), beavertail frame, Elite engraving on slide, front strap checkering, front cocking serrations, black hard-coat anodized frame, Nitron-finished slide, accessory rail.

Caliber: 9mm, .357 SIG, .40 S&W   Action: DA/SA, semi-auto   Barrel Length: 4"

OA Length: 7.4"   Wt.: 32 oz.

Sights: SIGLITE night sights   Capacity: 15, 12   Magazine: detachable box

Grips: one piece reduced reach ergo

D2C:	NIB	$1,025	Ex	$ 800	VG+	$570	Good	$ 515	LMP	$ 1,175
C2C:	Mint	$ 990	Ex	$ 725	VG+	$520	Good	$ 465	Fair	$ 240
Trade-In:	Mint	$ 730	Ex	$ 575	VG+	$400	Good	$ 360	Poor	$ 120

P229 PLATINUM ELITE REDBOOK CODE: RB-SS-H-229PLE

Stainless steel slide, black hard-coat anodized alloy frame with beavertail, (SRT) Short Reset Trigger, accessory rail, decocking lever, front cocking serrations, front-strap checkering.

Production: discontinued   Caliber: 9mm, .40 S&W   Action: DA/SA, semi-auto

Barrel Length: 4"   OA Length: 7.6"   Wt.: 32 oz.

Sights: adjustable SIGLITE combat night sights   Capacity: 12, 13

Magazine: detachable box   Grips: Hogue aluminum

D2C:	NIB	$1,130	Ex $	875	VG+	$630	Good $	565	LMP	$1,289
C2C:	Mint	$1,090	Ex $	800	VG+	$570	Good $	510	Fair	$ 260
Trade-In:	Mint $	810	Ex $	650	VG+	$450	Good $	400	Poor	$ 120

P229 SCORPION
NEW IN BOX

P229 SCORPION REDBOOK CODE: RB-SS-H-229SPN

Short Reset Trigger (SRT), beavertail grip, Flat Dark Earth coating on frame and slide, front cocking serrations, front-strap checkering, accessory rail, model with threaded barrel available.

Caliber: 9mm, .40 S&W   Action: DA/SA, semi-auto   Barrel Length: 4"

OA Length: 7.6"   Wt.: 32 oz.

Sights: SIGLITE night sights   Capacity: 10, 12, 15   Magazine: detachable box

Grips: Hogue Extreme G10

D2C:	NIB	$1,225	Ex $	950	VG+ $	680	Good $	615	LMP	$1,285
C2C:	Mint	$1,180	Ex $	850	VG+ $	620	Good $	555	Fair	$ 285
Trade-In:	Mint $	870	Ex $	700	VG+ $	480	Good $	430	Poor $	150

P229 SCT
EXCELLENT
Courtesy of Bud's Gun Shop

P229 SCT REDBOOK CODE: RB-SS-H-229SCT

Nitron-finished slide, black hard-coat anodized alloy frame, accessory rail, front cocking serrations, higher-capacity designed magazines.

Production: discontinued   Caliber: 9mm, .40 S&W   Action: DA/SA, semi-auto

Barrel Length: 4"   OA Length: 7.1"   Wt.: 32 oz.

Sights: TRUGLO TFO front, SIGLITE night rear   Capacity: 17, 14

Magazine: detachable box   Grips: black polymer factory

D2C:	NIB	$1,025	Ex $	800	VG+ $	570	Good $	515	LMP	$ 1,156
C2C:	Mint $	990	Ex $	725	VG+ $	520	Good $	465	Fair	$ 240
Trade-In:	Mint $	730	Ex $	575	VG+ $	400	Good $	360	Poor $	120

P230 REDBOOK CODE: RB-SS-H-P230XX

Blue or stainless finish, steel slide, aluminum alloy frame, decocking lever, magazine release in heel of butt, model P230 SL has stainless frame and slide and is worth a slight premium.

Production: 1977 - 1996   Caliber: .22 LR, .32 ACP, .380 ACP, 9mm Ultra

Action: DA/SA, semi-auto   Barrel Length: 3.6"   Weight: 18.5 oz., 23.6 oz. (stainless)

Sights: fixed   Capacity: 10 (.22), 8 (.32), 7 (.380 & 9mm)   Magazine: detachable box

Grips: black checkered polymer

D2C:	NIB $	540	Ex $	425	VG+	$300	Good $	270	LMP	$ 510
C2C:	Mint $	520	Ex $	375	VG+	$270	Good $	245	Fair	$ 125
Trade-In:	Mint $	390	Ex $	325	VG+	$220	Good $	190	Poor $	60

P232
NEW IN BOX

P232 REDBOOK CODE: RB-SS-H-P232XX

Nitron stainless steel slide, aluminum alloy black hard-coat anodized frame, snag-free edges, decocking lever, optional night sights, magazine release in heel of butt, models available in two-tone or stainless finish.

Production: 1997 - current   Caliber: .380 ACP   Action: DA/SA, semi-auto

Barrel Length: 3.6"   OA Length: 6.6"   Weight: 18.5 oz.

Sights: 3-dot contrast or SIGLITE night sights   Capacity: 7   Magazine: detachable box

Grips: black stippled polymer or Hogue rubber finger groove

D2C:	NIB	$ 605	Ex $ 475	VG+ $340	Good $ 305	LMP $ 649
C2C:	Mint $ 590	Ex $ 425	VG+ $ 310	Good $ 275	Fair $ 140
Trade-In:	Mint $ 430	Ex $ 350	VG+ $240	Good $ 215	Poor $ 90

P232 TWO-TONE
NEW IN BOX

P232 TWO-TONE REDBOOK CODE: RB-SS-H-P232TT

Natural stainless steel slide, aluminum alloy black hard-coat anodized frame, snag-free edges, decocking lever, models available in Nitron or stainless finish.

Caliber: .380 ACP   Action: DA/SA, semi-auto   Barrel Length: 3.6"   OA Length: 6.6"

Weight: 18.5 oz.   Sights: SIGLITE night sights   Capacity: 7   Magazine: detachable box

Grips: black stippled polymer or Hogue rubber finger groove

D2C:	NIB	$ 650	Ex $ 500	VG+ $365	Good $ 325	LMP $ 749
C2C:	Mint $ 630	Ex $ 450	VG+ $330	Good $ 295	Fair $ 150
Trade-In:	Mint $ 470	Ex $ 375	VG+ $260	Good $ 230	Poor $ 90

P232 STAINLESS
NEW IN BOX

P232 STAINLESS REDBOOK CODE: RB-SS-H-P232SS

Natural stainless steel slide and frame, snag-free edges, decocking lever, models available in Nitron or two-tone finish.

Caliber: .380 ACP   Action: DA/SA, semi-auto   Barrel Length: 3.6"   OA Length: 6.6"

Weight: 24 oz.   Sights: SIGLITE night sights   Capacity: 7   Magazine: detachable box

Grips: black stippled polymer or Hogue rubber finger groove

D2C:	NIB	$ 675	Ex $ 525	VG+ $375	Good $ 340	LMP $ 799
C2C:	Mint $ 650	Ex $ 475	VG+ $340	Good $ 305	Fair $ 160
Trade-In:	Mint $ 480	Ex $ 400	VG+ $270	Good $ 240	Poor $ 90

P238 TWO-TONE
NEW IN BOX

P238 TWO-TONE REDBOOK CODE: RB-SS-H-P238TT

Black anodized alloy beavertail style frame, natural stainless steel finish slide, optional night sights.

Caliber: .380 ACP   Action: SA, semi-auto   Barrel Length: 2.7"   OA Length: 5.5"

Weight: 15 oz.   Sights: contrast, SIGLITE night sights   Capacity: 6

Magazine: detachable box   Grips: fluted polymer

D2C:	NIB	$ 550	Ex $ 425	VG+ $ 310	Good $ 275	LMP $ 693
C2C:	Mint $ 530	Ex $ 400	VG+ $280	Good $ 250	Fair $ 130
Trade-In:	Mint $ 400	Ex $ 325	VG+ $220	Good $ 195	Poor $ 60

P238 NITRON
NEW IN BOX

P238 NITRON REDBOOK CODE: RB-SS-H-P238NI

Black anodized alloy beavertail style frame, Nitron finish slide, optional night sights.

Caliber: .380 ACP   Action: SA, semi-auto   Barrel Length: 2.7"   OA Length: 5.5"

Weight: 15 oz.   Sights: contrast, SIGLITE night sights   Capacity: 6

Magazine: detachable box   Grips: fluted polymer

D2C:	NIB	$ 550	Ex $ 425	VG+ $ 310	Good $ 275	LMP $ 679
C2C:	Mint $ 530	Ex $ 400	VG+ $280	Good $ 250	Fair $ 130
Trade-In:	Mint $ 400	Ex $ 325	VG+ $220	Good $ 195	Poor $ 60

SIG SAUER

P238 ROSEWOOD
NEW IN BOX

P238 ROSEWOOD REDBOOK CODE: RB-SS-H-P238RO

Nitron stainless steel slide, anodized alloy beavertail style frame, available with ambidextrous safety.

Production: 2012 - current Caliber: .380 ACP Action: SA, semi-auto

Barrel Length: 2.7" OA Length: 5.5" Weight: 15 oz. Sights: SIGLITE night sights

Capacity: 6 Magazine: detachable box Grips: rosewood

D2C:	NIB $ 575	Ex $ 450	VG+ $ 320	Good $ 290	LMP $ 752
C2C:	Mint $ 560	Ex $ 400	VG+ $ 290	Good $ 260	Fair $ 135
Trade-In:	Mint $ 410	Ex $ 325	VG+ $ 230	Good $ 205	Poor $ 60

P238 BLACKWOOD
NEW IN BOX
Courtesy of Bud's Gun Shop

P238 BLACKWOOD REDBOOK CODE: RB-SS-H-P238BW

Anodized alloy beavertail style frame, natural stainless steel slide, black hard-coat anodized frame, ambidextrous safety available.

Production: 2012 - current Caliber: .380 ACP Action: SA, semi-auto

Barrel Length: 2.7" OA Length: 5.5" Weight: 15 oz. Sights: SIGLITE night sights

Capacity: 6 Magazine: detachable box Grips: blackwood

D2C:	NIB $ 575	Ex $ 450	VG+ $ 320	Good $ 290	LMP $ 766
C2C:	Mint $ 560	Ex $ 400	VG+ $ 290	Good $ 260	Fair $ 135
Trade-In:	Mint $ 410	Ex $ 325	VG+ $ 230	Good $ 205	Poor $ 60

P238 RAINBOW
(TITANIUM)
NEW IN BOX

P238 RAINBOW (TITANIUM) REDBOOK CODE: RB-SS-H-P238RT

Rainbow Titanium slide and controls, beavertail style frame, black hard-coat anodized frame.

Production: 2012 - current Caliber: .380 ACP Action: SA, semi-auto

Barrel Length: 2.7" OA Length: 5.5" Weight: 15 oz. Sights: SIGLITE night sights

Capacity: 6 Magazine: detachable box Grips: rosewood

D2C:	NIB $ 575	Ex $ 450	VG+ $ 320	Good $ 290	LMP $ 752
C2C:	Mint $ 560	Ex $ 400	VG+ $ 290	Good $ 260	Fair $ 135
Trade-In:	Mint $ 410	Ex $ 325	VG+ $ 230	Good $ 205	Poor $ 60

P238 EQUINOX
NEW IN BOX

P238 EQUINOX REDBOOK CODE: RB-SS-H-P238EQ

Anodized alloy beavertail style frame, two-tone stainless steel slide, polished slide flats, nickel controls, available ambidextrous safety.

Production: 2012 - current Caliber: .380 ACP Action: SA, semi-auto

Barrel Length: 2.7" OA Length: 5.5" Weight: 15 oz. Sights: SIGLITE night sight rear, TRUGLO tritium fiber-optic front Capacity: 6 Magazine: detachable box Grips: wood

D2C:	NIB $ 600	Ex $ 475	VG+ $ 335	Good $ 300	LMP $ 780
C2C:	Mint $ 580	Ex $ 425	VG+ $ 300	Good $ 270	Fair $ 140
Trade-In:	Mint $ 430	Ex $ 350	VG+ $ 240	Good $ 210	Poor $ 60

P238 HD
NEW IN BOX

P238 HD REDBOOK CODE: RB-SS-H-P238HD

Natural stainless frame and slide finish, beavertail style frame, black accents.

Caliber: .380 ACP Action: SA, semi-auto Barrel Length: 2.7" OA Length: 5.5"

Weight: 20 oz. Sights: SIGLITE night sights Capacity: 6 Magazine: detachable box

Grips: G10 or wood

D2C:	NIB $ 600	Ex $ 475	VG+ $ 335	Good $ 300	LMP $ 786
C2C:	Mint $ 580	Ex $ 425	VG+ $ 300	Good $ 270	Fair $ 140
Trade-In:	Mint $ 430	Ex $ 350	VG+ $ 240	Good $ 210	Poor $ 60

P238 TACTICAL LASER
NEW IN BOX

P238 TACTICAL LASER   REDBOOK CODE: RB-SS-H-P238TL

Integrated laser, black hard-coat anodized frame, stainless slide finish, beavertail style frame.

Caliber: .380 ACP	Action: SA, semi-auto	Barrel Length: 2.7"	OA Length: 5.5"			
Weight: 15 oz.	Sights: SIGLITE night sights	Capacity: 6	Magazine: detachable box			
Grips: aluminum						
D2C:	NIB $ 775	Ex $600	VG+ $ 435	Good $ 390	LMP $ 829	
C2C:	Mint $ 750	Ex $550	VG+ $ 390	Good $ 350	Fair $ 180	
Trade-In:	Mint $ 560	Ex $450	VG+ $ 310	Good $ 275	Poor $ 90	

P238 SAS
NEW IN BOX

P238 SAS   REDBOOK CODE: RB-SS-H-P238SA

SIG Anti-Snag treatment, black hard-coat anodized frame, natural stainless steel slide, optional ambidextrous safety.

Caliber: .380 ACP	Action: SA, semi-auto	Barrel Length: 2.7"	OA Length: 5.5"			
Weight: 15 oz.	Sights: SIGLITE night sights	Capacity: 6	Magazine: detachable box			
Grips: custom wood						
D2C:	NIB $ 750	Ex $ 575	VG+ $420	Good $ 375	LMP $ 790	
C2C:	Mint $ 720	Ex $ 525	VG+ $380	Good $ 340	Fair $ 175	
Trade-In:	Mint $ 540	Ex $ 425	VG+ $300	Good $ 265	Poor $ 90	

P238 LADY
NEW IN BOX

P238 LADY   REDBOOK CODE: RB-SS-H-P238LA

Red Cerakote finish alloy frame, Nitron finished slide with scroll engraving and gold flower inlay, beavertail style frame.

Caliber: .380 ACP	Action: SA, semi-auto	Barrel Length: 2.7"	OA Length: 5.5"			
Weight: 15 oz.	Sights: SIGLITE night sights	Capacity: 6	Magazine: detachable box			
Grips: rosewood						
D2C:	NIB $ 575	Ex $ 450	VG+ $320	Good $ 290	LMP $ 752	
C2C:	Mint $ 560	Ex $ 400	VG+ $290	Good $ 260	Fair $ 135	
Trade-In:	Mint $ 410	Ex $ 325	VG+ $230	Good $ 205	Poor $ 60	

P238 EXTREME
NEW IN BOX

P238 EXTREME   REDBOOK CODE: RB-SS-H-P238EX

Nitron stainless steel slide, black hard-coat anodized frame, front cocking serrations, alloy beavertail style frame, X-Grip extended magazine.

Caliber: .380 ACP	Action: SA, semi-auto	Barrel Length: 2.7"	OA Length: 5.5"			
Weight: 15 oz.	Sights: SIGLITE night sights	Capacity: 7	Magazine: detachable box			
Grips: Hogue G-10						
D2C:	NIB $ 575	Ex $ 450	VG+ $320	Good $ 290	LMP $ 752	
C2C:	Mint $ 560	Ex $ 400	VG+ $290	Good $ 260	Fair $ 135	
Trade-In:	Mint $ 410	Ex $ 325	VG+ $230	Good $ 205	Poor $ 60	

P238 GAMBLER
NEW IN BOX

P238 GAMBLER   REDBOOK CODE: RB-SS-H-P238GA

Top of slide features a rendering of the "dead man's hand" in 24K gold inlay, "The Gambler" text featured on slide, black hard-coat anodized frame, Nitron slide finish, beavertail style grip.

Production: discontinued	Caliber: .380 ACP	Action: SA, semi-auto				
Barrel Length: 2.7"	OA Length: 5.5"	Weight: 15 oz.	Sights: SIGLITE night sights			
Capacity: 6	Magazine: detachable box	Grips: rosewood				
D2C:	NIB $ 575	Ex $ 450	VG+ $320	Good $ 290	LMP $ 752	
C2C:	Mint $ 560	Ex $ 400	VG+ $290	Good $ 260	Fair $ 135	
Trade-In:	Mint $ 410	Ex $ 325	VG+ $230	Good $ 205	Poor $ 60	

SIG SAUER

P238 DIAMOND PLATE
NEW IN BOX

P238 HDW
NEW IN BOX

P238 DESERT
NEW IN BOX

P238 SCORPION
NEW IN BOX

**P238 BLACK
DIAMOND PLATE**
NEW IN BOX

P238 DIAMOND PLATE REDBOOK CODE: RB-SS-H-P238DP

Stainless steel slide with diamond plate engraving, black hard-coat anodized frame, beavertail style frame.

Caliber: .380 ACP   Action: SA, semi-auto   Barrel Length: 2.7"   OA Length: 5.5"

Weight: 15 oz.   Sights: SIGLITE night sights   Capacity: 6   Magazine: detachable box

Grips: G-10

D2C:	NIB $	575	Ex $	450	VG+ $320	Good $	290	LMP $	752
C2C:	Mint $	560	Ex $	400	VG+ $290	Good $	260	Fair $	135
Trade-In:	Mint $	410	Ex $	325	VG+ $230	Good $	205	Poor $	60

P238 HDW REDBOOK CODE: RB-SS-H-P238HW

Stainless steel frame and slide, beavertail style frame.

Caliber: .380 ACP   Action: SA, semi-auto   Barrel Length: 2.7"   OA Length: 5.5"

Weight: 20 oz.   Sights: SIGLITE night sights   Capacity: 6   Magazine: detachable box

Grips: rosewood

D2C:	NIB $	600	Ex $	475	VG+ $335	Good $	300	LMP $	786
C2C:	Mint $	580	Ex $	425	VG+ $300	Good $	270	Fair $	140
Trade-In:	Mint $	430	Ex $	350	VG+ $240	Good $	210	Poor $	60

P238 DESERT REDBOOK CODE: RB-SS-H-P238DE

Anodized alloy frame with Light Tan finish, stainless steel slide with Desert Tan finish, beavertail style frame, extended magazine, optional ambidextrous model available.

Caliber: .380 ACP   Action: SA, semi-auto   Barrel Length: 2.7"   OA Length: 5.5"

Weight: 15 oz.   Sights: SIGLITE night sights   Capacity: 7   Magazine: detachable box

Grips: Hogue one-piece FDE rubber

D2C:	NIB $	600	Ex $	475	VG+ $ 335	Good $	300	LMP $	781
C2C:	Mint $	580	Ex $	425	VG+ $ 300	Good $	270	Fair $	140
Trade-In:	Mint $	430	Ex $	350	VG+ $ 240	Good $	210	Poor $	60

P238 SCORPION REDBOOK CODE: RB-SS-H-P238SC

Anodized alloy frame with Flat Dark Earth finish, stainless steel slide with Flat Dark Earth finish, beavertail style frame, ambidextrous safety, extended magazine.

Caliber: .380 ACP   Action: SA, semi-auto   Barrel Length: 2.7"   OA Length: 5.5"

Weight: 15 oz.   Sights: SIGLITE night sights   Capacity: 7   Magazine: detachable box

Grips: Hogue Piranha G-10

D2C:	NIB $	600	Ex $	475	VG+ $335	Good $	300	LMP $	795
C2C:	Mint $	580	Ex $	425	VG+ $300	Good $	270	Fair $	140
Trade-In:	Mint $	430	Ex $	350	VG+ $240	Good $	210	Poor $	60

P238 BLACK DIAMOND PLATE REDBOOK CODE: RB-SS-H-P238BD

Nitron finished stainless steel slide with diamond plate engraving, anodized alloy beavertail style frame, available with ambidextrous safety.

Caliber: .380 ACP   Action: SA, semi-auto   Barrel Length: 2.7"   OA Length: 5.5"

Weight: 15 oz.   Sights: SIGLITE night sights   Capacity: 6   Magazine: detachable box

Grips: black aluminum

D2C:	NIB $	600	Ex $	475	VG+ $335	Good $	300	LMP $	779
C2C:	Mint $	580	Ex $	425	VG+ $300	Good $	270	Fair $	140
Trade-In:	Mint $	430	Ex $	350	VG+ $240	Good $	210	Poor $	60

P238 ESP-NITRON
NEW IN BOX

P238 ESP-NITRON   REDBOOK CODE: RB-SS-H-P238ES

Polished and engraved Nitron stainless steel slide, black hard-coat anodized frame with beavertail style grip.

Caliber: .380 ACP   Action: SA, semi-auto   Barrel Length: 2.7"   OA Length: 5.5"

Weight: 15 oz.   Sights: SIGLITE night sights   Capacity: 6   Magazine: detachable box

Grips: Hogue Pink rubber finger groove

D2C:	NIB $ 600	Ex $ 475	VG+ $335	Good $ 300	LMP $ 779
C2C:	Mint $ 580	Ex $ 425	VG+ $300	Good $ 270	Fair $ 140
Trade-In:	Mint $ 430	Ex $ 350	VG+ $240	Good $ 210	Poor $ 60

P238 PEARL
NEW IN BOX

P238 PEARL   REDBOOK CODE: RB-SS-H-P238PE

Polished and engraved Nitron stainless steel slide, black hard-coat anodized frame with beavertail style grip.

Caliber: .380 ACP   Action: SA, semi-auto   Barrel Length: 2.7"   OA Length: 5.5"

Weight: 15 oz.   Sights: SIGLITE night sights   Capacity: 6   Magazine: detachable box

Grips: custom Pearlite

D2C:	NIB $ 600	Ex $475	VG+ $ 335	Good $ 300	LMP $ 779
C2C:	Mint $ 580	Ex $425	VG+ $300	Good $ 270	Fair $ 140
Trade-In:	Mint $ 430	Ex $350	VG+ $240	Good $ 210	Poor $ 60

P238 TRIBAL
NEW IN BOX

P238 TRIBAL   REDBOOK CODE: RB-SS-H-P238TB

Natural stainless slide with Nitron tribal pattern, black hard-coat anodized frame finish, beavertail style frame.

Caliber: .380 ACP   Action: SA, semi-auto   Barrel Length: 2.7"   OA Length: 5.5"

Weight: 15 oz.   Sights: SIGLITE night sights   Capacity: 6   Magazine: detachable box

Grips: custom tribal engraved aluminum

D2C:	NIB $ 600	Ex $ 475	VG+ $ 335	Good $ 300	LMP $ 752
C2C:	Mint $ 580	Ex $ 425	VG+ $ 300	Good $ 270	Fair $ 140
Trade-In:	Mint $ 430	Ex $ 350	VG+ $ 240	Good $ 210	Poor $ 60

P238 TRIBAL ROSEWOOD
NEW IN BOX

P238 TRIBAL ROSEWOOD   REDBOOK CODE: RB-SS-H-P238TR

Nickel plated stainless slide with Rainbow Titanium tribal pattern, black hard-coat anodized frame finish, beavertail style frame.

Caliber: .380 ACP   Action: SA, semi-auto   Barrel Length: 2.7"   OA Length: 5.5"

Weight: 15 oz.   Sights: SIGLITE night sights   Capacity: 6   Magazine: detachable box

Grips: custom rosewood

D2C:	NIB $ 750	Ex $ 575	VG+ $420	Good $ 375	LMP $ 766
C2C:	Mint $ 720	Ex $ 525	VG+ $380	Good $ 340	Fair $ 175
Trade-In:	Mint $ 540	Ex $ 425	VG+ $300	Good $ 265	Poor $ 90

P238 LIBERTY   REDBOOK CODE: RB-SS-H-P238LI

Limited production, black hard-coat anodized beavertail style frame, Nitron finished slide, rendering of Liberty Bell in 24k gold inlay, the text "We The People" adorns the slide.

Production: discontinued   Caliber: .380 ACP   Action: SA, semi-auto

Barrel Length: 2.7"   OA Length: 5.5"   Weight: 15 oz.   Sights: SIGLITE night sights

Capacity: 6   Magazine: detachable box   Grips: rosewood

D2C:	NIB $ 650	Ex $500	VG+ $365	Good $ 325	LMP
C2C:	Mint $ 630	Ex $450	VG+ $330	Good $ 295	Fair $ 150
Trade-In:	Mint $ 470	Ex $375	VG+ $260	Good $ 230	Poor $ 90

SIG SAUER

P239
NEW IN BOX

P239 REDBOOK CODE: RB-SS-H-P239XX

Black hard-coat anodized frame, Nitron slide finish, optional night sights, two-tone model available.

Caliber: 9mm, .40 S&W, .357 SIG Action: DA/SA, semi-auto Barrel Length: 3.6"				
OA Length: 6.6" Weight: 30 oz. Sights: contrast, SIGLITE night sights				
Capacity: 7, 8 Magazine: detachable box Grips: black polymer factory				

D2C:	NIB $ 750	Ex $575	VG+ $420	Good $ 375	LMP $ 993
C2C:	Mint $ 720	Ex $525	VG+ $380	Good $ 340	Fair $ 175
Trade-In:	Mint $ 540	Ex $425	VG+ $300	Good $ 265	Poor $ 90

P239 DAK REDBOOK CODE: RB-SS-H-P239DK

SIG's exclusive DAK trigger system, double strike capability, black hard anodized frame finish, Nitron slide finish.

Caliber: 9mm, .40 S&W, .357 SIG Action: DAK (DAO), semi-auto Barrel Length: 3.6"				
OA Length: 6.6" Weight: 30 oz. Sights: contrast, SIGLITE night sights				
Capacity: 7, 8 Magazine: detachable box Grips: black polymer factory				

D2C:	NIB $ 800	Ex $625	VG+ $445	Good $ 400	
C2C:	Mint $ 770	Ex $575	VG+ $400	Good $ 360	Fair $ 185
Trade-In:	Mint $ 570	Ex $450	VG+ $320	Good $ 280	Poor $ 90

P239 TACTICAL
NEW IN BOX

P239 TACTICAL REDBOOK CODE: RB-SS-H-P239TA

Short Reset Trigger (SRT), Nitron stainless steel slide, black hard-coat anodized frame, front cocking serrations, threaded barrel.

Caliber: 9mm Action: DA/SA, semi-auto Barrel Length: 4" OA Length: 7.6"				
Weight: 30 oz. Sights: SIGLITE night sights Capacity: 8 Magazine: detachable box				
Grips: black polymer factory				

D2C:	NIB $ 985	Ex $ 750	VG+ $ 550	Good $495	LMP $ 1,176
C2C:	Mint $ 950	Ex $700	VG+ $500	Good $445	Fair $ 230
Trade-In:	Mint $ 700	Ex $ 575	VG+ $390	Good $345	Poor $ 120

P239 SAS
NEW IN BOX

P239 SAS REDBOOK CODE: RB-SS-H-P239SA

Short Reset Trigger (SRT), SIG Anti-Snag dehorn treatment, Nitron finished slide, black hard-coat anodized frame, model available in two-tone finish. Increased value for Gen. 2 model.

Caliber: 9mm, .40 S&W, .357 SIG Action: DA/SA, semi-auto				
Barrel Length: 3.6" OA Length: 6.6" Weight: 30 oz. Sights: SIGLITE night sights				
Capacity: 7, 8 Magazine: detachable box Grips: black polymer factory				

D2C:	NIB $ 945	Ex $725	VG+ $ 525	Good $ 475	LMP $ 1,125
C2C:	Mint $ 910	Ex $675	VG+ $ 480	Good $ 430	Fair $ 220
Trade-In:	Mint $ 680	Ex $550	VG+ $ 370	Good $ 335	Poor $ 120

P239 SAS TWO-TONE
NEW IN BOX

P239 SAS TWO-TONE REDBOOK CODE: RB-SS-H-P239ST

Short Reset Trigger (SRT), SIG Anti-Snag dehorn treatment, natural stainless finish slide, black hard-coat anodized frame.

Caliber: 9mm, .40 S&W, .357 SIG Action: DA/SA, semi-auto Barrel Length: 3.6"				
OA Length: 6.6" Weight: 30 oz. Sights: SIGLITE night sights Capacity: 7, 8				
Magazine: detachable box Grips: black polymer factory				

D2C:	NIB $ 950	Ex $ 725	VG+ $530	Good $475	LMP $1,139
C2C:	Mint $ 920	Ex $ 675	VG+ $480	Good $430	Fair $ 220
Trade-In:	Mint $ 680	Ex $ 550	VG+ $380	Good $335	Poor $ 120

P239 RAINBOW
NEW IN BOX

P239 RAINBOW   REDBOOK CODE: RB-SS-H-P239RB

Rainbow Titanium stainless steel slide, Rainbow Titanium trigger and controls, black hard-coat anodized frame

Caliber: .40 S&W   Action: DA/SA, semi-auto   Barrel Length: 3.6"   OA Length: 6.6"
Weight: 30 oz.   Sights: SIGLITE night sights   Capacity: 7   Magazine: detachable box
Grips: black polymer factory

D2C:	NIB	$ 925	Ex	$725	VG+	$ 515	Good	$ 465	LMP	$ 1,119
C2C:	Mint	$ 890	Ex	$650	VG+	$470	Good	$ 420	Fair	$ 215
Trade-In:	Mint	$ 660	Ex	$525	VG+	$370	Good	$ 325	Poor	$ 120

P239 SCORPION
NEW IN BOX

P239 SCORPION   REDBOOK CODE: RB-SS-H-P239SC

Short Reset Trigger (SRT), Flat Dark Earth coating on slide and frame, front cocking serrations.

Caliber: 9mm, .40 S&W, .357 SIG   Action: DA/SA, semi-auto   Barrel Length: 3.6"
OA Length: 6.6"   Weight: 30 oz.   Sights: SIGLITE night sights   Capacity: 7, 8
Magazine: detachable box   Grips: Hogue G-10 Piranha

D2C:	NIB	$1,000	Ex	$ 775	VG+	$ 555	Good	$ 500	LMP	$ 1,219
C2C:	Mint	$ 960	Ex	$700	VG+	$500	Good	$ 450	Fair	$ 230
Trade-In:	Mint	$ 710	Ex	$ 575	VG+	$390	Good	$ 350	Poor	$ 120

P250 FULL SIZE
NEW IN BOX

P250 FULL SIZE   REDBOOK CODE: RB-SS-H-P250FS

Modular polymer frame, interchangeable grip frame sizes and caliber conversions, Nitron slide finish, accessory rail, night sights available, two-tone model available.

Caliber: 9mm, .357 SIG, .40 S&W, .45 ACP   Action: DAO, semi-auto   Barrel Length: 4.7"
OA Length: 8"   Weight: 30 oz.   Sights: contrast, SIGLITE night sights
Capacity: 10, 14, 17   Magazine: detachable box   Grips: black polymer

D2C:	NIB	$ 700	Ex	$ 550	VG+	$390	Good	$350	LMP	$ 642
C2C:	Mint	$ 680	Ex	$500	VG+	$350	Good	$ 315	Fair	$ 165
Trade-In:	Mint	$ 500	Ex	$400	VG+	$280	Good	$245	Poor	$ 90

P250 TACTICAL ALL TERRAIN DIGITAL
REDBOOK CODE: RB-SS-H-P250TA

Nitron finish slide, All Terrain Digital interchangeable polymer grip shell with stainless insert, threaded barrel.

Caliber: .380 ACP, 9mm, .357 SIG, .40 S&W, .45ACP   Action: DAO, semi-auto
Barrel Length: 4.8"   OA Length: 8"   Weight: 25 oz.   Sights: SIGLITE night sights
Capacity: 15, 13, 9   Magazine: detachable box   Grips: interchangeable polymer

D2C:	NIB	$ 750	Ex	$ 575	VG+	$420	Good	$ 375	LMP	$ 829
C2C:	Mint	$ 720	Ex	$ 525	VG+	$380	Good	$ 340	Fair	$ 175
Trade-In:	Mint	$ 540	Ex	$ 425	VG+	$300	Good	$ 265	Poor	$ 90

P250 COMPACT
NEW IN BOX

P250 COMPACT   REDBOOK CODE: RB-SS-H-P250CC

Modular polymer frame, interchangeable grip frame sizes and caliber conversions, Nitron slide finish, accessory rail, night sights available, two-tone model available, threaded barrel available with slight premium.

Caliber: .380 ACP, 9mm, .357 SIG, .40 S&W, .45ACP   Action: DAO, semi-auto
Barrel Length: 4"   OA Length: 7.2"   Weight: 25 oz.
Sights: contrast, SIGLITE night sights   Capacity: 15, 13, 9   Grips: black polymer

D2C:	NIB	$ 750	Ex	$ 575	VG+	$420	Good	$ 375	LMP	$ 642
C2C:	Mint	$ 720	Ex	$ 525	VG+	$380	Good	$ 340	Fair	$ 175
Trade-In:	Mint	$ 540	Ex	$ 425	VG+	$300	Good	$ 265	Poor	$ 90

P250 COMPACT DIGITAL CAMO   REDBOOK CODE: RB-SS-H-250CDC

Nitron finish slide, digital camo interchangeable polymer grip shell with stainless insert, accessory rail.

| Caliber: 9mm | Action: DAO, semi-auto | Barrel Length: 4" | OA Length: 7.2" |

| Weight: 25 oz. | Sights: SIGLITE night sights | Capacity: 15 | Magazine: detachable box |

Grips: digital camo polymer grip shell

D2C:	NIB $	775	Ex $600	VG+ $	435	Good $ 390		LMP $ 855	
C2C:	Mint $	750	Ex $ 550	VG+ $390		Good $ 350		Fair $ 180	
Trade-In:	Mint $	560	Ex $ 450	VG+ $ 310		Good $ 275		Poor $ 90	

P250 COMPACT DIAMOND PLATE (STAINLESS)

REDBOOK CODE: RB-SS-H-250CDS

Natural stainless slide finish with diamond plate engraving, modular black polymer frame, accessory rail.

| Caliber: 9mm | Action: DAO, semi-auto | Barrel Length: 4" | OA Length: 7.2" |

| Weight: 25 oz. | Sights: SIGLITE night sights | Capacity: 15 | Magazine: detachable box |

Grips: black polymer

D2C:	NIB $	550	Ex $ 425	VG+ $ 310		Good $ 275		LMP $ 667	
C2C:	Mint $	530	Ex $400	VG+ $280		Good $ 250		Fair $ 130	
Trade-In:	Mint $	400	Ex $ 325	VG+ $220		Good $ 195		Poor $ 60	

P250 COMPACT DIAMOND PLATE (STAINLESS)
NEW IN BOX

P250 COMPACT DIAMOND PLATE (BLACK)

REDBOOK CODE: RB-SS-H-250CDB

Black diamond plate engraving on slide, accessory rail, modular black polymer frame, Nitron slide finish.

| Caliber: 9mm | Action: DAO, semi-auto | Barrel Length: 4" | OA Length: 7.2" |

| Weight: 27 oz. | Sights: SIGLITE night sights | Capacity: 15 | Magazine: detachable box |

Grips: black polymer

D2C:	NIB $	550	Ex $ 425	VG+ $ 310		Good $ 275		LMP $ 667	
C2C:	Mint $	530	Ex $ 400	VG+ $280		Good $ 250		Fair $ 130	
Trade-In:	Mint $	400	Ex $ 325	VG+ $220		Good $ 195		Poor $ 60	

P250 SUBCOMPACT NITRON   REDBOOK CODE: RB-SS-H-250SCN

Modular polymer frame, interchangeable grip frame sizes and caliber conversions, Nitron slide finish, accessory rail, night sights available, two-tone model available.

| Caliber: .380 ACP, 9mm, .357 SIG, .40 S&W, .45ACP | Action: DAO, semi-auto |

| Barrel Length: 3.6" | OA Length: 6.7" | Weight: 25 oz. | Sights: contrast, SIGLITE |

| Night Sights | Capacity: 6, 10, 12 | Magazine: detachable box | Grips: black polymer |

D2C:	NIB $	530	Ex $ 425	VG+ $ 295		Good $ 265		LMP $ 641	
C2C:	Mint $	510	Ex $ 375	VG+ $270		Good $ 240		Fair $ 125	
Trade-In:	Mint $	380	Ex $300	VG+ $ 210		Good $ 190		Poor $ 60	

P250 SUBCOMPACT NITRON
NEW IN BOX

P250 SUBCOMPACT NITRON RAIL   REDBOOK CODE: RB-SS-H-250SNR

Modular black polymer frame, Nitron slide finish, accessory rail.

| Caliber: 9mm | Action: DAO, semi-auto | Barrel Length: 3.6" | OA Length: 6.7" |

| Weight: 25 oz. | Sights: SIGLITE night sights | Capacity: 12 | Magazine: detachable box |

Grips: black polymer

D2C:	NIB $	575	Ex $450	VG+ $320		Good $290		LMP $ 642	
C2C:	Mint $	560	Ex $400	VG+ $290		Good $ 260		Fair $ 135	
Trade-In:	Mint $	410	Ex $ 325	VG+ $230		Good $ 205		Poor $ 60	

P250 SUBCOMPACT NITRON RAIL
NEW IN BOX

**P250 SUBCOMPACT
NITRON .380**
NEW IN BOX

P250 SUBCOMPACT NITRON .380   REDBOOK CODE: RB-SS-H-250380

Modular black polymer frame, Nitron slide finish, 12-round flush fit magazine, 15-round extended magazine available, optional ambidextrous manual safety.

Caliber: .380 ACP Action: DAO, semi-auto Barrel Length: 3.6" OA Length: 6.7"									
Weight: 20 oz. Sights: SIGLITE night sights Capacity: 12, 15									
Magazine: detachable box Grips: black polymer									
D2C:	NIB	$	575	Ex	$ 450	VG+	$320	Good	$ 290	LMP	$ 666
C2C:	Mint	$	560	Ex	$ 400	VG+	$290	Good	$ 260	Fair	$ 135
Trade-In:	Mint	$	410	Ex	$ 325	VG+	$230	Good	$ 205	Poor	$ 60

P250 2SUM
NEW IN BOX

Courtesy of Bud's Gun Shop

P250 2SUM   REDBOOK CODE: RB-SS-H-2502SM

Includes modular full-size P250 plus all components to rapidly convert to subcompact, modular black polymer frame, Nitron slide finish, accessory rail on full-size frame, optional night sights available.

Caliber: 9mm, .40 S&W Action: DAO, semi-auto Barrel Length: 4.7" (full), 3.6"(sub)									
OA Length: 8" Weight: 30 oz. (full), 25 oz. (sub) Sights: contrast, SIGLITE Night									
Sights Capacity: 10, 12, 14, 17 Magazine: detachable box Grips: black polymer									
D2C:	NIB	$	750	Ex	$ 575	VG+	$420	Good	$ 375	LMP	$ 886
C2C:	Mint	$	720	Ex	$ 525	VG+	$380	Good	$ 340	Fair	$ 175
Trade-In:	Mint	$	540	Ex	$ 425	VG+	$300	Good	$ 265	Poor	$ 90

P290RS
NEW IN BOX

P290RS   REDBOOK CODE: RB-SS-H-290RSX

Customizable grip inserts, black polymer frame, Nitron slide finish, re-strike capable, extended magazine included, model available in two-tone finish.

Caliber: 9mm Action: DAO, semi-auto Barrel Length: 3" OA Length: 5.5"									
Weight: 21 oz. Sights: SIGLITE night sights Capacity: 6, 8									
Magazine: detachable box Grips: interchangeable polymer inserts									
D2C:	NIB	$	490	Ex	$375	VG+	$ 275	Good	$ 245	LMP	$ 570
C2C:	Mint	$	480	Ex	$350	VG+	$250	Good	$ 225	Fair	$ 115
Trade-In:	Mint	$	350	Ex	$275	VG+	$200	Good	$ 175	Poor	$ 60

**P290RS
TWO-TONE LASER**
NEW IN BOX

P290RS TWO-TONE LASER   REDBOOK CODE: RB-SS-H-290RTL

Integrated laser module, customizable grip inserts, black polymer frame, natural stainless slide finish, re-strike capable, extended magazine included.

Caliber: 9mm Action: DAO, semi-auto Barrel Length: 3" OA Length: 5.5"									
Weight: 21 oz. Sights: SIGLITE night sights Capacity: 6, 8									
Magazine: detachable box Grips: interchangeable polymer inserts									
D2C:	NIB	$	535	Ex	$ 425	VG+	$300	Good	$ 270	LMP	$ 685
C2C:	Mint	$	520	Ex	$ 375	VG+	$270	Good	$ 245	Fair	$ 125
Trade-In:	Mint	$	380	Ex	$300	VG+	$ 210	Good	$ 190	Poor	$ 60

**P290RS
TWO-TONE**
NEW IN BOX

P290RS TWO-TONE   REDBOOK CODE: RB-SS-H-290RTT

Customizable grip inserts, black polymer frame, natural stainless slide finish, re-strike capable, extended magazine included.

Caliber: 9mm Action: DAO, semi-auto Barrel Length: 3" OA Length: 2.9"									
Weight: 21 oz. Sights: SIGLITE night sights Capacity: 6, 8									
Magazine: detachable box Grips: interchangeable polymer inserts									
D2C:	NIB	$	530	Ex	$ 425	VG+	$295	Good	$ 265	LMP	$ 613
C2C:	Mint	$	510	Ex	$ 375	VG+	$270	Good	$ 240	Fair	$ 125
Trade-In:	Mint	$	380	Ex	$300	VG+	$ 210	Good	$ 190	Poor	$ 60

SIG SAUER

P290RS BLACK DIAMOND PLATE
NEW IN BOX
Courtesy of Bud's Gun Shop

P290RS RAINBOW
NEW IN BOX

P290RS ENHANCED
NEW IN BOX
Courtesy of Bud's Gun Shop

P290RS BLACK DIAMOND PLATE   REDBOOK CODE: RB-SS-H-290RBD

Nitron finished slide with diamond plate engraving, black polymer frame, customizable grip inserts, re-strike capable, extended magazine included.

Caliber: 9mm   Action: DAO, semi-auto   Barrel Length: 3"   OA Length: 5.5"
Weight: 21 oz.   Sights: SIGLITE night sights   Capacity: 6, 8
Magazine: detachable box   Grips: interchangeable polymer inserts

D2C:	NIB	$ 510	Ex $400	VG+	$285	Good $ 255	LMP	$ 595
C2C:	Mint $ 490	Ex $ 375	VG+	$260	Good $ 230	Fair	$ 120
Trade-In:	Mint $ 370	Ex $300	VG+	$200	Good $ 180	Poor	$ 60

P290RS RAINBOW   REDBOOK CODE: RB-SS-H-290RSR

Rainbow Titanium-coated slide and controls, black polymer frame, customizable grip inserts, re-strike capable, extended magazine included.

Caliber: 9mm   Action: DAO, semi-auto   Barrel Length: 3"   OA Length: 2.9"
Weight: 21 oz.   Sights: SIGLITE night sights   Capacity: 6, 8
Magazine: detachable box   Grips: interchangeable polymer inserts

D2C:	NIB $ 530	Ex $ 425	VG+ $ 295	Good $ 265	LMP $ 613
C2C:	Mint $ 510	Ex $ 375	VG+ $ 270	Good $ 240	Fair $ 125
Trade-In:	Mint $ 380	Ex $ 300	VG+ $ 210	Good $ 190	Poor $ 60

P290RS ENHANCED   REDBOOK CODE: RB-SS-H-290RSE

Enhanced customizable grip inserts, black polymer frame, Nitron slide finish, re-strike capable, extended magazine included.

Caliber: 9mm   Action: DAO, semi-auto   Barrel Length: 3"   OA Length: 5.5"
Weight: 21 oz.   Sights: SIGLITE night sights   Capacity: 6, 8
Magazine: detachable box   Grips: removable G-10 grip plates

D2C:	NIB $ 530	Ex $ 425	VG+ $ 295	Good $ 265	LMP $ 613
C2C:	Mint $ 510	Ex $ 375	VG+ $ 270	Good $ 240	Fair $ 125
Trade-In:	Mint $ 380	Ex $ 300	VG+ $ 210	Good $ 190	Poor $ 60

SP2340   REDBOOK CODE: RB-SS-H-SP2340

SIG's first polymer frame, blue slide, switchable between DA/SA and DAO, accessory rails on dust cover, decocking lever.

Production: 1998   Caliber: .357 SIG, .40 S&W   Action: DA/SA, DAO, semi-auto
Barrel Length: 4"   OA Length: 7.4"   Weight: 30 oz.   Sights: contrast, SIGLITE Night
Sights   Capacity: 10, 12   Magazine: detachable box   Grips: interchangeable polymer

D2C:	NIB $ 350	Ex $ 275	VG+ $195	Good $ 175	LMP $ 640
C2C:	Mint $ 340	Ex $ 250	VG+ $180	Good $ 160	Fair $ 85
Trade-In:	Mint $ 250	Ex $ 200	VG+ $140	Good $ 125	Poor $ 60

SP2009   REDBOOK CODE: RB-SS-H-SP2009

9mm variation of the SP2340, polymer frame, decocking lever.

Production: 1999 - 2005   Caliber: 9mm   Action: DA/SA, DAO, semi-auto
Barrel Length: 4"   OA Length: 7.4"   Weight: 27 oz.   Sights: contrast, SIGLITE Night
Sights   Capacity: 15   Magazine: detachable box   Grips: interchangeable polymer

D2C:	NIB $ 490	Ex $375	VG+ $ 275	Good $ 245	LMP $ 640
C2C:	Mint $ 480	Ex $350	VG+ $250	Good $ 225	Fair $ 115
Trade-In:	Mint $ 350	Ex $275	VG+ $200	Good $ 175	Poor $ 60

SIG SAUER

SP2022 NITRON
NEW IN BOX

SP2022 NITRON   REDBOOK CODE: RB-SS-H-2022NI

Black polymer frame, Nitron finished stainless slide, accessory rail, interchangeable grip sizes, available with slide mounted manual safety, optional night sights, model available with threaded barrel.

Caliber: 9mm   Action: DA/SA, semi-auto   Barrel Length: 4"   OA Length: 7.4"

Weight: 29 oz.   Sights: contrast, SIGLITE night sights   Capacity: 10, 12, 15

Magazine: detachable box   Grips: interchangeable polymer

	NIB	Ex	VG+	Good	LMP
D2C:	$ 475	Ex $ 375	VG+ $ 265	Good $ 240	LMP $ 641
C2C:	Mint $ 460	Ex $ 350	VG+ $ 240	Good $ 215	Fair $ 110
Trade-In:	Mint $ 340	Ex $ 275	VG+ $ 190	Good $ 170	Poor $ 60

SP2022 BLACK DIAMOND PLATE
NEW IN BOX

SP2022 BLACK DIAMOND PLATE   REDBOOK CODE: RB-SS-H-2022BD

Black polymer frame, Nitron finished slide with diamond plate engraving, accessory rail, interchangeable grip sizes.

Caliber: 9mm   Action: DA/SA, semi-auto   Barrel Length: 4"   OA Length: 7.4"

Weight: 29 oz.   Sights: SIGLITE night sights   Capacity: 15   Magazine: detachable box

Grips: interchangeable polymer

	NIB	Ex	VG+	Good	LMP
D2C:	$ 485	Ex $ 375	VG+ $270	Good $ 245	LMP $ 685
C2C:	Mint $ 470	Ex $ 350	VG+ $250	Good $ 220	Fair $ 115
Trade-In:	Mint $ 350	Ex $ 275	VG+ $190	Good $ 170	Poor $ 60

SP2022 DIAMOND PLATE
NEW IN BOX

SP2022 DIAMOND PLATE   REDBOOK CODE: RB-SS-H-2022DP

Black polymer frame, natural stainless slide with diamond plate engraving, accessory rail, interchangeable grip sizes.

Caliber: 9mm   Action: DA/SA, semi-auto   Barrel Length: 4"   OA Length: 7.4"

Weight: 29 oz.   Sights: SIGLITE night sights   Capacity: 15   Magazine: detachable box

Grips: interchangeable polymer

	NIB	Ex	VG+	Good	LMP
D2C:	$ 485	Ex $ 375	VG+ $ 270	Good $ 245	LMP $ 685
C2C:	Mint $ 470	Ex $ 350	VG+ $ 250	Good $ 220	Fair $ 115
Trade-In:	Mint $ 350	Ex $ 275	VG+ $ 190	Good $ 170	Poor $ 60

P938 BLACKWOOD
NEW IN BOX

P938 BLACKWOOD   REDBOOK CODE: RB-SS-H-P938BW

Black hard-coat anodized frame, microcompact size, natural stainless slide and controls, beavertail style frame, ambidextrous safety.

Caliber: 9mm   Action: SA, semi-auto   Barrel Length: 3"   OA Length: 5.9"

Weight: 16 oz.   Sights: SIGLITE night sights   Capacity: 6   Magazine: detachable box

Grips: custom blackwood

	NIB	Ex	VG+	Good	LMP
D2C:	$ 675	Ex $ 525	VG+ $ 375	Good $ 340	LMP $ 809
C2C:	Mint $ 650	Ex $ 475	VG+ $340	Good $ 305	Fair $ 160
Trade-In:	Mint $ 480	Ex $400	VG+ $270	Good $ 240	Poor $ 90

P938 EXTREME   REDBOOK CODE: RB-SS-H-P938EX

Black hard-coat anodized frame, microcompact size, Nitron stainless slide, beavertail style frame, ambidextrous safety, extended magazine.

Caliber: 9mm   Action: SA, semi-auto   Barrel Length: 3"   OA Length: 5.9"

Weight: 16 oz.   Sights: SIGLITE night sights   Capacity: 7   Magazine: detachable box

Grips: Hogue G-10 Extreme

	NIB	Ex	VG+	Good	LMP
D2C:	$ 690	Ex $ 525	VG+ $385	Good $ 345	LMP $ 823
C2C:	Mint $ 670	Ex $500	VG+ $350	Good $ 315	Fair $ 160
Trade-In:	Mint $ 490	Ex $400	VG+ $270	Good $ 245	Poor $ 90

P938 EXTREME
NEW IN BOX

P938 ROSEWOOD
NEW IN BOX

P938 ROSEWOOD REDBOOK CODE: RB-SS-H-P938RW

Black hard-coat anodized frame, microcompact size, Nitron stainless slide, beavertail style frame, ambidextrous safety.

Caliber: 9mm Action: SA, semi-auto Barrel Length: 3" OA Length: 5.9"

Weight: 16 oz. Sights: SIGLITE night sights Capacity: 6 Magazine: detachable box

Grips: rosewood

D2C:	NIB	$ 685	Ex	$ 525	VG+	$ 385	Good	$ 345	LMP	$ 795
C2C:	Mint	$ 660	Ex	$ 475	VG+	$ 350	Good	$ 310	Fair	$ 160
Trade-In:	Mint	$ 490	Ex	$ 400	VG+	$ 270	Good	$ 240	Poor	$ 90

P938 EQUINOX
NEW IN BOX

P938 EQUINOX REDBOOK CODE: RB-SS-H-P938EQ

Black hard-coat anodized frame, microcompact size, two-tone polished slide, stainless controls, beavertail style frame, ambidextrous safety.

Caliber: 9mm Action: SA, semi-auto Barrel Length: 3" OA Length: 5.9"

Weight: 16 oz. Sights: TRUGLO front and SIGLITE rear Capacity: 6

Magazine: detachable box Grips: Hogue black diamondwood

D2C:	NIB	$ 690	Ex	$ 525	VG+	$ 385	Good	$ 345	LMP	$ 823
C2C:	Mint	$ 670	Ex	$ 500	VG+	$ 350	Good	$ 315	Fair	$ 160
Trade-In:	Mint	$ 490	Ex	$ 400	VG+	$ 270	Good	$ 245	Poor	$ 90

P938 BLACK RUBBER GRIP
NEW IN BOX

P938 BLACK RUBBER GRIP REDBOOK CODE: RB-SS-H-P938BR

Black hard-coat anodized frame, microcompact size, Nitron stainless slide, beavertail style frame, ambidextrous safety, extended magazine.

Caliber: 9mm Action: SA, semi-auto Barrel Length: 3" OA Length: 5.9"

Weight: 16 oz. Sights: SIGLITE night sights Capacity: 7 Magazine: detachable box

Grips: Hogue black rubber

D2C:	NIB	$ 690	Ex	$ 525	VG+	$ 385	Good	$ 345	LMP	$ 823
C2C:	Mint	$ 670	Ex	$ 500	VG+	$ 350	Good	$ 315	Fair	$ 160
Trade-In:	Mint	$ 490	Ex	$ 400	VG+	$ 270	Good	$ 245	Poor	$ 90

P938 NIGHTMARE
NEW IN BOX

P938 NIGHTMARE REDBOOK CODE: RB-SS-H-P938NM

Black hard-coat anodized frame, microcompact size, Nitron stainless slide, stainless controls, beavertail style frame, ambidextrous safety.

Caliber: 9mm Action: SA, semi-auto Barrel Length: 3" OA Length: 5.9"

Weight: 16 oz. Sights: SIGLITE night sights Capacity: 6 Magazine: detachable box

Grips: Hogue black G-10

D2C:	NIB	$ 690	Ex	$ 525	VG+	$ 385	Good	$ 345	LMP	$ 823
C2C:	Mint	$ 670	Ex	$ 500	VG+	$ 350	Good	$ 315	Fair	$ 160
Trade-In:	Mint	$ 490	Ex	$ 400	VG+	$ 270	Good	$ 245	Poor	$ 90

P938 AG
NEW IN BOX

P938 AG REDBOOK CODE: RB-SS-H-P938AG

Black hard-coat anodized frame, microcompact size, natural stainless slide, stainless controls, beavertail style frame, ambidextrous safety.

Caliber: 9mm Action: SA, semi-auto Barrel Length: 3" OA Length: 5.9"

Weight: 16 oz. Sights: SIGLITE night sights Capacity: 6 Magazine: detachable box

Grips: black checkered aluminum

D2C:	NIB	$ 700	Ex	$ 550	VG+	$ 390	Good	$ 350	LMP	$ 823
C2C:	Mint	$ 680	Ex	$ 500	VG+	$ 350	Good	$ 315	Fair	$ 165
Trade-In:	Mint	$ 500	Ex	$ 400	VG+	$ 280	Good	$ 245	Poor	$ 90

P938 SAS
NEW IN BOX

P938 SAS REDBOOK CODE: RB-SS-H-P938SA

SIG Anti-Snag treatment, black hard-coat anodized frame, microcompact size, natural stainless slide, stainless controls, beavertail style frame, ambidextrous safety.

Caliber: 9mm Action: SA, semi-auto Barrel Length: 3" OA Length: 5.9"

Weight: 16 oz. Sights: SIGLITE night sights Capacity: 6 Magazine: detachable box

Grips: custom goncalo wood

D2C:	NIB	$ 725	Ex	$575	VG+	$405	Good	$ 365	LMP	$ 838
C2C:	Mint	$ 700	Ex	$525	VG+	$370	Good	$ 330	Fair	$ 170
Trade-In:	Mint	$ 520	Ex	$425	VG+	$290	Good	$ 255	Poor	$ 90

GSR REVOLUTION REDBOOK CODE: RB-SS-H-GSRREV

Black Nitron or matte stainless finish on a stainless steel slide and frame, accessory rail standard, match grade barrel, machined and hand-fitted, external extractor, name changed to Sig Sauer 1911 in 2006.

Production: 2004 - 2006 Caliber: .45 ACP Action: SA, semi-auto Barrel Length: 5"

OA Length: 8.7" Weight: 42 oz. Sights: Novak low mount Capacity: 8

Magazine: detachable box Grips: checkered wood or Ergo Grip XT Extreme Use grips

D2C:	NIB	$ 748	Ex $	575	VG+ $	420	Good $	375	LMP	$1,077
C2C:	Mint	$ 720	Ex $	525	VG+ $	380	Good $	340	Fair	$ 175
Trade-In:	Mint	$ 540	Ex $	425	VG+ $	300	Good $	265	Poor	$ 90

1911 NITRON
NEW IN BOX

1911 NITRON REDBOOK CODE: RB-SS-H-1911NI

Nitron finished stainless steel frame and slide, beavertail grip safety, match grade barrel, match grade hammer/sear and trigger, checkering on front strap and mainspring housing, de-horned frame and slide.

Production: 2006 - current Caliber: .45 ACP Action: SA, semi-auto Barrel Length: 5"

OA Length: 8.7" Weight: 42 oz. Sights: low profile night sights Capacity: 8

Magazine: detachable box Grips: custom wood

D2C:	NIB	$ 950	Ex	$725	VG+	$530	Good	$ 475	LMP	$1,099
C2C:	Mint	$ 920	Ex	$675	VG+	$480	Good	$ 430	Fair	$ 220
Trade-In:	Mint	$ 680	Ex	$550	VG+	$380	Good	$ 335	Poor	$ 120

1911 NITRON RAIL
NEW IN BOX

1911 NITRON RAIL REDBOOK CODE: RB-SS-H-1911NR

Nitron finished stainless steel frame and slide, beavertail grip safety, match grade barrel, match grade hammer/sear and trigger, checkering on front strap and mainspring housing, de-horned frame and slide, accessory rail.

Production: 2006 - current Caliber: .45 ACP Action: SA, semi-auto Barrel Length: 5"

OA Length: 8.7" Weight: 42 oz. Sights: low profile night sights Capacity: 8

Magazine: detachable box Grips: custom wood

D2C:	NIB	$ 980	Ex $	750	VG+	$545	Good	$ 490	LMP	$1,142
C2C:	Mint	$ 950	Ex $	700	VG+	$490	Good	$ 445	Fair	$ 230
Trade-In:	Mint	$ 700	Ex $	550	VG+	$390	Good	$ 345	Poor	$ 120

1911 TARGET NITRON REDBOOK CODE: RB-SS-H-1911TN

Nitron finished stainless steel frame and slide, beavertail grip safety, match grade barrel, match grade hammer/sear and trigger, checkering on front strap and mainspring housing, de-horned frame and slide.

Production: 2006 Caliber: .45 ACP Action: SA, semi-auto Barrel Length: 5"

OA Length: 8.7" Weight: 42 oz. Sights: adjustable target Capacity: 8

Magazine: detachable box Grips: rosewood custom

D2C:	NIB	$ 960	Ex $	750	VG+	$535	Good	$ 480	LMP	$1,113
C2C:	Mint	$ 930	Ex $	675	VG+	$480	Good	$ 435	Fair	$ 225
Trade-In:	Mint	$ 690	Ex $	550	VG+	$380	Good	$ 340	Poor	$ 120

SIG SAUER

1911 STAINLESS
NEW IN BOX

1911 STAINLESS   REDBOOK CODE: RB-SS-H-1911SS

Natural stainless steel frame and slide, beavertail grip safety, match grade barrel, match grade hammer/sear and trigger, checkering on front strap and mainspring housing, de-horned frame and slide.

Production: 2006   Caliber: .45 ACP   Action: SA, semi-auto   Barrel Length: 5"

OA Length: 8.7"   Weight: 42 oz.   Sights: low profile night sights   Capacity: 8

Magazine: detachable box   Grips: custom blackwood

D2C:	NIB $ 960	Ex $ 750	VG+ $ 535	Good $ 480	LMP $1,113
C2C:	Mint $ 930	Ex $ 675	VG+ $480	Good $ 435	Fair $ 225
Trade-In:	Mint $ 690	Ex $ 550	VG+ $380	Good $ 340	Poor $ 120

1911 STAINLESS RAIL
NEW IN BOX

1911 STAINLESS RAIL   REDBOOK CODE: RB-SS-H-1911SR

Natural stainless steel frame and slide, beavertail grip safety, match grade barrel, match grade hammer/sear and trigger, checkering on front strap and mainspring housing, de-horned frame and slide, accessory rail.

Production: 2006   Caliber: .45 ACP   Action: SA, semi-auto   Barrel Length: 5"

OA Length: 8.7"   Weight: 42 oz.   Sights: low profile night sights   Capacity: 8

Magazine: detachable box   Grips: custom wood

D2C:	NIB $1,000	Ex $ 775	VG+ $ 555	Good $ 500	LMP $1,156
C2C:	Mint $ 960	Ex $700	VG+ $500	Good $ 450	Fair $ 230
Trade-In:	Mint $ 710	Ex $ 575	VG+ $390	Good $ 350	Poor $ 120

1911 TARGET STAINLESS
NEW IN BOX

1911 TARGET STAINLESS   REDBOOK CODE: RB-SS-H-1911TS

Natural stainless steel frame and slide, beavertail grip safety, match grade barrel, match grade hammer/sear and trigger, checkering on front strap and mainspring housing, de-horned frame and slide.

Production: 2006   Caliber: .45 ACP   Action: SA, semi-auto   Barrel Length: 5"

OA Length: 8.7"   Weight: 42 oz.   Sights: adjustable target   Capacity: 8

Magazine: detachable box   Grips: black diamondwood

D2C:	NIB $1,000	Ex $ 775	VG+ $ 555	Good $ 500	LMP $1,142
C2C:	Mint $ 960	Ex $700	VG+ $500	Good $ 450	Fair $ 230
Trade-In:	Mint $ 710	Ex $ 575	VG+ $390	Good $ 350	Poor $ 120

1911 CARRY NITRON
NEW IN BOX

1911 CARRY NITRON   REDBOOK CODE: RB-SS-H-1911CN

Nitron finished stainless steel frame and slide, beavertail grip safety, match grade barrel, match grade hammer/sear and trigger, checkering on front strap and mainspring housing, de-horned frame and slide.

Production: 2006   Caliber: .45 ACP   Action: SA, semi-auto   Barrel Length: 4.2"

OA Length: 7.7"   Weight: 39 oz.   Sights: low profile night sights   Capacity: 8

Magazine: detachable box   Grips: custom rosewood

D2C:	NIB $ 980	Ex $ 750	VG+ $545	Good $ 490	LMP $ 1,128
C2C:	Mint $ 950	Ex $ 700	VG+ $490	Good $ 445	Fair $ 230
Trade-In:	Mint $ 700	Ex $ 550	VG+ $390	Good $ 345	Poor $ 120

1911 CARRY STAINLESS
NEW IN BOX

1911 CARRY STAINLESS   REDBOOK CODE: RB-SS-H-1911CS

Natural stainless steel frame and slide, beavertail grip safety, match grade barrel, match grade hammer/sear and trigger, checkering on front strap and mainspring housing, de-horned frame and slide.

Production: 2006 - current   Caliber: .45 ACP   Action: SA, semi-auto   Barrel Length: 4.2"

OA Length: 7.7"   Weight: 35 oz.   Sights: low profile night sights   Capacity: 8

Magazine: detachable box   Grips: custom gray diamondwood

D2C:	NIB	$1,000	Ex $	775	VG+	$ 555	Good $	500	LMP	$1,142
C2C:	Mint $	960	Ex $	700	VG+	$500	Good $	450	Fair	$ 230
Trade-In:	Mint $	710	Ex $	575	VG+	$390	Good $	350	Poor	$ 120

1911 XO BLACK
NEW IN BOX

1911 XO BLACK   REDBOOK CODE: RB-SS-H-1911XB

Black XO finished stainless steel frame and slide, beavertail grip safety, match grade barrel, match grade hammer/sear and trigger, checkering on front strap and mainspring housing, de-horned frame and slide.

Production: 2006   Caliber: .45 ACP   Action: SA, semi-auto   Barrel Length: 5"

OA Length: 8.7"   Weight: 42 oz.   Sights: low profile contrast   Capacity: 8

Magazine: detachable box   Grips: Ergo Grip XT Extreme-Use

D2C:	NIB	$ 980	Ex $	750	VG+	$545	Good $	490	LMP	$1,128
C2C:	Mint $	950	Ex $	700	VG+	$490	Good $	445	Fair	$ 230
Trade-In:	Mint $	700	Ex $	550	VG+	$390	Good $	345	Poor	$ 120

1911 XO STAINLESS   REDBOOK CODE: RB-SS-H-1911XS

Stainless XO finished stainless steel frame and slide, beavertail grip safety, match grade barrel, match grade hammer/sear and trigger, checkering on front strap and mainspring housing, de-horned frame and slide.

Production: 2006 - 2012   Caliber: .45 ACP   Action: SA, semi-auto   Barrel Length: 5"

OA Length: 8.7"   Weight: 40 oz.   Sights: Novak Contrast Combat   Capacity: 8

Magazine: detachable box   Grips: Ergo Grip XT Extreme-Use

D2C:	NIB	$ 980	Ex $	750	VG+	$545	Good $	490	LMP	$1,128
C2C:	Mint $	950	Ex $	700	VG+	$490	Good $	445	Fair	$ 230
Trade-In:	Mint $	700	Ex $	550	VG+	$390	Good $	345	Poor	$ 120

1911 TTT
NEW IN BOX

1911 TTT   REDBOOK CODE: RB-SS-H-1911TT

Nitron finished stainless steel slide, black controls, natural stainless frame, beavertail grip safety, match grade barrel, match grade hammer/sear and trigger, checkering on front strap and mainspring housing, de-horned frame and slide.

Production: 2006   Caliber: .45 ACP   Action: SA, semi-auto   Barrel Length: 5"

OA Length: 8.7"   Weight: 42 oz.   Sights: adjustable combat night sights   Capacity: 8

Magazine: detachable box   Grips: burled maple

D2C:	NIB	$1,025	Ex $	800	VG+	$ 570	Good $	515	LMP	$1,170
C2C:	Mint $	990	Ex $	725	VG+	$520	Good $	465	Fair	$ 240
Trade-In:	Mint $	730	Ex $	575	VG+	$400	Good $	360	Poor	$ 120

1911 STX
NEW IN BOX

1911 STX REDBOOK CODE: RB-SS-H-1911SX

Nitron finished stainless steel slide, natural stainless frame, beavertail grip safety, match grade barrel, match grade hammer/sear and trigger, checkering on front strap and mainspring housing, de-horned frame and slide, polished cocking serrations, flat top slide, stainless magwell.

Production: 2006 Caliber: .45 ACP Action: SA, semi-auto Barrel Length: 5"
OA Length: 8.7" Weight: 42 oz. Sights: adjustable combat night sights Capacity: 8
Magazine: detachable box Grips: burled maple

D2C:	NIB	$1,030	Ex $ 800	VG+ $ 575	Good $ 515	LMP $1,213
C2C:	Mint $ 990	Ex $ 725	VG+ $ 520	Good $ 465	Fair $ 240
Trade-In:	Mint $ 740	Ex $ 600	VG+ $ 410	Good $ 365	Poor $ 120

1911 COMPACT NITRON
NEW IN BOX

1911 COMPACT NITRON REDBOOK CODE: RB-SS-H-1911CT

Nitron finished stainless steel frame and slide, checkered front strap and mainspring housing, beavertail grip safety, shortened and de-horned slide and frame, match grade barrel.

Production: 2006 - discontinued Caliber: .45 ACP Action: SA, semi-auto
Barrel Length: 4.2" OA Length: 7.7" Weight: 36 oz. Sights: Novak night sights
Capacity: 7 Magazine: detachable box Grips: slim profile custom rosewood

D2C:	NIB	$1,025	Ex $ 800	VG+ $ 570	Good $ 515	LMP $1,200
C2C:	Mint $ 990	Ex $ 725	VG+ $ 520	Good $ 465	Fair $ 240
Trade-In:	Mint $ 730	Ex $ 575	VG+ $ 400	Good $ 360	Poor $ 120

1911 COMPACT STAINLESS REDBOOK CODE: RB-SS-H-1911SC

Stainless steel frame and slide, checkered front strap and mainspring housing, beavertail grip safety, shortened and de-horned slide and frame, match grade barrel.

Production: 2006 - discontinued Caliber: .45 ACP Action: SA, semi-auto
Barrel Length: 4.2" OA Length: 7.7" Weight: 36 oz. Sights: Novak night sights
Capacity: 7 Magazine: detachable box Grips: slim profile custom gray diamondwood

D2C:	NIB	$1,000	Ex $ 775	VG+ $ 555	Good $ 500	LMP $1,142
C2C:	Mint $ 960	Ex $ 700	VG+ $500	Good $ 450	Fair $ 230
Trade-In:	Mint $ 710	Ex $ 575	VG+ $390	Good $ 350	Poor $ 120

1911 COMPACT RCS NITRON
NEW IN BOX

1911 COMPACT RCS NITRON REDBOOK CODE: RB-SS-H-1911CR

SIG Anti-Snag treatment, Nitron finished slide and frame, beavertail grip safety, match grade barrel, match grade hammer/sear and trigger, checkering on front strap and mainspring housing, de-horned frame and slide.

Production: 2006 - current Caliber: .45 ACP Action: SA, semi-auto
Barrel Length: 4.2" OA Length: 7.7" Weight: 30 oz. Sights: low profile night sights
Capacity: 7 Magazine: detachable box Grips: slim profile custom rosewood

D2C:	NIB	$1,025	Ex $800	VG+ $570	Good $ 515	LMP $1,170
C2C:	Mint $ 990	Ex $ 725	VG+ $520	Good $ 465	Fair $ 240
Trade-In:	Mint $ 730	Ex $ 575	VG+ $400	Good $ 360	Poor $ 120

1911 RCS TWO-TONE
NEW IN BOX

1911 RCS TWO-TONE REDBOOK CODE: RB-SS-H-1911RT

SIG Anti-Snag treatment, stainless finished slide, Nitron finished frame, beavertail grip safety, match grade barrel, match grade hammer/sear and trigger, checkering on front strap and mainspring housing, de-horned frame and slide.

Production: 2006 Caliber: .45 ACP Action: SA, semi-auto Barrel Length: 4.2"
OA Length: 7.7" Weight: 30 oz. Sights: low profile night sights Capacity: 7
Magazine: detachable box Grips: slim profile custom rosewood

D2C:	NIB	$ 980	Ex $750	VG+ $545	Good $490	LMP $1,185
C2C:	Mint $ 950	Ex $700	VG+ $490	Good $ 445	Fair $ 230
Trade-In:	Mint $ 700	Ex $550	VG+ $390	Good $ 345	Poor $ 120

1911 C3
NEW IN BOX

1911 C3   REDBOOK CODE: RB-SS-H-1911C3

Natural stainless steel slide, lightweight black alloy frame, beavertail grip safety, match grade barrel, match grade hammer/sear and trigger, checkering on front strap and mainspring housing, de-horned frame and slide.

Production: 2008 - current   Caliber: .45 ACP   Action: SA, semi-auto   Barrel Length: 4.2"
OA Length: 7.7"   Weight: 30 oz.   Sights: low profile night sights   Capacity: 7
Magazine: detachable box   Grips: slim profile custom rosewood

D2C:	NIB	$ 975	Ex	$ 750	VG+	$ 545	Good	$ 490	LMP	$1,042
C2C:	Mint	$ 940	Ex	$ 675	VG+	$490	Good	$ 440	Fair	$ 225
Trade-In:	Mint	$ 700	Ex	$ 550	VG+	$390	Good	$ 345	Poor	$ 120

1911 C3+
NEW IN BOX

1911 C3+   REDBOOK CODE: RB-SS-H-1911CP

Natural stainless steel slide, lightweight black alloy frame, beavertail grip safety, match grade barrel, match grade hammer/sear and trigger, checkering on front strap and mainspring housing, de-horned frame and slide, stainless extended magwell.

Production: current   Caliber: .45 ACP   Action: SA, semi-auto   Barrel Length: 4.2"
OA Length: 7.7"   Weight: 30 oz.   Sights: low profile night sights   Capacity: 7
Magazine: detachable box   Grips: slim profile custom rosewood

D2C:	NIB	$ 975	Ex	$ 750	VG+	$ 545	Good	$ 490		
C2C:	Mint	$ 940	Ex	$ 675	VG+	$490	Good	$ 440	Fair	$ 225
Trade-In:	Mint	$ 700	Ex	$ 550	VG+	$390	Good	$ 345	Poor	$ 120

1911 C3-CTC-WD
NEW IN BOX

1911 C3-CTC-WD   REDBOOK CODE: RB-SS-H-1911CD

Natural stainless steel slide, lightweight black alloy frame, beavertail grip safety, match grade barrel, match grade hammer/sear and trigger, checkering on front strap and mainspring housing, de-horned frame and slide, stainless extended magwell.

Production: current   Caliber: .45 ACP   Action: SA, semi-auto   Barrel Length: 4.2"
OA Length: 7.7"   Weight: 30 oz.   Sights: low profile night sights   Capacity: 7
Magazine: detachable box   Grips: Crimson Trace burlwood Lasergrips

D2C:	NIB	$1,030	Ex	$ 800	VG+	$ 575	Good	$ 515	LMP	$1,285
C2C:	Mint	$ 990	Ex	$ 725	VG+	$ 520	Good	$ 465	Fair	$ 240
Trade-In:	Mint	$ 740	Ex	$ 600	VG+	$ 410	Good	$ 365	Poor	$ 120

1911 C3-CTC
NEW IN BOX

1911 C3-CTC   REDBOOK CODE: RB-SS-H-1911CC

Natural stainless steel slide, lightweight black alloy frame, beavertail grip safety, match grade barrel, match grade hammer/sear and trigger, checkering on front strap and mainspring housing, de-horned frame and slide, stainless extended magwell.

Production: current   Caliber: .45 ACP   Action: SA, semi-auto   Barrel Length: 4.2"
OA Length: 7.7"   Weight: 30 oz.   Sights: low profile night sights   Capacity: 7
Magazine: detachable box   Grips: Crimson Trace black Lasergrips

D2C:	NIB	$1,030	Ex	$800	VG+	$ 575	Good	$ 515	LMP	$1,285
C2C:	Mint	$ 990	Ex	$ 725	VG+	$520	Good	$ 465	Fair	$ 240
Trade-In:	Mint	$ 740	Ex	$600	VG+	$ 410	Good	$ 365	Poor	$ 120

1911 TACOPS
NEW IN BOX

1911 TACOPS   REDBOOK CODE: RB-SS-H-1911TA

Nitron finished stainless steel slide and frame, beavertail grip safety, match grade barrel, match grade hammer/sear and trigger, checkering on front strap and mainspring housing, de-horned frame and slide, ambidextrous safety, extended stainless magwell, integral accessory rail, includes four 8-round magazines.

Caliber: .45 ACP   Action: SA, semi-auto   Barrel Length: 5"   OA Length: 8.7"   Weight: 42 oz.
Sights: low profile night sights   Capacity: 8   Magazine: detachable box   Grips: Ergo Grip XT

D2C:	NIB	$1,015	Ex	$ 775	VG+	$ 565	Good	$ 510	LMP	$1,213
C2C:	Mint	$ 980	Ex	$ 725	VG+	$ 510	Good	$ 460	Fair	$ 235
Trade-In:	Mint	$ 730	Ex	$ 575	VG+	$400	Good	$ 360	Poor	$ 120

1911 TACOPS TB
NEW IN BOX

1911 TACOPS TB REDBOOK CODE: RB-SS-H-1911TB

Nitron finished stainless steel slide and frame, beavertail grip safety, match grade barrel, match grade hammer/sear and trigger, checkering on front strap and mainspring housing, de-horned frame and slide, ambidextrous safety, extended stainless magwell, integral accessory rail, includes four 8-round magazines, threaded barrel.

Caliber: .45 ACP   Action: SA, semi-auto   Barrel Length: 5.5"   OA Length: 8.75"

Weight: 42 oz.   Sights: low profile night sights   Capacity: 8   Magazine: detachable box

Grips: Ergo Grip XT

D2C:	NIB	$1,030	Ex	$800	VG+	$575	Good	$515	LMP	$1,285
C2C:	Mint	$990	Ex	$725	VG+	$520	Good	$465	Fair	$240
Trade-In:	Mint	$740	Ex	$600	VG+	$410	Good	$365	Poor	$120

1911 TACOPS CARRY
NEW IN BOX

1911 TACOPS CARRY REDBOOK CODE: RB-SS-H-1911TC

Nitron finished stainless steel slide and frame, beavertail grip safety, match grade barrel, match grade hammer/sear and trigger, checkering on front strap and mainspring housing, de-horned frame and slide, ambidextrous safety, extended stainless magwell, integral accessory rail, includes four 8-round magazines.

Production: current   Caliber: .45 ACP   Action: SA, semi-auto   Barrel Length: 4.2"

OA Length: 7.7"   Weight: 35 oz.   Sights: low profile night sights   Capacity: 8

Magazine: detachable box   Grips: Ergo Grip XT

D2C:	NIB	$1,020	Ex	$800	VG+	$570	Good	$510	LMP	$1,213
C2C:	Mint	$980	Ex	$725	VG+	$510	Good	$460	Fair	$235
Trade-In:	Mint	$730	Ex	$575	VG+	$400	Good	$360	Poor	$120

1911 TACOPS CARRY TB REDBOOK CODE: RB-SS-H-191TCT

Nitron finished stainless steel slide and frame, beavertail grip safety, match grade barrel, match grade hammer/sear and trigger, checkering on front strap and mainspring housing, de-horned frame and slide, ambidextrous safety, extended stainless magwell, integral accessory rail, includes four 8-round magazines, threaded barrel.

Production: current   Caliber: .45 ACP   Action: SA, semi-auto   Barrel Length: 4.7"

OA Length: 8.2"   Sights: low profile night sights   Capacity: 8

Magazine: detachable box   Grips: Ergo Grip XT

D2C:	NIB	$1,035	Ex	$800	VG+	$575	Good	$520	LMP	$1,285
C2C:	Mint	$1,000	Ex	$725	VG+	$520	Good	$470	Fair	$240
Trade-In:	Mint	$740	Ex	$600	VG+	$410	Good	$365	Poor	$120

1911 SCORPION
NEW IN BOX

1911 SCORPION REDBOOK CODE: RB-SS-H-191SCP

Cerakote Flat Dark Earth finished stainless steel frame and slide, beavertail grip safety, match grade barrel, match grade hammer/sear and trigger, checkering on front strap and mainspring housing, de-horned frame and slide, front cocking serrations, accessory rail.

Caliber: .45 ACP   Action: SA, semi-auto   Barrel Length: 5"   OA Length: 8.7"

Weight: 42 oz.   Sights: low profile night sights   Capacity: 8   Magazine: detachable box

Grips: Hogue Piranha G10

D2C:	NIB	$1,010	Ex	$775	VG+	$565	Good	$505	LMP	$1,213
C2C:	Mint	$970	Ex	$700	VG+	$510	Good	$455	Fair	$235
Trade-In:	Mint	$720	Ex	$575	VG+	$400	Good	$355	Poor	$120

1911 SCORPION TB
NEW IN BOX

1911 SCORPION TB   REDBOOK CODE: RB-SS-H-191SCT

Cerakote Flat Dark Earth finished stainless steel frame and slide, beavertail grip safety, match grade barrel, match grade hammer/sear and trigger, checkering on front strap and mainspring housing, de-horned frame and slide, front cocking serrations, accessory rail, threaded barrel.

Caliber: .45 ACP	Action: SA, semi-auto	Barrel Length: 5.5"	OA Length: 9.2"	
Weight: 43 oz.	Sights: low profile night sights	Capacity: 8	Magazine: detachable box	
Grips: Hogue Extreme Series G10 and magwell				

D2C:	NIB	$1,050	Ex $ 800	VG+ $ 585	Good $ 525	LMP	$1,285
C2C:	Mint	$1,010	Ex $ 725	VG+ $ 530	Good $ 475	Fair	$ 245
Trade-In:	Mint $ 750	Ex $ 600	VG+ $ 410	Good $ 370	Poor $ 120

1911 CARRY SCORPION
NEW IN BOX

1911 CARRY SCORPION   REDBOOK CODE: RB-SS-H-191CSC

Cerakote Flat Dark Earth finished stainless steel frame and slide, beavertail grip safety, match grade barrel, match grade hammer/sear and trigger, checkering on front strap and mainspring housing, de-horned frame and slide, front cocking serrations, accessory rail.

Production: current	Caliber: .45 ACP	Action: SA, semi-auto	Barrel Length: 4.2"	
OA Length: 7.7"	Weight: 35 oz.	Sights: low profile night sights	Capacity: 8	
Magazine: detachable box	Grips: Hogue G-10			

D2C:	NIB	$1,015	Ex $ 775	VG+ $ 565	Good $ 510	LMP	$1,213
C2C:	Mint	$ 980	Ex $ 725	VG+ $ 510	Good $ 460	Fair	$ 235
Trade-In:	Mint $ 730	Ex $ 575	VG+ $400	Good $ 360	Poor $ 120

1911 CARRY SCORPION TB
NEW IN BOX

1911 CARRY SCORPION TB

REDBOOK CODE: RB-SS-H-191CST

Cerakote Flat Dark Earth finished stainless steel frame and slide, beavertail grip safety, match grade barrel, match grade hammer/sear and trigger, checkering on front strap and mainspring housing, de-horned frame and slide, front cocking serrations, accessory rail, threaded barrel.

Production: current	Caliber: .45 ACP	Action: SA, semi-auto	Barrel Length: 4.7"	
OA Length: 8.2"	Weight: 37 oz.	Sights: low profile night sights	Capacity: 8	
Magazine: detachable box	Grips: Hogue Extreme Series G10 and magwell			

D2C:	NIB	$1,040	Ex $800	VG+ $580	Good $ 520	LMP	$1,285
C2C:	Mint	$1,000	Ex $ 725	VG+ $520	Good $ 470	Fair	$ 240
Trade-In:	Mint $ 740	Ex $600	VG+ $ 410	Good $ 365	Poor $ 120

1911 ULTRA COMPACT
NEW IN BOX

1911 ULTRA COMPACT   REDBOOK CODE: RB-SS-H-1911UC

Nitron stainless slide, lightweight black alloy frame, beavertail grip safety, match grade barrel, match grade hammer/sear and trigger, checkering on front strap and mainspring housing, de-horned frame and slide, two-tone finish available.

Production: 2011 - current	Caliber: .45 ACP	Action: SA, semi-auto	Barrel Length: 3.3"	
OA Length: 6.8"	Weight: 28 oz.	Sights: low profile night sights	Capacity: 7	
Magazine: detachable box	Grips: custom blackwood			

D2C:	NIB	$ 985	Ex $ 750	VG+ $550	Good $ 495	LMP	$1,142
C2C:	Mint	$ 950	Ex $700	VG+ $500	Good $ 445	Fair	$ 230
Trade-In:	Mint $ 700	Ex $ 575	VG+ $390	Good $ 345	Poor $ 120

1911 ULTRA COMPACT TWO-TONE
NEW IN BOX

1911 ULTRA COMPACT TWO-TONE

REDBOOK CODE: RB-SS-H-191UCT

Natural stainless steel slide, lightweight black alloy frame, beavertail grip safety, match grade barrel, match grade hammer/sear and trigger, checkering on front strap and mainspring housing, de-horned frame and slide.

Production: 2011 - current   Caliber: .45 ACP   Action: SA, semi-auto   Barrel Length: 3.3"
OA Length: 6.8"   Weight: 28 oz.   Sights: low profile night sights   Capacity: 7
Magazine: detachable box   Grips: custom rosewood

D2C:	NIB $ 925	Ex $ 725	VG+ $ 515	Good $ 465	LMP $1,156
C2C:	Mint $ 890	Ex $ 650	VG+ $470	Good $ 420	Fair $ 215
Trade-In:	Mint $ 660	Ex $ 525	VG+ $370	Good $ 325	Poor $ 120

1911 DESERT
NEW IN BOX

1911 DESERT   REDBOOK CODE: RB-SS-H-1911DE

Custom two-tone Cerakote Desert Tan slide and frame, beavertail grip safety, match grade barrel, match grade hammer/sear and trigger, checkering on front strap and mainspring housing, de-horned frame and slide.

Caliber: .45 ACP   Action: SA, semi-auto   Barrel Length: 5"   OA Length: 8.7"
Weight: 42 oz.   Sights: low profile night sights   Capacity: 8   Magazine: detachable box
Grips: Flat Dark Earth Ergo Grip XT

D2C:	NIB $ 805	Ex $ 625	VG+ $450	Good $ 405	LMP $1,070
C2C:	Mint $ 780	Ex $ 575	VG+ $ 410	Good $ 365	Fair $ 190
Trade-In:	Mint $ 580	Ex $ 475	VG+ $320	Good $ 285	Poor $ 90

1911 MAX
NEW IN BOX

1911 MAX   REDBOOK CODE: RB-SS-H-1911MX

Custom Max Michel edition, Nitron stainless slide, natural stainless frame, SIG flat trigger, external extractor, Dawson Ice Magwell, Koenig Speed Hammer, EGW sear and firing pin, forward cocking serrations.

Production: 2012 - current   Caliber: .45 ACP   Action: SA, semi-auto   Barrel Length: 5"
OA Length: 8.7"   Weight: 42 oz.   Sights: fiber optic front, adjustable rear   Capacity: 8
Magazine: detachable box   Grips: custom Hogue chainlink G-10

D2C:	NIB $ 940	Ex $ 725	VG+ $525	Good $ 470	LMP $1,713
C2C:	Mint $ 910	Ex $ 650	VG+ $470	Good $ 425	Fair $ 220
Trade-In:	Mint $ 670	Ex $ 550	VG+ $370	Good $ 330	Poor $ 120

1911 SPARTAN
NEW IN BOX

1911 SPARTAN   REDBOOK CODE: RB-SS-H-1911SP

Stainless steel slide and frame with oil rubbed bronze Nitron finish, 24kt gold inlay engraved Molon Labe on slide, beavertail grip safety, match grade barrel, match grade hammer/sear and trigger, checkering on front strap and mainspring housing, de-horned frame and slide.

Caliber: .45 ACP   Action: SA, semi-auto   Barrel Length: 5"   OA Length: 8.7"
Weight: 42 oz.   Sights: low profile night sights   Capacity: 8   Magazine: detachable box
Grips: Hogue Spartan

D2C:	NIB $ 975	Ex $ 750	VG+ $545	Good $ 490	LMP $1,356
C2C:	Mint $ 940	Ex $ 675	VG+ $490	Good $ 440	Fair $ 225
Trade-In:	Mint $ 700	Ex $ 550	VG+ $390	Good $ 345	Poor $ 120

1911 EXTREME
NEW IN BOX

1911 EXTREME REDBOOK CODE: RB-SS-H-1911EX

Nitron finished stainless steel frame and slide, flat trigger, extended magwell, ambidextrous safety, accessory rail, beavertail grip safety, match grade barrel, match grade hammer/sear and trigger, checkering on front strap and mainspring housing, de-horned frame and slide.

Caliber: .45 ACP   Action: SA, semi-auto   Barrel Length: 5"   OA Length: 8.7"

Weight: 42 oz.   Sights: low profile night sights   Capacity: 8   Magazine: detachable box

Grips: Hogue Extreme Piranha G-10

D2C:	NIB	$ 855	Ex $ 650	VG+ $ 475	Good $ 430	LMP $1,213
C2C:	Mint $ 830	Ex $ 600	VG+ $430	Good $ 385	Fair $ 200
Trade-In:	Mint $ 610	Ex $ 500	VG+ $340	Good $ 300	Poor $ 90

1911 FASTBACK NIGHTMARE
NEW IN BOX

1911 FASTBACK NIGHTMARE REDBOOK CODE: RB-SS-H-191FBN

Nitron finished stainless steel frame and slide, stainless controls, rounded Fastback frame, beavertail grip safety, match grade barrel, match grade hammer/sear and trigger, checkering on front strap and mainspring housing, de-horned frame and slide.

Production: 2012   Caliber: .45 ACP, .357 SIG   Action: SA, semi-auto   Barrel Length: 5"

OA Length: 8.7"   Weight: 42 oz.   Sights: low profile night sights   Capacity: 8, 9

Magazine: detachable box   Grips: black Hogue G-10 Fastback profile

D2C:	NIB $1,025	Ex $ 800	VG+ $ 570	Good $ 515	LMP $1,242
C2C:	Mint $ 990	Ex $ 725	VG+ $ 520	Good $ 465	Fair $ 240
Trade-In:	Mint $ 730	Ex $ 575	VG+ $ 400	Good $ 360	Poor $ 120

1911 FASTBACK CARRY NIGHTMARE
NEW IN BOX

1911 FASTBACK CARRY NIGHTMARE

REDBOOK CODE: RB-SS-H-191FCN

Nitron finished stainless steel frame and slide, stainless controls, rounded Fastback frame, beavertail grip safety, match grade barrel, match grade hammer/sear and trigger, checkering on front strap and mainspring housing, de-horned frame and slide.

Production: 2012   Caliber: .45 ACP, .357 SIG   Action: SA, semi-auto   Barrel Length: 4.2"

OA Length: 7.7"   Weight: 39 oz.   Sights: low profile night sights   Capacity: 8, 9

Magazine: detachable box   Grips: black Hogue G-10 Fastback profile

D2C:	NIB $1,025	Ex $800	VG+ $ 570	Good $ 515	LMP $1,242
C2C:	Mint $ 990	Ex $ 725	VG+ $520	Good $465	Fair $ 240
Trade-In:	Mint $ 730	Ex $ 575	VG+ $400	Good $ 360	Poor $ 120

1911 FASTBACK ROSEWOOD
NEW IN BOX

1911 FASTBACK ROSEWOOD

REDBOOK CODE: RB-SS-H-191FBR

Natural stainless slide, Nitron finished stainless steel frame, beavertail grip safety, match grade barrel, match grade hammer/sear and trigger, checkering on front strap and mainspring housing, de-horned frame and slide.

Production: 2012   Caliber: .45 ACP   Action: SA, semi-auto   Barrel Length: 5"

OA Length: 8.7"   Weight: 42 oz.   Sights: low profile night sights   Capacity: 8

Magazine: detachable box   Grips: Hogue rosewood double-diamond

D2C:	NIB $1,015	Ex $ 775	VG+ $565	Good $ 510	LMP $1,242
C2C:	Mint $ 980	Ex $ 725	VG+ $ 510	Good $ 460	Fair $235
Trade-In:	Mint $ 730	Ex $ 575	VG+ $400	Good $ 360	Poor $120

SIG SAUER

1911 FASTBACK CARRY ROSEWOOD

REDBOOK CODE: RB-SS-H-191FCR

Natural stainless slide, Nitron finished stainless steel frame, beavertail grip safety, match grade barrel, match grade hammer/sear and trigger, checkering on front strap and mainspring housing, de-horned frame and slide.

Production: 2012 - current	Caliber: .45 ACP	Action: SA, semi-auto	Barrel Length: 4.2"

OA Length: 7.7"	Weight: 39 oz.	Sights: low profile night sights	Capacity: 8

Magazine: detachable box	Grips: Hogue rosewood double-diamond		

D2C:	NIB	$1,005	Ex	$775	VG+	$560	Good	$505	LMP	$1,242
C2C:	Mint	$970	Ex	$700	VG+	$510	Good	$455	Fair	$235
Trade-In:	Mint	$720	Ex	$575	VG+	$400	Good	$355	Poor	$120

1911 TRADITIONAL REVERSE TWO-TONE
NEW IN BOX

1911 TRADITIONAL REVERSE TWO-TONE

REDBOOK CODE: RB-SS-H-191TRT

Natural stainless steel frame, Nitron coated stainless slide, black control levers, traditional round-top 1911 slide profile, front cocking serrations, skeletonized trigger, match grade barrel, checkering on front strap and mainspring housing.

Caliber: .45 ACP, .357 SIG	Action: SA, semi-auto	Barrel Length: 5"	OA Length: 8.7"

Weight: 42 oz.	Sights: low profile night sights	Capacity: 8, 9	

Magazine: detachable box	Grips: custom blackwood		

D2C:	NIB	$995	Ex	$775	VG+	$555	Good	$500	LMP	$1,142
C2C:	Mint	$960	Ex	$700	VG+	$500	Good	$450	Fair	$230
Trade-In:	Mint	$710	Ex	$575	VG+	$390	Good	$350	Poor	$120

1911 TRADITIONAL COMPACT STAINLESS
NEW IN BOX

1911 TRADITIONAL COMPACT STAINLESS

REDBOOK CODE: RB-SS-H-191TCS

Natural stainless steel slide and frame, traditional round-top 1911 slide profile, front cocking serrations, skeletonized trigger, match grade barrel, checkering on front strap and mainspring.

Production: current	Caliber: .45 ACP	Action: SA, semi-auto	Barrel Length: 4.2"

OA Length: 7.7"	Weight: 35.5 oz.	Sights: low profile night sights	Capacity: 7

Magazine: detachable box	Grips: custom blackwood		

D2C:	NIB	$990	Ex	$775	VG+	$550	Good	$495	LMP	$1,142
C2C:	Mint	$960	Ex	$700	VG+	$500	Good	$450	Fair	$230
Trade-In:	Mint	$710	Ex	$575	VG+	$390	Good	$350	Poor	$120

1911 TRADITIONAL MATCH ELITE STAINLESS
NEW IN BOX

1911 TRADITIONAL MATCH ELITE STAINLESS

REDBOOK CODE: RB-SS-H-191TMS

Natural stainless steel frame and slide, traditional round-top 1911 slide profile, front cocking serrations, skeletonized trigger, match grade barrel, checkering on front strap and mainspring.

Caliber: 9mm, .40 S&W	Action: SA, semi-auto	Barrel Length: 5"	OA Length: 8.7"

Weight: 42 oz.	Sights: adjustable target	Capacity: 10	Magazine: detachable box

Grips: custom wood			

D2C:	NIB	$985	Ex	$750	VG+	$550	Good	$495	LMP	$1,128
C2C:	Mint	$950	Ex	$700	VG+	$500	Good	$445	Fair	$230
Trade-In:	Mint	$700	Ex	$575	VG+	$390	Good	$345	Poor	$120

1911 TRADITIONAL TWO-TONE MATCH ELITE
NEW IN BOX

1911 TRADITIONAL TWO-TONE MATCH ELITE

REDBOOK CODE: RB-SS-H-191TTM

Natural stainless steel slide, Nitron coated stainless frame, traditional round-top 1911 slide profile, front cocking serrations, skeletonized trigger, match grade barrel, checkering on front strap and mainspring.

Caliber: .45 ACP   Action: SA, semi-auto   Barrel Length: 5"   OA Length: 8.7"
Weight: 42 oz.   Sights: low profile adjustable   Capacity: 8   Magazine: detachable box
Grips: custom wood

D2C:	NIB	$ 980	Ex $	750	VG+ $	545	Good $	490	LMP	$1,128
C2C:	Mint $	950	Ex $	700	VG+ $	490	Good $	445	Fair	$ 230
Trade-In:	Mint $	700	Ex $	550	VG+ $	390	Good $	345	Poor	$ 120

1911 TRADITIONAL TACOPS
NEW IN BOX

1911 TRADITIONAL TACOPS   REDBOOK CODE: RB-SS-H-191TTA

Nitron finished stainless steel frame and slide, ambidextrous safety, accessory rail, traditional round-top 1911 slide profile, front cocking serrations, skeletonized trigger, match grade barrel, checkering on front strap and mainspring.

Caliber: .45 ACP, .357 SIG   Action: SA, semi-auto   Barrel Length: 5"   OA Length: 8.7"
Weight: 42 oz.   Sights: low profile night sights   Capacity: 8, 9
Magazine: detachable box   Grips: black Magwell Ergo XT

D2C:	NIB	$1,005	Ex $	775	VG+ $	560	Good $	505	LMP	$1,213
C2C:	Mint $	970	Ex $	700	VG+ $	510	Good $	455	Fair	$ 235
Trade-In:	Mint $	720	Ex $	575	VG+ $	400	Good $	355	Poor	$ 120

1911 POW-MIA
EXCELLENT

1911 POW-MIA   REDBOOK CODE: RB-SS-H-191POW

Limited production run, stainless finished slide and frame, engraved dog tags and American flag on top of slide, barbed wire engraved on side of slide, includes custom POW-MIA Ka-Bar knife and Pelican Storm Case.

Caliber: .45 ACP   Action: SA, semi-auto   Barrel Length: 5"   OA Length: 8.7"
Weight: 42 oz.   Sights: SIGLITE night sights   Capacity: 8   Magazine: detachable box
Grips: custom Hogue POW-MIA

D2C:	NIB	$ 1,140	Ex $	875	VG+ $	635	Good $	570	LMP	$1,499
C2C:	Mint	$1,100	Ex $	800	VG+ $	570	Good $	515	Fair	$265
Trade-In:	Mint $	810	Ex $	650	VG+ $	450	Good $	400	Poor	$120

1911 BLACKWATER
EXCELLENT
Courtesy of Bud's Gun Shop

1911 BLACKWATER   REDBOOK CODE: RB-SS-H-1911BW

Limited production run, Nitron finished stainless slide and frame, integral accessory rail, ambidextrous safety, extended magwell, Blackwater logo engraving on slide.

Caliber: .45 ACP   Action: SA, semi-auto   Barrel Length: 5"   Sights: Novak night sights
Capacity: 8   Magazine: detachable box   Grips: custom Blackwater wood grips

D2C:	NIB	$1,200	Ex $	925	VG+	$670	Good $	600	LMP	$1,290
C2C:	Mint	$ 1,160	Ex $	850	VG+	$600	Good $	540	Fair	$ 280
Trade-In:	Mint $	860	Ex $	675	VG+	$470	Good $	420	Poor	$ 120

1911 PLATINUM ELITE CARRY   REDBOOK CODE: RB-SS-H-191PEC

Duo-tone natural stainless finish slide with polished flats, Nitron frame, stainless controls, checkered front strap.

Production: 2008 - discontinued   Caliber: .45 ACP   Action: SA, semi-auto
Barrel Length: 4.2"   OA Length: 7.7"   Weight: 35 oz.   Sights: Novak adjustable
night sights   Capacity: 8   Magazine: detachable box   Grips: aluminum

D2C:	NIB	$1,200	Ex $	925	VG+	$670	Good $	600	LMP	$1,275
C2C:	Mint	$ 1,160	Ex $	850	VG+	$600	Good $	540	Fair	$ 280
Trade-In:	Mint $	860	Ex $	675	VG+	$470	Good $	420	Poor	$ 120

1911-22
NEW IN BOX

1911-22 REDBOOK CODE: RB-SS-H-191122

Black slide and frame with PTFE finish, full-size 1911 dimensions, skeletonized trigger and hammer, grip safety, ambidextrous thumb safety.

Production: current   Caliber: .22 LR   Action: SA, semi-auto   Barrel Length: 5"
OA Length: 8.5"   Weight: 34 oz.   Sights: low profile contrast   Capacity: 10
Magazine: detachable box   Grips: rosewood

D2C:	NIB	$ 385	Ex	$300	VG+	$ 215	Good	$ 195	LMP	$ 419
C2C:	Mint	$ 370	Ex	$275	VG+	$200	Good	$ 175	Fair	$ 90
Trade-In:	Mint	$ 280	Ex	$ 225	VG+	$ 160	Good	$ 135	Poor	$ 60

1911-22 FLAT DARK EARTH
NEW IN BOX

1911-22 FLAT DARK EARTH REDBOOK CODE: RB-SS-H-19122F

Flat Dark Earth frame and slide, full-size 1911 dimensions, skeletonized trigger and hammer, grip safety, ambidextrous thumb safety.

Production: current   Caliber: .22 LR   Action: SA, semi-auto   Barrel Length: 5"
OA Length: 8.5"   Weight: 34 oz.   Sights: low profile contrast   Capacity: 10
Magazine: detachable box   Grips: walnut

D2C:	NIB	$ 405	Ex	$ 325	VG+	$ 225	Good	$ 205	LMP	$ 470
C2C:	Mint	$ 390	Ex	$300	VG+	$ 210	Good	$ 185	Fair	$ 95
Trade-In:	Mint	$ 290	Ex	$ 250	VG+	$ 160	Good	$ 145	Poor	$ 60

1911-22 OLIVE DRAB
NEW IN BOX

1911-22 OLIVE DRAB REDBOOK CODE: RB-SS-H-19122O

Olive Drab frame and slide, full-size 1911 dimensions, skeletonized trigger and hammer, grip safety, ambidextrous thumb safety.

Production: current   Caliber: .22 LR   Action: SA, semi-auto   Barrel Length: 5"
OA Length: 8.5"   Weight: 34 oz.   Sights: low profile contrast   Capacity: 10
Magazine: detachable box   Grips: rosewood

D2C:	NIB	$ 425	Ex	$ 325	VG+	$240	Good	$ 215	LMP	$ 470
C2C:	Mint	$ 410	Ex	$300	VG+	$220	Good	$ 195	Fair	$ 100
Trade-In:	Mint	$ 310	Ex	$ 250	VG+	$ 170	Good	$ 150	Poor	$ 60

1911-22 TWO-TONE OLIVE DRAB
REDBOOK CODE: RB-SS-H-19122T

Olive Drab slide and black frame, full-size 1911 dimensions, skeletonized trigger and hammer, grip safety, ambidextrous thumb safety

Production: 2013 - current   Caliber: .22 LR   Action: SA, semi-auto   Barrel Length: 5"
OA Length: 8.5"   Weight: 34 oz.   Sights: low profile contrast   Capacity: 10
Magazine: detachable box   Grips: rosewood

D2C:	NIB	$ 425	Ex	$ 325	VG+	$240	Good	$ 215	LMP	$ 470
C2C:	Mint	$ 410	Ex	$300	VG+	$220	Good	$ 195	Fair	$ 100
Trade-In:	Mint	$ 310	Ex	$ 250	VG+	$ 170	Good	$ 150	Poor	$ 60

1911-22 CAMO
NEW IN BOX

1911-22 CAMO REDBOOK CODE: RB-SS-H-19122C

Camo slide and frame, full-size 1911 dimensions, skeletonized trigger and hammer, grip safety, ambidextrous thumb safety.

Caliber: .22 LR   Action: SA, semi-auto   Barrel Length: 5"   OA Length: 8.5"
Weight: 34 oz.   Sights: low profile contrast   Capacity: 10   Magazine: detachable box
Grips: polymer camo

D2C:	NIB	$ 425	Ex	$ 325	VG+	$240	Good	$ 215	LMP	$ 518
C2C:	Mint	$ 410	Ex	$300	VG+	$220	Good	$ 195	Fair	$ 100
Trade-In:	Mint	$ 310	Ex	$ 250	VG+	$ 170	Good	$ 150	Poor	$ 60

MOSQUITO NEW IN BOX

MOSQUITO REDBOOK CODE: RB-SS-H-MOSQUI

Blue slide finish, black polymer frame, fixed barrel, blowback system, accessory rail, slide-mounted ambidextrous safety.

Production: current	Caliber: .22 LR	Action: DA/SA, semi-auto		Barrel Length: 4"	
OA Length: 7.2"	Weight: 25 oz.	Sights: adjustable rear	Capacity: 10		
Magazine: detachable box	Grips: black polymer				
D2C:	NIB $ 330	Ex $ 275	VG+ $185	Good $ 165	LMP $390
C2C:	Mint $ 320	Ex $ 250	VG+ $170	Good $ 150	Fair $ 80
Trade-In:	Mint $ 240	Ex $ 200	VG+ $130	Good $ 120	Poor $ 60

MOSQUITO PINK FINISH NEW IN BOX

MOSQUITO PINK FINISH REDBOOK CODE: RB-SS-H-MOSPNK

Pink finish over polymer frame, blue slide, fixed barrel, blowback system, accessory rail, slide-mounted ambidextrous safety.

Production: current	Caliber: .22 LR	Action: DA/SA, semi-auto		Barrel Length: 4"	
OA Length: 7.2"	Weight: 25 oz.	Sights: adjustable rear	Capacity: 10		
Magazine: detachable box	Grips: black polymer				
D2C:	NIB $ 270	Ex $225	VG+ $ 150	Good $ 135	LMP $405
C2C:	Mint $ 260	Ex $200	VG+ $ 140	Good $ 125	Fair $ 65
Trade-In:	Mint $ 200	Ex $ 175	VG+ $ 110	Good $ 95	Poor $ 30

MOSQUITO DESERT DIGITAL CAMO TB NEW IN BOX

MOSQUITO DESERT DIGITAL CAMO TB
REDBOOK CODE: RB-SS-H-MOSDDC

Desert Digital Camo frame and slide, polymer frame, fixed barrel, blowback system, accessory rail, slide-mounted ambidextrous safety, threaded barrel.

Caliber: .22 LR	Action: DA/SA, semi-auto	Barrel Length: 4"		OA Length: 8.3"	
Weight: 28 oz.	Sights: adjustable rear	Capacity: 10	Magazine: detachable box		
Grips: Desert Digital Camo					
D2C:	NIB $ 375	Ex $300	VG+ $ 210	Good $ 190	LMP $502
C2C:	Mint $ 360	Ex $275	VG+ $ 190	Good $ 170	Fair $ 90
Trade-In:	Mint $ 270	Ex $225	VG+ $ 150	Good $ 135	Poor $ 60

MOSQUITO FDE NEW IN BOX

MOSQUITO FDE REDBOOK CODE: RB-SS-H-MOSFDE

Flat Dark Earth finished frame and slide, fixed barrel, blowback system, accessory rail, slide-mounted ambidextrous safety, model with threaded barrel available.

Production: current	Caliber: .22 LR	Action: DA/SA, semi-auto		Barrel Length: 4"	
OA Length: 7.2"	Weight: 25 oz.	Sights: adjustable	Capacity: 10		
Magazine: detachable box	Grips: polymer				
D2C:	NIB $ 330	Ex $ 275	VG+ $ 185	Good $ 165	LMP $ 419
C2C:	Mint $ 320	Ex $ 250	VG+ $ 170	Good $ 150	Fair $ 80
Trade-In:	Mint $ 240	Ex $ 200	VG+ $ 130	Good $ 120	Poor $ 60

MOSQUITO OD GREEN NEW IN BOX

MOSQUITO OD GREEN REDBOOK CODE: RB-SS-H-MOSODG

OD Green finished frame and slide, fixed barrel, blowback system, accessory rail, slide-mounted ambidextrous safety, model with threaded barrel available.

Production: current	Caliber: .22 LR	Action: DA/SA, semi-auto		Barrel Length: 4"	
OA Length: 7.2"	Weight: 25 oz.	Sights: adjustable rear	Capacity: 10		
Magazine: detachable box	Grips: polymer				
D2C:	NIB $ 330	Ex $ 275	VG+ $ 185	Good $ 165	LMP $ 419
C2C:	Mint $ 320	Ex $ 250	VG+ $ 170	Good $ 150	Fair $ 80
Trade-In:	Mint $ 240	Ex $200	VG+ $ 130	Good $ 120	Poor $ 60

**MOSQUITO
CARBON FIBER**
NEW IN BOX

MOSQUITO CARBON FIBER   REDBOOK CODE: RB-SS-H-MOSCAR

Carbon Fiber finished frame and slide, fixed barrel, blowback system, accessory rail, slide mounted ambidextrous safety.

Caliber: .22 LR		Action: DA/SA, semi-auto		Barrel Length: 4"		OA Length: 7.2"				
Weight: 25 oz.		Sights: windage adjustable		Capacity: 10		Grips: polymer				
D2C:	NIB	$ 350	Ex	$ 275	VG+	$195	Good	$ 175	LMP	$469
C2C:	Mint	$ 340	Ex	$ 250	VG+	$180	Good	$ 160	Fair	$ 85
Trade-In:	Mint	$ 250	Ex	$200	VG+	$140	Good	$ 125	Poor	$ 60

MOSQUITO MULTICAM   REDBOOK CODE: RB-SS-H-MOSMUL

Multi-Cam finished frame and slide, fixed barrel, blowback system, accessory rail, slide mounted ambidextrous safety, model with threaded barrel available.

**MOSQUITO
MULTICAM**
NEW IN BOX

Production: current		Caliber: .22 LR		Action: DA/SA, semi-auto		Barrel Length: 4"				
OA Length: 7.2"		Weight: 25 oz.		Sights: adjustable rear		Capacity: 10				
Magazine: detachable box		Grips: MultiCam coated polymer								
D2C:	NIB	$ 350	Ex	$275	VG+	$ 195	Good	$ 175	LMP	$469
C2C:	Mint	$ 340	Ex	$250	VG+	$ 180	Good	$ 160	Fair	$ 85
Trade-In:	Mint	$ 250	Ex	$200	VG+	$ 140	Good	$ 125	Poor	$ 60

Smith & Wesson

Horace Smith and Daniel Wesson produced their first lever-action repeating pistol in 1852. After financial difficulties, they sold the majority of their company to Oliver Winchester, who would later form the Winchester Repeating Arms Company. On their second venture in 1856, Smith and Wesson patented a new, self-contained cartridge revolver, setting them on a course to become one of the most prominent and important manufacturers in the history of modern firearms. Since then, the company has produced many revolvers, pistols, and rifles and remained one of the largest companies in the weapons industry. The company is based in Springfield, Massachusetts.

**.32 DOUBLE ACTION
4TH MODEL**
VERY GOOD +

.32 DOUBLE ACTION 4TH MODEL   REDBOOK CODE: RB-RB-SW-H-32DA4M

Blue or nickel finish, top-break action, automatic shell extractor, rounded rear trigger guard, 6" barrel is rare, 8" and 10" barrel is very rare with increased value, ~239,600 manufactured, serial numbers: 43406-~283000.

Production: 1883 - 1909		Caliber: .32 S&W		Action: DA/SA, revolver						
Barrel Length: 3", 3.5", 6", 8", 10"		Sights: pinned round-blade front, notch-cut rear								
Capacity: 5		Grips: checkered hard rubber with S&W monograms								
D2C:	NIB	$ 510	Ex	$400	VG+	$285	Good	$ 255		
C2C:	Mint	$ 490	Ex	$ 375	VG+	$260	Good	$ 230	Fair	$120
Trade-In:	Mint	$ 370	Ex	$300	VG+	$200	Good	$ 180	Poor	$ 60

.32 DOUBLE ACTION 5TH MODEL   REDBOOK CODE: RB-SW-H-32DA5M

Blue or nickel finish, top-break action, automatic shell extractor, rounded trigger guard, 44,641 manufactured, serial numbers: ~283000-327645.

.32 DOUBLE ACTION 5TH MODEL
VERY GOOD +

Production: 1909 - 1919		Caliber: .32 S&W		Action: DA/SA, revolver						
Barrel Length: 3", 3.5", 6"		Sights: integral forged front, notch-cut rear								
Capacity: 5		Grips: checkered hard rubber with S&W monograms								
D2C:	NIB	$ 650	Ex	$ 500	VG+	$ 360	Good	$ 325		
C2C:	Mint	$ 630	Ex	$ 450	VG+	$ 330	Good	$ 295	Fair	$150
Trade-In:	Mint	$ 470	Ex	$ 375	VG+	$ 260	Good	$ 230	Poor	$ 90

**.32 SAFETY
HAMMERLESS 1ST MODEL**
NEW IN BOX

.32 SAFETY HAMMERLESS 1ST MODEL

REDBOOK CODE: RB-SW-H-32SH1M

Nickel or (rare) blue finish, known as the "Lemon Squeezer" or the "New Departure," top-break action, enclosed hammer, rounded trigger guard, push-down barrel release latch, rare 2" model known as "Bicycle Gun" has increased value, 6" barrel rare with increased value, 1⅜" barrel extremely rare with highly increased value, 91,417 manufactured, serial numbers: 1-91420.

Production: 1888 - 1902   Caliber: .32 S&W   Action: DAO, revolver

Barrel Length: 2", 3", 3.5", 6"   Sights: pinned round-blade front, notch-cut rear

Capacity: 5   Grips: checkered hard rubber with S&W monograms

D2C:	NIB	$ 540	Ex	$ 425	VG+	$300	Good	$ 270		
C2C:	Mint	$ 520	Ex	$ 375	VG+	$270	Good	$ 245	Fair	$ 125
Trade-In:	Mint	$ 390	Ex	$ 325	VG+	$220	Good	$ 190	Poor	$ 60

32 SAFETY HAMMERLESS 2ND MODEL
EXCELLENT

.32 SAFETY HAMMERLESS 2ND MODEL

REDBOOK CODE: RB-SW-H-32SH2M

Blue or nickel finish, rounded trigger guard, top-break action, changed to a lift-to-open T-shaped barrel latch, "32 S&W CTG" marking on barrel's left side, rare 2" model known as the "Bicycle Gun" has increased value, approximately 78,500 manufactured, serial numbers: 91418-170000.

Production: 1902 - 1909   Caliber: .32 S&W   Action: DAO, revolver

Barrel Length: 2", 3", 3.5", 6"   Sights: pinned blade front, notch-cut rear

Capacity: 5   Grips: checkered hard rubber with S&W monograms

D2C:	NIB	$ 700	Ex	$ 550	VG+	$385	Good	$ 350		
C2C:	Mint	$ 680	Ex	$ 500	VG+	$350	Good	$ 315	Fair	$ 165
Trade-In:	Mint	$ 500	Ex	$ 400	VG+	$280	Good	$ 245	Poor	$ 90

**.32 SAFETY
HAMMERLESS 3RD MODEL**
VERY GOOD +

.32 SAFETY HAMMERLESS 3RD MODEL

REDBOOK CODE: RB-SW-H-32SH3M

Blue or nickel finish, top-break action, lift-to-open T-shaped barrel latch, rounded trigger guard, round-butt frame, typically forged front sight but may be pinned on rare 2" model, 2" model known as the "Bicycle Gun" worth high premium, 1.5" barrel extremely rare with significant increased value, roughly 73,000 manufactured, serial numbers: roughly 170000-243000.

Production: 1909 - 1937   Caliber: .32 S&W   Action: DAO, revolver

Barrel Length: 1.5", 2", 3", 3.5", 6"   Sights: integral forged front, notch-cut rear

Capacity: 5   Grips: checkered hard rubber with S&W monograms

D2C:	NIB	$ 770	Ex	$ 600	VG+	$ 425	Good	$ 385		
C2C:	Mint	$ 740	Ex	$ 550	VG+	$ 390	Good	$ 350	Fair	$ 180
Trade-In:	Mint	$ 550	Ex	$ 450	VG+	$ 310	Good	$ 270	Poor	$ 90

.38 DOUBLE ACTION 3RD MODEL
EXCELLENT

.38 DOUBLE ACTION 3RD MODEL

REDBOOK CODE: RB-SW-H-38DA3M

Blue or nickel finish, top-break action, square-back trigger guard, round butt, no cylinder grooves, extended cylinder flutes, sideplate located on left side of frame. 203,700 manufactured. Serial numbers: 119001-322700. Known to be sold by S&W until at least 1905. Between 1892 and 1905, American Express Co. purchased 1000 models marked "AM. EX. Co." with 3-digit backstrap numbers and these models have increased value. Original 2" factory barrel is extremely rare and has increased value. 8" or 10" barrels rare and worth several times the standard value.

Production: 1884 - 1895   Caliber: .38 S&W   Action: DA/SA, revolver

Barrel Length: 2", 3.25", 4", 5", 6", 8", 10"   Sights: pinned round blade front, notch-cut rear   Capacity: 5   Grips: checkered hard rubber with S&W monograms

D2C:	NIB	$ 530	Ex	$ 425	VG+	$ 295	Good	$ 265		
C2C:	Mint	$ 510	Ex	$ 375	VG+	$270	Good	$ 240	Fair	$ 125
Trade-In:	Mint	$ 380	Ex	$300	VG+	$210	Good	$ 190	Poor	$ 60

SMITH & WESSON

.38 DOUBLE ACTION 4TH MODEL
VERY GOOD +

.38 DOUBLE ACTION 4TH MODEL   REDBOOK CODE: RB-SW-H-38DA4M

Blue or nickel finish, nickel model has blue trigger guard, top-break action, upgraded trigger and sear, increased value for pre-1898 models, 2.5" barrel extremely rare with significantly increased value, 6" barrel rare, 216,300 manufactured, serial numbers: 322700-539000

Production: 1895 - 1909   Caliber: .38 S&W   Action: DA/SA, revolver

Barrel Length: 2.5", 3.25", 4", 5", 6"   Sights: pinned round blade front, notch-cut rear or adjustable target sights   Capacity: 5   Grips: checkered hard rubber with S&W monograms or hard rubber target type with extension and square butt

D2C:	NIB $ 870	Ex $ 675	VG+ $480	Good $ 435		
C2C:	Mint $ 840	Ex $ 625	VG+ $440	Good $ 395	Fair $ 205	
Trade-In:	Mint $ 620	Ex $ 500	VG+ $340	Good $ 305	Poor $ 90	

.38 DOUBLE ACTION 5TH MODEL
VERY GOOD +

.38 DOUBLE ACTION 5TH MODEL   REDBOOK CODE: RB-SW-H-38DA5M

Blue or nickel finish, nickel finish has blue trigger guard, top-break action, target sights are rare with a square front sight, no patent markings on barrel, 15,000 manufactured, serial numbers: 539001-554080.

Production: 1909 - 1911   Caliber: .38 S&W   Action: DA/SA, revolver

Barrel Length: 1.5", 3.25", 4", 5", 6"   Sights: integral-forged round front, notch-cut rear or adjustable target sights (rare)   Capacity: 5   Grips: checkered hard rubber with S&W monograms or walnut target with extension (rare)

D2C:	NIB $ 520	Ex $ 400	VG+ $ 290	Good $ 260		
C2C:	Mint $ 500	Ex $ 375	VG+ $ 260	Good $ 235	Fair $ 120	
Trade-In:	Mint $ 370	Ex $ 300	VG+ $ 210	Good $ 185	Poor $ 60	

.38 DOUBLE ACTION PERFECTED MODEL
EXCELLENT

.38 DOUBLE ACTION PERFECTED MODEL

REDBOOK CODE: RB-SW-H-38DAPM

Blue or nickel finish, the last S&W break-open revolver produced, the only top-break model with trigger guard integral to the frame, the only S&W with a topstrap barrel release and a side thumb release, sideplate located on right side of frame, 59,400 manufactured, serial numbers: 1-59400.

Production: 1909 - 1920   Caliber: .38 S&W   Action: DA/SA, revolver

Barrel Length: 3.25", 4", 5", 6"   Sights: integral-forged round front, notch-cut rear or adjustable target sights (rare)   Capacity: 5   Grips: checkered hard rubber with S&W monograms or walnut target (rare)

D2C:	NIB $ 950	Ex $ 725	VG+ $525	Good $ 475		
C2C:	Mint $ 920	Ex $ 675	VG+ $480	Good $ 430	Fair $ 220	
Trade-In:	Mint $ 680	Ex $ 550	VG+ $380	Good $ 335	Poor $ 120	

.38 SAFETY HAMMERLESS 3RD MODEL
EXCELLENT

.38 SAFETY HAMMERLESS 3RD MODEL

REDBOOK CODE: RB-SW-H-38SH3M

Blue or nickel finish, top-break action, side plate located on left side, 2" and 6" barrels rare with high value, 73,500 manufactured, serial numbers: 42480-116010, 3rd Model changes include the barrel release catch protruding over the frame and pushing down to function. Cross pin was added to trigger guard. "USX" engraved models worth significant premium. "U.S." engraved Army test revolvers very rare and hold extremely substantial value.

Production: 1890 - 1898   Caliber: .38 S&W   Action: DAO, revolver

Barrel Length: 3.25", 4", 5", 6"   Sights: pinned round-blade front, notch-cut rear

Capacity: 5   Grips: checkered hard rubber with S&W monograms

D2C:	NIB $ 525	Ex $ 400	VG+ $290	Good $ 265		
C2C:	Mint $ 510	Ex $ 375	VG+ $270	Good $ 240	Fair $ 125	
Trade-In:	Mint $ 380	Ex $ 300	VG+ $ 210	Good $ 185	Poor $ 60	

**.38 SAFETY HAMMERLESS
4TH MODEL**
EXCELLENT

.38 SAFETY HAMMERLESS 4TH MODEL

REDBOOK CODE: RB-SW-H-38SH4M

Blue or nickel finish, nickel finish model has blue trigger guard and barrel latch, top-break action, lift-to-open T-type barrel latch, left side sideplate, 104,000 manufactured, serial numbers: 116003-~220000

Production: 1898 - 1907 Caliber: .38 S&W Action: DAO, revolver							
Barrel Length: 3.25", 4", 5", 6" Sights: pinned-round blade front, notch-cut rear							
Capacity: 5 Grips: checkered hard rubber with S&W monograms							
D2C:	NIB $ 475	Ex $ 375	VG+ $265	Good $ 240			
C2C:	Mint $ 460	Ex $ 350	VG+ $240	Good $ 215	Fair $ 110		
Trade-In:	Mint $ 340	Ex $ 275	VG+ $ 190	Good $ 170	Poor $ 60		

**.38 SAFETY
HAMMERLESS 5TH MODEL**
EXCELLENT

.38 SAFETY HAMMERLESS 5TH MODEL

REDBOOK CODE: RB-SW-H-38SH5M

Blue or nickel finish, nickel model has blue trigger guard and barrel latch, top-break action, lift-to-open T-shaped barrel latch, 41,500 manufactured, serial numbers: 220000-261495. Barrels have Smith & Wesson marking on right side, "38 S.&W. CTG" marking on left side. Some variations may have a pinned front sight. 2" model known as "Bicycle Gun" has increased value.

Production: 1907 - 1940 Caliber: .38 S&W Action: DAO, revolver							
Barrel Length: 1.5", 2", 3.25", 4", 5", 6" Sights: pinned round-blade front, notch-cut rear Capacity: 5 Grips: checkered hard rubber with or without S&W monograms or checkered walnut							
D2C:	NIB $ 530	Ex $ 425	VG+ $295	Good $ 265			
C2C:	Mint $ 510	Ex $ 375	VG+ $270	Good $ 240	Fair $ 125		
Trade-In:	Mint $ 380	Ex $300	VG+ $ 210	Good $ 190	Poor $ 60		

**.38 SINGLE ACTION
3RD MODEL**
EXCELLENT

.38 SINGLE ACTION 3RD MODEL (MODEL OF 1891)

REDBOOK CODE: RB-SW-H-38SA3M

Blue or nickel finish, top-break action, rounded trigger guard, flared sides on hammer spur, barrel typically marked "Model of 91," 26,850 manufactured, serial numbers: 1-28107. Available with exchangeable single shot barrels in .22, .32, and .38 calibers. Serial numbers include Mexican Model and Single Shot 1st Model.

Production: 1891 - 1911 Caliber: .38 S&W Action: SA, revolver							
Barrel Length: 3.25", 4", 5", 6" Sights: pinned round-blade front, notch-cut rear							
Capacity: 5 Grips: checkered hard rubber with S&W monograms or pearl with gold-plated S&W monograms							
D2C:	NIB $2,200	Ex $1,675	VG+ $1,210	Good $1,100			
C2C:	Mint $2,120	Ex $1,525	VG+ $1,100	Good $ 990	Fair $ 510		
Trade-In:	Mint $1,570	Ex $1,250	VG+ $860	Good $ 770	Poor $ 240		

**.38 SINGLE ACTION
MEXICAN MODEL**
GOOD

.38 SINGLE ACTION MEXICAN MODEL (MODEL OF 1891)

REDBOOK CODE: RB-SW-H-38SAMM

Blue or nickel finish, top-break action, single-shot exchangeable barrels were available, no frame extension around spur trigger, differences from 3rd Model include: flat sided hammer, non-integral spur trigger separate from frame, lack of half-cock notch on hammer, serial numbers run concurrent with 3rd Model, roughly 2,000 manufactured, serial numbers: 1-28107. Many forgeries in circulation. Conversion kits were available to convert 3rd Models to Mexican Models which are not original Mexican Models.

Production: 1891 - 1911 Caliber: .38 S&W Action: SA, revolver							
Barrel Length: 3.25", 4", 5", 6" Sights: pinned round-blade front, notch-cut rear							
Capacity: 5 Grips: checkered hard rubber with or without S&W monograms or checkered walnut							
D2C:	NIB $6,800	Ex $6,200	VG+ $5,310	Good $4,100			
C2C:	Mint $4,040	Ex $2,900	VG+ $2,100	Good $1,890	Fair $ 970		
Trade-In:	Mint $2,990	Ex $2,375	VG+ $1,640	Good $1,470	Poor $ 420		

**.44 DOUBLE ACTION
FIRST MODEL**
EXCELLENT

.44 DOUBLE ACTION FIRST MODEL (NEW MODEL NAVY NO. 3 REVOLVER) REDBOOK CODE: RB-SW-H-44DAFM

Blue or nickel finish, top-break action, round butt, ribbed barrel, target sight model has significantly higher value, 53,590 manufactured, serial numbers: 1-54670. The Wesson Favorite Model is included in serial number range but has much higher value. Other caliber options are very rare and worth high premium. 1 7/8" cylinder on early models and 1 9/16" on later models. Considered an antique as all frames of this model were manufactured no later than 1898.

Production: 1881 - 1913 Caliber: .44 S&W Russian Action: DA/SA, revolver

Barrel Length: 4", 5", 6", 6.5" Sights: service or target Capacity: 6

Grips: checkered hard rubber with S&W monograms or walnut

D2C:	NIB $3,800	Ex $2,900	VG+ $2,090	Good $1,900		
C2C:	Mint $3,650	Ex $2,625	VG+ $1,900	Good $1,710	Fair $ 875	
Trade-In:	Mint $2,700	Ex $2,150	VG+ $1,490	Good $1,330	Poor $ 390	

**.44 DOUBLE ACTION
FRONTIER**
EXCELLENT

.44 DOUBLE ACTION FRONTIER

REDBOOK CODE: RB-SW-H-44DAFR

Blue or nickel finish, top-break action, patent date markings dropped after 1900, 15,340 manufactured, serial numbers: 1-15340. Most later production models marked "44 Winchester Ctg." on left side of barrel. Considered an antique as all frames of this model were manufactured no later than 1898.

Production: 1886 - 1913 Caliber: .44-40 Win. (WCF) Action: DA/SA, revolver

Barrel Length: 4", 5", 6", 6.5" Sights: service or target Capacity: 6

Grips: checkered hard rubber with S&W monograms or walnut

D2C:	NIB $4,400	Ex $3,350	VG+ $2,420	Good $2,200		
C2C:	Mint $4,230	Ex $3,050	VG+ $2,200	Good $1,980	Fair $1,015	
Trade-In:	Mint $3,130	Ex $2,475	VG+ $1,720	Good $1,540	Poor $ 450	

**.38 WINCHESTER
DOUBLE ACTION**
VERY GOOD +

.38 WINCHESTER DOUBLE ACTION

REDBOOK CODE: RB-SW-H-38WNDA

Blue or nickel finish, top-break action, round butt, same configuration as .44 Double Action Frontier, 276 manufactured, serial numbers 1-276. "38 Winchester Ctg" marking on most barrels.

Production: 1900 - 1910 Caliber: .38-40 Win. (WCF) Action: DA/SA, revolver

Barrel Length: 4", 5", 6.5" Sights: service or target Capacity: 6

Grips: checkered hard rubber with S&W monograms or walnut

D2C:	NIB $7,100	Ex $5,400	VG+ $3,905	Good $3,550		
C2C:	Mint $6,820	Ex $4,900	VG+ $3,550	Good $3,195	Fair $1,635	
Trade-In:	Mint $5,050	Ex $4,000	VG+ $2,770	Good $2,485	Poor $ 720	

.22 LADYSMITH 1ST MODEL
GOOD

.22 LADYSMITH 1ST MODEL (MODEL 22 HAND EJECTOR)

REDBOOK CODE: RB-SW-H-22LS1M

Blue or nickel finish, small "M" frame, left side cylinder release on frame, round butt, 4,575 manufactured, serial numbers: 1-4575. Barrel lengths under 3" worth significant premium.

Production: 1902 - 1906 Caliber: .22 Long (.22 S&W) Action: DA/SA, revolver

Barrel Length: 2.25", 3", 3.5" Wt.: 9.5 oz.

Sights: round-blade front, notch-cut rear Capacity: 7 Grips: checkered hard rubber

D2C:	NIB $2,300	Ex $1,750	VG+ $1,265	Good $1,150		
C2C:	Mint $2,210	Ex $1,600	VG+ $1,150	Good $1,035	Fair $ 530	
Trade-In:	Mint $1,640	Ex $1,300	VG+ $900	Good $ 805	Poor $ 240	

SMITH & WESSON

.22 LADYSMITH 2ND MODEL
VERY GOOD +

.22 LADYSMITH 3RD MODEL
VERY GOOD +

.22/.32 KIT GUN
EXCELLENT

.22 LADYSMITH 2ND MODEL (MODEL 22 HAND EJECTOR)
REDBOOK CODE: RB-SW-H-22LS2M
Blue or nickel finish, rounded butt, cylinder locked and released by ejector rod, 9,374 manufactured, serial numbers: 4575-13950.

Production: 1906 - 1910 Caliber: .22 Long (.22 S&W) Action: DA/SA, revolver
Barrel Length: 3", 3.5" Sights: forged round blade front, notch-cut rear or target sights
Capacity: 7 Grips: hard rubber, pearl or ivory

D2C:	NIB	$2,600	Ex	$2,000	VG+	$1,430	Good	$1,300	
C2C:	Mint	$2,500	Ex	$1,800	VG+	$1,300	Good	$1,170	Fair $ 600
Trade-In:	Mint	$1,850	Ex	$1,475	VG+	$1,020	Good	$ 910	Poor $ 270

.22 LADYSMITH 3RD MODEL (MODEL 22 HAND EJECTOR)
REDBOOK CODE: RB-SW-H-22LS3M
Blue or nickel finish, small "M" frame, square butt, pinned barrel, ".22 S&W CTG" marked on barrel, ejector rod opens cylinder release, 12,203 manufactured, serial numbers: 13951-26154, 2.25" barrels have high value. 6" and 10" barrels have increased value. Target sights add significant value.

Production: 1910 - 1921 Caliber: .22 Long (.22 S&W) Action: DA/SA, revolver
Barrel Length: 2.25", 3", 3.5", 6", 10" Sights: fixed (standard model) or
pinned Paine front, adjustable rear (target model) Capacity: 7
Grips: smooth walnut with gold S&W monograms, pearl or ivory (rare)

D2C:	NIB	$2,100	Ex	$1,600	VG+	$1,155	Good	$1,050	
C2C:	Mint	$2,020	Ex	$1,450	VG+	$1,050	Good	$ 945	Fair $ 485
Trade-In:	Mint	$1,500	Ex	$1,200	VG+	$820	Good	$ 735	Poor $ 210

.22/.32 KIT GUN (PREWAR) REDBOOK CODE: RB-SW-H-2232KG
Blue or nickel finish, pinned barrel, built on "I" frame, serial numbers range between: 525670-536684.

Production: 1935 - 1941 Caliber: .22 LR Action: DA/SA, revolver Barrel Length: 4"
Wt.: 21 oz. Sights: Patridge or USRA front, adjustable rear Capacity: 6
Grips: checkered walnut with S&W monograms, round or extension type

D2C:	NIB	$5,400	Ex	$4,125	VG+	$2,970	Good	$2,700	
C2C:	Mint	$5,190	Ex	$3,750	VG+	$2,700	Good	$2,430	Fair $1,245
Trade-In:	Mint	$3,840	Ex	$3,025	VG+	$2,110	Good	$1,890	Poor $ 540

.22/.32 KIT GUN (POSTWAR) REDBOOK CODE: RB-SW-H-2232KP
Blue or nickel finish, built on "I" frame, hammer block, serial numbers range between: 536685-590000, 2" barrel has high value.

Production: 1946 - 1952 Caliber: .22 LR Action: DA/SA, revolver
Barrel Length: 2", 4" Sights: Patridge or USRA front, adjustable rear
Capacity: 6 Grips: checkered walnut with S&W monograms, small square or extension type

D2C:	NIB	$3,700	Ex	$2,825	VG+	$2,035	Good	$1,850	
C2C:	Mint	$3,560	Ex	$2,575	VG+	$1,850	Good	$1,665	Fair $ 855
Trade-In:	Mint	$2,630	Ex	$2,075	VG+	$1,450	Good	$1,295	Poor $ 390

MODEL OF 1953 .22/.32 KIT GUN
EXCELLENT

MODEL OF 1953 .22/.32 KIT GUN (PRE-MODEL 34)

REDBOOK CODE: RB-SW-H-PREM34

Blue or nickel finish, improved "I" frame, "J" frame on later models, pinned barrel, serial numbers range between: 101-135465, continued production in 1957 as the Model 34.

Production: 1953 - 1957		Caliber: .22 LR		Action: DA/SA, revolver					
Barrel Length: 2", 4"		Sights: ramped front, adjustable rear		Capacity: 6					
Grips: diamond checkered walnut									
D2C:	NIB	$ 1,150	Ex $ 875	VG+ $ 635	Good $ 575				
C2C:	Mint	$ 1,110	Ex $ 800	VG+ $ 580	Good $ 520	Fair $ 265			
Trade-In:	Mint $ 820		Ex $ 650	VG+ $ 450	Good $ 405	Poor $ 120			

.22/.32 KIT GUN AIRWEIGHT
EXCELLENT

.22/.32 KIT GUN AIRWEIGHT (PRE-MODEL 43)

REDBOOK CODE: RB-SW-H-PREM43

Blue or nickel (rare) finish, aluminum alloy "J" frame, no markings of "Airweight," serial numbers range between: 5000-135465, 2" barrel has increased value, nickel model is rare with high value, continued production in 1957 as the Model 43.

Production: 1955 - 1957		Caliber: .22 LR, .22 WMR		Action: DA/SA, revolver					
Barrel Length: 2", 3.5"		Sights: ramped front, adjustable rear		Capacity: 6					
Grips: diamond checkered walnut									
D2C:	NIB	$ 1,150	Ex $ 875	VG+ $ 635	Good $ 575				
C2C:	Mint	$ 1,110	Ex $ 800	VG+ $ 580	Good $ 520	Fair $ 265			
Trade-In:	Mint $ 820		Ex $ 650	VG+ $ 450	Good $ 405	Poor $ 120			

.32 HAND EJECTOR 1ST MODEL
EXCELLENT

.32 HAND EJECTOR 1ST MODEL (MODEL 1 OR MODEL OF 1896)

REDBOOK CODE: RB-SW-H-32HE1M

Blue or nickel finish, "I" frame, top strap cylinder stop, extractor rod releases cylinder, 19,712 manufactured, serial numbers: 1-19715, S&W's first swing-out cylinder, target model has increased value.

Production: 1896 - 1903		Caliber: .32 S&W Long		Action: DA/SA, revolver			
Barrel Length: 3.25", 4.25", 6"		Sights: pinned round-blade front, pinned rear or target sights (rare)		Capacity: 6		Grips: round butt hard rubber with S&W monogram or target extension type	
D2C:	NIB	$1,300	Ex $1,000	VG+ $ 715	Good $ 650		
C2C:	Mint	$1,250	Ex $ 900	VG+ $ 650	Good $ 585	Fair $ 300	
Trade-In:	Mint $ 930		Ex $ 750	VG+ $ 510	Good $ 455	Poor $ 150	

.32 HAND EJECTOR 2ND MODEL (MODEL OF 1903)

REDBOOK CODE: RB-SW-H-32HE2M

Blue or nickel finish, cylinder stop on bottomstrap of frame, pinned barrel, extractor rod locking lug, thumb-latch cylinder release on left side of frame, 19,425 manufactured, serial numbers: 1-19430, target model has a significantly increased value.

Production: 1903 - 1904		Caliber: .32 S&W Long		Action: DA/SA, revolver			
Barrel Length: 3.25", 4.25", 6"		Sights: forged round blade front, notch-cut rear or target sights		Capacity: 6		Grips: round butt hard rubber with S&W monogram or target extension type	
D2C:	NIB	$1,100	Ex $ 850	VG+ $ 605	Good $ 550		
C2C:	Mint	$1,060	Ex $ 775	VG+ $ 550	Good $ 495	Fair $ 255	
Trade-In:	Mint $ 790		Ex $ 625	VG+ $ 430	Good $ 385	Poor $ 120	

SMITH & WESSON

**.32 HAND EJECTOR
2ND MODEL**
VERY GOOD +

.32 HAND EJECTOR 2ND MODEL

(MODEL OF 1903 - 1ST CHANGE) REDBOOK CODE: RB-SW-H-32HE21

Blue or nickel finish, same as 2nd Model with a few internal changes, 31,700 manufactured, serial numbers between: 19426-51125.

Production: 1904 - 1906 Caliber: .32 S&W Long Action: DA/SA, revolver

Barrel Length: 3.25", 4.25", 6" Sights: forged round-blade front, notch-cut rear or target sights Capacity: 6 Grips: round butt hard rubber with S&W monogram or target extension type

D2C:	NIB	$1,100	Ex $	850	VG+	$605	Good $	550	
C2C:	Mint	$1,060	Ex $	775	VG+	$550	Good $	495	Fair $ 255
Trade-In:	Mint $	790	Ex $	625	VG+	$430	Good $	385	Poor $ 120

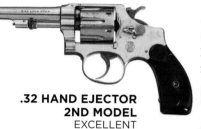

**.32 HAND EJECTOR
2ND MODEL**
EXCELLENT

.32 HAND EJECTOR 2ND MODEL

(MODEL OF 1903 - 2ND CHANGE) REDBOOK CODE: RB-SW-H-32HE22

Minor differences from 1st Change model, target model has increased value, 44,373 manufactured, serial numbers between: 51125-95500.

Production: 1906 - 1909 Caliber: .32 S&W Long Action: DA/SA, revolver

Barrel Length: 3.25", 4.25", 6" Sights: forged round-blade front, notch-cut rear or target sights Capacity: 6 Grips: round-butt hard rubber with S&W monogram or target extension type

D2C:	NIB	$1,100	Ex $	850	VG+ $	605	Good $	550	
C2C:	Mint	$1,060	Ex $	775	VG+ $	550	Good $	495	Fair $ 255
Trade-In:	Mint $	790	Ex $	625	VG+ $	430	Good $	385	Poor $ 120

.32 HAND EJECTOR 2ND MODEL

(MODEL OF 1903 - 3RD CHANGE) REDBOOK CODE: RB-SW-H-32HE23

Minor differences from 2nd Change model, target model has much higher value, 624 manufactured, serial numbers between: 95500-96125.

Production: 1909 - 1910 Caliber: .32 S&W Long Action: DA/SA, revolver

Barrel Length: 3.25", 4.25", 6" Sights: forged round-blade front, notch-cut rear or target sights Capacity: 6 Grips: round butt hard rubber with S&W monogram or target extension type

D2C:	NIB	$1,100	Ex $	850	VG+ $	605	Good $	550	
C2C:	Mint	$1,060	Ex $	775	VG+ $	550	Good $	495	Fair $ 255
Trade-In:	Mint $	790	Ex $	625	VG+ $	430	Good $	385	Poor $ 120

.32 HAND EJECTOR 2ND MODEL

(MODEL OF 1903 - 4TH CHANGE) REDBOOK CODE: RB-SW-H-32HE24

Minor differences from 3rd Change model, target model has increased value, 6,374 manufactured, serial numbers between: 96125-102500.

Production: 1910 Caliber: .32 S&W Long Action: DA/SA, revolver

Barrel Length: 3.25", 4.25", 6" Sights: forged round-blade front, notch-cut rear or target sights Capacity: 6 Grips: round butt hard rubber with S&W monogram or target extension type

D2C:	NIB	$1,100	Ex $	850	VG+ $	605	Good $	550	
C2C:	Mint	$1,060	Ex $	775	VG+ $	550	Good $	495	Fair $ 255
Trade-In:	Mint $	790	Ex $	625	VG+ $	430	Good $	385	Poor $ 120

**.32 HAND EJECTOR
2ND MODEL**
EXCELLENT

.32 HAND EJECTOR 2ND MODEL

(MODEL OF 1903 - 5TH CHANGE)   REDBOOK CODE: RB-SW-H-32HE25

Minor differences from 4th Change model, target model has much higher value, 160,499 manufactured, serial numbers between: 102500-263000.

Production: 1910 - 1917   Caliber: .32 S&W Long   Action: DA/SA, revolver

Barrel Length: 3.25", 4.25", 6"   Sights: forged round-blade front, notch-cut rear or target sights   Capacity: 6   Grips: round-butt hard rubber with S&W monogram or target extension type

D2C:	NIB	$1,100	Ex $ 850	VG+ $ 605	Good $ 550			
C2C:	Mint	$1,060	Ex $ 775	VG+ $ 550	Good $ 495	Fair $ 255		
Trade-In:	Mint $ 790		Ex $ 625	VG+ $ 430	Good $ 385	Poor $ 120		

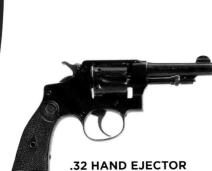

**.32 HAND EJECTOR
3RD MODEL**
VERY GOOD +

.32 HAND EJECTOR 3RD MODEL

REDBOOK CODE: RB-SW-H-32HE3M

Blue or nickel finish, "I" frame, hammer block arm, nearly identical to Model of 1903 5th Change, serial number range concurrent with .32 Regulation Police (Pre-War): 263000-536685, target model has a higher value.

Production: 1911 - 1942   Caliber: .32 S&W Long   Action: DA/SA, revolver

Barrel Length: 3.25", 4.25", 6"   Sights: round-blade front, square-notch rear

Capacity: 6   Grips: diamond checkered walnut or black rubber with S&W monograms

D2C:	NIB	$1,000	Ex $ 775	VG+ $550	Good $ 500		
C2C:	Mint $ 960		Ex $700	VG+ $500	Good $ 450	Fair $ 230	
Trade-In:	Mint $ 710		Ex $ 575	VG+ $390	Good $ 350	Poor $ 120	

.32 HAND EJECTOR
MINT

.32 HAND EJECTOR (POSTWAR) (PRE-MODEL 30)

REDBOOK CODE: RB-SW-H-PREM30

Blue or nickel finish, "I" frame, five-screw or improved four-screw frame, serial numbers run concurrent with .32 Regulation Police (Postwar): 536685-712955, continued as the Model 30 in 1957.

Production: 1946 - 1957   Caliber: .32 S&W Long   Action: DA/SA, revolver

Barrel Length: 2", 3.25", 4.25", 6"   Sights: round-blade front, square-notch rear

Capacity: 6   Grips: checkered walnut with S&W monograms

D2C:	NIB $ 750		Ex $ 575	VG+ $ 415	Good $ 375		
C2C:	Mint $ 720		Ex $ 525	VG+ $380	Good $ 340	Fair $ 175	
Trade-In:	Mint $ 540		Ex $ 425	VG+ $300	Good $ 265	Poor $ 90	

**.32 REGULATION
POLICE**
GOOD

.32 REGULATION POLICE (PREWAR)   REDBOOK CODE: RB-SW-H-32RPPW

Blue or nickel finish, "I" frame with five-screws, shouldered backstrap, new type of grip stock not interchangeable, serial numbers run concurrent with .32 Hand Ejector 3rd Model: ~260000-536000, target model has significantly higher value.

Production: 1917 - 1942   Caliber: .32 S&W Long   Action: DA/SA, revolver

Barrel Length: 3.25", 4.25", 6"   Sights: round-blade front, square-notch rear or target sights   Capacity: 6   Grips: diamond checkered walnut with S&W monograms

D2C:	NIB $ 810		Ex $625	VG+ $450	Good $ 405		
C2C:	Mint $ 780		Ex $ 575	VG+ $ 410	Good $ 365	Fair $ 190	
Trade-In:	Mint $ 580		Ex $ 475	VG+ $320	Good $ 285	Poor $ 90	

.32 REGULATION POLICE
VERY GOOD +

.32 REGULATION POLICE (POSTWAR) (PRE-MODEL 31)

REDBOOK CODE: RB-SW-H-PREM31

Blue or nickel finish, "I" frame with either 4 or 5 screw design, later changed to "J" frame, serial numbers concurrent with .32 Hand Ejector 3rd Model: 536000-712955, continued as the Model 31 in 1957.

Production: 1946 - 1957   Caliber: .32 S&W Long   Action: DA/SA, revolver

Barrel Length: 3.25", 4", 4.25", 6"   Sights: round-blade front, square-notch rear

Capacity: 6   Grips: diamond checkered walnut with S&W monograms

D2C:	NIB	$ 730	Ex	$ 575	VG+	$405	Good	$ 365		
C2C:	Mint	$ 710	Ex	$ 525	VG+	$370	Good	$ 330	Fair	$ 170
Trade-In:	Mint	$ 520	Ex	$ 425	VG+	$290	Good	$ 260	Poor	$ 90

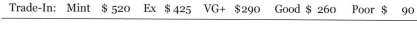

K-32 HAND EJECTOR (POSTWAR) (K32 MASTERPIECE)

(PRE-MODEL 16)   REDBOOK CODE: RB-SW-H-PREM16

"K" frame with five-screws, narrow rib barrel until 1950, serial numbers range in the K series, continued as the Model 16 in 1957.

Production: 1946 - 1957   Caliber: .32 S&W Long   Action: DA/SA, revolver

Barrel Length: 4", 6"   Sights: Patridge front, adjustable micrometer rear

Capacity: 6   Grips: diamond checkered walnut with S&W monograms

D2C:	NIB	$3,100	Ex	$2,375	VG+	$1,705	Good	$1,550		
C2C:	Mint	$2,980	Ex	$2,150	VG+	$1,550	Good	$1,395	Fair	$ 715
Trade-In:	Mint	$2,210	Ex	$1,750	VG+	$1,210	Good	$1,085	Poor	$ 330

.32 MILITARY & POLICE   REDBOOK CODE: RB-SW-H-32MIPO

"K" frame with five-screws, "Made in USA" marking on frame, serial numbers range in the .38 M&P series, 4,815 or so manufactured.

Production: 1948 - 1950   Caliber: .32 S&W Long   Action: DA/SA, revolver

Barrel Length: 2", 4", 5"   Sights: round-blade front, square-notch rear

Capacity: 6   Grips: diamond checkered walnut with S&W monograms

D2C:	NIB	$3,000	Ex	$2,300	VG+	$1,650	Good	$1,500		
C2C:	Mint	$2,880	Ex	$2,075	VG+	$1,500	Good	$1,350	Fair	$ 690
Trade-In:	Mint	$2,130	Ex	$1,700	VG+	$1,170	Good	$1,050	Poor	$ 300

.32-20 HAND EJECTOR - 1ST MODEL

REDBOOK CODE: RB-SW-H-3220H1

Blue or nickel finish, "K" frame with four-screws, yoke stop pin, pinned barrel without locking lug, ~5,310 manufactured, serial numbers: 1-5310, target model has significantly increased value.

Production: 1899 - 1902   Caliber: .32-20 Win. (WCF)   Action: DA/SA, revolver

Barrel Length: 4", 5", 6", 6.5"   Sights: round-blade front, square-notch rear

Capacity: 6   Grips: checkered hard rubber with S&W monograms or walnut

D2C:	NIB	$2,680	Ex	$2,050	VG+	$1,475	Good	$1,340		
C2C:	Mint	$2,580	Ex	$1,850	VG+	$1,340	Good	$1,210	Fair	$ 620
Trade-In:	Mint	$1,910	Ex	$1,525	VG+	$1,050	Good	$ 940	Poor	$ 270

.32-20 HAND EJECTOR - 2ND MODEL
VERY GOOD +

.32-20 HAND EJECTOR - 2ND MODEL (MODEL OF 1902)

REDBOOK CODE: RB-SW-H-3220H2

Blue or nickel finish, similar to 1st Model, added under barrel locking lug, knurled knob on extractor rod, 4,500 manufactured, serial numbers: 5312-9811, target model has significantly increased value.

Production: 1902 - 1905	Caliber: .32-20 Win. (WCF)	Action: DA/SA, revolver		
Barrel Length: 4", 5", 6.5"	Sights: round-blade front, square-notch rear			
Capacity: 6	Grips: checkered hard rubber with S&W monograms or walnut			
D2C:	NIB $1,950	Ex $1,500	VG+ $1,075	Good $ 975
C2C:	Mint $1,880	Ex $1,350	VG+ $980	Good $ 880	Fair $ 450
Trade-In:	Mint $1,390	Ex $1,100	VG+ $ 770	Good $ 685	Poor $ 210

.32-20 HAND EJECTOR - MODEL OF 1902 - 1ST CHANGE

REDBOOK CODE: RB-SW-H-322021

Blue or nickel finish, "K" frame with four-screws, changes from 2nd Model include square butt on walnut grip model and larger diameter barrel, enlarged area of frame where barrel attaches, 8,313 manufactured, serial numbers: 9812-18125, target model has significantly increased value.

Production: 1903 - 1905	Caliber: .32-20 Win. (WCF)	Action: DA/SA, revolver		
Barrel Length: 4", 5", 6.5"	Sights: round-blade front, square-notch rear			
Capacity: 6	Grips: checkered hard rubber with S&W monograms or walnut			
D2C:	NIB $1,800	Ex $1,375	VG+ $ 990	Good $ 900
C2C:	Mint $1,730	Ex $1,250	VG+ $ 900	Good $ 810	Fair $ 415
Trade-In:	Mint $1,280	Ex $1,025	VG+ $ 710	Good $ 630	Poor $ 180

.32-20 HAND EJECTOR - 3RD MODEL (MODEL OF 1905)

REDBOOK CODE: RB-SW-H-32203M

Similar to previous Model of 1902 in appearance, several internal changes including trigger guard screw, 4,300 manufactured, serial numbers: 18126-22426, target model has significantly increased value.

Production: 1905 - 1906	Caliber: .32-20 Win. (WCF)	Action: DA/SA, revolver		
Barrel Length: 4", 5", 6.5"	Sights: round-blade front, square-notch rear			
Capacity: 6	Grips: checkered hard rubber with S&W monograms or walnut			
D2C:	NIB $1,800	Ex $1,375	VG+ $ 990	Good $ 900
C2C:	Mint $1,730	Ex $1,250	VG+ $ 900	Good $ 810	Fair $ 415
Trade-In:	Mint $1,280	Ex $1,025	VG+ $ 710	Good $ 630	Poor $ 180

.32-20 HAND EJECTOR - 3RD MODEL

(MODEL OF 1905 - 1ST CHANGE) REDBOOK CODE: RB-SW-H-322031

Similar to previous Model of 1905 with several internal changes, patent date markings on barrel, rebound slide, ~11,073 manufactured, serial numbers concurrent with 2nd change and range between: 22427-33500, target model has significantly increased value.

Production: 1906 - 1909	Caliber: .32-20 Win. (WCF)	Action: DA/SA, revolver		
Barrel Length: 4", 5", 6", 6.5"	Sights: round-blade front, square-notch rear			
Capacity: 6	Grips: checkered hard rubber with S&W monograms or walnut			
D2C:	NIB $1,800	Ex $1,375	VG+ $ 990	Good $ 900
C2C:	Mint $1,730	Ex $1,250	VG+ $ 900	Good $ 810	Fair $ 415
Trade-In:	Mint $1,280	Ex $1,025	VG+ $ 710	Good $ 630	Poor $ 180

.32-20 HAND EJECTOR - 3RD MODEL

(MODEL OF 1905 - 2ND CHANGE) REDBOOK CODE: RB-SW-H-322032
Five-screw "K" frame, several internal changes from previous model, star extractor guide pins added, ~11,700 manufactured, serial numbers concurrent with 1st Change model range between: 33501-45200, target model has significantly increased value.

Production: 1906 - 1909 Caliber: .32-20 Win. (WCF) Action: DA/SA, revolver
Barrel Length: 4", 5", 6", 6.5" Sights: round-blade front, square-notch rear
Capacity: 6 Grips: checkered hard rubber with S&W monograms or walnut

D2C:	NIB	$1,800	Ex	$1,375	VG+	$ 990	Good	$ 900	
C2C:	Mint	$1,730	Ex	$1,250	VG+	$ 900	Good	$ 810	Fair $ 415
Trade-In:	Mint	$1,280	Ex	$1,025	VG+	$ 710	Good	$ 630	Poor $ 180

.32-20 HAND EJECTOR - 3RD MODEL

(MODEL OF 1905 - 3RD CHANGE) REDBOOK CODE: RB-SW-H-322033
Similar to previous model with internal changes, no patent dates until later production, ~20,500 manufactured, serial numbers range between: 45201-65700, target model has significantly increased value.

Production: 1909 - 1915 Caliber: .32-20 Win. (WCF) Action: DA/SA, revolver
Barrel Length: 4", 6" Sights: round-blade front, square-notch rear Capacity: 6
Grips: checkered hard rubber with S&W monograms or walnut

D2C:	NIB	$1,800	Ex	$1,375	VG+	$ 990	Good	$ 900	
C2C:	Mint	$1,730	Ex	$1,250	VG+	$ 900	Good	$ 810	Fair $ 415
Trade-In:	Mint	$1,280	Ex	$1,025	VG+	$ 710	Good	$ 630	Poor $ 180

.32-20 HAND EJECTOR - 3RD MODEL GOOD

.32-20 HAND EJECTOR - 3RD MODEL

(MODEL OF 1905 - 4TH CHANGE) REDBOOK CODE: RB-SW-H-322034
Slight internal variations from previous model, 78,985 manufactured, serial numbers range between: 65700-144684, target model has significantly increased value.

Production: 1915 - 1940 Caliber: .32-20 Win. (WCF) Action: DA/SA, revolver
Barrel Length: 4", 5", 6" Sights: round-blade front, square-notch rear
Capacity: 6 Grips: checkered hard rubber with S&W monograms or walnut

D2C:	NIB	$1,800	Ex	$1,375	VG+	$ 990	Good	$ 900	
C2C:	Mint	$1,730	Ex	$1,250	VG+	$ 900	Good	$ 810	Fair $ 415
Trade-In:	Mint	$1,280	Ex	$1,025	VG+	$ 710	Good	$ 630	Poor $ 180

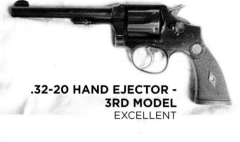
.32-20 HAND EJECTOR - 3RD MODEL EXCELLENT

.357 REGISTERED MAGNUM REDBOOK CODE: RB-SW-H-357RMG

Custom ordered and factory registered, blue or nickel finish, barrel lengths available in 1/4" increments, any combination of sights or grips and custom engraving available, counterbored cylinder, checkered topstrap, marked "S&W .357 Magnum" on right side, all Registered Magnums begin with "Reg" prefix before registration number, 5,500 manufactured, serial numbers range between the .44 Hand Ejector 3rd Model series, original box and original registration certificate must be present for full price at mint or excellent grades, many other custom options exist as this model was made-to-order with registration certificate.

Production: 1935 - 1939 Caliber: .357 Magnum Action: DA/SA, revolver
Barrel Length: 3.5", 4", 5", 6", 6.5", 8.75" or custom
Sights: King or Baughman front with any combination of six available main sights
Capacity: 6 Grips: checkered walnut, Magna, ivory, or other custom options

D2C:	NIB	$10,000	Ex	$9,200	VG+	$4,500	Good	$2,500	
C2C:	Mint	$9,500	Ex	$8,300	VG+	$4,300	Good	$2,400	Fair $1,500
Trade-In:	Mint	$7,500	Ex	$6,900	VG+	$3,100	Good	$1,750	Poor $ 900

.357 REGISTERED MAGNUM EXCELLENT

.357 MAGNUM PREWAR NON-REGISTERED

REDBOOK CODE: RB-SW-H-357MPN

Similar to Registered Magnum model without registration number marked in yoke, standardized barrel lengths, ~1,400 manufactured, custom barrel lengths have high value, boxes are same as Registered Magnum models, very rare nickel finish has increased value. Matching factory box must be present for mint grade value.

Production: 1938 - 1941 Caliber: .357 Magnum Action: DA/SA, revolver

Barrel Length: 3.5", 5", 6", 6.5", 8.375"   Sights: King or Baughman ramped front,

adjustable rear or custom options Capacity: 6 Grips: checkered walnut or custom

D2C:	NIB	$9,550	Ex	$7,275	VG+	$5,255	Good	$4,775	
C2C:	Mint	$9,170	Ex	$6,600	VG+	$4,780	Good	$4,300	Fair $2,200
Trade-In:	Mint	$6,790	Ex	$5,350	VG+	$3,730	Good	$3,345	Poor $ 960

.357 MAGNUM
EXCELLENT

.357 MAGNUM (PRE-MODEL 27) (POSTWAR)

REDBOOK CODE: RB-SW-H-PREM27

Similar to prewar model, blue or (rare) nickel finish, introduced a short-throw hammer, "N" frame with five-screws, continued as the Model 27 in 1957.

Production: 1950 - 1957 Caliber: .357 Magnum Action: DA/SA, revolver

Barrel Length: 3.5", 5", 6", 6.5", 8.375" Sights: King or Baughman ramped front,

adjustable micrometer rear or custom options Capacity: 6 Grips: checkered walnut

with S&W medallions or target

D2C:	NIB	$1,850	Ex	$1,425	VG+	$1,020	Good	$ 925	
C2C:	Mint	$1,780	Ex	$1,300	VG+	$930	Good	$ 835	Fair $ 430
Trade-In:	Mint	$1,320	Ex	$1,050	VG+	$730	Good	$ 650	Poor $ 210

HIGHWAY PATROLMAN (PRE-MODEL 28)

REDBOOK CODE: RB-SW-H-PREM28

Blue finish, "N" frame with five-screws, similar to Pre-Model 27 but with no serrations on barrel rib or rear sight and lacks highly polished finish, 6.5" barrel reduced to 6" on later models, serial numbers in "S" prefix series.

Production: 1954 - 1957 Caliber: .357 Magnum Action: DA/SA, revolver

Barrel Length: 4", 6", 6.5"   Sights: Baughman Quick Draw front, adjustable

micrometer rear Capacity: 6 Grips: Magna or diamond checkered walnut target grips

D2C:	NIB	$ 1,150	Ex	$ 875	VG+	$ 635	Good	$ 575	
C2C:	Mint	$ 1,110	Ex	$ 800	VG+	$ 580	Good	$ 520	Fair $ 265
Trade-In:	Mint	$ 820	Ex	$ 650	VG+	$ 450	Good	$ 405	Poor $ 120

.357 COMBAT MAGNUM (PRE-MODEL 19)

REDBOOK CODE: RB-SW-H-PREM19

Blue or nickel finish, built on "K" frame with four-screws, serial numbers have "K" series prefix, continued as the Model 19 in 1957.

Production: 1955 - 1957 Caliber: .357 Magnum Action: DA/SA, revolver

Barrel Length: 4" Sights: Baughman Quick Draw front, adjustable micrometer rear

Capacity: 6 Grips: diamond checkered walnut with S&W monograms

D2C:	NIB	$1,900	Ex	$1,450	VG+	$1,045	Good	$ 950	
C2C:	Mint	$1,830	Ex	$1,325	VG+	$ 950	Good	$ 855	Fair $ 440
Trade-In:	Mint	$1,350	Ex	$1,075	VG+	$ 750	Good	$ 665	Poor $ 210

**.38 MILITARY &
POLICE 1ST MODEL**
GOOD

.38 MILITARY & POLICE 1ST MODEL

(MODEL OF 1899) REDBOOK CODE: RB-SW-H-38MP1M

Blue or nickel finish, four-screw "K" series frame, no locking lug, walnut grip features impressed circle in the top portion, target model worth significant premium, 20,980 manufactured, serial numbers: 1-20975.

Production: 1899 - 1902 Caliber: .38 S&W Special, .38 Long Colt Action: DA/SA, revolver

Barrel Length: 4", 5", 6", 6.5" Sights: fixed service or adjustable target Capacity: 6

Grips: checkered hard rubber with S&W monograms or walnut with impressed circle

D2C:	NIB	$1,350	Ex	$1,050	VG+	$ 745	Good	$ 675	
C2C:	Mint	$1,300	Ex	$ 950	VG+	$ 680	Good	$ 610	Fair	$ 315
Trade-In:	Mint	$ 960	Ex	$ 775	VG+	$ 530	Good	$ 475	Poor	$ 150

**.38 MILITARY &
POLICE 2ND MODEL**
EXCELLENT

.38 MILITARY & POLICE 2ND MODEL

(MODEL OF 1902) REDBOOK CODE: RB-SW-H-38MP2M

Blue or nickel finish, four-screw "K" frame, added locking lug on barrel underside, enlarged extractor rod, 12,827 manufactured, serial numbers range: 20976-33800, target model worth significant premium.

Production: 1902 - 1903 Caliber: .38 S&W Special Action: DA/SA, revolver

Barrel Length: 4", 5", 6", 6.5" Sights: fixed service or adjustable target

Capacity: 6 Grips: checkered hard rubber with S&W monograms or checkered walnut

D2C:	NIB	$1,580	Ex	$1,225	VG+	$ 870	Good	$ 790	
C2C:	Mint	$1,520	Ex	$1,100	VG+	$ 790	Good	$ 715	Fair	$ 365
Trade-In:	Mint	$1,130	Ex	$ 900	VG+	$ 620	Good	$ 555	Poor	$ 180

**.38 MILITARY & POLICE
2ND MODEL - 1ST CHANGE**
VERY GOOD +

.38 MILITARY & POLICE 2ND MODEL - 1ST CHANGE

(MODEL OF 1902) REDBOOK CODE: RB-SW-H-38MP21

Blue or nickel finish, enlarged barrel and yoke with added dimensions to frame, 28,645 manufactured, serial numbers range: 33804-62449, target model worth significant premium.

Production: 1903 - 1905 Caliber: .38 S&W Special Action: DA/SA, revolver

Barrel Length: 4", 5", 6.5" Sights: fixed round-blade front, square-notch rear

or target sights Capacity: 6 Grips: checkered hard rubber with S&W monograms

or checkered walnut (round or square butt)

D2C:	NIB	$2,100	Ex	$1,600	VG+	$1,555	Good	$1,050	
C2C:	Mint	$2,020	Ex	$1,450	VG+	$1,050	Good	$ 945	Fair	$ 485
Trade-In:	Mint	$1,500	Ex	$1,200	VG+	$820	Good	$ 735	Poor	$ 210

**.38 MILITARY & POLICE
MODEL OF 1905**
GOOD

.38 MILITARY & POLICE MODEL OF 1905 (.38 HAND

EJECTOR M&P 3RD MODEL) REDBOOK CODE: RB-SW-H-38MP05

Blue or nickel finish, five-screw "K" frame, changes made to frame and cylinder stop cut, 10,800 manufactured, serial numbers range: 62450-73250, target model has higher value.

Production: 1905 - 1906 Caliber: .38 S&W Special Action: DA/SA, revolver

Barrel Length: 4", 5", 6.5" Sights: fixed round-blade front, square-notch rear

or target sights Capacity: 6 Grips: checkered hard rubber with S&W monograms

or checkered walnut (round or square butt)

D2C:	NIB	$1,500	Ex	$1,150	VG+	$825	Good	$ 750	
C2C:	Mint	$1,440	Ex	$1,050	VG+	$750	Good	$ 675	Fair	$ 345
Trade-In:	Mint	$1,070	Ex	$ 850	VG+	$590	Good	$ 525	Poor	$ 150

.38 MILITARY & POLICE MODEL OF 1905 - 1ST CHANGE

REDBOOK CODE: RB-SW-H-38MP51

Blue or nickel finish, five-screw "K" frame, several minor internal changes from previous model, serial numbers range between 1st and 2nd change: 73250-146900, target model has increased value.

Production: 1906 - ~ 1908   Caliber: .38 S&W Special   Action: DA/SA, revolver
Barrel Length: 4", 5", 6", 6.5"   Sights: fixed round-blade front, square-notch rear or target sights   Capacity: 6   Grips: checkered hard rubber with S&W monograms or checkered walnut (round or square butt)

D2C:	NIB	$1,350	Ex	$1,050	VG+	$ 745	Good	$ 675	
C2C:	Mint	$1,300	Ex	$ 950	VG+	$ 680	Good	$ 610	Fair $ 315
Trade-In:	Mint	$ 960	Ex	$ 775	VG+	$ 530	Good	$ 475	Poor $ 150

.38 MILITARY & POLICE MODEL OF 1905 - 2ND CHANGE

REDBOOK CODE: RB-SW-H-38MP52

Blue or nickel finish, "K" frame with five-screws, minor changes from 1st Change, overlapping serial numbers from 1st Change between: 73250-146900, target model has increased value.

Production: 1908 - 1909   Caliber: .38 S&W Special   Action: DA/SA, revolver
Barrel Length: 4", 5", 6", 6.5"   Sights: fixed round-blade front, square-notch rear or target sights   Capacity: 6   Grips: checkered hard rubber with S&W monograms or checkered walnut (round or square butt)

D2C:	NIB	$1,350	Ex	$1,050	VG+	$ 745	Good	$ 675	
C2C:	Mint	$1,300	Ex	$ 950	VG+	$ 680	Good	$ 610	Fair $ 315
Trade-In:	Mint	$ 960	Ex	$ 775	VG+	$ 530	Good	$ 475	Poor $ 150

.38 MILITARY & POLICE MODEL OF 1905 - 3RD CHANGE
GOOD

.38 MILITARY & POLICE MODEL OF 1905 - 3RD CHANGE

REDBOOK CODE: RB-SW-H-38MP53

Blue or nickel finish, "K" frame with five-screws, minor changes from 2nd Change, extractor rod knob changed to a solid piece, 94,800 manufactured, serial numbers range between: 146900-241705, target model has increased value.

Production: 1909 - 1915   Caliber: .38 S&W Special   Action: DA/SA, revolver
Barrel Length: 4", 6"   Sights: fixed round-blade front, square-notch rear or target sights   Capacity: 6   Grips: checkered hard rubber with S&W monograms or checkered walnut (round or square butt)

D2C:	NIB	$1,350	Ex	$1,050	VG+	$ 745	Good	$ 675	
C2C:	Mint	$1,300	Ex	$ 950	VG+	$ 680	Good	$ 610	Fair $ 315
Trade-In:	Mint	$ 960	Ex	$ 775	VG+	$ 530	Good	$ 475	Poor $ 150

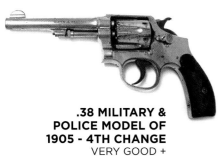

.38 MILITARY & POLICE MODEL OF 1905 - 4TH CHANGE
VERY GOOD +

.38 MILITARY & POLICE MODEL OF 1905 - 4TH CHANGE

REDBOOK CODE: RB-SW-H-38MP54

Blue or nickel finish, "K" frame with five-screws, few minor changes from 3rd Change, serial numbers range between: 241700-1000000, target model has higher value, 2" and square butt frame have significantly increased value, nickel finish has increased value.

Production: 1915 - 1942   Caliber: .38 S&W Special   Action: DA/SA, revolver
Barrel Length: 2", 4", 5", 6"   Sights: fixed round-blade front, square-notch rear or target sights   Capacity: 6   Grips: checkered hard rubber with S&W monograms or checkered walnut (round or square butt)

D2C:	NIB	$1,300	Ex	$1,000	VG+	$ 715	Good	$ 650	
C2C:	Mint	$1,250	Ex	$ 900	VG+	$ 650	Good	$ 585	Fair $ 300
Trade-In:	Mint	$ 930	Ex	$ 750	VG+	$ 510	Good	$ 455	Poor $ 150

**.38 MILITARY & POLICE
VICTORY MODEL**
GOOD

.38 MILITARY & POLICE VICTORY MODEL

REDBOOK CODE: RB-SW-H-38MPVM

Black Magic or sandblast blue finish, five-screw "K" frame, lanyard ring on butt, concurrent with .38/200 Model, serial numbers have "V" or "SV" prefix, rare 2" models have significantly increased value.

Production: 1942 - 1945   Caliber: .38 S&W Special   Action: DA/SA, revolver

Barrel Length: 2", 4"   Sights: fixed round-blade front, square-notch rear

Capacity: 6   Grips: smooth walnut

D2C:	NIB $ 925	Ex $ 725	VG+ $ 510	Good $ 465	
C2C:	Mint $ 890	Ex $ 650	VG+ $ 470	Good $ 420	Fair $ 215
Trade-In:	Mint $ 660	Ex $ 525	VG+ $ 370	Good $ 325	Poor $ 120

.38 MILITARY & POLICE
VERY GOOD +

.38 MILITARY & POLICE (POSTWAR) (PRE-MODEL 10)

REDBOOK CODE: SW-H-PREM10

Blue or nickel finish, five-screw "K" frame, serial numbers have "S" or "C" prefix, continued as the Model 10 in 1957.

Production: 1946 - 1957   Caliber: .38 S&W Special   Action: DA/SA, revolver

Barrel Length: 2", 4", 5", 6"   Sights: fixed round-blade or ramp front,
square-notch rear   Capacity: 6   Grips: diamond checkered walnut with S&W monograms

D2C:	NIB $ 975	Ex $ 750	VG+ $ 540	Good $ 490	
C2C:	Mint $ 940	Ex $ 675	VG+ $ 490	Good $ 440	Fair $ 225
Trade-In:	Mint $ 700	Ex $ 550	VG+ $ 390	Good $ 345	Poor $ 120

K-38 COMBAT MASTERPIECE
EXCELLENT

K-38 COMBAT MASTERPIECE (PRE-MODEL 15)

REDBOOK CODE: RB-SW-H-PREM15

Blue or nickel finish, five-screw target "K" frame, serial numbers have "K" prefix, continued as the Model 15 in 1957.

Production: 1949 - 1957   Caliber: .38 S&W Special   Action: DA/SA, revolver

Barrel Length: 4"   Sights: Baughman Quick Draw front, adjustable micrometer rear

Capacity: 6   Grips: diamond checkered walnut with S&W monograms

D2C:	NIB $ 850	Ex $ 650	VG+ $470	Good $ 425	
C2C:	Mint $ 820	Ex $ 600	VG+ $430	Good $ 385	Fair $200
Trade-In:	Mint $ 610	Ex $ 500	VG+ $340	Good $300	Poor $ 90

**.38 MILITARY &
POLICE AIRWEIGHT**
EXCELLENT

.38 MILITARY & POLICE AIRWEIGHT (PRE-MODEL 12)

REDBOOK CODE: RB-SW-H-PREM12

Five-screw alloy "KA" frame, blue or nickel finish, early alloy cylinders and later steel cylinders, continued as Model 12 in 1957.

Production: 1952 - 1957   Caliber: .38 S&W Special   Action: DA/SA, revolver

Barrel Length: 2", 4", 5", 6"   Sights: fixed-ramped front, square-notch rear

Capacity: 6   Grips: diamond checkered walnut with S&W monograms

D2C:	NIB $1,850	Ex $1,425	VG+ $1,020	Good $ 925	
C2C:	Mint $1,780	Ex $1,300	VG+ $ 930	Good $ 835	Fair $ 430
Trade-In:	Mint $1,320	Ex $1,050	VG+ $ 730	Good $ 650	Poor $ 210

USAF M13
(THE AIRCREWMAN)
VERY GOOD +
PICTURE IS A REPRODUCTION

.38 REGULATION POLICE
EXCELLENT

USAF M13 (THE AIRCREWMAN) REDBOOK CODE: RB-SW-H-M13AIR

Blue finish, five-screw "K" frame, "Property of U.S. Air Force" marked on backstrap, "REVOLVER, LIGHTWEIGHT, M13" marked on topstrap, serial numbers have a "C" prefix, very rare as most were destroyed, beware of fakes. Values listed represent actual historical sales.

Production: ~ 1952 - 1957   Caliber: .38 S&W Special   Action: DA/SA, revolver

Barrel Length: 2"   Sights: fixed-ramped front, square-notch rear

Capacity: 6   Grips: diamond checkered walnut with S&W monograms

D2C:	NIB –	Ex –	VG+ $ 5,390	Good –			
C2C:	Mint –	Ex –	VG+ $ 4,900	Good –	Fair –		
Trade-In:	Mint –	Ex –	VG+ $ 3,830	Good –	Poor $ 600		

.38 REGULATION POLICE (PREWAR) (PRE-MODEL 33)

REDBOOK CODE: RB-SW-H-PREM33

Blue finish, five-screw "I" frame with shouldered backstrap, walnut grips marked with patent date "Pat. June 5 1917," 54,474 manufactured, serial numbers: 1-54475, target model has significantly increased value, post-war model manufactured from 1949-1957, continued as the Model 33 in 1957.

Production: 1917 - 1940   Caliber: .38 S&W   Action: DA/SA, revolver

Barrel Length: 4"   Sights: round blade front, square-notch rear

Grips: new style diamond checkered walnut with S&W monograms

D2C:	NIB $1,530	Ex $1,175	VG+ $ 845	Good $ 765	
C2C:	Mint $1,470	Ex $1,075	VG+ $ 770	Good $ 690	Fair $ 355
Trade-In:	Mint $1,090	Ex $ 875	VG+ $ 600	Good $ 540	Poor $ 180

.38/.32 TERRIER (PREWAR) REDBOOK CODE: RB-SW-H-3832TP

Blue or nickel finish, five-screw "I" frame, "Smith & Wesson .38 S&W CTG" marked on left side of barrel, "Made in U.S.A." on right side of frame.

Production: 1936 - 1940   Caliber: .38 S&W   Action: DA/SA, revolver

Barrel Length: 2"   Sights: round-blade front, square-notch rear

Capacity: 5   Grips: hard rubber or checkered walnut with S&W monograms

D2C:	NIB $ 930	Ex $725	VG+ $515	Good $ 465	
C2C:	Mint $ 900	Ex $650	VG+ $470	Good $ 420	Fair $ 215
Trade-In:	Mint $ 670	Ex $525	VG+ $370	Good $ 330	Poor $ 120

.38/.32 TERRIER (POSTWAR) (PRE-MODEL 32)

REDBOOK CODE: RB-SW-H-PREM32

Blue or nickel finish, originally a five-screw and later a four or three-screw "I" frame, flat latch thumbpiece, serial numbers range in the .38/.22 Regulation Police range, continued as the Model 32 in 1957.

Production: 1948 - 1957   Caliber: .38 S&W   Action: DA/SA, revolver

Barrel Length: 2"   Sights: round-blade front, square-notch rear

Capacity: 5   Grips: hard rubber or checkered walnut with S&W monograms

D2C:	NIB $ 510	Ex $ 400	VG+ $ 285	Good $ 255	
C2C:	Mint $ 490	Ex $ 375	VG+ $ 260	Good $ 230	Fair $ 120
Trade-In:	Mint $ 370	Ex $ 300	VG+ $ 200	Good $ 180	Poor $ 60

.38 CHIEFS SPECIAL
VERY GOOD +

.38 CHIEFS SPECIAL (PRE-MODEL 36)

REDBOOK CODE: RB-SW-H-PREM36

Blue or nickel finish, originally a five-screw "J" frame but later four or three-screw, early standard thumbpiece later changed to flat thumbpiece, serial numbers range between: 1-786545, continued as the Model 36 in 1957, Chiefs Special Target (Pre-Model 50) has adjustable target sights and value premium.

Production: 1950 - 1957   Caliber: .38 S&W Special   Action: DA/SA, revolver

Barrel Length: 2", 3"   Sights: round blade (early) or ramped front, square-notch rear or target sights   Capacity: 5   Grips: diamond checkered walnut with S&W monograms

D2C:	NIB	$1,000	Ex $	775	VG+ $	550	Good $	500	
C2C:	Mint $	960	Ex $	700	VG+ $	500	Good $	450	Fair $ 230
Trade-In:	Mint $	710	Ex $	575	VG+ $	390	Good $	350	Poor $ 120

.38 CHIEFS SPECIAL AIRWEIGHT (PRE-MODEL 37)

REDBOOK CODE: RB-SW-H-PREM37

Blue or nickel finish, early five-screw and later four-screw alloy "J" frame, flat latch thumbpiece, original alloy cylinder later changed to steel, serial numbers range in Chiefs Special series.

Production: 1952 - 1957   Caliber: .38 S&W Special   Action: DA/SA, revolver

Barrel Length: 2", 3"   Sights: fixed-ramped front, square-notch rear

Capacity: 5   Grips: diamond checkered walnut with S&W monograms

D2C:	NIB $	940	Ex $	725	VG+ $	520	Good $	470	
C2C:	Mint $	910	Ex $	650	VG+ $	470	Good $	425	Fair $ 220
Trade-In:	Mint $	670	Ex $	550	VG+ $	370	Good $	330	Poor $ 120

BODYGUARD AIRWEIGHT (PRE-MODEL 38)

REDBOOK CODE: RB-SW-H-PREM38

Blue or nickel finish, four-screw alloy "J" frame, shrouded hammer with access for single action, steel cylinder, flat latch thumbpiece, serial numbers range in Chiefs Special series, 3" barrel rare and has increased value.

Production: 1955 - 1957   Caliber: .38 S&W Special   Action: DA/SA, revolver

Barrel Length: 2", 3"   Sights: fixed-ramped front, square-notch rear

Capacity: 5   Grips: diamond checkered walnut with S&W monograms

D2C:	NIB $	930	Ex $	725	VG+ $	515	Good $	465	
C2C:	Mint $	900	Ex $	650	VG+ $	470	Good $	420	Fair $ 215
Trade-In:	Mint $	670	Ex $	525	VG+ $	370	Good $	330	Poor $ 120

CENTENNIAL (PRE-MODEL 40)   REDBOOK CODE: RB-SW-H-PREM40

Blue or nickel finish, four-screw "J" frame, backstrap grip safety, concealed hammer, backstrap safety locking pin stored in grip, flat latch thumbpiece, serial numbers concurrent with Model 42 rage from: 1-30165, continued as the Model 40 in 1957.

Production: 1952 - 1957   Caliber: .38 S&W Special   Action: DAO, revolver

Barrel Length: 2"   Wt.: 19 oz.   Sights: fixed-ramped front, square-notch rear

Capacity: 5   Grips: diamond checkered walnut with S&W monograms or smooth high-horned wood

D2C:	NIB $	875	Ex $	675	VG+ $	485	Good $	440	
C2C:	Mint $	840	Ex $	625	VG+ $	440	Good $	395	Fair $ 205
Trade-In:	Mint $	630	Ex $	500	VG+ $	350	Good $	310	Poor $ 90

CENTENNIAL AIRWEIGHT (PRE-MODEL 42)
REDBOOK CODE: RB-SW-H-PREM42
Blue or nickel finish, lightweight alloy frame version of Centennial Model, serial numbers range concurrently with Centennial Model between: 1-30161.

Production: 1952 - 1957 Caliber: .38 S&W Special Action: DAO, revolver				
Barrel Length: 2" Wt.: 13 oz. Sights: fixed-ramped front, square-notch rear				
Capacity: 5 Grips: diamond checkered walnut with S&W monograms or smooth high-horned wood				
D2C:	NIB $ 950	Ex $ 725	VG+ $ 525	Good $ 475	
C2C:	Mint $ 920	Ex $ 675	VG+ $ 480	Good $ 430	Fair $ 220
Trade-In:	Mint $ 680	Ex $ 550	VG+ $ 380	Good $ 335	Poor $ 120

.38/44 HEAVY DUTY (PREWAR) REDBOOK CODE: RB-SW-H-3844HD
Blue or nickel finish, five-screw "N" frame, shrouded extractor rod, ~11,110 manufactured, serial numbers range in the .44 Hand Ejector 3rd Model series, 4" and 6.5" models worth significant premium.

Production: 1930 - 1941 Caliber: .38/44 S&W Special or .38 S&W Special				
Action: DA/SA, revolver Barrel Length: 4", 5", 6.5" Sights: round-blade front, square notch-cut rear Capacity: 6 Grips: diamond checkered walnut or Magna grips				
D2C:	NIB $2,350	Ex $1,800	VG+ $1,295	Good $1,175	
C2C:	Mint $2,260	Ex $1,625	VG+ $1,180	Good $1,060	Fair $ 545
Trade-In:	Mint $1,670	Ex $1,325	VG+ $ 920	Good $ 825	Poor $ 240

**.38/44 HEAVY DUTY
TRANSITION**
EXCELLENT

.38/44 HEAVY DUTY TRANSITION
(POSTWAR TRANSITIONAL) REDBOOK CODE: RB-SW-H-3844HT
Blue or nickel finish, five-screw "N" frame, hammer block, serial numbers have "S" prefix, 6.5" model is rare.

Production: 1946 - 1950 Caliber: .38/44 S&W Special or .38 S&W Special				
Action: DA/SA, revolver Barrel Length: 4", 5", 6.5" Sights: round-blade front, square notch-cut rear Capacity: 6 Grips: diamond checkered Magna grips				
D2C:	NIB $2,100	Ex $1,600	VG+ $1,155	Good $1,050	
C2C:	Mint $2,020	Ex $1,450	VG+ $1,050	Good $ 945	Fair $ 485
Trade-In:	Mint $1,500	Ex $1,200	VG+ $ 820	Good $ 735	Poor $ 210

**.38/44 HEAVY DUTY
MODEL OF 1950**
EXCELLENT

.38/44 HEAVY DUTY MODEL OF 1950
(PRE-MODEL 20) REDBOOK CODE: RB-SW-H-PREM20
Blue or nickel finish, five-screw "N" frame, serial numbers range with the "S" prefix, continued as the Model 20 in 1957.

Production: 1950 - 1957 Caliber: .38/44 S&W Special or .38 S&W Special				
Action: DA/SA, revolver Barrel Length: 4", 5", 6.5" Sights: round-blade front, square-notch cut rear Capacity: 6 Grips: diamond checkered Magna grips				
D2C:	NIB $1,350	Ex $1,050	VG+ $ 745	Good $ 675	
C2C:	Mint $1,300	Ex $ 950	VG+ $ 680	Good $ 610	Fair $ 315
Trade-In:	Mint $ 960	Ex $ 775	VG+ $ 530	Good $ 475	Poor $ 150

**.44 HAND EJECTOR
1ST MODEL**
VERY GOOD +

.44 HAND EJECTOR 1ST MODEL (NEW CENTURY, TRIPLE-LOCK, .44 MILITARY, MODEL OF 1908) REDBOOK CODE: RB-SW-H-44HE1M

Blue or nickel finish, five-screw "N" frame, third lock added at yoke, 15,375 manufactured, serial numbers rage: 1-15375 in the .44 HE series, serial number duplication occurs with the 455 MKII First Model from 1-5000, target models have increased value, special calibers other than the standard .44 S&W Special worth premium.

Production: 1907 - 1915 Caliber: .44 S&W Special, .44 S&W Russian, .44-40 Win., .45 LC, .450 Eley, .455 Mark II, .45 S&W Special, .38-40 Win., .22 LR

Action: DA/SA, revolver Barrel Length: 4", 5", 6.5", 7.5"

Sights: round-blade front, square-notch cut rear or target sights Capacity: 6

Grips: diamond checkered walnut with or without S&W medallions

D2C:	NIB $4,930	Ex $3,750	VG+ $2,715	Good $2,465	
C2C:	Mint $4,740	Ex $3,425	VG+ $2,470	Good $2,220	Fair $ 1,135
Trade-In:	Mint $3,510	Ex $2,775	VG+ $1,930	Good $1,730	Poor $ 510

**.44 HAND EJECTOR
2ND MODEL**
VERY GOOD +

.44 HAND EJECTOR 2ND MODEL REDBOOK CODE: RB-SW-H-44HE2M

Blue or nickel finish, five-screw "N" frame, similar to First Model but third lock feature removed and no shroud over extractor rod, 17,510 manufactured, serial numbers range in the .44 HE series, calibers other than the standard .44 S&W Special have significantly increased value.

Production: 1915 - 1940 Caliber: .44 S&W Special, .44-40 Win., .45 LC, .38-40 Win.

Action: DA/SA, revolver Barrel Length: 4", 5", 6.5" Sights: round-blade front, square-notch cut rear or target sights Capacity: 6 Grips: diamond checkered walnut with or without S&W medallions, later production Magna grips

D2C:	NIB $2,200	Ex $1,675	VG+ $1,210	Good $1,100	
C2C:	Mint $2,120	Ex $1,525	VG+ $1,100	Good $ 990	Fair $ 510
Trade-In:	Mint $1,570	Ex $1,250	VG+ $ 860	Good $ 770	Poor $ 240

.44 HAND EJECTOR 3RD MODEL (PREWAR)

(MODEL 1926 HAND EJECTOR) REDBOOK CODE: RB-SW-H-44HE3M

Five-screw "N" frame, blue or nickel finish, extractor shroud reintroduced, 4,976 manufactured, serial numbers range between: 28355-61412, target model has multiple times the value of standard 3rd Model.

Production: 1926 - 1941 Caliber: .44 S&W Special, .44-40 Win., .45 LC

Action: DA/SA, revolver Barrel Length: 4", 5", 6.5"

Sights: round-blade front, square-notch cut rear Capacity: 6

Grips: diamond checkered walnut with S&W monograms

D2C:	NIB $3,950	Ex $3,025	VG+ $2,175	Good $1,975	
C2C:	Mint $3,800	Ex $2,750	VG+ $1,980	Good $1,780	Fair $ 910
Trade-In:	Mint $2,810	Ex $2,225	VG+ $1,550	Good $1,385	Poor $ 420

.44 MAGNUM
EXCELLENT

.44 MAGNUM (PRE-MODEL 29) (5-SCREW) REDBOOK CODE: RB-SW-H-PREM29

"N" Target five-screw frame, blue or nickel finish, shrouded extractor rod, nickel finish rare and worth multiple times standard value, serial numbers range between: S130927-S167500 in the same range as the Model 29.

Production: 1955 - 1957 Caliber: .44 Magnum Action: DA/SA, revolver

Barrel Length: 4", 6.5" Sights: red insert target front, adjustable micrometer rear

Capacity: 6 Grips: diamond checkered goncalo alves target

D2C:	NIB $4,200	Ex $3,200	VG+ $2,310	Good $2,100	
C2C:	Mint $4,040	Ex $2,900	VG+ $2,100	Good $1,890	Fair $ 970
Trade-In:	Mint $2,990	Ex $2,375	VG+ $1,640	Good $1,470	Poor $ 420

SMITH & WESSON

.44 MAGNUM VERY GOOD +

.44 MAGNUM (PRE-MODEL 29)(4-SCREW)
REDBOOK CODE: RB-SW-H-PRE294
"N" Target four-screw frame, top sideplate screw removed, blue or nickel finish, 5" barrel models and nickel finished models worth higher premium.

Production: 1956 - 1958	Caliber: .44 Magnum	Action: DA/SA, revolver			
Barrel Length: 4", 6.5"	Sights: red-insert target front, adjustable micrometer rear				
Capacity: 6	Grips: diamond checkered goncalo alves target				
D2C:	NIB $2,650	Ex $2,025	VG+ $1,460	Good $1,325	
C2C:	Mint $2,550	Ex $1,850	VG+ $1,330	Good $1,195	Fair $ 610
Trade-In:	Mint $1,890	Ex $1,500	VG+ $1,040	Good $ 930	Poor $ 270

MODEL 10
EXCELLENT

MODEL 10 (.38 MILITARY & POLICE) REDBOOK CODE: RB-SW-H-MDL10X
"K" frame, original frame built with 5 screws and later 4 and 3 screws, continuation of the .38 M&P Model, blue or nickel finish, many variations exist for this model.

Production: 1957 - current	Caliber: .38 S&W Special, .38 S&W Special +P				
Action: DA/SA, revolver	Barrel Length: 2", 2.5", 3", 4", 5", 6"				
Sights: ramped-blade front, notch-cut rear	Capacity: 6				
Grips: diamond checkered walnut Magnas with S&W monograms					
D2C:	NIB $ 675	Ex $ 525	VG+ $ 375	Good $ 340	
C2C:	Mint $ 650	Ex $ 475	VG+ $340	Good $ 305	Fair $ 160
Trade-In:	Mint $ 480	Ex $ 400	VG+ $ 270	Good $ 240	Poor $ 90

MODEL 11 (.38/200 MILITARY & POLICE) REDBOOK CODE: RB-SW-H-MDL11X
Black Magic or sandblast blue finish, five-screw "K" frame, most with lanyard ring on butt, concurrent with Victory Model, serial numbers have "V" or "SV" prefix, continuation of the .38/200 British Service Revolver (K-200) which first began production in 1938, S&W changed to model names in 1957 and this model became the Model 11.

Production: 1957 - 1965	Caliber: .38 S&W (.38-200)	Action: DA/SA, revolver			
Barrel Length: 4", 5", 6"	Sights: fixed round-blade front, square-notch rear				
Capacity: 6	Grips: diamond checkered walnut with S&W monograms				
(commercial) or smooth walnut (military)					
D2C:	NIB $1,220	Ex $ 950	VG+ $ 675	Good $ 610	
C2C:	Mint $1,180	Ex $ 850	VG+ $ 610	Good $ 550	Fair $ 285
Trade-In:	Mint $ 870	Ex $ 700	VG+ $ 480	Good $ 430	Poor $ 150

MODEL 12
VERY GOOD +

MODEL 12 (.38 MILITARY & POLICE AIRWEIGHT) REDBOOK CODE: RB-SW-H-MDL12X
Five-screw alloy "KA" frame, blue or nickel finish, continuation of the .38 M&P Airweight after the 1957 model name change.

Production: 1957 - 1986	Caliber: .38 S&W Special	Action: DA/SA, revolver			
Barrel Length: 2", 4", 5", 6"	Sights: fixed-ramped front, square-notch rear				
Capacity: 6	Grips: diamond checkered walnut with S&W monograms				
D2C:	NIB $ 700	Ex $ 550	VG+ $ 385	Good $ 350	
C2C:	Mint $ 680	Ex $ 500	VG+ $ 350	Good $ 315	Fair $ 165
Trade-In:	Mint $ 500	Ex $ 400	VG+ $ 280	Good $ 245	Poor $ 90

MODEL 13
MINT

MODEL 13 (.357 MAGNUM MILITARY & POLICE HEAVY BARREL)
REDBOOK CODE: RB-SW-H-MDL13X
Blue or nickel finish, three-screw "K" frame, heavy barrel version of the Model 10, non-shrouded extractor rod, nickel model has slightly increased value.

Production: 1974 - 1998   Caliber: .357 Magnum   Action: DA/SA, revolver

Barrel Length: 3", 4"   Sights: fixed-ramped front, square-notch rear   Capacity: 6

Grips: diamond checkered walnut Magnas with S&W monograms

D2C:	NIB $	550	Ex $	425	VG+	$305	Good $	275		
C2C:	Mint $	530	Ex $	400	VG+	$280	Good $	250	Fair $	130
Trade-In:	Mint $	400	Ex $	325	VG+	$220	Good $	195	Poor $	60

MODEL 14
VERY GOOD +

MODEL 14 (K-38 TARGET MASTERPIECE)
REDBOOK CODE: RB-SW-H-MDL14X
Blue finish, target "K" frame, continuation of the K-38 Target Masterpiece after the name change in 1957.

Production: 1957 - 1982   Caliber: .38 S&W Special   Action: DA/SA, revolver

Barrel Length: 6", 8.375"   Sights: Patridge front, adjustable micrometer rear

Capacity: 6   Grips: diamond checkered walnut Magnas with S&W monograms

D2C:	NIB $	725	Ex $	575	VG+	$400	Good $	365		
C2C:	Mint $	700	Ex $	525	VG+	$370	Good $	330	Fair $	170
Trade-In:	Mint $	520	Ex $	425	VG+	$290	Good $	255	Poor $	90

MODEL 14 FULL LUG (K-38 TARGET MASTERPIECE)
REDBOOK CODE: RB-SW-H-MDL14F
Blue finish, target "K" frame, added a full-lug barrel.

Production: 1991 - 1999   Caliber: .38 S&W Special   Action: DA/SA, revolver

Barrel Length: 6"   Sights: Patridge front, adjustable micrometer rear

Capacity: 6   Grips: smooth-wood combat grips

D2C:	NIB	$630	Ex	$500	VG+	$350	Good $	315		
C2C:	Mint	$610	Ex	$450	VG+	$320	Good $	285	Fair $	145
Trade-In:	Mint	$450	Ex	$375	VG+	$250	Good $	225	Poor $	90

MODEL 14 SINGLE-ACTION (K-38 TARGET MASTERPIECE)
REDBOOK CODE: RB-SW-H-MDL14S
Same as Model 14 but single action only.

Production: 1961   Caliber: .38 S&W Special   Action: SA, revolver   Barrel Length: 6"

Sights: Patridge front, adjustable micrometer rear   Capacity: 6

Grips: diamond checkered walnut Magnas with S&W monograms

D2C:	NIB $	780	Ex $	600	VG+	$430	Good $	390		
C2C:	Mint $	750	Ex $	550	VG+	$390	Good $	355	Fair $	180
Trade-In:	Mint $	560	Ex $	450	VG+	$310	Good $	275	Poor $	90

MODEL 14 CLASSIC
VERY GOOD +

MODEL 14 CLASSIC   REDBOOK CODE: RB-SW-H-MDL14C
Newer reintroduction of the Model 14, blue finish.

Production: 2009 - 2011   Caliber: .38 S&W Special +P   Action: DA/SA, revolver

Barrel Length: 6"   Sights: Patridge front, adjustable micrometer rear

Capacity: 6   Grips: checkered walnut with S&W monograms

D2C:	NIB $	765	Ex $	600	VG+	$425	Good $	385		
C2C:	Mint $	740	Ex $	550	VG+	$390	Good $	345	Fair $	180
Trade-In:	Mint $	550	Ex $	450	VG+	$300	Good $	270	Poor $	90

SMITH & WESSON

MODEL 15
VERY GOOD +

MODEL 15 (K-38 COMBAT MASTERPIECE) REDBOOK CODE: RB-SW-H-MDL15X
Blue or nickel finish, originally built on five-screw target "K" frame, continued from the original K-38 Combat Masterpiece after the name change in 1957.

Production: 1957 - 1999   Caliber: .38 S&W Special   Action: DA/SA, revolver
Barrel Length: 2", 4", 6", 8.375"   Sights: Baughman Quick Draw front, adjustable micrometer rear   Capacity: 6   Grips: diamond checkered walnut Magnas with S&W monograms

D2C:	NIB	$ 680	Ex $	525	VG+ $	375	Good $	340		
C2C:	Mint $	660	Ex $	475	VG+ $	340	Good $	310	Fair $	160
Trade-In:	Mint $	490	Ex $	400	VG+ $	270	Good $	240	Poor $	90

MODEL 15-8 LEW HORTON HERITAGE SERIES
REDBOOK CODE: RB-SW-H-MDL158
Blue color case hardened 3-screw frame, round butt, S&W Performance Center logo, special Heritage Series box.

Production: 2001   Caliber: .38 S&W Special   Action: DA/SA, revolver   Capacity: 6
Barrel Length: 4"   Sights: ramped front, micrometer click rear   Grips: checkered wood

D2C:	NIB $	875	Ex $	675	VG+ $	485	Good $	440		
C2C:	Mint $	840	Ex $	625	VG+ $	440	Good $	395	Fair $	205
Trade-In:	Mint $	630	Ex $	500	VG+ $	350	Good $	310	Poor $	90

MODEL 15-9 LEW HORTON HERITAGE SERIES MCGIVERN
REDBOOK CODE: RB-SW-H-MDL159
Available in blue, nickel, or color case hardened finish. Model commemorates Ed McGivern's world speed records on sideplate.

Production: 2002   Caliber: .38 S&W Special   Action: DA/SA, revolver
Barrel Length: 6"   Sights: Patridge front with McGivern Gold bead, micrometer click rear   Grips: altamont checkered

D2C:	NIB $	975	Ex $	750	VG+ $	540	Good $	490		
C2C:	Mint $	940	Ex $	675	VG+ $	490	Good $	440	Fair $	225
Trade-In:	Mint $	700	Ex $	550	VG+ $	390	Good $	345	Poor $	120

MODEL 15 CLASSIC
NEW IN BOX

MODEL 15 CLASSIC (MASTERPIECE) REDBOOK CODE: RB-SW-H-MDL15C
Blue finish, medium frame, reintroduction of the Model 15.

Production: 2010 - 2011   Caliber: .38 S&W Special +P   Action: DA/SA, revolver
Barrel Length: 4"   Sights: fixed-ramped front, adjustable micrometer rear
Capacity: 6   Grips: diamond checkered wood with S&W medallions

D2C:	NIB $	810	Ex $	625	VG+ $	450	Good $	405		
C2C:	Mint $	780	Ex $	575	VG+ $	410	Good $	365	Fair $	190
Trade-In:	Mint $	580	Ex $	475	VG+ $	320	Good $	285	Poor $	90

MODEL 16
EXCELLENT

MODEL 16 (K-32 MASTERPIECE) REDBOOK CODE: RB-SW-H-MDL16X
Blue finish, five-screw target "K" frame, serial numbers have "K" prefix, continuation of the K-32 Masterpiece after the 1957 name change.

Production: 1957 - 1974   Caliber: .32 S&W Long   Action: DA/SA, revolver
Barrel Length: 6"   Sights: Patridge front, adjustable micrometer rear
Capacity: 6   Grips: diamond checkered walnut Magnas with S&W monograms

D2C:	NIB	$2,875	Ex	$2,200	VG+	$1,585	Good	$1,440		
C2C:	Mint	$2,760	Ex	$2,000	VG+	$1,440	Good	$1,295	Fair $	665
Trade-In:	Mint	$2,050	Ex	$1,625	VG+	$1,130	Good	$1,010	Poor $	300

MODEL 16-4
MASTERPIECE - FULL LUG
MINT

MODEL 16-4 MASTERPIECE - FULL LUG

REDBOOK CODE: RB-SW-H-MDL164
Blue finish, full-lug barrel version of the Model 16, roughly 8,800 manufactured.

Production: 1990 - 1993   Caliber: .32 H&R Magnum   Action: DA/SA, revolver

Barrel Length: 4", 6", 8.375"   Sights: Patridge front, adjustable micrometer rear

Capacity: 6   Grips: goncalo alves wood combat with S&W medallions

D2C:	NIB	$ 980	Ex $ 750	VG+ $540	Good $ 490		
C2C:	Mint	$ 950	Ex $ 700	VG+ $490	Good $ 445	Fair $ 230
Trade-In:	Mint	$ 700	Ex $ 550	VG+ $390	Good $ 345	Poor $ 120

MODEL 19
EXCELLENT

MODEL 19 (NO DASH)   REDBOOK CODE: RB-SW-H-MDL19X

Four-screw target K-frame, blue or nickel finish, high premium for nickel finish, six-groove backstrap, combat trigger, a continuation of the .357 Combat Magnum (Pre-Model 19) after the 1957 name change.

Production: 1957 - 1959   Caliber: .357 Magnum/.38 Special   Action: DA/SA, revolver

Barrel Length: 4"   Sights: Baughman Quick Draw front, micro adjustable rear

Capacity: 6   Grips: checkered walnut target with S&W monograms

D2C:	NIB	$1,050	Ex $ 800	VG+ $580	Good $ 525		
C2C:	Mint	$1,010	Ex $ 725	VG+ $530	Good $ 475	Fair $ 245
Trade-In:	Mint	$ 750	Ex $ 600	VG+ $410	Good $ 370	Poor $ 120

MODEL 25 CLASSIC
NEW IN BOX

MODEL 25 CLASSIC   REDBOOK CODE: RB-SW-H-MDL25C

Carbon steel construction, built on N-frame, bright-blue or nickel (discontinued) finish, a reintroduction of the Model 25, nickel finish brings an increased value.

Production: current   Caliber: .45 LC   Action: DA/SA, revolver   Barrel Length: 6.5"

OA Length: 12"   Sights: pinned Patridge front, micro adjustable rear   Capacity: 6

Grips: checkered square-butt walnut

D2C:	NIB	$ 865	Ex $ 675	VG+ $480	Good $ 435	LMP $1,009
C2C:	Mint	$ 840	Ex $ 600	VG+ $440	Good $ 390	Fair $ 200
Trade-In:	Mint	$ 620	Ex $ 500	VG+ $340	Good $ 305	Poor $ 90

MODEL 27
NEW IN BOX

MODEL 27   REDBOOK CODE: RB-SW-H-MDL27X

A continuation of the original ".357 Magnum" (see Pre-Model 27) after the 1957 name change, blue or nickel finish, built on the N-frame, checkered top strap, increased value for nickel finish or early 4-screw model.

Production: 1957 - 1994   Caliber: .357 Magnum   Action: DA/SA, revolver

Barrel Length: 3.5", 4", 5", 6", 6.5", 8.375"   Sights: Patridge or red ramp front, micro adjustable rear   Capacity: 6   Grips: diamond walnut Magnas or (later) goncalo alves

D2C:	NIB	$1,600	Ex $1,225	VG+ $880	Good $ 800		
C2C:	Mint	$1,540	Ex $1,125	VG+ $800	Good $ 720	Fair $ 370
Trade-In:	Mint	$ 1140	Ex $900	VG+ $630	Good $ 560	Poor $ 180

MODEL 27 CLASSICS
MINT

MODEL 27 CLASSICS   REDBOOK CODE: RB-SW-H-MDL27C

Bright-blue or nickel (discontinued) finish, carbon steel construction, nickel finish brings a slight premium.

Production: current   Caliber: .357 Magnum/.38 Special   Action: DA/SA, revolver

Barrel Length: 4", 6.5"   Sights: ramp front, micro adjustable rear   Capacity: 6

Grips: checkered walnut

D2C:	NIB	$ 860	Ex $ 675	VG+ $ 475	Good $430	LMP $1,019
C2C:	Mint	$ 830	Ex $600	VG+ $430	Good $390	Fair $ 200
Trade-In:	Mint	$ 620	Ex $500	VG+ $340	Good $305	Poor $ 90

SMITH & WESSON

MODEL 29 .44 MAGNUM (EARLY 4-SCREW)
REDBOOK CODE: RB-SW-H-ML294S
Blue or nickel finish, four-screw model, N frame, "S" serial prefix, 5" barrel worth multiple times standard value, nickel finish has increased premium, a continuation of the .44 Magnum (Pre-Model 29) after the 1957 model name change.

Production: 1957 - ~ 1961   Caliber: .44 Magnum   Action: DA/SA, revolver

Barrel Length: 4", 5", 6.5", 8.375"   Sights: red ramp front, adjustable micrometer rear

Capacity: 6   Grips: diamond checkered walnut target

D2C:	NIB $2,400	Ex $1,825	VG+ $1,320	Good $1,200	
C2C:	Mint $2,310	Ex $1,675	VG+ $1,200	Good $1,080	Fair $ 555
Trade-In:	Mint $ 1,710	Ex $1,350	VG+ $ 940	Good $ 840	Poor $ 240

MODEL 29-1   REDBOOK CODE: RB-SW-H-MDL291
Blue or nickel finish, most models have three-screw N-frame, 5" barrel is extremely rare, nickel finish has much higher value, this model is rare as the Model 29-2 quickly took its place.

Production: 1962   Caliber: .44 Magnum   Action: DA/SA, revolver

Barrel Length: 4", 5", 6.5", 8.375"   Sights: red ramp front, adjustable micrometer rear

Capacity: 6   Grips: diamond checkered walnut target

D2C:	NIB $3,900	Ex $2,975	VG+ $2,145	Good $1,950	
C2C:	Mint $3,750	Ex $2,700	VG+ $1,950	Good $1,755	Fair $ 900
Trade-In:	Mint $2,770	Ex $2,200	VG+ $1,530	Good $1,365	Poor $ 390

MODEL 29-2
EXCELLENT

MODEL 29-2   REDBOOK CODE: RB-SW-H-MDL292
3-screw N-frame, upgraded cylinder stop from previous model, 5" barrel very rare and worth multiple times standard value, nickel model has slightly higher value, this model was made famous in the film "Dirty Harry" starring Clint Eastwood, early "S" serial numbered models add approximately 80% premium.

Production: 1962 - 1982   Caliber: .44 Magnum   Action: DA/SA, revolver

Barrel Length: 4", 5", 6", 6.5", 8.375"   Sights: red ramp front, adjustable micrometer rear   Capacity: 6   Grips: checkered walnut with or without diamond pattern

D2C:	NIB $1,100	Ex $ 850	VG+ $ 605	Good $ 550	
C2C:	Mint $1,060	Ex $ 775	VG+ $ 550	Good $ 495	Fair $ 255
Trade-In:	Mint $ 790	Ex $ 625	VG+ $ 430	Good $ 385	Poor $ 120

MODEL 29-3 AND NEWER
EXCELLENT

MODEL 29-3 AND NEWER   REDBOOK CODE: RB-SW-H-ML2937
Built on the three-screw N-frame, blue or nickel finish, many variations and special editions exist that may influence value.

Production: 1982 - 1999   Caliber: .44 Magnum   Action: DA/SA, revolver

Barrel Length: 3", 4", 6", 6.5", 8.375"   Sights: red ramp front, adjustable micrometer rear   Capacity: 6   Grips: checkered walnut

D2C:	NIB $ 900	Ex $700	VG+ $ 495	Good $ 450	
C2C:	Mint $ 870	Ex $ 625	VG+ $450	Good $ 405	Fair $ 210
Trade-In:	Mint $ 640	Ex $ 525	VG+ $360	Good $ 315	Poor $ 90

MODEL 29 CLASSIC
NEW IN BOX

MODEL 29 CLASSIC   REDBOOK CODE: RB-SW-H-MDL29C

Blue or nickel finish, carbon steel frame and cylinder, modern reintroduction of the original Model 29, add roughly $170 for 6.5" barrel, slight premium for nickel.

Production: current   Caliber: .44 Magnum   Action: DA/SA, revolver
Barrel Length: 4", 6.5"   Wt.: 44 oz. (4"), 49 oz. (6.5")   Sights: red ramp front,
adjustable micrometer rear   Capacity: 6   Grips: checkered square-butt walnut

D2C:	NIB $ 860	Ex $ 675	VG+ $475	Good $ 430	LMP $ 999	
C2C:	Mint $ 830	Ex $ 600	VG+ $430	Good $ 390	Fair $ 200	
Trade-In:	Mint $ 620	Ex $ 500	VG+ $340	Good $ 305	Poor $ 90	

MODEL 36 CLASSIC
NEW IN BOX

MODEL 36 CLASSIC (.38 CHIEFS SPECIAL)

REDBOOK CODE: RB-SW-H-MDL36C
Carbon steel frame and cylinder, blue or discontinued nickel finish, nickel finish brings a slight premium.

Production: current   Caliber: .38 S&W Special +P   Action: DA/SA, revolver
Barrel Length: 1.875"   Wt.: 20 oz.
Sights: integral fixed   Capacity: 5   Grips: checkered wood

D2C:	NIB $ 650	Ex $ 500	VG+ $360	Good $ 325	LMP $ 749	
C2C:	Mint $ 630	Ex $ 450	VG+ $330	Good $ 295	Fair $ 150	
Trade-In:	Mint $ 470	Ex $ 375	VG+ $260	Good $ 230	Poor $ 90	

MODEL 57 CLASSICS
NEW IN BOX

MODEL 57 CLASSICS   REDBOOK CODE: RB-SW-H-MDL57C

Carbon steel construction, bright blue or discontinued nickel finish, 4" model no longer in production, classic thumbpiece, color-case wide-spur hammer and target trigger.

Production: current   Caliber: .41 Magnum   Action: DA/SA, revolver
Barrel Length: 4", 6"   Sights: red ramp front, micro adjustable rear
Capacity: 6   Grips: checkered square butt walnut

D2C:	NIB $ 865	Ex $ 675	VG+ $480	Good $ 435	LMP $ 1009	
C2C:	Mint $ 840	Ex $ 600	VG+ $440	Good $ 390	Fair $ 200	
Trade-In:	Mint $ 620	Ex $ 500	VG+ $340	Good $ 305	Poor $ 90	

MODEL 60
EXCELLENT

MODEL 60 (.38 SPECIAL CHIEFS SPECIAL)

REDBOOK CODE: RB-SW-H-MDL60S
Stainless steel round butt J-frame, S&W's first stainless steel revolver.

Production: 1965 - 1999   Caliber: .38 S&W Special   Action: DA/SA, revolver
Barrel Length: 2", 3"   Sights: ramp front, notch-cut rear   Capacity: 5
Grips: checkered walnut with S&W monograms

D2C:	NIB $ 575	Ex $ 450	VG+ $320	Good $ 290		
C2C:	Mint $ 560	Ex $ 400	VG+ $290	Good $ 260	Fair $ 135	
Trade-In:	Mint $ 410	Ex $ 325	VG+ $230	Good $ 205	Poor $ 60	

**MODEL 60 .357 MAGNUM
CHIEFS SPECIAL**
EXCELLENT

MODEL 60 .357 MAGNUM CHIEFS SPECIAL

(CURRENT MODEL)   REDBOOK CODE: RB-SW-H-MDL60M
Stainless steel frame and cylinder, built on J Magnum frame.

Production: 1996 - current   Caliber: .357 Magnum/.38 Special   Action: DA/SA, revolver
Barrel Length: 2.125", 3", 5"   Wt.: 23 oz., 23 oz.   Sights: fixed-ramp or night-
sight front, adjustable rear (3" model)   Capacity: 5   Grips: synthetic or wood

D2C:	NIB $ 630	Ex $ 500	VG+ $350	Good $ 315	LMP $ 729	
C2C:	Mint $ 610	Ex $ 450	VG+ $320	Good $ 285	Fair $ 145	
Trade-In:	Mint $ 450	Ex $ 375	VG+ $250	Good $ 225	Poor $ 90	

MODEL 64
NEW IN BOX

MODEL 64   REDBOOK CODE: RB-SW-H-MDL64X

Stainless steel K-frame, early round butt and later changed to square butt, three-screw frame.

Production: current   Caliber: .38 S&W Special +P   Action: DA/SA, revolver

Barrel Length: 2", 3", 4"   Wt.: 36 oz.   Sights: integral front, fixed rear

Capacity: 6   Grips: synthetic or checkered walnut

D2C:	NIB	$ 590	Ex $ 450	VG+	$325	Good $ 295	LMP $ 689		
C2C:	Mint $ 570	Ex $ 425	VG+	$300	Good $ 270	Fair $ 140			
Trade-In:	Mint $ 420	Ex $ 350	VG+	$240	Good $ 210	Poor $ 60			

MODEL 327
NEW IN BOX

MODEL 327   REDBOOK CODE: RB-SW-H-M327XX

Scandium-alloy frame, titanium-alloy cylinder and barrel shroud, Performance Center color-case trigger with overstop, Performance Center tuned action.

Production: 2008 - current   Caliber: .357 Magnum/.38 S&W Special +P

Action: DA/SA, revolver   Barrel Length: 2"   Wt.: 22 oz.

Sights: red ramp front, notch-cut rear   Capacity: 8   Grips: wood grips with finger grooves

D2C:	NIB $ 1,150	Ex $ 875	VG+ $ 635	Good $ 575	LMP $1,269	
C2C:	Mint $ 1,110	Ex $ 800	VG+ $ 580	Good $ 520	Fair $ 265	
Trade-In:	Mint $ 820	Ex $ 650	VG+ $ 450	Good $ 405	Poor $ 120	

MODEL 327 PD
NEW IN BOX

MODEL 327 PD   REDBOOK CODE: RB-SW-H-M327PD

Scandium-alloy frame, titanium-alloy cylinder, matte black finish.

Production: 2008 - 2009   Caliber: .357 Magnum/.38 S&W Special +P

Action: DA/SA, revolver   Barrel Length: 4"   Wt.: 24 oz.   Sights: pinned red HI-VIZ front, adjustable v-notch rear   Capacity: 8   Grips: wood or rubber grip included

D2C:	NIB $ 760	Ex $ 600	VG+ $ 420	Good $ 380		
C2C:	Mint $ 730	Ex $ 525	VG+ $ 380	Good $ 345	Fair $ 175	
Trade-In:	Mint $ 540	Ex $ 450	VG+ $ 300	Good $ 270	Poor $ 90	

**MODEL 327
NIGHT GUARD**
NEW IN BOX

MODEL 327 NIGHT GUARD   REDBOOK CODE: RB-SW-H-M327NG

Stainless steel cylinder, scandium alloy frame, matte black finish.

Production: 2008 - 2012   Caliber: .357 Magnum/.38 S&W Special +P

Action: DA/SA, revolver   Barrel Length: 2.5"   Wt.: 28 oz.

Sights: XS Sights 24/7 tritium front, fixed rear   Capacity: 8   Grips: synthetic

D2C:	NIB $ 925	Ex $ 725	VG+ $ 510	Good $ 465		
C2C:	Mint $ 890	Ex $ 650	VG+ $ 470	Good $ 420	Fair $ 215	
Trade-In:	Mint $ 660	Ex $ 525	VG+ $ 370	Good $ 325	Poor $ 120	

MODEL 327 TRR8
NEW IN BOX

MODEL 327 TRR8   REDBOOK CODE: RB-SW-H-M327TR

Scandium-alloy frame, stainless steel cylinder, Performance Center color-case trigger with overstop, Performance Center tuned action, equipment rails included.

Production: 2006 - current   Caliber: .357 Magnum/.38 S&W Special +P

Action: DA/SA, revolver   Barrel Length: 5"   Wt.: 35 oz.

Sights: interchangeable front, adjustable v-notch rear   Capacity: 8   Grips: synthetic

D2C:	NIB $ 1,150	Ex $ 875	VG+ $ 635	Good $ 575	LMP $1,289	
C2C:	Mint $ 1,110	Ex $ 800	VG+ $ 580	Good $ 520	Fair $ 265	
Trade-In:	Mint $ 820	Ex $ 650	VG+ $ 450	Good $ 405	Poor $ 120	

MODEL 329 PD
NEW IN BOX

MODEL 329 PD   REDBOOK CODE: RB-SW-H-M329PD

Scandium-alloy frame, titanium-alloy cylinder, matte black finish.

Production: current   Caliber: .44 Magnum, .44 S&W Special   Action: DA/SA, revolver
Barrel Length: 4"   Wt.: 25 oz.   Sights: pinned red HI-VIZ front,
adjustable v-notch rear   Capacity: 6   Grips: wood or rubber grip included

D2C:	NIB	$1,030	Ex $ 800	VG+ $570	Good $ 515	LMP $ 1,159
C2C:	Mint $ 990	Ex $ 725	VG+ $520	Good $ 465	Fair $ 240
Trade-In:	Mint $ 740	Ex $ 600	VG+ $410	Good $ 365	Poor $ 120

MODEL 329 PD ALASKA BACKPACKER
NEW IN BOX

MODEL 329 PD ALASKA BACKPACKER

REDBOOK CODE: RB-SW-H-M329AB

Two-tone finish, scandium-alloy frame, stainless steel cylinder, laser-engraved grizzly bear on sideplate, TALO exclusive.

Production: current   Caliber: .44 Magnum, .44 S&W Special   Action: DA/SA, revolver
Barrel Length: 2.5"   Wt.: 30 oz.
Sights: Patridge gold bead front, adjustable v-notch rear   Capacity: 6   Grips: synthetic

D2C:	NIB $ 950	Ex $ 725	VG+ $525	Good $ 475	
C2C:	Mint $ 920	Ex $ 675	VG+ $480	Good $ 430	Fair $ 220
Trade-In:	Mint $ 680	Ex $ 550	VG+ $380	Good $ 335	Poor $ 120

MODEL 329 NIGHT GUARD
NEW IN BOX

MODEL 329 NIGHT GUARD   REDBOOK CODE: RB-SW-H-M329NG

Scandium alloy frame, stainless steel cylinder, matte black finish, large frame.

Production: ~2008 - discontinued   Caliber: .44 Magnum/.44 S&W Special
Action: DA/SA, revolver   Barrel Length: 2.5"   Wt.: 30 oz.
Sights: XS Sights 24/7 tritium front, fixed rear   Capacity: 6   Grips: synthetic

D2C:	NIB $ 910	Ex $ 700	VG+ $505	Good $ 455	
C2C:	Mint $ 880	Ex $ 650	VG+ $460	Good $ 410	Fair $ 210
Trade-In:	Mint $ 650	Ex $ 525	VG+ $360	Good $ 320	Poor $ 120

MODEL 331 AIRLITE TI CHIEFS SPECIAL
MINT

MODEL 331 AIRLITE TI CHIEFS SPECIAL

REDBOOK CODE: RB-SW-H-M331AT

Three-screw aluminum alloy J-magnum frame, titanium cylinder, atomic symbol for titanium marked on sideplate, aluminum alloy barrel shroud and yoke.

Production: 1999 - 2003   Caliber: .32 H&R Magnum   Action: DA/SA, revolver
Barrel Length: 1.87"   Wt.: 12 oz.   Sights: fixed forged front, notch-cut rear
Capacity: 6   Grips: Uncle Mike's Boot Grip or Dymondwood grip

D2C:	NIB $ 700	Ex $ 550	VG+ $385	Good $ 350	
C2C:	Mint $ 680	Ex $ 500	VG+ $350	Good $ 315	Fair $ 165
Trade-In:	Mint $ 500	Ex $ 400	VG+ $280	Good $ 245	Poor $ 90

MODEL 332 AIRLITE TI CENTENNIAL

REDBOOK CODE: RB-SW-H-M332AT

Three-screw aluminum alloy J-magnum frame, titanium cylinder, atomic symbol for titanium marked on sideplate, aluminum alloy barrel shroud and yoke, hammerless.

Production: 1999 - 2003   Caliber: .32 H&R Magnum   Action: DAO, revolver
Barrel Length: 1.87"   Wt.: 12 oz.   Sights: fixed forged front, notch-cut rear
Capacity: 6   Grips: Uncle Mike's Boot Grip or Dymondwood grip

D2C:	NIB $ 650	Ex $ 500	VG+ $360	Good $ 325	
C2C:	Mint $ 630	Ex $ 450	VG+ $330	Good $ 295	Fair $ 150
Trade-In:	Mint $ 470	Ex $ 375	VG+ $260	Good $ 230	Poor $ 90

MODEL 337 AIRLITE TI CHIEFS SPECIAL

REDBOOK CODE: RB-SW-H-M337AT

Three-screw aluminum alloy J-magnum frame, titanium cylinder, atomic symbol for titanium marked on sideplate, aluminum alloy barrel shroud and yoke, target Kit Gun version worth a slight premium.

Production: 1998 - 2004	Caliber: .38 S&W Special +P	Action: DA/SA, revolver			
Barrel Length: 1.87"	Wt.: 11 oz.	Sights: pinned fixed front, notch-cut rear			
Capacity: 5	Grips: wood or rubber				
D2C:	NIB $ 700	Ex $ 550	VG+ $385	Good $ 350	
C2C:	Mint $ 680	Ex $ 500	VG+ $350	Good $ 315	Fair $ 165
Trade-In:	Mint $ 500	Ex $ 400	VG+ $280	Good $ 245	Poor $ 90

MODEL 337 PD AIRLITE TI CHIEFS SPECIAL

REDBOOK CODE: RB-SW-H-M337PD

Three-screw aluminum alloy J-magnum frame, titanium cylinder, atomic symbol for titanium marked on sideplate, aluminum alloy barrel shroud and yoke, configuration similar to Model 337 but with black finish and upgraded sight and grip.

Production: 2000 - 2004	Caliber: .38 S&W Special +P	Action: DA/SA, revolver			
Barrel Length: 1.87"	Wt.: 11 oz.	Sights: red ramp front, notch-cut rear			
Capacity: 5	Grips: Hogue Bantam rubber				
D2C:	NIB $ 700	Ex $ 550	VG+ $385	Good $ 350	
C2C:	Mint $ 680	Ex $ 500	VG+ $350	Good $ 315	Fair $ 165
Trade-In:	Mint $ 500	Ex $ 400	VG+ $280	Good $ 245	Poor $ 90

MODEL 340 AIRLITE SC CENTENNIAL
NEW IN BOX

MODEL 340 AIRLITE SC CENTENNIAL

REDBOOK CODE: RB-SW-H-M340AS

Scandium alloy J-magnum frame with three-screws, titanium cylinder, matte stainless gray finish.

Production: 2001 - discontinued	Caliber: .357 Magnum	Action: DAO, revolver			
Barrel Length: 1.87"	Wt.: 12 oz.	Sights: pinned black ramp front, notch-cut rear			
Capacity: 5	Grips: Hogue Bantam rubber				
D2C:	NIB $ 850	Ex $ 650	VG+ $470	Good $ 425	
C2C:	Mint $ 820	Ex $ 600	VG+ $430	Good $ 385	Fair $ 200
Trade-In:	Mint $ 610	Ex $ 500	VG+ $340	Good $ 300	Poor $ 90

MODEL 340 PD AIRLITE SC CENTENNIAL
NEW IN BOX

MODEL 340 PD AIRLITE SC CENTENNIAL

REDBOOK CODE: RB-SW-H-M340PD

Scandium-alloy J-magnum frame with three-screws, titanium cylinder, similar to Model 340 but with black finish and red sight.

Production: 2001 - discontinued	Caliber: .357 Magnum	Action: DAO, revolver			
Barrel Length: 1.87"	Wt.: 12 oz.	Sights: red ramp front, notch-cut rear			
Capacity: 5	Grips: Hogue Bantam rubber				
D2C:	NIB $ 875	Ex $ 675	VG+ $485	Good $ 440	
C2C:	Mint $ 840	Ex $ 625	VG+ $440	Good $ 395	Fair $ 205
Trade-In:	Mint $ 630	Ex $ 500	VG+ $350	Good $ 310	Poor $ 90

MODEL 438 BODYGUARD
NEW IN BOX

MODEL 438 BODYGUARD   REDBOOK CODE: RB-SW-H-M438BG

Aluminum alloy J-frame, stainless barrel and cylinder, shrouded accessible hammer, matte black finish.

Caliber: .38 S&W Special +P   Action: DA/SA, revolver   Barrel Length: 1.875"

Wt.: 15 oz.   Sights: integral front, fixed rear   Capacity: 5   Grips: synthetic

D2C:	NIB $ 430	Ex $ 350	VG+ $240	Good $215					
C2C:	Mint $ 420	Ex $ 300	VG+ $220	Good $195	Fair $100				
Trade-In:	Mint $ 310	Ex $ 250	VG+ $170	Good $155	Poor $60				

MODEL 442 AIRWEIGHT
NEW IN BOX

MODEL 442 AIRWEIGHT   REDBOOK CODE: RB-SW-H-M442AW

Aluminum alloy J-frame, carbon steel cylinder, enclosed hammer, matte black finish.

Production: current   Caliber: .38 S&W Special +P   Action: DAO, revolver

Barrel Length: 1.875"   Wt.: 15 oz.   Sights: blade front, fixed rear

Capacity: 5   Grips: synthetic

D2C:	NIB $ 430	Ex $ 350	VG+ $240	Good $ 215	LMP $ 469
C2C:	Mint $ 420	Ex $ 300	VG+ $220	Good $ 195	Fair $ 100
Trade-In:	Mint $ 310	Ex $ 250	VG+ $170	Good $ 155	Poor $ 60

MODEL 442 MOON CLIP
NEW IN BOX

MODEL 442 MOON CLIP (PRO SERIES)

REDBOOK CODE: RB-SW-H-M442PS

Aluminum alloy J-frame, stainless steel cylinder, cut for moon clips, enclosed hammer, matte black finish.

Production: current   Caliber: .38 S&W Special +P   Action: DAO, revolver

Barrel Length: 1.875"   Wt.: 15 oz.   Sights: integral front, fixed rear

Capacity: 5   Grips: synthetic

D2C:	NIB $ 460	Ex $ 350	VG+ $255	Good $ 230	LMP $ 499
C2C:	Mint $ 450	Ex $ 325	VG+ $230	Good $ 210	Fair $ 110
Trade-In:	Mint $ 330	Ex $ 275	VG+ $180	Good $ 165	Poor $ 60

MODEL 460V
NEW IN BOX

MODEL 460V   REDBOOK CODE: RB-SW-H-M460VX

Stainless steel X-frame, stainless cylinder, gain-twist rifling, removable compensator.

Production: current   Caliber: .460 S&W Magnum/.454 Casull/.45 LC

Action: DA/SA, revolver   Barrel Length: 5"   Wt.: 61 oz.

Sights: red ramp front, adjustable rear   Capacity: 5   Grips: synthetic

D2C:	NIB $1,180	Ex $ 900	VG+ $650	Good $ 590	LMP $1,369
C2C:	Mint $1,140	Ex $ 825	VG+ $590	Good $ 535	Fair $ 275
Trade-In:	Mint $ 840	Ex $ 675	VG+ $470	Good $ 415	Poor $ 120

MODEL 500
NEW IN BOX

MODEL 500   REDBOOK CODE: RB-SW-H-MDL500

Stainless steel X-frame, stainless cylinder, 2 removable muzzle compensators, subtract $70 for 6.5" model.

Production: 2003 - current   Caliber: .500 S&W Magnum   Action: DA/SA, revolver

Barrel Length: 4", 6.5", 8.375"   Wt.: 56 oz., 61 oz., 73 oz.

Sights: red ramp front, adjustable rear   Capacity: 5   Grips: synthetic

D2C:	NIB $1,180	Ex $ 900	VG+ $650	Good $ 590	LMP $1,369
C2C:	Mint $1,140	Ex $ 825	VG+ $590	Good $ 535	Fair $ 275
Trade-In:	Mint $ 840	Ex $ 675	VG+ $470	Good $ 415	Poor $ 120

MODEL 586
VERY GOOD +

MODEL 586 REDBOOK CODE: RB-SW-H-MDL586

Blue or nickel finish, three-screw target L-frame, available in a "+" version that holds 7 rounds, add a slight premium for nickel finish.

Production: 1980 - 1999   Caliber: .357 Magnum/.38 Special +P   Action: DA/SA, revolver

Barrel Length: 2.5", 3", 4", 6", 8.375"   Sights: Baughman or red ramp front, adjustable rear   Capacity: 6, 7   Grips: goncalo alves or synthetic

D2C:	NIB	$ 640	Ex $ 500	VG+ $ 355	Good $ 320				
C2C:	Mint $ 620	Ex $ 450	VG+ $320	Good $ 290	Fair $ 150				
Trade-In:	Mint $ 460	Ex $ 375	VG+ $250	Good $ 225	Poor $ 90				

MODEL 586 CLASSICS
NEW IN BOX

MODEL 586 CLASSICS   REDBOOK CODE: RB-SW-H-M586CD

Blue finish, carbon steel construction, a reintroduction of the original 586.

Production: 2012 - current   Caliber: .357 Magnum/.38 Special +P

Action: DA/SA, revolver   Barrel Length: 4", 6"   Wt.: 40 oz.

Sights: red ramp front, adjustable rear   Capacity: 6   Grips: checkered wood

D2C:	NIB $ 760	Ex $ 600	VG+ $420	Good $ 380	LMP $ 839	
C2C:	Mint $ 730	Ex $ 525	VG+ $380	Good $ 345	Fair $ 175	
Trade-In:	Mint $ 540	Ex $ 450	VG+ $300	Good $ 270	Poor $ 90	

MODEL 625 JM
NEW IN BOX

MODEL 625 JM (JERRY MICULEK DESIGN)

REDBOOK CODE: RB-SW-H-M625JM

Stainless steel frame and cylinder, "Miculek style" .265 inch wide grooved speed trigger, low-reflection bead-blast finish, includes moon clips, designed in collaboration with Jerry Miculek.

Production: current   Caliber: .45 ACP   Action: DA/SA, revolver

Barrel Length: 4"   Wt.: 40 oz.   Sights: gold bead Patridge, adjustable rear

Capacity: 6   Grips: Jerry Miculek wood

D2C:	NIB $ 880	Ex $ 675	VG+ $485	Good $ 440	LMP $ 979	
C2C:	Mint $ 850	Ex $ 625	VG+ $440	Good $ 400	Fair $ 205	
Trade-In:	Mint $ 630	Ex $ 500	VG+ $350	Good $ 310	Poor $ 90	

**MODEL 625 JM
PERFORMANCE CENTER**
NEW IN BOX

MODEL 625 JM PERFORMANCE CENTER

REDBOOK CODE: RB-SW-H-625JMP

Performance Center tuned action and trigger, custom teardrop hammer, chamfered charge holes, deep-cut broached rifling, stainless frame and cylinder.

Production: 2005 - current   Caliber: .45 ACP   Action: DA/SA, revolver

Barrel Length: 4"   Wt.: 42 oz.   Sights: gold-bead front, adjustable rear

Capacity: 6   Grips: Altamont red, white, and blue laminated

D2C:	NIB $ 999	Ex $ 775	VG+ $550	Good $ 500	LMP $1,079	
C2C:	Mint $ 960	Ex $ 700	VG+ $500	Good $ 450	Fair $ 230	
Trade-In:	Mint $ 710	Ex $ 575	VG+ $390	Good $ 350	Poor $ 120	

MODEL 627 (MODEL OF 1989) REDBOOK CODE: RB-SW-H-MDL627

Similar to the Model 27 but with stainless-steel construction, unfluted cylinder, limited production, the Model 627-0 has upgraded stop notches and is far more common than the rare original no-dash 627.

Production: 1989 Caliber: .357 Magnum/.38 Special +P					
Action: DA/SA, revolver Barrel Length: 5.5" Wt.: 51 oz.					
Sights: blade front, adjustable rear Capacity: 6 Grips: goncalo alves combat					
D2C:	NIB $ 875	Ex $ 675	VG+ $485	Good $ 440	
C2C:	Mint $ 840	Ex $ 625	VG+ $440	Good $ 395	Fair $ 205
Trade-In:	Mint $ 630	Ex $ 500	VG+ $350	Good $ 310	Poor $ 90

MODEL 627 - PERFORMANCE CENTER
NEW IN BOX

MODEL 627 - PERFORMANCE CENTER (170210)

REDBOOK CODE: RB-SW-H-627PC5

Performance Center tuned action, stainless frame and cylinder, chrome-flashed teardrop hammer, chrome trigger, and trigger stop.

Production: 2003 - current Caliber: .357 Magnum/.38 Special +P Action: DA/SA,					
revolver Barrel Length: 5" Wt.: 44 oz. Sights: gold-bead front, adjustable rear					
Capacity: 8 Grips: wood with synthetic included					
D2C:	NIB $1,140	Ex $ 875	VG+ $ 630	Good $ 570	LMP $1,249
C2C:	Mint $1,100	Ex $ 800	VG+ $ 570	Good $ 515	Fair $ 265
Trade-In:	Mint $ 810	Ex $ 650	VG+ $ 450	Good $ 400	Poor $ 120

MODEL 627 - PERFORMANCE CENTER
NEW IN BOX

MODEL 627 - PERFORMANCE CENTER (170133)

REDBOOK CODE: RB-SW-H-627PC2

Stainless steel frame and cylinder, chrome-flashed custom tear-drop hammer and trigger with stop, Performance Center action with ball detent lock-up, cut for moon clips.

Production: 1999 - current Caliber: .357 Magnum/.38 Special +P Action: DA/SA, revolver					
Barrel Length: 2.62" Wt.: 38 oz. Sights: dovetail red ramp front, adjustable white-					
outline rear Capacity: 8 Grips: wood					
D2C:	NIB $ 960	Ex $ 750	VG+ $530	Good $ 480	LMP $1,049
C2C:	Mint $ 930	Ex $ 675	VG+ $480	Good $ 435	Fair $ 225
Trade-In:	Mint $ 690	Ex $ 550	VG+ $380	Good $ 340	Poor $ 120

MODEL 627-3 V-COMP
NEW IN BOX

MODEL 627-3 V-COMP REDBOOK CODE: RB-SW-H-M627VC

Two-tone finish, stainless steel frame and cylinder, Performance Center tuned action, chrome hammer and trigger with stop, removable barrel compensator.

Production: current Caliber: .357 Magnum/.38 Special +P Action: DA/SA, revolver					
Barrel Length: 5" Wt.: 47 oz. Sights: adjustable orange dovetail front, adjustable rear					
Capacity: 8 Grips: synthetic					
D2C:	NIB $1,350	Ex $1,050	VG+ $ 745	Good $ 675	LMP $1,509
C2C:	Mint $1,300	Ex $ 950	VG+ $ 680	Good $ 610	Fair $ 315
Trade-In:	Mint $ 960	Ex $ 775	VG+ $ 530	Good $ 475	Poor $ 150

MODEL 627 PRO SERIES
NEW IN BOX

MODEL 627 PRO SERIES   REDBOOK CODE: RB-SW-H-M627PS

Stainless-steel frame and cylinder, matte stainless finish, chamfered charge holes, custom barrel with recessed precision crown, bossed mainspring.

Production: 2008 - current Caliber: .357 Magnum/.38 Special +P Action: DA/SA, revolver				
Barrel Length: 4" Wt.: 41 oz. Sights: interchangeable front, adjustable rear				
Capacity: 8 Grips: synthetic				
D2C:	NIB $ 880	Ex $ 675	VG+ $ 485	Good $ 440	LMP $ 999
C2C:	Mint $ 850	Ex $ 625	VG+ $ 440	Good $ 400	Fair $ 205
Trade-In:	Mint $ 630	Ex $ 500	VG+ $ 350	Good $ 310	Poor $ 90

MODEL 629
NEW IN BOX

MODEL 629   REDBOOK CODE: RB-SW-H-MDL629

Similar to the Model 29 but with all stainless steel construction, built on N-frame, polished stainless finish.

Production: current Caliber: .44 Magnum/.44 S&W Special Action: DA/SA, revolver				
Barrel Length: 3", 4", 6", 8.375" Wt.: 41 oz. Sights: red ramp front, micro adjustable				
rear Capacity: 6 Grips: goncalo alves or synthetic				
D2C:	NIB $ 820	Ex $ 625	VG+ $ 455	Good $ 410	
C2C:	Mint $ 790	Ex $ 575	VG+ $ 410	Good $ 370	Fair $ 190
Trade-In:	Mint $ 590	Ex $ 475	VG+ $ 320	Good $ 290	Poor $ 90

MODEL 629 CLASSIC
NEW IN BOX

MODEL 629 CLASSIC   REDBOOK CODE: RB-SW-H-M629CL

Large N-frame, stainless steel frame and cylinder, satin stainless finish, full-lug barrel, interchangeable front sights, chamfered cylinder, drilled and tapped for scope mounts.

Production: current Caliber: .44 Magnum/.44 S&W Special Action: DA/SA, revolver				
Barrel Length: 5", 6.5", 8.375" Wt.: 48 oz. Sights: red ramp front, adjustable rear				
Capacity: 6 Grips: Hogue synthetic				
D2C:	NIB $ 840	Ex $ 650	VG+ $ 465	Good $ 420	LMP $ 989
C2C:	Mint $ 810	Ex $ 600	VG+ $ 420	Good $ 380	Fair $ 195
Trade-In:	Mint $ 600	Ex $ 475	VG+ $ 330	Good $ 295	Poor $ 90

**MODEL 629
CLASSIC DELUXE**
NEW IN BOX

MODEL 629 CLASSIC DELUXE (CLASSIC DX)

REDBOOK CODE: RB-SW-H-M629CD

Similar to the Model 629 Classic but includes both synthetic and wood grips, 5 interchangeable front sights.

Production: discontinued Caliber: .44 Magnum/.44 S&W Special Action: DA/SA, revolver				
Barrel Length: 6.5", 8.375" Wt.: 48 oz. Sights: red ramp front, adjustable rear				
Capacity: 6 Grips: Hogue synthetic, or wood				
D2C:	NIB $ 850	Ex $ 650	VG+ $470	Good $ 425	
C2C:	Mint $ 820	Ex $ 600	VG+ $430	Good $ 385	Fair $ 200
Trade-In:	Mint $ 610	Ex $ 500	VG+ $340	Good $ 300	Poor $ 90

MODEL 629 - PERFORMANCE CENTER (170135)

REDBOOK CODE: RB-SW-H-629PFC

Performance Center action with ball detent lock-up, stainless frame and cylinder, chrome-flashed tear-drop hammer and trigger with stop, matte stainless finish.

Production: current Caliber: .44 Magnum/.44 S&W Special Action: DA/SA, revolver				
Barrel Length: 2.62" Wt.: 40 oz. Sights: red ramp front, adjustable rear				
Capacity: 6 Grips: wood				
D2C:	NIB $ 950	Ex $ 725	VG+ $525	Good $ 475	LMP $1,079
C2C:	Mint $ 920	Ex $ 675	VG+ $480	Good $ 430	Fair $ 220
Trade-In:	Mint $ 680	Ex $ 550	VG+ $380	Good $ 335	Poor $ 120

MODEL 632 PRO SERIES
NEW IN BOX

MODEL 632 PRO SERIES   REDBOOK CODE: RB-SW-H-M632PS

Matte silver finish, small J-frame, stainless steel frame and cylinder, enclosed hammer, fluted barrel.

Production: 2009 - discontinued   Caliber: .327 Federal Magnum   Action: DAO, revolver

Barrel Length: 2.12"   Wt.: 23 oz.   Sights: front and rear tritium night sights

Capacity: 6   Grips: synthetic

D2C:	NIB $	720	Ex $	550	VG+	$400	Good $	360		
C2C:	Mint $	700	Ex $	500	VG+	$360	Good $	325	Fair $	170
Trade-In:	Mint $	520	Ex $	425	VG+	$290	Good $	255	Poor $	90

MODEL 632 CARRY COMP
NEW IN BOX

MODEL 632 CARRY COMP (POWERPORT)
REDBOOK CODE: RB-SW-H-632CCP

Matte black finish, small J-frame, stainless steel frame and cylinder, full-lug barrel, Powerport-ported barrel, exposed hammer.

Production: 2009 - discontinued   Caliber: .327 Federal Magnum   Action: DA/SA, revolver

Barrel Length: 3"   Wt.: 25 oz.   Sights: ramp front, adjustable rear

Capacity: 6   Grips: synthetic

D2C:	NIB $	730	Ex $	575	VG+	$405	Good $	365	LMP $	899
C2C:	Mint $	710	Ex $	525	VG+	$370	Good $	330	Fair $	170
Trade-In:	Mint $	520	Ex $	425	VG+	$290	Good $	260	Poor $	90

MODEL 637
NEW IN BOX

MODEL 637   REDBOOK CODE: RB-SW-H-MDL637

Aluminum alloy J-frame with three-screws, stainless cylinder, matte silver finish. This is an upgraded reintroduction of the 1989 production model.

Production: 1996 - current   Caliber: .38 S&W Special +P   Action: DA/SA, revolver

Barrel Length: 1.87", 2"   Wt.: 15 oz.   Sights: integral front, fixed rear

Capacity: 5   Grips: synthetic

D2C:	NIB $	435	Ex $	350	VG+	$240	Good $	220	LMP $	469
C2C:	Mint $	420	Ex $	325	VG+	$220	Good $	200	Fair $	105
Trade-In:	Mint $	310	Ex $	250	VG+	$170	Good $	155	Poor $	60

MODEL 640 (16390)
NEW IN BOX

MODEL 640 (16390)   REDBOOK CODE: RB-SW-H-640390

Small J-frame, satin stainless finish, stainless steel frame and cylinder, internal hammer.

Production: current   Caliber: .357 Magnum/.38 S&W Special +P   Action: DAO, revolver

Barrel Length: 2.125"   Wt.: 23 oz.   Sights: blade front, fixed rear

Capacity: 5   Grips: synthetic

D2C:	NIB $	640	Ex $	500	VG+	$355	Good $	320	LMP $	729
C2C:	Mint $	620	Ex $	450	VG+	$320	Good $	290	Fair $	150
Trade-In:	Mint $	460	Ex $	375	VG+	$250	Good $	225	Poor $	90

MODEL 640 PRO SERIES
NEW IN BOX

MODEL 640 PRO SERIES (178043)   REDBOOK CODE: RB-SW-H-640P43

Satin stainless finish, stainless steel frame and cylinder, small J-frame, cylinder cut for moon clips.

Production: discontinued   Caliber: .357 Magnum/.38 S&W Special +P   Action: DAO, revolver

Barrel Length: 2.125"   Wt.: 23 oz.   Sights: blade front, fixed rear

Capacity: 5   Grips: synthetic

D2C:	NIB $	650	Ex $	500	VG+	$360	Good $	325		
C2C:	Mint $	630	Ex $	450	VG+	$330	Good $	295	Fair $	150
Trade-In:	Mint $	470	Ex $	375	VG+	$260	Good $	230	Poor $	90

**MODEL 640
PRO SERIES**
NEW IN BOX

**MODEL 642 CT
AIRWEIGHT**
NEW IN BOX

**MODEL 642 PRO
SERIES AIRWEIGHT**
NEW IN BOX

**MODEL 642
LADYSMITH AIRWEIGHT**
NEW IN BOX

MODEL 649
NEW IN BOX

MODEL 640 PRO SERIES (178044) REDBOOK CODE: RB-SW-H-640P44

Stainless steel frame and cylinder, satin stainless finish, small J-frame, cylinder cut for moon clips, internal hammer.

Production: current	Caliber: .357 Magnum / .38 S&W Special +P		Action: DAO, revolver			
Barrel Length: 2.125" Wt.: 23 oz.		Sights: front and rear tritium night sights				
Capacity: 5 Grips: synthetic						
D2C:	NIB $ 730	Ex $ 575	VG+ $405	Good $ 365	LMP $ 839	
C2C:	Mint $ 710	Ex $ 525	VG+ $370	Good $ 330	Fair $ 170	
Trade-In:	Mint $ 520	Ex $ 425	VG+ $290	Good $ 260	Poor $ 90	

MODEL 642 CT AIRWEIGHT REDBOOK CODE: RB-SW-H-M642CT

Aluminum alloy frame, stainless steel cylinder and barrel, matte silver finish, small J-frame.

Production: current	Caliber: .38 S&W Special +P		Action: DAO, revolver			
Barrel Length: 1.87" Wt.: 15 oz.		Sights: integral front, fixed rear				
Capacity: 5 Grips: Crimson Trace Lasergrips						
D2C:	NIB $ 585	Ex $ 450	VG+ $325	Good $ 295	LMP $ 699	
C2C:	Mint $ 570	Ex $ 425	VG+ $300	Good $ 265	Fair $ 135	
Trade-In:	Mint $ 420	Ex $ 350	VG+ $230	Good $ 205	Poor $ 60	

MODEL 642 PRO SERIES AIRWEIGHT

REDBOOK CODE: RB-SW-H-M642PS

Aluminum alloy frame, stainless steel cylinder and barrel, matte silver finish, small J-frame, cylinder cut for moon clips.

Production: current	Caliber: .38 S&W Special +P		Action: DAO, revolver			
Barrel Length: 1.87" Wt.: 15 oz.		Sights: integral front, fixed rear				
Capacity: 5 Grips: synthetic						
D2C:	NIB $ 450	Ex $ 350	VG+ $250	Good $ 225	LMP $ 499	
C2C:	Mint $ 440	Ex $ 325	VG+ $230	Good $ 205	Fair $ 105	
Trade-In:	Mint $ 320	Ex $ 275	VG+ $180	Good $ 160	Poor $ 60	

MODEL 642 LADYSMITH AIRWEIGHT

REDBOOK CODE: RB-SW-H-M642LS

Aluminum alloy frame, small round butt J-frame, internal hammer, matte silver finish, smooth target trigger.

Production: current	Caliber: .38 S&W Special +P		Action: DAO, revolver			
Barrel Length: 1.87" Wt.: 15 oz.		Sights: integral front, fixed rear				
Capacity: 5 Grips: Dymondwood						
D2C:	NIB $ 450	Ex $ 350	VG+ $250	Good $ 225	LMP $ 499	
C2C:	Mint $ 440	Ex $ 325	VG+ $230	Good $ 205	Fair $ 105	
Trade-In:	Mint $ 320	Ex $ 275	VG+ $180	Good $ 160	Poor $ 60	

MODEL 649 (.38 S&W SPECIAL BODYGUARD)

REDBOOK CODE: RB-SW-H-ML6493

Stainless steel J-frame, satin stainless finish, shrouded accessible trigger, discontinued but reintroduced as a .357 Magnum.

Production: 1985 - 1998	Caliber: .38 S&W Special		Action: DA/SA, revolver			
Barrel Length: 2", 2.125" Wt.: 20 oz.		Sights: ramp front, notch rear				
Capacity: 5 Grips: checkered walnut or synthetic						
D2C:	NIB $ 525	Ex $ 400	VG+ $290	Good $ 265		
C2C:	Mint $ 510	Ex $ 375	VG+ $270	Good $ 240	Fair $ 125	
Trade-In:	Mint $ 380	Ex $ 300	VG+ $210	Good $ 185	Poor $ 60	

MODEL 649
NEW IN BOX

MODEL 649 (.357 MAGNUM BODYGUARD)
REDBOOK CODE: RB-SW-H-MDL649
Satin stainless finish, stainless steel frame and cylinder, small magnum J-frame, shrouded accessible hammer.

Production: current Caliber: .357 Magnum/.38 S&W Special +P Action: DA/SA, revolver					
Barrel Length: 2.12" Wt.: 23 oz. Sights: blade front, fixed rear Capacity: 5					
Grips: synthetic					
D2C:	NIB $ 660	Ex $ 525	VG+ $ 365	Good $ 330	LMP $ 729
C2C:	Mint $ 640	Ex $ 475	VG+ $ 330	Good $ 300	Fair $ 155
Trade-In:	Mint $ 470	Ex $ 375	VG+ $ 260	Good $ 235	Poor $ 90

MODEL 686
NEW IN BOX

MODEL 686   REDBOOK CODE: RB-SW-H-M686XX
Stainless steel frame and cylinder, L-frame, satin stainless finish, similar to the Model 586, many variations exist of earlier productions.

Production: current Caliber: .357 Magnum/.38 Special +P Action: DA/SA, revolver					
Barrel Length: 2.5", 4", 6", 8.37" Wt.: 35 oz. (2.5") Sights: red ramp front, adjustable					
white outline rear Capacity: 6 Grips: synthetic or wood					
D2C:	NIB $ 740	Ex $ 575	VG+ $ 410	Good $ 370	LMP $ 829
C2C:	Mint $ 720	Ex $ 525	VG+ $ 370	Good $ 335	Fair $ 175
Trade-In:	Mint $ 530	Ex $ 425	VG+ $ 290	Good $ 260	Poor $ 90

MODEL 686 PLUS
NEW IN BOX

MODEL 686 PLUS (PRO SERIES)   REDBOOK CODE: RB-SW-H-ML686P
Stainless steel frame and cylinder, L-frame, satin stainless finish, cylinder cut for moon clips, plus model with 7-shot cylinder.

Production: 1996 - current Caliber: .357 Magnum/.38 Special +P Action: DA/SA, revolver					
Barrel Length: 2.5", 4", 6" Wt.: 34 oz. (2.5") Sights: red ramp front, adjustable white					
outline rear Capacity: 7 Grips: synthetic or wood					
D2C:	NIB $ 765	Ex $ 600	VG+ $ 425	Good $ 385	LMP $ 849
C2C:	Mint $ 740	Ex $ 550	VG+ $ 390	Good $ 345	Fair $ 180
Trade-In:	Mint $ 550	Ex $ 450	VG+ $ 300	Good $ 270	Poor $ 90

MODEL 686 SSR
NEW IN BOX

MODEL 686 SSR (PRO SERIES)   REDBOOK CODE: RB-SW-H-686SSR
Stainless steel frame and barrel, medium frame, chamfered charge holes, custom barrel with recessed crown, bossed mainspring.

Production: current Caliber: .357 Magnum/.38 Special +P Action: DA/SA, revolver					
Barrel Length: 4" Wt.: 38 oz. Sights: interchangeable front, adjustable rear					
Capacity: 6 Grips: ergonomic wood					
D2C:	NIB $ 860	Ex $ 675	VG+ $ 475	Good $ 430	LMP $ 999
C2C:	Mint $ 830	Ex $ 600	VG+ $ 430	Good $ 390	Fair $ 200
Trade-In:	Mint $ 620	Ex $ 500	VG+ $ 340	Good $ 305	Poor $ 90

MODEL M&P R8
NEW IN BOX

MODEL M&P R8   REDBOOK CODE: RB-SW-H-MPR8XX
Scandium-alloy frame, stainless steel cylinder, Performance Center color case trigger with overstop, Performance Center tuned action, variant of the Model 327, accessory rail.

Production: current Caliber: .357 Magnum/.38 Special +P Action: DA/SA, revolver					
Barrel Length: 5" Wt.: 36 oz. Sights: interchangeable front, adjustable v-notch rear					
Capacity: 8 Grips: synthetic					
D2C:	NIB $1,180	Ex $ 900	VG+ $ 650	Good $ 590	LMP $1,289
C2C:	Mint $1,140	Ex $ 825	VG+ $ 590	Good $ 535	Fair $ 275
Trade-In:	Mint $ 840	Ex $ 675	VG+ $ 470	Good $ 415	Poor $ 120

SMITH & WESSON

M&P 340
NEW IN BOX

M&P 340   REDBOOK CODE: RB-SW-H-MP340X

Scandium-alloy J-Frame, stainless steel cylinder, matte black finish, available with or without internal lock system.

Production: 2007 - current Caliber: .357 Magnum +P Action: DAO, revolver						
Barrel Length: 1.87" Wt.: 13 oz. Sights: XS Sights 24/7 Tritium night sights front,						
integral u-notch rear Capacity: 5 Grips: synthetic textured						
D2C:	NIB $ 775	Ex $ 600	VG+ $430	Good $ 390	LMP $ 869
C2C:	Mint $ 750	Ex $ 550	VG+ $390	Good $ 350	Fair $ 180
Trade-In:	Mint $ 560	Ex $ 450	VG+ $ 310	Good $ 275	Poor $ 90

M&P 340 CT
NEW IN BOX

M&P 340 CT   REDBOOK CODE: RB-SW-H-MP340C

Scandium-alloy J-Frame, stainless steel cylinder, matte black finish, available with or without internal lock system.

Production: 2007 - current Caliber: .357 Magnum +P Action: DAO, revolver						
Barrel Length: 1.87" Wt.: 13 oz. Sights: XS Sights 24/7 Tritium Night Sights front,						
integral u-notch rear Capacity: 5 Grips: Crimson Trace Lasergrips						
D2C:	NIB $ 985	Ex $ 750	VG+ $545	Good $ 495	LMP $ 1129
C2C:	Mint $ 950	Ex $ 700	VG+ $500	Good $ 445	Fair $ 230
Trade-In:	Mint $ 700	Ex $ 575	VG+ $390	Good $ 345	Poor $ 120

M&P 360
NEW IN BOX

M&P 360   REDBOOK CODE: RB-SW-H-MP360X

Scandium-alloy J-Frame, stainless steel PVD coated cylinder, matte black finish.

Production: 2007 - 2012 Caliber: .357 Magnum +P Action: DA/SA, revolver						
Barrel Length: 1.87", 3" Wt.: 13 oz., 15 oz. Sights: XS Sights 24/7 Tritium						
night sights front, integral u-notch rear Capacity: 5 Grips: synthetic textured						
D2C:	NIB $ 760	Ex $ 600	VG+ $420	Good $ 380	
C2C:	Mint $ 730	Ex $ 525	VG+ $380	Good $ 345	Fair $ 175
Trade-In:	Mint $ 540	Ex $ 450	VG+ $300	Good $ 270	Poor $ 90

GOVERNOR
NEW IN BOX

GOVERNOR   REDBOOK CODE: RB-SW-H-GOVERN

Scandium-alloy frame, stainless PVD cylinder, matte black finish, includes moon clips, add $250 for Crimson Trace Lasergips model.

Production: 2011 - current Caliber: .45 ACP, .45 LC Action: DA/SA, revolver						
Barrel Length: 2.75" Wt.: 30 oz. Sights: dovetailed Tritium night sight front, fixed notch-						
cut rear Capacity: 6 Grips: synthetic						
D2C:	NIB $ 760	Ex $ 600	VG+ $420	Good $ 380	LMP $ 869
C2C:	Mint $ 730	Ex $ 525	VG+ $380	Good $ 345	Fair $ 175
Trade-In:	Mint $ 540	Ex $ 450	VG+ $300	Good $ 270	Poor $ 90

MODEL 59   REDBOOK CODE: RB-SW-H-MDL59X

Alloy frame, steel slide, blue or nickel finish, most feature serrated front and rear grip straps, models with smooth front/rear grip straps are rare and worth a high premium.

Production: 1971 - 1982 Caliber: 9mm Action: DA/SA, semi-auto						
Barrel Length: 4" Sights: fixed Capacity: 14 Magazine: detachable box						
Grips: checkered plastic						
D2C:	NIB $ 580	Ex $ 450	VG+ $320	Good $ 290	
C2C:	Mint $ 560	Ex $ 425	VG+ $290	Good $ 265	Fair $ 135
Trade-In:	Mint $ 420	Ex $ 325	VG+ $230	Good $ 205	Poor $ 60

MODEL 61 ESCORT
MINT

MODEL 61 ESCORT REDBOOK CODE: RB-SW-H-MDL61E

Blue or nickel finish, manual thumb safety, serial number range: B1-B65440.

Production: 1970 - 1974 Caliber: .22 LR Action: SA, semi-auto

Barrel Length: 2.12" Wt.: 14 oz. Sights: fixed-ramped front, dovetail rear

Capacity: 5 Magazine: detachable box Grips: checkered plastic

D2C:	NIB $ 460	Ex $ 350	VG+ $ 255	Good $ 230		
C2C:	Mint $ 450	Ex $ 325	VG+ $ 230	Good $ 210	Fair $ 110
Trade-In:	Mint $ 330	Ex $ 275	VG+ $ 180	Good $ 165	Poor $ 60

MODEL 457
EXCELLENT

MODEL 457 REDBOOK CODE: RB-SW-H-ML457C

Alloy frame, carbon steel slide, matte black finish, external safety/decocker, bobbed hammer, also available in a stainless model.

Production: 1996 - discontinued Caliber: .45 ACP Action: DA/SA, semi-auto

Barrel Length: 3.75" Wt.: 29 oz. Sights: 3-dot Capacity: 7

Magazine: detachable box Grips: one-piece synthetic

D2C:	NIB $ 485	Ex $ 375	VG+ $ 270	Good $ 245		
C2C:	Mint $ 470	Ex $ 350	VG+ $ 250	Good $ 220	Fair $ 115
Trade-In:	Mint $ 350	Ex $ 275	VG+ $ 190	Good $ 170	Poor $ 60

MODEL 469 VERY GOOD +

MODEL 469 (MINI) REDBOOK CODE: RB-SW-H-ML469M

Alloy frame, carbon steel slide, matte blue or nickel (rare) finish, bobbed hammer, ambidextrous safety.

Production: 1983 - 1988 Caliber: 9mm Action: DA/SA, semi-auto

Barrel Length: 3.5" Wt.: 26 oz. Sights: ramp front, drift adjustable rear

Capacity: 12 Magazine: detachable box Grips: black synthetic

D2C:	NIB $ 460	Ex $ 350	VG+ $ 255	Good $ 230		
C2C:	Mint $ 450	Ex $ 325	VG+ $ 230	Good $ 210	Fair $ 110
Trade-In:	Mint $ 330	Ex $ 275	VG+ $ 180	Good $ 165	Poor $ 60

MODEL 908
VERY GOOD +

MODEL 908 REDBOOK CODE: RB-SW-H-ML908C

Steel slide, compact alloy frame, bobbed hammer, external safety.

Production: 1996 - discontinued Caliber: 9mm Action: DA/SA, semi-auto

Barrel Length: 3.5" Wt.: 25 oz. Sights: 3-dot fixed Capacity: 8

Magazine: detachable box Grips: black synthetic

D2C:	NIB $ 520	Ex $ 400	VG+ $ 290	Good $ 260		
C2C:	Mint $ 500	Ex $ 375	VG+ $ 260	Good $ 235	Fair $ 120
Trade-In:	Mint $ 370	Ex $ 300	VG+ $ 210	Good $ 185	Poor $ 60

MODEL 908S REDBOOK CODE: RB-SW-H-M908SC

An all stainless steel version of the Model 908, bobbed hammer, external safety.

Production: discontinued Caliber: 9mm Action: DA/SA, semi-auto

Barrel Length: 3.5" Wt.: 25 oz. Sights: 3-dot fixed Capacity: 8

Magazine: detachable box Grips: black synthetic

D2C:	NIB $ 520	Ex $ 400	VG+ $ 290	Good $ 260		
C2C:	Mint $ 500	Ex $ 375	VG+ $ 260	Good $ 235	Fair $ 120
Trade-In:	Mint $ 370	Ex $ 300	VG+ $ 210	Good $ 185	Poor $ 60

SMITH & WESSON

MODEL 3913 COMPACT MINT
Courtesy of Bud's Gun Shop

MODEL 3913 COMPACT REDBOOK CODE: RB-SW-H-M3913C

Alloy frame, stainless steel slide, stainless finish, ambidextrous safety/decocking lever.

Production: 1999 Caliber: 9mm Action: DA/SA, semi-auto Barrel Length: 3.5"										
Wt.: 25 oz. Sights: fixed Capacity: 8 Magazine: detachable box Grips: black synthetic										
D2C:	NIB $	510	Ex $	400	VG+ $	285	Good $	255		
C2C:	Mint $	490	Ex $	375	VG+ $	260	Good $	230	Fair $	120
Trade-In:	Mint $	370	Ex $	300	VG+ $	200	Good $	180	Poor $	60

MODEL 3914 COMPACT REDBOOK CODE: RB-SW-H-3914CT

Blue finish, alloy frame, carbon steel slide, bobbed hammer, similar to Model 3913.

Production: 1996 Caliber: 9mm Action: DA/SA, semi-auto										
Barrel Length: 3.5" Wt.: 25 oz. Sights: fixed Capacity: 8										
Magazine: detachable box Grips: black synthetic										
D2C:	NIB $	510	Ex $	400	VG+ $	285	Good $	255		
C2C:	Mint $	490	Ex $	375	VG+ $	260	Good $	230	Fair $	120
Trade-In:	Mint $	370	Ex $	300	VG+ $	200	Good $	180	Poor $	60

MODEL 3953 COMPACT REDBOOK CODE: RB-SW-H-3953CS

Aluminum alloy frame, stainless steel slide, external safety/decocking lever.

Production: 1999 Caliber: 9mm Action: DAO, semi-auto										
Barrel Length: 3.5" Wt.: 24 oz. Sights: fixed 3-dot Capacity: 8										
Magazine: detachable box Grips: black synthetic										
D2C:	NIB $	495	Ex $	400	VG+ $	275	Good $	250		
C2C:	Mint $	480	Ex $	350	VG+ $	250	Good $	225	Fair $	115
Trade-In:	Mint $	360	Ex $	300	VG+ $	200	Good $	175	Poor $	60

MODEL 4006
EXCELLENT
Courtesy of Bud's Gun Shop

MODEL 4006 REDBOOK CODE: RB-SW-H-4006SS

Stainless steel slide and frame, ambidextrous safety, add roughly $100 for night sights.

Caliber: .40 S&W Action: DA/SA, semi-auto Barrel Length: 4"										
Sights: windage and elevation adjustable rear Capacity: 11 Magazine: detachable box										
Grips: one-piece wraparound										
D2C:	NIB $	530	Ex $	425	VG+ $	295	Good $	265		
C2C:	Mint $	510	Ex $	375	VG+ $	270	Good $	240	Fair $	125
Trade-In:	Mint $	380	Ex $	300	VG+ $	210	Good $	190	Poor $	60

MODEL 4006 TSW REDBOOK CODE: RB-SW-H-4006TS

Stainless steel slide, alloy frame, satin stainless finish with black controls, tactical accessory rail.

Caliber: .40 S&W Action: DA/SA, semi-auto Barrel Length: 4" Sights: 3-dot Novak										
Capacity: 10 Magazine: detachable box Grips: one-piece wraparound										
D2C:	NIB $	670	Ex $	525	VG+ $	370	Good $	335		
C2C:	Mint $	650	Ex $	475	VG+ $	340	Good $	305	Fair $	155
Trade-In:	Mint $	480	Ex $	400	VG+ $	270	Good $	235	Poor $	90

MODEL 4013 COMPACT REDBOOK CODE: RB-SW-H-4013CS
Aluminum alloy frame, stainless steel slide, ambidextrous safety/decocking lever.

Production: 1991 - ~1995 Caliber: .40 S&W Action: DA/SA, semi-auto					
Barrel Length: 3.5" Sights: fixed 3-dot Capacity: 8					
Magazine: detachable box Grips: black synthetic					
D2C:	NIB $ 610	Ex $ 475	VG+ $340	Good $ 305	
C2C:	Mint $ 590	Ex $ 425	VG+ $ 310	Good $ 275	Fair $ 145
Trade-In:	Mint $ 440	Ex $ 350	VG+ $240	Good $ 215	Poor $ 90

MODEL 4014 COMPACT REDBOOK CODE: RB-SW-H-4014CO
Aluminum alloy frame, carbon steel slide, blue finish, ambidextrous safety/decocking lever.

Production: 1991 - 1993 Caliber: .40 S&W Action: DA/SA, semi-auto					
Barrel Length: 3.5" Sights: fixed 3-dot Capacity: 8					
Magazine: detachable box Grips: black synthetic					
D2C:	NIB $ 580	Ex $ 450	VG+ $320	Good $ 290	
C2C:	Mint $ 560	Ex $ 425	VG+ $290	Good $ 265	Fair $ 135
Trade-In:	Mint $ 420	Ex $ 325	VG+ $230	Good $ 205	Poor $ 60

MODEL 4053 COMPACT REDBOOK CODE: RB-SW-H-4053CS
Aluminum alloy frame, stainless slide, similar to the Model 4013 but double-action only.

Production: 1991 - discontinued Caliber: .40 S&W Action: DAO, semi-auto					
Barrel Length: 3.5" Wt.: 28 oz. Sights: 3-dot fixed Capacity: 8					
Magazine: detachable box Grips: black synthetic					
D2C:	NIB $ 580	Ex $ 450	VG+ $320	Good $ 290	
C2C:	Mint $ 560	Ex $ 425	VG+ $290	Good $ 265	Fair $ 135
Trade-In:	Mint $ 420	Ex $ 325	VG+ $230	Good $ 205	Poor $ 60

MODEL 4516 COMPACT
EXCELLENT

MODEL 4516 COMPACT REDBOOK CODE: RB-SW-H-4516CS
Compact stainless steel frame and slide, bobbed hammer, ambi safety.

Production: 1990 Caliber: .45 ACP Action: DA/SA, semi-auto					
Barrel Length: 3.75" Wt.: 35 oz. Sights: 3-dot fixed Capacity: 7					
Magazine: detachable box Grips: black synthetic					
D2C:	NIB $ 550	Ex $ 425	VG+ $305	Good $ 275	
C2C:	Mint $ 530	Ex $ 400	VG+ $280	Good $ 250	Fair $ 130
Trade-In:	Mint $ 400	Ex $ 325	VG+ $220	Good $ 195	Poor $ 60

MODEL CS9 CHIEFS SPECIAL
NEW IN BOX

MODEL CS9 CHIEFS SPECIAL REDBOOK CODE: RB-SW-H-CS9CSP
Aluminum alloy frame, stainless steel slide in blue or stainless finish, external safety.

Production: 1998 - discontinued Caliber: 9mm Action: DA/SA, semi-auto					
Barrel Length: 3" Wt.: 21 oz. Sights: 3-dot fixed Capacity: 7					
Magazine: detachable box Grips: Hogue-wraparound rubber					
D2C:	NIB $ 625	Ex $ 475	VG+ $345	Good $ 315	
C2C:	Mint $ 600	Ex $ 450	VG+ $320	Good $ 285	Fair $ 145
Trade-In:	Mint $ 450	Ex $ 350	VG+ $250	Good $ 220	Poor $ 90

SMITH & WESSON

SIGMA MODEL SW9F
EXCELLENT

SIGMA MODEL SW9F   REDBOOK CODE: RB-SW-H-SMSW9F

Polymer frame, blue carbon steel slide, stainless barrel, optional tritium night sights.

Production: 1994 - 1998   Caliber: 9mm   Action: striker-fired, semi-auto

Barrel Length: 4.5"   Wt.: 26 oz.   Sights: fixed white 3-dot or tritium night sights

Capacity: 10, 17   Magazine: detachable box   Grips: polymer grip frame

D2C:	NIB $ 450	Ex $ 350	VG+ $ 250	Good $ 225		
C2C:	Mint $ 440	Ex $ 325	VG+ $ 230	Good $ 205	Fair $ 105
Trade-In:	Mint $ 320	Ex $ 275	VG+ $ 180	Good $ 160	Poor $ 60

SIGMA MODEL SW9C   REDBOOK CODE: RB-SW-H-SMSW9C

Polymer frame, blue carbon steel slide, stainless barrel, optional tritium night sights.

Production: 1995 - 1998   Caliber: 9mm   Action: striker-fired, semi-auto

Barrel Length: 4"   Wt.: 25 oz.   Sights: fixed white 3-dot or tritium night sights

Capacity: 10, 17   Magazine: detachable box   Grips: polymer grip frame

D2C:	NIB $ 460	Ex $ 350	VG+ $ 255	Good $ 230	
C2C:	Mint $ 450	Ex $ 325	VG+ $ 230	Good $ 210	Fair $ 110
Trade-In:	Mint $ 330	Ex $ 275	VG+ $ 180	Good $ 165	Poor $ 60

SIGMA MODEL SW9M   REDBOOK CODE: RB-SW-H-SMSW9M

Polymer frame, carbon steel slide with black finish, ultra-compact design.

Production: 1996 - 1998   Caliber: 9mm   Action: striker-fired, semi-auto

Barrel Length: 3.25"   Wt.: 18 oz.   Sights: post front, fixed channel rear

Capacity: 7   Magazine: detachable box   Grips: polymer grip frame

D2C:	NIB $ 475	Ex $ 375	VG+ $ 265	Good $ 240	
C2C:	Mint $ 460	Ex $ 350	VG+ $ 240	Good $ 215	Fair $ 110
Trade-In:	Mint $ 340	Ex $ 275	VG+ $ 190	Good $ 170	Poor $ 60

SIGMA MODEL SW9V   REDBOOK CODE: RB-SW-H-MSW9V

Gray or black polymer frame, stainless finish on slide.

Production: 1996 - 1998   Caliber: 9mm   Action: striker-fired, semi-auto

Barrel Length: 4"   Wt.: 25 oz.   Sights: fixed white 3-dot

Capacity: 10, 16   Magazine: detachable box   Grips: polymer grip frame

D2C:	NIB $ 400	Ex $ 325	VG+ $ 220	Good $ 200	
C2C:	Mint $ 390	Ex $ 300	VG+ $ 200	Good $ 180	Fair $ 95
Trade-In:	Mint $ 290	Ex $ 225	VG+ $ 160	Good $ 140	Poor $ 60

**SIGMA MODEL
SW40F**
EXCELLENT

SIGMA MODEL SW40F   REDBOOK CODE: RB-SW-H-SMSW4F

Polymer frame, blue carbon steel slide, stainless barrel, optional tritium night sights.

Production: 1994 - 1998   Caliber: .40 S&W   Action: striker-fired, semi-auto

Barrel Length: 4.5"   Wt.: 26 oz.   Sights: fixed white 3-dot or tritium night sights

Capacity: 10, 15   Magazine: detachable box   Grips: polymer grip frame

D2C:	NIB $ 400	Ex $ 325	VG+ $ 220	Good $ 200	
C2C:	Mint $ 390	Ex $ 300	VG+ $ 200	Good $ 180	Fair $ 95
Trade-In:	Mint $ 290	Ex $ 225	VG+ $ 160	Good $ 140	Poor $ 60

SIGMA MODEL SW40C REDBOOK CODE: RB-SW-H-SMSW4C

Polymer frame, blue carbon steel slide, stainless barrel, optional tritium night sights.

Production: 1995 - 1998 Caliber: .40 S&W Action: striker-fired, semi-auto
Barrel Length: 4" Wt.: 25 oz. Sights: fixed white 3-dot or tritium night sights
Capacity: 10, 14 Magazine: detachable box Grips: polymer grip frame

D2C:	NIB $ 410	Ex $ 325	VG+ $230	Good $ 205		
C2C:	Mint $ 400	Ex $ 300	VG+ $210	Good $ 185	Fair $ 95	
Trade-In:	Mint $ 300	Ex $ 250	VG+ $160	Good $ 145	Poor $ 60	

SIGMA MODEL SW380 REDBOOK CODE: RB-SW-H-SMSW38

Black polymer frame, blue-steel slide.

Production: 1995 - 2000 Caliber: .380 ACP Action: striker-fired, semi-auto
Barrel Length: 3" Wt.: 14 oz. Sights: post front, fixed channel rear
Capacity: 6 Magazine: detachable box Grips: polymer grip frame

	NIB/Mint	Ex	VG+	Good	Fair/Poor	
D2C:	NIB $ 299	Ex $ 250	VG+ $165	Good $ 150		
C2C:	Mint $ 290	Ex $ 225	VG+ $150	Good $ 135	Fair $ 70	
Trade-In:	Mint $ 220	Ex $ 175	VG+ $120	Good $ 105	Poor $ 30	

SIGMA MODEL SW380
EXCELLENT
Courtesy of Bud's Gun Shop

ENHANCED SIGMA MODEL SW9G REDBOOK CODE: RB-SW-H-ESSW9G

Enhanced Sigma Series, NATO green polymer frame, stainless slide with Melonite coating, accessory groove on frame, model SW9GVE has stainless finish slide.

Production: 2001 - 2003 Caliber: 9mm Action: striker-fired, semi-auto
Barrel Length: 4" Wt.: 25 oz. Sights: tritium front, 2-dot rear
Capacity: 10, 16 Magazine: detachable box Grips: green polymer grip frame

	NIB/Mint	Ex	VG+	Good	Fair/Poor	
D2C:	NIB $ 360	Ex $ 275	VG+ $200	Good $ 180		
C2C:	Mint $ 350	Ex $ 250	VG+ $180	Good $ 165	Fair $ 85	
Trade-In:	Mint $ 260	Ex $ 225	VG+ $150	Good $ 130	Poor $ 60	

ENHANCED SIGMA MODEL SW9E REDBOOK CODE: RB-SW-H-ESSW9E

Enhanced Sigma Series, polymer frame, stainless slide with Melonite coating, accessory groove on frame.

Production: 1999 - 2002 Caliber: 9mm Action: striker-fired, semi-auto
Barrel Length: 4" Wt.: 25 oz. Sights: tritium night sights
Capacity: 10, 16 Magazine: detachable box Grips: polymer grip frame

	NIB/Mint	Ex	VG+	Good	Fair/Poor	
D2C:	NIB $ 360	Ex $ 275	VG+ $200	Good $ 180		
C2C:	Mint $ 350	Ex $ 250	VG+ $180	Good $ 165	Fair $ 85	
Trade-In:	Mint $ 260	Ex $ 225	VG+ $150	Good $ 130	Poor $ 60	

ENHANCED SIGMA MODEL SW9VE REDBOOK CODE: RB-SW-H-ESSW9V

Stainless steel slide and barrel, textured polymer frame, two-tone finish, accessory rail, SDT trigger system, rear slide serrations.

Production: 1999 - current Caliber: 9mm Action: striker-fired, semi-auto
Barrel Length: 4" Wt.: 23 oz. Sights: fixed white 3-dot Capacity: 10, 16
Magazine: detachable box Grips: polymer grip frame

D2C:	NIB $ 325	Ex $ 250	VG+ $ 180	Good $ 165	LMP $ 379	
C2C:	Mint $ 320	Ex $ 225	VG+ $ 170	Good $ 150	Fair $ 75	
Trade-In:	Mint $ 240	Ex $ 200	VG+ $ 130	Good $ 115	Poor $ 60	

ENHANCED SIGMA MODEL SW9VE
EXCELLENT

ENHANCED SIGMA MODEL SW40E REDBOOK CODE: RB-SW-H-ESSW4E
Enhanced Sigma Series, polymer frame, stainless slide with Melonite coating, accessory groove on frame.

Production: 1999 - 2002 Caliber: .40 S&W Action: striker-fired, semi-auto					
Barrel Length: 4" Wt.: 24 oz. Sights: tritium night sights					
Capacity: 10, 14 Magazine: detachable box Grips: polymer grip frame					
D2C:	NIB $ 340	Ex $ 275	VG+ $ 190	Good $ 170	
C2C:	Mint $ 330	Ex $ 250	VG+ $ 170	Good $ 155	Fair $ 80
Trade-In:	Mint $ 250	Ex $ 200	VG+ $ 140	Good $ 120	Poor $ 60

ENHANCED SIGMA MODEL SW40G REDBOOK CODE: RB-SW-H-ESSW4G
Enhanced Sigma Series, NATO green polymer frame, stainless slide with Melonite coating or stainless finish, accessory groove on frame.

Production: 2001 - discontinued Caliber: .40 S&W Action: striker-fired, semi-auto					
Barrel Length: 4" Wt.: 25 oz. Sights: tritium front, 2-dot rear Capacity: 10, 14					
Magazine: detachable box Grips: green polymer grip frame					
D2C:	NIB $ 340	Ex $ 275	VG+ $ 190	Good $ 170	
C2C:	Mint $ 330	Ex $ 250	VG+ $ 170	Good $ 155	Fair $ 80
Trade-In:	Mint $ 250	Ex $ 200	VG+ $ 140	Good $ 120	Poor $ 60

ENHANCED SIGMA MODEL SW40VE REDBOOK CODE: SW-H-ESSW4V
Stainless steel slide and barrel, textured polymer frame, two-tone finish, accessory rail, SDT trigger system, front and rear serrations.

Production: 1999 - current Caliber: .40 S&W Action: striker-fired, semi-auto					
Barrel Length: 4" Wt.: 23 oz. Sights: fixed white 3-dot					
Capacity: 10, 14 Magazine: detachable box Grips: polymer grip frame					
D2C:	NIB $ 325	Ex $ 250	VG+ $180	Good $ 165	LMP $ 379
C2C:	Mint $ 320	Ex $ 225	VG+ $170	Good $ 150	Fair $ 75
Trade-In:	Mint $ 240	Ex $ 200	VG+ $130	Good $ 115	Poor $ 60

ENHANCED SIGMA MODEL SW40VE
NEW IN BOX

SW99 REDBOOK CODE: RB-SW-H-MLSW99
Stainless steel slide and barrel with Melonite finish, decocking lever on slide, polymer frame, similar to Walther P99 as this model is a joint collaboration with S&W and Walther.

Production: 1999 - 2004 Caliber: 9mm, .40 S&W, .45 ACP Action: striker-fired, semi-auto					
Barrel Length: 4", 4.125", 4.25" Wt.: 25 oz. Sights: white-dot front, adjustable rear					
Capacity: 9, 10, 12, 16 Magazine: detachable box Grips: polymer grip frame with interchangeable backstraps					
D2C:	NIB $ 545	Ex $ 425	VG+ $300	Good $ 275	
C2C:	Mint $ 530	Ex $ 400	VG+ $280	Good $ 250	Fair $ 130
Trade-In:	Mint $ 390	Ex $ 325	VG+ $220	Good $ 195	Poor $ 60

SW99
EXCELLENT

SW99 COMPACT REDBOOK CODE: RB-SW-H-SW99CO
Compact size, stainless steel slide and barrel with Melonite finish, decocking lever on slide, polymer frame.

Production: 2003 - 2004 Caliber: 9mm, .40 S&W Action: striker-fired, semi-auto					
Barrel Length: 3.5" Wt.: 23 oz. Sights: white-dot front, adjustable rear					
Capacity: 8, 9 Magazine: detachable box Grips: polymer grip frame with interchangeable backstraps					
D2C:	NIB $ 545	Ex $ 425	VG+ $300	Good $ 275	
C2C:	Mint $ 530	Ex $ 400	VG+ $280	Good $ 250	Fair $ 130
Trade-In:	Mint $ 390	Ex $ 325	VG+ $220	Good $ 195	Poor $ 60

SD40 VE
NEW IN BOX

SD40 VE   REDBOOK CODE: RB-SW-H-SD40VE
Stainless steel slide and barrel, polymer frame, SDT (Self Defense Trigger), accessory rail, textured grip and backstrap, two-tone finish, front and rear serrations.

Production: current   Caliber: .40 S&W   Action: striker-fired, semi-auto

Barrel Length: 4"   Wt.: 23 oz.   Sights: white-dot front, fixed 2-dot rear

Capacity: 10, 14   Magazine: detachable box   Grips: polymer grip frame

D2C:	NIB $ 340	Ex $ 275	VG+ $ 190	Good $ 170	LMP $ 379
C2C:	Mint $ 330	Ex $ 250	VG+ $ 170	Good $ 155	Fair $ 80
Trade-In:	Mint $ 250	Ex $ 200	VG+ $ 140	Good $ 120	Poor $ 60

SD9 VE
NEW IN BOX

SD9 VE   REDBOOK CODE: RB-SW-H-MSD9VE
Stainless steel slide and barrel, polymer frame, SDT (Self Defense Trigger), accessory rail, textured grip and backstrap, two-tone finish, front and rear serrations.

Production: current   Caliber: 9mm   Action: striker-fired, semi-auto

Barrel Length: 4"   Wt.: 23 oz.   Sights: white-dot front, fixed 2-dot rear

Capacity: 10, 16   Magazine: detachable box   Grips: polymer grip frame

D2C:	NIB $ 340	Ex $ 275	VG+ $ 190	Good $ 170	LMP $ 379
C2C:	Mint $ 330	Ex $ 250	VG+ $ 170	Good $ 155	Fair $ 80
Trade-In:	Mint $ 250	Ex $ 200	VG+ $ 140	Good $ 120	Poor $ 60

BODYGUARD .380
NEW IN BOX

BODYGUARD .380   REDBOOK CODE: RB-SW-H-BGD380
Integrated INSIGHT laser, polymer frame, stainless slide with matte black finish.

Production: 2010 - current   Caliber: .380 ACP   Action: DAO, semi-auto

Barrel Length: 2.75"   Wt.: 12 oz.   Sights: fixed front, drift adjustable rear

Capacity: 6   Magazine: detachable box   Grips: polymer grip frame

D2C:	NIB $ 389	Ex $ 300	VG+ $ 215	Good $ 195	LMP $ 419
C2C:	Mint $ 380	Ex $ 275	VG+ $200	Good $ 180	Fair $ 90
Trade-In:	Mint $ 280	Ex $ 225	VG+ $ 160	Good $ 140	Poor $ 60

M&P SHIELD
NEW IN BOX

M&P SHIELD   REDBOOK CODE: RB-SW-H-MPSHLD
Lightweight polymer frame, corrosion resistant coated stainless slide and barrel, thumb safety.

Production: 2012 - current   Caliber: 9mm, .40 S&W   Action: striker-fired, semi-auto

Barrel Length: 3.1"   Wt.: 19 oz.   Sights: fixed white-dot front, white 2-dot rear

Capacity: 6 (9mm), 7, or 8 (.40 cal.)   Magazine: detachable box

Grips: textured polymer grip frame

D2C:	NIB $ 429	Ex $ 350	VG+ $240	Good $ 215	LMP $ 449
C2C:	Mint $ 420	Ex $ 300	VG+ $220	Good $ 195	Fair $ 100
Trade-In:	Mint $ 310	Ex $ 250	VG+ $ 170	Good $ 155	Poor $ 60

M&P45
NEW IN BOX

M&P45   REDBOOK CODE: RB-SW-H-MP45XX
Full size black or Dark Earth Brown polymer frame, stainless steel slide and barrel with black finish, ambidextrous slide stop, reversible magazine catch, optional thumb safety, accessory rail, model with Crimson Trace Lasergrips adds $230 to retail value.

Production: 2007 - current   Caliber: .45 ACP   Action: striker-fired, semi-auto

Barrel Length: 4", 4.5"   Wt.: 28 oz.   Sights: white-dot front, low-profile carry rear

Capacity: 10   Magazine: detachable box   Grips: polymer grip frame with interchangeable grip sizes

D2C:	NIB $ 510	Ex $ 400	VG+ $ 285	Good $ 255	LMP $ 599
C2C:	Mint $ 490	Ex $ 375	VG+ $ 260	Good $ 230	Fair $ 120
Trade-In:	Mint $ 370	Ex $ 300	VG+ $ 200	Good $ 180	Poor $ 60

SMITH & WESSON

M&P45C
NEW IN BOX

M&P45C   REDBOOK CODE: RB-SW-H-MP45CX

Compact polymer frame in black or Dark Earth Brown, stainless slide and barrel, ambidextrous slide stop, reversible magazine catch, optional thumb safety, accessory rail.

Production: 2008 - current   Caliber: .45 ACP   Action: striker-fired, semi-auto

Barrel Length: 4"   Wt.: 26 oz.   Sights: white-dot front, low-profile carry rear

Capacity: 8   Magazine: detachable box

Grips: polymer grip frame with interchangeable grip sizes

D2C:	NIB	$510	Ex	$400	VG+	$285	Good $255	LMP	$599
C2C:	Mint	$490	Ex	$375	VG+	$260	Good $ 230	Fair	$120
Trade-In:	Mint	$370	Ex	$300	VG+	$200	Good $ 180	Poor	$ 60

M&P9
NEW IN BOX

M&P9   REDBOOK CODE: RB-SW-H-MP9XXX

Full-size black polymer frame, stainless slide and barrel, ambidextrous slide stop, reversible magazine catch, optional thumb safety, accessory rail, optional "Carry and Range Kit" adds $40 to LMP, slight premium for TALO exclusive nickel boron coated model.

Production: 2006 - current   Caliber: 9mm   Action: striker-fired, semi-auto

Barrel Length: 4.25"   Wt.: 24-26 oz.   Sights: white-dot front, low-profile carry rear or optional tritium night sights   Capacity: 10, 17

Magazine: detachable box   Grips: polymer grip frame with interchangeable grip sizes

D2C:	NIB	$ 489	Ex $	375	VG+	$270	Good $ 245	LMP	$ 569
C2C:	Mint	$ 470	Ex $	350	VG+	$250	Good $ 225	Fair	$ 115
Trade-In:	Mint	$ 350	Ex $	275	VG+	$200	Good $ 175	Poor	$ 60

M&P9 PRO SERIES
NEW IN BOX

M&P9 PRO SERIES   REDBOOK CODE: RB-SW-H-MP9PRO

Full size black polymer frame, stainless steel slide and barrel with black finish, reduced 4-5.5 lb. trigger, ambidextrous slide stop, reversible magazine catch, accessory rail.

Production: 2008 - current   Caliber: 9mm   Action: striker-fired, semi-auto

Barrel Length: 4.25", 5"   Wt.: 24-26 oz.   Sights: Novak green fiber-optic front, Novak reduced glare rear (5" model) or front and rear night sights (4.25" model)

Capacity: 10, 17   Magazine: detachable box

Grips: polymer grip frame with interchangeable grip sizes

D2C:	NIB	$ 615	Ex $	475	VG+ $	340	Good $ 310	LMP	$ 669
C2C:	Mint	$ 600	Ex $	425	VG+ $	310	Good $ 280	Fair	$ 145
Trade-In:	Mint	$ 440	Ex $	350	VG+ $	240	Good $ 220	Poor	$ 90

M&P9C
NEW IN BOX

M&P9C   REDBOOK CODE: RB-SW-H-MP9CXX

Compact polymer frame, stainless slide and barrel, ambidextrous slide stop, reversible magazine catch, optional thumb safety, optional magazine safety, accessory rail, Crimson Trace Lasergrips model adds $240 to retail price.

Production: current   Caliber: 9mm   Action: striker-fired, semi-auto

Barrel Length: 3.5"   Wt.: 22 oz.   Sights: white-dot front, low-profile carry rear

Capacity: 10, 12   Magazine: detachable box

Grips: polymer grip frame with interchangeable grip sizes

D2C:	NIB	$ 510	Ex $ 400	VG+	$ 285	Good $ 255	LMP	$ 569
C2C:	Mint	$ 490	Ex $ 375	VG+	$ 260	Good $ 230	Fair	$ 120
Trade-In:	Mint	$ 370	Ex $ 300	VG+	$ 200	Good $ 180	Poor	$ 60

M&P40 NEW IN BOX

M&P40   REDBOOK CODE: RB-SW-H-MP40XX

Full-size black polymer frame, stainless slide and barrel, ambidextrous slide stop, reversible magazine catch, optional thumb safety, accessory rail.

Production: current   Caliber: .40 S&W   Action: striker-fired, semi-auto

Barrel Length: 4.25"   Wt.: 24.25 oz.

Sights: white-dot front, low-profile carry rear or optional tritium night sights   Capacity: 10, 15

Magazine: detachable box   Grips: polymer grip frame with interchangeable grip sizes

D2C:	NIB $ 489	Ex $ 375	VG+ $270	Good $ 245	LMP $ 569
C2C:	Mint $ 470	Ex $ 350	VG+ $250	Good $ 225	Fair $ 115
Trade-In:	Mint $ 350	Ex $ 275	VG+ $200	Good $ 175	Poor $ 60

M&P40 PRO SERIES NEW IN BOX

M&P40 PRO SERIES   REDBOOK CODE: RB-SW-H-MP40PS

Full-size black polymer frame, stainless steel slide and barrel with black finish, reduced 4-5.5 lb. trigger, ambidextrous slide stop, reversible magazine catch, accessory rail.

Production: 2008 - current   Caliber: .40 S&W   Action: striker-fired, semi-auto

Barrel Length: 4.25", 5"   Wt.: 24-26 oz.

Sights: Novak green fiber-optic front, Novak reduced glare rear (5" model) or front and rear night sights (4.25" model)   Capacity: 15

Magazine: detachable box   Grips: polymer grip frame with interchangeable grip sizes

D2C:	NIB $615	Ex $475	VG+ $340	Good $310	LMP $669
C2C:	Mint $600	Ex $425	VG+ $310	Good $280	Fair $145
Trade-In:	Mint $440	Ex $350	VG+ $240	Good $220	Poor $90

M&P40C NEW IN BOX

M&P40C   REDBOOK CODE: RB-SW-H-MP40CX

Compact polymer frame, stainless slide and barrel, ambidextrous slide stop, reversible magazine catch, optional thumb safety, accessory rail.

Production: current   Caliber: .40 S&W   Action: striker-fired, semi-auto

Barrel Length: 3.5"   Wt.: 22 oz.   Sights: steel-ramp dovetail mount front, Novak lo-mount carry rear (tritium sights optional)   Capacity: 10

Magazine: detachable box   Grips: polymer grip frame with interchangeable grip sizes

D2C:	NIB $ 510	Ex $ 400	VG+ $285	Good $ 255	LMP $ 569
C2C:	Mint $ 490	Ex $ 375	VG+ $260	Good $ 230	Fair $ 120
Trade-In:	Mint $ 370	Ex $ 300	VG+ $200	Good $ 180	Poor $ 60

M&P40 VTAC NEW IN BOX

M&P40 VTAC   REDBOOK CODE: RB-SW-H-MP40VT

Flat Dark Earth PVD coated stainless slide, Flat Dark Earth frame, ambidextrous slide stop, reversible magazine catch, accessory rail.

Production: ~2011 - current   Caliber: .40 S&W   Action: striker-fired, semi-auto

Barrel Length: 4.25"   Wt.: 24 oz.   Sights: VTAC Warrior front and rear

Capacity: 15   Magazine: detachable box   Grips: Flat Dark Earth polymer grip frame

D2C:	NIB $ 720	Ex $ 550	VG+ $400	Good $ 360	LMP $ 779
C2C:	Mint $ 700	Ex $ 500	VG+ $360	Good $ 325	Fair $ 170
Trade-In:	Mint $ 520	Ex $ 425	VG+ $290	Good $ 255	Poor $ 90

SMITH & WESSON

M&P357
NEW IN BOX

M&P357   REDBOOK CODE: RB-SW-H-MP357X

Full-size black polymer frame, stainless slide and barrel, ambidextrous slide stop, reversible magazine catch, optional thumb safety, accessory rail.

Production: 2010   Caliber: .357 Sig.   Action: striker-fired, semi-auto

Barrel Length: 4.25"   Wt.: 25 oz.   Sights: steel-ramp dovetail mount front,

Novak lo-mount carry rear (tritium sights optional)   Capacity: 10, 15

Magazine: detachable box   Grips: polymer grip frame with interchangeable grip sizes

D2C:	NIB $ 580	Ex $ 450	VG+ $320	Good $ 290	LMP $ 727
C2C:	Mint $ 560	Ex $ 425	VG+ $290	Good $ 265	Fair $ 135
Trade-In:	Mint $ 420	Ex $ 325	VG+ $230	Good $ 205	Poor $ 60

M&P357C
NEW IN BOX

M&P357C   REDBOOK CODE: RB-SW-H-MP357C

Compact polymer frame, stainless slide and barrel, ambidextrous slide stop, reversible magazine catch, optional thumb safety, accessory rail.

Production: 2007 - 2010   Caliber: .357 Sig.   Action: striker-fired, semi-auto

Barrel Length: 3.5"   Wt.: 22 oz.   Sights: steel ramp dovetail mount front,

Novak lo-mount carry rear (tritium sights optional)   Capacity: 10

Magazine: detachable box   Grips: polymer grip frame with interchangeable grip sizes

D2C:	NIB $ 580	Ex $ 450	VG+ $ 320	Good $ 290	LMP $ 727
C2C:	Mint $ 560	Ex $ 425	VG+ $ 290	Good $ 265	Fair $ 135
Trade-In:	Mint $ 420	Ex $ 325	VG+ $ 230	Good $ 205	Poor $ 60

M&P22
NEW IN BOX

M&P22   REDBOOK CODE: RB-SW-H-MP22XX

Full-size black polymer frame, alloy slide, ambidextrous slide stop, reversible magazine catch, ambidextrous thumb safety, accessory rail.

Production: 2011 - current   Caliber: .22 LR   Action: striker-fired, semi-auto

Barrel Length: 4.1"   Wt.: 24 oz.   Sights: drift adjustable front, click adjustable rear

Capacity: 10, 12   Magazine: detachable box   Grips: polymer grip frame

D2C:	NIB $ 375	Ex $ 300	VG+ $ 210	Good $ 190	LMP $ 419
C2C:	Mint $ 360	Ex $ 275	VG+ $ 190	Good $ 170	Fair $ 90
Trade-In:	Mint $ 270	Ex $ 225	VG+ $ 150	Good $ 135	Poor $ 60

**M&P9 PRO
SERIES C.O.R.E.**
NEW IN BOX

M&P9 PRO SERIES C.O.R.E.   REDBOOK CODE: RB-SW-H-MP9PSC

Full-size polymer frame, black-coated stainless barrel, includes Pro Series features, C.O.R.E. (Competition Optics Ready Equipment) slide is cut to accept Trijicon RMR, Leupold Delta Point, Jpoint, Doctor, C-More STS, and Insight MRDS optics.

Production: 2013 - current   Caliber: 9mm   Action: striker-fired, semi-auto

Barrel Length: 4.25"   Wt.: 24 oz.   Sights: white-dot dovetail front, fixed 2-dot rear

Capacity: 17   Magazine: detachable box

Grips: polymer grip frame with enhanced textured interchangeable grip sizes

D2C:	NIB $ 675	Ex $ 525	VG+ $375	Good $ 340	LMP $ 729
C2C:	Mint $ 650	Ex $ 475	VG+ $340	Good $ 305	Fair $ 160
Trade-In:	Mint $ 480	Ex $ 400	VG+ $270	Good $ 240	Poor $ 90

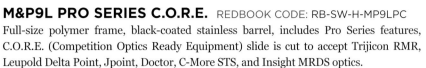

M&P9L PRO SERIES C.O.R.E. NEW IN BOX

M&P9L PRO SERIES C.O.R.E. REDBOOK CODE: RB-SW-H-MP9LPC

Full-size polymer frame, black-coated stainless barrel, includes Pro Series features, C.O.R.E. (Competition Optics Ready Equipment) slide is cut to accept Trijicon RMR, Leupold Delta Point, Jpoint, Doctor, C-More STS, and Insight MRDS optics.

Production: 2013 - current   Caliber: 9mm   Action: striker-fired, semi-auto

Barrel Length: 5"   Wt.: 26 oz.   Sights: white dot dovetail front, fixed 2-dot rear

Capacity: 17   Magazine: detachable box

Grips: polymer grip frame with enhanced textured interchangeable grip sizes

D2C:	NIB	$ 675	Ex $ 525	VG+ $ 375	Good $ 340	LMP $ 729
C2C:	Mint $ 650	Ex $ 475	VG+ $340	Good $ 305	Fair $ 160
Trade-In:	Mint $ 480	Ex $ 400	VG+ $270	Good $ 240	Poor $ 90

M&P40 PRO SERIES C.O.R.E. NEW IN BOX

M&P40 PRO SERIES C.O.R.E. REDBOOK CODE: RB-SW-H-MP40PC

Full size polymer frame, black coated stainless barrel, includes Pro Series features, C.O.R.E. (Competition Optics Ready Equipment) slide is cut to accept Trijicon RMR, Leupold Delta Point, Jpoint, Doctor, C-More STS, and Insight MRDS optics.

Production: 2013 - current   Caliber: .40 S&W   Action: striker-fired, semi-auto

Barrel Length: 4.25"   Wt.: 24 oz.   Sights: white-dot dovetail front, fixed 2-dot rear

Capacity: 15   Magazine: detachable box

Grips: polymer grip frame with enhanced textured interchangeable grip sizes

D2C:	NIB	$ 675	Ex $ 525	VG+ $ 375	Good $ 340	LMP $ 729
C2C:	Mint $ 650	Ex $ 475	VG+ $340	Good $ 305	Fair $ 160
Trade-In:	Mint $ 480	Ex $ 400	VG+ $270	Good $ 240	Poor $ 90

M&P40L PRO SERIES C.O.R.E. NEW IN BOX

M&P40L PRO SERIES C.O.R.E. REDBOOK CODE: RB-SW-H-MP40LC

Full-size polymer frame, black-coated stainless barrel, includes Pro Series features, C.O.R.E. (Competition Optics Ready Equipment) slide is cut to accept Trijicon RMR, Leupold Delta Point, Jpoint, Doctor, C-More STS, and Insight MRDS optics.

Production: 2013 - current   Caliber: .40 S&W   Action: striker-fired, semi-auto

Barrel Length: 5"   Wt.: 26 oz.   Sights: white-dot dovetail front, fixed 2-dot rear

Capacity: 15   Magazine: detachable box

Grips: polymer grip frame with enhanced textured interchangeable grip sizes

D2C:	NIB	$ 675	Ex $ 525	VG+ $ 375	Good $ 340	LMP $ 729
C2C:	Mint $ 650	Ex $ 475	VG+ $340	Good $ 305	Fair $ 160
Trade-In:	Mint $ 480	Ex $ 400	VG+ $270	Good $ 240	Poor $ 90

SW1911 NEW IN BOX

SW1911 REDBOOK CODE: RB-SW-H-SW1911

Black-finished stainless steel slide and frame, external extractor, single slide serrations, GI-spec recoil guide, optional target sights available, model with stainless finish available, add a slight premium for model with accessory rail, model also available without firing pin block.

Production: 2003 - current   Caliber: .45 ACP   Action: SA, semi-auto   Barrel Length: 5"

Wt.: 40 oz.   Sights: white dot front, low profile carry rear or adjustable target sights

Capacity: 8   Magazine: detachable box   Grips: laminated double-diamond walnut grips with silver S&W medallions or synthetic grips

D2C:	NIB	$ 899	Ex $ 700	VG+ $ 495	Good $ 450	LMP $1,039
C2C:	Mint $ 870	Ex $ 625	VG+ $ 450	Good $ 405	Fair $ 210
Trade-In:	Mint $ 640	Ex $ 525	VG+ $ 360	Good $ 315	Poor $ 90

SMITH & WESSON

SMITH & WESSON

SW1911 - CRIMSON TRACE REDBOOK CODE: SW-H-1911CT

Stainless frame and slide, non-reflective finish, external extractor, full-length guide rod.

Production: 2005 - discontinued Caliber: .45 ACP Action: SA, semi-auto

Barrel Length: 5" Wt.: 40 oz. Sights: white dot front, low profile carry rear

Capacity: 8 Magazine: detachable box

Grips: "Exclusive" Olive Drab Crimson Trace Lasergrip

D2C:	NIB	$1,020	Ex	$800	VG+	$565	Good	$510	LMP	$1,454
C2C:	Mint	$980	Ex	$725	VG+	$510	Good	$480	Fair	$235
Trade-In:	Mint	$730	Ex	$575	VG+	$400	Good	$360	Poor	$120

**SW1911 DK -
CHAMPION SERIES -
DOUG KOENIG**
EXCELLENT

SW1911 DK - CHAMPION SERIES - DOUG KOENIG

(PERFORMANCE CENTER) REDBOOK CODE: RB-SW-H-11DKPC

Full-size stainless steel frame and slide, two-tone finish, oversize external extractor, competition magazine well extension, 3.5-4 lbs. trigger pull, 30 LPI front strap checkering, designed in collaboration with Doug Koenig, Performance Center styled slide with two-tone finish, Doug Koenig Competition Speed Hammer, Competition Vertical Speed Trigger, hand lapped frame and slide, hand-fitted barrel and bushing.

Production: 2005 Caliber: .38 Super Action: SA, semi-auto Barrel Length: 5"

Wt.: 43 oz. Sights: fixed post front, adjustable target rear Capacity: 10

Magazine: detachable box Grips: Micarta grips with logo

D2C:	NIB	$2,240	Ex	$1,725	VG+	$1,235	Good	$1,120	LMP	$2,694
C2C:	Mint	$2,160	Ex	$1,550	VG+	$1,120	Good	$1,010	Fair	$520
Trade-In:	Mint	$1,600	Ex	$1,275	VG+	$880	Good	$785	Poor	$240

**SW1911 DK -
CHAMPION SERIES -
DOUG KOENIG**
EXCELLENT

SW1911 DK - CHAMPION SERIES - DOUG KOENIG

REDBOOK CODE: RB-SW-H-1911DK

Stainless steel frame, carbon steel slide, two-tone finish, oversized magwell, ambidextrous safety, Doug Koenig Speed Hammer, flat competition speed trigger, external extractor.

Production: 2005 - discontinued Caliber: .45 ACP Action: SA, semi-auto

Barrel Length: 5" Wt.: 42 oz. Sights: black blade front, adjustable target rear

Capacity: 8 Magazine: detachable box

Grips: checkered rosewood with silver S&W medallions

D2C:	NIB	$1,100	Ex	$850	VG+	$605	Good	$550	LMP	$1,309
C2C:	Mint	$1,060	Ex	$775	VG+	$550	Good	$495	Fair	$255
Trade-In:	Mint	$790	Ex	$625	VG+	$430	Good	$385	Poor	$120

**SW1911 -
PRO SERIES .45**
NEW IN BOX

SW1911 - PRO SERIES .45 REDBOOK CODE: RB-SW-H-11PC45

Two-tone stainless slide and frame, hand-polished feed ramp, precision-crowned muzzle, 30 LPI front-strap checkering, ambidextrous safety, 4-4.5 lb. trigger pull.

Production: 2008 - current Caliber: .45 ACP Action: SA, semi-auto

Barrel Length: 5" Wt.: 41 oz. Sights: Novak fiber-optic front,

adjustable fiber-optic rear Capacity: 8 Magazine: detachable box

Grips: checkered wood

D2C:	NIB	$1,260	Ex	$975	VG+	$695	Good	$630	LMP	$1,379
C2C:	Mint	$1,210	Ex	$875	VG+	$630	Good	$570	Fair	$290
Trade-In:	Mint	$900	Ex	$725	VG+	$500	Good	$445	Poor	$150

SW1911 - PRO SERIES 9MM
NEW IN BOX

SW1911 - PRO SERIES 9MM   REDBOOK CODE: RB-SW-H-11PC9M

Stainless steel frame and slide, hand-polished integral feed ramp, precision crowned muzzle, oversized external extractor, full-length guide rod, ambidextrous safety, extended magwell, 30 LPI front strap checkering, stoned hammer and sear, 4-5.5 trigger pull.

Production: 2008 - current   Caliber: 9mm   Action: SA, semi-auto

Barrel Length: 5"   Wt.: 41 oz.   Sights: dovetail white-dot front, fixed white 2-dot rear or adjustable target sights   Capacity: 10   Magazine: detachable box

Grips: checkered wood

D2C:	NIB	$1,385	Ex	$1,075	VG+	$765	Good	$695	LMP $1,489
C2C:	Mint	$1,330	Ex	$975	VG+	$700	Good	$625	Fair $320
Trade-In:	Mint	$990	Ex	$800	VG+	$550	Good	$485	Poor $150

SW1911 - PRO SERIES SUB COMPACT
NEW IN BOX

SW1911 - PRO SERIES SUB COMPACT

REDBOOK CODE: RB-SW-H-11PSSC
Scandium alloy frame, stainless steel slide, oversized external extractor, full-length guide rod.

Production: 2009 - current   Caliber: .45 ACP   Action: SA, semi-auto

Barrel Length: 3"   Wt.: 27 oz.   Sights: dovetail white-dot front, fixed white 2-dot rear

Capacity: 7   Magazine: detachable box   Grips: checkered synthetic

D2C:	NIB	$985	Ex	$750	VG+	$545	Good	$495	LMP $1,159
C2C:	Mint	$950	Ex	$700	VG+	$500	Good	$445	Fair $230
Trade-In:	Mint	$700	Ex	$575	VG+	$390	Good	$345	Poor $120

SW1911 - 100TH ANNIVERSARY SPECIAL
NEW IN BOX

SW1911 - 100TH ANNIVERSARY SPECIAL

REDBOOK CODE: RB-SW-H-11100A
Engraved glass bead finish, stainless steel frame and slide, glass top presentation case, special serial number range: JBP0000-JBP0500.

Production: 2011 - current   Caliber: .45 ACP   Action: SA, semi-auto

Barrel Length: 5"   Wt.: 40 oz.   Sights: white-dot front, low-profile carry rear

Capacity: 8   Magazine: detachable box

Grips: diamond checkered wood with S&W monograms

D2C:	NIB	$1,125	Ex	$875	VG+	$620	Good	$565	LMP $1,235
C2C:	Mint	$1,080	Ex	$800	VG+	$570	Good	$510	Fair $260
Trade-In:	Mint	$800	Ex	$650	VG+	$440	Good	$395	Poor $120

SW1911 - E-SERIES
NEW IN BOX

SW1911 - E-SERIES   REDBOOK CODE: RB-SW-H-1911ES

Stainless steel slide and frame, satin stainless finish, precision-fit trigger, chamfered and recessed muzzle, "fish scale" scalloped slide serrations, tactical accessory rail model add $400.

Production: 2011 - current   Caliber: .45 ACP   Action: SA, semi-auto

Barrel Length: 5"   Wt.: 40 oz.   Sights: white dot front, white 2-dot rear

Capacity: 8   Magazine: detachable box   Grips: wooden laminate E-Series

D2C:	NIB	$830	Ex	$650	VG+	$460	Good	$415	LMP $919
C2C:	Mint	$800	Ex	$575	VG+	$420	Good	$375	Fair $195
Trade-In:	Mint	$590	Ex	$475	VG+	$330	Good	$295	Poor $90

SMITH & WESSON

SW1911 SC - E-SERIES
NEW IN BOX

SW1911 SC - E-SERIES   REDBOOK CODE: RB-SW-H-11SCES

Scandium-alloy frame with black anodized finish, stainless steel slide, oversized extractor, combat ejection port, round-butt frame, titanium firing pin, precision trigger, recessed muzzle.

Production: 2011 - current   Caliber: .45 ACP   Action: SA, semi-auto

Barrel Length: 4.25"   Wt.: 30 oz.   Sights: tritium night sights

Capacity: 8   Magazine: detachable box   Grips: wooden-laminate E-Series

D2C:	NIB	$1,230	Ex $ 950	VG+ $680	Good $ 615	LMP $1,369
C2C:	Mint	$1,190	Ex $ 850	VG+ $620	Good $ 555	Fair $ 285
Trade-In:	Mint	$ 880	Ex $ 700	VG+ $480	Good $ 435	Poor $ 150

SW1911 CT - E-SERIES, CRIMSON TRACE LASERGRIPS
NEW IN BOX

SW1911 CT - E-SERIES, CRIMSON TRACE LASERGRIPS

REDBOOK CODE: RB-SW-H-11ESCT

Stainless steel slide and frame, satin stainless finish, precision-fit trigger, chamfered and recessed muzzle, "fish scale" scalloped slide serrations.

Production: 2011 - current   Caliber: .45 ACP   Action: SA, semi-auto

Barrel Length: 5"   Wt.: 40 oz.   Sights: white dot front, white 2-dot rear

Capacity: 8   Magazine: detachable box   Grips: Crimson Trace Lasergrips

D2C:	NIB	$ 985	Ex $ 750	VG+ $545	Good $ 495	LMP $1,089
C2C:	Mint	$ 950	Ex $ 700	VG+ $500	Good $ 445	Fair $ 230
Trade-In:	Mint	$ 700	Ex $ 575	VG+ $390	Good $ 345	Poor $ 120

SW1911 - COMPACT ES - EXTENDED SLIDE
NEW IN BOX

SW1911 - COMPACT ES - EXTENDED SLIDE

REDBOOK CODE: RB-SW-H-11CESE

Scandium alloy frame, stainless steel slide, two-tone finish, compact frame with extended slide.

Production: 2009 - current   Caliber: .45 ACP   Action: SA, semi-auto

Barrel Length: 4.25"   Wt.: 30 oz.   Sights: white-dot front, low-profile carry rear

Capacity: 7   Magazine: detachable box   Grips: diamond checkered wood

D2C:	NIB	$1,020	Ex $ 800	VG+ $565	Good $ 510	LMP $1,139
C2C:	Mint	$ 980	Ex $ 725	VG+ $510	Good $ 460	Fair $ 235
Trade-In:	Mint	$ 730	Ex $ 575	VG+ $400	Good $ 360	Poor $ 120

SW1911 PC - 170344
NEW IN BOX

SW1911 PC - 170344   REDBOOK CODE: RB-SW-H-11PC44

Scandium-alloy round butt (bobtail) frame, stainless steel slide, two-tone finish, Performance Center action job, 3.5-4 lb. trigger pull, throated barrel, precision crowned muzzle, polished feed ramp, Briley Spherical barrel bushing, 30 lpi checkering, ambidextrous frame safety, slide-ported lightening cuts.

Production: 2012 - current   Caliber: .45 ACP   Action: SA, semi-auto

Barrel Length: 4.25"   Wt.: 30 oz.   Sights: white dot front, low profile carry rear

Capacity: 8   Magazine: detachable box   Grips: G10 custom wood

D2C:	NIB	$1,389	Ex $1,075	VG+ $765	Good $ 695	LMP $1,539
C2C:	Mint	$1,340	Ex $ 975	VG+ $700	Good $ 630	Fair $ 320
Trade-In:	Mint	$ 990	Ex $ 800	VG+ $550	Good $ 490	Poor $ 150

SW1911 PC - 170343
NEW IN BOX

SW1911 PC - 170343   REDBOOK CODE: RB-SW-H-11PC43

Stainless slide and frame, glass bead finish, Performance Center action job, 3.5-4 lb. trigger pull, throated barrel, precision-crowned muzzle, polished feed ramp, Briley Spherical barrel bushing, 30 lpi checkering, ambidextrous frame safety, slide-ported lightening cuts.

Production: 2012 - current   Caliber: .45 ACP   Action: SA, semi-auto

Barrel Length: 5" Wt.: 40 oz.   Sights: black-post front, adjustable target rear

Capacity: 8   Magazine: detachable box   Grips: G10 custom wood

D2C:	NIB $1,389	Ex $1,075	VG+ $ 765	Good $ 695	LMP $1,539
C2C:	Mint $1,340	Ex $ 975	VG+ $ 700	Good $ 630	Fair $ 320
Trade-In:	Mint $ 990	Ex $ 800	VG+ $ 550	Good $ 490	Poor $ 150

SW1911 PD
NEW IN BOX

SW1911 PD   REDBOOK CODE: RB-SW-H-1911PD

Scandium-alloy frame, stainless steel slide, black finish, Commander configuration, external extractor, model available with accessory rail, available with Crimson Trace Lasergrips, Gunsite Edition has increased value.

Production: 2004 - 2011   Caliber: .45 ACP   Action: SA, semi-auto

Barrel Length: 4.25", 5" Wt.: 30 oz., 33 oz.   Sights: white-dot front, low profile carry rear   Capacity: 8   Magazine: detachable box   Grips: checkered wood

D2C:	NIB $ 980	Ex $ 750	VG+ $ 540	Good $ 490	LMP $1,109
C2C:	Mint $ 950	Ex $ 700	VG+ $ 490	Good $ 445	Fair $ 230
Trade-In:	Mint $ 700	Ex $ 550	VG+ $ 390	Good $ 345	Poor $ 120

Springfield Armory (Springfield, Inc.)

The original Springfield Armory traces its origins back to 1777, when George Washington ordered its creation to store munitions for the budding U.S. Military. By 1794, the armory began producing muskets for the U.S. government and would continue producing legendary firearms, such as the M1911 and the M1 Garand, until the Government shut down the facility in 1968. In 1974, Robert Reese licensed the Springfield Armory name and established the Springfield Armory, Inc., manufacturing many of the original armory's designs and traditional weapons. Today, the company is based in Geneseo, Illinois and manufactures a wide variety of celebrated pistols and rifles.

1911-A1 STANDARD (PRE-90S)   REDBOOK CODE: RB-RB-SA-H-M1911A

Similar to the original M1911 A1, the company's standard model 1911. Parkerized, blue, or bi-tone finish.

Production: 1985 - 1990   Caliber: 9mm, .38 Super, .45 ACP, 10mm

Action: SA, semi-auto   Barrel Length: 5"   Sights: fixed

Capacity: 7, 9, 10   Magazine: detachable box   Grips: checkered walnut

D2C:	NIB $ 710	Ex $ 550	VG+ $ 395	Good $ 355	
C2C:	Mint $ 690	Ex $ 500	VG+ $ 360	Good $ 320	Fair $ 165
Trade-In:	Mint $ 510	Ex $ 400	VG+ $ 280	Good $ 250	Poor $ 90

1911-A1 COMPACT (PRE-90S) REDBOOK CODE: RB-SA-H-1911AC
A compact variant of the 1911-A1. Shortened barrel, blue or bi-tone finish, barrel compensator.

Production: 1990 Caliber: .45 ACP Action: SA, semi-auto Barrel Length: 4.5"									
Wt.: 37 oz. Sights: 3-dot Capacity: 7 Magazine: detachable box Grips: checkered walnut									
D2C:	NIB	$ 820	Ex $ 625	VG+ $ 455	Good $ 410				
C2C:	Mint $ 790	Ex $ 575	VG+ $ 410	Good $ 370	Fair $ 190				
Trade-In:	Mint $ 590	Ex $ 475	VG+ $ 320	Good $ 290	Poor $ 90				

1911-A1 COMMANDER (PRE-90S) REDBOOK CODE: RB-SA-H-COMMAN
Blue, parkerized, or bi-tone finish, Commander hammer, shortened version of the full-size 1911-A1, discontinued due to a lawsuit from Colt over the Commander name.

Production: discontinued Caliber: .45 ACP Action: SA, semi-auto							
Barrel Length: 3.625" Sights: 3-dot Capacity: 8							
Magazine: detachable box Grips: checkered walnut							
D2C:	NIB	$ 840	Ex $ 650	VG+ $ 465	Good $ 420		
C2C:	Mint $ 810	Ex $ 600	VG+ $ 420	Good $ 380	Fair $ 195		
Trade-In:	Mint $ 600	Ex $ 475	VG+ $ 330	Good $ 295	Poor $ 90		

1911-A1 COMBAT COMMANDER (PRE-90S)
REDBOOK CODE: RB-SA-H-COMCOM
Bobbed hammer, blue, parkerized, or bi-tone finish, shortened version of the full-size 1911-A1, discontinued due to a lawsuit from Colt over the Combat Commander name.

Production: 1988 - discontinued Caliber: .45 ACP Action: SA, semi-auto							
Barrel Length: 4.25" Sights: fixed Capacity: 7							
Magazine: detachable box Grips: checkered walnut							
D2C:	NIB	$ 860	Ex $ 675	VG+ $ 475	Good $ 430		
C2C:	Mint $ 830	Ex $ 600	VG+ $ 430	Good $ 390	Fair $ 200		
Trade-In:	Mint $ 620	Ex $ 500	VG+ $ 340	Good $ 305	Poor $ 90		

1911-A1 DEFENDER (PRE-90S) REDBOOK CODE: RB-SA-H-DEFEND
Similar to the Standard 1911-A1, blue or parkerized finish, steel frame and slide, beveled magwell, extended thumb safety, bobbed hammer.

Production: late 1980s - 1990 Caliber: .45 ACP Action: SA, semi-auto							
Barrel Length: 5" Wt.: 40 oz. Sights: fixed combat Capacity: 8							
Magazine: detachable box Grips: checkered walnut							
D2C:	NIB	$ 790	Ex $ 625	VG+ $ 435	Good $ 395		
C2C:	Mint $ 760	Ex $ 550	VG+ $ 400	Good $ 360	Fair $ 185		
Trade-In:	Mint $ 570	Ex $ 450	VG+ $ 310	Good $ 280	Poor $ 90		

1911-A1 COMPACT (90'S EDITION) REDBOOK CODE: RB-SA-H-90ECOM
Steel or alloy frame option, blue, parkerized, or bi-tone finish, slight premium for lightweight alloy model.

Caliber: .45 ACP Action: SA, semi-auto Barrel Length: 4" Sights: 3-dot							
Capacity: 7 Magazine: detachable box Grips: checkered walnut							
D2C:	NIB	$ 650	Ex $ 500	VG+ $ 360	Good $ 325		
C2C:	Mint $ 630	Ex $ 450	VG+ $ 330	Good $ 295	Fair $ 150		
Trade-In:	Mint $ 470	Ex $ 375	VG+ $ 260	Good $ 230	Poor $ 90		

**1911-A1 DEFENDER
FACTORY COMP**
VERY GOOD +

1911-A1 DEFENDER FACTORY COMP   REDBOOK CODE: RB-SA-H-PDPDEF

Duo-tone finish, fitted compensator on end of slide, beveled magwell, lightweight speed trigger, slight premium for target-sight model.

Production: discontinued   Caliber: .38 Super, .45 ACP   Action: SA, semi-auto

Barrel Length: 4.25", 5"   Sights: adjustable target or fixed combat   Capacity: 8, 10

Magazine: detachable box   Grips: checkered walnut

D2C:	NIB	$ 800	Ex $ 625	VG+ $ 440	Good $ 400		
C2C:	Mint $ 770	Ex $ 575	VG+ $ 400	Good $ 360	Fair $ 185
Trade-In:	Mint $ 570	Ex $ 450	VG+ $ 320	Good $ 280	Poor $ 90

1911-A1 DEFENDER LOADED LIGHTWEIGHT

REDBOOK CODE: RB-SA-H-LODEFL

Similar to the M1911-A1 Defender, but noted for its improved frame design, lightweight components, ambidextrous design, lowered ejection port.

Production: late 1980s - early 1990s   Caliber: .45 ACP   Action: SA, semi-auto

Barrel Length: 5"   Sights: dovetail or adjustable combat-style   Capacity: 8

Magazine: detachable box   Grips: synthetic or wood

D2C:	NIB $ 850	Ex $ 650	VG+ $ 470	Good $ 425	
C2C:	Mint $ 820	Ex $ 600	VG+ $ 430	Good $ 385	Fair $ 200
Trade-In:	Mint $ 610	Ex $ 500	VG+ $ 340	Good $ 300	Poor $ 90

1911-A1 CHAMPION
NEW IN BOX

1911-A1 CHAMPION (CUSTOM LOADED)   REDBOOK CODE: RB-SA-H-CHAMPI

A somewhat more compact variant of the M1911A1. Noted for its shortened slide and barrel. Parkerized, blue, or stainless finishes, less value without custom loaded features.

Production: early 1990s - early 2000s   Caliber: .45 ACP   Action: SA, semi-auto

Barrel Length: 4"   Sights: 3-dot   Capacity: 7   Magazine: detachable box

Grips: checkered walnut

D2C:	NIB $ 900	Ex $ 700	VG+ $ 495	Good $ 450	
C2C:	Mint $ 870	Ex $ 625	VG+ $ 450	Good $ 405	Fair $ 210
Trade-In:	Mint $ 640	Ex $ 525	VG+ $ 360	Good $ 315	Poor $ 90

**1911-A1 CHAMPION
LIGHTWEIGHT**
EXCELLENT

1911-A1 CHAMPION LIGHTWEIGHT (CUSTOM LOADED)

REDBOOK CODE: RB-SA-H-CHAMPX

A lightweight variant of the Champion, noted for its aluminum alloy frame and accessory rail.

Production: mid-2000s - mid-2010s   Caliber: .45 ACP   Action: SA, semi-auto

Barrel Length: 4"   Sights: military-style adjustable   Capacity: 7

Magazine: detachable box   Grips: checkered walnut

D2C:	NIB $ 920	Ex $ 700	VG+ $ 510	Good $ 460	
C2C:	Mint $ 890	Ex $ 650	VG+ $ 460	Good $ 415	Fair $ 215
Trade-In:	Mint $ 660	Ex $ 525	VG+ $ 360	Good $ 325	Poor $ 120

1911-A1 CHAMPION COMP   REDBOOK CODE: RB-SA-H-90ECCX

Beveled magwell, compensator, blue finish.

Caliber: .45 ACP   Action: SA, semi-auto   Barrel Length: 4"

Sights: 3-dot   Capacity: 7   Magazine: detachable box   Grips: checkered walnut

D2C:	NIB $1,080	Ex $ 825	VG+ $ 595	Good $ 540	
C2C:	Mint $ 1040	Ex $ 750	VG+ $ 540	Good $ 490	Fair $ 250
Trade-In:	Mint $ 770	Ex $ 625	VG+ $ 430	Good $ 380	Poor $ 120

SPRINGFIELD ARMORY

1911-A1 FACTORY COMP
EXCELLENT

1911-A1 MICRO COMPACT GI
NEW IN BOX

1911-A1 MICRO COMPACT CUSTOM LOADED
NEW IN BOX

1911-A1 MICRO COMPACT LIGHTWEIGHT
NEW IN BOX

1911-A1 GI
EXCELLENT

1911-A1 FACTORY COMP REDBOOK CODE: RB-SA-H-FACCOM

Similar to other M1911A1 models. Noted for its barrel compensator, lengthened thumb safety, and enhanced trigger.

Production: late 1990s - early 2000s Caliber: .45 ACP, .38 Super Action: SA, semi-auto
Barrel Length: 5.7" Sights: adjustable rear Capacity: 7, 10
Magazine: detachable box Grips: checkered walnut

D2C:	NIB	$1,200	Ex $ 925	VG+ $ 660	Good $ 600			
C2C:	Mint	$1,160	Ex $ 850	VG+ $ 600	Good $ 540	Fair	$ 280	
Trade-In:	Mint	$ 860	Ex $ 675	VG+ $ 470	Good $ 420	Poor	$ 120	

1911-A1 MICRO COMPACT GI REDBOOK CODE: RB-SA-H-MICCGI

Similar features of the full-size GI model, parkerized finish, forged steel slide and frame, dual-spring recoil system with full-length guide rod.

Production: 2004 - discontinued Caliber: .45 ACP Action: SA, semi-auto
Barrel Length: 3" Wt.: 33 oz. Sights: low-profile military Capacity: 6
Magazine: detachable box Grips: U.S. engraved hardwood

D2C:	NIB	$ 700	Ex $ 550	VG+ $ 385	Good $ 350			
C2C:	Mint	$ 680	Ex $ 500	VG+ $ 350	Good $ 315	Fair	$ 165	
Trade-In:	Mint	$ 500	Ex $ 400	VG+ $ 280	Good $ 245	Poor	$ 90	

1911-A1 MICRO COMPACT CUSTOM LOADED

REDBOOK CODE: RB-SA-H-MICCOM

Stainless steel slide with steel frame. Armory Kote finishes in bi-tone stainless/black or OD Green. Ambidextrous safety, carry bevel treatment, dual-spring recoil system, less value without custom loaded features, slight premium for light-rail model.

Production: discontinued Caliber: .45 ACP Action: SA, semi-auto
Barrel Length: 3" Wt.: 32 oz. Sights: 3-dot tritium Capacity: 6
Magazine: detachable box Grips: thinline cocobolo

D2C:	NIB	$1,075	Ex $ 825	VG+ $ 595	Good $ 540			
C2C:	Mint	$1,040	Ex $ 750	VG+ $ 540	Good $ 485	Fair	$ 250	
Trade-In:	Mint	$ 770	Ex $ 625	VG+ $ 420	Good $ 380	Poor	$ 120	

1911-A1 MICRO COMPACT LIGHTWEIGHT

(CUSTOM LOADED) REDBOOK CODE: RB-SA-H-MICCLI

Features a lightweight aluminum frame. Armory Kote finishes in either bi-tone stainless/black or OD Green. Ambidextrous thumb safety, dual-spring recoil system, carry bevel treatment, less value without custom loaded features, slight premium for light rail model.

Production: early 2000s Caliber: .45 ACP Action: SA, semi-auto
Barrel Length: 3" Wt.: 24 oz. Sights: 3-dot tritium Capacity: 6
Magazine: detachable box Grips: thinline cocobolo

D2C:	NIB	$1,075	Ex $ 825	VG+ $ 595	Good $ 540			
C2C:	Mint	$1,040	Ex $ 750	VG+ $ 540	Good $ 485	Fair	$ 250	
Trade-In:	Mint	$ 770	Ex $ 625	VG+ $ 420	Good $ 380	Poor	$ 120	

1911-A1 GI REDBOOK CODE: RB-SA-H-1911GI

Available in parkerized, stainless, or OD Green finishes. Forged frame and slide, spur hammer, vertical slide serrations, titanium firing pin.

Caliber: .45 ACP Action: SA, semi-auto Barrel Length: 5"
OA Length: 8.5" Wt.: 36 oz. Sights: fixed military-style
Capacity: 7 Magazine: detachable box Grips: walnut or black polymer

D2C:	NIB	$ 540	Ex $ 425	VG+ $ 300	Good $ 270			
C2C:	Mint	$ 520	Ex $ 375	VG+ $ 270	Good $ 245	Fair	$ 125	
Trade-In:	Mint	$ 390	Ex $ 325	VG+ $ 220	Good $ 190	Poor	$ 60	

1911-A1 GI HIGH CAP
NEW IN BOX

1911-A1 GI HIGH CAP   REDBOOK CODE: RB-SA-H-1911GH

Wide frame high-capacity variation, forged steel frame and slide, parkerized finish, spur hammer, vertical slide serrations, titanium firing pin.

Caliber: .45 ACP   Action: SA, semi-auto   Barrel Length: 5"

OA Length: 8.5"   Wt.: 40 oz.   Sights: fixed military-style

Capacity: 13   Magazine: detachable box   Grips: black polymer

D2C:	NIB $ 600	Ex $ 475	VG+ $ 330	Good $ 300		
C2C:	Mint $ 580	Ex $ 425	VG+ $ 300	Good $ 270	Fair $ 140
Trade-In:	Mint $ 430	Ex $ 350	VG+ $ 240	Good $ 210	Poor $ 60

1911-A1 SUPER TUNED CHAMPION   REDBOOK CODE: RB-SA-H-SUTUCH

Blue or parkerized finish, skeletonized trigger, super tuned by Springfield's Custom Shop.

Production: discontinued   Caliber: .45 ACP   Action: SA, semi-auto

Barrel Length: 4"   Wt.: 36 oz.   Sights: Novak low mount   Capacity: 7

Magazine: detachable box   Grips: checkered hardwood

D2C:	NIB $ 900	Ex $ 700	VG+ $ 495	Good $ 450		
C2C:	Mint $ 870	Ex $ 625	VG+ $ 450	Good $ 405	Fair $ 210
Trade-In:	Mint $ 640	Ex $ 525	VG+ $ 360	Good $ 315	Poor $ 90

**1911-A1
ULTRA COMPACT**
EXCELLENT

1911-A1 ULTRA COMPACT (CUSTOM LOADED)

REDBOOK CODE: RB-SA-H-ULCOCL

Stainless steel or bi-tone finish, match grade barrel, lightweight speed trigger, beveled magwell, reduced price without custom loaded features.

Caliber: .45 ACP   Action: SA, semi-auto   Barrel Length: 3.5"   Wt.: 32 oz.

Sights: 3-dot fixed   Capacity: 6   Magazine: detachable box   Grips: rubber

D2C:	NIB $1,010	Ex $ 775	VG+ $ 560	Good $ 505		
C2C:	Mint $ 970	Ex $ 700	VG+ $ 510	Good $ 455	Fair $ 235
Trade-In:	Mint $ 720	Ex $ 575	VG+ $ 400	Good $ 355	Poor $ 120

1911-A1 ULTRA COMPACT MIL-SPEC   REDBOOK CODE: RB-SA-H-ULCOMS

Similar to the standard Ultra Compact model but with black parkerized or blue finish.

Caliber: .45 ACP   Action: SA, semi-auto   Barrel Length: 3.5"   Wt.: 32 oz.

Sights: 3-dot fixed   Capacity: 6   Magazine: detachable box

D2C:	NIB $ 810	Ex $ 625	VG+ $ 450	Good $ 405		
C2C:	Mint $ 780	Ex $ 575	VG+ $ 410	Good $ 365	Fair $ 190
Trade-In:	Mint $ 580	Ex $ 475	VG+ $ 320	Good $ 285	Poor $ 90

1911-A1 ULTRA COMPACT HIGH CAPACITY

REDBOOK CODE: RB-SA-H-ULCOHC

Parkerized or stainless steel finish, wide high capacity frame, stainless model with ported barrel has increased value.

Caliber: .45 ACP   Action: SA, semi-auto   Barrel Length: 3.5"   Wt.: 31 oz.

Sights: Novak, fixed   Capacity: 10   Magazine: detachable box   Grips: black polymer

D2C:	NIB $ 1050	Ex $ 800	VG+ $ 580	Good $ 525		
C2C:	Mint $ 1010	Ex $ 725	VG+ $ 530	Good $ 475	Fair $ 245
Trade-In:	Mint $ 750	Ex $ 600	VG+ $ 410	Good $ 370	Poor $ 120

SPRINGFIELD ARMORY

1911-A1 ULTRA COMPACT V10 PORTED
EXCELLENT

1911-A1 ULTRA COMPACT V10 PORTED

REDBOOK CODE: RB-SA-H-ULCOVP

Stainless or bi-tone finish, similar to Ultra Compact model but with ported barrel, less value without custom loaded features.

Caliber: .45 ACP Action: SA, semi-auto Barrel Length: 3.5" Wt.: 32 oz.

Sights: Novak, fixed Capacity: 6 Magazine: detachable box Grips: checkered

D2C:	NIB	$ 950	Ex $	725	VG+ $	525	Good $	475	
C2C:	Mint $	920	Ex $	675	VG+ $	480	Good $	430	Fair $ 220
Trade-In:	Mint $	680	Ex $	550	VG+ $	380	Good $	335	Poor $ 120

1911-A1 TRP
NEW IN BOX

1911-A1 TRP (TACTICAL RESPONSE PISTOL)

REDBOOK CODE: RB-SA-H-TRPLRA

Black Armory Kote or stainless finish, forged steel or stainless steel frame and slide, 2-piece full-length guide rod, 2-piece extended magwell.

Production: current Caliber: .45 ACP Action: SA, semi-auto Barrel Length: 5"

OA Length: 8.6" Wt.: 42 oz. Sights: fixed combat 3-dot tritium

Capacity: 7 Magazine: detachable box Grips: G10 composite

D2C:	NIB	$1,420	Ex $1,100	VG+	$785	Good $	710		
C2C:	Mint $1,370	Ex $1,000	VG+	$710	Good $	640	Fair $ 330		
Trade-In:	Mint $1,010	Ex $ 800	VG+	$560	Good $	500	Poor $ 150		

1911-A1 TRP LIGHT RAIL
NEW IN BOX

1911-A1 TRP LIGHT RAIL (OPERATOR)

REDBOOK CODE: RB-SA-H-TRPOPR

Forged steel frame and slide with Black Armory Kote, stainless match bull barrel (bushingless), Picatinny accessory rail, 2-piece full-length guide rod, lightweight speed trigger.

Production: current Caliber: .45 ACP Action: SA, semi-auto Barrel Length: 5"

OA Length: 8.6" Wt.: 45 oz. Sights: adjustable 3-dot tritium

Capacity: 7 Magazine: detachable box Grips: G10 composite

D2C:	NIB	$1,530	Ex $1,175	VG+ $	845	Good $	765		
C2C:	Mint $1,470	Ex $1,075	VG+ $	770	Good $	690	Fair $ 355		
Trade-In:	Mint $1,090	Ex $ 875	VG+ $	600	Good $	540	Poor $ 180		

1911-A1 LOADED
NEW IN BOX

1911-A1 LOADED (PARKERIZED OR STAINLESS)

REDBOOK CODE: RB-SA-H-LOADED

Forged steel frame and slide with parkerized finish or optional stainless model, 2-piece full-length guide rod, lightweight speed trigger, beveled magwell, slight premium for stainless model.

Production: current Caliber: .45 ACP, 9mm Action: SA, semi-auto

Barrel Length: 5" Wt.: 40 oz. Sights: 3-dot combat, tritium

Capacity: 7 Magazine: detachable box Grips: checkered cocobolo

D2C:	NIB	$ 840	Ex $ 650	VG+	$465	Good $	420		
C2C:	Mint $ 810	Ex $ 600	VG+	$420	Good $	380	Fair $ 195		
Trade-In:	Mint $ 600	Ex $ 475	VG+	$330	Good $	295	Poor $ 90		

1911-A1 LOADED TARGET
NEW IN BOX

1911-A1 LOADED TARGET (STAINLESS)

REDBOOK CODE: RB-SA-H-LFS45X

Forged parkerized steel or forged stainless steel construction, front and rear cocking serrations, Carry Bevel treatment, delta lightweight hammer, titanium firing pin, loaded chamber indicator, beavertail grip safety.

Production: current Caliber: .45 ACP, 9mm Action: SA, semi-auto

Barrel Length: 5" Wt.: 40 oz. Sights: 3-dot low-profile combat or adjustable target

Capacity: 7, 9 Magazine: detachable box Grips: checkered cocobolo

D2C:	NIB $	860	Ex $	675	VG+ $	475	Good $	430	
C2C:	Mint $	830	Ex $	600	VG+ $	430	Good $	390	Fair $ 200
Trade-In:	Mint $	620	Ex $	500	VG+ $	340	Good $	305	Poor $ 90

1911-A1 MIL-SPEC
NEW IN BOX

1911-A1 MIL-SPEC (PARKERIZED OR STAINLESS)

REDBOOK CODE: RB-SA-H-MSFSPA

Forged steel slide and frame, parkerized finish, GI-style recoil system, beveled magwell, lowered and flared ejection port, polished feed ramp, throated stainless barrel, angled slide serrations, add roughly $70 for stainless model.

Production: current Caliber: .45 ACP Action: SA, semi-auto Barrel Length: 5"

OA Length: 8.6" Wt.: 39 oz. Sights: 3-dot combat Capacity: 7

Magazine: detachable Grips: cocobolo or black synthetic

D2C:	NIB $	630	Ex $	500	VG+ $	350	Good $	315	
C2C:	Mint $	610	Ex $	450	VG+ $	320	Good $	285	Fair $ 145
Trade-In:	Mint $	450	Ex $	375	VG+ $	250	Good $	225	Poor $ 90

1911-A1 RANGE OFFICER
NEW IN BOX

1911-A1 RANGE OFFICER REDBOOK CODE: RB-SA-H-RO45XX

Parkerized finish, frame-mounted and grip safety, skeletonized trigger, steel slide and frame.

Production: current Caliber: .45 ACP Action: SA, semi-auto Barrel Length: 5"

Wt.: 40 oz. Sights: low-profile adjustable target Capacity: 7, 9

Magazine: detachable box Grips: cocobolo, checkered

D2C:	NIB $	810	Ex $	625	VG+ $	450	Good $	405	
C2C:	Mint $	780	Ex $	575	VG+ $	410	Good $	365	Fair $ 190
Trade-In:	Mint $	580	Ex $	475	VG+ $	320	Good $	285	Poor $ 90

1911-A1 TROPHY MATCH
NEW IN BOX

1911-A1 TROPHY MATCH REDBOOK CODE: RB-SA-H-TRPHYX

Stainless steel slide and frame, lightweight speed trigger, 2-piece full-length guide rod, extended magwell, 20 lpi frontstrap checkering.

Production: current Caliber: .45 ACP, .40 S&W Action: SA, semi-auto

Barrel Length: 5" Wt.: 41 oz. Sights: adjustable target Capacity: 7

Magazine: detachable box Grips: checkered cocobolo

D2C:	NIB $	1,290	Ex $	1,000	VG+ $	710	Good $	645	
C2C:	Mint $	1,240	Ex $	900	VG+ $	650	Good $	585	Fair $ 300
Trade-In:	Mint $	920	Ex $	725	VG+ $	510	Good $	455	Poor $ 150

1911-A1 LOADED MC OPERATOR
NEW IN BOX

Courtesy of Bud's Gun Shop

1911-A1 LOADED MC OPERATOR

REDBOOK CODE: RB-SA-H-LDMCOP

Forged steel slide with Black Armory Kote, forged steel frame with Olive Drab Armory Kote, integral accessory rail, military-style guide rod.

Production: current Caliber: .45 ACP Action: SA, semi-auto Barrel Length: 5"

Wt.: 43 oz. Sights: tritium 3-dot Capacity: 7 Grips: Pachmayr wraparound

D2C:	NIB $	1090	Ex $	850	VG+ $	600	Good $	545	
C2C:	Mint $	1050	Ex $	775	VG+ $	550	Good $	495	Fair $ 255
Trade-In:	Mint $	780	Ex $	625	VG+ $	430	Good $	385	Poor $ 120

1911-A1 LIGHTWEIGHT OPERATOR
NEW IN BOX

1911-A1 LIGHTWEIGHT OPERATOR REDBOOK CODE: RB-SA-H-LWOXXX

A lightweight variant of the Operator. Accessory rail, full-length guide rod, aluminum frame, steel slide.

Production: current	Caliber: .45 ACP	Action: SA, semi-auto	Barrel Length: 5"
OA Length: 7.5"	Sights: tritium 3-dot	Capacity: 7	Magazine: detachable box
Grips: cocobolo hardwood			

D2C:	NIB $1,010	Ex $775	VG+ $560	Good $505	
C2C:	Mint $970	Ex $700	VG+ $510	Good $455	Fair $235
Trade-In:	Mint $720	Ex $575	VG+ $400	Good $355	Poor $120

1911-A1 LIGHTWEIGHT CHAMPION OPERATOR
NEW IN BOX

1911-A1 LIGHTWEIGHT CHAMPION OPERATOR

REDBOOK CODE: RB-SA-H-LWCOXX

Fully-supported ramped bull barrel (bushingless), forged lightweight alloy frame with black hardcoat anodized finish, forged steel slide with Black Armory Kote finish, accessory rail, dual-spring recoil system, lightweight speed trigger, beveled magwell.

Production: current	Caliber: .45 ACP	Action: SA, semi-auto	Barrel Length: 4"
Wt.: 31 oz.	Sights: fixed combat 3-dot tritium	Capacity: 7	Magazine: detachable box
Grips: cocobolo hardwood			

D2C:	NIB $920	Ex $700	VG+ $510	Good $460	
C2C:	Mint $890	Ex $650	VG+ $460	Good $415	Fair $215
Trade-In:	Mint $660	Ex $525	VG+ $360	Good $325	Poor $120

1911-A1 EMP
NEW IN BOX

1911-A1 EMP REDBOOK CODE: RB-SA-H-EMPMP1

Forged steel or alloy frame with black hard-coat anodized finish, forged stainless slide with satin stainless finish, stainless fully-supported ramped bull barrel, dual-spring recoil system.

Production: current	Caliber: 9mm, .40 S&W, .45 GAP	Action: SA, semi-auto
Barrel Length: 3"	Wt.: 27 oz.	Sights: combat 3-dot tritium Capacity: 8, 9
Magazine: detachable	Grips: thinline cocobolo or G10	

D2C:	NIB $1,090	Ex $850	VG+ $600	Good $545	
C2C:	Mint $1,050	Ex $775	VG+ $550	Good $495	Fair $255
Trade-In:	Mint $780	Ex $625	VG+ $430	Good $385	Poor $120

1911-A1 EXPERT REDBOOK CODE: RB-SA-H-EXPERT

Triple-port compensator, match barrel, beveled magwell, bi-tone finish.

Caliber: .45 ACP	Action: SA, semi-auto	Sights: Bo-Mar adjustable rear	Capacity: 7
Magazine: detachable box	Grips: Pachmayr wraparound	Barrel Length: 5"	

D2C:	NIB $1,750	Ex $1,350	VG+ $965	Good $875	
C2C:	Mint $1,680	Ex $1,225	VG+ $880	Good $790	Fair $405
Trade-In:	Mint $1,250	Ex $1,000	VG+ $690	Good $615	Poor $180

OMEGA
EXCELLENT

OMEGA REDBOOK CODE: RB-SA-H-OMEGA1

Blue, parkerized, or stainless finish, frame-mounted and grip safety, some with caliber conversion kits, have very high premium, choice of ported or unported barrel.

Production: late 1980s - early 1990s	Caliber: .38 Super, .45 ACP, 10mm
Action: SA, semi-auto	Barrel Length: 5", 6" Sights: adjustable
Capacity: 7	Magazine: detachable box Grips: Pachmayr, wood, or synthetic

D2C:	NIB $1,050	Ex $800	VG+ $580	Good $525	
C2C:	Mint $1,010	Ex $725	VG+ $530	Good $475	Fair $245
Trade-In:	Mint $750	Ex $600	VG+ $410	Good $370	Poor $120

P9
VERY GOOD +

P9   REDBOOK CODE: RB-SA-H-P9XXXX
Similar design to the CZ-75 or Browning Hi-Power. Blue or parkerized finish. Stainless steel model has slightly higher value.

Production: mid-1980s - early 1990s Caliber: 9mm, 9x21mm, .40 S&W, .45 ACP
Action: DA/SA, semi-auto Barrel Length: 4.7" Wt.: 35 oz. Sights: blade front, adjustable rear Capacity: 11, 15 Magazine: detachable box Grips: synthetic or wood

D2C:	NIB $ 625	Ex $ 475	VG+ $ 345	Good $ 315		
C2C:	Mint $ 600	Ex $ 450	VG+ $ 320	Good $ 285	Fair $ 145	
Trade-In:	Mint $ 450	Ex $ 350	VG+ $ 250	Good $ 220	Poor $ 90	

P9 COMPACT
NEW IN BOX

P9 COMPACT   REDBOOK CODE: RB-SA-H-P9COMP
A compact variant of the P9, frame-mounted safety, blue or stainless finish.

Production: early 1990s Caliber: 9mm, .40 S&W Action: DA/SA, semi-auto
Barrel Length: 3.6" Sights: blade front, adjustable rear Capacity: 10, 13
Magazine: detachable box Grips: synthetic or wood

D2C:	NIB $ 570	Ex $ 450	VG+ $ 315	Good $ 285		
C2C:	Mint $ 550	Ex $ 400	VG+ $ 290	Good $ 260	Fair $ 135	
Trade-In:	Mint $ 410	Ex $ 325	VG+ $ 230	Good $ 200	Poor $ 60	

XD FULL SIZE
NEW IN BOX

XD FULL SIZE   REDBOOK CODE: RB-SA-H-XDXXXX
Full-size frame, accessory rail, grip safety, Melonite-finished fully-supported ramped barrel. Polymer frame in black or OD Green, Dark Earth, or optional bi-tone finish with stainless slide.

Production: early 2000s - current Caliber: .45 ACP, 9mm, .40 S&W, .357 Sig, .45 GAP
Action: striker-fired, semi-auto Barrel Length: 4", 5"
Wt.: 28-33 oz. Sights: 3-dot dovetail Capacity: 12, 13, 16 Magazine: detachable box
Grips: polymer grip frame

D2C:	NIB $ 570	Ex $ 450	VG+ $ 315	Good $ 285		
C2C:	Mint $ 550	Ex $ 400	VG+ $ 290	Good $ 260	Fair $ 135	
Trade-In:	Mint $ 410	Ex $ 325	VG+ $ 230	Good $ 200	Poor $ 60	

XD COMPACT
NEW IN BOX

XD COMPACT   REDBOOK CODE: RB-SA-H-XDC45X
Compact polymer frame with full-size steel slide, grip safety, accessory rail. Available in black, bi-tone, Dark Earth, or OD Green. Includes X-Tension magazine holding 13 rounds.

Production: late 2000s - current Caliber: .45 ACP Action: striker-fired, semi-auto
Barrel Length: 4", 5" Wt.: 29 or 32 oz. Sights: 3-dot dovetail
Capacity: 10, (13 w/extension) Magazine: detachable box Grips: polymer grip frame

D2C:	NIB $ 560	Ex $ 450	VG+ $ 310	Good $ 280		
C2C:	Mint $ 540	Ex $ 400	VG+ $ 280	Good $ 255	Fair $ 130	
Trade-In:	Mint $ 400	Ex $ 325	VG+ $ 220	Good $ 200	Poor $ 60	

XD SUB-COMPACT
NEW IN BOX

XD SUB-COMPACT   REDBOOK CODE: RB-SA-H-XDSC9M
Grip safety, accessory rail, Melonite fully-supported ramped barrel, loaded chamber indicator, striker-status indicator. Polymer frame in black or OD Green or optional bi-tone finish with stainless slide. Includes a Mag X-Tension extended magazine holding 12 (.40) and 16 (9mm) rounds.

Production: early 2000s - current Caliber: 9mm, .40 S&W Action: striker-fired, semi-auto
Barrel Length: 3" Wt.: 26 oz. Sights: 3-dot dovetail Capacity: 9, 13
Magazine: detachable box Grips: polymer grip frame

D2C:	NIB $ 560	Ex $ 450	VG+ $ 310	Good $ 280		
C2C:	Mint $ 540	Ex $ 400	VG+ $ 280	Good $ 255	Fair $ 130	
Trade-In:	Mint $ 400	Ex $ 325	VG+ $ 220	Good $ 200	Poor $ 60	

XD (M) FULL SIZE
NEW IN BOX

XD (M) COMPACT
NEW IN BOX

**XD(M)
COMPETITION SERIES**
EXCELLENT

XD-S 3.3"
NEW IN BOX

XD-S 4"
NEW IN BOX

XD(M) FULL SIZE   REDBOOK CODE: RB-SA-H-XDMXXX

An upgraded and enhanced version of the XD, enhanced contours and textures, interchangeable backstraps, match grade barrel, accessory rail, black or bi-tone finish, includes Mag X-Tension extended magazine.

Production: 2007 - current Caliber: 9mm, .40 S&W, .45 ACP Action: striker-fired, semi-auto
Barrel Length: 3.8", 4.5" Wt.: 29-31 oz.   Sights: 3-dot low profile
Capacity: 13, 16, 19 Magazine: detachable box Grips: polymer grip frame

D2C:	NIB $ 610	Ex $ 475	VG+ $ 340	Good $ 305		
C2C:	Mint $ 590	Ex $ 425	VG+ $ 310	Good $ 275	Fair $ 145
Trade-In:	Mint $ 440	Ex $ 350	VG+ $ 240	Good $ 215	Poor $ 90

XD(M) COMPACT   REDBOOK CODE: RB-SA-H-XDMCOM

A compact variant of the XD(M), an upgraded and enhanced version of the XD, enhanced contours and textures, interchangeable backstraps, match grade barrel, accessory rail, black or bi-tone finish, includes Mag X-Tension extended magazine.

Production: 2007 - current Caliber: 9mm, .40 S&W, .45 ACP Action: striker-fired, semi-auto
Barrel Length: 3.8" Wt.: 27-28 oz.   Sights: 3-dot low profile Capacity: 13, 11, 9
Magazine: detachable box Grips: polymer grip frame

D2C:	NIB $ 610	Ex $ 475	VG+ $ 340	Good $ 305		
C2C:	Mint $ 590	Ex $ 425	VG+ $ 310	Good $ 275	Fair $ 145
Trade-In:	Mint $ 440	Ex $ 350	VG+ $ 240	Good $ 215	Poor $ 90

XD(M) COMPETITION SERIES   REDBOOK CODE: RB-SA-H-XDMCSX

Includes the same features of the XD(M) but with extended barrel and slide with lightening cut on the slide, upgraded sights.

Production: current Caliber: 9mm, .40 S&W, .45 ACP Action: striker-fired, semi-auto
Barrel Length: 5.25" Wt.: 29-32 oz.   Sights: fiber-optic front, adjustable target rear
Capacity: 19, 16, 13 Magazine: detachable box Grips: polymer grip frame

D2C:	NIB $ 750	Ex $ 575	VG+ $ 415	Good $ 375		
C2C:	Mint $ 720	Ex $ 525	VG+ $ 380	Good $ 340	Fair $ 175
Trade-In:	Mint $ 540	Ex $ 425	VG+ $ 300	Good $ 265	Poor $ 90

XD-S 3.3"   REDBOOK CODE: RB-SA-H-XDSXXX

Forged steel slide with Melonite finish, polymer frame, accessory rail, USA Trigger System, grip safety, loaded chamber indicator, carrying case, available with a stainless slide, X-Tension magazines available.

Production: 2012 - current Caliber: .45 ACP, 9mm Action: striker-fired, semi-auto
Barrel Length: 3.3" Wt.: 21.5-23 oz.   Sights: fiber-optic front, dovetail rear
Capacity: 5, 7 Magazine: detachable box Grips: polymer grip frame

D2C:	NIB $ 510	Ex $ 400	VG+ $ 285	Good $ 255		
C2C:	Mint $ 490	Ex $ 375	VG+ $ 260	Good $ 230	Fair $ 120
Trade-In:	Mint $ 370	Ex $ 300	VG+ $ 200	Good $ 180	Poor $ 60

XD-S 4"   REDBOOK CODE: RB-SA-H-XDS1XX

Forged steel slide with Melonite finish, polymer frame, accessory rail, USA Trigger System, grip safety, loaded chamber indicator, carrying case, available with stainless slide.

Production: 2014 - current Caliber: 9mm Action: striker-fired, semi-auto
Barrel Length: 4" Wt.: 25 oz.   Sights: fiber-optic front, dovetail rear
Capacity: 7, 9 (with X-Tension) Magazine: detachable box Grips: polymer grip frame

D2C:	NIB $ 520	Ex $ 400	VG+ $ 290	Good $ 260		
C2C:	Mint $ 500	Ex $ 375	VG+ $ 260	Good $ 235	Fair $ 120
Trade-In:	Mint $ 370	Ex $ 300	VG+ $ 210	Good $ 185	Poor $ 60

STI International

Located in Georgetown, Texas, STI specializes in custom M1911-style pistols. The company manufactures a host of parts and custom kits for the M1911 and their 2011 pistols and became the first to incorporate a modular polymer frame into a high-capacity, M1911-style pistol. STI builds most of their pistols to customer specifications.

DUTY ONE
NEW IN BOX
Courtesy of Bud's Gun Shop

DUTY ONE (ORIGINAL MODEL) REDBOOK CODE: RB-RB-SI-H-DTYONE

1911 single stack frame, tactical rail, undercut trigger guard, 30 lpi front strap checkering, flat top slide with front and rear cocking serrations, ramped bull barrel, beavertail grip safety, ambidextrous thumb safeties, RecoilMaster guide rod, blue finish, this model was replaced by the Duty One 5.0.

Production: discontinued   Caliber: 9mm, .40 S&W, .45 ACP   Action: SA, semi-auto
Barrel Length: 5.01"   OA Length: 8.5"   Wt.: 32 oz.   Sights: ramped front,
fixed rear   Capacity: 8, 9   Magazine: detachable box   Grips: rosewood or G10

D2C:	NIB	$1,230	Ex	$950	VG+	$680	Good	$615	
C2C:	Mint	$1,190	Ex	$850	VG+	$620	Good	$555	Fair $285
Trade-In:	Mint	$880	Ex	$700	VG+	$480	Good	$435	Poor $150

DUTY ONE 3.0
NEW IN BOX

DUTY ONE 3.0 REDBOOK CODE: RB-SI-H-DTYON3

Officers-size 1911 frame with lightened integral rail, grid-style front strap checkering, flat-top slide with custom grid-style front and rear cocking serrations, ramped bull barrel, beavertail grip safety, RecoilMaster guide rod, matte blue finish.

Production: 2013 - current   Caliber: 9mm, .40 S&W, .45 ACP   Action: SA, semi-auto
Barrel Length: 3.24"   OA Length: 6.5"   Wt.: 32 oz.   Sights: ramped front, ledge-style rear
Capacity: 6, 7, 8   Magazine: detachable box   Grips: STI G10 grip panels

D2C:	NIB	$1,384	Ex	$1,075	VG+	$ 765	Good	$ 695	LMP $1,384
C2C:	Mint	$1,330	Ex	$ 975	VG+	$ 700	Good	$ 625	Fair $ 320
Trade-In:	Mint	$ 990	Ex	$ 800	VG+	$ 540	Good	$ 485	Poor $ 150

DUTY ONE 4.0
NEW IN BOX

DUTY ONE 4.0 REDBOOK CODE: RB-SI-H-DTYON4

1911 frame with lightened integral rail, grid-style front strap checkering, flat-top slide with custom grid-style front and rear cocking serrations, ramped bushing barrel, high-ride beavertail grip safety, RecoilMaster guide rod, matte blue finish.

Production: 2013 - current   Caliber: 9mm, .40 S&W, .45 ACP   Action: SA, semi-auto
Barrel Length: 4.37"   OA Length: 8"   Wt.: 36 oz.   Sights: ramped front, ledge style rear
Capacity: 8, 9   Magazine: detachable box   Grips: STI G10 grip panels

D2C:	NIB	$1,384	Ex	$1,075	VG+	$ 765	Good	$ 695	LMP $1,384
C2C:	Mint	$1,330	Ex	$ 975	VG+	$ 700	Good	$ 625	Fair $ 320
Trade-In:	Mint	$ 990	Ex	$ 800	VG+	$ 540	Good	$ 485	Poor $ 150

DUTY ONE 5.0
NEW IN BOX

DUTY ONE 5.0 REDBOOK CODE: RB-SI-H-DTYON5

1911 frame with lightened integral rail, grid-style front strap checkering, flat-top slide with custom grid-style front and rear cocking serrations, ramped bushing barrel, high-ride beavertail grip safety, RecoilMaster guide rod, matte blue finish, updated design to the original Duty One.

Production: 2013 - current   Caliber: 9mm, .40 S&W, .45 ACP   Action: SA, semi-auto
Barrel Length: 5.11"   OA Length: 8.5"   Wt.: 37 oz.   Sights: ramped front, ledge style rear
Capacity: 8, 9   Magazine: detachable box   Grips: STI G10 grip panels

D2C:	NIB	$1,384	Ex	$1,075	VG+	$ 765	Good	$ 695	LMP $1,384
C2C:	Mint	$1,330	Ex	$ 975	VG+	$ 700	Good	$ 625	Fair $ 320
Trade-In:	Mint	$ 990	Ex	$ 800	VG+	$ 540	Good	$ 485	Poor $ 150

STI INTERNATIONAL

ELEKTRA
NEW IN BOX

ELEKTRA REDBOOK CODE: RB-SI-H-ELECTR

1911 Officer length forged aluminum frame, stippled front strap, round-top stainless slide with STI Sabertooth cocking serrations, integral sight channel, ramped bull barrel, beavertail grip safety, RecoilMaster guide rod, several color schemes available, Cerakoted with sides of slide polished.

Production: current   Caliber: 9mm, .45 ACP   Action: SA, semi-auto

Barrel Length: 3.24"   OA Length: 7"   Wt.: 24 oz.

Sights: lowered inset Heinie Straight Eight 2-dot tritium   Capacity: 6, 8

Magazine: detachable box   Grips: pearl style

D2C:	NIB	$1,472	Ex	$1,125	VG+	$810	Good	$740	LMP $1,472
C2C:	Mint	$1,420	Ex	$1,025	VG+	$740	Good	$665	Fair $340
Trade-In:	Mint	$1,050	Ex	$825	VG+	$580	Good	$520	Poor $150

ESCORT
EXCELLENT
Courtesy of Bud's Gun Shop

ESCORT REDBOOK CODE: RB-SI-H-ESCORT

1911 Officer length forged aluminum frame, stippled front strap, round top with rear cocking serrations, ramped bull barrel, beavertail grip safety, RecoilMaster guide rod, titanium strut, Cerakoted frame, blue slide.

Production: current   Caliber: 9mm, .40 S&W, .45 ACP   Action: SA, semi-auto

Barrel Length: 3.24"   OA Length: 7"   Wt.: 23 oz.   Sights: competition front, 3-dot white TAS adjustable rear   Capacity: 6, 7, 8   Magazine: detachable box   Grips: cocobolo

D2C:	NIB	$1,233	Ex	$950	VG+	$680	Good	$620	LMP $1,233
C2C:	Mint	$1,190	Ex	$875	VG+	$620	Good	$555	Fair $285
Trade-In:	Mint	$880	Ex	$700	VG+	$490	Good	$435	Poor $150

GI
NEW IN BOX

GI REDBOOK CODE: RB-SI-H-GIXXXX

1911 Government steel frame, classic slide with rear cocking serrations, bald front strap, beavertail grip safety, thumb safety, matte blue finish.

Production: 2013 - current   Caliber: .45 ACP   Action: SA, semi-auto

Barrel Length: 5.11"   OA Length: 8.5"   Wt.: 35 oz.   Sights: GI steel

Capacity: 7   Magazine: detachable box   Grips: checkered wood

D2C:	NIB	$874	Ex	$675	VG+	$485	Good	$440	LMP $874
C2C:	Mint	$840	Ex	$625	VG+	$440	Good	$395	Fair $205
Trade-In:	Mint	$630	Ex	$500	VG+	$350	Good	$310	Poor $90

GUARDIAN
NEW IN BOX

GUARDIAN REDBOOK CODE: RB-SI-H-GUARDI

1911 Officer fame, stippled front strap, stainless steel flat top slide, rear cocking serrations, ramped bull barrel, beavertail grip safety, RecoilMaster guide rod, blue finish frame, stainless finish slide with polished sides.

Production: current   Caliber: 9mm, .40 S&W, .45 ACP   Action: SA, semi-auto

Barrel Length: 3.96"   OA Length: 7.5"   Wt.: 32 oz.   Sights: 3-dot white TAS adjustable sights

Capacity: 6, 7, 8   Magazine: detachable box   Grips: cocobolo

D2C:	NIB	$1,181	Ex	$900	VG+	$650	Good	$595	LMP $1,181
C2C:	Mint	$1,140	Ex	$825	VG+	$600	Good	$535	Fair $275
Trade-In:	Mint	$840	Ex	$675	VG+	$470	Good	$415	Poor $120

LAWMAN 3.0
NEW IN BOX

LAWMAN 3.0 REDBOOK CODE: RB-SI-H-LAWMN3

Classic 1911 style, Officer length, forged aluminum frame, checkered front strap and mainspring housing, carbon steel slide with round top, front and rear cocking serrations, ramped bull barrel, beavertail grip safety, RecoilMaster guide rod, blue slide with polished sides of slide, Cerakoted frame.

Production: current Caliber: 9mm, .45 ACP Action: SA, semi-auto

Barrel Length: 3.24" OA Length: 6.5" Wt.: 25 oz.

Sights: ramped front, TAS adjustable rear Capacity: 6, 8 Magazine: detachable box

Grips: STI G10 grip panels

D2C:	NIB $1,534	Ex $1,175	VG+ $ 845	Good $ 770	LMP $1,534
C2C:	Mint $1,480	Ex $1,075	VG+ $ 770	Good $ 695	Fair $ 355
Trade-In:	Mint $1,090	Ex $ 875	VG+ $ 600	Good $ 540	Poor $ 180

LAWMAN 4.0 & 5.0
EXCELLENT

LAWMAN 4.0 & 5.0 REDBOOK CODE: RB-SI-H-LAWM45

Classic 1911 style, Government length (5"), Commander length (4"), forged steel frame with checkered front strap and mainspring housing, front and rear cocking serrations, ramped bushing barrel, beavertail grip safety, full-length guide rod, blue finish with polished sides of slide, Cerakoted frames with optional color schemes.

Production: current Caliber: 9mm, .45 ACP Action: SA, semi-auto

Barrel Length: 4.26", 5.11" OA Length: 8.5", 7.75" Wt.: 39 oz., 36 oz.

Sights: ramped front, TAS adjustable rear Capacity: 8, 9 Magazine: detachable box

Grips: STI G10 grip panels

D2C:	NIB $ 1,514	Ex $1,175	VG+ $ 835	Good $ 760	LMP $ 1,514
C2C:	Mint $1,460	Ex $1,050	VG+ $ 760	Good $ 685	Fair $ 350
Trade-In:	Mint $1,080	Ex $ 850	VG+ $ 600	Good $ 530	Poor $ 180

RANGER II
EXCELLENT

RANGER II REDBOOK CODE: RB-SI-H-RNGER2

1911 Government frame, stippled front strap, rear cocking serrations, ramped bull barrel, beavertail grip safety, RecoilMaster guide rod, matte blue finish, custom options available.

Production: current Caliber: 9mm, .40 S&W, .45 ACP Action: SA, semi-auto

Barrel Length: 4.26" OA Length: 7.75" Wt.: 34 oz. Sights: ramped front, Heinie fixed rear Capacity: 8, 9 Magazine: detachable box Grips: cocobolo

D2C:	NIB $ 1,181	Ex $ 900	VG+ $ 650	Good $ 595	LMP $ 1,181
C2C:	Mint $ 1,140	Ex $ 825	VG+ $ 600	Good $ 535	Fair $ 275
Trade-In:	Mint $ 840	Ex $ 675	VG+ $ 470	Good $ 415	Poor $ 120

RANGER III
NEW IN BOX

RANGER III REDBOOK CODE: RB-SI-H-RNGER3

Similar to Ranger II but has (SFS) Safety Fast Shooting system, 1911 Government frame, stippled front strap, rear cocking serrations, ramped bull barrel, beavertail grip safety, RecoilMaster guide rod, matte blue finish, custom options available.

Production: current Caliber: 9mm, .40 S&W, .45 ACP Action: SA, semi-auto

Barrel Length: 4.26" OA Length: 7.75" Wt.: 34 oz.

Sights: ramped front, Heinie fixed rear Capacity: 8, 9 Magazine: detachable box

Grips: cocobolo

D2C:	NIB $ 1,341	Ex $1,025	VG+ $ 740	Good $ 675	LMP $ 1,341
C2C:	Mint $1,290	Ex $ 950	VG+ $ 680	Good $ 605	Fair $ 310
Trade-In:	Mint $ 960	Ex $ 775	VG+ $ 530	Good $ 470	Poor $ 150

STI INTERNATIONAL

SENTINEL PREMIER
EXCELLENT

SENTINEL PREMIER REDBOOK CODE: RB-SI-H-SNTNLP

Hard-chromed with polished sided slide, 1911 Government-sized frame, 30 lpi front strap checkering, blended steel magwell, front and rear cocking serrations, crowned and ramped bushing barrel, beavertail grip safety, RecoilMaster guide rod.

Production: current   Caliber: .45 ACP   Action: SA, semi-auto   Barrel Length: 5.11"

OA Length: 8.5"   Wt.: 38 oz.   Sights: ramped tritium front, tritium TAS adjustable rear

Capacity: 8   Magazine: detachable box   Grips: STI G10 grip panels

D2C:	NIB $2,370	Ex $1,825	VG+ $1,305	Good $1,185	LMP $2,370
C2C:	Mint $2,280	Ex $1,650	VG+ $1,190	Good $1,070	Fair $ 550
Trade-In:	Mint $1,690	Ex $1,350	VG+ $ 930	Good $ 830	Poor $ 240

SENTRY
NEW IN BOX

SENTRY REDBOOK CODE: RB-SI-H-SENTRY

1911 Government frame, checkered front strap, flat top with front and rear cocking serrations, ramped bushing barrel, beavertail grip safety, ambidextrous thumb safeties, RecoilMaster guide rod, blue finish with polished sides of slide.

Production: current   Caliber: 9mm, .40 S&W, .45 ACP   Action: SA, semi-auto

Barrel Length: 5.11"   OA Length: 8.5"   Wt.: 37 oz.   Sights: Dawson fiber-optic front,

adjustable rear   Capacity: 8, 9   Magazine: detachable box   Grips: cocobolo

D2C:	NIB $1,753	Ex $1,350	VG+ $ 965	Good $ 880	LMP $1,753
C2C:	Mint $1,690	Ex $1,225	VG+ $ 880	Good $ 790	Fair $ 405
Trade-In:	Mint $1,250	Ex $1,000	VG+ $ 690	Good $ 615	Poor $ 180

SHADOW
VERY GOOD +

SHADOW REDBOOK CODE: RB-SI-H-SHADOW

Aluminum alloy 1911 officer frame, steel slide, sight channel cut into slide to lower sights, ramped bull barrel, stippled front strap, bobbed beavertail grip, RecoilMaster dual recoil spring, matte black Cerakote coating, stippled polymer mainspring housing, undercut trigger guard.

Production: current   Caliber: 9mm, .40 S&W, .45 ACP   Action: SA, semi-auto

Barrel Length: 3.24"   OA Length: 7"   Wt.: 23 oz.   Sights: lowered inset Heinie

Straight Eight 2-dot tritium   Capacity: 6, 7, 8   Magazine: detachable box

Grips: black G10 Micarta with STI logo

D2C:	NIB $1,472	Ex $1,125	VG+ $ 810	Good $ 740	LMP $1,472
C2C:	Mint $1,420	Ex $1,025	VG+ $ 740	Good $ 665	Fair $ 340
Trade-In:	Mint $1,050	Ex $ 825	VG+ $ 580	Good $ 520	Poor $ 150

SPARTAN III & IV
NEW IN BOX

SPARTAN III & IV REDBOOK CODE: RB-SI-H-SPRN34

1911 Commander steel frame, classic slide with rear cocking serrations, chromed non-ramped bull barrel, beavertail grip safety, full-length guide rod, parkerized finish.

Production: current   Caliber: 9mm, .45 ACP   Action: SA, semi-auto

Barrel Length: 3.24", 4.26"   Wt.: 33 oz.   Sights: competition front, fixed rear

Capacity: 8, 9   Magazine: detachable box

Grips: hand-checkered double-diamond mahogany

D2C:	NIB $ 754	Ex $ 575	VG+ $ 415	Good $ 380	LMP $ 754
C2C:	Mint $ 730	Ex $ 525	VG+ $ 380	Good $ 340	Fair $ 175
Trade-In:	Mint $ 540	Ex $ 425	VG+ $ 300	Good $ 265	Poor $ 90

SPARTAN V
NEW IN BOX

SPARTAN V REDBOOK CODE: RB-SI-H-SPRTN5

1911 Government steel frame, classic slide with rear cocking serrations, chromed non-ramped bull barrel, beavertail grip safety, full-length guide rod, parkerized finish.

Caliber: 9mm, .45 ACP Action: SA, semi-auto Barrel Length: 5.11" OA Length: 8.5"

Wt.: 35 oz. Sights: fiber-optic front, LPA adjustable rear Capacity: 8, 9

Magazine: detachable box Grips: hand-checkered double-diamond mahogany

D2C:	NIB $ 754	Ex $ 575	VG+ $ 415	Good $ 380	LMP $ 754
C2C:	Mint $ 730	Ex $ 525	VG+ $ 380	Good $ 340	Fair $ 175
Trade-In:	Mint $ 540	Ex $ 425	VG+ $ 300	Good $ 265	Poor $ 90

TROJAN 5.0
VERY GOOD +

TROJAN 5.0 REDBOOK CODE: RB-SI-H-TROJN5

Matte blue finish, Government steel slide, front and rear cocking serrations, stippled front strap, bushing barrel, .40 and 9mm have ramped barrel, beavertail grip safety, one-piece full-length guide rod.

Caliber: 9mm, .40 S&W, .40 Super, .45 ACP Action: SA, semi-auto

Barrel Length: 5" OA Length: 8.5" Wt.: 36 oz. Sights: competition front, adjustable rear

Capacity: 8 (.45 ACP) Magazine: detachable box Grips: rosewood

D2C:	NIB $ 1222	Ex $ 950	VG+ $ 675	Good $ 615	
C2C:	Mint $ 1180	Ex $ 850	VG+ $ 620	Good $ 550	Fair $ 285
Trade-In:	Mint $ 870	Ex $ 700	VG+ $ 480	Good $ 430	Poor $ 150

APEIRO
NEW IN BOX

APEIRO REDBOOK CODE: RB-SI-H-APEIRO

STI modular steel 2011 frame, stainless steel slide with STI Sabertooth cocking serrations, ramped Schuemann island-style bull barrel, high-ride grip safety, ambidextrous thumb safeties, RecoilMaster guide rod, stainless finished slide with polished sides, blue frame.

Production: 2009 - current Caliber: 9mm, .40 S&W, .45 ACP Action: SA, semi-auto

Barrel Length: 5.01" OA Length: 8.69" Wt.: 40 oz.

Sights: Dawson fiber-optic front, adjustable rear Capacity: 11, 14, 17, 20, 22, 26

Magazine: detachable box Grips: modular polymer grip

D2C:	NIB $2,934	Ex $2,250	VG+ $1,615	Good $1,470	LMP $2,934
C2C:	Mint $2,820	Ex $2,025	VG+ $1,470	Good $1,325	Fair $ 675
Trade-In:	Mint $2,090	Ex $1,650	VG+ $1,150	Good $1,030	Poor $ 300

NITRO 10
NEW IN BOX

NITRO 10 REDBOOK CODE: RB-SI-H-NTRO10

1911 Government frame, 'Nitro 10' front strap, classic slide with 'Nitro' front and rear cocking serrations, ramped bull barrel, beavertail grip safety, one-piece steel guide rod, matte blue finish.

Production: 2014 - current Caliber: 10mm Action: SA, semi-auto

Barrel Length: 5" OA Length: 8.77" Wt.: 39 oz.

Sights: competition front, fixed ledge-style rear Capacity: 8

Magazine: detachable box Grips: Nitro cocobolo grip panels

D2C:	NIB $ 1,514	Ex $1,175	VG+ $ 835	Good $ 760	LMP $ 1,514
C2C:	Mint $1,460	Ex $1,050	VG+ $ 760	Good $ 685	Fair $ 350
Trade-In:	Mint $1,080	Ex $ 850	VG+ $ 600	Good $ 530	Poor $ 180

STI INTERNATIONAL

EAGLE 5.0
NEW IN BOX

EAGLE 5.0 REDBOOK CODE: RB-SI-H-EAGLE5

STI modular steel 2011 frame, classic round-top slide, front and rear cocking serrations, ramped bushing barrel, beavertail grip safety, ambidextrous thumb safeties, one-piece steel guide rod, blue finish with polished sides of slide, custom options available.

Production: current   Caliber: 9mm, 9x23, .357 SIG, .38 Super, .40 S&W, .40 Super, .45 ACP, .45HP, 10mm   Action: SA, semi-auto   Barrel Length: 5.11"

OA Length: 8.77"   Wt.: 35 oz.   Sights: competition front, adjustable rear

Capacity: 11, 14, 17, 20, 22, 26   Magazine: detachable box   Grips: modular polymer grip

D2C:	NIB $2,123	Ex $1,625	VG+ $1,170	Good $1,065	LMP $2,123
C2C:	Mint $2,040	Ex $1,475	VG+ $1,070	Good $ 960	Fair $ 490
Trade-In:	Mint $1,510	Ex $1,200	VG+ $ 830	Good $ 745	Poor $ 240

EDGE
EXCELLENT
Courtesy of Bud's Gun Shop

EDGE REDBOOK CODE: RB-SI-H-EDGEXX

Fully supported and ramped bull barrel, STI's modular steel 2011 frame, front and rear cocking serrations, aluminum magwell, beavertail grip safety, ambidextrous thumb safeties, RecoilMaster guide rod, extended magazines available, many custom options.

Production: current   Caliber: 9mm, .38 Super, .40 S&W, .45 ACP

Action: SA, semi-auto   Barrel Length: 5.01"   OA Length: 8.5"   Wt.: 38 oz.

Sights: competition front or Dawson fiber-optic front, adjustable rear

Capacity: 11, 14, 17, 20, 22, 26   Magazine: detachable box   Grips: modular polymer grip

D2C:	NIB $2,180	Ex $1,675	VG+ $1,200	Good $1,090	LMP $2,180
C2C:	Mint $2,100	Ex $1,525	VG+ $1,090	Good $ 985	Fair $ 505
Trade-In:	Mint $1,550	Ex $1,225	VG+ $ 860	Good $ 765	Poor $ 240

EXECUTIVE
EXCELLENT

EXECUTIVE REDBOOK CODE: RB-SI-H-EXECUT

STI modular steel 2011 frame, round-top slide with front and rear cocking serrations, beavertail grip safety, ambidextrous thumb safeties, RecoilMaster guide rod, hard chrome finish with polished sides and black color inlay, custom options available.

Production: current   Caliber: .40 S&W   Action: SA, semi-auto   Barrel Length: 5"

OA Length: 8.6"   Wt.: 40 oz.   Sights: Dawson fiber-optic front, adjustable rear

Capacity: 14, 17, 22   Magazine: detachable box   Grips: gray modular polymer grip

D2C:	NIB $2,520	Ex $1,925	VG+ $1,390	Good $1,260	LMP $2,638
C2C:	Mint $2,420	Ex $1,750	VG+ $1,260	Good $1,135	Fair $ 580
Trade-In:	Mint $1,790	Ex $1,425	VG+ $990	Good $ 885	Poor $ 270

LEGEND
EXCELLENT

LEGEND REDBOOK CODE: RB-SI-H-LEGEND

STI modular steel 2011 frame, tri-top slide with 30 lpi and Sabertooth cocking serrations, ramped bull barrel, beavertail grip safety, ambi thumb safeties, RecoilMaster guide rod, hard chromed slide with black inlay, blue frame.

Production: current   Caliber: 9mm, .40 S&W, .45 ACP   Action: SA, semi-auto

Barrel Length: 5.01"   OA Length: 8.67"   Wt.: 39 oz.

Sights: Dawson fiber-optic front, adjustable rear   Capacity: 11, 14, 17, 20, 22, 26

Magazine: detachable box   Grips: modular polymer grip

D2C:	NIB $2,886	Ex $2,200	VG+ $1,590	Good $1,445	LMP $2,886
C2C:	Mint $2,780	Ex $2,000	VG+ $1,450	Good $1,300	Fair $ 665
Trade-In:	Mint $2,050	Ex $1,625	VG+ $1,130	Good $1,015	Poor $ 300

PERFECT 10
EXCELLENT

PERFECT 10 REDBOOK CODE: RB-SI-H-PRFC10

STI 2011 modular polymer frame with aluminum magwell, classic round-top slide with front and rear cocking serrations, ramped bull barrel, beavertail grip safety, ambi thumb safeties, two-piece steel guide rod, blue finish.

Production: current   Caliber: 10mm   Action: SA, semi-auto   Barrel Length: 6"

OA Length: 9.5"   Wt.: 41 oz.   Sights: competition front, adjustable rear   Capacity: 14 ,17

Magazine: detachable box   Grips: modular polymer grip

D2C:	NIB	$2,520	Ex	$1,925	VG+	$1,390	Good	$1,260	LMP $2,621
C2C:	Mint	$2,420	Ex	$1,750	VG+	$1,260	Good	$1,135	Fair $ 580
Trade-In:	Mint	$1,790	Ex	$1,425	VG+	$ 990	Good	$ 885	Poor $ 270

TACTICAL 3.0 & 4.0
EXCELLENT

TACTICAL 3.0 & 4.0 REDBOOK CODE: RB-SI-H-TACT34

STI modular aluminum 2011 frame with integral rail, 4140 carbon steel slide, front and rear cocking serrations, ramped bull barrel, beavertail grip safety, ambi thumb safeties, RecoilMaster guide rod, blue slide, black Cerakoted frame.

Production: current   Caliber: 9mm, .40 S&W, .45 ACP   Action: SA, semi-auto

Barrel Length: 3.75", 4.26"   Wt.: 30 oz., 34 oz.   Sights: ramped front, ledge-style rear

Capacity: 11, 14, 17, 20, 22, 26   Magazine: detachable box   Grips: modular polymer grip

D2C:	NIB	$2,144	Ex	$1,650	VG+	$1,180	Good	$1,075	LMP $2,144
C2C:	Mint	$2,060	Ex	$1,500	VG+	$1,080	Good	$ 965	Fair $ 495
Trade-In:	Mint	$1,530	Ex	$1,225	VG+	$840	Good	$ 755	Poor $ 240

TACTICAL 5.0
NEW IN BOX

TACTICAL 5.0 REDBOOK CODE: RB-SI-H-TACT50

STI modular steel 2011 frame with integral rail, 4140 carbon steel slide, front and rear cocking serrations, ramped bull barrel, beavertail grip safety, ambi thumb safeties, RecoilMaster guide rod, blue slide and frame.

Production: current   Caliber: 9mm, .40 S&W, .45 ACP   Action: SA, semi-auto

Barrel Length: 5.01"   OA Length: 8.5"   Wt.: 34 oz.

Sights: ramped front, ledge style rear   Capacity: 11, 14, 17, 20, 22, 26

Magazine: detachable box   Grips: modular polymer grip

D2C:	NIB	$2,144	Ex	$1,650	VG+	$1,180	Good	$1,075	LMP $2,144
C2C:	Mint	$2,060	Ex	$1,500	VG+	$1,080	Good	$ 965	Fair $ 495
Trade-In:	Mint	$1,530	Ex	$1,225	VG+	$ 840	Good	$ 755	Poor $ 240

TACTICAL 4.15
NEW IN BOX

TACTICAL 4.15 REDBOOK CODE: RB-SI-H-TAC415

STI modular steel 2011 double-stack frame with integral rail, flat-top slide with serrations, ramped bull barrel, RecoilMaster guide rod, matte blue finish, ambi thumb safety, beavertail grip safety.

Production: 2004 - current   Caliber: 9mm, .40 S&W, .45 ACP   Action: SA, semi-auto

Barrel Length: 4.15"   OA Length: 7.75"   Wt.: 34 oz.

Sights: ramped front, Heinie fixed rear   Capacity: 10, 12, 15   Magazine: detachable box

Grips: modular polymer grip

D2C:	NIB	$1,999	Ex	$1,525	VG+	$1,100	Good	$1,000	LMP $1,999
C2C:	Mint	$1,920	Ex	$1,400	VG+	$1,000	Good	$ 900	Fair $ 460
Trade-In:	Mint	$1,420	Ex	$1,125	VG+	$ 780	Good	$ 700	Poor $ 210

TACTICAL SS 3.0 & 4.0   REDBOOK CODE: RB-SI-H-TACS34

Forged steel frame with integral tactical rail, 30 lpi font strap checkering, flat-top slide with front and rear cocking serrations, ramped bull barrel, beavertail grip safety, ambi thumb safeties, RecoilMaster guide rod, matte blue finish.

Production: 2013 - current   Caliber: 9mm, .40 S&W, .45 ACP   Action: SA, semi-auto

Barrel Length: 3.96", 4.26" Wt.: 34 oz., 38 oz.   Sights: ramped front, ledge-style rear

Capacity: 6, 7, 8, 9 Magazine: detachable box   Grips: STI G10 grip panels

D2C:	NIB $1,940	Ex $1,475	VG+ $1,070	Good $ 970	LMP $1,940
C2C:	Mint $1,870	Ex $1,350	VG+ $ 970	Good $ 875	Fair $ 450
Trade-In:	Mint $1,380	Ex $1,100	VG+ $ 760	Good $ 680	Poor $ 210

TACTICAL SS 5.0   REDBOOK CODE: RB-SI-H-TACSS5

Forged steel frame with integral tactical rail, 30 lpi font strap checkering, flat-top slide with front and rear cocking serrations, ramped bull barrel, beavertail grip safety, ambi thumb safeties, RecoilMaster guide rod, matte blue finish.

Production: 2013 - current   Caliber: 9mm, .40 S&W, .45 ACP   Action: SA, semi-auto

Barrel Length: 5.01" OA Length: 8.75" Wt.: 41 oz.   Sights: ramped front, ledge style rear

Capacity: 8, 9 Magazine: detachable box   Grips: STI G10 grip panels

D2C:	NIB $1,940	Ex $1,475	VG+ $1,070	Good $ 970	LMP $1,940
C2C:	Mint $1,870	Ex $1,350	VG+ $ 970	Good $ 875	Fair $ 450
Trade-In:	Mint $1,380	Ex $1,100	VG+ $ 760	Good $ 680	Poor $ 210

VIP
NEW IN BOX

VIP   REDBOOK CODE: RB-SI-H-VIPXXX

STI 2011 modular steel or aluminum grip and frame, stainless slide with rear cocking serrations, stainless ramped bull barrel, beavertail grip safety, RecoilMaster guide rod.

Production: current   Caliber: 9mm, .40 S&W, .45 ACP   Action: SA, semi-auto

Barrel Length: 3.96" OA Length: 6.75" Wt.: 30 oz. (steel), 26 oz. (aluminum)

Sights: ramped front, fixed rear Capacity: 10, 12, 15 Magazine: detachable box

Grips: modular polymer grip

D2C:	NIB $1,690	Ex $1,300	VG+ $ 930	Good $ 845	LMP $ 1,775
C2C:	Mint $1,630	Ex $1,175	VG+ $ 850	Good $ 765	Fair $ 390
Trade-In:	Mint $1,200	Ex $ 950	VG+ $ 660	Good $ 595	Poor $ 180

OFF DUTY   REDBOOK CODE: RB-SI-H-OFFDTY

Steel Commander frame, stippled front strap, steel slide with rear cocking serrations, beveled magwell, integrated rail, ramped bull barrel, beavertail grip safety, RecoilMaster guide rod, blue finish, custom options available.

Production: 2009 - discontinued   Caliber: 9mm, .45 ACP   Action: SA, semi-auto

Barrel Length: 3" OA Length: 7" Wt.: 31.3 oz.   Sights: 3-dot adjustable

Capacity: 6, 7 Magazine: detachable box   Grips: checkered cocobolo wood

D2C:	NIB $ 1,120	Ex $ 875	VG+ $ 620	Good $ 560	
C2C:	Mint $1,080	Ex $ 775	VG+ $ 560	Good $ 505	Fair $ 260
Trade-In:	Mint $ 800	Ex $ 650	VG+ $ 440	Good $ 395	Poor $ 120

ROGUE REDBOOK CODE: RB-SI-H-ROGUEX

Officer forged aluminum frame, stippled front strap, lightened stainless slide with cocking serrations, bull barrel, beavertail grip safety, RecoilMaster guide rod, blue slide, Duracoated frame.

Production: discontinued Caliber: 9mm Action: SA, semi-auto
Barrel Length: 3" OA Length: 6" Wt.: 21 oz. Sights: integral fixed
Capacity: 7 Magazine: detachable box Grips: checkered cocobolo wood

D2C:	NIB	$1,024	Ex $	800	VG+ $	565	Good $	515	LMP $1,024
C2C:	Mint $	990	Ex $	725	VG+ $	520	Good $	465	Fair $ 240
Trade-In:	Mint $	730	Ex $	575	VG+ $	400	Good $	360	Poor $ 120

TOTAL ECLIPSE
EXCELLENT
Courtesy of Bud's Gun Shop

TOTAL ECLIPSE REDBOOK CODE: RB-SI-H-TTLECL

STI 2011 modular aluminum frame, classic slide with rear cocking serrations and integral sight channel, ramped bull barrel, beavertail grip safety, RecoilMaster guide rod, black Cerakoted.

Caliber: 9mm, .40 S&W, .45 ACP Action: SA, semi-auto
Barrel Length: 3.24" OA Length: 7" Wt.: 22 oz. Sights: lowered inset 2-dot tritium
Capacity: 7, 8, 12 Magazine: detachable box Grips: modular polymer grip

D2C:	NIB	$1,685	Ex	$1,300	VG+ $	930	Good $	845	LMP $1,870
C2C:	Mint	$1,620	Ex	$1,175	VG+ $	850	Good $	760	Fair $ 390
Trade-In:	Mint	$1,200	Ex $	950	VG+ $	660	Good $	590	Poor $ 180

BLS9 & BLS40 REDBOOK CODE: RB-SI-H-BLS940

Government length steel frame, high rise cut under trigger guard, stippled front strap, front and rear slide serrations, ramped linkless barrel, no bushing, high-rise grip safety, matte blue finish.

Caliber: 9mm, .40 S&W Action: SA, semi-auto Barrel Length: 3.4" OA Length: 7"
Height: 5.25" Wt.: 30 oz. Sights: integral front, Heinie Low Mount rear Capacity: 8, 9
Magazine: detachable box Grips: rosewood

D2C:	NIB $	720	Ex $	550	VG+ $	400	Good $	360	
C2C:	Mint $	700	Ex $	500	VG+ $	360	Good $	325	Fair $ 170
Trade-In:	Mint $	520	Ex $	425	VG+ $	290	Good $	255	Poor $ 90

LS9 & LS40
EXCELLENT

LS9 & LS40 REDBOOK CODE: RB-SI-H-LS9X40

Steel single stack frame, high rise cut under trigger guard, stippled front strap, exterior extractor, front and rear slide serrations, ramped linkless barrel, no bushing, matte blue finish.

Production: discontinued Caliber: 9mm, .40 S&W Action: SA, semi-auto
Barrel Length: 3.4" OA Length: 7" Wt.: 28 oz. Sights: slide integral front,
Heine Low Mount rear Capacity: 6, 7 Magazine: detachable box
Grips: checkered cocobolo wood

D2C:	NIB	$880	Ex	$675	VG+	$485	Good	$440	
C2C:	Mint $	850	Ex	$625	VG+	$440	Good	$400	Fair $ 205
Trade-In:	Mint $	630	Ex	$500	VG+	$350	Good	$310	Poor $ 90

FALCON 3.9 REDBOOK CODE: RB-SI-H-FALCON

STI modular steel or aluminum 2011 frame, polymer grip, classic slide with rear cocking serrations, stainless ramped bull barrel, beavertail grip safety, ambidextrous thumb safety, blue finish.

Production: discontinued Caliber: .40 S&W, .45 ACP Action: SA, semi-auto
Barrel Length: 3.9" Wt.: 25 oz. (steel), 30 oz. (aluminum) Sights: STI fixed
Capacity: 10 Magazine: detachable box Grips: modular polymer grip

D2C:	NIB	$1,100	Ex $	850	VG+ $	605	Good $	550	
C2C:	Mint	$1,060	Ex $	775	VG+ $	550	Good $	495	Fair $ 255
Trade-In:	Mint $	790	Ex $	625	VG+ $	430	Good $	385	Poor $ 120

STI INTERNATIONAL

DUTY CT   REDBOOK CODE: RB-SI-H-DUTYCT

Matte blue finish, accessory rail, 30 lpi front strap checkering, ramped bull barrel, beavertail grip safety, ambidextrous thumb safeties, RecoilMaster guide rod.

Production: discontinued   Caliber: 9mm, .40 S&W, .45 ACP   Action: SA, semi-auto
Barrel Length: 4.15"   OA Length: 7.75"   Wt.: 36 oz.   Sights: ramped front, Heinie fixed rear   Capacity: 8, 9   Magazine: detachable box   Grips: checkered cocobolo wood

D2C:	NIB	$1,350	Ex	$1,050	VG+	$ 745	Good	$ 675	LMP $1,440
C2C:	Mint	$1,300	Ex	$ 950	VG+	$ 680	Good	$ 610	Fair $ 315
Trade-In:	Mint	$ 960	Ex	$ 775	VG+	$ 530	Good	$ 475	Poor $ 150

TRUSIGHT   REDBOOK CODE: RB-SI-H-TRUSGT

STI modular steel frame with polymer grip, rear slide serrations, ramped bull barrel with expansion chamber, beavertail grip safety, ambidextrous thumb safeties, RecoilMaster guide rod, matte blue finish, magwell.

Production: 2006 - discontinued   Caliber: 9mm, .40 S&W, .45 ACP
Action: SA, semi-auto   Barrel Length: 4.15"   Wt.: 36 oz.
Sights: Dawson fiber-optic front, adjustable rear   Capacity: 11, 14, 17, 20, 22, 26
Magazine: detachable box   Grips: modular polymer grip

D2C:	NIB	$1,700	Ex	$1,300	VG+	$ 935	Good	$ 850	
C2C:	Mint	$1,640	Ex	$1,175	VG+	$ 850	Good	$ 765	Fair $ 395
Trade-In:	Mint	$1,210	Ex	$ 975	VG+	$ 670	Good	$ 595	Poor $ 180

LEGACY   REDBOOK CODE: RB-SI-H-LEGACY

PVD finish on all parts, Government slide and frame, rear cocking serrations, checkered front strap, beavertail grip safety, ambidextrous thumb safeties, ramped bushing barrel.

Production: 2006 - discontinued   Caliber: .45 ACP   Action: SA, semi-auto
Barrel Length: 5"   OA Length: 8.5"   Wt.: 37 oz.   Sights: ramped front, adjustable rear
Capacity: 8   Magazine: detachable box   Grips: smooth cocobolo wood

D2C:	NIB	$1,450	Ex	$1,125	VG+	$ 800	Good	$ 725	
C2C:	Mint	$1,400	Ex	$1,025	VG+	$ 730	Good	$ 655	Fair $ 335
Trade-In:	Mint	$1,030	Ex	$ 825	VG+	$ 570	Good	$ 510	Poor $ 150

LEGACY
EXCELLENT

Steyr

Founded in 1864 by Josef Werndl, Steyr Arms has changed hands several times during its history, including being a part of the Österreichische Waffenfabriksgesellschaft and Steyr Werke AG / Steyr-Daimler-Puch AG conglomerate. The company currently manufactures their weapons in Steyr, Austria, making an array of various pistols, rifles, assault rifles, and sub-machine guns. Steyr Arms Inc. is the U.S. subsidiary of Steyr Mannlicher.

STEYR

MODEL GB   REDBOOK CODE: RB-SX-H-XXXXXX

Originally manufactured by L.E.S., steel frame, slide-mounted safety, blue finish. Slight premium for later military variations.

Production: mid-1970s - 1988   Caliber: 9mm   Action: DA/SA, semi-auto
Barrel Length: 5.3"   OA Length: 8.5"   Sights: fixed   Capacity: 18   Grips: black synthetic

D2C:	NIB	$ 700	Ex	$ 550	VG+	$ 385	Good	$ 350	
C2C:	Mint	$ 680	Ex	$ 500	VG+	$ 350	Good	$ 315	Fair $ 165
Trade-In:	Mint	$ 500	Ex	$ 400	VG+	$ 280	Good	$ 245	Poor $ 90

MODEL GB
EXCELLENT

MODEL M SERIES
NEW IN BOX

MODEL M SERIES REDBOOK CODE: RB-SX-H-XXXXX2
Polymer frame, frame-mounted and trigger safety, matte black finish.

Production: late 1999 - early 2000s Caliber: 9mm, 9x21mm, .40 S&W, .357 Sig					
Action: striker-fired, semi-auto Barrel Length: 4" Sights: fixed					
Capacity: 10, 12, 14, 17 Magazine: detachable box Grips: black synthetic					
D2C:	NIB $ 540	Ex $ 425	VG+ $ 300	Good $ 270	
C2C:	Mint $ 520	Ex $ 375	VG+ $ 270	Good $ 245	Fair $ 125
Trade-In:	Mint $ 390	Ex $ 325	VG+ $ 220	Good $ 190	Poor $ 60

M-A1 SERIES
NEW IN BOX

MODEL SPP/TMP REDBOOK CODE: RB-SX-H-XXXXX1
Tactical, Uzi-style design. Black finish, blowback action, polymer frame, may feature front grip. TMP = military version. SPP = civilian version.

Production: early 1990s - early 2000s Caliber: 9mm					
Action: SA, semi-auto Barrel Length: 5" Sights: fixed					
Capacity: 15, 20, 30 Magazine: detachable box Grips: black synthetic					
D2C:	NIB $1,200	Ex $ 925	VG+ $ 660	Good $ 600	
C2C:	Mint $1,160	Ex $ 850	VG+ $ 600	Good $ 540	Fair $ 280
Trade-In:	Mint $ 860	Ex $ 675	VG+ $ 470	Good $ 420	Poor $ 120

M-A1 SERIES REDBOOK CODE: RB-SX-H-XXXXX3
Similar to the Model M, accessory rail, polymer frame, matte finish.

Production: current Caliber: 9mm, .40 S&W Action: striker-fired, semi-auto					
Barrel Length: 3.5", 4" Sights: fixed Capacity: 12, 17 Magazine: detachable box					
Grips: black synthetic					
D2C:	NIB $ 515	Ex $ 400	VG+ $ 285	Good $ 260	
C2C:	Mint $ 500	Ex $ 375	VG+ $ 260	Good $ 235	Fair $ 120
Trade-In:	Mint $ 370	Ex $ 300	VG+ $ 210	Good $ 185	Poor $ 60

Sturm, Ruger & Company

William B. Ruger and Alexander M. Sturm founded the Sturm, Ruger & Co., Inc. in 1949, producing .22 caliber pistols in their Southport, Connecticut machine shop. This pistol, known as the Ruger Standard model, became the most successful .22 semi-automatic in history. This success helped the company design and manufacture other firearms, most of which found great commercial success. Ruger currently manufactures an extensive line of pistols, revolvers, rifles, and shotguns from its facilities in Newport, New Hampshire and Prescott, Arizona.

SINGLE SIX
(FLAT GATE MODEL)
EXCELLENT

SINGLE SIX (FLAT GATE MODEL) REDBOOK CODE: RB-RU-H-SNGSFM
Early production Single Six models were referred to as "flat gates." These models feature a flat cartridge loading gate and black checkered hard-rubber grips. Serial numbers under 2000 may increase this model's value up to 30 percent.

Production: 1953 - 1957 Caliber: .22 LR, .22 WMR Action: SA, revolver					
Barrel Length: 5.5" Sights: fixed Capacity: 6 Grips: rubber					
D2C:	NIB $ 880	Ex $ 675	VG+ $ 485	Good $ 440	
C2C:	Mint $ 850	Ex $ 625	VG+ $ 440	Good $ 400	Fair $ 205
Trade-In:	Mint $ 630	Ex $ 500	VG+ $ 350	Good $ 310	Poor $ 90

STEYR

STURM, RUGER & CO.

SINGLE SIX
(CONTOURED GATE MODEL)
GOOD

SINGLE SIX (CONTOURED GATE MODEL)

REDBOOK CODE: RB-RU-H-SNGSCM

Contoured or "round gate" models were introduced in the late 1950s. These models replaced the flat gate models of the late 1950s. Factory engraved models are in a different price category.

Production: late 1950s - discontinued Caliber: .22 LR, .22 WMR Action: SA, revolver
Barrel Length: 4.625", 5.5", 9.5" Sights: fixed Capacity: 6

Grips: rubber or walnut

D2C:	NIB $ 620	Ex $ 475	VG+ $ 345	Good $ 310		
C2C:	Mint $ 600	Ex $ 450	VG+ $ 310	Good $ 280	Fair $ 145
Trade-In:	Mint $ 450	Ex $ 350	VG+ $ 250	Good $ 220	Poor $ 90

SINGLE SIX CONVERTIBLE
EXCELLENT

SINGLE SIX CONVERTIBLE   REDBOOK CODE: RB-RU-H-SNGSCO

Blue finish, interchangeable cylinders, black checkered hard-rubber grips, also available in stainless finish. Similar to the Colt Single Action Army.

Production: early 1960s - discontinued Caliber: .22 LR/.22 WMR Action: SA, revolver
Barrel Length: 5.5", 6.5", 9.5" Sights: fixed Capacity: 6 Grips: rubber

D2C:	NIB $ 680	Ex $ 525	VG+ $ 375	Good $ 340		
C2C:	Mint $ 660	Ex $ 475	VG+ $ 340	Good $ 310	Fair $ 160
Trade-In:	Mint $ 490	Ex $ 400	VG+ $ 270	Good $ 240	Poor $ 90

**LIGHTWEIGHT
SINGLE SIX**
GOOD

LIGHTWEIGHT SINGLE SIX   REDBOOK CODE: RB-RU-H-LWSSIX

Tri-color or blue finish, aluminum alloy frame, alloy or steel cylinder, rollmark on cylinder frame. Serial numbers began around 200,000.

Production: 1956 - discontinued Caliber: .22 LR, .22 WMR Action: SA, revolver
Barrel Length: 4.6" Sights: fixed Capacity: 6 Grips: rubber or walnut

D2C:	NIB $1,300	Ex $1,000	VG+ $ 715	Good $ 650		
C2C:	Mint $1,250	Ex $900	VG+ $ 650	Good $ 585	Fair $ 300
Trade-In:	Mint $ 930	Ex $750	VG+ $ 510	Good $ 455	Poor $ 150

**COLORADO
CENTENNIAL
SINGLE SIX**
EXCELLENT

COLORADO CENTENNIAL SINGLE SIX

REDBOOK CODE: RB-RU-H-COCNSS

Polished blue finish, 15,000 produced, special display box included. Special engravings, commemorating the Colorado Centennial.

Production: 1975 Caliber: .22 LR, .22 WMR Action: SA, revolver
Barrel Length: 6.5" Sights: blade front, adjustable rear Capacity: 6 Grips: walnut

D2C:	NIB $ 500	Ex $ 400	VG+ $ 275	Good $ 250		
C2C:	Mint $ 480	Ex $ 350	VG+ $ 250	Good $ 225	Fair $ 115
Trade-In:	Mint $ 360	Ex $ 300	VG+ $ 200	Good $ 175	Poor $ 60

SINGLE SIX 50TH ANNIVERSARY   REDBOOK CODE: RB-RU-H-SS50AN

Blue steel finish, limited edition, gold rollmark, red Ruger medallions, red case included.

Production: 2003 only Caliber: .22 LR, .22 WMR Action: SA, revolver
Barrel Length: 4.6" Sights: fixed Capacity: 6 Grips: cocobolo

D2C:	NIB $ 450	Ex $ 350	VG+ $ 250	Good $ 225		
C2C:	Mint $ 440	Ex $ 325	VG+ $ 230	Good $ 205	Fair $ 105
Trade-In:	Mint $ 320	Ex $ 275	VG+ $ 180	Good $ 160	Poor $ 60

**BLACKHAWK FLATTOP
.44 MAGNUM**
VERY GOOD +

BLACKHAWK FLATTOP .44 MAGNUM

REDBOOK CODE: RB-RU-H-BLFLAT

Blue finish, flat-top steel cylinder frame, slight premium for 7.5" barrel models, nearly double value for 10" barrel models.

Production: 1956 - 1963 Caliber: .44 Magnum Action: SA, revolver					
Barrel Length: 6.5", 7.5", 10" Sights: adjustable rear Capacity: 6 Grips: walnut					
D2C:	NIB $1,400	Ex $1,075	VG+ $ 770	Good $ 700	
C2C:	Mint $1,350	Ex $ 975	VG+ $ 700	Good $ 630	Fair $ 325
Trade-In:	Mint $1,000	Ex $ 800	VG+ $ 550	Good $ 490	Poor $ 150

**BLACKHAWK FLATTOP
.357 MAGNUM**
EXCELLENT

BLACKHAWK FLATTOP .357 MAGNUM

REDBOOK CODE: RB-RU-H-FLATXX

Blue finish, larger version of the Single Six model, earlier revolvers equipped with rubber grips, slightly less for 6.5" barrel models, nearly double value for 10" barrel models.

Production: 1955 - 1962 Caliber: .357 Magnum Action: SA, revolver					
Barrel Length: 4.6", 6.5", 10" Sights: adjustable rear Capacity: 6					
Grips: checkered hard rubber or walnut					
D2C:	NIB $1,300	Ex $1,000	VG+ $ 715	Good $ 650	
C2C:	Mint $1,250	Ex $ 900	VG+ $ 650	Good $ 585	Fair $ 300
Trade-In:	Mint $ 930	Ex $ 750	VG+ $ 510	Good $ 455	Poor $ 150

BLACKHAWK
EXCELLENT

BLACKHAWK REDBOOK CODE: RB-RU-H-BLACKH

Blue finish, equipped with a hooded rear sight, very high premium for original factory brass grip frame.

Production: late 1950s Caliber: .357 Magnum, .41 Magnum, .45 LC, .30 Carbine					
Action: SA, revolver Barrel Length: 4.6", 6.5"					
Sights: ramp front, adjustable rear Capacity: 6 Grips: walnut					
D2C:	NIB $ 950	Ex $ 725	VG+ $ 525	Good $ 475	
C2C:	Mint $ 920	Ex $ 675	VG+ $ 480	Good $ 430	Fair $ 220
Trade-In:	Mint $ 680	Ex $ 550	VG+ $ 380	Good $ 335	Poor $ 120

BLACKHAWK CONVERTIBLE REDBOOK CODE: RB-RU-H-BLACON

Interchangeable cylinders, blue finish, hooded rear sight.

Production: discontinued Caliber: .357 Magnum, .45 LC Action: SA, revolver					
Barrel Length: 4.6", 6.5" Sights: ramp front, adjustable rear Capacity: 6					
Grips: walnut					
D2C:	NIB $ 700	Ex $ 550	VG+ $ 385	Good $ 350	
C2C:	Mint $ 680	Ex $ 500	VG+ $ 350	Good $ 315	Fair $ 165
Trade-In:	Mint $ 500	Ex $ 400	VG+ $ 280	Good $ 245	Poor $ 90

SUPER BLACKHAWK
(OLD MODEL)
EXCELLENT

SUPER BLACKHAWK (OLD MODEL) REDBOOK CODE: RB-RU-H-SUPBLA

Blue finish, unfluted cylinder, rear square trigger guard, additional price for revolvers with brass grip frames, original wood presentation case brings a high premium.

Production: 1959 - 1972 Caliber: .44 Magnum Action: SA, revolver					
Barrel Length: 7.5" Wt.: 48 oz. Sights: adjustable Capacity: 6 Grips: walnut					
D2C:	NIB $ 750	Ex $ 575	VG+ $ 415	Good $ 375	
C2C:	Mint $ 720	Ex $ 525	VG+ $ 380	Good $ 340	Fair $ 175
Trade-In:	Mint $ 540	Ex $ 425	VG+ $ 300	Good $ 265	Poor $ 90

STURM, RUGER & CO.

BEARCAT
(OLD MODEL)
EXCELLENT

BEARCAT (OLD MODEL) REDBOOK CODE: RB-RU-H-OBEARC

Alloy frame, anodized, brass colored alloy trigger guard, blue finish, unfluted cylinder, music wire coil springs, this model had plastic grips until early 1960s.

Production: 1958 - 1970	Caliber: .22 LR	Action: SA, revolver			
Barrel Length: 4" Sights: fixed Capacity: 6 Grips: walnut					
D2C:	NIB $ 675	Ex $ 525	VG+ $ 375	Good $ 340	
C2C:	Mint $ 650	Ex $ 475	VG+ $ 340	Good $ 305	Fair $ 160
Trade-In:	Mint $ 480	Ex $ 400	VG+ $ 270	Good $ 240	Poor $ 90

SUPER BEARCAT
(OLD MODEL)
EXCELLENT

SUPER BEARCAT (OLD MODEL) REDBOOK CODE: RB-RU-H-OSBEAR

Blue finish, brass colored aluminum or blue trigger guard, music wire coil springs.

Production: early 1970s - 1974	Caliber: .22 LR	Action: SA, revolver			
Barrel Length: 4" Sights: fixed Capacity: 6 Grips: walnut					
D2C:	NIB $ 700	Ex $ 550	VG+ $ 385	Good $ 350	
C2C:	Mint $ 680	Ex $ 500	VG+ $ 350	Good $ 315	Fair $ 165
Trade-In:	Mint $ 500	Ex $ 400	VG+ $ 280	Good $ 245	Poor $ 90

LCRX
(5430)
NEW IN BOX

LCRX (5430) REDBOOK CODE: RB-RU-H-LC5430

Polymer frame, exposed hammer, grip peg, matte black synergistic finish, IonBond Diamondback cylinder finish, stainless steel barrel, includes a soft case, also available with Crimson Trace Lasergrips.

Production: current	Caliber: .38 Special +P	Action: DA/SA, revolver			
Barrel Length: 1.87" Wt.: 13 oz. Sights: replaceable, pinned ramp front,					
U-Notch Integral rear Capacity: 5 Grips: Hogue Tamer Monogrip					
D2C:	NIB $ 430	Ex $ 350	VG+ $ 240	Good $ 215	LMP $ 529
C2C:	Mint $ 420	Ex $ 300	VG+ $ 220	Good $ 195	Fair $ 100
Trade-In:	Mint $ 310	Ex $ 250	VG+ $ 170	Good $ 155	Poor $ 60

LCR
(5413)
EXCELLENT

LCR (5413) REDBOOK CODE: RB-RU-H-LC5413

Polymer frame, matte black, synergistic finish, IonBond Diamondback cylinder finish, grip peg, includes a soft case, distinguished by Crimson Trace Lasergrips.

Production: current	Caliber: .22 LR	Action: DAO, revolver			
Barrel Length: 1.87" Wt.: 14 oz. Sights: replaceable, pinned ramp front,					
U-Notch Integral rear Capacity: 8 Grips: Crimson Trace Lasergrips					
D2C:	NIB $ 630	Ex $ 500	VG+ $ 350	Good $ 315	LMP $ 799
C2C:	Mint $ 610	Ex $ 450	VG+ $ 320	Good $ 285	Fair $ 145
Trade-In:	Mint $ 450	Ex $ 375	VG+ $ 250	Good $ 225	Poor $ 90

LCR
(5414)
NEW IN BOX

LCR (5414) REDBOOK CODE: RB-RU-H-LC5414

Polymer frame, matte black, synergistic finish, IonBond Diamondback cylinder finish, grip peg, includes soft case, distinguished by Hogue Tamer Monogrips, also available with Crimson Trace Lasergrips.

Production: current	Caliber: .22 WMRF	Action: DAO, revolver			
Barrel Length: 1.87" Wt.: 17 oz. Sights: replaceable, pinned ramp front,					
U-Notch Integral rear Capacity: 6 Grips: Hogue Tamer Monogrip					
D2C:	NIB $ 440	Ex $ 350	VG+ $ 245	Good $ 220	LMP $ 529
C2C:	Mint $ 430	Ex $ 325	VG+ $ 220	Good $ 200	Fair $ 105
Trade-In:	Mint $ 320	Ex $ 250	VG+ $ 180	Good $ 155	Poor $ 60

LCR
(5450)
NEW IN BOX

LCR (5450) REDBOOK CODE: RB-RU-H-LC5450

Polymer frame, IonBond Diamondback finish, stainless steel barrel, grip peg, includes a soft case, distinguished by Hogue Tamer Monogrips, also available with Crimson Trace Lasergrips.

Production: current Caliber: .357 Magnum Action: DAO, revolver

Barrel Length: 1.87" Wt.: 17 oz. Sights: replaceable, pinned ramp front,

U-Notch Integral rear Capacity: 5 Grips: Hogue Tamer Monogrip

D2C:	NIB	$ 498	Ex $ 400	VG+ $ 275	Good $ 250	LMP $ 599
C2C:	Mint $ 480	Ex $ 350	VG+ $ 250	Good $ 225	Fair $ 115
Trade-In:	Mint $ 360	Ex $ 300	VG+ $ 200	Good $ 175	Poor $ 60

LCR
(5451)
NEW IN BOX

LCR (5451) REDBOOK CODE: RB-RU-H-LC5451

Polymer frame, IonBond Diamondback finish, grip peg, includes a soft case, distinguished by Crimson Trace Lasergrips.

Production: current Caliber: .357 Magnum Action: DAO, revolver

Barrel Length: 1.87" Wt.: 17 oz. Sights: replaceable, pinned ramp front,

U-Notch Integral rear Capacity: 5 Grips: Crimson Trace Lasergrips

D2C:	NIB	$ 680	Ex $ 525	VG+ $ 375	Good $ 340	LMP $ 879
C2C:	Mint $ 660	Ex $ 475	VG+ $ 340	Good $ 310	Fair $ 160
Trade-In:	Mint $ 490	Ex $ 400	VG+ $ 270	Good $ 240	Poor $ 90

LCR
(5401)
NEW IN BOX

LCR (5401) REDBOOK CODE: RB-RU-H-LC5401

Polymer frame, matte black, synergistic finish, IonBond Diamondback cylinder finish, stainless steel barrel, grip peg, includes a soft case, distinguished by Hogue Tamer Monogrips, also available with Crimson Trace Lasergrips.

Production: current Caliber: .38 Special +P Action: DAO, revolver

Barrel Length: 1.87" Wt.: 13 oz. Sights: replaceable, pinned ramp front,

U-Notch Integral rear Capacity: 5 Grips: Hogue Tamer Monogrip

D2C:	NIB	$ 440	Ex $ 350	VG+ $ 245	Good $ 220	LMP $ 529
C2C:	Mint $ 430	Ex $ 325	VG+ $ 220	Good $ 200	Fair $ 105
Trade-In:	Mint $ 320	Ex $ 250	VG+ $ 180	Good $ 155	Poor $ 60

LCR
(5402)
NEW IN BOX

LCR (5402) REDBOOK CODE: RB-RU-H-LC5402

Polymer frame, matte black, synergistic finish, IonBond Diamondback cylinder finish, stainless steel barrel, grip peg, includes a soft case, distinguished by Crimson Trace Lasergrips.

Production: current Caliber: .38 Special +P Action: DAO, revolver

Barrel Length: 1.87" Wt.: 13 oz. Sights: replaceable, pinned ramp front,

U-Notch Integral rear Capacity: 5 Grips: Crimson Trace Lasergrips

D2C:	NIB	$ 630	Ex $ 500	VG+ $ 350	Good $ 315	LMP $ 799
C2C:	Mint $ 610	Ex $ 450	VG+ $ 320	Good $ 285	Fair $ 145
Trade-In:	Mint $ 450	Ex $ 375	VG+ $ 250	Good $ 225	Poor $ 90

SP101
(5718)
NEW IN BOX

Courtesy of Bud's Gun Shop

SP101 (5718) REDBOOK CODE: RB-RU-H-SP5718

Satin stainless finish, transfer-bar safety, steel frame, triple-locking cylinder.

Production: current Caliber: .357 Magnum Action: DA/SA, revolver

Barrel Length: 2.25" Wt.: 26 oz. Sights: black ramp front, fixed rear

Capacity: 5 Grips: black rubber

D2C:	NIB	$ 497	Ex $ 400	VG+ $ 275	Good $ 250	LMP $ 659
C2C:	Mint $ 480	Ex $ 350	VG+ $ 250	Good $ 225	Fair $ 115
Trade-In:	Mint $ 360	Ex $ 300	VG+ $ 200	Good $ 175	Poor $ 60

SP101
(5720)
NEW IN BOX

SP101 (5720) REDBOOK CODE: RB-RU-H-SP5720
Satin stainless finish, double-action only model, triple-locking cylinder.

Production: current	Caliber: .357 Magnum	Action: DAO, revolver			
Barrel Length: 2.25" Wt.: 25 oz.	Sights: black ramp front, fixed rear				
Capacity: 5 Grips: black rubber					
D2C:	NIB $ 497	Ex $ 400	VG+ $ 275	Good $ 250	LMP $ 659
C2C:	Mint $ 480	Ex $ 350	VG+ $ 250	Good $ 225	Fair $ 115
Trade-In:	Mint $ 360	Ex $ 300	VG+ $ 200	Good $ 175	Poor $ 60

SP101
(5771)
NEW IN BOX

SP101 (5771) REDBOOK CODE: RB-RU-H-SP5771
Satin stainless finish, engraved wood in black rubber grips, triple-locking cylinder, distinguished by fiber-optic front sight.

Production: current	Caliber: .357 Magnum	Action: DA/SA, revolver			
Barrel Length: 4.2" Wt.: 30 oz.	Sights: fiber-optic front, adjustable rear				
Capacity: 5 Grips: black rubber with engraved wood					
D2C:	NIB $ 560	Ex $ 450	VG+ $ 310	Good $ 280	LMP $ 699
C2C:	Mint $ 540	Ex $ 400	VG+ $ 280	Good $ 255	Fair $ 130
Trade-In:	Mint $ 400	Ex $ 325	VG+ $ 220	Good $ 200	Poor $ 60

SP101
(5737)
NEW IN BOX

SP101 (5737) REDBOOK CODE: RB-RU-H-SP5737
Satin stainless finish, triple-locking cylinder.

Production: current	Caliber: .38 Special +P	Action: DA/SA, revolver			
Barrel Length: 2.25" Wt.: 26 oz.	Sights: black ramp front, fixed rear				
Capacity: 5 Grips: black rubber					
D2C:	NIB $ 497	Ex $ 400	VG+ $ 275	Good $ 250	LMP $ 659
C2C:	Mint $ 480	Ex $ 350	VG+ $ 250	Good $ 225	Fair $ 115
Trade-In:	Mint $ 360	Ex $ 300	VG+ $ 200	Good $ 175	Poor $ 60

GP100 MATCH CHAMPION
NEW IN BOX

GP100 MATCH CHAMPION (1754) REDBOOK CODE: RB-RU-H-GP1754
Standard model, satin stainless finish, slab-sided, half-lug barrel with "Match Champion" engraving, triple-locking cylinder, transfer-bar safety, contoured cylinder.

Production: current	Caliber: .357 Magnum	Action: DA/SA, revolver			
Barrel Length: 4.2" Wt.: 38 oz.	Sights: fiber-optic front, fixed rear				
Capacity: 6 Grips: Hogue Stippled Hardwood					
D2C:	NIB $ 740	Ex $ 575	VG+ $ 410	Good $ 370	LMP $ 899
C2C:	Mint $ 720	Ex $ 525	VG+ $ 370	Good $ 335	Fair $ 175
Trade-In:	Mint $ 530	Ex $ 425	VG+ $ 290	Good $ 260	Poor $ 90

GP100
(1702)
NEW IN BOX

GP100 (1702) REDBOOK CODE: RB-RU-H-GP1702
Standard model, blue finish, triple-locking cylinder, full-shrouded barrel, transfer-bar safety.

Production: current	Caliber: .357 Magnum	Action: DA/SA, revolver			
Barrel Length: 4" Wt.: 40 oz.	Sights: ramp front, adjustable rear				
Capacity: 6 Grips: Hogue Monogrip					
D2C:	NIB $ 560	Ex $ 450	VG+ $ 310	Good $ 280	LMP $ 699
C2C:	Mint $ 540	Ex $ 400	VG+ $ 280	Good $ 255	Fair $ 130
Trade-In:	Mint $ 400	Ex $ 325	VG+ $ 220	Good $ 200	Poor $ 60

GP100
(1704)
NEW IN BOX

GP100 (1704) REDBOOK CODE: RB-RU-H-GP1704

Standard model, blue finish, triple-locking cylinder, full-shrouded barrel, transfer-bar safety. Similar to GP100 1702 model but with a longer barrel.

Production: current Caliber: .357 Magnum Action: DA/SA, revolver

Barrel Length: 6" Wt.: 45 oz. Sights: ramp front, adjustable rear

Capacity: 6 Grips: Hogue Monogrip

D2C:	NIB $	570	Ex $	450	VG+ $	315	Good $	285	LMP $ 699
C2C:	Mint $	550	Ex $	400	VG+ $	290	Good $	260	Fair $ 135
Trade-In:	Mint $	410	Ex $	325	VG+ $	230	Good $	200	Poor $ 60

GP100
(1705)
NEW IN BOX

GP100 (1705) REDBOOK CODE: RB-RU-H-GP1705

Standard model, satin stainless finish, triple-locking cylinder, full-shrouded barrel, transfer-bar safety.

Production: current Caliber: .357 Magnum Action: DA/SA, revolver

Barrel Length: 4" Wt.: 40 oz. Sights: ramp front, adjustable rear

Capacity: 6 Grips: Hogue Monogrip

D2C:	NIB $	620	Ex $	475	VG+ $	345	Good $	310	LMP $ 759
C2C:	Mint $	600	Ex $	450	VG+ $	310	Good $	280	Fair $ 145
Trade-In:	Mint $	450	Ex $	350	VG+ $	250	Good $	220	Poor $ 90

GP100
(1707)
NEW IN BOX

GP100 (1707) REDBOOK CODE: RB-RU-H-GP1707

Similar to GP100 1705 model but with a longer barrel. Standard model, Satin stainless finish, triple-locking cylinder, full-shrouded barrel, transfer-bar safety.

Production: current Caliber: .357 Magnum Action: DA/SA, revolver

Barrel Length: 6" Wt.: 45 oz. Sights: ramp front, adjustable rear

Capacity: 6 Grips: Hogue Monogrip

D2C:	NIB $	620	Ex $	475	VG+ $	345	Good $	310	LMP $ 759
C2C:	Mint $	600	Ex $	450	VG+ $	310	Good $	280	Fair $ 145
Trade-In:	Mint $	450	Ex $	350	VG+ $	250	Good $	220	Poor $ 90

GP100
(1715)
NEW IN BOX

GP100 (1715) REDBOOK CODE: RB-RU-H-GP1715

Satin stainless finish, triple-locking cylinder, full-shrouded barrel, transfer-bar safety.

Production: current Caliber: .357 Magnum Action: DA/SA, revolver

Barrel Length: 3" Wt.: 36 oz. Sights: ramp front, fixed rear Capacity: 6

Grips: Hogue Monogrip

D2C:	NIB $	570	Ex $	450	VG+ $	315	Good $	285	LMP $ 729
C2C:	Mint $	550	Ex $	400	VG+ $	290	Good $	260	Fair $ 135
Trade-In:	Mint $	410	Ex $	325	VG+ $	230	Good $	200	Poor $ 60

GP100 (1753) REDBOOK CODE: RB-RU-H-GP1753

Blue finish, transfer-bar safety, triple-locking cylinder, Talo distributor exclusive.

Production: current Caliber: .357 Magnum Action: DA/SA, revolver

Barrel Length: 3" Wt.: 36 oz. Sights: gold bead Novak front, lo-mount Novak rear

Capacity: 6 Grips: black rubber with wood insert

D2C:	NIB $	690	Ex $	525	VG+ $	380	Good $	345	
C2C:	Mint $	670	Ex $	500	VG+ $	350	Good $	315	Fair $ 160
Trade-In:	Mint $	490	Ex $	400	VG+ $	270	Good $	245	Poor $ 90

STURM, RUGER & CO.

REDHAWK
MINT

REDHAWK STAINLESS
NEW IN BOX

SUPER REDHAWK
MINT

SUPER REDHAWK ALASKAN
NEW IN BOX

REDHAWK REDBOOK CODE: RB-RU-H-REDHAK

Blue finish, large frame, shrouded ejector rod, also offered in stainless steel finish.

Production: 1979 - early 2010s Caliber: .357 Magnum, .41 Magnum, .44 Magnum, .45 LC

Action: DA/SA, revolver Barrel Length: 4", 5.5", 7.5"

Sights: fixed, some with fiber-optic front Capacity: 6 Grips: walnut or rubber

D2C:	NIB $ 820	Ex $ 625	VG+ $ 455	Good $ 410			
C2C:	Mint $ 790	Ex $ 575	VG+ $ 410	Good $ 370	Fair $ 190
Trade-In:	Mint $ 590	Ex $ 475	VG+ $ 320	Good $ 290	Poor $ 90

REDHAWK STAINLESS REDBOOK CODE: RB-RU-H-REDHKS

A stainless steel variant of the Redhawk.

Production: discontinued Caliber: .357 Magnum, .41 Magnum, .44 Magnum, .45 LC

Action: DA/SA, revolver Barrel Length: 4", 5.5", 7.5"

Sights: ramped front, adjustable rear Capacity: 6 Grips: synthetic, Hogue, or wood

D2C:	NIB $ 890	Ex $ 700	VG+ $ 490	Good $ 445	
C2C:	Mint $ 860	Ex $ 625	VG+ $ 450	Good $ 405	Fair $ 205
Trade-In:	Mint $ 640	Ex $ 500	VG+ $ 350	Good $ 315	Poor $ 90

SUPER REDHAWK (5501) (5505)

REDBOOK CODE: RB-RU-H-SR5501, RB-RU-H-SR5505

Satin stainless finish, dual chambering, scope system, transfer-bar safety.

Production: late 1980s - current Caliber: .44 Magnum, .454 Casull, .480 Ruger

Action: DA/SA, revolver Barrel Length: 7.5" Wt.: 53 oz. Sights: ramp front, adjustable rear Capacity: 6 Grips: Hogue Tamer Monogrip

D2C:	NIB $ 854	Ex $ 650	VG+ $ 470	Good $ 430	LMP $1,049
C2C:	Mint $ 820	Ex $ 600	VG+ $ 430	Good $ 385	Fair $ 200
Trade-In:	Mint $ 610	Ex $ 500	VG+ $ 340	Good $ 300	Poor $ 90

SUPER REDHAWK ALASKAN (5303) (5301)

REDBOOK CODE: RB-RU-H-RA5303, RB-RU-H-RA5301

Satin stainless finish, dual chambering, triple-locking cylinder, transfer-bar safety.

Production: current Caliber: .44 Magnum, .454 Casull Action: DA/SA, revolver

Barrel Length: 2.5" Wt.: 45 oz. Sights: ramp front, adjustable rear

Capacity: 6 Grips: Hogue Tamer Monogrip

D2C:	NIB $ 865	Ex $ 675	VG+ $ 480	Good $ 435	LMP $1,079
C2C:	Mint $ 840	Ex $ 600	VG+ $ 440	Good $ 390	Fair $ 200
Trade-In:	Mint $ 620	Ex $ 500	VG+ $ 340	Good $ 305	Poor $ 90

SUPER REDHAWK ALASKAN (5517) REDBOOK CODE: RB-RU-H-RA5517

Similar to 5301 model, but with a longer barrel. Satin stainless finish, dual chambering, triple-locking cylinder, transfer-bar safety.

Production: current Caliber: .454 Casull Action: DA/SA, revolver

Barrel Length: 5" Wt.: 47 oz. Sights: ramp front, adjustable rear

Capacity: 6 Grips: Hogue Tamer Monogrip

D2C:	NIB $ 940	Ex $ 725	VG+ $ 520	Good $ 470	LMP $ 1,115
C2C:	Mint $ 910	Ex $ 650	VG+ $ 470	Good $ 425	Fair $ 220
Trade-In:	Mint $ 670	Ex $ 550	VG+ $ 370	Good $ 330	Poor $ 120

STURM, RUGER & CO.

NEW BEARCAT 50TH ANNIVERSARY

REDBOOK CODE: RB-RU-H-NBC50A

Distinguished by gold trigger guard and gold accents on cylinder, blue finish, transfer-bar safety.

Production: 2008	Caliber: .22 LR	Action: SA, revolver						
Barrel Length: 4"	Sights: blade front, notch rear	Capacity: 6	Grips: Cocobolo					
D2C:	NIB $ 600	Ex $ 475	VG+ $ 330	Good $ 300				
C2C:	Mint $ 580	Ex $ 425	VG+ $ 300	Good $ 270	Fair $ 140			
Trade-In:	Mint $ 430	Ex $ 350	VG+ $ 240	Good $ 210	Poor $ 60			

NEW BEARCAT (0912)
NEW IN BOX

NEW BEARCAT (0912) REDBOOK CODE: RB-RU-H-NB0912

Standard model, blue finish, transfer-bar safety, decoration on cylinder.

Production: current	Caliber: .22 LR	Action: SA, revolver	Barrel Length: 4.2"		
Wt.: 24 oz.	Sights: blade front, integral notch rear	Capacity: 6	Grips: hardwood		
D2C:	NIB $ 485	Ex $ 375	VG+ $ 270	Good $ 245	LMP $ 569
C2C:	Mint $ 470	Ex $ 350	VG+ $ 250	Good $ 220	Fair $ 115
Trade-In:	Mint $ 350	Ex $ 275	VG+ $ 190	Good $ 170	Poor $ 60

NEW BEARCAT (0913)
NEW IN BOX

NEW BEARCAT (0913) REDBOOK CODE: RB-RU-H-NB0913

Standard model, satin stainless finish, transfer-bar safety, unique engravings on cylinder.

Production: current	Caliber: .22 LR	Action: SA, revolver	Barrel Length: 4.2"		
Wt.: 24 oz.	Sights: blade front, integral notch rear	Capacity: 6	Grips: hardwood		
D2C:	NIB $ 520	Ex $ 400	VG+ $ 290	Good $ 260	LMP $ 619
C2C:	Mint $ 500	Ex $ 375	VG+ $ 260	Good $ 235	Fair $ 120
Trade-In:	Mint $ 370	Ex $ 300	VG+ $ 210	Good $ 185	Poor $ 60

NEW MODEL SINGLE-TEN (8100)
NEW IN BOX

NEW MODEL SINGLE-TEN (8100) REDBOOK CODE: RB-RU-H-ST8100

Satin stainless finish, transfer-bar safety, distinguished by Williams adjustable sights.

Production: current	Caliber: .22 LR	Action: SA, revolver	Barrel Length: 5.5"		
Wt.: 38 oz.	Sights: Williams adjustable fiber-optic	Capacity: 10			
Grips: hardwood gunfighter					
D2C:	NIB $ 530	Ex $ 425	VG+ $ 295	Good $ 265	LMP $ 639
C2C:	Mint $ 510	Ex $ 375	VG+ $ 270	Good $ 240	Fair $ 125
Trade-In:	Mint $ 380	Ex $ 300	VG+ $ 210	Good $ 190	Poor $ 60

NEW MODEL SINGLE-NINE (8150)
NEW IN BOX

NEW MODEL SINGLE-NINE (8150) REDBOOK CODE: RB-RU-H-SN8150

Similar to the New Model Single-Ten, but has a longer barrel and 9-shot magazine. Satin stainless finish, transfer-bar safety, distinguished by Williams adjustable sights.

Production: current	Caliber: .22 WMRF	Action: SA, revolver	Barrel Length: 6.5"		
Wt.: 39 oz.	Sights: Williams adjustable fiber-optic	Capacity: 9			
Grips: hardwood gunfighter					
D2C:	NIB $ 530	Ex $ 425	VG+ $ 295	Good $ 265	LMP $ 639
C2C:	Mint $ 510	Ex $ 375	VG+ $ 270	Good $ 240	Fair $ 125
Trade-In:	Mint $ 380	Ex $ 300	VG+ $ 210	Good $ 190	Poor $ 60

NEW MODEL SINGLE-SIX CONVERTIBLE (0621)

REDBOOK CODE: RB-RU-H-SC0621
Alloy steel, blue finish, transfer-bar safety, extra cylinder option.

Production: current Caliber: .22 LR Action: SA, revolver   Barrel Length: 5.5"

Wt.: 33 oz. Sights: ramp front, adjustable rear   Capacity: 6

Grips: black checkered hard rubber

D2C:	NIB $	470	Ex $	375	VG+ $	260	Good $	235	LMP $ 569
C2C:	Mint $	460	Ex $	325	VG+ $	240	Good $	215	Fair $ 110
Trade-In:	Mint $	340	Ex $	275	VG+ $	190	Good $	165	Poor $ 60

NEW MODEL SINGLE-SIX CONVERTIBLE (0623)
NEW IN BOX

NEW MODEL SINGLE-SIX CONVERTIBLE (0623)

REDBOOK CODE: RB-RU-H-SC0623
Alloy steel, blue finish, transfer-bar safety, extra cylinder option.

Production: current Caliber: .22 LR Action: SA, revolver   Barrel Length: 4.6"

Wt.: 32 oz. Sights: ramp front, adjustable rear   Capacity: 6

Grips: black checkered hard rubber

D2C:	NIB $	470	Ex $	375	VG+ $	260	Good $	235	LMP $ 569
C2C:	Mint $	460	Ex $	325	VG+ $	240	Good $	215	Fair $ 110
Trade-In:	Mint $	340	Ex $	275	VG+ $	190	Good $	165	Poor $ 60

NEW MODEL SINGLE-SIX CONVERTIBLE (0625)
NEW IN BOX

NEW MODEL SINGLE-SIX CONVERTIBLE (0625)

REDBOOK CODE: RB-RU-H-SC0625
Satin stainless finish, transfer-bar safety, extra cylinder option.

Production: current Caliber: .22 LR Action: SA, revolver   Barrel Length: 5.5"

Wt.: 39 oz. Sights: ramp front, adjustable rear   Capacity: 6   Grips: hardwood

D2C:	NIB $	520	Ex $	400	VG+ $	290	Good $	260	LMP $ 639
C2C:	Mint $	500	Ex $	375	VG+ $	260	Good $	235	Fair $ 120
Trade-In:	Mint $	370	Ex $	300	VG+ $	210	Good $	185	Poor $ 60

NEW MODEL SINGLE-SIX CONVERTIBLE (0626)
EXCELLENT

NEW MODEL SINGLE-SIX CONVERTIBLE (0626)

REDBOOK CODE: RB-RU-H-SC0626
Satin stainless finish, transfer-bar safety, extra cylinder option.

Production: current Caliber: .22 LR Action: SA, revolver   Barrel Length: 6.5"

Wt.: 40 oz. Sights: ramp front, adjustable rear   Capacity: 6   Grips: hardwood

D2C:	NIB $	530	Ex $	425	VG+ $	295	Good $	265	LMP $ 639
C2C:	Mint $	510	Ex $	375	VG+ $	270	Good $	240	Fair $ 125
Trade-In:	Mint $	380	Ex $	300	VG+ $	210	Good $	190	Poor $ 60

NEW MODEL SINGLE-SIX CONVERTIBLE (0646)
NEW IN BOX

NEW MODEL SINGLE-SIX CONVERTIBLE (0646)

REDBOOK CODE: RB-RU-H-SC0646
Blue finish, alloy steel, transfer-bar safety, extra cylinder option.

Production: current Caliber: .22 LR Action: SA, revolver   Barrel Length: 6.5"

Wt.: 35 oz. Sights: blade front, fixed rear   Capacity: 6

Grips: black checkered hard rubber

D2C:	NIB $	460	Ex $	350	VG+ $	255	Good $	230	LMP $ 569
C2C:	Mint $	450	Ex $	325	VG+ $	230	Good $	210	Fair $ 110
Trade-In:	Mint $	330	Ex $	275	VG+ $	180	Good $	165	Poor $ 60

NEW MODEL SINGLE-SIX HUNTER (0662)
NEW IN BOX

NEW MODEL SINGLE-SIX HUNTER (0662)

REDBOOK CODE: RB-RU-H-SH0662

Hunter model, satin stainless finish, black laminate grips, transfer-bar safety, extra cylinder option.

Production: current   Caliber: .22 LR   Action: SA, revolver   Barrel Length: 7.5"

Wt.: 45 oz.   Sights: bead front, adjustable rear   Capacity: 6   Grips: black laminate

D2C:	NIB	$ 610	Ex $ 475	VG+ $ 340	Good $ 305	LMP $ 799
C2C:	Mint	$ 590	Ex $ 425	VG+ $ 310	Good $ 275	Fair $ 145
Trade-In:	Mint	$ 440	Ex $ 350	VG+ $ 240	Good $ 215	Poor $ 90

NEW MODEL SINGLE-SIX .17 HMR
NEW IN BOX

NEW MODEL SINGLE-SIX .17 HMR (0661)

REDBOOK CODE: RB-RU-H-HMR061

Alloy steel, blue finish, transfer-bar safety.

Production: current   Caliber: .17 HMR   Action: SA, revolver   Barrel Length: 6.5"

Wt.: 35 oz.   Sights: ramp front, adjustable rear   Capacity: 6

Grips: black checkered hard rubber

D2C:	NIB	$ 440	Ex $ 350	VG+ $ 245	Good $ 220	LMP $ 569
C2C:	Mint	$ 430	Ex $ 325	VG+ $ 220	Good $ 200	Fair $ 105
Trade-In:	Mint	$ 320	Ex $ 250	VG+ $ 180	Good $ 155	Poor $ 60

NEW MODEL BLACKHAWK
NEW IN BOX

NEW MODEL BLACKHAWK   REDBOOK CODE: RB-RU-H-BL0505

Alloy steel, blue finish, aluminum grip frame, transfer-bar safety.

Production: current   Caliber: .30 Carbine , .357 Magnum, .44 Special, .45 Colt

Action: SA, revolver   Barrel Length: 4.6", 5.5", 7.5"   Wt.: 46 oz.

Sights: ramp front, adjustable rear   Capacity: 6   Grips: black checkered hard rubber

D2C:	NIB	$ 490	Ex $ 375	VG+ $ 270	Good $ 245	LMP $ 609
C2C:	Mint	$ 480	Ex $ 350	VG+ $ 250	Good $ 225	Fair $ 115
Trade-In:	Mint	$ 350	Ex $ 275	VG+ $ 200	Good $ 175	Poor $ 60

NEW MODEL BLACKHAWK (5232)
NEW IN BOX

NEW MODEL BLACKHAWK (5232)

REDBOOK CODE: RB-RU-H-BL5232

Blue finish, alloy steel grip frame, transfer-bar safety.

Production: current   Caliber: .44 Special   Action: SA, revolver

Barrel Length: 4.6"   Wt.: 42 oz.   Sights: ramp front, adjustable rear

Capacity: 6   Grips: black checkered

D2C:	NIB	$ 490	Ex $ 375	VG+ $ 270	Good $ 245	LMP $ 609
C2C:	Mint	$ 480	Ex $ 350	VG+ $ 250	Good $ 225	Fair $ 115
Trade-In:	Mint	$ 350	Ex $ 275	VG+ $ 200	Good $ 175	Poor $ 60

NEW MODEL BLACKHAWK (0445)
EXCELLENT

NEW MODEL BLACKHAWK (0445)   REDBOOK CODE: RB-RU-H-BL0445

Blue finish, aluminum grip frame, transfer-bar safety.

Production: current   Caliber: .45 Colt   Action: SA, revolver

Barrel Length: 4.62"   Wt.: 39 oz.   Sights: ramp front, adjustable rear

Capacity: 6   Grips: black checkered hard rubber

D2C:	NIB	$ 490	Ex $ 375	VG+ $ 270	Good $ 245	LMP $ 609
C2C:	Mint	$ 480	Ex $ 350	VG+ $ 250	Good $ 225	Fair $ 115
Trade-In:	Mint	$ 350	Ex $ 275	VG+ $ 200	Good $ 175	Poor $ 60

STURM, RUGER & CO.

NEW MODEL BLACKHAWK (0465)
EXCELLENT

NEW MODEL BLACKHAWK (0465) REDBOOK CODE: RB-RU-H-BL0465

Alloy steel, blue finish, aluminum grip frame, transfer-bar safety.

Production: current Caliber: .45 LC Action: SA, revolver Barrel Length: 5.5"

Wt.: 40 oz. Sights: ramp front, adjustable rear Capacity: 6

Grips: black checkered hard rubber

D2C:	NIB $ 490	Ex $ 375	VG+ $ 270	Good $ 245	LMP $ 609
C2C:	Mint $ 480	Ex $ 350	VG+ $ 250	Good $ 225	Fair $ 115
Trade-In:	Mint $ 350	Ex $ 275	VG+ $ 200	Good $ 175	Poor $ 60

NEW MODEL BLACKHAWK STAINLESS (0309)
NEW IN BOX

NEW MODEL BLACKHAWK STAINLESS (0309)

REDBOOK CODE: RB-RU-H-BL0309

Satin stainless finish, stainless steel grip frame, transfer-bar safety.

Production: current Caliber: .357 Magnum Action: SA, revolver

Barrel Length: 4.6" Wt.: 46 oz. Sights: ramp front, adjustable rear

Capacity: 6 Grips: hardwood

D2C:	NIB $ 599	Ex $ 475	VG+ $ 330	Good $ 300	LMP $ 729
C2C:	Mint $ 580	Ex $ 425	VG+ $ 300	Good $ 270	Fair $ 140
Trade-In:	Mint $ 430	Ex $ 350	VG+ $ 240	Good $ 210	Poor $ 60

NEW MODEL BLACKHAWK CONVERTIBLE (0446)
NEW IN BOX

NEW MODEL BLACKHAWK CONVERTIBLE (0446)

REDBOOK CODE: RB-RU-H-BL0446

Alloy steel, blue finish, transfer-bar safety.

Production: current Caliber: .45 LC Action: SA, revolver Barrel Length: 4.6"

Wt.: 39 oz. Sights: ramp front, adjustable rear Capacity: 6

Grips: black checkered hard rubber

D2C:	NIB $ 560	Ex $ 450	VG+ $ 310	Good $ 280	LMP $ 679
C2C:	Mint $ 540	Ex $ 400	VG+ $280	Good $ 255	Fair $ 130
Trade-In:	Mint $ 400	Ex $ 325	VG+ $220	Good $ 200	Poor $ 60

NEW MODEL BLACKHAWK BISLEY (0831)
NEW IN BOX

NEW MODEL BLACKHAWK BISLEY (0831)

REDBOOK CODE: RB-RU-H-BL0831

Blue finish, alloy-steel grip frame, transfer-bar safety.

Production: current Caliber: .44 Magnum Action: SA, revolver Barrel Length: 7.5"

Wt.: 50 oz. Sights: ramp front, adjustable rear Capacity: 6 Grips: hardwood

D2C:	NIB $ 640	Ex $ 500	VG+ $ 355	Good $ 320	LMP $ 799
C2C:	Mint $ 620	Ex $ 450	VG+ $ 320	Good $ 290	Fair $ 150
Trade-In:	Mint $ 460	Ex $ 375	VG+ $ 250	Good $ 225	Poor $ 90

NEW MODEL SUPER BLACKHAWK
NEW IN BOX

NEW MODEL SUPER BLACKHAWK

REDBOOK CODE: RB-RU-H-SB0804

Satin stainless or blue finish, unfluted cylinder.

Production: current Caliber: .44 Magnum Action: SA, revolver Barrel Length: 5.5"

Wt.: 48 oz. Sights: ramp front, adjustable rear Capacity: 6 Grips: hardwood

D2C:	NIB $ 590	Ex $ 450	VG+ $ 325	Good $ 295	LMP $ 739
C2C:	Mint $ 570	Ex $ 425	VG+ $ 300	Good $ 270	Fair $ 140
Trade-In:	Mint $ 420	Ex $ 350	VG+ $ 240	Good $ 210	Poor $ 60

OLD MODEL BEARCAT
EXCELLENT

NEW MODEL SUPER BLACKHAWK (0813)
NEW IN BOX

NEW MODEL SUPER BLACKHAWK HUNTER
(0860)
NEW IN BOX

NEW MODEL SUPER BLACKHAWK BISLEY HUNTER
(0662)
NEW IN BOX

NEW MODEL SUPER BLACKHAWK (0806)

REDBOOK CODE: RB-RU-H-SB0806
Satin stainless finish, unfluted cylinder.

Production: current Caliber: .44 Magnum Action: SA, revolver Barrel Length: 10.5"
Wt.: 55 oz. Sights: ramp front, adjustable rear Capacity: 6 Grips: hardwood

	NIB	Ex	VG+	Good	LMP	Fair	Poor
D2C:	$ 615	$ 475	$ 340	$ 310	$ 769		
C2C:	Mint $ 600	$ 425	$ 310	$ 280		$ 145	
Trade-In:	Mint $ 440	$ 350	$ 240	$ 220			$ 90

NEW MODEL SUPER BLACKHAWK (0813)

REDBOOK CODE: RB-RU-H-SB0813
Stainless or blue finish, unfluted cylinder, distinguished by its 4.6" barrel.

Production: current Caliber: .44 Magnum , .45 LC Action: SA, revolver Barrel Length: 4.6"
Wt.: 45 oz. Sights: ramp front, adjustable rear Capacity: 6 Grips: hardwood

	NIB	Ex	VG+	Good	LMP	Fair	Poor
D2C:	$ 590	$ 450	$ 325	$ 295	$ 739		
C2C:	Mint $ 570	$ 425	$ 300	$ 270		$ 140	
Trade-In:	Mint $ 420	$ 350	$ 240	$ 210			$ 60

NEW MODEL SUPER BLACKHAWK HUNTER (0860)

REDBOOK CODE: RB-RU-H-SB0860
Satin stainless finish, ribbed barrel, unfluted cylinder, transfer-bar safety.

Production: current Caliber: .44 Magnum Action: SA, revolver Barrel Length: 7.5"
Wt.: 52 oz. Sights: ramp front, adjustable rear Capacity: 6 Grips: black laminate

	NIB	Ex	VG+	Good	LMP	Fair	Poor
D2C:	$ 685	$ 525	$ 380	$ 345	$ 859		
C2C:	Mint $ 660	$ 475	$ 350	$ 310		$ 160	
Trade-In:	Mint $ 490	$ 400	$ 270	$ 240			$ 90

NEW MODEL SUPER BLACKHAWK BISLEY HUNTER (0862)

REDBOOK CODE: RB-RU-H-SB0862
Satin stainless finish, similar to the New Model Super Blackhawk Hunter.

Production: current Caliber: .44 Magnum Action: SA, revolver Barrel Length: 7.5"
Wt.: 52 oz. Sights: ramp front, adjustable rear Capacity: 6 Grips: black laminate

	NIB	Ex	VG+	Good	LMP	Fair	Poor
D2C:	$ 685	$ 525	$ 380	$ 345	$ 859		
C2C:	Mint $ 660	$ 475	$ 350	$ 310		$ 160	
Trade-In:	Mint $ 490	$ 400	$ 270	$ 240			$ 90

BISLEY REDBOOK CODE: RB-RU-H-BISLEY

Engraved cylinder, optional fluted or unfluted cylinder, blue finish, curved trigger.

Production: early 1980s - discontinued Caliber: .22 LR, .32 H&R Magnum,
.357 Magnum, .41 Magnum, .44 Magnum, .45 LC Action: SA, revolver
Barrel Length: 6.5", 7.5" Sights: adjustable Capacity: 6
Grips: smooth wood with medallions

	NIB	Ex	VG+	Good	Fair	Poor
D2C:	$ 480	$ 375	$ 265	$ 240		
C2C:	Mint $ 470	$ 350	$ 240	$ 220	$ 115	
Trade-In:	Mint $ 350	$ 275	$ 190	$ 170		$ 60

STURM, RUGER & CO.

BISLEY VAQUERO
(5107)
MINT

BISLEY VAQUERO REDBOOK CODE: RB-RU-H-BISVAQ

Blue case-colored or stainless finish, longer grip frame, slight premium for ivory grips.

Production: late 1990s - discontinued Caliber: .44 Magnum, .45 LC Action: SA, revolver
Barrel Length: 5.5" Sights: fixed Capacity: 6 Grips: rosewood or ivory

D2C:	NIB $	760	Ex $	600	VG+ $	420	Good $	380	
C2C:	Mint $	730	Ex $	525	VG+ $	380	Good $	345	Fair $ 175
Trade-In:	Mint $	540	Ex $	450	VG+ $	300	Good $	270	Poor $ 90

VAQUERO
EXCELLENT

VAQUERO REDBOOK CODE: RB-RU-H-ORGVAQ

Blue color cased or stainless finish, similar to the New Model Blackhawk but with fixed sights.

Production: 1993 - 2005 Caliber: .357 Magnum, .44 WCF, .44 Magnum, .45 LC
Action: SA, revolver   Barrel Length: 4.6", 5.5", 7.5" Sights: fixed Capacity: 6
Grips: walnut or faux ivory

D2C:	NIB $	540	Ex $	425	VG+ $	300	Good $	270	
C2C:	Mint $	520	Ex $	375	VG+ $	270	Good $	245	Fair $ 125
Trade-In:	Mint $	390	Ex $	325	VG+ $	220	Good $	190	Poor $ 60

VAQUERO BIRD'S HEAD
EXCELLENT

VAQUERO BIRD'S HEAD REDBOOK CODE: RB-RU-H-VBHXXX

Blue color cased or stainless polished finish, steel frame, distinguished by birds-head grip.

Production: early 2000s-discontinued Caliber: .357 Magnum, .45 LC
Action: SA, revolver Barrel Length: 3.75" Sights: fixed Capacity: 6
Grips: black Micarta or ivory

D2C:	NIB $	580	Ex $	450	VG+ $	320	Good $	290	
C2C:	Mint $	560	Ex $	425	VG+ $	290	Good $	265	Fair $ 135
Trade-In:	Mint $	420	Ex $	325	VG+ $	230	Good $	205	Poor $ 60

**NEW MODEL
VAQUERO BLUE**
(5107)
NEW IN BOX

NEW MODEL VAQUERO BLUE (5107) REDBOOK CODE: RB-RU-H-VB5107

Alloy steel, blue finish, beveled cylinder, mid-size frame, transfer-bar safety, internal lock, ejector rod head. Sometimes referred to as "NM" for "New Model."

Production: current Caliber: .357 Magnum, .45 LC Action: SA, revolver
Barrel Length: 4.6" Wt.: 43 oz. Sights: fixed Capacity: 6 Grips: hardwood

D2C:	NIB $	585	Ex $	450	VG+ $	325	Good $	295	LMP $ 739
C2C:	Mint $	570	Ex $	425	VG+ $	300	Good $	265	Fair $ 135
Trade-In:	Mint $	420	Ex $	350	VG+ $	230	Good $	205	Poor $ 60

NM VAQUERO STAINLESS
(5109)
NEW IN BOX
Courtesy of Bud's Gun Shop

NM VAQUERO STAINLESS (5109) REDBOOK CODE: RB-RU-H-VS5109

High-gloss stainless finish, mid-sized, transfer-bar safety, internal lock, ejector rod head.

Production: current Caliber: .357 Magnum, .45 LC Action: SA, revolver
Barrel Length: 4.6", 5.5" Wt.: 43 oz. Sights: fixed Capacity: 6 Grips: hardwood

D2C:	NIB $	585	Ex $	450	VG+ $	325	Good $	295	LMP $ 739
C2C:	Mint $	570	Ex $	425	VG+ $	300	Good $	265	Fair $ 135
Trade-In:	Mint $	420	Ex $	350	VG+ $	230	Good $	205	Poor $ 60

NM VAQUERO BISLEY
(5130)
MINT

NM VAQUERO BISLEY (5130) REDBOOK CODE: RB-RU-H-VB5130

High-gloss stainless finish, distinguished by ivory grips, transfer-bar safety, internal lock, ejector rod head. "NM" for "New Model"

Production: current Caliber: .357 Magnum, .45 LC Action: SA, revolver
Barrel Length: 5.5" Wt.: 45 oz. Sights: fixed Capacity: 6 Grips: simulated ivory

D2C:	NIB $	640	Ex $	500	VG+ $	355	Good $	320	LMP $ 809
C2C:	Mint $	620	Ex $	450	VG+ $	320	Good $	290	Fair $ 150
Trade-In:	Mint $	460	Ex $	375	VG+ $	250	Good $	225	Poor $ 90

**NEW MODEL VAQUERO
SASS** (5133)
NEW IN BOX

NEW MODEL VAQUERO SASS (5133)
REDBOOK CODE: RB-RU-H-SA5133
Comes as a 2-gun set with consecutive serial numbers and pricing includes both guns. High-gloss stainless finish, Reverse Indexing Pawl, transfer-bar safety, internal lock, large ejector rod head.

Production: current Caliber: .357 Magnum Action: SA, revolver					
Barrel Length: 4.6", 5.5" Wt.: 45 oz. Sights: fixed Capacity: 6 Grips: hardwood					
D2C:	NIB $1,360	Ex $1,050	VG+ $ 750	Good $ 680	LMP $1,618
C2C:	Mint $1,310	Ex $ 950	VG+ $ 680	Good $ 615	Fair $ 315
Trade-In:	Mint $ 970	Ex $ 775	VG+ $ 540	Good $ 480	Poor $ 150

NEW MODEL VAQUERO SASS (5134)
REDBOOK CODE: RB-RU-H-SA5134
Comes as a 2-gun set with consecutive serial numbers and pricing includes both guns. High-gloss stainless finish, Reverse Indexing Pawl, transfer-bar safety, internal lock, large ejector rod head.

Production: current Caliber: .45 LC Action: SA, revolver					
Barrel Length: 5.5" Wt.: 40 oz. Sights: fixed Capacity: 6 Grips: hardwood					
D2C:	NIB $1,360	Ex $1,050	VG+ $ 750	Good $ 680	LMP $1,618
C2C:	Mint $1,310	Ex $ 950	VG+ $ 680	Good $ 615	Fair $ 315
Trade-In:	Mint $ 970	Ex $ 775	VG+ $ 540	Good $ 480	Poor $ 150

SPEED SIX
MINT

SPEED SIX (MODEL 207, 208, 209) REDBOOK CODE: RB-RU-H-SSM207
Round butt, blue finish, bobbed hammer version available, significant premium for military marked versions, premium for stainless models.

Production: 1973 - discontinued Caliber: .357 Magnum, .38 Special, 9mm					
Action: DA/SA, revolver Barrel Length: 2.75", 4" Sights: fixed					
Capacity: 6 Grips: checkered walnut or synthetic					
D2C:	NIB $ 530	Ex $ 425	VG+ $ 295	Good $ 265	
C2C:	Mint $ 510	Ex $ 375	VG+ $ 270	Good $ 240	Fair $ 125
Trade-In:	Mint $ 380	Ex $ 300	VG+ $ 210	Good $ 190	Poor $ 60

SECURITY SIX
(MODEL 117)
EXCELLENT

SECURITY SIX (MODEL 117) REDBOOK CODE: RB-RU-H-SSM117
Square butt, blue or stainless finish, also available with fixed sights.

Production: 1970 - 1985 Caliber: .357 Magnum Action: DA/SA, revolver					
Barrel Length: 2.75", 4", 6" Sights: adjustable Capacity: 6 Grips: checkered walnut					
D2C:	NIB $ 550	Ex $ 425	VG+ $ 305	Good $ 275	
C2C:	Mint $ 530	Ex $ 400	VG+ $ 280	Good $ 250	Fair $ 130
Trade-In:	Mint $ 400	Ex $ 325	VG+ $ 220	Good $ 195	Poor $ 60

**POLICE SERVICE
SIX MODEL**
MINT

POLICE SERVICE SIX MODEL REDBOOK CODE: RB-RU-H-PSS107
Blue or stainless finish, square butt, few 6" models produced so 6" models have higher value.

Production: 1972 - 1988 Caliber: .357 Magnum, .38 Special, 9mm					
Action: DA/SA, revolver Barrel Length: 2.75", 4", 6" Sights: fixed					
Capacity: 6 Grips: walnut or synthetic					
D2C:	NIB $ 450	Ex $ 350	VG+ $ 250	Good $ 225	
C2C:	Mint $ 440	Ex $ 325	VG+ $ 230	Good $ 205	Fair $ 105
Trade-In:	Mint $ 320	Ex $ 270	VG+ $ 180	Good $ 180	Poor $ 60

STURM, RUGER & CO.

STANDARD MODEL
VERY GOOD +

MARK I TARGET
EXCELLENT

MARK II STANDARD
EXCELLENT

MARK II TARGET
MINT

STANDARD MODEL REDBOOK CODE: RB-RU-H-RESTDX

Serial numbers range between approximately 1 and 35,000. The first Ruger pistol ever produced. Models with Red Eagle logos in the grips were manufactured pre-1950 and are of note.

Production: 1949 - 1982   Caliber: .22 LR   Action: SA, semi-auto

Barrel Length: 4.75", 6" (post-1954)   Sights: fixed   Capacity: 9

Magazine: detachable box   Grips: hard rubber

D2C:	NIB $	375	Ex $	300	VG+ $	210	Good $	190	
C2C:	Mint $	360	Ex $	275	VG+ $	190	Good $	170	Fair $ 90
Trade-In:	Mint $	270	Ex $	225	VG+ $	150	Good $	135	Poor $ 60

MARK I TARGET REDBOOK CODE: RB-RU-H-M1TRXX

Blue finish, additional price for red medallion and walnut grips, early models distinguished by Red Eagle grips and worth a premium. Also available with tapered or bull barrels.

Production: 1951 - 1982   Caliber: .22 LR   Action: SA, semi-auto

Barrel Length: 5.25", 5.5", 6.87"   Sights: adjustable   Capacity: 10

Magazine: detachable box   Grips: rubber or checkered walnut

D2C:	NIB $	470	Ex $	375	VG+ $	260	Good $	235	
C2C:	Mint $	460	Ex $	325	VG+ $	240	Good $	215	Fair $ 110
Trade-In:	Mint $	340	Ex $	275	VG+ $	190	Good $	165	Poor $ 60

MARK II STANDARD REDBOOK CODE: RB-RU-H-M2STXX

Many aftermarket options available. Blue or brushed satin stainless finish, grooved trigger, blowback action, frame-mounted safety, equipped with plastic case and lock.

Production: 1982 - 2005   Caliber: .22 LR   Action: SA, semi-auto

Barrel Length: 4.75", 6"   Sights: blade front, adjustable rear

Capacity: 10   Magazine: detachable box   Grips: synthetic or wood

D2C:	NIB $	320	Ex $	250	VG+ $	180	Good $	160	
C2C:	Mint $	310	Ex $	225	VG+ $	160	Good $	145	Fair $ 75
Trade-In:	Mint $	230	Ex $	200	VG+ $	130	Good $	115	Poor $ 60

MARK II TARGET REDBOOK CODE: RB-RU-H-M2TRXX

Blue or stainless finish, steel frame, blowback action, fame-mounted safety, bulled or tapered barrel.

Production: late 1980s - discontinued   Caliber: .22 LR   Action: SA, semi-auto

Barrel Length: 5.25", 6.875"   Sights: adjustable   Capacity: 10

Magazine: detachable box   Grips: rubber or checkered walnut

D2C:	NIB $	420	Ex $	325	VG+ $	235	Good $	210	
C2C:	Mint $	410	Ex $	300	VG+ $	210	Good $	190	Fair $ 100
Trade-In:	Mint $	300	Ex $	250	VG+ $	170	Good $	150	Poor $ 60

MARK III HUNTER REDBOOK CODE: RB-RU-H-M322HU

Stainless or blue finish, fluted barrel, interchangeable inserts.

Production: early 2000s - discontinued   Caliber: .22 LR   Action: SA, semi-auto

Barrel Length: 4.5", 6.875"   Sights: fiber-optic front, adjustable rear

Capacity: 10   Magazine: detachable box   Grips: black

D2C:	NIB $	450	Ex $	350	VG+ $	250	Good $	225	
C2C:	Mint $	440	Ex $	325	VG+ $	230	Good $	205	Fair $ 105
Trade-In:	Mint $	320	Ex $	275	VG+ $	180	Good $	160	Poor $ 60

MARK III STANDARD (10104) REDBOOK CODE: RB-RU-H-M10104
Blue finish, tapered barrel, frame-mounted safety.

Production: current Caliber: .22 LR Action: SA, semi-auto Barrel Length: 4.75"					
Wt.: 35 oz. Sights: fixed Capacity: 10 Magazine: detachable box Grips: checkered black					
D2C:	NIB $ 320	Ex $ 250	VG+ $ 180	Good $ 160	LMP $ 389
C2C:	Mint $ 310	Ex $ 225	VG+ $ 160	Good $ 145	Fair $ 75
Trade-In:	Mint $ 230	Ex $ 200	VG+ $ 130	Good $ 115	Poor $ 60

MARK III STANDARD (10104)
NEW IN BOX

MARK III STANDARD (10105) REDBOOK CODE: RB-RU-H-M10105
A longer barrel variant of the Mark III Standard 10104. Blue finish, tapered barrel, frame-mounted safety.

Production: current Caliber: .22 LR Action: SA, semi-auto Barrel Length: 6"					
Wt.: 37 oz. Sights: fixed Capacity: 10 Magazine: detachable box Grips: checkered black					
D2C:	NIB $ 320	Ex $ 250	VG+ $ 180	Good $ 160	LMP $ 389
C2C:	Mint $ 310	Ex $ 225	VG+ $ 160	Good $ 145	Fair $ 75
Trade-In:	Mint $ 230	Ex $ 200	VG+ $ 130	Good $ 115	Poor $ 60

MARK III STANDARD (10105)
NEW IN BOX
Courtesy of Bud's Gun Shop

MARK III TARGET (10101) REDBOOK CODE: RB-RU-H-M10101
A target variant of the Mark III Standard. Noted for its scope mounts and adjustable rear sight. Blue finish, bull barrel, frame-mounted safety, chamber indicator.

Production: current Caliber: .22 LR Action: SA, semi-auto Barrel Length: 5.5"					
Wt.: 42 oz. Sights: fixed front, adjustable rear, scope mount Capacity: 10					
Magazine: detachable box Grips: checkered synthetic					
D2C:	NIB $ 389	Ex $ 300	VG+ $ 215	Good $ 195	LMP $ 459
C2C:	Mint $ 380	Ex $ 275	VG+ $ 200	Good $ 180	Fair $ 90
Trade-In:	Mint $ 280	Ex $ 225	VG+ $ 160	Good $ 140	Poor $ 60

MARK III TARGET (10101)
NEW IN BOX

MARK III TARGET (10103) REDBOOK CODE: RB-RU-H-M10103
A stainless, target variant of the Mark III Standard. Bull barrel, chamber indicator, frame-mounted safety.

Production: current Caliber: .22 LR Action: SA, semi-auto Barrel Length: 5.5"					
Wt.: 42 oz. Sights: fixed front, adjustable rear Capacity: 10 Magazine: detachable box					
Grips: checkered					
D2C:	NIB $ 470	Ex $ 375	VG+ $ 260	Good $ 235	LMP $ 569
C2C:	Mint $ 460	Ex $ 325	VG+ $ 240	Good $ 215	Fair $ 110
Trade-In:	Mint $ 340	Ex $ 275	VG+ $ 190	Good $ 165	Poor $ 60

MARK III TARGET (10103)
NEW IN BOX

MARK III TARGET (10159) REDBOOK CODE: RB-RU-H-M10159
A target variant of the Mark III Standard, noted for its finger-molded grips. Blue finish, frame-mounted safety, loaded chamber indicator.

Production: current Caliber: .22 LR Action: SA, semi-auto Barrel Length: 5.5"					
Wt.: 43 oz. Sights: fixed front, adjustable rear Capacity: 10 Magazine: detachable box					
Grips: target-style, finger-molded laminate wood					
D2C:	NIB $ 430	Ex $ 350	VG+ $ 240	Good $ 215	LMP $ 509
C2C:	Mint $ 420	Ex $ 300	VG+ $ 220	Good $ 195	Fair $ 100
Trade-In:	Mint $ 310	Ex $ 250	VG+ $ 170	Good $ 155	Poor $ 60

MARK III TARGET (10159)
NEW IN BOX

STURM, RUGER & CO.

MARK III HUNTER (10118)
NEW IN BOX

MARK III HUNTER (10118) REDBOOK CODE: RB-RU-H-H10118

A hunter variant of the Mark III Standard, noted for its fluted barrel and fiber-optic sights. Stainless finish, frame-mounted safety.

Production: current Caliber: .22 LR Action: SA, semi-auto Barrel Length: 6.88"

Wt.: 41 oz. Sights: fiber-optic front, adjustable rear Capacity: 10 Magazine: detachable box

Grips: brown laminate wood

D2C:	NIB $ 530	Ex $ 425	VG+ $ 295	Good $ 265	LMP $ 679
C2C:	Mint $ 510	Ex $ 375	VG+ $ 270	Good $ 240	Fair $ 125
Trade-In:	Mint $ 380	Ex $ 300	VG+ $ 210	Good $ 190	Poor $ 60

MARK III HUNTER (10160)
NEW IN BOX

MARK III HUNTER (10160) REDBOOK CODE: RB-RU-H-H10160

A hunter variant of the Mark III Standard, noted for its fluted barrel, finger-molded grips, and fiber-optic sights. Stainless finish, frame-mounted finish.

Production: current Caliber: .22 LR Action: SA, semi-auto Barrel Length: 6.88"

Wt.: 44 oz. Sights: fiber-optic front, adjustable rear Capacity: 10

Magazine: detachable box Grips: brown laminate wood, thumb-molded grip

D2C:	NIB $ 580	Ex $ 450	VG+ $ 320	Good $ 290	LMP $ 729
C2C:	Mint $ 560	Ex $ 425	VG+ $ 290	Good $ 265	Fair $ 135
Trade-In:	Mint $ 420	Ex $ 325	VG+ $ 230	Good $ 205	Poor $ 60

MARK III COMPETITION (10112)
NEW IN BOX

MARK III COMPETITION (10112) REDBOOK CODE: RB-RU-H-M10112

A competition-style variant of the Mark III Standard, noted for its flat-sided barrel and adjustable sights. Stainless finish, frame-mounted safety.

Production: current Caliber: .22 LR Action: SA, semi-auto Barrel Length: 6.88"

Wt.: 45 oz. Sights: fixed front, adjustable rear Capacity: 10 Magazine: detachable box

Grips: checkered brown laminate

D2C:	NIB $ 600	Ex $ 475	VG+ $ 330	Good $ 300	LMP $ 659
C2C:	Mint $ 580	Ex $ 425	VG+ $ 300	Good $ 270	Fair $ 140
Trade-In:	Mint $ 430	Ex $ 350	VG+ $ 240	Good $ 210	Poor $ 60

P85
VERY GOOD +

P85 REDBOOK CODE: RB-RU-H-P85XXX

Matte black or stainless finish, alloy frame, ambidextrous safety, name changed in 1991 to P85 Mark II.

Production: 1987 - discontinued Caliber: 9mm Action: DA/SA, semi-auto

Barrel Length: 4.5" Sights: fixed Capacity: 15 Magazine: detachable box

Grips: synthetic

D2C:	NIB $ 450	Ex $ 350	VG+ $ 250	Good $ 225	
C2C:	Mint $ 440	Ex $ 325	VG+ $ 230	Good $ 205	Fair $ 105
Trade-In:	Mint $ 320	Ex $ 275	VG+ $ 180	Good $ 160	Poor $ 60

P89
EXCELLENT

P89 REDBOOK CODE: RB-RU-H-P89XXX

Blue finish, manual safety, firing pin block, resembles Ruger's P85 model.

Production: 1991 - discontinued Caliber: 9mm Action: DA/SA, semi-auto

Barrel Length: 4.5" Sights: fixed Capacity: 10, 15 Magazine: detachable box

Grips: synthetic

D2C:	NIB $ 380	Ex $ 300	VG+ $ 210	Good $ 190	
C2C:	Mint $ 370	Ex $ 275	VG+ $ 190	Good $ 175	Fair $ 90
Trade-In:	Mint $ 270	Ex $ 225	VG+ $ 150	Good $ 135	Poor $ 60

P90
EXCELLENT

P95
NEW IN BOX

STURM, RUGER & CO.

P90   REDBOOK CODE: RB-RU-H-P90XXX
Blue finish, ambidextrous safety, aluminum alloy frame, steel slide.

Production: 1991 - discontinued	Caliber: .45 ACP	Action: DA/SA, semi-auto			
Barrel Length: 4.5"	Sights: fixed	Capacity: 8	Grips: black synthetic		

D2C:	NIB $ 400	Ex $ 325	VG+ $ 220	Good $ 200	
C2C:	Mint $ 390	Ex $ 300	VG+ $ 200	Good $ 180	Fair $ 95
Trade-In:	Mint $ 290	Ex $ 225	VG+ $ 160	Good $ 140	Poor $ 60

P95   REDBOOK CODE: RB-RU-H-P95XXX
Blue or stainless finish and slide, decocker only, polymer grip frame, also available in a DAO version.

Production: mid-1990s - discontinued	Caliber: 9mm	Action: DA/SA, semi-auto			
Barrel Length: 4"	Sights: fixed, 3-dot	Capacity: 15			
Magazine: detachable box	Grips: polymer				

D2C:	NIB $ 325	Ex $ 250	VG+ $ 180	Good $ 165	
C2C:	Mint $ 320	Ex $ 225	VG+ $ 170	Good $ 150	Fair $ 75
Trade-In:	Mint $ 240	Ex $ 200	VG+ $ 130	Good $ 115	Poor $ 60

KP93DC   REDBOOK CODE: RB-RU-H-KP93XX
Compact version, decocker, stainless steel finish, aluminum frame.

Production: early 1990s - discontinued	Caliber: 9mm	Action: DA/SA, semi-auto			
Barrel Length: 3.9"	Sights: fixed	Capacity: 10, 15	Magazine: detachable box		
Grips: checkered black plastic					

D2C:	NIB $ 485	Ex $ 375	VG+ $ 270	Good $ 245	
C2C:	Mint $ 470	Ex $ 350	VG+ $ 250	Good $ 220	Fair $ 115
Trade-In:	Mint $ 350	Ex $ 275	VG+ $ 190	Good $ 170	Poor $ 60

KP345   REDBOOK CODE: RB-RU-H-KP345X
Internal lock, duo-tone finish, loaded chamber indicator, magazine disconnect.

Production: early 2000s - discontinued	Caliber: .45 ACP	Action: DA/SA, semi-auto			
Barrel Length: 4.5"	Sights: 3-dot	Capacity: 8			
Magazine: detachable box	Grips: black polymer checkered				

D2C:	NIB $ 460	Ex $ 350	VG+ $ 255	Good $ 230	
C2C:	Mint $ 450	Ex $ 325	VG+ $ 230	Good $ 210	Fair $ 110
Trade-In:	Mint $ 330	Ex $ 275	VG+ $ 180	Good $ 165	Poor $ 60

LCP (3730)
NEW IN BOX

LCP (3730)   REDBOOK CODE: RB-RU-H-L3730X
Noted for its brushed stainless slide. Nylon grip frame, compact design, six-grooved barrel. Laser Max and Crimson Trace mounted-laser models also available.

Production: 2014 - current	Caliber: .380 ACP	Action: DAO, semi-auto			
Barrel Length: 2.75"	Sights: fixed				
Capacity: 6	Magazine: detachable box	Grips: nylon			

D2C:	NIB $ 368	Ex $ 300	VG+ $ 205	Good $ 185	LMP $ 429
C2C:	Mint $ 360	Ex $ 275	VG+ $ 190	Good $ 170	Fair $ 85
Trade-In:	Mint $ 270	Ex $ 225	VG+ $ 150	Good $ 130	Poor $ 60

LCP (3701)
NEW IN BOX

LCP (3701) REDBOOK CODE: RB-RU-H-L3701X

A matte black variant of the LCP. Nylon grip frame, compact design, six-grooved barrel.

Production: current Caliber: .380 ACP Action: DAO, semi-auto

Barrel Length: 2.75" Sights: fixed

Capacity: 6 Magazine: detachable box Grips: nylon

D2C:	NIB $ 320	Ex $ 250	VG+ $ 180	Good $ 160	LMP $ 379
C2C:	Mint $ 310	Ex $ 225	VG+ $ 160	Good $ 145	Fair $ 75
Trade-In:	Mint $ 230	Ex $ 200	VG+ $ 130	Good $ 115	Poor $ 60

LCP (3713)
NEW IN BOX

LCP (3713) REDBOOK CODE: RB-RU-H-L3713X

A variant of the LCP featuring Crimson Trace laser sights. Steel slide, polymer frame, matte black finish.

Production: current Caliber: .380 ACP Action: DAO, semi-auto

Barrel Length: 2.75" Sights: fixed, Crimson Trace lasers Capacity: 6

Magazine: detachable box Grips: nylon

D2C:	NIB $ 460	Ex $ 350	VG+ $ 255	Good $ 230	LMP $ 559
C2C:	Mint $ 450	Ex $ 325	VG+ $ 230	Good $ 210	Fair $ 110
Trade-In:	Mint $ 330	Ex $ 275	VG+ $ 180	Good $ 165	Poor $ 60

LCP (3718)
NEW IN BOX

LCP (3718) REDBOOK CODE: RB-RU-H-L3718X

A variant of the LCP featuring LazerMax laser sights. Steel slide, polymer frame, matte black finish.

Production: current Caliber: .380 ACP Action: DAO, semi-auto

Barrel Length: 2.75" Sights: fixed, LaserMax laser sights Capacity: 6

Magazine: detachable box Grips: nylon

D2C:	NIB $ 380	Ex $ 300	VG+ $ 210	Good $ 190	LMP $ 449
C2C:	Mint $ 370	Ex $ 275	VG+ $ 190	Good $ 175	Fair $ 90
Trade-In:	Mint $ 270	Ex $ 225	VG+ $ 150	Good $ 135	Poor $ 60

LC380 (3231)
NEW IN BOX

LC380 (3231) REDBOOK CODE: RB-RU-H-LC3231

A slightly larger variant of the LCP. Noted for its LaserMax laser sights. Steel slide, polymer frame, matte black finish.

Production: current Caliber: .380 ACP Action: DAO, semi-auto

Barrel Length: 3.12" Wt.: 18 oz. Sights: fixed front, adjustable rear and

LaserMax laser sights Capacity: 7 Magazine: detachable box Grips: nylon

D2C:	NIB $ 440	Ex $ 350	VG+ $ 245	Good $ 220	LMP $ 529
C2C:	Mint $ 430	Ex $ 325	VG+ $ 220	Good $ 200	Fair $ 105
Trade-In:	Mint $ 320	Ex $ 250	VG+ $ 180	Good $ 155	Poor $ 60

LC380 (3219)
NEW IN BOX

LC380 (3219) REDBOOK CODE: RB-RU-H-LC3219

A slightly larger variant of the LCP. Compact design, polymer frame, steel slide, matte black finish.

Production: current Caliber: .380 ACP Action: DAO, semi-auto Barrel Length: 3.12"

Wt.: 17 oz. Sights: fixed front, adjustable rear, 3-dot Capacity: 7 Magazine: detachable box

Grips: nylon

D2C:	NIB $ 360	Ex $ 275	VG+ $ 200	Good $ 180	LMP $ 449
C2C:	Mint $ 350	Ex $ 250	VG+ $ 180	Good $ 165	Fair $ 85
Trade-In:	Mint $ 260	Ex $ 225	VG+ $ 150	Good $ 130	Poor $ 60

LC9 (3200)
NEW IN BOX

LC9 (3200) REDBOOK CODE: RB-RU-H-LC3200

Steel slide, polymer frame, matte black finish, compact design.

Production: current Caliber: 9mm Action: DAO, semi-auto Barrel Length: 3.12"

Wt.: 18 oz. Sights: 3-dot, adjustable Capacity: 7 Magazine: detachable box

Grips: nylon

D2C:	NIB $ 370	Ex $ 300	VG+ $ 205	Good $ 185	LMP $ 449
C2C:	Mint $ 360	Ex $ 275	VG+ $ 190	Good $ 170	Fair $ 90
Trade-In:	Mint $ 270	Ex $ 225	VG+ $ 150	Good $ 130	Poor $ 60

SR9C COMPACT (3313)
NEW IN BOX

SR9C COMPACT (3313) REDBOOK CODE: RB-RU-H-SC3313

Stainless or black slide. Compact design, black nylon grip frame, reversible backstrap, accessory rail.

Production: current Caliber: 9mm Action: striker-fired, semi-auto

Barrel Length: 3.4" Wt.: 23 oz. Sights: 3-dot, adjustable

Capacity: 17 Magazine: detachable box Grips: nylon

D2C:	NIB $ 430	Ex $ 350	VG+ $ 240	Good $ 215	LMP $ 529
C2C:	Mint $ 420	Ex $ 300	VG+ $ 220	Good $ 195	Fair $ 100
Trade-In:	Mint $ 310	Ex $ 250	VG+ $ 170	Good $ 155	Poor $ 60

SR9C COMPACT (3316)
NEW IN BOX

SR9C COMPACT (3316) REDBOOK CODE: RB-RU-H-SC3316

A variant of the SR9C, noted for its 10-round magazine. Compact design, black nylon grip frame, reversible backstrap, accessory rail.

Production: current Caliber: 9mm Action: striker-fired, semi-auto

Barrel Length: 3.4" Wt.: 23 oz. Sights: 3-dot, adjustable

Capacity: 10 Magazine: detachable box Grips: nylon

D2C:	NIB $ 430	Ex $ 350	VG+ $ 240	Good $ 215	LMP $ 529
C2C:	Mint $ 420	Ex $ 300	VG+ $ 220	Good $ 195	Fair $ 100
Trade-In:	Mint $ 310	Ex $ 250	VG+ $ 170	Good $ 155	Poor $ 60

SR9 (3301)
NEW IN BOX

SR9 (3301) REDBOOK CODE: RB-RU-H-S93301

Stainless steel slide, nylon grip frame, accessory rail, reversible backstrap, frame-mounted and trigger safety.

Production: current Caliber: 9mm Action: striker-fired, semi-auto

Barrel Length: 4.1" Wt.: 26 oz. Sights: 3-dot, adjustable

Capacity: 17 Magazine: detachable box Grips: nylon

D2C:	NIB $ 430	Ex $ 350	VG+ $ 240	Good $ 215	LMP $ 529
C2C:	Mint $ 420	Ex $ 300	VG+ $ 220	Good $ 195	Fair $ 100
Trade-In:	Mint $ 310	Ex $ 250	VG+ $ 170	Good $ 155	Poor $ 60

SR9 (3321)
NEW IN BOX

SR9 (3321) REDBOOK CODE: RB-RU-H-S3321X

Like the SR9 3301, but noted for its black slide. Black nylon grip frame, accessory rail, reversible backstrap, frame-mounted and trigger safety.

Production: current Caliber: 9mm Action: striker-fired, semi-auto

Barrel Length: 4.14" Wt.: 26 oz. Sights: 3-dot, adjustable

Capacity: 17 Magazine: detachable box Grips: nylon

D2C:	NIB $ 430	Ex $ 350	VG+ $ 240	Good $ 215	LMP $ 529
C2C:	Mint $ 420	Ex $ 300	VG+ $ 220	Good $ 195	Fair $ 100
Trade-In:	Mint $ 310	Ex $ 250	VG+ $ 170	Good $ 155	Poor $ 60

SR40C
NEW IN BOX

SR40
NEW IN BOX

SR40 (3472)
NEW IN BOX

SR45
NEW IN BOX
Courtesy of Bud's Gun Shop

SR1911 (6700)
NEW IN BOX

SR40C REDBOOK CODE: RB-RU-H-S3477X

A .40 S&W variant of the SR9. Black nitride or stainless slide, black nylon grip frame, compact design, accessory rail, reversible backstrap, has frame-mounted and trigger safety.

Production: current Caliber: .40 S&W Action: striker-fired, semi-auto

Barrel Length: 3.5" Wt.: 23 oz. Sights: 3-dot, adjustable

Capacity: 15 Magazine: detachable box Grips: nylon

D2C:	NIB $	430	Ex $	350	VG+ $	240	Good $	215	LMP $	529
C2C:	Mint $	420	Ex $	300	VG+ $	220	Good $	195	Fair $	100
Trade-In:	Mint $	310	Ex $	250	VG+ $	170	Good $	155	Poor $	60

SR40 REDBOOK CODE: RB-RU-H-S3470X

A larger variant of the SR40c. Stainless steel or black slide, black nylon grip frame, accessory rail, reversible backstrap, has frame-mounted and trigger safety.

Production: current Caliber: .40 S&W Action: striker-fired, semi-auto

Barrel Length: 4.14" Wt.: 27 oz. Sights: 3-dot, adjustable

Capacity: 15 Magazine: detachable box Grips: nylon

D2C:	NIB $	430	Ex $	350	VG+ $	240	Good $	215	LMP $	529
C2C:	Mint $	420	Ex $	300	VG+ $	220	Good $	195	Fair $	100
Trade-In:	Mint $	310	Ex $	250	VG+ $	170	Good $	155	Poor $	60

SR40 (3472) REDBOOK CODE: RB-RU-H-S3472X

A smaller capacity variant of the SR40 3470. Stainless slide, black nylon grip frame, accessory rail, reversible backstrap, frame-mounted and trigger safety.

Production: current Caliber: .40 S&W Action: striker-fired, semi-auto

Barrel Length: 4.14" Wt.: 27 oz. Sights: 3-dot, adjustable

Capacity: 10 Magazine: detachable box Grips: nylon

D2C:	NIB $	430	Ex $	350	VG+ $	240	Good $	215	LMP $	529
C2C:	Mint $	420	Ex $	300	VG+ $	220	Good $	195	Fair $	100
Trade-In:	Mint $	310	Ex $	250	VG+ $	170	Good $	155	Poor $	60

SR45 REDBOOK CODE: RB-RU-H-S3800X

Similar to the SR9 and SR40. Black or stainless slide finish, black nylon grip frame, accessory rail, reversible backstrap, loaded chamber indicator, has frame-mounted and trigger safety.

Production: current Caliber: .45 ACP Action: striker-fired, semi-auto

Barrel Length: 4.5" Wt.: 30 oz. Sights: 3-dot, adjustable

Capacity: 10 Magazine: detachable box Grips: nylon

D2C:	NIB $	430	Ex $	350	VG+ $	240	Good $	215	LMP $	529
C2C:	Mint $	420	Ex $	300	VG+ $	220	Good $	195	Fair $	100
Trade-In:	Mint $	310	Ex $	250	VG+ $	170	Good $	155	Poor $	60

SR1911 (6700) REDBOOK CODE: RB-RU-H-191100

M1911-style design. Low-glare stainless slide and frame, skeletonized trigger and hammer, frame-mounted and grip safety, sold with two magazines.

Production: current Caliber: .45 ACP Action: SA, semi-auto   Barrel Length: 5"

Wt.: 40 oz. Sights: Novak, adjustable   Capacity: 8 Magazine: detachable box

Grips: checkered hardwood

D2C:	NIB $	720	Ex $	550	VG+ $	400	Good $	360	LMP $	829
C2C:	Mint $	700	Ex $	500	VG+ $	360	Good $	325	Fair $	170
Trade-In:	Mint $	520	Ex $	425	VG+ $	290	Good $	255	Poor $	90

SR1911 (6702)
NEW IN BOX

SR1911 (6702) REDBOOK CODE: RB-RU-H-191102
Indicative of the M1911 Commander. Low-glare stainless slide and frame, skeletonized trigger and hammer, frame-mounted and grip safety.

Production: current Caliber: .45 ACP Action: SA, semi-auto

Barrel Length: 4.25" Wt.: 36 oz. Sights: Novak, adjustable 3-dot

Capacity: 7 Magazine: detachable box Grips: checkered hardwood

		NIB		Ex		VG+		Good		LMP	
D2C:	NIB $	720	Ex $	550	VG+ $	400	Good $	360	LMP $	829
C2C:	Mint $	700	Ex $	500	VG+ $	360	Good $	325	Fair $	170
Trade-In:	Mint $	520	Ex $	425	VG+ $	290	Good $	255	Poor $	90

SR22 (3600)
NEW IN BOX

SR22 (3600) REDBOOK CODE: RB-RU-H-S3600X
Black anodized slide, black polymer grip frame, accessory rail, frame-mounted safety.

Production: current Caliber: .22 LR Action: DA/SA, semi-auto

Barrel Length: 3.5" Wt.: 18 oz. Sights: 3-dot, adjustable

Capacity: 10 Magazine: detachable box Grips: polymer

D2C:	NIB $	340	Ex $	275	VG+ $	190	Good $	170	LMP $	399
C2C:	Mint $	330	Ex $	250	VG+ $	170	Good $	155	Fair $	80
Trade-In:	Mint $	250	Ex $	200	VG+ $	140	Good $	120	Poor $	60

SR22 (3604)
NEW IN BOX

SR22 (3604) REDBOOK CODE: RB-RU-H-S3604X
A threaded-barrel variant of the SR22 (3600). Black polymer grip frame, accessory rail, frame-mounted safety.

Production: current Caliber: .22 LR Action: DA/SA, semi-auto

Barrel Length: 3.5" Wt.: 18 oz. Sights: 3-dot, adjustable

Capacity: 10 Magazine: detachable box Grips: polymer

D2C:	NIB $	370	Ex $	300	VG+ $	205	Good $	185	LMP $	439
C2C:	Mint $	360	Ex $	275	VG+ $	190	Good $	170	Fair $	90
Trade-In:	Mint $	270	Ex $	225	VG+ $	150	Good $	130	Poor $	60

SR22 (3607)
NEW IN BOX

SR22 (3607) REDBOOK CODE: RB-RU-H-S3607X
A silver-slide variant of the SR22 (3600). Black polymer grip frame, accessory rail, frame-mounted safety.

Production: current Caliber: .22 LR Action: DA/SA, semi-auto

Barrel Length: 3.5" Wt.: 18 oz. Sights: 3-dot, adjustable

Capacity: 10 Magazine: detachable box Grips: polymer

D2C:	NIB $	360	Ex $	275	VG+ $	200	Good $	180	LMP $	419
C2C:	Mint $	350	Ex $	250	VG+ $	180	Good $	165	Fair $	85
Trade-In:	Mint $	260	Ex $	225	VG+ $	150	Good $	130	Poor $	60

22/45 TARGET (10107)
NEW IN BOX

22/45 TARGET (10107) REDBOOK CODE: RB-RU-H-10107T
Target model, loaded ßchamber indicator, bull barrel, blue finish, molded grip panels, Zytel polymer grip frame.

Production: current Caliber: .22 LR Action: SA, semi-auto Barrel Length: 4"

Wt.: 32 oz. Sights: fixed front, adjustable rear Capacity: 10 Magazine: detachable box

Grips: polymer

D2C:	NIB $	280	Ex $	225	VG+ $	155	Good $	140	LMP $	359
C2C:	Mint $	270	Ex $	200	VG+ $	140	Good $	130	Fair $	65
Trade-In:	Mint $	200	Ex $	175	VG+ $	110	Good $	100	Poor $	30

22/45 TARGET (10158)
NEW IN BOX

22/45 LITE (3903)
NEW IN BOX

22/45 LITE (3906)
NEW IN BOX

22/45 THREADED BARREL (10149)
NEW IN BOX

22/45 THREADED BARREL (10150)
NEW IN BOX

22/45 TARGET (10158) REDBOOK CODE: RB-RU-H-10158T

Target model, bull barrel, blue finish, loaded chamber indicator, equipped with replaceable black laminate grip panels and Zytel polymer grip frame.

Production: current Caliber: .22 LR Action: SA, semi-auto Barrel Length: 5.5"

Wt.: 33 oz. Sights: fixed front, adjustable rear Capacity: 10 Magazine: detachable box

Grips: black laminate

D2C:	NIB $	320	Ex $	250	VG+ $	180	Good $	160	LMP $ 399
C2C:	Mint $	310	Ex $	225	VG+ $	160	Good $	145	Fair $ 75
Trade-In:	Mint $	230	Ex $	200	VG+ $	130	Good $	115	Poor $ 60

22/45 LITE (3903) REDBOOK CODE: RB-RU-H-223903

A lightweight variant of the 22/45, noted for its fluted and threaded barrel. Zytel polymer grip frame, frame-mounted safety, blue finish.

Production: current Caliber: .22 LR Action: SA, semi-auto Barrel Length: 4.4"

Wt.: 23 oz. Sights: fixed front, adjustable rear Capacity: 10 Magazine: detachable box

Grips: black rubber

D2C:	NIB $	410	Ex $	325	VG+ $	230	Good $	205	LMP $ 499
C2C:	Mint $	400	Ex $	300	VG+ $	210	Good $	185	Fair $ 95
Trade-In:	Mint $	300	Ex $	250	VG+ $	160	Good $	145	Poor $ 60

22/45 LITE (3906) REDBOOK CODE: RB-RU-H-223906

A lightweight variant of the 22/45, noted riveted barrel shroud. Zytel polymer grip frame, stainless slide, frame-mounted safety.

Production: 2014 - current Caliber: .22 LR Action: SA, semi-auto

Barrel Length: 4.4" Wt.: 23 oz. Sights: fixed front, adjustable rear

Capacity: 10 Magazine: detachable box Grips: black rubber

D2C:	NIB $	410	Ex $	325	VG+ $	230	Good $	205	LMP $ 499
C2C:	Mint $	400	Ex $	300	VG+ $	210	Good $	185	Fair $ 95
Trade-In:	Mint $	300	Ex $	250	VG+ $	160	Good $	145	Poor $ 60

22/45 THREADED BARREL (10149)

REDBOOK CODE: RB-RU-H-221014

Threaded barrel, Zytel polymer grip frame, scope mount, accessory rail, blue finish, frame-mounted safety.

Production: current Caliber: .22 LR Action: SA, semi-auto Barrel Length: 4.5"

Wt.: 32 oz. Sights: scope mounts Capacity: 10 Magazine: detachable box

Grips: polymer

D2C:	NIB $	370	Ex $	300	VG+ $	205	Good $	185	LMP $ 449
C2C:	Mint $	360	Ex $	275	VG+ $	190	Good $	170	Fair $ 90
Trade-In:	Mint $	270	Ex $	225	VG+ $	150	Good $	130	Poor $ 60

22/45 THREADED BARREL (10150) REDBOOK CODE: RB-RU-H-221015

Unlike the 22/45 Threaded Barrel 10149, this model doesn't feature an accessory rail or scope mount. Threaded barrel, matte blue finish, Zytel polymer grip frame.

Production: current Caliber: .22 LR Action: SA, semi-auto Barrel Length: 4.5"

Wt.: 32 oz. Sights: fixed front, adjustable rear Capacity: 10 Magazine: detachable box

Grips: polymer

D2C:	NIB $	370	Ex $	300	VG+ $	205	Good $	185	LMP $ 449
C2C:	Mint $	360	Ex $	275	VG+ $	190	Good $	170	Fair $ 90
Trade-In:	Mint $	270	Ex $	225	VG+ $	150	Good $	130	Poor $ 60

Taurus

Headquartered in Porto Alegre, Brazil, Forjas Taurus began manufacturing its first firearm in 1941. Taurus soon purchased the Beretta plant in Sao Paulo along with the designs and tooling for their Model 92. The company began incorporating their unique designs into M92 tooling and thus created the well known PT-92 and PT-99 product line. When Taurus ventured into the US market, they introduced a lifetime warranty on every weapon, which dramatically increased sales. Taurus now has a US subsidiary, Taurus USA, and manufactures a wide range of revolvers, pistols, and rifles.

MODEL 905
NEW IN BOX

MODEL 905 REDBOOK CODE: RB-TA-H-905XXX

Small frame, blue finish, Taurus Security System, transfer-bar safety, Taurus Stellar Clips, also available in stainless steel finish for an additional price.

Production: current Caliber: 9mm Action: DA/SA, revolver Barrel Length: 2"
Wt.: 22.2 oz. Sights: fixed Capacity: 5 Grips: rubber

D2C:	NIB $	430	Ex $	350	VG+ $240	Good $	215	LMP $	480
C2C:	Mint $	420	Ex $	300	VG+ $220	Good $	195	Fair $	100
Trade-In:	Mint $	310	Ex $	250	VG+ $ 170	Good $	155	Poor $	60

MINI REVOLVER
NEW IN BOX

MINI REVOLVER (M380IBULB) REDBOOK CODE: RB-TA-H-M380IB

Lightweight, small frame, blue finish, transfer-bar safety, stainless steel finish also available for an additional price.

Production: current Caliber: .380 ACP Action: DAO, revolver Barrel Length: 1.75"
Wt.: 15.5 oz. Sights: fixed front, adjustable rear Capacity: 5 Grips: rubber

D2C:	NIB $	365	Ex $	300	VG+ $205	Good $	185	LMP $	433
C2C:	Mint $	360	Ex $	275	VG+ $ 190	Good $	165	Fair $	85
Trade-In:	Mint $	260	Ex $	225	VG+ $ 150	Good $	130	Poor $	60

MODEL 94
NEW IN BOX

MODEL 94 REDBOOK CODE: RB-TA-H-94XXXX

Small frame, extended ejector rod, Taurus Security System, transfer-bar safety, blue or stainless steel finish for an additional price, optional 9-round cylinder and optional Crimson Trace Lasergrips on all small frame revolvers.

Production: current Caliber: .22 LR, .22 Magnum Action: DA/SA, revolver
Barrel Length: 2", 4", 5" Wt.: Sights: fixed front, adjustable rear Capacity: 9
Grips: rubber

D2C:	NIB $	380	Ex $	300	VG+ $ 210	Good $	190	LMP	$438
C2C:	Mint $	370	Ex $	275	VG+ $ 190	Good $	175	Fair $	90
Trade-In:	Mint $	270	Ex $	225	VG+ $ 150	Good $	135	Poor $	60

MODEL 94 ULTRA-LITE
NEW IN BOX

MODEL 94 ULTRA-LITE REDBOOK CODE: RB-TA-H-94ULXX

Small frame, extended ejector rod, Taurus Security System, transfer-bar safety, stainless steel finish, optional 9-round cylinder.

Production: current Caliber: .22 S/L/LR Action: DA/SA, revolver
Barrel Length: 2" Wt.: 18 oz. Sights: fixed front, adjustable rear
Capacity: 9 Grips: rubber

D2C:	NIB $	415	Ex $	325	VG+ $230	Good $	210	LMP $	513
C2C:	Mint $	400	Ex $	300	VG+ $ 210	Good $	190	Fair $	100
Trade-In:	Mint $	300	Ex $	250	VG+ $ 170	Good $	150	Poor $	60

TAURUS

MODEL 941
NEW IN BOX

MODEL 941   REDBOOK CODE: RB-TA-H-941XXX

Blue finish, transfer-bar safety, Taurus Security System, also available in stainless steel finish for an additional price, choice of 9-round cylinder.

Production: current   Caliber: .22 Magnum   Action: DA/SA, revolver

Barrel Length: 2", 4"   Wt.: 24-27 oz.   Sights: fixed front, adjustable rear

Capacity: 8   Grips: rubber

D2C:	NIB $	410	Ex $	325	VG+ $	230	Good $	205	LMP $	465
C2C:	Mint $	400	Ex $	300	VG+ $	210	Good $	185	Fair $	95
Trade-In:	Mint $	300	Ex $	250	VG+ $	160	Good $	145	Poor $	60

MODEL 941 ULTRA-LITE
NEW IN BOX

MODEL 941 ULTRA-LITE   REDBOOK CODE: RB-TA-H-941UTL

Matte stainless steel or blue finish, transfer-bar safety, Taurus Security System, extended ejector rod, add roughly $30 for stainless model.

Production: current   Caliber: .22 Magnum   Action: DA/SA, revolver

Barrel Length: 2"   Wt.: 18.5 oz.   Sights: front fixed, adjustable rear

Capacity: 8   Grips: rubber

D2C:	NIB $	440	Ex $	350	VG+ $	245	Good $	220	LMP $	499
C2C:	Mint $	430	Ex $	325	VG+ $	220	Good $	200	Fair $	105
Trade-In:	Mint $	320	Ex $	250	VG+ $	180	Good $	155	Poor $	60

MODEL 605
(605B2)
NEW IN BOX

MODEL 605 (605B2)   REDBOOK CODE: RB-TA-H-605B2X

Small frame, blue finish, Taurus Security System, transfer-bar safety, also available in stainless steel finish.

Production: current   Caliber: .357 Magnum   Action: DA/SA, revolver

Barrel Length: 2"   Wt.: 24 oz.   Sights: fixed   Capacity: 5   Grips: rubber

D2C:	NIB $	370	Ex $	300	VG+ $	205	Good $	185	LMP $	456
C2C:	Mint $	360	Ex $	275	VG+ $	190	Good $	170	Fair $	90
Trade-In:	Mint $	270	Ex $	225	VG+ $	150	Good $	130	Poor $	60

DT REVOLVER
(605PLYDTB2)
NEW IN BOX

DT REVOLVER (605PLYDTB2)   REDBOOK CODE: RB-TA-H-DTREVO

Lightweight polymer frame and cover, blue finish, optional stainless model available.

Production: discontinued   Caliber: .357 Magnum   Action: DA/SA, revolver

Barrel Length: 2"   Wt.: 20.2 oz.   Sights: fixed, fiber-optic   Capacity: 5   Grips: rubber

D2C:	NIB $	360	Ex $	275	VG+ $	200	Good $	180	LMP $	445
C2C:	Mint $	350	Ex $	250	VG+ $	180	Good $	165	Fair $	85
Trade-In:	Mint $	260	Ex $	225	VG+ $	150	Good $	130	Poor $	60

PROTECTOR MODEL 651   REDBOOK CODE: RB-TA-H-651PMX

Small frame, blue finish, Taurus Security System, transfer-bar safety, also available in Shadow Gray Titanium finish and stainless steel finish for an additional price.

Production: discontinued   Caliber: .357 Magnum   Action: DA/SA, revolver

Barrel Length: 2"   Wt.: 25 oz.   Sights: fixed   Capacity: 5   Grips: rubber

D2C:	NIB $	373	Ex $	300	VG+ $	210	Good $	190	LMP $	477
C2C:	Mint $	360	Ex $	275	VG+ $	190	Good $	170	Fair $	90
Trade-In:	Mint $	270	Ex $	225	VG+ $	150	Good $	135	Poor $	60

MODEL 445 REDBOOK CODE: RB-TA-H-445XX1

Compact frame, blue or stainless steel finish, transfer-bar safety, factory ported barrel on some models, Taurus Security System.

Production: discontinued Caliber: .44 Special Action: DA/SA, revolver					
Barrel Length: 2" Wt.: 28 oz. Sights: fixed Capacity: 5 Grips: rubber					
D2C:	NIB $ 289	Ex $ 225	VG+ $ 160	Good $ 145	
C2C:	Mint $ 280	Ex $ 200	VG+ $ 150	Good $ 135	Fair $ 70
Trade-In:	Mint $ 210	Ex $ 175	VG+ $ 120	Good $ 105	Poor $ 30

MODEL 445 WITH CONCEALED HAMMER

REDBOOK CODE: RB-TA-H-445XX2

Concealed hammer, compact frame, blue or stainless steel finish, transfer-bar safety, factory ported barrel, Taurus Security System.

Production: discontinued Caliber: .44 Special Action: DAO, revolver					
Barrel Length: 2" Wt.: 28 oz. Sights: fixed Capacity: 5 Grips: rubber					
D2C:	NIB $ 289	Ex $ 225	VG+ $ 160	Good $ 145	
C2C:	Mint $ 280	Ex $ 200	VG+ $ 150	Good $ 135	Fair $ 70
Trade-In:	Mint $ 210	Ex $ 175	VG+ $ 120	Good $ 105	Poor $ 30

MODEL 82 REDBOOK CODE: RB-TA-H-82SECX

Custom DA/SA trigger, blue finish, Taurus Security System, transfer-bar safety, similar to the Model 65.

Production: current Caliber: .38 Special +P Action: DA/SA, revolver					
Barrel Length: 3", 4" Wt.: 36.5 oz. Sights: fixed Capacity: 6					
Grips: custom-molded rubber					
D2C:	NIB $ 390	Ex $ 300	VG+ $ 215	Good $ 195	LMP $ 471
C2C:	Mint $ 380	Ex $ 275	VG+ $200	Good $ 180	Fair $ 90
Trade-In:	Mint $ 280	Ex $ 225	VG+ $ 160	Good $ 140	Poor $ 60

MODEL 85 (85B2FS) REDBOOK CODE: RB-TA-H-85B2FS

Small frame, blue finish, Taurus Security System, transfer-bar safety.

Production: current Caliber: .38 Special +P Action: DA/SA, revolver					
Barrel Length: 2" Wt.: 21 oz. Sights: fixed Capacity: 5 Grips: rubber					
D2C:	NIB $ 330	Ex $ 275	VG+ $ 185	Good $ 165	LMP $ 397
C2C:	Mint $ 320	Ex $ 250	VG+ $ 170	Good $ 150	Fair $ 80
Trade-In:	Mint $ 240	Ex $ 200	VG+ $ 130	Good $ 120	Poor $ 60

MODEL 85 (85B2FS)
NEW IN BOX

MODEL 85 (85SHC) REDBOOK CODE: RB-TA-H-85XXX1

Small frame, Shadow Gray Titanium finish, all titanium construction, transfer-bar safety, Taurus Security System, additional finishes and fiber-optic front sights available.

Production: discontinued Caliber: .38 Special +P Action: DA/SA, revolver					
Barrel Length: 2" Wt.: 17 oz. Sights: fixed Capacity: 5 Grips: rubber					
D2C:	NIB $ 580	Ex $ 450	VG+ $320	Good $ 290	LMP $ 650
C2C:	Mint $ 560	Ex $ 425	VG+ $290	Good $ 265	Fair $ 135
Trade-In:	Mint $ 420	Ex $ 325	VG+ $230	Good $ 205	Poor $ 60

MODEL 85 (85PLYB2FS) REDBOOK CODE: RB-TA-H-85XXX2
Small frame, blue finish, Taurus Security System, transfer-bar safety.

Production: current Caliber: .38 Special +P Action: DA/SA, revolver
Barrel Length: 2.5" Wt.: 18.2 oz. Sights: fixed, red fiber-optic front
Capacity: 5 Grips: rubber

D2C:	NIB $	360	Ex $	275	VG+	$200	Good $	180	LMP $ 397
C2C:	Mint $	350	Ex $	250	VG+	$180	Good $	165	Fair $ 85
Trade-In:	Mint $	260	Ex $	225	VG+	$150	Good $	130	Poor $ 60

MODEL 85 (85SS2ULFS)
NEW IN BOX

MODEL 85 (85SS2ULFS) REDBOOK CODE: RB-TA-H-85XXX3
Matte stainless steel finish, smooth trigger, transfer-bar safety, Taurus Security System, ultra-light revolver.

Production: current Caliber: .38 Special +P Action: DA/SA, revolver
Barrel Length: 2" Wt.: 17 oz. Sights: fixed front Capacity: 5 Grips: rubber

D2C:	NIB $	375	Ex $	300	VG+	$210	Good $	190	LMP $ 443
C2C:	Mint $	360	Ex $	275	VG+	$190	Good $	170	Fair $ 90
Trade-In:	Mint $	270	Ex $	225	VG+	$150	Good $	135	Poor $ 60

MODEL 85 (85BPP2)
NEW IN BOX

MODEL 85 (85BPP2) REDBOOK CODE: RB-TA-H-85XXX4
Blue finish, small frame, transfer-bar safety, Taurus Security System.

Production: discontinued Caliber: .38 Special +P Action: DA/SA, revolver
Barrel Length: 2" Wt.: 21 oz. Sights: fixed Capacity: 5 Grips: pink mother of pearl

D2C:	NIB $	430	Ex $	350	VG+	$240	Good $	215	LMP $ 498
C2C:	Mint $	420	Ex $	300	VG+	$220	Good $	195	Fair $ 100
Trade-In:	Mint $	310	Ex $	250	VG+	$170	Good $	155	Poor $ 60

85 VIEW
(85VTA)
NEW IN BOX

85 VIEW (85VTA) REDBOOK CODE: RB-TA-H-85VIEW
Clear-view Lexan sideplate, matte stainless or pink finish, Taurus Security System, small ultra-lightweight frame, contoured grips.

Production: 2014 - current Caliber: .38 Special Action: DAO, revolver
Barrel Length: 1.41" Wt.: 9 oz. Sights: fixed Capacity: 5 Grips: rubber

D2C:	NIB $	550	Ex $	425	VG+	$305	Good $	275	LMP $ 599
C2C:	Mint $	530	Ex $	400	VG+	$280	Good $	250	Fair $ 130
Trade-In:	Mint $	400	Ex $	325	VG+	$220	Good $	195	Poor $ 60

CIA MODEL 650
(650B2)
NEW IN BOX

CIA MODEL 650 (650B2) REDBOOK CODE: RB-TA-H-CI6501
Blue finish, transfer-bar safety, Taurus Security System.

Production: current Caliber: .357 Magnum/.38 Special Action: DAO, revolver
Barrel Length: 2" Wt.: 24 oz. Sights: fixed Capacity: 5 Grips: rubber

D2C:	NIB $	430	Ex $	350	VG+	$240	Good $	215	LMP $ 512
C2C:	Mint $	420	Ex $	300	VG+	$220	Good $	195	Fair $ 100
Trade-In:	Mint $	310	Ex $	250	VG+	$170	Good $	155	Poor $ 60

CIA MODEL 650 (650SS2) REDBOOK CODE: RB-TA-H-CI6502
Small frame, stainless steel finish, transfer-bar safety, Taurus Security System.

Production: discontinued Caliber: .357 Magnum/.38 Special Action: DAO, revolver
Barrel Length: 2" Wt.: 24 oz. Sights: fixed Capacity: 5 Grips: rubber

D2C:	NIB $	449	Ex $	350	VG+	$250	Good $	225	LMP $ 555
C2C:	Mint $	440	Ex $	325	VG+	$230	Good $	205	Fair $ 105
Trade-In:	Mint $	320	Ex $	275	VG+	$180	Good $	160	Poor $ 60

CIA MODEL 850
(850B2UL)
NEW IN BOX

CIA MODEL 850 (850B2UL)   REDBOOK CODE: RB-TA-H-CI8501
Small frame, blue finish, transfer-bar safety, Taurus Security System.

Production: current Caliber: .38 Special +P Action: DAO, revolver
Barrel Length: 2" Sights: fixed Capacity: 5 Grips: rubber

D2C:	NIB	$ 399	Ex $ 325	VG+ $220	Good $ 200	LMP $ 474
C2C:	Mint $ 390	Ex $ 300	VG+ $200	Good $ 180	Fair $ 95
Trade-In:	Mint $ 290	Ex $ 225	VG+ $160	Good $ 140	Poor $ 60

CIA MODEL 850   REDBOOK CODE: RB-TA-H-CI8502
Small frame, Shadow Gray Titanium finish, transfer-bar safety, enclosed hammer, also available in blue and stainless steel finish.

Production: discontinued Caliber: .38 Special +P Action: DAO, revolver
Barrel Length: 2" Wt.: 16 oz. Sights: fixed Capacity: 5 Grips: rubber

D2C:	NIB $ 410	Ex $ 325	VG+ $230	Good $ 205	LMP $ 666
C2C:	Mint $ 400	Ex $ 300	VG+ $ 210	Good $ 185	Fair $ 95
Trade-In:	Mint $ 300	Ex $ 250	VG+ $160	Good $ 145	Poor $ 60

CIA MODEL 850 "BLUE PEARL"   REDBOOK CODE: RB-TA-H-CI8503
Small frame, Spectrum Blue Titanium finish, transfer bar safety, Taurus Security System, enclosed hammer.

Production: discontinued Caliber: .38 Special +P Action: DAO, revolver
Barrel Length: 2" Wt.: 21 oz. Sights: fixed Capacity: 5 Grips: mother-of-pearl

D2C:	NIB $ 430	Ex $ 350	VG+ $240	Good $ 215	LMP $ 698
C2C:	Mint $ 420	Ex $ 300	VG+ $220	Good $ 195	Fair $ 100
Trade-In:	Mint $ 310	Ex $ 250	VG+ $170	Good $ 155	Poor $ 60

990 TRACKER STAINLESS   REDBOOK CODE: RB-TA-H-990TR1
Taurus Security System, transfer-bar safety, matte stainless finish, extended ejector rod, 6.5" barrel has vent rib.

Production: current Caliber: .22 LR Action: DA/SA, revolver
Barrel Length: 4", 6.5" Wt.: 38-44 oz.
Sights: fixed front, adjustable rear Capacity: 9 Grips: rubber

D2C:	NIB $ 440	Ex $ 350	VG+ $245	Good $ 220	LMP $ 555
C2C:	Mint $ 430	Ex $ 325	VG+ $220	Good $ 200	Fair $ 105
Trade-In:	Mint $ 320	Ex $ 250	VG+ $180	Good $ 155	Poor $ 60

990 TRACKER STAINLESS
NEW IN BOX

990 TRACKER BLUE   REDBOOK CODE: RB-TA-H-990TR2
Blue finish, transfer bar, Taurus Security System, extended ejector rod, 6.5" barrel has vent rib.

Production: discontinued Caliber: .22 LR Action: DA/SA, revolver
Barrel Length: 4", 6.5" Wt.: 38-44 oz. Sights: fixed front, adjustable rear
Capacity: 9 Grips: rubber

D2C:	NIB $ 390	Ex $ 300	VG+ $ 215	Good $ 195	LMP $ 492
C2C:	Mint $ 380	Ex $ 275	VG+ $200	Good $ 180	Fair $ 90
Trade-In:	Mint $ 280	Ex $ 225	VG+ $160	Good $ 140	Poor $ 60

990 TRACKER BLUE
NEW IN BOX

991 TRACKER (991SS4)   REDBOOK CODE: RB-TA-H-991TR1
Taurus Security System, transfer-bar safety, matte stainless steel finish, extended ejector rod, also available with a 6" barrel.

Production: current Caliber: .22 Magnum Action: DA/SA, revolver Barrel Length: 4"
Wt.: 38 oz. Sights: fixed front, adjustable rear Capacity: 9 Grips: rubber

D2C:	NIB $ 460	Ex $ 350	VG+ $255	Good $ 230	LMP $ 555
C2C:	Mint $ 450	Ex $ 325	VG+ $230	Good $ 210	Fair $ 110
Trade-In:	Mint $ 330	Ex $ 275	VG+ $180	Good $ 165	Poor $ 60

TAURUS

TAURUS

991 TRACKER
NEW IN BOX

991 TRACKER (991B4) REDBOOK CODE: RB-TA-H-991TR2

Blue finish, extended ejector rod, transfer-bar safety, Taurus Security System, also available with a 6.5" barrel.

Production: discontinued Caliber: .22 Magnum Action: DA/SA, revolver

Barrel Length: 4" Wt.: 38 oz. Sights: fixed front, adjustable rear

Capacity: 9 Grips: rubber

D2C:	NIB $ 410	Ex $ 325	VG+ $230	Good $ 205	LMP $ 492
C2C:	Mint $ 400	Ex $ 300	VG+ $210	Good $ 185	Fair $ 95
Trade-In:	Mint $ 300	Ex $ 250	VG+ $160	Good $ 145	Poor $ 60

992 TRACKER (992B4) REDBOOK CODE: RB-TA-H-992TR1

Blue finish, Taurus Security System, transfer-bar safety, removable cylinder, transforms from .22 LR to .22 Magnum, also available with 6.5" barrel.

Production: current Caliber: .22 LR Action: DA/SA, revolver

Barrel Length: 4" Sights: fixed front, adjustable rear Capacity: 9 Grips: rubber

D2C:	NIB $ 499	Ex $ 400	VG+ $ 275	Good $ 250	LMP $ 590
C2C:	Mint $ 480	Ex $ 350	VG+ $250	Good $ 225	Fair $ 115
Trade-In:	Mint $ 360	Ex $ 300	VG+ $200	Good $ 175	Poor $ 60

992 TRACKER (992B6)
NEW IN BOX

992 TRACKER (992B6) REDBOOK CODE: RB-TA-H-992TR2

Taurus Security System, transforms from .22 LR to .22 Mag., transfer-bar safety, removable cylinder, blue finish, stainless model available for slight premium.

Production: current Caliber: .22 LR Action: DA/SA, revolver Barrel Length: 6.5"

Wt.: 55 oz. Sights: fixed front, adjustable rear Capacity: 9 Grips: rubber

D2C:	NIB $ 499	Ex $ 400	VG+ $ 275	Good $ 250	LMP $ 590
C2C:	Mint $ 480	Ex $ 350	VG+ $250	Good $ 225	Fair $ 115
Trade-In:	Mint $ 360	Ex $ 300	VG+ $200	Good $ 175	Poor $ 60

627 TRACKER REDBOOK CODE: RB-TA-H-627TRX

Factory tuned and ported barrel with gas expansion chamber, vent rib for optional scope mount base on 6.5" model, Taurus Security System, transfer-bar safety, matte stainless finish. Add $125 for total titanium model.

Production: current Caliber: .357 Magnum Action: DA/SA, revolver

Barrel Length: 4", 6.5" Wt.: 28.8 oz., 40.8 oz. Sights: fixed front, adjustable rear

Capacity: 7 Grips: rubber

D2C:	NIB $ 560	Ex $ 450	VG+ $ 310	Good $ 280	LMP $ 670
C2C:	Mint $ 540	Ex $ 400	VG+ $280	Good $ 255	Fair $ 130
Trade-In:	Mint $ 400	Ex $ 325	VG+ $220	Good $ 200	Poor $ 60

44 TRACKER
(44TRACKER4B)
NEW IN BOX

44 TRACKER (44TRACKER4B) REDBOOK CODE: RB-TA-H-44TRXX

Blue steel finish, factory tuned ported barrel, Taurus Security, transfer safety bar, smooth trigger, also available in matte stainless steel finish for an additional price.

Production: current Caliber: .44 Magnum Action: DA/SA, revolver

Barrel Length: 4" Wt.: 34 oz. Sights: fixed front, adjustable rear

Capacity: 5 Grips: rubber

D2C:	NIB $ 530	Ex $ 425	VG+ $295	Good $ 265	LMP $ 645
C2C:	Mint $ 510	Ex $ 375	VG+ $270	Good $ 240	Fair $ 125
Trade-In:	Mint $ 380	Ex $ 300	VG+ $210	Good $ 190	Poor $ 60

44 TRACKER
(44TRACKER4SS)
NEW IN BOX

44 TRACKER (44TRACKER4SS) REDBOOK CODE: RB-TA-H-44TRSS

Matte stainless finish, ported barrel, Taurus Security System, smooth trigger, transfer-bar safety.

Production: current Caliber: .44 Magnum Action: DA/SA, revolver

Barrel Length: 4" Wt.: 34 oz. Sights: fixed front, adjustable rear

Capacity: 5 Grips: rubber

D2C:	NIB $	570	Ex $	450	VG+ $	315	Good $	285	LMP $ 693
C2C:	Mint $	550	Ex $	400	VG+ $	290	Good $	260	Fair $ 135
Trade-In:	Mint $	410	Ex $	325	VG+ $	230	Good $	200	Poor $ 60

MODEL 617 (617SS2) REDBOOK CODE: RB-TA-H-617XX1

Stainless steel finish, compact frame, Taurus Security System, transfer-bar safety.

Production: current Caliber: .357 Magnum Action: DA/SA, revolver

Barrel Length: 2" Wt.: 28.3 oz. Sights: fixed Capacity: 7 Grips: rubber

D2C:	NIB $	475	Ex $	375	VG+ $	265	Good $	240	LMP $ 560
C2C:	Mint $	460	Ex $	350	VG+ $	240	Good $	215	Fair $ 110
Trade-In:	Mint $	340	Ex $	275	VG+ $	190	Good $	170	Poor $ 60

MODEL 617 (617CHB) REDBOOK CODE: RB-TA-H-617XX2

Blue finish, smooth trigger, compact frame, concealed hammer, Taurus Security System, transfer-bar safety.

Production: discontinued Caliber: .357 Magnum Action: DA/SA, revolver

Barrel Length: 2" Wt.: 28 oz. Sights: fixed Capacity: 7 Grips: rubber

D2C:	NIB $	430	Ex $	350	VG+ $	240	Good $	215
C2C:	Mint $	420	Ex $	300	VG+ $	220	Good $	195	Fair $ 100
Trade-In:	Mint $	310	Ex $	250	VG+ $	170	Good $	155	Poor $ 60

MODEL 617 (617SH2C) REDBOOK CODE: RB-TA-H-617XX3

Shadow Gray Titanium finish, transfer-bar safety, compact frame, Taurus Security System, also available in blue finish.

Production: discontinued Caliber: .357 Magnum Action: DA/SA, revolver

Barrel Length: 2" Wt.: 20 oz. Sights: fixed Capacity: 7 Grips: rubber

D2C:	NIB $	650	Ex $	500	VG+ $	360	Good $	325	LMP $ 707
C2C:	Mint $	630	Ex $	450	VG+ $	330	Good $	295	Fair $ 150
Trade-In:	Mint $	470	Ex $	375	VG+ $	260	Good $	230	Poor $ 90

MODEL 617 (617SSC) REDBOOK CODE: RB-TA-H-617XX4

Stainless steel finish, compact frame, Taurus Security System, transfer-bar safety.

Production: discontinued Caliber: .357 Magnum Action: DA/SA, revolver

Barrel Length: 2" Wt.: 28 oz. Sights: fixed Capacity: 7 Grips: rubber

D2C:	NIB $	420	Ex $	325	VG+ $	235	Good $	210
C2C:	Mint $	410	Ex $	300	VG+ $	210	Good $	190	Fair $ 100
Trade-In:	Mint $	300	Ex $	250	VG+ $	170	Good $	150	Poor $ 60

THE JUDGE (4510TKR-3B)   REDBOOK CODE: RB-TA-H-JUDGX1

Fires both .410 GA. 2.5-inch shotshell and .45 Long Colt cartridges, extended ejector rod, Taurus Security System, transfer-bar safety, blue finish.

Production: current   Caliber: .45 LC / .410 Ga. (2.5" chamber)   Action: DA/SA, revolver
Barrel Length: 3"   Wt.: 29 oz.   Sights: fixed red fiber-optic front, fixed rear   Capacity: 5
Grips: rubber

D2C:	NIB $ 510	Ex $ 400	VG+ $ 285	Good $ 255	LMP $ 620
C2C:	Mint $ 490	Ex $ 375	VG+ $ 260	Good $ 230	Fair $ 120
Trade-In:	Mint $ 370	Ex $ 300	VG+ $ 200	Good $ 180	Poor $ 60

THE JUDGE
(4510TKR-3SS)
NEW IN BOX

THE JUDGE (4510TKR-3SS)   REDBOOK CODE: RB-TA-H-JUDGX2

Fires both .410 GA. 2.5-inch shotshell and .45 Long Colt cartridges, extended ejector rod, Taurus Security System, transfer-bar safety, stainless finish.

Production: current   Caliber: .45 LC / .410 Ga. (2.5" chamber)   Action: DA/SA, revolver
Barrel Length: 3"   Wt.: 29 oz.   Sights: fixed red fiber-optic front, fixed rear
Capacity: 5   Grips: rubber

D2C:	NIB $ 545	Ex $ 425	VG+ $300	Good $ 275	LMP $ 653
C2C:	Mint $ 530	Ex $ 400	VG+ $280	Good $ 250	Fair $ 130
Trade-In:	Mint $ 390	Ex $ 325	VG+ $220	Good $ 195	Poor $ 60

THE JUDGE (4510TRACKERB)   REDBOOK CODE: RB-TA-H-JUDGX3

Fires both .410 GA. 2.5-inch shotshell and .45 Long Colt cartridges, extended ejector rod, Taurus Security System, transfer-bar safety, blue finish.

Production: current   Caliber: .45 LC / .410 Ga. (2.5" chamber)   Action: DA/SA, revolver
Barrel Length: 6.5"   Wt.: 32 oz.   Sights: fixed red fiber-optic front, fixed rear
Capacity: 5   Grips: rubber

D2C:	NIB $ 498	Ex $ 400	VG+ $ 275	Good $ 250	LMP $ 607
C2C:	Mint $ 480	Ex $ 350	VG+ $250	Good $ 225	Fair $ 115
Trade-In:	Mint $ 360	Ex $ 300	VG+ $200	Good $ 175	Poor $ 60

THE JUDGE (4510TRACKERSS)   REDBOOK CODE: RB-TA-H-JUDGX4

Fires both .410 GA. 2.5-inch shotshell and .45 Long Colt cartridges, extended ejector rod, Taurus Security System, transfer-bar safety, stainless finish.

Production: current   Caliber: .45 LC / .410 Ga. (2.5" chamber)   Action: DA/SA, revolver
Barrel Length: 6.5"   Wt.: 32 oz.   Sights: fixed red fiber-optic front, fixed rear
Capacity: 5   Grips: rubber

D2C:	NIB $ 545	Ex $ 425	VG+ $300	Good $ 275	LMP $ 653
C2C:	Mint $ 530	Ex $ 400	VG+ $280	Good $ 250	Fair $ 130
Trade-In:	Mint $ 390	Ex $ 325	VG+ $220	Good $ 195	Poor $ 60

THE JUDGE (4510TKR-3BMAG)   REDBOOK CODE: RB-TA-H-JUDGX5

Fires both .410 GA. 3-inch Magnum shotshell and .45 Long Colt cartridges, extended ejector rod, Taurus Security System, transfer-bar safety, blue finish.

Production: current   Caliber: .45 LC / .410 Ga. (3" Mag. chamber)   Action: DA/SA, revolver
Barrel Length: 3"   Wt.: 37 oz.   Sights: fixed red fiber-optic front, fixed rear
Capacity: 5   Grips: rubber

D2C:	NIB $ 560	Ex $ 450	VG+ $ 310	Good $ 280	LMP $ 668
C2C:	Mint $ 540	Ex $ 400	VG+ $280	Good $ 255	Fair $ 130
Trade-In:	Mint $ 400	Ex $ 325	VG+ $220	Good $ 200	Poor $ 60

THE JUDGE (4510TKR-3MAG)   REDBOOK CODE: RB-TA-H-JUDGX6

Fires both .410 GA. 3-inch Magnum shotshell and .45 Long Colt cartridges, extended ejector rod, Taurus Security System, transfer-bar safety, stainless finish.

Production: current Caliber: .45 LC / .410 Ga. (3" Mag. chamber) Action: DA/SA, revolver
Barrel Length: 3" Wt.: 37 oz.   Sights: fixed red fiber-optic front, fixed rear
Capacity: 5 Grips: rubber

D2C:	NIB	$	615	Ex $	475	VG+ $	340	Good $	310	LMP $	716
C2C:	Mint	$	600	Ex $	425	VG+ $	310	Good $	280	Fair $	145
Trade-In:	Mint	$	440	Ex $	350	VG+ $	240	Good $	220	Poor $	90

THE JUDGE (4510B6MAG)   REDBOOK CODE: RB-TA-H-JUDGX7

Fires both .410 GA. 3-inch Magnum shotshell and .45 Long Colt cartridges, extended ejector rod, Taurus Security System, transfer-bar safety, blue finish.

Production: current Caliber: .45 LC / .410 Ga. (3" Mag. chamber) Action: DA/SA, revolver
Barrel Length: 6.5" Sights: fixed red fiber-optic front Capacity: 5 Grips: rubber

D2C:	NIB	$	545	Ex $	425	VG+ $	300	Good $	275	LMP $	653
C2C:	Mint	$	530	Ex $	400	VG+ $	280	Good $	250	Fair $	130
Trade-In:	Mint	$	390	Ex $	325	VG+ $	220	Good $	195	Poor $	60

THE JUDGE (4510SS6MAG)   REDBOOK CODE: RB-TA-H-JUDGX8

Fires both .410 GA. 3-inch Magnum shotshell and .45 Long Colt cartridges, extended ejector rod, Taurus Security System, transfer-bar safety, stainless finish.

Production: current Caliber: .45 LC / .410 Ga. (3" Mag. chamber) Action: DA/SA, revolver
Barrel Length: 6.5" Sights: fixed red fiber-optic front, fixed rear
Capacity: 5 Grips: rubber

D2C:	NIB	$	615	Ex $	475	VG+ $	340	Good $	310	LMP $	716
C2C:	Mint	$	600	Ex $	425	VG+ $	310	Good $	280	Fair $	145
Trade-In:	Mint	$	440	Ex $	350	VG+ $	240	Good $	220	Poor $	90

THE JUDGE (4510PD-3SS)   REDBOOK CODE: RB-TA-H-JUDGX9

Fires both .410 GA. 2.5-inch shotshell and .45 Long Colt cartridges, extended ejector rod, Taurus Security System, transfer-bar safety, reduced profile hammer, stainless finish.

Production: current Caliber: .45 LC / .410 Ga. (2.5" chamber) Action: DA/SA, revolver
Barrel Length: 2.5" Wt.: 28 oz.   Sights: fixed red fiber-optic front
Capacity:5 Grips: rubber

D2C:	NIB	$	545	Ex $	425	VG+ $	300	Good $	275	LMP $	653
C2C:	Mint	$	530	Ex $	400	VG+ $	280	Good $	250	Fair $	130
Trade-In:	Mint	$	390	Ex $	325	VG+ $	220	Good $	195	Poor $	60

THE JUDGE (4510PD-3B)
NEW IN BOX

THE JUDGE (4510PD-3B)   REDBOOK CODE: RB-TA-H-JUDG10

Fires both .410 GA. 2.5-inch shotshell and .45 Long Colt cartridges, extended ejector rod, Taurus Security System, transfer-bar safety, reduced profile hammer, blue finish.

Production: current Caliber: .45 LC / .410 Ga. (2.5" chamber) Action: DA/SA, revolver
Barrel Length: 2.5" Wt.: 28 oz. Sights: fixed red fiber-optic front
Capacity: 5 Grips: rubber

D2C:	NIB	$	499	Ex $	400	VG+ $	275	Good $	250	LMP $	607
C2C:	Mint	$	480	Ex $	350	VG+ $	250	Good $	225	Fair $	115
Trade-In:	Mint	$	360	Ex $	300	VG+ $	200	Good $	175	Poor $	60

TAURUS

PUBLIC DEFENDER POLYMER
(THE JUDGE)
(4510PLY-SS2)
NEW IN BOX

PUBLIC DEFENDER POLYMER (THE JUDGE)

(4510PLY-SS2)   REDBOOK CODE: RB-TA-H-JUDG11

Fires both .410 GA. 2.5-inch shotshell and .45 Long Colt cartridges, light polymer body frame and cover, extended ejector rod, Taurus Security System, transfer-bar safety, stainless steel or blue finish.

Production: current Caliber: .45 LC / .410 Ga. (2.5" chamber) Action: DA/SA, revolver

Barrel Length: 2.5" Wt.: 23 oz. Sights: fixed red fiber-optic front, adjustable rear

Capacity: 5 Grips: rubber

D2C:	NIB $ 545	Ex $ 425	VG+ $300	Good $ 275	LMP $ 653	
C2C:	Mint $ 530	Ex $ 400	VG+ $280	Good $ 250	Fair $ 130	
Trade-In:	Mint $ 390	Ex $ 325	VG+ $220	Good $ 195	Poor $ 60	

PUBLIC DEFENDER POLYMER (THE JUDGE)(4510PLYFS)

REDBOOK CODE: RB-TA-H-PDXXX1

Small frame, Taurus Security System, blue finish, transfer bar, extended ejector rod.

Production: current Caliber: .45 LC / .410 Ga. (2.5" chamber) Action: DA/SA, revolver

Barrel Length: 2" Wt.: 27 oz. Sights: fixed red fiber-optic front

Capacity: 5 Grips: rubber

D2C:	NIB $ 460	Ex $ 350	VG+ $ 255	Good $ 230	LMP $ 514	
C2C:	Mint $ 450	Ex $ 325	VG+ $ 230	Good $ 210	Fair $ 110	
Trade-In:	Mint $ 330	Ex $ 275	VG+ $ 180	Good $ 165	Poor $ 60	

PUBLIC DEFENDER (THE JUDGE) (4510PD-3TI)

REDBOOK CODE: RB-TA-H-PDXXX3

Blue finish, medium frame, transfer-bar safety, carbon steel or stainless steel frame, optional titanium cylinder.

Production: discontinued Caliber: .45 LC/.410 Ga. Action: DA/SA, revolver

Barrel Length: 2.5" Wt.: 26 oz. Sights: fixed red fiber-optic front

Capacity: 5 Grips: rubber

D2C:	NIB $ 599	Ex $ 475	VG+ $ 330	Good $ 300	LMP $ 648	
C2C:	Mint $ 580	Ex $ 425	VG+ $ 300	Good $ 270	Fair $ 140	
Trade-In:	Mint $ 430	Ex $ 350	VG+ $ 240	Good $ 210	Poor $ 60	

MODEL 65
(65B4)
NEW IN BOX

MODEL 65 (65B4)   REDBOOK CODE: RB-TA-H-65XXXX

Medium frame, Taurus Security System, transfer-bar safety, blue finish, also available in stainless steel finish for an additional price.

Production: current Caliber: .357 Magnum & .38 Special +P Action: DA/SA, revolver

Barrel Length: 4" Wt.: 38 oz. Sights: fixed Capacity: 6 Grips: rubber

D2C:	NIB $ 415	Ex $ 325	VG+ $ 230	Good $ 210	LMP $ 488	
C2C:	Mint $ 400	Ex $ 300	VG+ $ 210	Good $ 190	Fair $ 100	
Trade-In:	Mint $ 300	Ex $ 250	VG+ $ 170	Good $ 150	Poor $ 60	

MODEL 66
NEW IN BOX

MODEL 66   REDBOOK CODE: RB-TA-H-66XXXX

Medium frame, Taurus Security System, transfer-bar safety, blue or stainless finish.

Production: current Caliber: .357 Magnum & .38 Special +P Action: DA/SA, revolver

Barrel Length: 4", 6" Wt.: 38-43.2 oz. Sights: fixed front, adjustable rear

Capacity: 7 Grips: rubber

D2C:	NIB $ 450	Ex $ 350	VG+ $ 250	Good $ 225	LMP $ 543	
C2C:	Mint $ 440	Ex $ 325	VG+ $ 230	Good $ 205	Fair $ 105	
Trade-In:	Mint $ 320	Ex $ 275	VG+ $ 180	Good $ 160	Poor $ 60	

MODEL 608
(608SS4/6)
NEW IN BOX

MODEL 608 (608SS4/6) REDBOOK CODE: RB-TA-H-608XX1

Forged-hammer, ported barrel, Taurus Security System, transfer-bar safety, stainless finish.

Production: current Caliber: .357 Magnum Action: DA/SA, revolver

Barrel Length: 4", 6.5" Wt.: 44 oz., 51 oz. Sights: fixed front, adjustable rear

Capacity: 8 Grips: rubber

D2C:	NIB	$ 570	Ex $ 450	VG+ $ 315	Good $ 285	LMP $ 688
C2C:	Mint $ 550	Ex $ 400	VG+ $ 290	Good $ 260	Fair $ 135
Trade-In:	Mint $ 410	Ex $ 325	VG+ $ 230	Good $ 200	Poor $ 60

MODEL 608
NEW IN BOX

MODEL 608 (608SS8) REDBOOK CODE: RB-TA-H-608XX2

Matte stainless steel finish, smooth trigger, transfer-bar safety, Taurus Security System, ported barrel.

Production: discontinued Caliber: .357 Magnum Action: DA/SA, revolver

Barrel Length: 8.375" Wt.: 56 oz. Sights: fixed front, adjustable rear

Capacity: 8 Grips: rubber

D2C:	NIB $ 630	Ex $ 500	VG+ $ 350	Good $ 315	LMP $ 703
C2C:	Mint $ 610	Ex $ 450	VG+ $ 320	Good $ 285	Fair $ 145
Trade-In:	Mint $ 450	Ex $ 375	VG+ $ 250	Good $ 225	Poor $ 90

MODEL 44
(STAINLESS)
NEW IN BOX

MODEL 44 (STAINLESS) REDBOOK CODE: RB-TA-H-44XXXX

Large frame, vent rib for optional scope mount base on 6.5" model, ported barrel, Taurus Security System, transfer-bar safety, matte stainless finish.

Production: current Caliber: .44 Magnum Action: DA/SA, revolver

Barrel Length: 4", 6.5", 8.375" Wt.: 45 oz., 52 oz., 57 oz.

Sights: fixed front, adjustable rear Capacity: 6 Grips: soft rubber

D2C:	NIB $ 610	Ex $ 475	VG+ $ 340	Good $ 305	LMP $726
C2C:	Mint $ 590	Ex $ 425	VG+ $ 310	Good $ 275	Fair $145
Trade-In:	Mint $ 440	Ex $ 350	VG+ $ 240	Good $ 215	Poor $ 90

MODEL 416 RAGING BULL
NEW IN BOX

MODEL 416 RAGING BULL REDBOOK CODE: RB-TA-H-416RBX

Large frame, stainless steel finish, ported barrel, transfer-bar safety, double lock-up cylinder, Taurus Security System.

Production: discontinued Caliber: .41 Magnum Action: DA/SA, revolver

Barrel Length: 6.5", 8.375" Wt.: 53 oz. Sights: Patridge front, adjustable rear

Capacity: 6 Grips: Raging Bull cushioned insert grip

D2C:	NIB $ 655	Ex $ 500	VG+ $ 365	Good $ 330	LMP $ 780
C2C:	Mint $ 630	Ex $ 475	VG+ $ 330	Good $ 295	Fair $ 155
Trade-In:	Mint $ 470	Ex $ 375	VG+ $ 260	Good $ 230	Poor $ 90

**MODEL 444
RAGING BULL**
NEW IN BOX

MODEL 444 RAGING BULL (444B6/SS6)(444B8/SS8)

REDBOOK CODE: RB-TA-H-444RB1

Double lock-up cylinder, factory tuned porting with gas expansion chamber, Taurus Security System, transfer-bar safety, blue or matte stainless finish.

Production: current Caliber: .44 Magnum Action: DA/SA, revolver

Barrel Length: 6.5", 8.375" Wt.: 53-63 oz. Sights: Patridge front, adjustable rear

Capacity: 6 Grips: Raging Bull-cushioned insert grip

D2C:	NIB $ 620	Ex $ 475	VG+ $ 345	Good $ 310	LMP $ 752
C2C:	Mint $ 600	Ex $ 450	VG+ $ 310	Good $ 280	Fair $ 145
Trade-In:	Mint $ 450	Ex $ 350	VG+ $ 250	Good $ 220	Poor $ 90

MODEL 444 MULTI
(444MULTI)
NEW IN BOX

**MODEL 454
RAGING BULL**
(454B6/8)
NEW IN BOX

**MODEL 454
RAGING BULL**
(45455)
NEW IN BOX

**MODEL 513 RAGING
JUDGE ULTRA-LITE**
NEW IN BOX

**MODEL 513 RAGING
JUDGE MAGNUM**
NEW IN BOX

MODEL 444 MULTI (444MULTI) REDBOOK CODE: RB-TA-H-444MBS

UltraLite Titanium in either blue or stainless finish, multi-alloy construction, transfer-bar safety, double lock-up cylinder, Taurus Security System.

Production: current   Caliber: .44 Magnum   Action: DA/SA, revolver
Barrel Length: 2.25", 4"   Wt.: 28 oz.   Sights: red fiber-optic front, fixed rear
Capacity: 6   Grips: Raging Bull cushioned insert grip

D2C:	NIB $ 650	Ex $ 500	VG+ $ 360	Good $ 325	LMP $ 799
C2C:	Mint $ 630	Ex $ 450	VG+ $ 330	Good $ 295	Fair $ 150
Trade-In:	Mint $ 470	Ex $ 375	VG+ $ 260	Good $ 230	Poor $ 90

MODEL 454 RAGING BULL (454B6/8)

REDBOOK CODE: RB-TA-H-454RBX

Large frame, double lock-up cylinder, ported barrel, Taurus Security System, transfer-bar safety, blue finish.

Production: discontinued   Caliber: .454 Casull   Action: DA/SA, revolver
Barrel Length: 6.5", 8.375"   Wt.: 53-63 oz.   Sights: Patridge front, adjustable rear
Capacity: 5   Grips: Raging Bull cushioned insert grip

D2C:	NIB $ 775	Ex $ 600	VG+ $ 430	Good $ 390	LMP $1,016
C2C:	Mint $ 750	Ex $ 550	VG+ $ 390	Good $ 350	Fair $ 180
Trade-In:	Mint $ 560	Ex $ 450	VG+ $ 310	Good $ 275	Poor $ 90

MODEL 454 RAGING BULL (454SS)

REDBOOK CODE: RB-TA-H-454RB1

Large frame, matte stainless steel finish, ported barrel, double lockup cylinder, Taurus Security System, transfer-bar safety.

Production: current   Caliber: .454 Casull   Action: DA/SA, revolver
Barrel Length: 2.25", 5", 6.5", 8.375"   Wt.: 48-63 oz.
Sights: Patridge front, adjustable rear   Capacity: 5   Grips: Raging Bull cushioned insert grip

D2C:	NIB $ 850	Ex $ 650	VG+ $ 470	Good $ 425	LMP $1,070
C2C:	Mint $ 820	Ex $ 600	VG+ $ 430	Good $ 385	Fair $ 200
Trade-In:	Mint $ 610	Ex $ 500	VG+ $ 340	Good $ 300	Poor $ 90

MODEL 513 RAGING JUDGE ULTRA-LITE

REDBOOK CODE: RB-TA-H-513RJ2

Large frame, Taurus Security System, transfer-bar safety, extended ejector rod, double lock-up cylinder, ultra-lite revolver, also available in stainless steel finish for an additional price.

Production: discontinued   Caliber: .45 LC/.410 Ga.   Action: DA/SA, revolver
Barrel Length: 3"   Wt.: 41 oz.   Sights: fixed red fiber-optic front, fixed rear
Capacity: 7   Grips: Raging Bull cushioned insert grip

D2C:	NIB $ 760	Ex $ 600	VG+ $ 420	Good $ 380	LMP $ 983
C2C:	Mint $ 730	Ex $ 525	VG+ $ 380	Good $ 345	Fair $ 175
Trade-In:	Mint $ 540	Ex $ 450	VG+ $ 300	Good $ 270	Poor $ 90

MODEL 513 RAGING JUDGE MAGNUM

REDBOOK CODE: RB-TA-H-513RJM

Double lock-up cylinder, extended ejector rod, Taurus Security System, transfer-bar safety, cushion insert grip, blue or stainless finish.

Production: current   Caliber: .454 Casull/.45 LC/.410 GA   Action: DA/SA, revolver
Barrel Length: 3", 6"   Wt.: 60-73 oz.   Sights: fixed red fiber-optic front, fixed rear
Capacity: 7   Grips: Raging Bull cushioned insert grip

D2C:	NIB $ 815	Ex $ 625	VG+ $ 450	Good $ 410	LMP $1,012
C2C:	Mint $ 790	Ex $ 575	VG+ $ 410	Good $ 370	Fair $ 190
Trade-In:	Mint $ 580	Ex $ 475	VG+ $ 320	Good $ 290	Poor $ 90

TAURUS

RAGING BULL 500 (500MSS10) REDBOOK CODE: RB-TA-H-RB500X
Matte stainless steel finish, factory barrel porting, transfer-bar safety, Taurus Security System, large frame.

Production: discontinued Caliber: .500 Magnum Action: DA/SA, revolver					
Barrel Length: 2.25", 4", 6.5", 10" Wt.: 68-72 oz. Sights: fixed front, adjustable rear					
Capacity: 5 Grips: Raging Bull cushioned-insert grip					
D2C:	NIB $ 930	Ex $ 725	VG+ $ 515	Good $ 465	LMP $ 934
C2C:	Mint $ 900	Ex $ 650	VG+ $ 470	Good $ 420	Fair $ 215
Trade-In:	Mint $ 670	Ex $ 525	VG+ $ 370	Good $ 330	Poor $ 120

TRACKER MODEL 425 REDBOOK CODE: RB-TA-H-425TRA
Shadow Gray Titanium or Ultrafine Titanium S/S finish, transfer-bar safety, compact frame, ported barrel.

Production: discontinued Caliber: .41 Magnum Action: DA/SA, revolver					
Barrel Length: 4" 6.5" Wt.: 28 oz. Sights: fixed front, adjustable rear					
Capacity: 5 Grips: rubber					
D2C:	NIB $ 525	Ex $ 400	VG+ $ 290	Good $ 265	LMP $ 796
C2C:	Mint $ 510	Ex $ 375	VG+ $ 270	Good $ 240	Fair $ 125
Trade-In:	Mint $ 380	Ex $ 300	VG+ $ 210	Good $ 185	Poor $ 60

GAUCHO REDBOOK CODE: RB-TA-H-GAUCHO
Sundance Blue, case hardened or stainless steel finish, large frame, transfer-bar safety, includes a "Four Click" cocking action and Taurus Security System.

Production: discontinued Caliber: .45 LC, .44-40, .38/.357 Action: SA, revolver					
Barrel Length: 4.75", 5.5", 7.5" Wt.: 38 oz. Sights: fixed					
Capacity: 6 Grips: checkered					
D2C:	NIB $ 410	Ex $ 325	VG+ $ 230	Good $ 205	LMP $ 520
C2C:	Mint $ 400	Ex $ 300	VG+ $ 210	Good $ 185	Fair $ 95
Trade-In:	Mint $ 300	Ex $ 250	VG+ $ 160	Good $ 145	Poor $ 60

PT111 MILLENNIUM G2 REDBOOK CODE: RB-TA-H-111MG2
Blue finish, accessory rail, loaded chamber indicator, Taurus Security System, fixed white-dot front, 2-dot adjustable rear, sub compact, trigger safety, option of blue/matte stainless slide.

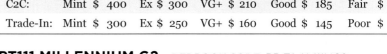

Production: current Caliber: 9mm Action: DA/SA, semi-auto					
Barrel Length: 3.2" Wt.: 22 oz. Sights: 3 white-dot, adjustable					
Capacity: 12 Magazine: detachable box Grips: black polymer					
D2C:	NIB $ 385	Ex $ 300	VG+ $ 215	Good $ 195	LMP $ 434
C2C:	Mint $ 370	Ex $ 275	VG+ $ 200	Good $ 175	Fair $ 90
Trade-In:	Mint $ 280	Ex $ 225	VG+ $ 160	Good $ 135	Poor $ 60

PT111 MILLENNIUM G2
NEW IN BOX

PT140 MILLENNIUM G2 REDBOOK CODE: RB-TA-H-140MG2
Blue finish, loaded chamber indicator, ramped barrel, Taurus Security System, accessory rail, trigger safety, option of blue/matte stainless slide.

Production: current Caliber: .40 S&W Action: DA/SA, semi-auto					
Barrel Length: 3.2" Wt.: 22 oz. Sights: 3 white dot, adjustable					
Capacity: 10 Magazine: detachable box Grips: black					
D2C:	NIB $ 385	Ex $ 300	VG+ $ 215	Good $ 195	LMP $ 434
C2C:	Mint $ 370	Ex $ 275	VG+ $ 200	Good $ 175	Fair $ 90
Trade-In:	Mint $ 280	Ex $ 225	VG+ $ 160	Good $ 135	Poor $ 60

PT140 MILLENNIUM G2
NEW IN BOX

TAURUS

709 SLIM
(709FS)
NEW IN BOX

709 SLIM (709FS) REDBOOK CODE: RB-TA-H-709SLM

Blue or matte stainless steel finish, loaded chamber indicator, "Strike Two" trigger system, Taurus Security System, Crimson Trace Laserguard option.

Production: current Caliber: 9mm Action: striker-fired, semi-auto

Barrel Length: 3" Wt.: 20 oz. Sights: fixed Capacity: 7 Magazine: detachable box

Grips: checkered polymer

D2C:	NIB $	320	Ex $	250	VG+ $	180	Good $	160	LMP $403
C2C:	Mint $	310	Ex $	225	VG+ $	160	Good $	145	Fair $ 75
Trade-In:	Mint $	230	Ex $	200	VG+ $	130	Good $	115	Poor $ 60

709 SLIM
(709SS)
NEW IN BOX

709 SLIM (709SS) REDBOOK CODE: RB-TA-H-709SSX

Matte stainless steel finish, "Strike Two" trigger capability, compact frame, also available in blue finish.

Production: current Caliber: 9mm Action: striker-fired, semi-auto

Barrel Length: 3" Wt.: 20 oz. Sights: fixed Capacity: 7

Magazine: detachable box Grips: checkered polymer

D2C:	NIB $	340	Ex $	275	VG+ $	190	Good $	170	LMP $ 503
C2C:	Mint $	330	Ex $	250	VG+ $	170	Good $	155	Fair $ 80
Trade-In:	Mint $	250	Ex7$	200	VG+ $	140	Good $	120	Poor $ 60

740 SLIM
(740FS)
NEW IN BOX

740 SLIM (740FS) REDBOOK CODE: RB-TA-H-740SLM

Blue or stainless steel finish, loaded chamber indicator, "Strike Two" trigger system, Taurus Security System, Crimson Trace Laserguard option.

Production: current Caliber: .40 S&W Action: striker-fired, semi-auto

Barrel Length: 3.20" Wt.: 20 oz. Sights: adjustable Capacity: 6

Magazine: detachable box Grips: polymer

D2C:	NIB $	320	Ex $	250	VG+ $	180	Good $	160	LMP $403
C2C:	Mint $	310	Ex $	225	VG+ $	160	Good $	145	Fair $ 75
Trade-In:	Mint $	230	Ex $	200	VG+ $	130	Good $	115	Poor $ 60

PT-22
(22BGR)
NEW IN BOX

PT-22 (22BGR) REDBOOK CODE: RB-TA-H-22XXX1

Blue and gold finish, Taurus Security System, gold accents.

Production: current Caliber: .22 LR Action: DAO, semi-auto Barrel Length: 2.75"

Wt.: 12 oz. Sights: fixed Capacity: 8 Magazine: detachable box Grips: rosewood

D2C:	NIB $	269	Ex $	225	VG+ $	150	Good $	135	LMP $ 299
C2C:	Mint $	260	Ex $	200	VG+ $	140	Good $	125	Fair $ 65
Trade-In:	Mint $	200	Ex $	175	VG+ $	110	Good $	95	Poor $ 30

PT-22
(22BR)
NEW IN BOX

PT-22 (22BR) REDBOOK CODE: RB-TA-H-22XXX2

Blue finish, tip-up barrel, Taurus Security System, blowback action.

Production: current Caliber: .22 LR Action: DAO, semi-auto Barrel Length: 2.75"

Wt.: 12 oz. Sights: fixed Capacity: 8 Magazine: detachable box Grips: rosewood

D2C:	NIB $	247	Ex $	200	VG+ $	140	Good $	125	LMP $ 276
C2C:	Mint $	240	Ex $	175	VG+ $	130	Good $	115	Fair $ 60
Trade-In:	Mint $	180	Ex $	150	VG+ $	100	Good $	90	Poor $ 30

PT-22
(22BSYN)
NEW IN BOX

PT-22
(22NGR)
NEW IN BOX

PT-22
(22PLY)
NEW IN BOX

PT-22
(22PLYSS)
NEW IN BOX

PT-1911
(1911B-1)
NEW IN BOX

PT-1911
(1911B-9)
NEW IN BOX

PT-22 (22BSYN)   REDBOOK CODE: RB-TA-H-22XXX3
Blue finish, tip-up barrel, Taurus Security System, blowback action.

Production: current Caliber: .22 LR Action: DAO, semi-auto Barrel Length: 2.75"
Sights: fixed Capacity: 8 Magazine: detachable box Grips: black synthetic

D2C:	NIB $ 239	Ex $ 200	VG+ $ 135	Good $ 120	LMP $ 268
C2C:	Mint $ 230	Ex $ 175	VG+ $ 120	Good $ 110	Fair $ 55
Trade-In:	Mint $ 170	Ex $ 150	VG+ $ 100	Good $ 85	Poor $ 30

PT-22 (22NGR)   REDBOOK CODE: RB-TA-H-22XXX6
Tip-up barrel, Taurus Security System, blowback action, gold accents, stainless finish, special edition.

Production: current Caliber: .22 LR Action: DAO, semi-auto Barrel Length: 2.75"
Wt.: 12 oz. Sights: fixed Capacity: 8 Magazine: detachable box Grips: rosewood

D2C:	NIB $ 269	Ex $ 225	VG+ $ 150	Good $ 135	LMP $ 299
C2C:	Mint $ 260	Ex $ 200	VG+ $ 140	Good $ 125	Fair $ 65
Trade-In:	Mint $ 200	Ex $ 175	VG+ $ 110	Good $ 95	Poor $ 30

PT-22 (22PLY)   REDBOOK CODE: RB-TA-H-22XXX9
Tip-up barrel, Taurus Security System, blowback action, blue steel finish, polymer frame.

Production: current Caliber: .22 LR Action: DAO, semi-auto Barrel Length: 2.33"
Wt.: 12 oz. Sights: fixed Capacity: 8 Magazine: detachable box Grips: polymer

D2C:	NIB $ 247	Ex $ 200	VG+ $ 140	Good $ 125	LMP $ 276
C2C:	Mint $ 240	Ex $ 175	VG+ $ 130	Good $ 115	Fair $ 60
Trade-In:	Mint $ 180	Ex $ 150	VG+ $ 100	Good $ 90	Poor $ 30

PT-22 (22PLYSS)   REDBOOK CODE: RB-TA-H-22XX10
Tip-up barrel, Taurus Security System, blowback action, matte stainless steel finish, polymer frame.

Production: current Caliber: .22 LR Action: DAO, semi-auto Barrel Length: 2.33"
Wt.: 12 oz. Sights: fixed Capacity: 8 Magazine: detachable box Grips: polymer

D2C:	NIB $ 263	Ex $ 200	VG+ $ 145	Good $ 135	LMP $ 293
C2C:	Mint $ 260	Ex $ 200	VG+ $ 140	Good $ 120	Fair $ 65
Trade-In:	Mint $ 190	Ex $ 150	VG+ $ 110	Good $ 95	Poor $ 30

PT-1911 (1911B-1)   REDBOOK CODE: RB-TA-H-1911X1
Blue finish, forged slide and frame, accessory rail, target hammer, Taurus Security System, skeletonized trigger, beveled magwell.

Production: current Caliber: .45 ACP Action: SA, semi-auto Barrel Length: 5"
Wt.: 40 oz. Sights: Novak Capacity: 8 Magazine: detachable box
Grips: checkered black

D2C:	NIB $ 635	Ex $ 500	VG+ $ 350	Good $ 320	LMP $ 834
C2C:	Mint $ 610	Ex $ 450	VG+ $ 320	Good $ 290	Fair $ 150
Trade-In:	Mint $ 460	Ex $ 375	VG+ $ 250	Good $ 225	Poor $ 90

PT-1911 (1911B-9)   REDBOOK CODE: RB-TA-H-1911X2
Skeletonized trigger, blue steel finish, target hammer, flared ejection port, beveled magwell.

Production: current Caliber: 9mm Action: SA, semi-auto Barrel Length: 5"
Wt.: 40 oz. Sights: Novak Capacity: 9 Magazine: detachable box
Grips: checkered black

D2C:	NIB $ 549	Ex $ 425	VG+ $ 305	Good $ 275	LMP $ 718
C2C:	Mint $ 530	Ex $ 400	VG+ $ 280	Good $ 250	Fair $ 130
Trade-In:	Mint $ 390	Ex $ 325	VG+ $ 220	Good $ 195	Poor $ 60

TAURUS

PT-1911
(1911B-BHW)
NEW IN BOX

PT-1911 (1911B-BHW)   REDBOOK CODE: RB-TA-H-1911X3
Blue finish, skeletonized trigger, beveled magwell, target hammer, flared ejection port.

Production: discontinued   Caliber: .45 ACP   Action: SA, semi-auto

Barrel Length: 5"   Wt.: 38 oz.   Sights: Novak   Capacity: 8

Magazine: detachable box   Grips: walnut

D2C:	NIB $ 615	Ex $ 475	VG+ $ 340	Good $ 310	LMP $ 803	
C2C:	Mint $ 600	Ex $ 425	VG+ $ 310	Good $ 280	Fair $ 145	
Trade-In:	Mint $ 440	Ex $ 350	VG+ $ 240	Good $ 220	Poor $ 90	

PT-1911
(1911DT)
NEW IN BOX

PT-1911 (1911DT)   REDBOOK CODE: RB-TA-H-1911X4
Skeletonized trigger, blue steel, duo-tone finish, forged slide and frame, target hammer, beveled magwell, flared ejection port.

Production: current   Caliber: .45 ACP   Action: SA, semi-auto   Barrel Length: 5"

Wt.: 38 oz.   Sights: Novak   Capacity: 8   Magazine: detachable box

Grips: checkered black

D2C:	NIB $ 640	Ex $ 500	VG+ $ 355	Good $ 320	LMP $ 866
C2C:	Mint $ 620	Ex $ 450	VG+ $ 320	Good $ 290	Fair $ 150
Trade-In:	Mint $ 460	Ex $ 375	VG+ $ 250	Good $ 225	Poor $ 90

PT-1911
(1911FS)
NEW IN BOX

PT-1911 (1911FS)   REDBOOK CODE: RB-TA-H-1911X5
Skeletonized trigger, blue steel finish, forged slide and frame, Taurus Security System, target hammer, beveled magwell, flared ejection port.

Production: current   Caliber: .45 ACP   Action: SA, semi-auto   Barrel Length: 5"

Wt.: 38 oz.   Sights: Novak   Capacity: 8   Magazine: detachable box

Grips: checkered black

D2C:	NIB $ 520	Ex $ 400	VG+ $ 290	Good $ 260	LMP $ 684
C2C:	Mint $ 500	Ex $ 375	VG+ $ 260	Good $ 235	Fair $ 120
Trade-In:	Mint $ 370	Ex $ 300	VG+ $ 210	Good $ 185	Poor $ 60

PT-1911
(1911DT)
NEW IN BOX

PT-1911 (1911SS)   REDBOOK CODE: RB-TA-H-1911X6
Stainless steel finish, Taurus Security System, forged slide and frame, skeletonized trigger, beveled magwell, target hammer, flared ejection port. Available with accessory rail on 1911AR model.

Production: current   Caliber: .45 ACP   Action: SA, semi-auto   Barrel Length: 5"

Wt.: 38 oz.   Sights: Novak   Capacity: 8   Magazine: detachable box

Grips: checkered black or pearl

D2C:	NIB $ 660	Ex $ 525	VG+ $ 365	Good $ 330	LMP $ 906
C2C:	Mint $ 640	Ex $ 475	VG+ $ 330	Good $ 300	Fair $ 155
Trade-In:	Mint $ 470	Ex $ 375	VG+ $ 260	Good $ 235	Poor $ 90

PT-1911
(1911SS-9)
NEW IN BOX

PT-1911 (1911SS-9)   REDBOOK CODE: RB-TA-H-1911X8
Stainless steel finish, skeletonized trigger, target hammer, beveled magwell, flared ejection port.

Production: current   Caliber: 9mm   Action: SA, semi-auto   Barrel Length: 5"

Wt.: 40 oz.   Sights: Novak   Capacity: 9   Magazine: detachable box

Grips: checkered black

D2C:	NIB $ 630	Ex $ 500	VG+ $ 350	Good $ 315	LMP $ 825
C2C:	Mint $ 610	Ex $ 450	VG+ $ 320	Good $ 285	Fair $ 145
Trade-In:	Mint $ 450	Ex $ 375	VG+ $ 250	Good $ 225	Poor $ 90

PT-1911
(1911SSBHW)
NEW IN BOX

PT-1911 (1911SSBHW)   REDBOOK CODE: RB-TA-H-1911X9

Stainless finish, target hammer, beveled magwell, skeletonized trigger, flared ejection port.

Production: current Caliber: .45 ACP Action: SA, semi-auto Barrel Length: 5"									
Wt.: 38 oz. Sights: Genuine Novak Capacity: 8									
Magazine: detachable box Grips: walnut									
D2C:	NIB	$ 729	Ex	$575	VG+	$405	Good	$365	LMP	$ 975
C2C:	Mint	$700	Ex	$525	VG+	$370	Good	$330	Fair	$ 170
Trade-In:	Mint	$520	Ex	$425	VG+	$290	Good	$260	Poor	$ 90

PT-25
(25BR)
NEW IN BOX

PT-25 (25BR)   REDBOOK CODE: RB-TA-H-25XXX1

Blue finish, tip-up barrel, Taurus Security System, match grade trigger.

Production: current Caliber: .25 ACP Action: DAO, semi-auto Barrel Length: 2.75"									
Wt.: 12 oz. Sights: fixed Capacity: 9 Magazine: detachable box Grips: rosewood									
D2C:	NIB	$ 259	Ex	$ 200	VG+	$ 145	Good	$ 130	LMP	$ 282
C2C:	Mint	$ 250	Ex	$ 200	VG+	$ 130	Good	$ 120	Fair	$ 60
Trade-In:	Mint	$ 190	Ex	$ 150	VG+	$ 110	Good	$ 95	Poor	$ 30

PT-25
(25NGR)
NEW IN BOX

PT-25 (25NGR)   REDBOOK CODE: RB-TA-H-25XXX4

Nickel and gold finish, gold accents, tip-up barrel, Taurus Security System.

Production: current Caliber: .25 ACP Action: DAO, semi-auto Barrel Length: 2.75"									
Wt.: 12 oz. Sights: fixed Capacity: 9 Magazine: detachable box Grips: rosewood									
D2C:	NIB	$ 270	Ex	$ 225	VG+	$ 150	Good	$ 135	LMP	$ 304
C2C:	Mint	$ 260	Ex	$ 200	VG+	$ 140	Good	$ 125	Fair	$ 65
Trade-In:	Mint	$ 200	Ex	$ 175	VG+	$ 110	Good	$ 95	Poor	$ 30

PT-25
(25PLY)
NEW IN BOX

PT-25 (25PLY)   REDBOOK CODE: RB-TA-H-25XXX6

Tip-up barrel, Taurus Security System, blowback operated, black finish.

Production: current Caliber: .25 ACP Action: DAO, semi-auto Barrel Length: 2.375"									
Wt.: 12 oz. Sights: fixed Capacity: 9 Magazine: detachable box Grips: polymer									
D2C:	NIB	$ 240	Ex	$ 200	VG+	$ 135	Good	$ 120	LMP	$ 276
C2C:	Mint	$ 240	Ex	$ 175	VG+	$ 120	Good	$ 110	Fair	$ 60
Trade-In:	Mint	$ 180	Ex	$ 150	VG+	$ 100	Good	$ 85	Poor	$ 30

PT-25
(25PLYSS)
NEW IN BOX

PT-25 (25PLYSS)   REDBOOK CODE: RB-TA-H-25XXX7

Tip-up barrel, Taurus Security System, blowback action, blue or matte stainless finish.

Production: current Caliber: .25 ACP Action: DAO, semi-auto Barrel Length: 2.33"									
Wt.: 11 oz. Sights: fixed Capacity: 9 Magazine: detachable box Grips: polymer									
D2C:	NIB	$ 260	Ex	$ 200	VG+	$ 145	Good	$ 130	LMP	$ 293
C2C:	Mint	$ 250	Ex	$ 200	VG+	$ 130	Good	$ 120	Fair	$ 60
Trade-In:	Mint	$ 190	Ex	$ 150	VG+	$ 110	Good	$ 95	Poor	$ 30

PT-738 TCP
NEW IN BOX

PT-738 TCP   REDBOOK CODE: RB-TA-H-738TCP

Blue steel, matte stainless or black stainless slide, pink polymer frame option, Taurus Security System, loaded chamber indicator, choice of Crimson Trace Laserguard.

Production: current Caliber: .380 ACP Action: DAO, semi-auto									
Barrel Length: 2.84" Wt.: 10 oz. Sights: fixed Capacity: 6									
Magazine: detachable box Grips: checkered polymer									
D2C:	NIB	$ 220	Ex	$ 175	VG+	$ 125	Good	$ 110	LMP	$ 254
C2C:	Mint	$ 220	Ex	$ 175	VG+	$ 110	Good	$ 100	Fair	$ 55
Trade-In:	Mint	$ 160	Ex	$ 125	VG+	$ 90	Good	$ 80	Poor	$ 30

TAURUS

PT-738 FS
NEW IN BOX

PT-738 FS   REDBOOK CODE: RB-TA-H-738FSX

Blue steel finish, compact polymer frame, loaded chamber indicator, Taurus Security System.

Production: current Caliber: .380 ACP Action: DAO, semi-auto Barrel Length: 2.84"

Wt.: 10 oz. Sights: fixed Capacity: 6 Magazine: detachable box Grips: checkered polymer

D2C:	NIB $	220	Ex $	175	VG+ $	125	Good $	110	LMP $ 254
C2C:	Mint $	220	Ex $	175	VG+ $	110	Good $	100	Fair $ 55
Trade-In:	Mint $	160	Ex $	125	VG+ $	90	Good $	80	Poor $ 30

PT-38S   REDBOOK CODE: RB-TA-H-38SXXX

Stainless steel finish, loaded chamber indicator, Taurus Security System, forged steel slide, 3-position ambidextrous safety.

Production: discontinued Caliber: .38 Super Action: DA/SA, semi-auto

Barrel Length: 4.25" Wt.: 30 oz. Sights: fixed 1-dot front, fixed 2-dot rear

Capacity: 10 Magazine: detachable box Grips: checkered rubber

D2C:	NIB $	530	Ex $	425	VG+ $	295	Good $	265	LMP $ 674
C2C:	Mint $	510	Ex $	375	VG+ $	270	Good $	240	Fair $ 125
Trade-In:	Mint $	380	Ex $	300	VG+ $	210	Good $	190	Poor $ 60

PT-58   REDBOOK CODE: RB-TA-H-58XXXX

Blue finish, smooth trigger, alloy frame, hammer-forged steel slide, 3-position ambidextrous safety, also available in stainless finish for slight premium.

Production: discontinued Caliber: .380 ACP Action: DAO, semi-auto

Barrel Length: 3.25" Wt.: 18.7 oz. Sights: fixed Capacity: 19

Magazine: detachable box Grips: rubber

D2C:	NIB $	375	Ex $	300	VG+ $	210	Good $	190	LMP $ 650
C2C:	Mint $	360	Ex $	275	VG+ $	190	Good $	170	Fair $ 90
Trade-In:	Mint $	270	Ex $	225	VG+ $	150	Good $	135	Poor $ 60

PT-92
NEW IN BOX

PT-92   REDBOOK CODE: RB-TA-H-92BGR

Loaded chamber indicator, Taurus Security System, 3-position ambidextrous safety/decocker, available in blue, stainless or nickel finishes, slight premium for stainless model, some models with Picatinny rail, some models with gold accents and pearl or wood grips.

Production: discontinued Caliber: 9mm Action: DA/SA, semi-auto

Barrel Length: 5" Wt.: 34 oz. Sights: fixed 3-dot Capacity: 10, 17

Magazine: detachable box Grips: checkered rubber

D2C:	NIB $	480	Ex $	375	VG+ $	265	Good $	240	
C2C:	Mint $	470	Ex $	350	VG+ $	240	Good $	220	Fair $ 115
Trade-In:	Mint $	350	Ex $	275	VG+ $	190	Good $	170	Poor $ 60

PT-99
NEW IN BOX

PT-99   REDBOOK CODE: RB-TA-H-99XXXX

Stainless steel or blue finish, hammer decocker, loaded chamber indicator, ambidextrous safety.

Production: discontinued Caliber: 9mm Action: DA/SA, semi-auto

Barrel Length: 5" Wt.: 34 oz. Sights: fixed dot front, adjustable rear

Capacity: 17 Magazine: detachable box Grips: checkered rubber

D2C:	NIB $	499	Ex $	400	VG+ $	275	Good $	250	LMP $ 605
C2C:	Mint $	480	Ex $	350	VG+ $	250	Good $	225	Fair $ 115
Trade-In:	Mint $	360	Ex $	300	VG+ $	200	Good $	175	Poor $ 60

PT-100 REDBOOK CODE: RB-TA-H-100XXX

Stainless steel finish, loaded chamber indicator, decocker lever, 3-position safety decocker, accessory rail, Taurus Security System.

Production: discontinued Caliber: .40 S&W Action: DA/SA, semi-auto

Barrel Length: 5" Wt.: 34 oz. Sights: fixed 2-dot rear Capacity: 11

Magazine: detachable box Grips: checkered rubber

D2C:	NIB	$ 502	Ex $ 400	VG+ $ 280	Good $ 255	LMP $ 648
C2C:	Mint $ 490	Ex $ 350	VG+ $ 260	Good $ 230	Fair $ 120
Trade-In:	Mint $ 360	Ex $ 300	VG+ $ 200	Good $ 180	Poor $ 60

PT-101
EXCELLENT

PT-101 REDBOOK CODE: RB-TA-H-101XXX

Stainless steel finish, ambidextrous 3-position safety, forged-alloy frame.

Production: discontinued Caliber: .40 S&W Action: DA/SA, semi-auto

Barrel Length: 5" Wt.: 34 oz. Sights: adjustable Capacity: 11

Magazine: detachable box Grips: checkered rubber

D2C:	NIB $ 475	Ex $ 375	VG+ $ 265	Good $ 240	LMP $ 605
C2C:	Mint $ 460	Ex $ 350	VG+ $ 240	Good $ 215	Fair $ 110
Trade-In:	Mint $ 340	Ex $ 275	VG+ $ 190	Good $ 170	Poor $ 60

PT-111
MILLENNIUM PRO
NEW IN BOX

PT-111 MILLENNIUM PRO REDBOOK CODE: RB-TA-H-111XXX

Blue finish, compact frame, smooth trigger, loaded chamber indicator.

Production: discontinued Caliber: 9mm Action: DA/SA, semi-auto

Barrel Length: 3.25" Wt.: 20 oz. Sights: Heinie front, Straight Eight rear

Capacity: 12 Magazine: detachable box Grips: checkered polymer

D2C:	NIB $ 360	Ex $ 275	VG+ $ 200	Good $ 180	LMP $ 467
C2C:	Mint $ 350	Ex $ 250	VG+ $ 180	Good $ 165	Fair $ 85
Trade-In:	Mint $ 260	Ex $ 225	VG+ $ 150	Good $ 130	Poor $ 60

PT-132
MILLENNIUM PRO
NEW IN BOX
Courtesy of Bud's Gun Shop

PT-132 MILLENNIUM PRO REDBOOK CODE: RB-TA-H-132XXX

Blue finish, loaded chamber indicator, also available in stainless steel finish for an additional price.

Production: discontinued Caliber: .32 ACP Action: DAO, semi-auto

Barrel Length: 3.25" Wt.: 20 oz. Sights: Heinie front, Straight Eight rear

Capacity: 10 Magazine: detachable box Grips: checkered polymer

D2C:	NIB $ 337	Ex $ 275	VG+ $ 190	Good $ 170	LMP $ 441
C2C:	Mint $ 330	Ex $ 250	VG+ $ 170	Good $ 155	Fair $ 80
Trade-In:	Mint $ 240	Ex $ 200	VG+ $ 140	Good $ 120	Poor $ 60

PT-138
MILLENNIUM PRO
NEW IN BOX

PT-138 MILLENNIUM PRO REDBOOK CODE: RB-TA-H-138PXX

Blue finish, loaded chamber indicator, compact frame, ambidextrous finger index memory pads.

Production: discontinued Caliber: .380 ACP Action: DA/SA, semi-auto

Barrel Length: 3.25" Wt.: 20 oz. Sights: Heinie front, Straight Eight rear

Capacity: 12 Magazine: detachable box Grips: checkered polymer

D2C:	NIB $ 350	Ex $ 275	VG+ $ 195	Good $ 175	LMP $ 441
C2C:	Mint $ 340	Ex $ 250	VG+ $ 180	Good $ 160	Fair $ 85
Trade-In:	Mint $ 250	Ex $ 200	VG+ $ 140	Good $ 125	Poor $ 60

**PT-140
MILLENNIUM PRO**
NEW IN BOX

PT-140 MILLENNIUM PRO   REDBOOK CODE: RB-TA-H-140PXX

Blue steel finish, compact frame, Taurus Security System, loaded chamber indicator, smooth trigger, also available in matte stainless steel for an additional price.

Production: discontinued   Caliber: .40 S&W   Action: DA/SA, semi-auto

Barrel Length: 3.25"   Wt.: 20 oz.   Sights: Heinie front, Straight Eight rear

Capacity: 10   Magazine: detachable box   Grips: checkered polymer

D2C:	NIB $	370	Ex $	300	VG+ $	205	Good $	185	LMP $ 483
C2C:	Mint $	360	Ex $	275	VG+ $	190	Good $	170	Fair $ 90
Trade-In:	Mint $	270	Ex $	225	VG+ $	150	Good $	130	Poor $ 60

**PT-145
MILLENNIUM PRO**
NEW IN BOX

PT-145 MILLENNIUM PRO   REDBOOK CODE: RB-TA-H-145PXX

Blue steel or stainless finish, compact frame, Taurus Security System, loaded chamber indicator, accessory rail, also available in matte stainless steel for an additional price.

Production: discontinued   Caliber: .45 ACP   Action: DA/SA, semi-auto

Barrel Length: 3.25"   Wt.: 22 oz.   Sights: Heinie front, Straight Eight rear

Capacity: 10   Magazine: detachable box   Grips: checkered polymer

D2C:	NIB $	370	Ex $	300	VG+ $	205	Good $	185	LMP $ 483
C2C:	Mint $	360	Ex $	275	VG+ $	190	Good $	170	Fair $ 90
Trade-In:	Mint $	270	Ex $	225	VG+ $	150	Good $	130	Poor $ 60

**PT-745 MILLENNIUM
PRO COMPACT**
NEW IN BOX

PT-745 MILLENNIUM PRO COMPACT   REDBOOK CODE: RB-TA-H-745PXX

Blue finish, compact frame, smooth trigger, loaded chamber indicator, Taurus Security System.

Production: discontinued   Caliber: .45 ACP   Action: DA/SA, semi-auto

Barrel Length: 3.25"   Wt.: 21 oz.   Sights: fixed 3-dot   Capacity: 6

Magazine: detachable box   Grips: checkered polymer

D2C:	NIB $	370	Ex $	300	VG+ $	205	Good $	185	LMP $ 483
C2C:	Mint $	360	Ex $	275	VG+ $	190	Good $	170	Fair $ 90
Trade-In:	Mint $	270	Ex $	225	VG+ $	150	Good $	130	Poor $ 60

**PT-809C
(809BC)**
NEW IN BOX

PT-809C (809BC)   REDBOOK CODE: RB-TA-H-809BCX

Blue finish, "Strike Two" trigger capability, external hammer, loaded chamber indicator, Taurus Security System, ambidextrous thumb rests and finger index memory pad.

Production: current   Caliber: 9mm   Action: DA/SA, semi-auto

Barrel Length: 3.625"   Sights: Novak   Capacity: 17

Magazine: detachable box   Grips: polymer

D2C:	NIB $	375	Ex $	300	VG+ $	210	Good $	190	LMP $ 486
C2C:	Mint $	360	Ex $	275	VG+ $	190	Good $	170	Fair $ 90
Trade-In:	Mint $	270	Ex $	225	VG+ $	150	Good $	135	Poor $ 60

**PT-809
(809B)**
NEW IN BOX

PT-809 (809B)   REDBOOK CODE: RB-TA-H-809BXX

Black Tennifer finish, "Strike Two" trigger capability, external hammer, ambidextrous 3-position safety and decocker, ambidextrous magazine release, Picatinny accessory rail, Taurus Security System, ambidextrous thumb rests and finger index memory pad.

Production: current   Caliber: 9mm   Action: DA/SA, semi-auto   Barrel Length: 4"

Wt.: 30 oz.   Sights: Novak   Capacity: 17   Magazine: detachable box   Grips: checkered polymer

D2C:	NIB $	375	Ex $	300	VG+ $	210	Good $	190	LMP $ 486
C2C:	Mint $	360	Ex $	275	VG+ $	190	Good $	170	Fair $ 90
Trade-In:	Mint $	270	Ex $	225	VG+ $	150	Good $	135	Poor $ 60

PT-840C
(840BC)
NEW IN BOX

PT-840C (840BC)   REDBOOK CODE: RB-TA-H-840BCX

Blue finish, "Strike Two" trigger capability, loaded chamber indicator, Taurus Security System, ambidextrous thumb rests and finger index memory pad.

Production: current   Caliber: .40 S&W   Action: DA/SA, semi-auto

Barrel Length: 3.625"   Sights: Novak   Capacity: 15   Magazine: detachable box

Grips: polymer

D2C:	NIB $	375	Ex $	300	VG+ $	210	Good $	190	LMP $	486
C2C:	Mint $	360	Ex $	275	VG+ $	190	Good $	170	Fair $	90
Trade-In:	Mint $	270	Ex $	225	VG+ $	150	Good $	135	Poor $	60

PT-840
(840B)
NEW IN BOX

PT-840 (840B)   REDBOOK CODE: RB-TA-H-840BXX

Black Tennifer finish, "Strike Two" trigger capability, loaded chamber indicator, Taurus Security System, external hammer, decocker, accessory rail, Taurus Security System, ambidextrous thumb rests and finger index memory pad.

Production: current   Caliber: .40 S&W   Action: DA/SA, semi-auto

Barrel Length: 4"   Wt.: 30 oz.   Sights: Novak   Capacity: 15

Magazine: detachable box   Grips: checkered polymer with interchangeable backstraps

D2C:	NIB $	375	Ex $	300	VG+ $	210	Good $	190	LMP $	486
C2C:	Mint $	360	Ex $	275	VG+ $	190	Good $	170	Fair $	90
Trade-In:	Mint $	270	Ex $	225	VG+ $	150	Good $	135	Poor $	60

PT-845
NEW IN BOX

PT-845   REDBOOK CODE: RB-TA-H-845XXX

Stainless steel or black finish, medium frame, smooth trigger, ambidextrous 3-position safety, "Strike Two" capability.

Production: discontinued   Caliber: .45 ACP   Action: DA/SA, semi-auto

Barrel Length: 4"   Wt.: 28 oz.   Sights: fixed 3-dot   Capacity: 12

Magazine: detachable box   Grips: checkered polymer

D2C:	NIB $	390	Ex $	300	VG+ $	215	Good $	195	LMP $	498
C2C:	Mint $	380	Ex $	275	VG+ $	200	Good $	180	Fair $	90
Trade-In:	Mint $	280	Ex $	225	VG+ $	160	Good $	140	Poor $	60

PT-909
NEW IN BOX

PT-909   REDBOOK CODE: RB-TA-H-909XXX

Blue steel finish, medium frame, smooth trigger, alloy frame, decocker lever, 3-position ambidextrous safety.

Production: discontinued   Caliber: 9mm   Action: DA/SA, semi-auto

Barrel Length: 4"   Wt.: 28 oz.   Sights: fixed 3-dot   Capacity: 17

Magazine: detachable box   Grips: hard rubber

D2C:	NIB $	470	Ex $	375	VG+ $	260	Good $	235	LMP $	615
C2C:	Mint $	460	Ex $	325	VG+ $	240	Good $	215	Fair $	110
Trade-In:	Mint $	340	Ex $	275	VG+ $	190	Good $	165	Poor $	60

PT-911
NEW IN BOX

PT-911   REDBOOK CODE: RB-TA-H-911XXX

Blue or stainless finish, medium frame, ambidextrous 3-position safety.

Production: discontinued   Caliber: 9mm   Action: DA/SA, semi-auto

Barrel Length: 4"   Wt.: 28 oz.   Sights: fixed 3-dot   Capacity: 15

Magazine: detachable box   Grips: rubber

D2C:	NIB $	475	Ex $	375	VG+ $	265	Good $	240	LMP $	615
C2C:	Mint $	460	Ex $	350	VG+ $	240	Good $	215	Fair $	110
Trade-In:	Mint $	340	Ex $	275	VG+ $	190	Good $	170	Poor $	60

PT-917C "COMPACT PLUS"   REDBOOK CODE: RB-TA-H-917CXX

Blue or stainless steel finish, medium frame, 3-position frame-mounted ambidextrous safety.

Production: discontinued   Caliber: 9mm   Action: DA/SA, semi-auto

Barrel Length: 4"   Wt.: 32 oz.   Sights: fixed 3-dot   Capacity: 20

Magazine: detachable box   Grips: rubber

D2C:	NIB $ 450	Ex $ 350	VG+ $ 250	Good $ 225	LMP $ 589
C2C:	Mint $ 440	Ex $ 325	VG+ $ 230	Good $ 205	Fair $ 105
Trade-In:	Mint $ 320	Ex $ 275	VG+ $ 180	Good $ 160	Poor $ 60

PT-938   REDBOOK CODE: RB-TA-H-938XXX

Blue finish, Taurus Security System, loaded chamber indicator, firing pin block.

Production: discontinued   Caliber: .380 ACP   Action: DA/SA, semi-auto

Barrel Length: 3"   Sights: fixed 3-dot

Capacity: 15   Magazine: detachable box   Grips: rubber

D2C:	NIB $ 375	Ex $ 300	VG+ $ 210	Good $ 190	
C2C:	Mint $ 360	Ex $ 275	VG+ $ 190	Good $ 170	Fair $ 90
Trade-In:	Mint $ 270	Ex $ 225	VG+ $ 150	Good $ 135	Poor $ 60

PT-938
NEW IN BOX

PT-940   REDBOOK CODE: RB-TA-H-940XXX

Blue finish or stainless at additional price, loaded chamber indicator, Taurus Security System.

Production: discontinued   Caliber: .40 S&W   Action: DA/SA, semi-auto

Barrel Length: 3.625"   Wt.: 28 oz.   Sights: fixed 3-dot   Capacity: 10

Magazine: detachable box   Grips: checkered rubber, pearl, or rosewood

D2C:	NIB $ 485	Ex $ 375	VG+ $ 270	Good $ 245	LMP $ 615
C2C:	Mint $ 470	Ex $ 350	VG+ $ 250	Good $ 220	Fair $ 115
Trade-In:	Mint $ 350	Ex $ 275	VG+ $ 190	Good $ 170	Poor $ 60

PT-940
NEW IN BOX

PT-945   REDBOOK CODE: RB-TA-H-945XXX

Blue finish, loaded chamber indicator, Taurus Security System, manual safety.

Production: discontinued   Caliber: .45 ACP   Action: DA/SA, semi-auto

Barrel Length: 4.25"   Wt.: 30 oz.   Sights: fixed 3-dot   Capacity: 8

Magazine: detachable box   Grips: checkered rubber

D2C:	NIB $ 540	Ex $ 425	VG+ $ 300	Good $ 270	LMP $ 658
C2C:	Mint $ 520	Ex $ 375	VG+ $ 270	Good $ 245	Fair $ 125
Trade-In:	Mint $ 390	Ex $ 325	VG+ $ 220	Good $ 190	Poor $ 60

PT-957   REDBOOK CODE: RB-TA-H-957XXX

Stainless steel finish, smooth trigger, rubber grips, chamber loader indicator, factory ported barrel.

Production: discontinued   Caliber: .357 SIG   Action: DA/SA, semi-auto

Barrel Length: 3.625"   Wt.: 28 oz.   Sights: fixed 3-dot   Capacity: 10

Magazine: detachable box   Grips: rubber

D2C:	NIB $ 395	Ex $ 325	VG+ $ 220	Good $ 200	LMP $ 523
C2C:	Mint $ 380	Ex $ 275	VG+ $ 200	Good $ 180	Fair $ 95
Trade-In:	Mint $ 290	Ex $ 225	VG+ $ 160	Good $ 140	Poor $ 60

PT2011 DT INTEGRAL
NEW IN BOX

PT2011 DT HYBRID 9MM
NEW IN BOX

**PT2011 DT HYBRID
.40 CAL**
(2011H40SS)
NEW IN BOX

PT 24/7 PRO
NEW IN BOX

PT 24/7 G2
(24/7-G29B-17)
NEW IN BOX

PT2011 DT INTEGRAL REDBOOK CODE: RB-TA-H-2011DT

Loaded chamber indicator, removable backstraps, matte stainless steel finish, Taurus Security System, trigger safety, manual safety.

Production: discontinued Caliber: 9mm, .380 ACP, .40 S&W Action: DA/SA, semi-auto
Barrel Length: 3.2" Wt.: 21 oz. Sights: adjustable rear Capacity: 13
Magazine: detachable box Grips: polymer

D2C:	NIB	$ 455	Ex $ 350	VG+ $ 255	Good $ 230	LMP $ 588
C2C:	Mint	$ 440	Ex $ 325	VG+ $ 230	Good $ 205	Fair $ 105
Trade-In:	Mint	$ 330	Ex $ 275	VG+ $ 180	Good $ 160	Poor $ 60

PT2011 DT HYBRID 9MM REDBOOK CODE: RB-TA-H-DTHY9M

Blue or matte stainless finish, manual safety, loaded chamber indicator, Taurus Security System, "Strike Two" trigger, ambidextrous thumb rests, and finger index memory pads.

Production: current Caliber: 9mm Action: DA/SA, semi-auto Barrel Length: 3.2"
Wt.: 24 oz. Sights: adjustable rear Capacity: 13 Magazine: detachable box
Grips: polymer

D2C:	NIB	$ 480	Ex $ 375	VG+ $ 265	Good $ 240	LMP $ 605
C2C:	Mint	$ 470	Ex $ 350	VG+ $ 240	Good $ 220	Fair $ 115
Trade-In:	Mint	$ 350	Ex $ 275	VG+ $ 190	Good $ 170	Poor $ 60

PT2011 DT HYBRID .40 CAL (2011H40SS)
REDBOOK CODE: RB-TA-H-DTHY40

Blue or matte stainless finish, manual safety, Taurus Security System, "Strike Two" trigger, ambidextrous thumb rests and finger index memory pads.

Production: current Caliber: .40 S&W Action: DA/SA, semi-auto
Barrel Length: 3.2" Wt.: 24 oz. Sights: adjustable rear Capacity: 11
Magazine: detachable box Grips: polymer

D2C:	NIB	$ 480	Ex $ 375	VG+ $ 265	Good $ 240	LMP $ 605
C2C:	Mint	$ 470	Ex $ 350	VG+ $ 240	Good $ 220	Fair $ 115
Trade-In:	Mint	$ 350	Ex $ 275	VG+ $ 190	Good $ 170	Poor $ 60

PT 24/7 PRO REDBOOK CODE: RB-TA-H-24PROX

Stainless steel finish, also available in blue or duo-tone finish, loaded chamber indicator, accessory rail, Taurus Security System, either compact, medium or long slide sizes.

Production: discontinued Caliber: 9mm, .40 S&W, .45 ACP Action: DA/SA, semi-auto
Barrel Length: 3,25", 4", 5.5" Wt.: 28.2 oz. Sights: Heinie front, Straight-8 rear
Capacity: 10 Magazine: detachable box Grips: rubber grip overlay

D2C:	NIB	$ 420	Ex $ 325	VG+ $ 235	Good $ 210	LMP $ 492
C2C:	Mint	$ 410	Ex $ 300	VG+ $ 210	Good $ 190	Fair $ 100
Trade-In:	Mint	$ 300	Ex $ 250	VG+ $ 170	Good $ 150	Poor $ 60

PT 24/7 G2 (24/7-G29B-17) REDBOOK CODE: RB-TA-H-24G2X1

"Strike Two" trigger capability, contoured thumb rests, trigger safety, vertical loaded chamber indicator, Picatinny accessory rail, ambidextrous magazine release, ambidextrous manual decocker/safety, firing pin cocking indicator, firing pin block, ambidextrous slide stop, Taurus Security System, blue finish.

Production: current Caliber: 9mm Action: DA/SA, semi-auto Barrel Length: 4.2"
Wt.: 28 oz. Sights: fixed front, adjustable rear Capacity: 17 Magazine: detachable box
Grips: checkered polymer

D2C:	NIB	$ 459	Ex $ 350	VG+ $ 255	Good $ 230	LMP $ 528
C2C:	Mint	$ 450	Ex $ 325	VG+ $ 230	Good $ 210	Fair $ 110
Trade-In:	Mint	$ 330	Ex $ 275	VG+ $ 180	Good $ 165	Poor $ 60

TAURUS

PT 24/7 G2
(24/7-G29SS-17)
NEW IN BOX

PT 24/7 G2 (24/7-G29SS-17)   REDBOOK CODE: RB-TA-H-24G2X2

"Strike Two" trigger capability, contoured thumb rests, trigger safety, vertical loaded chamber indicator, Picatinny accessory rail, ambidextrous magazine release, ambidextrous manual decocker/safety, firing pin cocking indicator, firing pin block, ambidextrous slide stop, Taurus Security System, stainless finish.

Production: current Caliber: 9mm Action: DA/SA, semi-auto Barrel Length: 4.2"
Wt.: 28 oz. Sights: fixed front, adjustable rear Capacity: 17 Magazine: detachable box
Grips: checkered polymer

D2C:	NIB $ 475	Ex $ 375	VG+ $ 265	Good $ 240	LMP $ 543
C2C:	Mint $ 460	Ex $ 350	VG+ $ 240	Good $ 215	Fair $ 110
Trade-In:	Mint $ 340	Ex $ 275	VG+ $ 190	Good $ 170	Poor $ 60

PT 24/7 G2
(24/7-G240B-15)
NEW IN BOX

PT 24/7 G2 (24/7-G240B-15)   REDBOOK CODE: RB-TA-H-24G2X3

"Strike Two" trigger capability, contoured thumb rests, trigger safety, vertical loaded chamber indicator, Picatinny accessory rail, ambidextrous magazine release, ambidextrous manual decocker/safety, firing pin cocking indicator, firing pin block, ambidextrous slide stop, Taurus Security System, blue finish.

Production: current Caliber: .40 S&W Action: DA/SA, semi-auto
Barrel Length: 4.2" Wt.: 28 oz. Sights: fixed front, adjustable rear
Capacity: 15 Magazine: detachable box Grips: checkered polymer

D2C:	NIB $ 459	Ex $ 350	VG+ $ 255	Good $ 230	LMP $ 528
C2C:	Mint $ 450	Ex $ 325	VG+ $ 230	Good $ 210	Fair $ 110
Trade-In:	Mint $ 330	Ex $ 275	VG+ $ 180	Good $ 165	Poor $ 60

PT 24/7 G2
(24/7-G240SS-15)
NEW IN BOX

PT 24/7 G2 (24/7-G240SS-15)   REDBOOK CODE: RB-TA-H-24G2X4

"Strike Two" trigger capability, contoured thumb rests, trigger safety, vertical loaded chamber indicator, Picatinny accessory rail, ambidextrous magazine release, ambidextrous manual decocker/safety, firing pin cocking indicator, firing pin block, ambidextrous slide stop, Taurus Security System, stainless finish.

Production: current Caliber: .40 S&W Action: DA/SA, semi-auto
Barrel Length: 4.2" Wt.: 28 oz. Sights: fixed front, adjustable rear
Capacity: 15 Magazine: detachable box Grips: checkered polymer

D2C:	NIB $ 475	Ex $ 375	VG+ $ 265	Good $ 240	LMP $ 543
C2C:	Mint $ 460	Ex $ 350	VG+ $ 240	Good $ 215	Fair $ 110
Trade-In:	Mint $ 340	Ex $ 275	VG+ $ 190	Good $ 170	Poor $ 60

PT 24/7 G2
(24/7-G245B-12)
NEW IN BOX

PT 24/7 G2 (24/7-G245B-12)   REDBOOK CODE: RB-TA-H-24G2X5

"Strike Two" trigger capability, contoured thumb rests, trigger safety, vertical loaded chamber indicator, Picatinny accessory rail, ambidextrous magazine release, ambidextrous manual decocker/safety, firing pin cocking indicator, firing pin block, ambidextrous slide stop, Taurus Security System, blue finish.

Production: current Caliber: .45 ACP Action: DA/SA, semi-auto
Barrel Length: 4.2" Wt.: 28 oz. Sights: fixed front, adjustable rear
Capacity: 12 Magazine: detachable box Grips: checkered polymer

D2C:	NIB $ 459	Ex $ 350	VG+ $ 255	Good $ 230	LMP $ 528
C2C:	Mint $ 450	Ex $ 325	VG+ $ 230	Good $ 210	Fair $ 110
Trade-In:	Mint $ 330	Ex $ 275	VG+ $ 180	Good $ 165	Poor $ 60

PT 24/7 G2
(24/7-G245SS-12)
NEW IN BOX

PT 24/7 G2 (24/7-G245SS-12) REDBOOK CODE: RB-TA-H-24G2X6

"Strike Two" trigger capability, contoured thumb rests, trigger safety, vertical loaded chamber indicator, Picatinny accessory rail, ambidextrous magazine release, ambidextrous manual decocker/safety, firing pin cocking indicator, firing pin block, ambidextrous slide stop, Taurus Security System, stainless finish.

Production: current Caliber: .45 ACP Action: DA/SA, semi-auto									
Barrel Length: 4.2" Wt.: 28 oz. Sights: fixed front, adjustable rear									
Capacity: 12 Magazine: detachable box Grips: checkered polymer									
D2C:	NIB $	475	Ex $375	VG+ $	265	Good $	240	LMP $	543
C2C:	Mint $	460	Ex $350	VG+ $	240	Good $	215	Fair $	110
Trade-In:	Mint $	340	Ex $275	VG+ $	190	Good $	170	Poor $	60

PT 24/7 G2
(24/7-G29BC-17)
NEW IN BOX

PT 24/7 G2 (24/7-G29BC-17) REDBOOK CODE: RB-TA-H-24G2X7

"Strike Two" trigger capability, shortened grip frame, contoured thumb rests, trigger safety, vertical loaded chamber indicator, Picatinny accessory rail, ambidextrous magazine release, ambidextrous manual decocker/safety, firing pin cocking indicator, firing pin block, ambidextrous slide stop, Taurus Security System, blue finish.

Production: current Caliber: 9mm Action: DA/SA, semi-auto									
Barrel Length: 3.5" Wt.: 27 oz. Sights: fixed front, adjustable rear									
Capacity: 17 Magazine: detachable box Grips: checkered polymer									
D2C:	NIB $	459	Ex $ 350	VG+ $	255	Good $	230	LMP $	528
C2C:	Mint $	450	Ex $ 325	VG+ $	230	Good $	210	Fair $	110
Trade-In:	Mint $	330	Ex $ 275	VG+ $	180	Good $	165	Poor $	60

PT 24/7 G2
(24/7-G29SSC-17)
NEW IN BOX

PT 24/7 G2 C (24/7-G29SSC-17) REDBOOK CODE: RB-TA-H-24G2X8

Compact model, "Strike Two" trigger capability, shortened grip frame, contoured thumb rests, trigger safety, vertical loaded chamber indicator, Picatinny accessory rail, ambidextrous magazine release, ambidextrous manual decocker/safety, firing pin cocking indicator, firing pin block, ambidextrous slide stop, Taurus Security System, stainless finish.

Production: current Caliber: 9mm Action: DA/SA, semi-auto									
Barrel Length: 3.5" Wt.: 27 oz. Sights: fixed front, adjustable rear									
Capacity: 17 Magazine: detachable box Grips: checkered polymer									
D2C:	NIB $	475	Ex $ 375	VG+ $	265	Good $	240	LMP $	543
C2C:	Mint $	460	Ex $ 350	VG+ $	240	Good $	215	Fair $	110
Trade-In:	Mint $	340	Ex $ 275	VG+ $	190	Good $	170	Poor $	60

PT 24/7 G2
(24/7-G240BC-15)
NEW IN BOX

PT 24/7 G2 (24/7-G240BC-15) REDBOOK CODE: RB-TA-H-24G2X9

Compact model, "Strike Two" trigger capability, shortened grip frame, contoured thumb rests, trigger safety, vertical loaded chamber indicator, Picatinny accessory rail, ambidextrous magazine release, ambidextrous manual decocker/safety, firing pin cocking indicator, firing pin block, ambidextrous slide stop, Taurus Security System, blue finish.

Production: current Caliber: .40 S&W Action: DA/SA, semi-auto									
Barrel Length: 3.5" Wt.: 27 oz. Sights: fixed front, adjustable rear									
Capacity: 15 Magazine: detachable box Grips: checkered polymer									
D2C:	NIB $	459	Ex $ 350	VG+ $	255	Good $	230	LMP $	528
C2C:	Mint $	450	Ex $ 325	VG+ $	230	Good $	210	Fair $	110
Trade-In:	Mint $	330	Ex $ 275	VG+ $	180	Good $	165	Poor $	60

PT 24/7 G2
(24/7-G240SSC-15)
NEW IN BOX

PT 24/7 G2 (24/7-G240SSC-15) REDBOOK CODE: RB-TA-H-24G210

"Strike Two" trigger capability, shortened grip frame, contoured thumb rests, trigger safety, vertical loaded chamber indicator, Picatinny accessory rail, ambidextrous magazine release, ambidextrous manual decocker/safety, firing pin cocking indicator, firing pin block, ambidextrous slide stop, Taurus Security System, stainless finish.

Production: current Caliber: .40 S&W Action: DA/SA, semi-auto

Barrel Length: 3.5" Wt.: 27 oz. Sights: fixed front, adjustable rear

Capacity: 15 Magazine: detachable box Grips: checkered polymer

D2C:	NIB $	475	Ex $	375	VG+ $	265	Good $ 240	LMP $	543
C2C:	Mint $	460	Ex $	350	VG+ $	240	Good $ 215	Fair $	110
Trade-In:	Mint $	340	Ex $	275	VG+ $	190	Good $ 170	Poor $	60

PT 24/7 G2
(24/7-G245BC-12)
NEW IN BOX

PT 24/7 G2 (24/7-G245BC-12) REDBOOK CODE: RB-TA-H-24G211

"Strike Two" trigger capability, shortened grip frame, contoured thumb rests, trigger safety, vertical loaded chamber indicator, Picatinny accessory rail, ambidextrous magazine release, ambidextrous manual decocker/safety, firing pin cocking indicator, firing pin block, ambidextrous slide stop, Taurus Security System, blue finish.

Production: current Caliber: .45 ACP Action: DA/SA, semi-auto

Barrel Length: 3.5" Wt.: 27 oz. Sights: fixed front, adjustable rear

Capacity: 12 Magazine: detachable box Grips: checkered polymer

D2C:	NIB $	459	Ex $	350	VG+ $	255	Good $ 230	LMP $	528
C2C:	Mint $	450	Ex $	325	VG+ $	230	Good $ 210	Fair $	110
Trade-In:	Mint $	330	Ex $	275	VG+ $	180	Good $ 165	Poor $	60

PT 24/7 G2
(24/7-G245SSC-12)
NEW IN BOX

PT 24/7 G2 (24/7-G245SSC-12) REDBOOK CODE: RB-TA-H-24G212

Compact model, "Strike Two" trigger capability, loaded chamber indicator, accessory rail, slide-mounted safety, firing pin cocking indicator, Taurus Security System, stainless finish.

Production: current Caliber: .45 ACP Action: DA/SA, semi-auto

Barrel Length: 3.5" Wt.: 27 oz. Sights: fixed front, adjustable rear

Capacity: 12 Magazine: detachable box Grips: checkered polymer

D2C:	NIB $	475	Ex $	375	VG+ $	265	Good $ 240	LMP $	543
C2C:	Mint $	460	Ex $	350	VG+ $	240	Good $ 215	Fair $	110
Trade-In:	Mint $	340	Ex $	275	VG+ $	190	Good $ 170	Poor $	60

UA Arms

UA Arms—previously Uselton Arms—started building custom M1911 handguns in 1999 for which they are likely most recognized. In 2006, the company introduced the Warrior Mountain Lite bolt-action rifle, which can fire, among others, 7.82/.308 Lazzeroni Warbird rounds—a unique, long-range and high-velocity caliber. In 2011, the company introduced its Integrated Aluminum (IA) M1911 pistols, which bonds stainless steel and aluminum to form their lightweight slide and frame. The company is based in Franklin, Tennessee.

M1911 COMPACT CLASSIC COMPANION
NEW IN BOX

M1911 COMPACT CLASSIC COMPANION

REDBOOK CODE: RB-UA-H-CCCXXX

Brushed stainless steel, beavertail grip safety, front and rear cocking serrations, stainless steel parts, solid or 3-hole trigger, lowered ejection port.

Production: 2001 - discontinued Caliber: .45 ACP Action: SA, semi-auto

Barrel Length: 3.5" Wt.: 34 oz. Sights: fiber-optic front, adjustable Novak rear

Capacity: 7 Magazine: detachable Grips: G10 grenade grips with UA medallion

D2C:	NIB $	2,899	Ex $	2,225	VG+ $	1,595	Good $1,450	LMP $	2,899
C2C:	Mint $	2,790	Ex $	2,025	VG+ $	1,450	Good $1,305	Fair $	670
Trade-In:	Mint $	2,060	Ex $	1,625	VG+ $	1,140	Good $1,015	Poor $	300

COMPACT MATCH
NEW IN BOX

COMPACT MATCH   REDBOOK CODE: RB-UA-H-COMMAT

Brushed stainless steel, beavertail grip safety, front and rear cocking serrations, stainless steel parts, solid or 3-hole trigger, lowered ejection port.

Production: 2012 - current   Caliber: .45 ACP   Action: SA, semi-auto
Barrel Length: 3.5"   Wt.: 34 oz.   Sights: fiber-optic front, adjustable Novak rear
Capacity: 8   Magazine: detachable   Grips: G10 grenade grips with UA medallion

D2C:	NIB	$ 2,799	Ex	$2,150	VG+	$1,540	Good	$1,400	LMP	$2,799
C2C:	Mint	$2,690	Ex	$1,950	VG+	$1,400	Good	$1,260	Fair	$ 645
Trade-In:	Mint	$1,990	Ex	$1,575	VG+	$1,100	Good	$980	Poor	$ 300

COMMANDER MATCH
NEW IN BOX

COMMANDER MATCH   REDBOOK CODE: RB-UA-H-COMMM

Brushed stainless steel, beavertail grip safety, front and rear cocking serrations, stainless steel parts, solid or 3-hole trigger, lowered ejection port.

Production: 2011 - current   Caliber: .45 ACP   Action: SA, semi-auto
Barrel Length: 4.25"   Wt.: 34 oz.   Sights: fiber-optic front, adjustable Novak rear
Capacity: 8   Magazine: detachable   Grips: G10 grenade grips with UA medallion

D2C:	NIB	$2,799	Ex	$2,150	VG+	$1,540	Good	$1,400	LMP	$2,799
C2C:	Mint	$2,690	Ex	$1,950	VG+	$1,400	Good	$1,260	Fair	$ 645
Trade-In:	Mint	$1,990	Ex	$1,575	VG+	$1,100	Good	$ 980	Poor	$ 300

IA 3.5 PUG COMPACT TAC WITH RAIL
NEW IN BOX

IA 3.5 PUG COMPACT TAC WITH RAIL
REDBOOK CODE: RB-UA-H-IA35CT

IA frame and S/S slide, UA Armor coated in standard H-152 stainless color, high-ride beavertail grip safety, front and rear cocking serrations, stainless steel parts, solid or 3-hole trigger, lowered ejection port, includes monogrammed UA soft pistol case.

Production: 2013 - current   Caliber: .45 ACP   Action: SA, semi-auto
Barrel Length: 3.5"   Wt.: 28 oz.   Sights: fiber-optic front, Novak rear
Capacity: 8   Magazine: detachable   Grips: G10 grenade grips with UA medallion

D2C:	NIB	$3,499	Ex	$2,675	VG+	$1,925	Good	$1,750	LMP	$3,499
C2C:	Mint	$3,360	Ex	$2,425	VG+	$1,750	Good	$1,575	Fair	$ 805
Trade-In:	Mint	$2,490	Ex	$1,975	VG+	$1,370	Good	$1,225	Poor	$ 360

Walther

With over a century of manufacturing and designing quality firearms, Walther imports and distributes firearms to the U.S. through the Walther Arms, Inc. name. The company states that Carl Walther and his sons designed their first semi-automatic pistol in 1908, making them one of the earliest developers of such a handgun. In the early 1990s, Walther merged with Umarex Sportwaffen GmbH & Co KG. Prior to the creation of Walther Arms, Inc., Smith & Wesson distributed Walther firearms in the United States. Now, Smith & Wesson continues to manufacture the PPK for Walther, while Walther manufactures the M&P22 for S&W, a somewhat unorthodox business practice.

PPQ
NEW IN BOX

PPQ   REDBOOK CODE: RB-WA-H-PPQXXX

Matte black finish, integral safety, loaded chamber indicator, ambidextrous magazine release and slide lock.

Production: current   Caliber: 9mm, .40 S&W   Action: DA/SA, semi-auto
Barrel Length: 4"   Sights: adjustable rear   Capacity: 15, 12 (9mm), 17, 14 (.40 S&W)
Magazine: detachable box   Grips: polymer grip frame with interchageable inserts

D2C:	NIB	$ 556	Ex	$ 425	VG+	$ 310	Good	$ 280		
C2C:	Mint	$ 540	Ex	$ 400	VG+	$ 280	Good	$ 255	Fair	$ 130
Trade-In:	Mint	$ 400	Ex	$ 325	VG+	$ 220	Good	$ 195	Poor	$ 60

WALTHER

PPQ M2
NEW IN BOX

PPQ M2   REDBOOK CODE: RB-WA-H-PPQM2X

Tennifer coated slide and barrel, front and rear slide serrations, ambidextrous slide stop, ambidextrous magazine release, 2 drop safeties and a firing pin block, checkered trigger guard, 1913 Picatinny rail, quick-defense trigger, available threaded barrel on Navy model, optional metal self-illumination and metal tritium night sights.

Production: 2013 - current   Caliber: 9mm, .40 S&W   Action: striker-fired, semi-auto
Barrel Length: 4", 4.1", 4.6", 5"   Wt.: 25 oz.   Sights: low-profile 3-dot polymer adjustable
Capacity: 15 (9mm), 11 (.40 cal), 15/17 (Navy)   Magazine: detachable box
Grips: polymer grip frame with interchageable inserts

D2C:	NIB $	580	Ex $	450	VG+ $	320	Good $	290	
C2C:	Mint $	560	Ex $	425	VG+ $	290	Good $	265	Fair $ 135
Trade-In:	Mint $	420	Ex $	325	VG+ $	230	Good $	205	Poor $ 60

P99
NEW IN BOX

P99   REDBOOK CODE: RB-WA-H-P99XXX

Various finishes, polymer frame, steel slide, decocker mechanism, compact design, loaded chamber indicator.

Production: 1997 - late 1990s   Caliber: 9mm, .40 S&W   Action: striker-fired, semi-auto
Barrel Length: 4", 4.2"   Sights: interchangeable front, adjustable rear
Capacity: 16 (9mm), 12 (.40 S&W)   Magazine: detachable box   Grips: polymer

D2C:	NIB $	599	Ex $	475	VG+ $	330	Good $	300	
C2C:	Mint $	580	Ex $	425	VG+ $	300	Good $	270	Fair $ 140
Trade-In:	Mint $	430	Ex $	350	VG+ $	240	Good $	210	Poor $ 60

P99 QA   REDBOOK CODE: RB-WA-H-P99QAX

A variant of the P99, denoted by its modified trigger and internal striker. Steel slide, polymer frame, Tennifer treated.

Production: early 2000s   Caliber: 9mm, .40 S&W   Action: striker-fired, semi-auto
Barrel Length: 4"   Sights: adjustable rear   Capacity: 12, 16
Magazine: detachable box   Grips: polymer

D2C:	NIB $	599	Ex $	475	VG+ $	330	Good $	300	
C2C:	Mint $	580	Ex $	425	VG+ $	300	Good $	270	Fair $ 140
Trade-In:	Mint $	430	Ex $	350	VG+ $	240	Good $	210	Poor $ 60

P99 AS
NEW IN BOX

P99 AS   REDBOOK CODE: RB-WA-H-P99ASX

Tennifer coated slide and barrel, rear slide serrations, paddle-style ambidextrous magazine release, 2 drop safeties and firing pin block , checkered trigger guard, Picatinny rail, AS (anti-stress) trigger, optional self-illumination and tritium night sights.

Production: late 1990s - current   Caliber: 9mm, .40 S&W   Action: striker-fired, semi-auto
Barrel Length: 4"   Wt.: 24 oz.   Sights: low profile, 3-dot polymer combat
Capacity: 15 (9mm), 12 (.40 cal)   Magazine: detachable box   Grips: polymer grip frame with interchageable inserts

D2C:	NIB $	599	Ex $	475	VG+ $	330	Good $	300	
C2C:	Mint $	580	Ex $	425	VG+ $	300	Good $	270	Fair $ 140
Trade-In:	Mint $	430	Ex $	350	VG+ $	240	Good $	210	Poor $ 60

P99 AS COMPACT REDBOOK CODE: RB-WA-H-P99ASC

Tennifer coated slide and barrel, compact model, rear slide serrations, paddle-style ambidextrous magazine release, 2 drop safeties and a firing pin block, checkered trigger guard, Picatinny rail, AS (anti-stress) trigger, optional self-illumination and tritium night sights.

P99 AS COMPACT
NEW IN BOX

Production: late 1990s - current Caliber: 9mm, .40 S&W Action: striker-fired, semi-auto
Barrel Length: 3.5" Wt.: 21 oz. Sights: low profile, 3-dot polymer combat
Capacity: 10 (9mm), 8 (.40 cal) Magazine: detachable box Grips: polymer grip frame with interchageable inserts

D2C:	NIB	$ 599	Ex $ 475	VG+ $ 330	Good $ 300				
C2C:	Mint $ 580	Ex $ 425	VG+ $ 300	Good $ 270	Fair $ 140				
Trade-In:	Mint $ 430	Ex $ 350	VG+ $ 240	Good $ 210	Poor $ 60				

WALTHER

PPS REDBOOK CODE: RB-WA-H-PPSXXX

Tennifer coated slide and barrel, rear slide serrations, paddle-style ambidextrous magazine release, loaded chamber viewport, red cocking indicator, no-snag slide stop, Picatinny rail, AS (anti-stress) trigger, optional metal self-illumination and metal tritium night sights.

PPS
NEW IN BOX

Production: 2007 - current Caliber: 9mm, .40 S&W Action: striker-fired, semi-auto
Barrel Length: 3.2" Wt.: 21 oz. Sights: low profile, 3-dot polymer combat
Capacity: 5, 6, 7, 8 Magazine: detachable box Grips: polymer grip frame with interchangeable inserts

D2C:	NIB $ 560	Ex $ 450	VG+ $ 310	Good $ 280	
C2C:	Mint $ 540	Ex $ 400	VG+ $ 280	Good $ 255	Fair $ 130
Trade-In:	Mint $ 400	Ex $ 325	VG+ $ 220	Good $ 200	Poor $ 60

PPX REDBOOK CODE: RB-WA-H-PPXXXX

Tennifer coated slide and barrel, front and rear slide serrations, ambidextrous slide stop, ambidextrous magazine release button, loaded chamber viewport, checkered trigger guard, bobbed hammer, 1913 Picatinny rail, black or stainless finish available, optional threaded barrel on SD model.

PPX
NEW IN BOX

Production: 2013 - current Caliber: 9mm, .40 S&W Action: striker-fired, semi-auto
Barrel Length: 4", 4.6" Wt.: 27 oz. Sights: low profile, 3-dot polymer combat
Capacity: 16 (9mm), 14 (.40 cal) Magazine: detachable box Grips: polymer grip frame

D2C:	NIB $ 375	Ex $ 300	VG+ $ 210	Good $ 190	
C2C:	Mint $ 360	Ex $ 275	VG+ $ 190	Good $ 170	Fair $ 90
Trade-In:	Mint $ 270	Ex $ 225	VG+ $ 150	Good $ 135	Poor $ 60

PPK REDBOOK CODE: RB-WA-H-PPKXXX

Rear slide serrations, manual safety, internal slide stop, beavertail extension, loaded chamber viewport, wave-cut top strap, blue or stainless finish.

PPK
NEW IN BOX

Production: current Caliber: .380 ACP Action: DA/SA, semi-auto
Barrel Length: 3.3" Wt.: 22 oz. Sights: fixed steel Capacity: 6, 7
Magazine: detachable box Grips: checkered panels

D2C:	NIB $ 575	Ex $ 450	VG+ $320	Good $ 290	
C2C:	Mint $ 560	Ex $ 400	VG+ $290	Good $ 260	Fair $ 135
Trade-In:	Mint $ 410	Ex $ 325	VG+ $230	Good $ 205	Poor $ 60

PPK/S
NEW IN BOX

PPK/S .22
NEW IN BOX

PPK/E
NEW IN BOX

PK 380
NEW IN BOX

PPK/S   REDBOOK CODE: RB-WA-H-PPKSXX

"S" sport version with longer grip, rear slide serrations, manual safety, internal slide stop, beavertail extension, loaded chamber viewport, wave-cut top strap, duo-tone, blue or stainless finish.

Production: current   Caliber: .380 ACP   Action: DA/SA, semi-auto

Barrel Length: 3.3"   Wt.: 22 oz.   Sights: fixed   Capacity: 7   Magazine: detachable box

Grips: plastic

D2C:	NIB	$ 575	Ex $ 450	VG+ $ 320	Good $ 290		
C2C:	Mint $ 560	Ex $ 400	VG+ $ 290	Good $ 260	Fair $ 135
Trade-In:	Mint $ 410	Ex $ 325	VG+ $ 230	Good $ 205	Poor $ 60

PPK/S .22   REDBOOK CODE: RB-WA-H-PPKS22

"S" sport version with longer grip, rear slide serrations, manual safety, internal slide stop, beavertail extension, wave-cut top strap, blue or stainless finish.

Production: 2013 - current   Caliber: .22 LR   Action: DA/SA, semi-auto

Barrel Length: 3.3"   Wt.: 24 oz.   Sights: fixed steel   Capacity: 10

Magazine: detachable box   Grips: checkered panels

D2C:	NIB $ 380	Ex $ 300	VG+ $ 210	Good $ 190	
C2C:	Mint $ 370	Ex $ 275	VG+ $ 190	Good $ 175	Fair $ 90
Trade-In:	Mint $ 270	Ex $ 225	VG+ $ 150	Good $ 135	Poor $ 60

PPK/E   REDBOOK CODE: RB-WA-H-PPKEXX

Made by FEG, marked as a Walther. A variant of the PPK, however, the parts are not interchangeable. Blue finish, frame-mounted safety, compact design.

Production: 2000 - current   Caliber: .380 ACP, .32 ACP, .22 LR   Action: DA/SA, semi-auto

Barrel Length: 3.4"   Sights: fixed steel   Capacity: 7, 8   Magazine: detachable box

Grips: plastic

D2C:	NIB $ 540	Ex $ 425	VG+ $ 300	Good $ 270	
C2C:	Mint $ 520	Ex $ 375	VG+ $ 270	Good $ 245	Fair $ 125
Trade-In:	Mint $ 390	Ex $ 325	VG+ $ 220	Good $ 190	Poor $ 60

PK 380   REDBOOK CODE: RB-WA-H-PK380X

Tennifer coated slide and barrel, loaded chamber viewport, rear slide serrations, internal slide stop, manual safety, Picatinny rail, ambidextrous slide stop, paddle-style ambidextrous magazine release, nickel or matte black slide, optional laser set.

Production: 2009 - current   Caliber: .380 ACP   Action: DA/SA, semi-auto

Barrel Length: 3.6"   Wt.: 20 oz.   Sights: low profile, 3-dot polymer combat

Capacity: 8   Magazine: detachable box   Grips: Ergonomic Walther Grip

D2C:	NIB $ 370	Ex $ 300	VG+ $ 205	Good $ 185	
C2C:	Mint $ 360	Ex $ 275	VG+ $ 190	Good $ 170	Fair $ 90
Trade-In:	Mint $ 270	Ex $ 225	VG+ $ 150	Good $ 130	Poor $ 60

MODEL P88   REDBOOK CODE: RB-WA-H-P88XXX

Double-stack magazine, blue finish, steel slide, black finish, slide-mounted safety, ambidextrous decocker mechanism.

Production: 1988 - late 1990s   Caliber: 9mm, .22 LR   Action: DA/SA, semi-auto

Barrel Length: 4"   Wt.: 20 oz.   Sights: fixed   Capacity: 15   Magazine: detachable box

Grips: black polymer or deluxe wood

D2C:	NIB $1,750	Ex $1,350	VG+ $ 965	Good $ 875	
C2C:	Mint $1,680	Ex $1,225	VG+ $ 880	Good $ 790	Fair $ 405
Trade-In:	Mint $1,250	Ex $1,000	VG+ $ 690	Good $ 615	Poor $ 180

MODEL P88 COMPACT
EXCELLENT

MODEL P88 COMPACT REDBOOK CODE: RB-WA-H-P88COM

Short recoil, auto-firing pin safety, blade front sight and adjustable notch rear sight, black finish.

Production: 1992 Caliber: 9mm Action: DA/SA, semi-auto Barrel Length: 3.8"					
Sights: fixed Capacity: 14 Magazine: detachable box Grips: polymer					
D2C:	NIB $1,750	Ex $1,350	VG+ $ 965	Good $ 875	
C2C:	Mint $1,680	Ex $1,225	VG+ $ 880	Good $ 790	Fair $ 405
Trade-In:	Mint $1,250	Ex $1,000	VG+ $ 690	Good $ 615	Poor $ 180

MODEL P5
EXCELLENT

MODEL P5 REDBOOK CODE: RB-WA-H-P5XXXX

Black matte high-polish finish, decocker lever, short recoil operated, aluminum alloy frame.

Production: mid-1970's - discontinued Caliber: 9mm, 9x21, 7.65 Luger					
Action: DA/SA, semi-auto Barrel Length: 3.5" Sights: blade front, notch rear					
Capacity: 8 Magazine: detachable box Grips: black plastic					
D2C:	NIB $1,800	Ex $1,375	VG+ $ 990	Good $ 900	
C2C:	Mint $1,730	Ex $1,250	VG+ $ 900	Good $ 810	Fair $ 415
Trade-In:	Mint $1,280	Ex $1,025	VG+ $ 710	Good $ 630	Poor $ 180

MODEL P5 COMPACT
EXCELLENT

MODEL P5 COMPACT REDBOOK CODE: RB-WA-H-P5COMP

Shorter barrel version of the P5, black matte high-polish finish, decocker lever, short -recoil operated, aluminum alloy frame.

Production: mid-1970s - discontinued Caliber: 9mm, 9x21, 7.65 Luger					
Action: DA/SA, semi-auto Barrel Length: 3.1" Sights: blade front, notch rear					
Capacity: 8 Magazine: detachable box Grips: black plastic					
D2C:	NIB $1,850	Ex $1,425	VG+ $1,020	Good $ 925	
C2C:	Mint $1,780	Ex $1,300	VG+ $ 930	Good $ 835	Fair $ 430
Trade-In:	Mint $1,320	Ex $1,050	VG+ $ 730	Good $ 650	Poor $ 210

P22
NEW IN BOX

P22 REDBOOK CODE: RB-WA-H-P22XXX

Loaded chamber indicator, front and rear slide serrations, external slide stop, manual safety, magazine disconnect safety, firing pin block safety, Picatinny rail, paddle-style ambidextrous magazine release, threaded barrel, multiple finishes and sights available.

Production: 2003 - current Caliber: .22 LR Action: DA/SA, semi-auto					
Barrel Length: 3.4", 5" Wt.: 20 oz. Sights: low profile, 3-dot polymer combat					
Capacity: 10 Magazine: detachable box					
Grips: Ergonomic Walther Grip with interchangeable inserts					
D2C:	NIB $ 340	Ex $ 275	VG+ $ 190	Good $ 170	
C2C:	Mint $ 330	Ex $ 250	VG+ $ 170	Good $ 155	Fair $ 80
Trade-In:	Mint $ 250	Ex $ 200	VG+ $ 140	Good $ 120	Poor $ 60

P22 TARGET
NEW IN BOX

P22 TARGET REDBOOK CODE: RB-WA-H-P22TAR

Tactical design, slide-mounted safety, ambidextrous design, polymer frame, extended target-style barrel.

Production: current Caliber: .22 LR Action: DA/SA, semi-auto					
Barrel Length: 5" Wt.: 20 oz. Sights: 3-dot adjustable Capacity: 10					
Magazine: detachable box Grips: polymer					
D2C:	NIB $ 410	Ex $ 325	VG+ $ 230	Good $ 205	
C2C:	Mint $ 400	Ex $ 300	VG+ $ 210	Good $ 185	Fair $ 95
Trade-In:	Mint $ 300	Ex $ 250	VG+ $ 160	Good $ 145	Poor $ 60

TPH
(GERMAN MFG.)
EXCELLENT

TPH (GERMAN MFG.) REDBOOK CODE: RB-WA-H-TPHXXX

Silver or blue finish, aluminum alloy frame, compact design, slide-mounted safety, engraved model available for an additional price.

Production: late 1960s - discontinued	Caliber: .22 LR., .25 ACP				
Action: DA/SA, semi-auto	Barrel Length: 2.8"	Sights: blade front, adjustable rear			
Capacity: 6	Magazine: detachable box	Grips: black plastic			
D2C:	NIB $1,600	Ex $1,225	VG+ $880	Good $800	
C2C:	Mint $1,540	Ex $1,125	VG+ $800	Good $720	Fair $370
Trade-In:	Mint $1,140	Ex $900	VG+ $630	Good $560	Poor $180

Wilson Combat

Located in Berryville, Arkansas, Wilson Combat manufactures custom M1911s, AR-style rifles, and a line of shotguns. Bill Wilson began customizing pistols in 1977, which he soon grew into a full-scale business. The company later bought Scattergun Technologies and now manufactures combat shotguns, and more recently, AR-15 rifles. The company continues to manufacture all of the parts for their M1911s in-house.

CLASSIC SUPERGRADE
NEW IN BOX

CLASSIC SUPERGRADE REDBOOK CODE: WC.H.CSXXXX

Full-size stainless frame, polished blue steel slide, front and rear slide serrations, beavertail grip safety, ambi thumb safety, magwell, fully-machined Bullet Proof parts.

Production: current	Caliber: .45 ACP, 10mm, .40 S&W, .38 Super, 9mm				
Action: SA, semi-auto	Barrel Length: 5"	Wt.: 37 oz.	Sights: ramp front, adjustable rear		
Capacity: 8	Magazine: detachable box	Grips: cocobolo double-diamond			
D2C:	NIB $5,150	Ex $3,925	VG+ $2,835	Good $2,575	LMP $5,195
C2C:	Mint $4,950	Ex $3,575	VG+ $2,580	Good $2,320	Fair $1,185
Trade-In:	Mint $3,660	Ex $2,900	VG+ $2,010	Good $1,805	Poor $540

TACTICAL SUPERGRADE
NEW IN BOX

TACTICAL SUPERGRADE REDBOOK CODE: RB-WC-H-TSXXXX

Full-size steel frame, 30 lpi high cut checkering on front strap, beavertail grip safety, ambi thumb safety, magwell, carbon steel slide, stainless barrel and bushing, fully-machined Bullet Proof parts.

Production: current	Caliber: .45 ACP, 10mm, .40 S&W, .38 Super, 9mm				
Action: SA, semi-auto	Barrel Length: 5"	Wt.: 37 oz.			
Sights: Battlesight with tritium front	Capacity: 8	Grips: G10 Starburst			
D2C:	NIB $5,200	Ex $3,975	VG+ $2,860	Good $2,600	LMP $5,045
C2C:	Mint $5,000	Ex $3,600	VG+ $2,600	Good $2,340	Fair $1,200
Trade-In:	Mint $3,700	Ex $2,925	VG+ $2,030	Good $1,820	Poor $540

TACTICAL SUPERGRADE PROFESSIONAL
REDBOOK CODE: RB-WC-H-TSPXXX

Carbon steel frame and slide, stainless cone barrel, full-length guide rod with reverse plug, beavertail grip safety, ambi thumb safety, magwell, fully-machined Bullet Proof parts.

**TACTICAL SUPERGRADE
PROFESSIONAL,** NEW IN BOX

Production: current	Caliber: .45 ACP, 9mm, .38 Super	Action: SA, semi-auto			
Barrel Length: 4"	Wt.: 39 oz.	Sights: Battlesight with white-outline tritium front			
Capacity: 8	Magazine: detachable box	Grips: G10 Starburst			
D2C:	NIB $4,995	Ex $3,800	VG+ $2,750	Good $2,500	LMP $5,045
C2C:	Mint $4,800	Ex $3,450	VG+ $2,500	Good $2,250	Fair $1,150
Trade-In:	Mint $3,550	Ex $2,800	VG+ $1,950	Good $1,750	Poor $510

TACTICAL SUPERGRADE COMPACT
NEW IN BOX

TACTICAL SUPERGRADE COMPACT

REDBOOK CODE: RB-WC-H-TSCXXX

Compact carbon steel frame and slide, stainless cone barrel, front strap checkering, beavertail grip safety, ambi thumb safety, magwell, fully-machined Bullet Proof parts.

Production: current Caliber: .45 ACP, 9mm, .38 Super Action: SA, semi-auto
Barrel Length: 4" Wt.: 36 oz. Sights: Battlesight with tritium front Capacity: 7
Magazine: detachable box Grips: G10 Starburst

D2C:	NIB	$4,995	Ex	$3,800	VG+	$2,750	Good	$2,500	LMP	$5,045
C2C:	Mint	$4,800	Ex	$3,450	VG+	$2,500	Good	$2,250	Fair	$1,150
Trade-In:	Mint	$3,550	Ex	$2,800	VG+	$1,950	Good	$1,750	Poor	$510

ULTRALIGHT CARRY
NEW IN BOX

ULTRALIGHT CARRY   REDBOOK CODE: RB-WC-H-ULCXXX

Full-size aluminum round butt frame, contoured magwell, concealment beavertail grip safety, tactical thumb safety, heavy-chamfer on bottom of slide, carry and ball endmill cuts, fully-machined Bullet Proof parts.

Production: current Caliber: .45 ACP, .38 Super, 9mm Action: SA, semi-auto
Barrel Length: 5" Wt.: 33 oz. Sights: Battlesight with fiber-optic front
Capacity: 8 Magazine: detachable box Grips: G10 Starburst

D2C:	NIB	$3,475	Ex	$2,650	VG+	$1,915	Good	$1,740	LMP	$3,650
C2C:	Mint	$3,340	Ex	$2,400	VG+	$1,740	Good	$1,565	Fair	$800
Trade-In:	Mint	$2,470	Ex	$1,950	VG+	$1,360	Good	$1,220	Poor	$360

ULTRALIGHT CARRY COMPACT
EXCELLENT

ULTRALIGHT CARRY COMPACT   REDBOOK CODE: RB-WC-H-ULCCOM

Compact-size aluminum round butt frame, contoured magwell, concealment beavertail grip safety, tactical thumb safety, heavy-chamfer on bottom of slide, carry and ball endmill cuts, fully-machined Bullet Proof parts.

Production: current Caliber: .45 ACP, .38 Super, 9mm Action: SA, semi-auto
Barrel Length: 4" Wt.: 26 oz. Sights: Battlesight with fiber-optic front
Capacity: 7 Magazine: detachable box Grips: G10 Starburst

D2C:	NIB	$3,475	Ex	$2,650	VG+	$1,915	Good	$1,740	LMP	$3,650
C2C:	Mint	$3,340	Ex	$2,400	VG+	$1,740	Good	$1,565	Fair	$800
Trade-In:	Mint	$2,470	Ex	$1,950	VG+	$1,360	Good	$1,220	Poor	$360

X-TAC
NEW IN BOX

X-TAC   REDBOOK CODE: RB-WC-H-XTACXX

Full-size carbon steel frame and slide, black parkerized finish, contoured magwell, X-Tac frontstrap and rear cocking serration treatment, stainless match barrel and bushing, beavertail grip safety, thumb safety.

Production: current Caliber: .45 ACP Action: SA, semi-auto Barrel Length: 5"
Wt.: 38 oz. Sights: Battlesight with fiber-optic front Capacity: 8 Magazine: detachable box
Grips: G10 Starburst

D2C:	NIB	$2,800	Ex	$2,150	VG+	$1,540	Good	$1,400	LMP	$2,760
C2C:	Mint	$2,690	Ex	$1,950	VG+	$1,400	Good	$1,260	Fair	$645
Trade-In:	Mint	$1,990	Ex	$1,575	VG+	$1,100	Good	$980	Poor	$300

X-TAC COMPACT
NEW IN BOX

X-TAC COMPACT   REDBOOK CODE: RB-WC-H-XTACCO

Compact carbon steel round butt frame, carbon steel slide, black parkerized finish, contoured magwell, X-Tac frontstrap and rear cocking serration treatment, stainless match grade cone barrel, beavertail grip safety, thumb safety.

Production: current Caliber: .45 ACP Action: SA, semi-auto Barrel Length: 4"
Wt.: 34 oz. Sights: Battlesight with fiber-optic front Capacity: 7 Magazine: detachable box
Grips: G10 Starburst

D2C:	NIB	$2,850	Ex	$2,175	VG+	$1,570	Good	$1,425	LMP	$2,785
C2C:	Mint	$2,740	Ex	$1,975	VG+	$1,430	Good	$1,285	Fair	$660
Trade-In:	Mint	$2,030	Ex	$1,600	VG+	$1,120	Good	$1,000	Poor	$300

CQB
NEW IN BOX

CQB ELITE
NEW IN BOX

CQB TACTICAL LE
NEW IN BOX

CQB LIGHT-RAIL LIGHTWEIGHT
NEW IN BOX

**CQB LIGHT-RAIL LIGHTWEIGHT
PROFESSIONAL**
NEW IN BOX

CQB REDBOOK CODE: RB-WC-H-CQBXXX

Full-size carbon steel frame and slide, beavertail grip safety, thumb safety, contoured magwell, high-cut checkered front strap, stainless match barrel and bushing.

Production: current Caliber: .45 ACP, 10mm, .40 S&W, .38 Super, 9mm
Action: SA, semi-auto Barrel Length: 5" Wt.: 37 oz. Sights: Battlesight with fiber-optic front Capacity: 8 Magazine: detachable box Grips: G10 Starburst

D2C:	NIB	$2,650	Ex	$2,025	VG+	$1,460	Good	$1,325	LMP $2,685
C2C:	Mint	$2,550	Ex	$1,850	VG+	$1,330	Good	$1,195	Fair $ 610
Trade-In:	Mint	$1,890	Ex	$1,500	VG+	$1,040	Good	$ 930	Poor $ 270

CQB ELITE REDBOOK CODE: RB-WC-H-CQBELI

Full-size carbon steel frame and slide, Bullet Proof parts and controls, beavertail grip safety, magwell, stainless match barrel and bushing.

Production: current Caliber: .45 ACP, 10mm, .40 S&W, .38 Super, 9mm
Action: SA, semi-auto Barrel Length: 5" Wt.: 37 oz.
Sights: Battlesight with fiber-optic front Capacity: 8 Magazine: detachable box
Grips: G10 diagonal flat bottom

D2C:	NIB	$3,250	Ex	$2,475	VG+	$1,790	Good	$1,625	LMP $3,425
C2C:	Mint	$3,120	Ex	$2,250	VG+	$1,630	Good	$1,465	Fair $ 750
Trade-In:	Mint	$2,310	Ex	$1,825	VG+	$1,270	Good	$1,140	Poor $ 330

CQB TACTICAL LE REDBOOK CODE: RB-WC-H-CQBTAC

Full-size stainless steel frame, carbon steel slide, integral light rail, Speed-Chute magwell, beavertail grip safety, thumb safety, stainless bushingless cone barrel.

Production: current Caliber: .45 ACP, 10mm, .40 S&W, .38 Super, 9mm
Action: SA, semi-auto Barrel Length: 5" Wt.: 40 oz.
Sights: Battlesight with fiber-optic front Capacity: 8 Magazine: detachable box
Grips: G10 diagonal-flat bottom

D2C:	NIB	$2,925	Ex	$2,225	VG+	$1,610	Good	$1,465	LMP $ 3,115
C2C:	Mint	$2,810	Ex	$2,025	VG+	$1,470	Good	$1,320	Fair $ 675
Trade-In:	Mint	$2,080	Ex	$1,650	VG+	$1,150	Good	$1,025	Poor $ 300

CQB LIGHT-RAIL LIGHTWEIGHT REDBOOK CODE: RB-WC-H-CQBLRL

Full-sized aluminum frame, carbon steel slide, integral light rail, contoured magwell, beavertail grip safety, thumb safety, stainless match barrel with bushing.

Production: current Caliber: .45 ACP, 9mm, .38 Super Action: SA, semi-auto
Barrel Length: 5" Wt.: 37 oz. Sights: Battlesight with fiber-optic front
Capacity: 8 Magazine: detachable box Grips: G10 Starburst

D2C:	NIB	$3,130	Ex	$2,400	VG+	$1,725	Good	$1,565	LMP $3,145
C2C:	Mint	$3,010	Ex	$2,175	VG+	$1,570	Good	$1,410	Fair $ 720
Trade-In:	Mint	$2,230	Ex	$1,775	VG+	$1,230	Good	$1,100	Poor $ 330

CQB LIGHT-RAIL LIGHTWEIGHT PROFESSIONAL

REDBOOK CODE: RB-WC-H-CQBLRP

Shorter professional-size aluminum frame, integral light rail, beavertail grip safety, thumb safety, contoured magwell, carbon steel slide, stainless match grade cone barrel, optional bobtail.

Production: current Caliber: .45 ACP, 9mm, .38 Super Action: SA, semi-auto
Barrel Length: 4" Wt.: 28 oz. Sights: Battlesight with fiber-optic front
Capacity: 8 Magazine: detachable box Grips: G10 Starburst

D2C:	NIB	$3,150	Ex	$2,400	VG+	$1,735	Good	$1,575	LMP $3,205
C2C:	Mint	$3,030	Ex	$2,175	VG+	$1,580	Good	$1,420	Fair $ 725
Trade-In:	Mint	$2,240	Ex	$1,775	VG+	$1,230	Good	$1,105	Poor $ 330

CQB COMPACT
NEW IN BOX

CQB COMPACT REDBOOK CODE: RB-WC-H-CQBCOM

Compact frame, carbon steel frame and slide, contoured magwell, stainless match grade cone barrel, beavertail grip safety, thumb safety, high premium for stainless or 9mm custom options.

Production: current Caliber: .45 ACP, 9mm, .38 Super Action: SA, semi-auto
Barrel Length: 4" Wt.: 36 oz. Sights: Battlesight with fiber-optic front
Capacity: 7 Magazine: detachable box Grips: G10 Starburst

D2C:	NIB	$2,785	Ex	$2,125	VG+	$1,535	Good	$1,395	LMP $2,890
C2C:	Mint	$2,680	Ex	$1,925	VG+	$1,400	Good	$1,255	Fair $ 645
Trade-In:	Mint	$1,980	Ex	$1,575	VG+	$1,090	Good	$ 975	Poor $ 300

CQB LIGHT-RAIL LIGHTWEIGHT COMPACT
NEW IN BOX

CQB LIGHT-RAIL LIGHTWEIGHT COMPACT

REDBOOK CODE: RB-WC-H-CQBLRC

Compact aluminum round-butt frame, carbon steel slide, integral light rail, beavertail grip safety, thumb safety, stainless match grade cone barrel.

Production: current Caliber: .45 ACP, 9mm, .38 Super Action: SA, semi-auto
Barrel Length: 4" Wt.: 27 oz. Sights: Battlesight with fiber-optic front
Capacity: 8 Magazine: detachable box Grips: G10 Starburst

D2C:	NIB	$3,195	Ex	$2,450	VG+	$1,760	Good	$1,600	LMP $3,280
C2C:	Mint	$3,070	Ex	$2,225	VG+	$1,600	Good	$1,440	Fair $ 735
Trade-In:	Mint	$2,270	Ex	$1,800	VG+	$1,250	Good	$1,120	Poor $ 330

PROFESSIONAL
NEW IN BOX

PROFESSIONAL REDBOOK CODE: RB-WC-H-PROFXX

Shorter professional-size carbon steel frame, carbon steel slide, contoured magwell, beavertail grip safety, thumb safety, stainless match grade cone barrel.

Production: current Caliber: .45 ACP, 9mm, .38 Super Action: SA, semi-auto
Barrel Length: 4" Wt.: 36 oz. Sights: Battlesight with fiber-optic front
Capacity: 8 Magazine: detachable box Grips: G10 Starburst

D2C:	NIB	$2,900	Ex	$2,225	VG+	$1,595	Good	$1,450	LMP $2,920
C2C:	Mint	$2,790	Ex	$2,025	VG+	$1,450	Good	$1,305	Fair $ 670
Trade-In:	Mint	$2,060	Ex	$1,625	VG+	$1,140	Good	$1,015	Poor $ 300

PROFESSIONAL LIGHTWEIGHT
NEW IN BOX

PROFESSIONAL LIGHTWEIGHT REDBOOK CODE: RB-WC-H-PROFLW

Professional-size aluminum frame, carbon steel slide, beavertail grip safety, thumb safety, contoured magwell, stainless match grade cone barrel, optional bobtail.

Production: current Caliber: .45 ACP, 9mm, .38 Super Action: SA, semi-auto
Barrel Length: 4" Wt.: 28 oz. Sights: Battlesight with fiber-optic front
Capacity: 8 Magazine: detachable box Grips: G10 Starburst

D2C:	NIB	$2,950	Ex	$2,250	VG+	$1,625	Good	$1,475	LMP $3,090
C2C:	Mint	$2,840	Ex	$2,050	VG+	$1,480	Good	$1,330	Fair $ 680
Trade-In:	Mint	$2,100	Ex	$1,675	VG+	$1,160	Good	$1,035	Poor $ 300

CLASSIC
NEW IN BOX

CLASSIC REDBOOK CODE: RB-WC-H-CLASSX

Full-size carbon steel frame and slide, contoured magwell, beavertail grip safety, thumb safety, stainless match grade barrel and bushing.

Production: current Caliber: .45 ACP, 10mm, .40 S&W, .38 Super, 9mm
Action: SA, semi-auto Barrel Length: 5" Wt.: 38 oz.
Sights: ramp front, adjustable rear Capacity: 8 Magazine: detachable box
Grips: cocobolo double-diamond

D2C:	NIB	$3,100	Ex	$2,375	VG+	$1,705	Good	$1,550	LMP $3,030
C2C:	Mint	$2,980	Ex	$2,150	VG+	$1,550	Good	$1,395	Fair $ 715
Trade-In:	Mint	$2,210	Ex	$1,750	VG+	$1,210	Good	$1,085	Poor $ 330

WILSON COMBAT

TACTICAL ELITE
NEW IN BOX

TACTICAL ELITE REDBOOK CODE: RB-WC-H-TACELI

Full-size carbon steel frame and slide, Bullet Proof beavertail grip safety and ambi thumb safety, magazine well, stainless heavy flanged cone barrel, full-length guide rod with reverse plug.

Production: current Caliber: .45 ACP, 10mm, .40 S&W, .38 Super, 9mm

Action: SA, semi-auto Barrel Length: 5" Wt.: 40 oz. Sights: Battlesight with

fiber-optic front Capacity: 8 Magazine: detachable box Grips: G10 Starburst

D2C:	NIB	$3,600	Ex	$2,750	VG+	$1,980	Good	$1,800	LMP $3,650
C2C:	Mint	$3,460	Ex	$2,500	VG+	$1,800	Good	$1,620	Fair $ 830
Trade-In:	Mint	$2,560	Ex	$2,025	VG+	$1,410	Good	$1,260	Poor $ 360

ELITE PROFESSIONAL
NEW IN BOX

ELITE PROFESSIONAL REDBOOK CODE: RB-WC-H-ELIPRO

Professional-sized carbon steel frame and slide, Bullet Proof beavertail grip safety and ambi thumb safety, magwell, stainless heavy-flanged cone barrel.

Production: current Caliber: .45 ACP, 9mm, .38 Super Action: SA, semi-auto

Barrel Length: 4" Wt.: 37 oz. Sights: Battlesight with fiber-optic front

Capacity: 8 Magazine: detachable box Grips: G10 diagonal, flat bottom

D2C:	NIB	$3,475	Ex	$2,650	VG+	$1,915	Good	$1,740	LMP $3,650
C2C:	Mint	$3,340	Ex	$2,400	VG+	$1,740	Good	$1,565	Fair $ 800
Trade-In:	Mint	$2,470	Ex	$1,950	VG+	$1,360	Good	$1,220	Poor $ 360

STEALTH
NEW IN BOX

STEALTH REDBOOK CODE: RB-WC-H-STEALT

Compact frame, carbon steel frame and slide, Bullet Proof beavertail grip safety and thumb safety, contoured magwell, stainless heavy-flanged cone barrel.

Production: current Caliber: .45 ACP, 9mm, .38 Super Action: SA, semi-auto

Barrel Length: 4" Wt.: 36 oz. Sights: Battlesight with fiber-optic front

Capacity: 7 Magazine: detachable box Grips: G10 Starburst

D2C:	NIB	$3,400	Ex	$2,600	VG+	$1,870	Good	$1,700	LMP $3,535
C2C:	Mint	$3,270	Ex	$2,350	VG+	$1,700	Good	$1,530	Fair $ 785
Trade-In:	Mint	$2,420	Ex	$1,925	VG+	$1,330	Good	$1,190	Poor $ 360

CARRY COMP
NEW IN BOX

CARRY COMP REDBOOK CODE: RB-WC-H-CCOMPX

Compact frame, contoured magazine well, carbon steel frame and slide, black Armor-Tuff finish, Bullet Proof beavertail grip safety and thumb safety, stainless compensated barrel.

Production: current Caliber: .45 ACP Action: SA, semi-auto Barrel Length: 4"

Wt.: 36 oz. Sights: Battlesight with fiber-optic front Capacity: 7 Grips: G10 Starburst

D2C:	NIB	$3,800	Ex	$2,900	VG+	$2,090	Good	$1,900	LMP $3,765
C2C:	Mint	$3,650	Ex	$2,625	VG+	$1,900	Good	$1,710	Fair $ 875
Trade-In:	Mint	$2,700	Ex	$2,150	VG+	$1,490	Good	$1,330	Poor $ 390

SENTINEL
NEW IN BOX

SENTINEL REDBOOK CODE: RB-WC-H-SENTXX

Sub-compact carbon steel frame and slide, Bullet Proof beavertail grip safety and thumb safety, contoured magwell, carry cuts, stainless cone barrel, full-length guide rod.

Production: current Caliber: 9mm Action: SA, semi-auto Barrel Length: 3.6"

Wt.: 31 oz. Sights: Battlesight with fiber-optic front Capacity: 8

Magazine: detachable box Grips: G10 Slimline

D2C:	NIB	$3,100	Ex	$2,375	VG+	$1,705	Good	$1,550	LMP $3,310
C2C:	Mint	$2,980	Ex	$2,150	VG+	$1,550	Good	$1,395	Fair $ 715
Trade-In:	Mint	$2,210	Ex	$1,750	VG+	$1,210	Good	$1,085	Poor $ 330

SUPER SENTINEL
NEW IN BOX

SUPER SENTINEL   REDBOOK CODE: RB-WC-H-SUPSEN

Sub-compact aluminum frame, carbon steel slide, contoured magazine well, carry cuts, fluted barrel/chamber, full-length guide rod, stainless cone barrel.

Production: current Caliber: .38 Super Action: SA, semi-auto Barrel Length: 3.6"
Wt.: 25 oz.   Sights: Battlesight with fiber-optic front Capacity: 8
Magazine: detachable box Grips: G10 Slimline

D2C:	NIB	$3,750	Ex	$2,850	VG+	$2,065	Good	$1,875	LMP	$3,875
C2C:	Mint	$3,600	Ex	$2,600	VG+	$1,880	Good	$1,690	Fair	$ 865
Trade-In:	Mint	$2,670	Ex	$2,100	VG+	$1,470	Good	$1,315	Poor	$ 390

MS. SENTINEL
NEW IN BOX

MS. SENTINEL   REDBOOK CODE: RB-WC-H-MSSENT

Sub-compact aluminum frame, carbon steel slide, Bullet Proof controls, countersunk slide stop, heavy-machine chamfer on bottom of slide, stainless cone barrel, fluted barrel and chamber, carry and ball endmill cuts.

Production: current Caliber: 9mm Action: SA, semi-auto Barrel Length: 3.6"
Wt.: 27 oz.   Sights: Battlesight with fiber-optic front Capacity: 8
Magazine: detachable box Grips: cocobolo

D2C:	NIB	$3,750	Ex	$2,850	VG+	$2,065	Good	$1,875	LMP	$3,875
C2C:	Mint	$3,600	Ex	$2,600	VG+	$1,880	Good	$1,690	Fair	$ 865
Trade-In:	Mint	$2,670	Ex	$2,100	VG+	$1,470	Good	$1,315	Poor	$ 390

ULTRALIGHT CARRY SENTINEL
NEW IN BOX

ULTRALIGHT CARRY SENTINEL   REDBOOK CODE: RB-WC-H-ULCSEN

Sub-compact size aluminum round-butt frame, contoured magwell, concealment beavertail grip safety, tactical thumb safety, fluted barrel and chamber, carry and ball endmill cuts, fully-machined Bullet Proof parts.

Production: current Caliber: 9mm Action: SA, semi-auto Barrel Length: 3.6"
Wt.: 25 oz.   Sights: Battlesight with fiber-optic front Capacity: 8
Magazine: detachable box Grips: G10 Starburst

D2C:	NIB	$3,750	Ex	$2,850	VG+	$2,065	Good	$1,875	LMP	$3,875
C2C:	Mint	$3,600	Ex	$2,600	VG+	$1,880	Good	$1,690	Fair	$ 865
Trade-In:	Mint	$2,670	Ex	$2,100	VG+	$1,470	Good	$1,315	Poor	$ 390

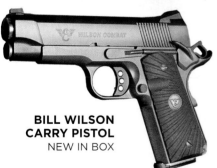

BILL WILSON CARRY PISTOL
NEW IN BOX

BILL WILSON CARRY PISTOL   REDBOOK CODE: RB-WC-H-BWCPIS

Compact carbon steel frame with gray Armor-Tuff finish, carbon steel slide with black Armor-Tuff finish, Bullet Proof beavertail grip and thumb safety, round-butt frame, contoured magwell, carry cuts, stainless cone barrel.

Production: current Caliber: .45 ACP Action: SA, semi-auto Barrel Length: 4"
Wt.: 35 oz.   Sights: Battlesight with fiber-optic front Capacity: 7
Magazine: detachable box Grips: G10 Starburst

D2C:	NIB	$2,900	Ex	$2,225	VG+	$1,595	Good	$1,450	LMP	$3,205
C2C:	Mint	$2,790	Ex	$2,025	VG+	$1,450	Good	$1,305	Fair	$ 670
Trade-In:	Mint	$2,060	Ex	$1,625	VG+	$1,140	Good	$1,015	Poor	$ 300

SPEC-OPS 9
NEW IN BOX

SPEC-OPS 9   REDBOOK CODE: RB-WC-H-SPECOP

Mid-size polymer frame with stainless steel rail insert, thin profile carbon steel slide, Bullet Proof thumb safety and beavertail grip safety, Spec-Ops ultralight hammer, countersunk slide stop, contoured magwell, carry and ball endmill cuts.

Production: current Caliber: 9mm Action: SA, semi-auto Barrel Length: 4.5"
Wt.: 30 oz.   Sights: fiber-optic front, Spec-Ops low-profile rear Capacity: 16
Magazine: detachable box Grips: polymer grip frame

D2C:	NIB	$ 2,175	Ex	$1,675	VG+	$1,200	Good	$1,090	LMP	$2,285
C2C:	Mint	$2,090	Ex	$1,525	VG+	$1,090	Good	$980	Fair	$ 505
Trade-In:	Mint	$1,550	Ex	$1,225	VG+	$850	Good	$765	Poor	$ 240

HUNTER
NEW IN BOX

HUNTER REDBOOK CODE: RB-WC-H-HUNTER

Full-size carbon steel frame and slide, countersunk slide stop, Bullet Proof beavertail grip safety and thumb safety, contoured magwell, Crimson Trace Lasergrips, single-port compensated barrel.

Production: current Caliber: .460 Rowland, 10mm Action: SA, semi-auto

Barrel Length: 5.5" Wt.: 40 oz. Sights: ramp front, adjustable rear Capacity: 7

Magazine: detachable box Grips: Crimson Trace Lasergrips

D2C:	NIB	$4,100	Ex	$3,125	VG+	$2,255	Good	$2,050	LMP $4,100
C2C:	Mint	$3,940	Ex	$2,850	VG+	$2,050	Good	$1,845	Fair $ 945
Trade-In:	Mint	$2,920	Ex	$2,300	VG+	$1600	Good	$1,435	Poor $ 420

PROTECTOR
NEW IN BOX

PROTECTOR REDBOOK CODE: RB-WC-H-PROTEC

Full-size carbon steel frame and slide, Bullet Proof beavertail grip safety and thumb safety, contoured magwell, stainless barrel and bushing.

Production: current Caliber: .45 ACP, 10mm, .40 S&W, .38 Super, 9mm

Action: SA, semi-auto Barrel Length: 5" Wt.: 38 oz. Sights: Battlesight with

fiber-optic front Capacity: 8 Magazine: detachable box Grips: G10 Starburst

D2C:	NIB	$2,800	Ex	$2,150	VG+	$1,540	Good	$1,400	LMP $2,920
C2C:	Mint	$2,690	Ex	$1,950	VG+	$1,400	Good	$1,260	Fair $ 645
Trade-In:	Mint	$1,990	Ex	$1,575	VG+	$1,100	Good	$ 980	Poor $ 300

1996A2
EXCELLENT

1996A2 REDBOOK CODE: RB-WC-H-1996A2

Bullet Proof controls, beavertail grip safety, thumb safety, blue finish, many upgrades were available to the base model that affect pricing.

Production: 1996 - discontinued Caliber: .45 ACP Action: SA, semi-auto

Barrel Length: 5" Wt.: 38 oz. Sights: 3-dot Capacity: 8 Magazine: detachable box

Grips: cocobolo

D2C:	NIB	$1,625	Ex	$1,250	VG+	$ 895	Good	$ 815	
C2C:	Mint	$1,560	Ex	$1,125	VG+	$ 820	Good	$ 735	Fair $ 375
Trade-In:	Mint	$1,160	Ex	$ 925	VG+	$ 640	Good	$ 570	Poor $ 180

CONTEMPORARY CLASSIC CENTENNIAL

REDBOOK CODE: RB-WC-H-CCCXXX

Limited 100 serial numbers commemorating 100 years of the 1911, serial number range: JMB001-JMB100, full-size carbon steel frame and slide, hand-fit carbon match grade barrel and bushing, walnut presentation case.

Production: 2011 - discontinued Caliber: .45 ACP Action: SA, semi-auto

Barrel Length: 5" Wt.: 37 oz. Sights: Battlesight with Gold bead front

Capacity: 8 Magazine: detachable box

Grips: presentation French walnut double-diamond

D2C:	NIB	$3,785	Ex	$2,900	VG+	$2,085	Good	$1,895	LMP $3,995
C2C:	Mint	$3,640	Ex	$2,625	VG+	$1,900	Good	$1,705	Fair $ 875
Trade-In:	Mint	$2,690	Ex	$2,125	VG+	$1,480	Good	$1,325	Poor $ 390

APPENDIX A: MANUFACTURERS

Accu-Tek Firearms (Excel Industries, Inc.)
accu-tekfirearms.com
Corporate Office & Manufacturing Facility
1601 Fremont Ct.
Ontario California 91761
Phone: 909-947-4867

Armscor USA and Rock Island Armory
us.armscor.com
150 North Smart Way
Pahrump, NV 89060
Phone: 775-537-1444

American Tactical Imports (ATI)
americantactical.us
100 Airpark Dr.
Rochester, NY 14624
Phone: 800-290-0065

Beretta USA
berettausa.com
Phone: 800-929-2901

Bersa (Eagle Imports, Inc.)
bersa.com
1750 Brielle Ave, Unit B-1
Wanamassa, NJ 07712
Phone: 732-493-0333

Bond Arms
bondarms.com
P.O. Box 1296
Granbury, TX 76048
Phone: 817-573-4445

Cabot Guns
cabotgun.com
799 N. Pike Road
Cabot, PA 16023
Phone: 724-602-4431

Charter Arms
charterfirearms.com
18 Brewster Lane
Shelton, CT 06484
Phone: 203-922-1652

Christensen Arms
christensenarms.com
P.O. Box 240
550 N. Cemetery Rd.
Gunnison, UT 84634
Phone: 435-528-7999

Colt (Colt's Manufacturing Company)
coltsmfg.com
P.O. Box 118
Hartford, CT 06141
Phone: 800-241-2485/860-232-4489

CZ-USA
cz-usa.com
P.O. Box 171073
Kansas City, KS 66117-0073
Phone: 913-321-1811

Diamondback Firearms, LLC
diamondbackfirearms.com
4135 Pine Tree PL
Cocoa, FL 32926
Phone: 877-997-6774

EAA (European American Armory)
eaacorp.com
P.O. Box 560746
Rockledge, FL 32956-0746
Phone: 321-639-4842

Ed Brown Products, Inc.
edbrown.com
P.O. Box 492
Perry, MO 63462
Phone: 573-565-3261

FMK Firearms
fmkfirearms.com
P.O. Box 1358
Placentia, CA 92871
Phone: 714-630-0658

FNH USA, LLC (FN Herstal)
fnhusa.com
P.O. Box 9424
McLean, VA 22102
Phone: 703-288-3500

Glock, Inc.
glock.com
6000 Highlands Parkway
Smyrna, GA 30082
Phone: 770-432-1202

Heckler & Koch
hecklerkoch-usa.com
5675 Transport Boulevard
Columbus, GA 31907 USA
Phone: 706-568-1906

Hi-Point Firearms
hi-pointfirearms.com

Infinity Firearms
sviguns.com
71229 Interstate 20
Gordon, TX 76453
800-928-1911/972-513-1911

APPENDIX A: MANUFACTURERS

Kahr Arms
kahr.com
130 Goddard Memorial Drive
Worcester, MA 01603
Phone: 508-795-3919

Kel-Tec (Kel-Tec CNC Industries Inc.)
keltecweapons.com
1505 Cox Road
Cocoa, FL 32926
Phone: 321-631-0068

Kimber Manufacturing
kimberamerica.com
30 Lower Valley Road
Kalispell, MT 59901
Phone: 888-243-4522

Les Baer Custom, Inc.
lesbaer.com
1804 Iowa Drive
LeClaire, IA 52753
Phone: 563-289-2126

Magnum Research
magnumresearch.com
12602 33rd Avenue S.W.
Pillager, MN 56473
Phone: 508-635-4273

Nighthawk Custom
nighthawkcustom.com
1306 W. Trimble Rd.
Berryville, AR 72616
Phone: 870-423-4867

Para USA, LLC
para-usa.com
10620 Southern Loop Blvd.
Pineville, NC 28134-7381
Phone: 704-930-7600

Remington Arms Company, LLC
remington.com
870 Remington Drive
P.O. Box 700
Madison, NC 27025-0700
Phone: 800-243-9700

Republic Forge
republicforge.com
820 S. Industrial
P.O. Box 774
Perryton, TX 79070
Phone: 806-648-1911

Rossi
rossiusa.com
16175 N.W. 49th Avenue
Miami, FL 33014

Ruger (Sturm, Ruger & Co.)
ruger.com
411 Sunapee Street
Newport, NH 03773
Phone: 603-865-2442

SIG Sauer
sigsauer.com
72 Pease Boulevard
Newington, NH 03801
Phone: 603-610-3000

Smith & Wesson
smith-wesson.com
2100 Roosevelt Avenue
Springfield, MA 01104
Phone: 800-331-0852

Springfield Armory USA
springfield-armory.com
420 West Main Street
Geneseo, IL 61254
Phone: 800-680-6866

Steyr Arms (Steyr Mannlicher)
steyrarms.com
P.O. Box 840
Trussville, AL 35173
Phone: 205-655-8299

STI International, Inc.
stiguns.com
114 Halmar Cove
Georgetown, TX 78628
Phone: 512-819-0656

Taurus International Manufacturing Inc.
taurususa.com
16175 N.W. 49th Avenue
Miami, FL 33014
Phone: 305-624-1115

UA Arms (Uselton Arms Inc.)
uaarms.com
Franklin, TN 37064
Phone: 615-970-9555

Walther Arms
waltherarms.com
7700 Chad Colley Blvd
Fort Smith, AR 72916
Phone: 479-242-8500

Wilson Combat
wilsoncombat.com
2234 CR 719
Berryville, AR 72616
Phone: 800-955-4856

APPENDIX B: ORGANIZATIONS

A Girl and A Gun

A ladies only organization established by women shooters for women shooters with a passion for pistol, rifle, and shotgun sports. The organization's mission is "to educate and encourage women about firearms usage and safety and to promote women's shooting interest and participation in the competitive shooting sports."
agirlandagunclub.com

America's 1st Freedom

An official journal of the National Rifle Association that is focused on the constitutional right to bear arms. America's 1st Freedom features the latest news concerning Second Amendment freedoms.
nranews.com/americas1stfreedom

American Handgunner

A magazine devoted to handguns, hunting, competition shooting, tactical knives, and shooting-related activities that features reviews on guns, knives, ammunition, shooting gear, historical articles, self-defense, and gun rights information.
americanhandgunner.com

American Rifleman

A firearms publication owned by the National Rifle Association. The publication includes information on guns, newsletters, reviews, guides, galleries, videos, blogs, training tips, and top stories on guns.
americanrifleman.org

Bud's Gun Shop

One of the largest and most trusted online gun stores, known for their fair prices and wide selection.
budsgunshop.com

Collectors Firearms

A gun auction website with one of the largest collections of militaria, uniforms, weapons, and memorabilia. Collectors Firearms features all kinds of antique and modern vintage firearms.
collectorsfirearms.com

Civilian Marksmanship Program (CMP)

"The Civilian Marksmanship Program is a national organization dedicated to training and educating U.S. citizens in responsible uses of firearms and air guns through gun safety training, marksmanship training, and competitions." With an emphasis on youth, the CMP Mission promotes firearm safety and marksmanship training.
odcmp.com

Combat Focus Shooting

A program that features handgun courses and instruction with a focus on real world defense scenarios. This shooting program is "designed to help the student become a more efficient shooter in the context of a dynamic critical incident."
combatfocusshooting.com

Concealed Carry Magazine

A concealed and carry publication featuring crucial advice for armed citizens to better protect themselves. Includes gun and gear reviews, life-saving training tips, and answers to concealed carry questions written by leading experts on self-defense and concealed carry. *Concealed Carry Magazine* is a publication of the United States Concealed Carry Association.
concealedcarrymagazine.com

Gun Broker

"The World's Largest Online Auction of Firearms and Accessories," Gun Broker provides a secure and safe way to purchase guns and hunting and shooting accessories while promoting responsible gun ownership.
gunbroker.com

International Defensive Pistol Association (IDPA)

"The International Defensive Pistol Association is the governing body of a shooting sport that simulates self-defense scenarios and real life encounters." Founded in 1996, the IDPA was formed to appeal to shooters worldwide. The organization has more than 22,000 members, representing 50 countries.
idpa.com

International Practical Shooting Confederation (IPSC)

– "The IPSC was established to promote, maintain, improve, and advance the sport of IPSC shooting, to safeguard its principles, and to regulate its conduct worldwide in order to cultivate safe, recreational use."
ipsc.org

National Rifle Association of America (NRA)

An influential American lobbying group and large supporter of the Second Amendment. Known as "America's longest standing civil rights organization." Formed in 1871 by Union veterans Col. William C. Church and Gen. George Wingate. The NRA sponsors marksmanship events and publishes firearm-based magazines such as *American Rifleman, American Hunter, America's 1st Freedom, Shooting Illustrated, Shooting Sports USA* and *NRA Insights*.
nra.org

APPENDIX B: ORGANIZATIONS

National Shooting Sports Foundation (NSSF)
Known as "the trade association for the firearms industry." The NSSF promotes, protects, and preserves hunting and shooting sports.
nssf.org

NRA Insights
An NRA publication geared towards young shooters. The publication includes stories, gun safety, games, videos, tips, and pointers.
nrainsights.org

Rock Island Auction
One of the largest and most important firearm auction houses, specializing in antique and collectible models from all eras.
rockislandauction.com

Second Amendment Foundation (SAF)
Strong supporter of Second Amendment rights. Promotes firearm rights through educational and legal action programs designed to inform the public about the gun control debate.
saf.org

Shooting Illustrated
An NRA publication that highlights firearm news, weekly polls, tips, feature stories, blogs, videos, galleries, and firearm-related gear.
shootingillustrated.com

Shooting Sports USA
An NRA publication that focuses on competition shooting news.
nrapublications.org/index.php/shooting-sports-usa

Shooting for Women Alliance (SFWA)
A nonprofit organization dedicated to educate women and youth worldwide about personal defense, firearms safety, conservation, and enjoyment of the shooting sports.
shootingforwomenalliance.com

Sporting Classics
Formed 1981, *Sporting Classics* is a bi-monthly magazine that consistently publishes some of the best hunting, fishing, and outdoor writing and photography.
sportingclassics.com

Springfield Armory Museum
From 1777 to 1968, the Springfield Armory Museum was the primary manufacturing center for U.S. Military-issued firearms. Courtesy of Wikipedia.com: "The Springfield Armory National Historic Site commemorates the critical role of the nation's first armory by preserving and interpreting the world's largest historic U.S. military small arms collection, along with historic archives, buildings, and landscapes."
nps.gov/spar/index.htm

Tactical-Life
Owned by Harris Publications, Tactical-Life is an umbrella website for the following publications: *Tactical Weapons, Guns & Weapons for Law Enforcement, Special Weapons for Military & Police, Rifle Firepower, Combat Handguns, Tactical Knives, Guns of the Old West,* and *The New Pioneer.*
tactical-life.com

The Sportsman Channel
A television channel designed for outdoor enthusiasts with a focus on hunting, shooting, and fishing for entertainment and educational purposes. Known as "the leader in outdoor TV for the American sportsman."
thesportsmanchannel.com

United States Practical Shooting Association (USPSA) – "The premier competitive shooting organization in the world." The USPSA site offers a club finder, articles for competitors, a rule book, match announcements, and top news about the organization.
uspsa.org

USA Carry
A leading concealed carry online resource featuring concealed carry articles, news, and training. USA Carry also features a directory where users can discover firearm instructors, gun shops, ranges, and gunsmiths.
usacarry.com

Women & Guns
A firearms publication for women that provides information on firearms, self-defense, articles, events seminars, and training information for women.
womenshooters.com

BIBLIOGRAPHY & WORKS REFERENCED

American-firearms.com

Armalite.com

Auto-ordnance.com

Berettausa.com

Bersa.com

Bluebookofgunvalues.com

Budsgunshop.com

Browning.com

Cabotgun.com

Carpenteri, Stephen D. Shooter's Bible (104th Ed.) New York, NY: Skyhorse Pub., 2012. Print.

Charterfirearms.com

Charlesdaly-us.com

Cheaperthandirt.com

Chiappafirearms.com

Cimarron-firearms.com

Cobrapistols.net

Collectorsfirearms.com

Colt.com

Coltautos.com

Coltfever.com

Comanchepistols.com

Coolgunsite.com

Coonaninc.com

Cornell, Joseph Madden., and Dan Shideler. Standard Catalog of Browning Firearms. Iola, WI: Gun Digest, 2008. Print.

Cva.com

Cz-usa.com

Eaacorp.com

Edbrown.com

Excelarms.com

Ezell, Edward C. Small Arms of the World. 12th ed. New York: Stackpole, 1993.

Firearmspriceguide.com

Fjestad, S. P. Blue Book of Gun Values (34th Ed.). Minneapolis, MN: Blue Book Publications, 2013. Print.

Flayderman, Norm. Flayderman's Guide to Antique American Firearms (9th Ed.). Iola, WI: Gun Digest.

Freedomarms.com

Forgottenweapons.com

Fnhusa.com

Genitron.com

Girsan.com.tr/en

Gunauction.com

Gunbroker.com

Gundersonmilitaria.com

Haemmerli.info

Heritagemfg.com

Hi-pointfirearms.com

Highstandard.com

Horstheld.com

Hr1871.com

icollector.com

Iverjohnsonarms.com

Jimenezarmsinc.com

Kahr.com

Kahrtalk.com

Keltecweapons.com

Kimberamerica.com

Korthusa.com

Lee, Jerry. 2014 Standard Catalog of Firearms: The Collector's Price and Reference Guide (24th Ed.) Iola, WI: Gun Digest, 2014. Print.

Lee, Jerry. Gun Digest (68th Ed.), 2014. Chicago: Gun Digest, 2014. Print.

Legacysports.com

Lugerforum.com

Model1911A1.com

Momoneypawn.com

Nighthawkcustom.com

Northamericanarms.com

Olyarms.com

Para-usa.com

Phoenixinvestmentarms.com

Proofhouse.com

Relicman.com

Remington.com

Rezzguns.com

Rockislandauction.com

Rockriverarms.com

Rossiusa.com

Ruger.com

Sightm1911.com

Sigsauer.com

Slickguns.com

Smith-wesson.com

Springfield-armory.com

Stiguns.com

Steyrarms.com

Tarr, James, and Rick Sapp. Standard Catalog of Colt Firearms (2nd Ed.). Iola, WI: Gun Digest, 2013. Print.

Thefirearmblog.com

Thefiringline.com

Taurususa.com

Unblinkingeye.com

Us.armscor.com

Useltonarms.com

Whichgun.com

Wikipedia.com

PHOTOGRAPHY CREDITS

The following organizations and individuals have contributed photography for this book. We greatly appreciate their assistance.

Bob Adams: Beretta 951

Adamsguns.com: CZ 38

Beretta

BudsGunShop.com

EAA

Cabot Guns

Charter Arms

Chiappa Firearms

Christensen Arms

Cimarron Firearms

Cobra Pistols

CollectorsFirearms.com

Colt.com

Cva.com

C.Z. U.S.A.

Ed Brown Products, Inc.

F.N. Herstal

Glock

Heckler & Koch

Hi-Point Firearms

Heritage Manufacturing

icollector.com

Kahr Arms

Kimber America

Legacy Sports

Nighthawk Custom

Para USA

Remington Arms

Rezz Guns

Rock Island Auction Company

Rossi Firearms

SIG Sauer

Smith & Wesson

Springfield Amory

STI, International

Sturm, Ruger & Co., Inc.

Taurus International

Umarex USA

UA Arms

Walther Arms, Inc.

Wikipedia.com